PRINCIPLES OF STELLAR EVOLUTION AND NUCLEOSYNTHESIS

Principles of Stellar Evolution and Nucleosynthesis

With a new Preface

Donald D. Clayton

The University of Chicago Press
Chicago and London

The University of Chicago Press, Chicago 60637
The University of Chicago Press, Ltd., London

 Published 1968
University of Chicago Press edition 1983
Printed in the United States of America

90 89 88 87 86 85 84 83   1 2 3 4 5

Library of Congress Cataloging in Publication Data

Clayton, Donald D.
Principles of stellar evolution and nucleosynthesis.

Reprint. Originally published: New York: McGraw-Hill, [1968]
Includes bibliographical references and index.
1. Stars—Evolution. 2. Nucleosynthesis. I. Title.
QB806.C55 1983 521′.58 83–5106
ISBN 0–226–10952–6
ISBN 0–226–10953–4 (pbk.)

CONTENTS

PREFACE (1983)

The Preface that I wrote fifteen years ago for the first edition of this textbook is still applicable to the specific role that it plays in the education of astrophysicists. Were that not so, a reprint edition would not be as useful as many of my colleagues appear to believe that it will be. Some have urged me to write instead an expanded and updated version; but most have reinforced my own opinion that the understanding of principles needed by students remains about what it was. That being the case, we have chosen to keep the cost accessible to students by offering this reprint edition.

Of course, scientific research itself has increased by leaps and bounds, both in stellar evolution and in nucleosynthesis. The number of research papers and the complexity of their results dwarf what was known fifteen years ago. But this textbook was not conceived to be a review paper then, and neither need it be today. What we choose instead is to provide in this Preface a brief commentary with selected references to recent literature. These will be useful to students seeking to bridge the text to recent research. In making these remarks, I strive for brevity. The recent references that I selected update those topics that are addressed explicitly or implicitly in the first edition. Rather than trying to be fairly complete in the references, I have singled out some that will help the student continue study of a particular topic, either because they are readable accounts, or especially influential papers, or reviews possessing a good reference list. In regard to stellar evolution, the *Annual Reviews of Astronomy and Astrophysics* are so useful that they are designated explicitly in this Preface (*Ann. Rev.*) as well as listed in the reference list at the end. The many IAU Symposia published by D. Reidel are also excellent sources to today's stellar evolution problems. One book alone, *Essays in Nuclear Astrophysics*, C. A. Barnes, D. D. Clayton, and D. N. Schramm, eds. (Cambridge University Press, Cambridge 1982), contains such a collection of authoritative and interestingly written articles as to serve as a major reference source for nuclear astrophysics. Many references will be to its contents. It is identified in this text for convenience as *EssNA*. Most instructors will, in any case, have their own references and supplementary material that fit their own emphases for the course.

Knowledge of the abundances of the elements in the solar system has greatly improved and continues to provide the basic impetus for nucleosynthesis theory. This includes improved knowledge of the abundances in meteorites, in the sun, and in the solar wind. Two outstanding recent compilations are those of Cameron (1982, in *EssNA*) and of Anders and Ebihara (1982), whose tables also indicate the nuclear processes believed to be responsible for the synthesis of the abundance of each isotope. A thrilling development was the discovery in the 1970s of small variations in isotopic composition between differing meteoritic samples. The differences strongly point to validation of key ideas of nucleosynthesis theory, ideas that had heretofore been accepted on faith; specifically, that the solar system abundances comprise a mixture of nuclei with different nucleosynthesis histories. I have myself taken pains (Clayton 1979) to demonstrate that these isotopic variations are not random but support the expectations of nucleosynthesis theory. Their interpretation gives birth to a new field of astronomy based on the chemical history of presolar dust (Clayton 1981). Significant observations of overabundances of iron (Kirshner and Oke 1975) and of the products of explosive oxygen burning (Chevalier and Kirshner 1979) in young supernova remnants have provided a long-sought demonstration of the significance of supernovae to nucleosynthesis (Woosley and Weaver 1982a, 1982b, in *EssNA*).

The discovery of pulsars was not anticipated by the text material. The subsequent attention to neutron stars emphasized that the equation of state of matter at very high densities, up to and beyond the density of nuclear matter on earth (2×10^{14} g/cm^3), is an essential part of the problem of stellar evolution. The pressure in the idealized (but unrealistic) case of a perfect neutron gas (degenerate) requires only that its mass M_N replace the electron mass M_e in chapter 2; however, that analysis gives a maximum "Chandrasekhar mass" of only 0.7 $M_\odot$ for a neutron star. It is the imperfect nature of the *nn* interaction (repulsion) that raises the maximum mass to about 2.5 $M_\odot$. Baym and Pethick (1979, *Ann. Rev.*) provide one good description of the physics of this imperfect gas and associated uncertainties in it with special emphasis on the structure of neutron stars. Canuto (1974, 1975, *Ann. Rev.*) provides a more thorough review of the physical principles of the equation of state itself. The problem at subnuclear densities is obfuscated by the complicated clustering of nucleons into rather exotic neutron-rich nuclei and by the degree of electron capture as the material becomes more neutron rich. Careful treatment of that density regime is required not only for the outer parts of the neutron star but also for the problem of presupernova – core collapse (e.g., Bethe 1982, in *EssNA*). A direct measure of the radii of neutron stars may be provided by a model of x-ray bursts based on the thermonuclear explosions on the surfaces of neutron stars (Woosley and Tamm 1976; van Paradijs 1979).

The problem of stellar opacity has been reviewed by Carson (1976, *Ann. Rev.*). The difficult problem of energy transport by convection is usefully discussed by Spiegel (1971, 1972, *Ann. Rev.*). These two problems remain at the heart of many aspects of stellar structure and evolution, and there is still no assurance that they are being correctly treated. Barkat (1975, *Ann. Rev.*) discusses the neutrino emitting processes in stellar interiors along with a discussion of their influence on the various stages of stellar evolution. Bahcall and Davis (1982, in *EssNA*) present an engaging review of the solar neutrino problem.

Many improvements in thermonuclear reaction rates have outdated the limited numbers tabulated in the text. Three reviews by Fowler, Caughlan, and Zimmerman (1967, 1975, *Ann. Rev.*) and Harris et al. (1983, *Ann. Rev.*) have assumed major roles as thermonuclear data evaluations. Hydrogen-burning reactions are discussed also by Kavanagh (1982, in *EssNA*) and by Rodney and Rolfs (1982, in *EssNA*). More advanced thermonuclear stages are described by Barnes (1982, in *EssNA*) and by Tombrello, Koonin, and Flanders (1982, in *EssNA*). These sources will lead to almost all of the nuclear data needed for the energy generation aspects of stellar evolution.

The more advanced stages of thermonuclear burning require a physics treatment slightly different than that in chapter 4, because the very high temperatures result in energies that are significant fractions of the Coulomb barrier and because they usually proceed through a large number of compound-nucleus resonances whose properties cannot all be measured. It therefore becomes necessary to calculate many thermonuclear rates with the aid of a model of thermonuclear reactions and from statistical data about the numbers and properties of such resonances. Such statistical theories usually employ the Hauser-Feshbach formulation of compound resonances. The validity of the results is measured by the degree (quite impressively good) to which such calculations reproduce the reactions where the cross section is well measured. Woosley et al. (1978) and Holmes et al. (1976) present a large and useful compilation of such calculations. Barnes (1982, in *EssNA*) also summarizes the theory nicely and shows how it compares with experiment. The most readable account was prepared by W. A. Fowler (1974) for the 1973 George Darwin Lecture of the RAS. His dedication to obtaining the correct thermonuclear rates provided the thrust for this approach to high-temperature reactions, and his account of it is especially to be recommended for those trying to understand for the first time the differences with chapter 4. Ward and Fowler (1980) discuss the thermalization of long-lived isomeric states that are not in thermal equilibrium at low temperatures.

The original material in chapter 5 is now inadequate on the nuclear problems of carbon burning, oxygen burning, neon burning, and silicon burning. What was missing then was a lot more nuclear data and the computer codes for reaction networks capable of handling many reactions and their inverses. There are a number of good discussions of these: carbon burning (Pardo, Couch, and Arnett 1974; Couch and Arnett 1975); neon burning (Arnett 1974a); oxygen burning (Woosley, Arnett, and Clayton 1972, 1973; Arnett 1974b); silicon burning (Woosley, Arnett, and Clayton 1973; Arnett 1977). Silicon burning approaches nuclear statistical equilibrium in the manner introduced in Chapter 7, with later quantitative treatments by Woosley, Arnett, and Clayton (1973), Hainebach et al. (1974), and Arnett (1977). Despite some subsequent improvements in the nuclear data, these papers are very clear in their displays of the nature of the burning, the setting of the burning, and the nucleosynthesis that accompanies it. The numerical technique for solving the simultaneous nonlinear differential equations for the rates of change of the abundances was devised and described by Arnett and Truran (1969). See also section 2 of Woosley, Arnett, and Clayton 1973 for further clarification.

Following these papers, it has become conventional, by the way, to let the variable Y_i or $Y(i)$ designate the abundance of species i in units of moles per gram. Then the number density $N(i) = \rho N_{AVO} Y(i)$, and $Y(i)$ clearly is equal to $X(i)/A(i)$, the ratio of mass fraction $X(i)$ to atomic weight $A(i)$. I note this here because of the now widespread (and commendable) use of this notation. Notice that in a simple expansion of matter without nuclear reactions or mixing the abundance $Y(i)$ remains constant, whereas the abundance $N(i)$ is proportional to ρ.

There is a large literature in stellar structure and evolution. But I make no attempt here to review it; rather, in the spirit of the principles of the subject, I choose to point the student to a few major works. In the same year that this textbook was published, another monograph by Cox and Giuli (1968) appeared, containing in two volumes a more exhaustive (though perhaps somewhat more demanding) treatment of the physics of stellar structure. Perhaps the best educational format today is to have these excellent volumes on reserve to elaborate specific sections of this text when desired. The second edition of *Galactic Astronomy* (Mihalas and Binney 1982) does an admirable job of briefly reviewing stellar evolution (in chapter 3) and, more important, of relating the many problems of stellar evolution to the general problems of the Galaxy. This book is, in other words, to be highly recommended to the student seeking interactions between stellar evolution and galactic structure and evolution.

The reviews by Iben (1967, 1974 *Ann. Rev.*) remain excellent quantitative explanations of the applications of stellar-structure theory to the important early phases of stellar evolution. A series of papers by Arnett referred to above describes clearly the advanced burning stages of the helium cores. The evolution of the entire massive stars through carbon burning is quantitatively described by Lamb et al. (1976). Woosley et al. (1978) presented the complete evolution through all six stages of nuclear burning and core collapse of 15 $M_\odot$ and 25 $M_\odot$ stars. Particularly useful papers in evolution through the asymptotic giant branch are Iben 1982 and Becker and Iben 1979, 1980. The related fascinating mystery of the carbon stars is addressed by Iben (1981); and the whole question of the last phases of evolution of low and intermediate mass stars ($< 8\ M_\odot$) is reviewed by Iben and Renzini (1983, *Ann. Rev.*). These few papers are singled out for the particularly relevant way in which they extend the principles of this textbook. The series of IAU Symposia are also rich sources for the student. In particular, *Physical Processes in Red Giants* (Iben and

Renzini 1981), *Stellar Instability and Evolution* (Ledoux 1974), *Chemical and Dynamical Evolution of Our Galaxy* (Basinska-Grzesik and Mayor 1977) and *Effects of Mass Loss in Stellar Evolution* (Chiosi and Stalio 1981) contain a host of stimulating treatments of problems in stellar evolution.

One particular problem deserving special mention is the thermal instability in evolved stars with two burning shells (He and H). This instability, which was discovered by Schwarzschild and Härm (1965), has led to very rich interplay between the principles of stellar evolution and of nucleosynthesis. It is this interplay that makes it noteworthy in the context of this text. In particular, Iben's work (1975a,b; 1976) is rich in its treatment of the thermal instability and in its relation to the *s*-process of nucleosynthesis. Going beyond these, Iben's (1978) study of thermal oscillations during carbon burning is typically thorough and physically illuminating.

Many problems concern stellar rotation. Dicke (1970, *Ann. Rev.*) and Gilman (1974, *Ann. Rev.*) discuss the issues and evidence for the sun. General features of stellar rotation are reviewed by Strittmatter (1969, *Ann. Rev.*), and the evolution of rotating stars is discussed by Fricke and Kippenhann (1972, *Ann. Rev.*).

One very important problem that is also related to the problem of stellar rotation is that of star formation. It is fairly well established that stars form in molecular clouds when they reach a density so great that self-gravitation takes over, leading to an extended period of almost free fall onto a growing hydrostatic core. Hydrodynamic models now exist (e.g., Larson 1973, *Ann. Rev.;* Woodward 1978, *Ann. Rev.;* Bodenheimer and Black 1978). The major issue is whether spherical symmetry is a good approximation as matter falls onto a growing core, or whether the angular momentum barrier will require an intermediate disk phase for redistributing the angular momentum. Most of the calculations are of spherical protostar collapse, and the problem is brilliantly described in a series of papers by Stahler, Shu, and Tamm (1980a,b; 1981) and by Winkler and Newman (1980a,b). Such calculations are, however, irrelevant if rotation forces the collapse in another direction. The general theory of accretion disks is laid out by Pringle (1981), but there are not yet many calculations of the growth of stars in such a way. But the structure of such prestellar accretion disks has become a new branch of stellar evolution, as described most clearly in papers by Lin (1981) and Lin and Bodenheimer (1982). Another issue is the first optical appearance of protostars. Where on the almost vertical Hayashi track does the spherically symmetric collapse first appear, and how does that compare to the T-Tauri phenomena? What does star formation look like if stars are instead grown via a disk? This has become perhaps ths most important problem in stellar evolution because of the crucial way in which it relates to the larger issue of the evolution of the phases of the interstellar medium and the exciting derivative issue of the formation of the solar system.

Interest in white dwarfs (Weidemann 1968, *Ann. Rev.;* Liebert 1980, *Ann. Rev.)* has increased over the years as questions beyond that of their structure have come to the fore. Ostriker (1971, *Ann. Rev.*) reviews many modern developments in their theory. Salpeter (1971, *Ann. Rev.*) summarizes the arguments for their being the central stars of planetary nebulae. Important questions for stellar evolution hinge on the evolutionary state of the preplanetary nebula stars and its connection to the mass and composition of the white dwarfs. Another line of development involves binary systems, wherein mass transfer onto the dwarf leading to a thermonuclear runaway in the hydrogen-rich skin has become the accepted model of the common nova (Gallagher and Starrfield 1978, *Ann. Rev.;* Truran 1982, in *EssNA*). A related fascination is the Type I supernova explosions, some of which may be triggered when a white dwarf mass surpasses the Chandrasekhar limit owing to either electron capture or accretion, or both. This has led to a lot of study of an initially degenerate carbon-detonation model (Arnett 1969). See, for example, the proceedings of the Texas Workshop on Type I supernovae (Wheeler 1980) or the book *Supernovae* (Rees and Stoneham 1982) or Trimble's work (1982, 1983). These books are excellent literature on the supernova explosion itself, which is the single most important event in stellar evolution and nucleosynthesis. Arnett (1982a, in *EssNA,* 1982b) reviews general arguments about the supernova phenomenon. Bethe (1982, in *EssNA*) describes the physics relevant to the collapse models of supernovae. Woosley and Weaver (1982a, 1982b, in *EssNA*) review the nucleosynthesis expected in massive supernovae. And I (Clayton, 1982, in *EssNA*) describe the hope of confirming nucleosynthesis theory by detecting gamma rays from the freshly synthesized radioactive nuclei expected to be synthesized in these explosions. Many other chapters in Rees and Stoneham's 1982 study address the many astrophysical problems related to supernovae. Hillebrandt (1978) reviews the relation of supernova hydrodynamics to the *r*-process of nucleosynthesis. Let it be said quite frankly that the supernova involves every single principle of stellar physics, and because it is also the primary source of heavy atomic nuclei, its central role in astrophysics is appreciated ever more fully.

Chapter 7 remains a useful introduction to the nucleosynthesis of the heavy elements, even though many improvements and clarifications have occurred. More modern *e*-process calculations are presented by Hainebach et al. 1974. The *s*-process is reviewed by Ulrich (1982, in *EssNA*), who played a leading role in the development of the idea that successive neutron-liberating pulses followed by mixing within a single star produce the exponential distribution of neutron fluences that characterizes solar abundances. The neutron-liberating reactions in these pulses are either $^{13}C(\alpha,n)^{16}O$, which requires some episodic mixing of hydrogen into a pulsationally unstable helium-burning shell (Iben and Renzini 1982), or the $^{22}Ne(\alpha,n)^{25}Mg$ reaction, which requires the high helium-burning-shell temperatures of moderately massive stars. The detailed problems associated with the required mixing are not yet fully solved. Especially to be recommended for this problem are the many papers of the Illinois group—Iben, Truran, and colleagues—as referenced above and in Ulrich's review. The interplay between stellar models and the nuclear physics has been very rewarding. The latest and best compendium of cross-section data and the good support it gives to the exponential-distribution model is the work of Käppeler et al. (1982). A major analysis of the mathematics of branching competition between neutron capture and beta decay is to be found in Ward, Newman, and Clayton 1976.

The *r*-process of heavy-element nucleosynthesis is updated in three excellent recent reviews (Schramm 1982, in *EssNA;* Hillebrandt 1978; Klapdor et al. 1981). Four related questions are dominating modern research: (1) Where in stellar evolution do the sites for intense ($n_n \gtrsim 10^{20}$ cm^{-3}) bursts of free neutrons occur? (2) Is it the neutron-separation energies, the beta-decay rates, or the neutron-capture cross sections along the capture path in neutron-rich nuclei that are dominating the abundance pattern? (3) Are the seed nuclei iron or are they heavier nuclei that have been produced in excess by an *s*-process during the prior evolution of the star? (4) How does the freeze-out during expansion set the final abundances? The view taken in chapter 7 was that the temperature and free-neutron densities were so high during the explosion that neutron-separation energies do regulate the flow and that the waiting-point approximation is a physically valid one. Important improvements to the beta-decay rates are described by Klapdor et. al. 1981. The alternative now receiving much study is that relatively low-temperature shock waves through the helium shell can liberate neutrons that drive *s*-process seed through neutron-magic, neutron-rich nuclei in circumstances where the (n,γ) cross sections dominate. See Cameron, Cowan, and Truran (1983) for a provocative study of this process and of the controversy over the "waiting-point approximation" within the context of helium-burning shells hit by a supernova shock wave. This is an interesting and important controversy that may be resolvable within a few years' time. The studies of presupernova structure strongly influence the problem through their control of allowable shock structures and of the pre-explosion nuclear evolution. It has also become apparent that the origins of rare neutron-rich nuclei in brief explosive neutron bursts (Howard et al. 1972; Wefel et al. 1981) are a key diagnostic aspect of such neutron bursts, many of which are too weak to produce the *r*-process peaks in the heavy elements.

Schramm (1982, in *EssNA*) and Thielemann, Metzinger, and Klapdor (1983) also review readably the implications from the *r*-process nucleosynthesis of long-lived radioactivities. The decays of $^{235,\,238}$U, and ^{187}Re continue to show that nucleosynthesis (*r*-process) began between 10 and 16 Gyr ago; but it has still not been possible to derive the exact answer with high precision. This age stands comfortably with those of the globular clusters and of the universe based on the Hubble expansion, but the possibility of improvement or of rude shocks remains. The discoveries and interpretation of the shorter lived extinct radioactivities ^{22}Na, ^{26}Al, ^{41}Ca, ^{107}Pd, ^{129}I, ^{135}Ba, and ^{244}Pu (Wasserburg and Papanastassiou 1982, in *EssNA*) have been very active and controversial (Schramm 1982 in *EssNA;* Clayton 1979). The major controversy revolves around (1) whether these nuclei were actually alive at one time in the solar system bodies wherein their daughters are now found, or whether the parent-daughter correlations now found in meteorites have been inherited from a much larger correlation present in interstellar dust, and (2) if they were alive in solar system objects, what those abundances imply for the conditions and causes of the formation of the solar system (Clayton 1979). I have advanced an astrophysical model (Clayton 1983) for the free-decay interval, prior to formation of the solar system, based upon the rate of mixing of fresh nucleosynthesis products into the dense phases of the interstellar medium. I cannot resist author's license to remark that the last paragraph of this book has been abundantly borne out.

The p-process of nucleosynthesis of the proton-rich, neutron-shielded heavy nuclei received almost no treatment in chapter 7; and, indeed, it has remained much the least studied of the nucleosynthesis mechanisms. It is very clear that elucidation of the explosive stars and their prior stellar evolution is to be found in the p-abundances, however. The original concept of rapid proton capture in a hot proton-rich environment was evaluated again by Audouze and Truran (1975), with good general fit to the observed p-abundances, but some severe points of difficulty remain. A compelling site for such a process has not been located, but hydrogen thermonuclear eruptions on the skin of a white dwarf, the novae (Truran 1982, in *EssNA*), is a possibility. Woosley and Howard (1978) made a thorough study of a separate process, a photo-ejection process in which (γ,n) and (γ,α) reactions strip down the normal heavy elements to proton-rich isotopes. I believe that this process is the more believable of the two, but the undecided question of stellar site remains elusive. Käppeler et al. 1982 also provides a useful discussion of this problem.

An ultimate goal of this science remains that of interpreting the full range of stellar abundances within the natural evolution of stars and stellar systems. There is a wide current literature for these problems, which can best be addressed according to the interest of the instructor. Suffice it here to say that the types of problems include, for example, (1) abundance gradients in galaxies, (2) the problem of the paucity of metal-poor dwarfs, and (3) the underabundance patterns in younger stellar systems such as the Magellanic Clouds. Problems of this type affect much larger issues in astrophysics than the scope of this textbook, however. See chapter 4 of Mihalas and Binney (1981) and Basinska-Grzesik and Mayor (1977).

It is my hope that these brief remarks will help this textbook remain as useful to new students as it was to those studying the subject during the 1970s. I thank those good friends and excellent colleagues, Dave Arnett, Willy Fowler, Icko Iben, and Stan Woosley, for their helpful assistance in improving the review aspects of this Preface.

REFERENCES

Anders, E., and Ebihara, M. 1982. *Geochim. Cosmochim, Acta* **46:** 2363.
Arnett, W. D. 1969. *Ap. Space Sci.* **5:** 180.
———. 1974a. *Ap. J.* **193:** 169.
———. 1974b. *Ap. J.* **194:** 373.
———. 1977. *Ap. J. Suppl.* **35:** 145.
———. 1982a. In *Essays in Nuclear Astrophysics.*
———. 1982b. In *Supernovae,* ed. M. J. Rees and R. J. Stoneham, p. 221. Dordrecht: Reidel.
Arnett, W. D., and Truran, J. W. 1969. *Ap. J.* **157:** 339.
Audouze, J., and Truran, J. W. 1975. *Ap. J.* **202:** 204.
Bahcall, J. N., and Davis, R. 1982. In *Essays in Nuclear Astrophysics.*

Barkat, Z. 1975. *Ann. Rev. Astron. Astrophys.* **13:** 45.
Barnes, C. A. 1982. In *Essays in Nuclear Astrophysics.*
Basinska-Grzesik, E., and Mayor, M., eds. 1977. *Chemical and Dynamical Evolution of Our Galaxy.* I.A.U. Colloquium no. 45. Geneva: Observatoire de Genève.
Baym, G., and Pethick, C. 1979. *Ann. Rev. Astron. Astrophys.* **17:** 415.
Becker, S. A., and Iben, I. 1979. *Ap. J.* **232:** 831.
———. 1980. *Ap. J.* **237:** 111.
Bethe, H. 1982. In *Essays in Nuclear Astrophysics.*
Bodenheimer, P., and Black, D. C. 1978. In *Protostars and Planets,* ed. T. Gehrels, p. 288. Tucson: Univ. Arizona Press.
Cameron, A. G. W. 1982. In *Essays in Nuclear Astrophysics.*
Cameron, A. G. W.; Cowan, J. J.; and Truran, J. W. 1983. *Astrophys. and Space Sci.* In press.
Canuto, V. 1974. *Ann. Rev. Astron. Astrophys.* **12:** 167.
———. 1975. *Ann. Rev. Astron. Astrophys.* **13:** 335.
Carson, T. R. 1976. *Ann. Rev. Astron. Astrophys.* **14:** 95.
Chevalier, R., and Kirshner, R. 1979. *Ap. J.* **233:** 154.
Chiosi, C., and Stalio, R., eds. 1981. *Effects of Mass Loss in Stellar Evolution.* Dordrecht: D. Reidel.
Clayton, D. D. 1979. *Space Sci. Rev.* **24:** 147.
———. 1981. *Quar. J. Roy. Astron. Soc.* **23:** 174.
———. 1982. In *Essays in Nuclear Astrophysics.*
———. 1983. In *Ap. J.* **268:** 381.
Couch, R. G., and Arnett, W. D. 1975. *Ap. J.* **196:** 791.
Cox, J. P., and Guili, R. T. 1968. *Principles of Stellar Structure.* New York: Gordon and Breach.
Dicke, R. H. 1970. *Ann. Rev. Astron. Astrophys.* **8:** 297.
Fowler, W. A. 1974. *Quar. J. Roy. Astron. Soc.* **15:** 82.
Fowler, W. A.; Caughlan, G. R.; and Zimmerman, B. A. 1967. *Ann. Rev. Astron. Astrophys.* **5:** 525.
———. 1975. *Ann. Rev. Astron. Astrophys.* **13:** 69.
Fricke, K., and Kippenhann, R. 1972. *Ann. Rev. Astron, Astrophys.* **10:** 45.
Gallagher, J. S., and Starrfield, S. 1978. *Ann. Rev. Astron. Astrophys.* **16:** 171.
Gilman, P. A. 1974. *Ann. Rev. Astron. Astrophys.* **12:** 47.
Hainebach, K. L.; Clayton, D. D.; Arnett, W. D.; and Woosley, S. E. 1974. *Ap. J.* **193:** 157.
Harris, M. J.; Fowler, W. A.; Caughlan, G. R.; and Zimmerman, B. A. 1983. *Ann. Rev. Astron. Astrophys.* **21.** In press.
Hillebrandt, W. 1978. *Spa. Sci. Rev.* **21:** 639.
Holmes, J. A.; Woosley, S. E.; Fowler, W. A.; and Zimmerman, B. A. 1976. *Atomic and Nuclear Data Tables.* **18:** 305.
Howard, W. M.; Arnett, W. D.; Clayton, D. D.; and Woosley, S. E. 1972. *Ap. J.* **175:** 201.

Iben, I. 1967. *Ann. Rev. Astron. Astrophys.* **5:** 571.
———. 1974. *Ann. Rev. Astron. Astrophys.* **12:** 215.
———. 1975a. *Ap. J.* **196:** 525.
———. 1975b. *Ap. J.* **196:** 549.
———. 1976. *Ap. J.* **208:** 165.
———. 1978. *Ap. J.* **226:** 966.
———. 1981. *Ap. J.* **246:** 278.
———. 1982. *Ap. J.* **260:** 821.
Iben, I., and Renzini, A., eds. 1981. *Physical Processes in Red Giants.* Dordrecht: D. Riedel.
———. 1982. *Ap. J. (Letters)* **263:** L23.
———. 1983. *Ann. Rev. Astron. Astrophys.* **21.** In press.
Käppeler, F.; Beer, H.; Wisshak, K.; Clayton, D. D.; Macklin, R. L.; and Ward, R. A. 1982. *Ap. J.* **257:** 821.
Kavanagh, R. W. 1982. In *Essays in Nuclear Astrophysics.*
Kirshner, R., and Oke, J. 1975. *Ap. J.* **200:** 574.
Klapdor, H. V.; Oda, T.; Metzinger, J.; Hillebrandt, W.; and Thielemann, F.-K. 1981. *Zeitschrift f. Phys. A.* **299**: 213.
Lamb, S.; Iben, I,; and Howard, W. M. 1976. *Ap. J.* **207:** 209.
Larson, R. B. 1973. *Ann. Rev. Astron. Astrophys.* **11:** 219.
Ledoux, P., ed. 1974. *Stellar Instability and Evolution.* Dordrecht: D. Reidel.
Liebert, J. 1980. *Ann. Rev. Astron. Astrophys.* **18:** 363.
Lin, D. N. C. 1981. *Ap. J.* **246:** 972.
Lin, D. N. C., and Bodenheimer, P. 1982. *Ap. J.* **262:** 768.
Mihalas, D., and Binney, J. P. 1982. *Galactic Astronomy,* 2d edition. San Francisco: W. H. Freeman.
Ostriker, J. P. 1971. *Ann. Rev. Astron. Astrophys.* **9:** 353.
Pardo, R. C.; Couch, R. G.; and Arnett, W. D. 1974. *Ap. J.* **191:** 711.
Pringle, J. E. 1981. *Ann. Rev. Astron. Astrophys.* **19:** 137.
Rees, M. J., and Stoneham, R. J. 1982. *Supernovae.* Dordrecht: D. Reidel.
Rodney, W. S., and Rolfs, C. 1982. In *Essays in Nuclear Astrophysics.*
Salpeter, E. E. 1971. *Ann. Rev. Astron. Astrophys.* **9:** 127.
Schramm, D. N. 1982. In *Essays in Nuclear Astrophysics.*
Schwarzchild, M., and Härm, R. 1965. *Ap. J.* **142:** 855.
Spiegel, E. A. 1971. *Ann. Rev. Astron. Astrophys.* **9:** 323.
———. 1972. *Ann. Rev. Astron. Astrophys.* **10:** 261.
Stahler, S. W.; Shu, F. H.; and Tamm, R. E. 1980a. *Ap. J.* **241:** 637.
———. 1980b. *Ap. J.* **242:** 226.
———. 1981. *Ap. J.* **248:** 727.
Strittmatter, P. A. 1969. *Ann. Rev. Astron. Astrophys.* **7:** 665.
Thielemann, F.-K.; Metzinger, J.; and Klapdor, H. V. 1983. *Zeitschrift f. Phys. A.* **309:** 301.

Tombrello, T. A.; Koonin, S. E.; and Flanders, B. A. 1982. In *Essays in Nuclear Astrophysics*.
Trimble, V. 1982. *Revs. Mod. Phys.* **54:** 1183.
———. 1983. *Revs. Mod. Phys.* In press.
Truran, J. W. 1982. In *Essays in Nuclear Astrophysics*.
Ulrich, R. K. 1982. In *Essays in Nuclear Astrophysics*.
van Paradijs, J. 1979. *Ap. J.* **234:** 609.
Ward, R. A., and Fowler, W. A. 1980. *Ap. J.* **238:** 266.
Ward, R. A.; Newman, M. J.; and Clayton, D. D. 1976. *Ap. J. Suppl.* **31:** 33.
Wasserburg, G. J., and Papanastassiou, D. A. 1982. In *Essays in Nuclear Astrophysics*.
Weaver, T. A.; Zimmerman, A.; and Woosley, S. E. 1978. *Ap. J.* **225:** 1021.
Wefel, J. P.; Schramm, D. N.; Blake, J. B.; and Pridmore-Brown, D. 1981. *Ap. J. Suppl.* **45:** 565.
Weidemann, V. 1968. *Ann. Rev. Astron. Astrophys.* **6:** 351.
Wheeler, J. C., ed. 1980. *Type I Supernovae*. Austin: University of Texas.
Winkler, K. H., and Newman, M. J. 1980a. *Ap. J.* **236:** 201.
———. 1980b. *Ap. J.* **238:** 311.
Woodward, P. R. 1978. *Ann. Rev. Astron. Astrophys.* **16:** 585.
Woosley, S. E.; Arnett, W. D.; and Clayton, D. D. 1972. *Ap. J.* **175:** 731.
———. 1973. *Ap. J. Suppl.* **26:** 231.
Woosley, S. E.; Fowler, W. A.; Holmes, J. A.; and Zimmerman, B. A. 1978. *Atomic Data and Nuclear Data Tables* **22:** 371.
Woosley. S. E., and Howard, W. M. 1978. *Ap. J. Suppl.* **36:** 285.
Woosley, S. E., and Tamm, R. E. 1976. *Nature* **263:** 101.
Woosley, S. E., and Weaver, T. A. 1982a. In *Supernovae*, ed. M. J. Rees and R. J. Stoneham, p. 79. Dordrecht: D. Reidel.
Woosley, S. E., and Weaver, T. A. 1982b. In *Essays in Nuclear Astrophysics*.

PREFACE (1968)

The combined sciences of stellar evolution and nucleosynthesis comprise one of the most active and intriguing areas in modern science. The advent of the space age has been accompanied by an increased interest in the physics of natural phenomena, and foremost among them is the star. Since the entire realm of astrophysics is moderated to some degree by our concepts of the structure and evolution of stars and the synthesis of the elements, the intensifying interest in the physical principles of those subjects is readily understandable. In 1963 I initiated at Rice University a graduate-level course covering the physical principles of the discipline, and this book is primarily the result of having taught that course for three years. It seems to me that the subject is sufficiently well-founded to warrant a textbook devoted to those physical principles.

I have concentrated in this book on those ideas that will best enable the student to turn with understanding to the astrophysical literature in the field. In the selection of the subject matter I have been guided by my students and by my own experience as a student. I have forgone detailed results and specialized interpretations in favor of the physics that underlies such efforts. I have attempted to limit the length of the book to subject matter that can be introduced in one year's time, although it is neither possible nor desirable to reproduce its entire contents in the classroom.

Material has been included that will be familiar to many senior physics students, either for the sake of continuity or because I deemed the particular subjects worthy of a brief review. Some material has been adapted from other sources, but much of it is original. My purpose at all times has been to cultivate the physical thinking that will best accelerate the assimilation of more advanced treatments. I have tried to make the book self-contained, with no pretense of completeness.

The prerequisites for a graduate-level course in a division of physics, astronomy, or space science are a senior introduction to modern physics and graduate quantum mechanics, although I have arranged material so as to make concurrent enrollment in the latter adequate. The basic thermodynamic and astronomical concepts have been introduced as needed within each discussion. I regard the result as an outline and discussion of the basic physical principles, the details of which may be filled in by the student as his understanding matures. I have not attempted to write a book for specialists; I try to make my own course profitable to students in associated disciplines who may never do research in stellar evolution and nucleosynthesis. The problems and answers I have placed within the text have been found helpful in assimilating the material.

One additional word of explanation seems advisable. The book has a strong orientation toward nuclear physics, a fact that might be interpreted as reflecting my own specialized training. There is, however, a better reason for that emphasis; viz., nuclear physics is the thread that unites the discipline. My personal motivation has been to clarify the interplay of astronomy and nuclear physics in the mind of the student. The region of overlap between those disciplines is called *nuclear astrophysics*.

The book is divided into seven chapters, each of which is to be regarded as an introduction to its subject matter. The chapters are not totally independent; I have included selected comments in each discussion relating that subject to the overall context of stellar evolution. I have included as footnotes to the text those references which may serve the readers as an introduction to the literature. Many of the figures have been borrowed from important publications on the subject matter, and those sources are indicated in the legends. I have been arbitrary in this selection process, choosing either major review articles or specific research papers that extend a subject introduced in the text in a particularly meaningful way. For a complete bibliography the reader may consult:

Kuchowicz, B.: "Nuclear Astrophysics: A Bibliographic Survey," Nuclear Information Center, Warsaw, 1965.

Langer, E., M. Herz, and J. P. Cox: "Recent Work on Stellar Interiors: A Bibliography of Material Published between 1958 and Mid-1966," *Joint Inst. Lab. Astrophys. Rept.* 88, 1966.

I naturally hope that this textbook will be useful for advanced courses with a slightly different emphasis from my own. The book is expandable at almost any point according to the interests of the instructor and the level of the class. I myself have found that, once the basic ideas were available in note form for my students, I was able to advantageously extend the treatment without overextending the student; I hope my book may have this usefulness elsewhere.

I am grateful to the administration of Rice University for allowing me the freedom to develop the course on which this book is based, to the U.S. Air Force

Office of Scientific Research for supporting the associated research program under grant no. AFOSR 855–65, and to the Alfred P. Sloan Foundation for a Research Fellowship during the year 1966–1967, when the final version was written. I also wish to thank Prof. W. A. Fowler and Prof. F. Hoyle, who, respectively, extended to me the hospitality of the California Institute of Technology, Pasadena, and of the Institute of Theoretical Astronomy, Cambridge, during 1966–1967. I am indebted to Prof. Peter B. Shaw for many helpful discussions about the presentation of material and to Prof. Icko Iben, Jr., for a clarifying review of Chap. 6.

I wish to express my special gratitude to Richard Feynman and Marshal Wrubel. They gave me personal encouragement to complete a difficult task.

This book bears a double dedication: to the members of my family, who have cheerfully borne the considerable investment of my personal time, and to William Alfred Fowler, whose influence as my own teacher can be found on almost every page.

Donald D. Clayton

Twinkle, twinkle, little star,
How I wonder what you are,
Up above the world so high,
Like a diamond in the sky.

JANE TAYLOR
"Rhymes for the Nursery"
(1806)

chapter 1

A PHYSICAL INTRODUCTION TO THE STARS

Over the course of the years man's wonderings about the stars have grown and crystallized into a physical theory. Today the ideas regarding the nature of stars and their evolution provide a frame of reference used for interpreting much of the universe. It seems that all of astrophysics, ranging in proximity from the earth and its origins to the farthest reaches of the observable universe, is moderated in part by our conception of a star. An outline of these relationships is indicated schematically in Fig. 1-1. The many scientific specialties that are related to the various functions of stars are evident there.

The subject received its factual food for growth from the astronomer, who studied not only individual stars but also the hierarchy of geometrical groupings in which they are seen—star clusters, galaxies, and clusters of galaxies. Not only did this study lead to knowledge about star clusters, but the properties of the stars within a given cluster, taken as a group, indicated surprising new features of individual stars, viz., how they evolve. The improved understanding of individual clusters that followed from the development of a science of stellar evolution has made it possible to use them as indicators of the structure and evolution of our galaxy and galaxies in general. These ideas provide the observational framework for cosmology, the study of the large-scale properties of the universe.

The key to the history of the stellar interior was found in the nuclear laboratory. The conceptual framework there revolves around the sequence of nuclear reactions that occur in the interiors of stars and the three major functions of those

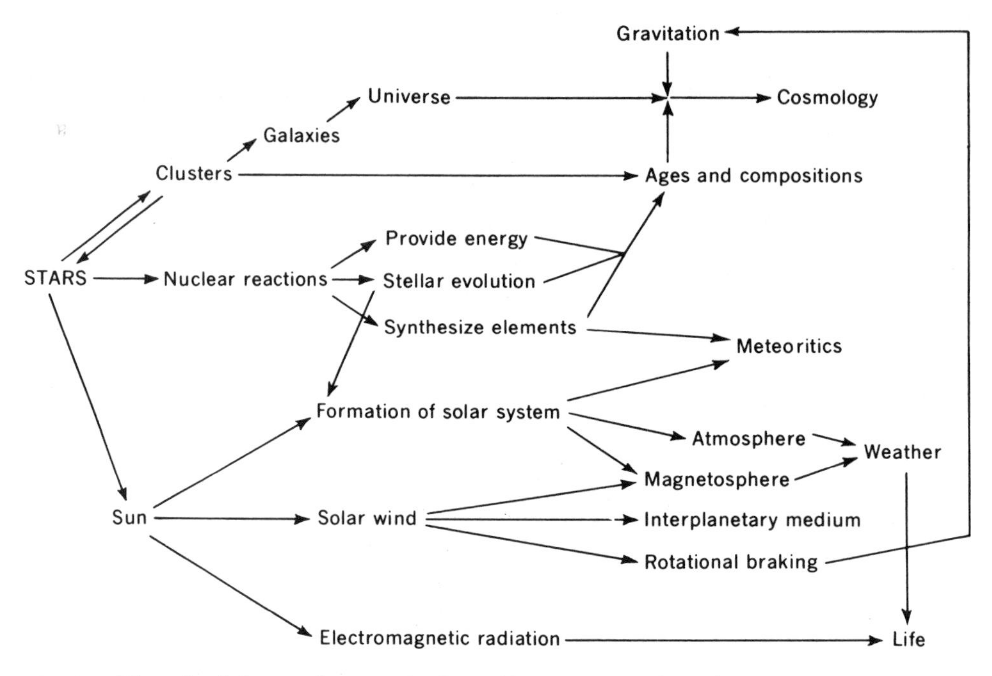

Fig. 1-1 The role of the star in astrophysics. Almost every subject in astrophysics and space science is influenced by our conception of the structure and evolution of the stars.

reactions: (1) they provide, like a giant nuclear reactor, the internal power that allows the stars to shine for long periods of time without cooling off; (2) they synthesize heavy elements from lighter ones; and (3) they determine the evolution of the star. The proper sequence of reactions was and is being found by careful laboratory study of nuclear reactions, the early demonstrations of which prompted Eddington to exclaim in 1920, "And what is possible in the Cavendish Laboratory may not be too difficult in the sun." The measured peculiarities of nuclear structure enter into the quantitative theories of stellar evolution and nucleosynthesis, with which we may hope to compute the age of a given cluster and the abundances of elements created by the stars within it.

A general theory of stellar evolution allows (and demands) a much more sophisticated picture of the sun than was permissible previously. Because the sun is the dominating feature of the solar system, all of our immediate environment has been shaped by the sun and the events attending its formation. In the physics of its initial condensation are to be found the principles for understanding the solar system as we know it. The formation of meteorites and the evolution of the atmospheres and oceans of the planets were determined by those early events. The electromagnetic radiation to which the earth is constantly exposed along with the ever-expanding solar corona or solar wind results in countless phenomena studied by geophysicists. Even the problem of life itself is linked with the problem of the origin and evolution of a star. Small wonder then that the principles in terms of which stars are interpreted are sought by increasingly larger numbers of scientists in these many interrelated fields.

This first chapter introduces those physical properties of stars which will be basic to our subject and provides an introduction to the main body of the book.

1-1 LUMINOSITY

The most striking observational fact about stars is their enormous range in apparent brightness. This fact is obvious just from looking at the night sky. The brightest star in the sky, Sirius, is 2 billion times as bright as the faintest star that can be seen with modern telescopes. Today even the untrained observer knows that a large part of the variation in brightness is due to a large range of distances of the various stars from us. Astronomers at one time considered the possibility that all stars had the same luminosity as the sun and tried to derive the distances of stars from how bright they appeared to be. This working hypothesis, although a good idea at the time, led to a number of inconsistencies. For instance, at many places in our own galaxy, clusters of stars can be seen that are well isolated from surrounding matter. The stars of such clusters are all in approximately the same location; i.e., the distance between the stars appears to be much less than the distance from the earth to the cluster. The only sensible thing to assume is that all such stars are the same distance away. One such example, perhaps the most famous, is the cluster called the Pleiades. Its brightest star, quite visible to the eye, is a blue star with a luminosity that is 0.4 times that of the North Star. Its faintest star, visible only by large telescopes, is a red star 1 million times fainter than the North Star. This variation must be an intrinsic luminosity difference. Observations of many other clusters of stars show that the optical luminosity of stars varies roughly between one-millionth and 1 million times the luminosity of the sun.

$$10^{-6}L_\odot < L < 10^{6}L_\odot$$

Because the science of observing the stars is so old, the historical paths traced to the modern definitions of luminosity are replete with ancient traditions and nomenclature. Systematic observations of the stars were made at least 4,000 years ago in China. By 2000 B.C. the Chinese had constructed a working calendar based on the apparent motions of the planets. There is a story that two astronomers, Hi and Ho, were executed for failing to predict an eclipse, but modern authorities suspect that their "crime" was actually one of carelessness in the preparation of the official calendar.

The first significant conceptual development of astronomical science was made by the Greeks, in particular Hipparchus, who introduced several advanced concepts to astronomy during his working years, 160 to 127 B.C. in Rhodes and Alexandria. His life testament was a catalog of about 1,000 stars, which remained the standard reference of the sky for 16 centuries. He assigned to these stars six categories of visual brightness, which are now called *magnitudes*. The brightest stars were placed in the *first magnitude*, whereas those at the limit of detectability by the unaided eye were of the *sixth magnitude*. The classification system has survived to the present day with a quantitative definition of the magnitudes.

In this book the *luminosity* of a star will be defined as the total power required of the star to sustain the energy efflux from a large surface surrounding the star. The energy flux is primarily of three types, photons, neutrinos, and mass loss. Photons constitute the traditional luminosity, and it is this form which is generally of most importance, inasmuch as photons are usually the dominant luminosity mechanism and the only form that is generally observable. In most cases these photons come directly from a thin surface layer of the star called the *photosphere*, although, in principle, they must include photons radiated by circumstellar matter deriving power for emission from the central star, e.g., fluorescence.

Neutrino luminosity is today unmeasurable and must be calculated from a model of the structure of the star. These neutrinos are basically from two different sources: (1) nuclear reactions in the interior and (2) a *weak interaction* of photons with matter. In either case the neutrinos produced escape immediately from the star and result in a form of luminosity that is so far invisible.

Mass loss from a star requires power to raise the matter from its gravitational binding in the photosphere and disperse it to large distances. Although a wide variety of stars are now known to lose mass at greatly differing rates, most mass loss has not yet been observed. The power output of the star is represented by the sum of these three luminosities:

$$L = L_\gamma + L_\nu + L_{\dot{m}} \tag{1-1}$$

The last two forms of luminosity will be discussed later. For the moment we turn our attention to the generally measurable photon luminosity.

PHOTON LUMINOSITY

The photon energy flux from a star is measured with the aid of a telescope and a detector of known sensitivity. The conversion of the observed energy flux into the photon luminosity of the star also requires a measurement of the distance to the star. Research in stellar distances began with nearby stars, whose distances could be determined by methods of angular parallax. Opposite points in the earth's orbit about the sun provide a broad base line for viewing a given star against the field of stars at infinity. The angular displacement of the star relative to the direction of the fixed field stars can then be measured if it is greater than the angular resolution of the instrument, as is illustrated in Fig. 1-2. This measurement determines the angle subtended at the star by the radius of the earth's orbit; this angle is called the *parallax* of the star.

Stellar parallaxes are very small, being only about 1 sec of arc for the nearest stars, and thus require careful work with a precise instrument for their measurement. The first successful measurements took place in about 1838, when Bessel (in Germany), Henderson (at the Cape of Good Hope), and Struve (in Russia) detected the parallaxes of the stars 61 Cygni, Alpha Centauri, and Vega, respectively.

The radius of the earth's orbit and the parallax of the star determine the distance to that star. The nearest star, Alpha Centauri, produces 1.52 sec of arc of angular displacement to the total diameter of the earth's orbit. Therefore, it has a parallax of 0.76 sec of arc. The distance to a star having 1 sec of parallax is defined as 1 parsec (pc). (One must take into account the direction of the star relative to the plane of the earth's orbit.) One parsec is equal to 3.086×10^{18} cm, or 3.262 light-years. The distances to about 7,000 stars have been determined by this method. Only for about 700 stars, however, are the parallaxes large enough (about 0.05 sec of arc or more) to be measured with a precision of 10 percent or better. There are other ways of determining the distances to farther stars, but the parallax method provided the starting point for determining astronomical distance scales.

It often happens that relative distances can be determined more accurately

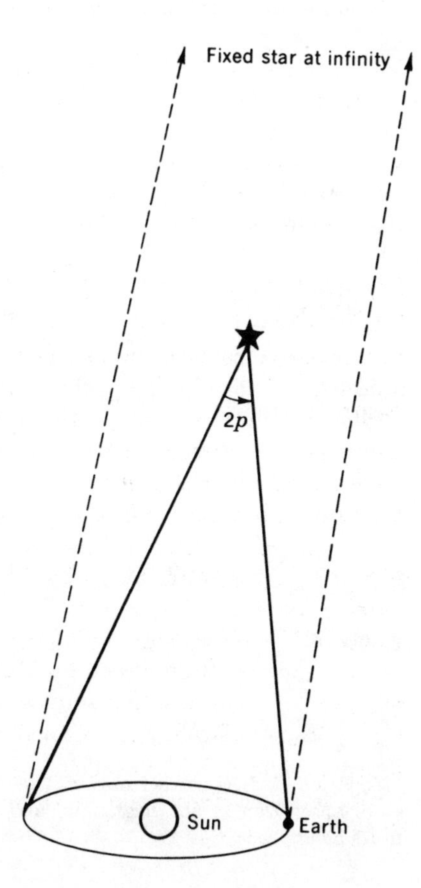

Fig. 1-2 The parallax of a star is a direct measure of its distance from the solar system. The angular displacement from season to season of a nearby star relative to the direction of distant field stars is inversely proportional to the distance to the star.

than absolute distances, as whenever groups of stars can be observed, such as those located in globular star clusters or in open galactic clusters, like the Pleiades, in which we know from the spatial arrangement of the stars that they are approximately the same distance from us. In those cases, very accurate measurements of *the ratio of apparent brightnesses* can be made which are often preferable to those obtained by applying individual corrections for the distances to the stars.

Once the distance to a star has been determined, its photon luminosity can be calculated from its apparent brightness. The energy received at the earth from the star can be corrected for the inverse-square-distance relationship, thus determining the absolute rate of energy output of the star. It is only necessary for this end to make an additional assumption concerning that fraction of radiant energy flux recorded in the measurement of the apparent brightness. This last requirement comes about because part of the energy flux is absorbed in the earth's atmosphere (or in interstellar space) before reaching the detector, and, in general, part of the energy reaching a given detector will not be recorded by it. Furthermore, the several different types of detectors in common use have different spectral sensitivities.

The photon luminosity of the star is given by

$$L_\gamma = 4\pi R_0 \int_0^\infty F_\lambda \, d\lambda \tag{1-2}$$

where $F_\lambda \, d\lambda$ is the net outgoing energy flux at wavelength λ in the interval $d\lambda$ from a surface of radius R surrounding the stellar atmosphere. If the earth is a distance r from the star, the incident energy flux on the top of the earth's atmosphere is[1]

$$f_\lambda = \left(\frac{R}{r}\right)^2 F_\lambda \tag{1-3}$$

It is clearly from a measurement of $r^2 f_\lambda$ at the earth that the corresponding product $R^2 F_\lambda$ can be determined.

The determination of f_λ is the basic observational problem in astronomy. It is not f_λ that is directly measured but rather the response of a specific type of detector *after* f_λ has been degraded by the earth's atmosphere. The apparent brightness of a star may be expressed as

$$b = \pi a^2 \int_0^\infty f_\lambda A_\lambda(z) R_\lambda \, d\lambda \tag{1-4}$$

where πa^2 = collecting area of telescope
$A_\lambda(z)$ = fraction of energy transmitted through earth's atmosphere at wavelength λ and zenith angle z; $A_\lambda(z) \approx e^{k_\lambda \sec z}$
R_λ = efficiency for recording photon of wavelength λ in detector

[1] Equation (1-3) is correct only in the absence of interstellar absorption. For many distant stars absorption is not negligible, and an appropriate correction for interstellar extinction must be applied.

In practice, the dependence on collecting area is removed by measuring brightness ratios of stars. Atmospheric extinction presents a more complicated problem, both because of the extensive observational research required to determine A_λ and because it is not constant either in time or from place to place. The method of dealing with this problem consists of the careful calibration of the spectral intensity of a set of standard stars, which are then compared to stars being observed. A_λ must be measured absolutely for the standard stars, however, and this is done by observing the rate of change of extinction with zenith angle. The measurements of A_λ indicate the nature of one problem in stellar luminosity. The earth's atmosphere absorbs almost all radiations with wavelengths shorter than about 3000 Å and most radiations longer than about 10,000 Å. The region between these limits is called the *optical window*. The maximum sensitivity of the human eye (4500 to 7500 Å) lies squarely in the middle of the optical window.[1] It is largely from the radiant flux in the optical window (to date!) that the rate of energy output of a star must be deduced, a process requiring a correction for the energy absorbed by the earth's atmosphere.

The ancients, particularly Hipparchus, set the tradition for the magnitude scale. Modern measurements of stellar brightness have revealed that their magnitude scale is nearly logarithmic in apparent brightness,[2] and in such a way that a difference of 5 magnitudes corresponded to nearly a factor of 100 in apparent brightness. To preserve the value of astronomical records and tradition, it was decided by international agreement that the ancient nomenclature should be retained but with the precise definition

$$\frac{b_1}{b_2} = 100^{\frac{1}{5}(m_2 - m_1)} = 2.512^{m_2 - m_1} \qquad (1\text{-}5)$$

where m_2 and m_1 are the apparent magnitudes of two stars of apparent brightness b_1 and b_2. It will be noted that a difference of 1 magnitude (1^m) corresponds to a brightness ratio of 2.512 in the sense that the larger magnitude is the smaller brightness. An alternative form for Eq. (1-5) is[3]

$$m_2 - m_1 = 2.500 \log \frac{b_1}{b_2} \qquad (1\text{-}6)$$

Although it is generally true that a magnitude difference corresponds to a brightness ratio in the sense of Eq. (1-6), there exist several different magnitude scales corresponding to the various types of detectors used in making the brightness comparisons. The ancient scale was, of course, based on the human eye, and the corresponding *visual magnitude* is designated by m_v. With the advent of

[1] It may seem remarkable that the world would be dark if our eyes were not so constructed as to operate best at those frequencies. What possible explanations may account for this coincidence?

[2] The human eye has a nearly logarithmic subjective response to radiant energy flux.

[3] Astronomers conventionally measure all logarithms to the base 10. We shall follow that custom throughout this book, using ln for $\log_e$.

photographic techniques, however, there arose a brightness scale based on the photochemical effect of starlight on standard emulsions. Since photographic plates respond to a much bluer part of the spectrum (3700 to 5000 Å) than the human eye (4500 to 7500 Å), the brightness ratio of two stars of differing temperatures does not have the same value in the two systems. The *photographic magnitude* is designated by m_{pg}. Two other magnitude scales designed to approximate m_v by using detectors having response similar to the eye are m_{pv}, the *photovisual magnitude,* measured with a photographic plate plus yellow filters, and V, the *photoelectric visual magnitude,* measured by a combination of photocathode and filters. Equations like (1-5) and (1-6) can be written for each of these magnitude systems, and for others as well.

Equation (1-6) defines only the differences in two magnitudes but is clearly equivalent to

$$m = -2.5 \log b + c \tag{1-7}$$

where c is the zero-point constant that depends on the system of magnitudes used as well as on the units of brightness. The zero point of the m_v scale was essentially set by the historical definition. The photographic scale is set such that $m_{pg} = m_v$ for a special type[1] of star, A0, but m_{pg} will differ slightly from m_v for other stellar types. Similar relationships exist for normalizing the zero point of each scale.

For an absolute comparison of intrinsic brightness it is common to discuss the magnitudes the stars would have if they were all at the same geocentric distance. The *absolute magnitude* of a star is defined as its magnitude viewed from a distance of 10 pc. From the inverse-square law it follows that the ratio of the absolute brightness B to the apparent brightness b is, for a star at a distance r (in parsecs),

$$\frac{B}{b} = \left(\frac{r}{10}\right)^2 \tag{1-8}$$

From Eq. (1-7) it then follows that the difference between the absolute and apparent magnitudes is

$$M - m = 5 - 5 \log r \tag{1-9}$$

This equation and the concept associated with it are used in each system of magnitude scales. By use of the distance modulus [Eq. (1-9)], absolute comparisons between stars can be made.

For our purposes the most significant physical quantity is the rate at which photon energy is radiated from the star. In terms of the language of magnitudes, the luminosity is measured by a quantity called the *absolute bolometric magnitude,* which is the absolute magnitude the star would have if the detector were able to respond to its entire radiant spectrum. It has already been noted that this quan-

[1] The spectral types will be discussed later.

tity is not amenable to direct physical measurement because a considerable portion of the energy arriving from a star may be absorbed in the earth's atmosphere and not be measured visibly, on photographic plates, or with photoelectric instruments. To obtain the total energy radiated from a star therefore requires making what is called a *bolometric correction.* The bolometric correction adds to the received energy from the star that amount of energy which is believed to be absorbed in the earth's atmosphere or is otherwise unmeasured by the detector. Although it might not seem so, this correction can be made. To first approximation, the emission spectrum of stars resembles that of a blackbody having a temperature equal to the surface temperature of the star. By measuring the spectrum of the star over the visible range of frequencies, one can determine the temperature of the surface of the star and then calculate what fraction of the energy output lies outside the range of atmospheric transmission. For the hottest stars, the greatest intensity of visible radiation lies in the blue and decreases, gradually becoming very dim, in the red. For these stars, most of the energy may be radiated in the ultraviolet. For most stars, the peak in the spectrum will lie somewhere in the visible region. The sun, for instance, has its maximum radiation intensity near 5000 Å, which is in the yellow. Cool stars will appear very red to the eye, the bulk of their energy being radiated in the infrared.

In terms of the notation used earlier, the bolometric correction (BC) will be defined as

$$\begin{aligned} \mathrm{BC} &= 2.500 \log \frac{\text{incident energy flux}}{\text{recorded energy flux}} \\ &= 2.500 \log \frac{\int f_\lambda \, d\lambda}{\int f_\lambda A_\lambda R_\lambda \, d\lambda} \end{aligned} \tag{1-10}$$

Although at first this may seem a cumbersome definition, Eq. (1-10) is defined to be of the same form as Eq. (1-6) in order that the BC will be a *magnitude increment.*

Problem 1-1: Show that $m - \mathrm{BC} = -2.500 \log \int f_\lambda \, d\lambda + \text{const.}$

Of course, the actual surface of the star is not a blackbody, since many factors may cause a deviation from local thermodynamic equilibrium on the surface of the star. However, even the deviations of the continuous spectrum from that of a blackbody spectrum are made understandable by modern calculations of the model atmospheres of stars. By understanding why the deviations occur, it is still possible to calculate a complete continuous spectrum from the observed continuous spectrum. The important theoretical problem is to determine what fraction of the energy flux lies outside that part of the star's radiation which is recorded by the instrument and therefrom to compute an absolute bolometric magnitude of the star.

In short, then, astronomers measure the apparent magnitude, spectrum, and distance of a star. The absolute magnitude of the star is computed by mentally viewing it from a distance of 10 pc. This absolute magnitude is then converted to the absolute bolometric magnitude of the star by correcting for the amount of radiant energy lying outside that portion received by the detecting instrument. If the bolometric correction is designated by BC and the absolute visual magnitude by M_v, then the absolute bolometric magnitude is defined by[1]

$$M_b = M_v - \text{BC} \tag{1-11}$$

We then have

$$M_b = -2.5 \log L + c \tag{1-12}$$

where L is the luminosity and c is a constant that depends upon the units used in expressing the luminosity. The luminosity might, for instance, be expressed in ergs per second, in which case $c = 88.70$. Or we may rewrite the equation, expressing the luminosity in units of the luminosity of the sun,[2]

$$M_b = -2.5 \log \frac{L}{L_\odot} + 4.72 \tag{1-13}$$

It will be apparent from this last equation that the absolute bolometric magnitude of the sun is +4.72. If the luminosity is different from that of the sun, the bolometric magnitude changes on a logarithmic scale in such a way that the most luminous stars have the smallest magnitudes. In fact, if a star is very bright, its absolute bolometric magnitude becomes less than zero. Equation (1-12) may be inverted to read

$$L = L_\odot \times 10^{(4.72-M_b)/2.5} \tag{1-14}$$

where

$$L_\odot = 3.90 \pm 0.04 \times 10^{33} \text{ ergs/sec}$$

This equation will be useful for understanding the language of astronomy. In resumé, M_b is a number determined observationally on a comparative basis. To convert this quantity to an absolute luminosity requires comparison with something whose luminosity is well known, i.e., the sun. Further elaboration of these points can be found in any textbook on astronomy.

Problem 1-2: The naked eye can see only stars brighter than $m_v = 6.0$. Assuming that no bolometric correction is required for the sun, how far away could it be seen by the eye?
Ans: 56 light-years.

[1] The minus sign is arbitrarily inserted in Eq. (1-11) so that the BC will be a positive quantity. This convention is not uniformly followed, however.

[2] C. W. Allen, "Astrophysical Quantities," University of London Press, Ltd., London, 1963.

Problem 1-3: The 200-in. telescope on Mount Palomar has a light-gathering diameter roughly 10^3 times that of the human eye. How far away could the sun be seen with the aid of such a telescope?

NEUTRINO LUMINOSITY

The mean free path of a neutrino in water is 1 billion times greater than the radius of the sun. Because they interact so weakly with matter, neutrinos produced in a stellar interior almost always escape from the star without further interaction. The only commonly discussed exception to this general rule occurs for matter at extremely high density and temperature such as may be found in the imploded core of a supernova. Because of this lack of interaction, neutrino emission provides a heat loss directly from the stellar interior. As such, the neutrino luminosity plays a logically different role in stellar evolution than photon luminosity or mass-loss luminosity. In the latter two types of luminosity, the interior energy must be transported to the stellar surface before being emitted, and the rate of energy transport is proportional to the temperature gradient in the star. The neutrino losses, however, represent a localized depletion of the thermal store of energy; they are emitted directly from the interior point into space without the need of interactive transport from the interior to the surface.

Because of its unique role in stellar evolution, neutrino luminosity is treated separately from the other (surface) luminosities. The energy of those neutrinos emitted in nuclear reactions is simply subtracted from the total energy released by the reaction in determining the effective magnitude of the thermonuclear energy sources. The energy of those neutrinos created by the weak interaction of photons with matter, on the other hand, is regarded as an instantaneous localized heat sink insofar as the structure of the star is concerned. By this treatment the observable properties of neutrino emission on stellar structure are deployed at their source.

Another peculiar aspect of the neutrino luminosity is that it is unobserved. Unlike the photon luminosity, it is not determined from observations as one of the requirements to be explained by stellar models. Quite the contrary; the neutrino luminosity must be calculated from the model of the star in question, and the observable tests of the correctness of the neutrino luminosity are to be found in the effects it has upon the observable evolution of the stars. Certain stages of stellar evolution are found to depend critically upon the inner structure of the star through the detailed physics of the neutrino-production mechanisms. In this way knowledge of the neutrino-emitting processes is leading to sensitive checks of understanding of stellar structure rather than to unobservable ambiguities, as might at first seem the case.

It should be added that very extensive experimental arrangements are being used in attempts to detect cosmic neutrinos, and hopes are high for the ultimate future of neutrino astronomy. But the difficulty in observing discrete sources other than the sun (which by its proximity produces the most easily detectable flux) is very great, and it remains to be seen what can be accomplished when the full power of scientific technology is directed toward this problem.

MASS-LOSS LUMINOSITY

Many different types of stars are observed to lose mass from their surfaces. Some of these, like supernovas, explosively eject a large amount of mass with high energy, characteristically masses comparable to the solar mass being blasted off with velocities measured in thousands of kilometers per second. Such short-duration single events are called *catastrophic mass loss* because they result in a rapid and violent change in the structure of the star. The associated theoretical problems are of such a special type that they require special treatment and will not be considered further here. Another type of star, the planetary nebula, shows evidence of a very large amount of mass loss, but it is still uncertain whether the loss occurred in a catastrophic event or whether the surrounding nebula is a stagnated accumulation of matter continuously ejected over a relatively long time scale. In certain other types, specifically red giants and T Tauri stars, matter can be observed to be flowing away from the star with velocities of tens or hundreds of kilometers per second in amounts as great as $10^{-7}M_\odot$/year.[1] But for the vast majority of stars, the escaping matter cannot be seen at all with ground-based telescopes, because it is probably a nearly collisionless plasma in a high degree of ionization relative to the photosphere from which it came. Ultraviolet astronomy from space may be the key to the mass-loss problem.

As the best known of all stars, the sun provides an example of the type of nearly invisible phenomenon that may be almost universal. The solar photosphere is the opaque boundary of the sun from which nearly all the light originates with a continuous spectrum similar to a blackbody heated to nearly 6000°K. A variety of nonthermal processes originating in the photosphere and chromosphere produce a hot and tenuous overlying layer called the *corona*. The ionized gas of this overlying layer is linked by magnetic fields that contribute to the complicated plasma phenomena observed in the solar chromosphere. Hydromagnetic waves steepen into shock waves in the outer layers. The dissipation and eventual thermalization of these waves provide mechanical heating of the gas and result in the phenomenon of the corona. At a distance of about $2R_\odot$ the plasma has a thermal temperature of $1 - 2 \times 10^6$ °K, more than two orders of magnitude greater than the photospheric temperature. This plasma is not contained by the sun but continually expands into space. Heated from below, the plasma accelerates as it moves outward, eventually moving faster than the Alfvén plasma speed (commonly called *supersonic*), after which point it expands freely into the vacuum. Various space probes have observed this plasma streaming past the earth's orbit at velocities between 300 and 1,000 km/sec with an internal temperature of a few times 10^5 °K and an average density of about 10 amu/cm^3.

Problem 1-4: Assuming at the earth a characteristic velocity of 400 km/sec and density of 10 amu/cm^3 for the solar wind, calculate the rate of mass loss for the sun.
Ans: $0.4 \times 10^{-13}M_\odot$/year.

[1] R. Weymann, *Ann. Rev. Astron. Astrophys.*, **1**:97 (1963) is a good review of mass loss in stars.

Since the solar wind and outer solar corona are nearly invisible (except in the ultraviolet), the analog stellar winds in stars in general may be optically undetectable. Nor can stellar winds be computed for other stars, because the situation is not even well enough understood in the sun to permit computation of the properties of the corona (or even its existence) and the rate of the wind. In spite of this lack of fundamental understanding, the power required to maintain a stellar wind can be expressed in terms of its properties.

In translating each gram of photospheric material, designated by the subscript e ($\Omega_G = -GM/R,\ T = T_e,\ v = 0$), to its condition at large distance ($\Omega_G = 0,\ T = T_\infty,\ v = v_\infty$), energy is required to overcome the gravitational binding, to heat the plasma to its asymptotic temperature, to alter the degree of ionization, and to produce the asymptotic kinetic energy. The corresponding luminosity is

$$L_{\dot{m}} = \left[G\frac{M}{R} + (U_\infty - U_e) + \tfrac{1}{2}v_\infty{}^2 \right] \frac{dM}{dt} \tag{1-15}$$

where U is the internal energy per gram (see Chap. 2) and dM/dt is the rate at which mass crosses an appropriate circumstellar sphere. For ionized matter $U = \frac{3}{2}kT \times$ (number of particles per gram).

Problem 1-5: Consider the example of the sun, again assuming $v_\infty = 400$ km/sec and $T_\infty = 3 \times 10^5$ °K. What are the magnitudes of the three terms in the mass-loss luminosity associated with the rate of solar mass loss computed in the previous problem? (Neglect ionization.)
Ans: 5×10^{27}, 1×10^{26}, 2×10^{27} ergs/sec.

Since the optical luminosity is $L_\odot = 3.9 \times 10^{33}$ ergs/sec, the total mass-loss luminosity of the sun is only about $2 \times 10^{-6} L_\odot$. This is certainly a negligible number for the sun in that it is much less than the uncertainty in the optical luminosity, but it would be unsafe to extend that conclusion to all stars, especially since mass-loss rates as great as 10^7 times the solar rate have been observed.

It should be emphasized here that the phenomenon being discussed is not described by hyperbolic orbits of individual masses with velocities in excess of the stellar escape velocity. At least in the sun the mass is pushed away hydrodynamically, the fluid properties of an otherwise almost collisionless gas being provided by the magnetic field permeating the plasma. Because of its high conductivity, the magnetic-field lines are "frozen into" the plasma, and all parts of the plasma expand collectively. As proof of this it need only be noted that the solar wind accelerates rather slowly for the first few solar radii and achieves its peak velocity at distances far from the solar photosphere. The magnetic-field energy and the energy of associated hydromagnetic waves have been omitted from Eq. (1-15). In principle they should be included, but it seems most likely that at large distances they will be unimportant.

The overall point to be made is that the necessary power for mass loss must be derived from the body of the star. If the surface of the star is not to cool off, this power must be generated in the bulk of the star and transported to the surface, in exact analogy to the situation for the photon luminosity. In this sense it would be appropriate to define a bolometric correction for stellar winds which could be applied to the apparent magnitude to convert it into a total surface power. Furthermore, the effect of this added luminosity on stellar evolution may be appreciable. Because of the extra power demand, the star will evolve faster with a stellar wind than it would with only the photon luminosity to demand power. The actual role of stellar winds in evolution is unknown, however, largely because of the paucity of information on the magnitude of mass-loss rates.

1-2 STELLAR TEMPERATURES

The physical concept of temperature is linked to the concept of thermal equilibrium. A given mechanical system has many possible configurations corresponding to the available distribution of the energies among its subsystems. For instance, the many configurations of a gas can be described in terms of the numbers of particles of given energies moving in given directions. Among all these configurations there is one that is *most probable*, the configuration of thermal equilibrium. That configuration is computable by the techniques of statistical mechanics. For a gas, the most probable configuration depends upon the nature of the gas particles, which for elementary particles, fall into three classes: (1) in the classical limit we may have identical but distinguishable particles, (2) identical but indistinguishable particles of half-integral spin angular momentum, e.g., electron, neutrino, proton, (3) identical but indistinguishable particles of integral spin angular momentum, e.g., photon, He^4 nuclei, π mesons.

If the number of gas particles of energy ϵ is designated by $n(\epsilon)$ and the number of possible particle states of energy ϵ by $g(\epsilon)$, the most probable configurations corresponding to the three types of gases are[1]

$$n(\epsilon) = \frac{g(\epsilon)}{e^{\alpha+\epsilon/kT} + 0} \qquad \text{Maxwell-Boltzmann statistics} \tag{1-16a}$$

$$n(\epsilon) = \frac{g(\epsilon)}{e^{\alpha+\epsilon/kT} + 1} \qquad \text{Fermi-Dirac statistics} \tag{1-16b}$$

$$n(\epsilon) = \frac{g(\epsilon)}{e^{\alpha+\epsilon/kT} - 1} \qquad \text{Einstein-Bose statistics} \tag{1-16c}$$

The parameter α is a real number whose value depends upon the density of particles. If at any given temperature $n(\epsilon)$ is the number density of particles of energy ϵ, then $\int n(\epsilon)\, d\epsilon$ (which is a function of α) must give the total particle density. For small density, α is a large positive number, whereas for very high density, α is a large negative number. These formulas are arrived at by con-

[1] For a physically oriented discussion and derivation of these three types of statistics, the reader is referred to R. B. Leighton, "Principles of Modern Physics," chap. 10, McGraw-Hill Book Company, New York, 1959.

sidering the various ways of arranging a fixed number of particles in individual energy states such that the total energy of the gas is conserved.

A qualification of major importance occurs for photons, which are zero-mass bosons. Since a photon has energy $h\nu$, it takes vanishingly small amounts of energy to create photons of vanishingly small frequency. This has the effect that only the total energy of the photon gas, not the number of photons, must be conserved in establishing the most probable configuration for a photon gas. For this reason, α does not appear in the photon distribution (or, alternatively, $\alpha = 0$). Thus

$$n(\epsilon) = \frac{g(\epsilon)}{e^{\epsilon/kT} - 1} \qquad \text{for photons} \tag{1-16c'}$$

In each type of gas, the number of particles of energy ϵ is proportional to the number of distinct particle states of energy ϵ. Simple physical consequences of the distribution laws can best be seen by considering the ratio $n(\epsilon)/g(\epsilon)$, which is the intrinsic probability that a state of energy ϵ is occupied. This probability, sometimes called the *occupation index*, will be designated by $P(\epsilon)$. Then, respectively,

$$P(\epsilon) = e^{-\alpha}e^{-\epsilon/kT} \qquad \text{Maxwell-Boltzmann} \tag{1-17a}$$

$$P(\epsilon) = \frac{1}{e^{\alpha+\epsilon/kT} + 1} \qquad \text{Fermi-Dirac} \tag{1-17b}$$

$$P(\epsilon) = \frac{1}{e^{\epsilon/kT} - 1} \qquad \text{photons} \tag{1-17c}$$

The behavior of these occupation indices (Fig. 1-3) is interesting. The Maxwell-Boltzmann distribution is a pure exponential in the energy for all temperatures and all values of α. The associated probability $P(\epsilon)$ that any state be occupied depends on both temperature and density and may, therefore, have any possible value for given (ϵ,T) depending on the value of the parameter α. Actually, however, Maxwell-Boltzmann statistics are valid for real particles (which are really indistinguishable) only when α is a large positive number, i.e., when the occupation is much smaller than unity.

The occupation index of the Fermi-Dirac system, on the other hand, never exceeds unity. This limit is an expression of the Pauli exclusion principle for fermions, which says that no more than one identical particle of half-integral spin may be in the same state. This limitation was built into the statistical argument that led to the distribution function. It is apparent, however, that if the temperature and density fall into a domain for which α is a large positive number, the Fermi-Dirac statistics reduce to Maxwell-Boltzmann statistics. The major practical application of the Fermi-Dirac statistics in stars lies in computing the pressure of an electron gas, which the reader will find in Chap. 2.

For a gas of Einstein-Bose particles there is a tendency for large occupation indices to occur at low energies, depending on the size of α. For the special case

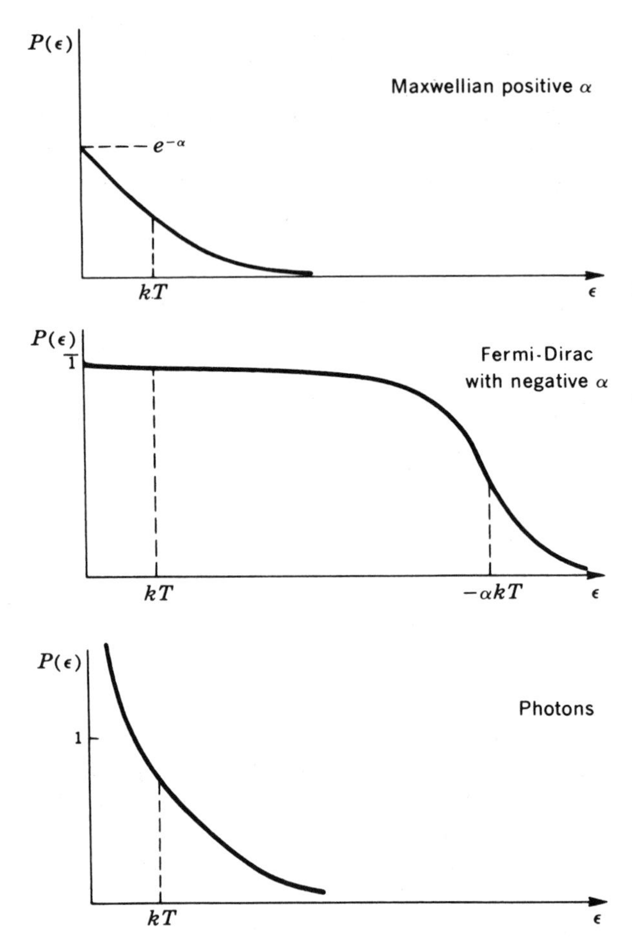

Fig. 1-3 Schematic occupation index. The Maxwell-Boltzmann distribution, which is a pure exponential curve at all temperatures and all values of ϵ, is valid only if $\alpha \gg 0$. The Fermi-Dirac distribution has, because of the Pauli exclusion principle, an upper bound of unity on the occupation index. This upper bound is approached at low energy and low temperature (or high density). The Einstein-Bose distribution for photons has a large occupation index for $\epsilon \ll kT$ and an exponentially decreasing index for $\epsilon \gg kT$

of a photon gas ($\alpha = 0$), $P(\epsilon)$ increases without bound at zero energy. This effect also has a quantum-mechanical analog, viz., that the mutual wave functions of bosons must be symmetric in the pair of particles. This symmetry has the physical effect of causing bosons to cluster into the same state. One would say that a boson prefers to be in a state already populated by other bosons, an effect that is in some sense the opposite of the exclusion principle for fermions. For the specific case of a photon gas of prescribed total energy, the distribution therefore favors a very large number of low-energy photons. For energies much greater than kT, on the other hand, the occupation index changes asymptotically to an exponential in the energy.

The use of these statistical-equilibrium distributions in the calculation of the properties of a gas requires counting the states available to free particles. The number of gas particles in each free state of given energy is a product of the occupation index and the number of free states having the given energy. In the continuum of states, of course, all energies are possible, because the energy becomes a continuous variable for free particles, so that it is common to talk of the *density of free states* $g(\epsilon)$, defined such that $g(\epsilon)\,d\epsilon$ is the number of states per unit volume of energy ϵ in the range $d\epsilon$. It is somewhat easier, moreover, to discuss the density of states in momentum space rather than in energy space. The analogous density of states $g(\mathbf{p})$ is defined such that $g(\mathbf{p})\,dp_x\,dp_y\,dp_z$ is the number of momentum states per unit volume with momentum $\mathbf{p}$ in the range $dp_x\,dp_y\,dp_z$. It is useful to think of this as the number of states per unit volume for which the set of momentum vectors terminates within a differential momentum volume $dp_x\,dp_y\,dp_z$ at the end of $\mathbf{p}$, as illustrated in Fig. 1-4. It is somewhat difficult to grasp the meaning of the density of momentum states for the continuum of free particles because any value of the momentum is possible for a free particle. Yet the number of *independent* free-particle states per unit volume is a finite quantity which may be calculated most easily in terms of a convenient representation of continuum states. The physical applications of the density of continuum states are almost countless, and a sketch of their enumeration follows.

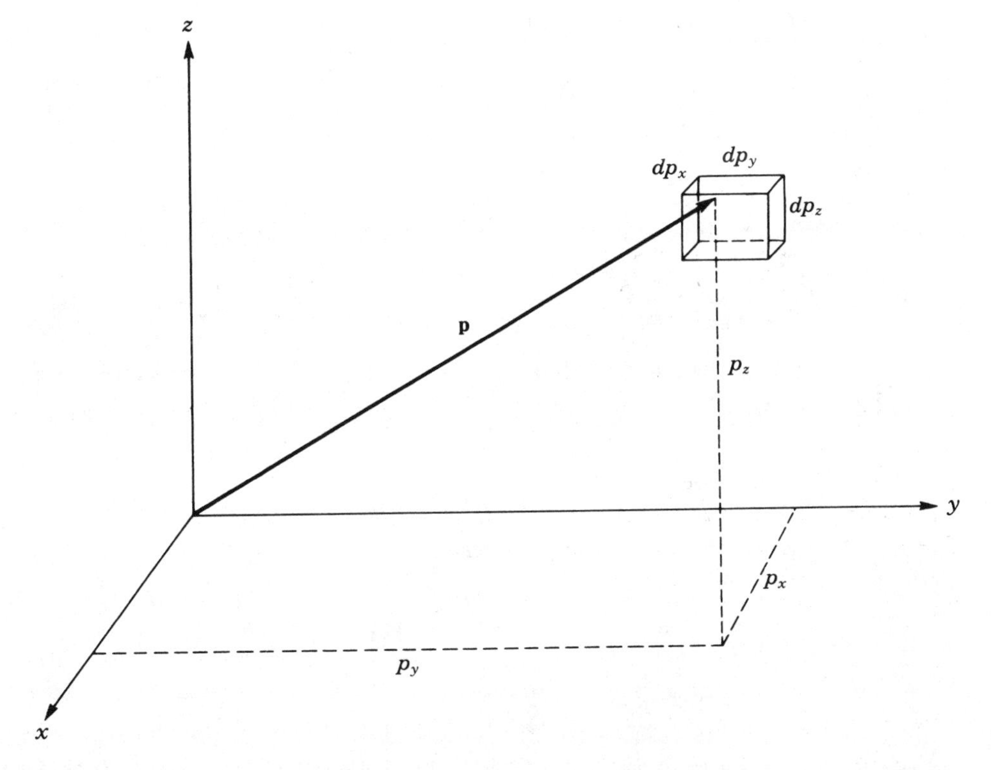

Fig. 1-4 A differential volume $dp_x\,dp_y\,dp_z$ about the point $\mathbf{p}$ in momentum space. The density of states $g(\mathbf{p})$ is defined such that $g(\mathbf{p})\,dp_x\,dp_y\,dp_z$ is the number of denumerable momentum states per unit volume having momentum $\mathbf{p}$ in the range $dp_x\,dp_y\,dp_z$.

The conceptual difficulty stems from the propensity to think classically—that the position and momentum of the particles are simultaneously definable. Yet from elementary wave mechanics we know that particles are describable by wave packets possessing characteristic spreads Δx and Δp_x in the conjugate coordinates such that the product of the uncertainties is greater than $h/2\pi$—the so-called *uncertainty principle* of Heisenberg. If a particle is localized in space, it posesses a spread in the expectation values of its momenta determinable from the Fourier transform of the spatial wave function.[1] The actual wave function of the free particle is expressible as a linear combination (or integral) over the plane-wave momentum eigenstates. Since the momentum operator is $\mathbf{p}_{\text{op}} = h/2\pi i\boldsymbol{\nabla}$, the state of a particle with a specific expectation value for, say, the x component p_x of the momentum is described by the factor $\psi \propto \exp\,[(2\pi i/h)p_x x]$. Imagine space to be divided into an infinite lattice of rectangular volumes of linear dimensions a, b, and c. It can be shown that the number of distinguishable free-particle states can be counted by demanding that the wave functions be periodic in this lattice (the so-called *periodic boundary condition*).[2] Then, by the Fourier theorem, the wave functions can all be expanded in terms of the momentum eigenfunctions

$$\psi \propto \exp\left[2\pi i\left(\frac{lx}{a} + \frac{my}{b} + \frac{nz}{c}\right)\right] \tag{1-18}$$

where l, m, and n are integers. The x component of the momentum is $p_x = hl/a$. In the next momentum eigenstate of the complete set, the value of p_x must be increased by the discrete step $\delta p_x = h/a$. Evidently the number of otherwise identical states with p_x lying in the interval Δp_x near p_x is given by the number of discrete momentum steps that can be fitted into the interval Δp_x:

$$\text{Number in } \Delta p_x = \frac{\Delta p_x}{\delta p_x} = \frac{a}{h}\,\Delta p_x \tag{1-19}$$

The total number of states in the box lying in the differential momentum volume $\Delta p_x\,\Delta p_y\,\Delta p_z$ about any value of p is then

$$\begin{aligned}\Delta N &= \left(\frac{a}{h}\,\Delta p_x\right)\left(\frac{b}{h}\,\Delta p_y\right)\left(\frac{c}{h}\,\Delta p_z\right)\\ &= \frac{1}{h^3}\,\Delta p_x\,\Delta p_y\,\Delta p_z\,(abc)\end{aligned} \tag{1-20}$$

Since abc is the volume of the box in which the states are counted and $\Delta p_x\,\Delta p_y\,\Delta p_z$ is the volume of momentum space in which the states lie, the factor h^3 represents the volume of a continuum state in six-dimensional phase space. The number of states *per unit volume* with momentum $\mathbf{p}$ in the range $dp_x\,dp_y\,dp_z$

[1] E. Merzbacher, "Quantum Mechanics," chap. 2, John Wiley & Sons, Inc., New York, 1961.

[2] The reader may find informative A. Messiah, "Quantum Mechanics," vol. 1, sec. 11, p. 190, John Wiley & Sons, Inc., New York, 1961.

is then

$$g(\mathbf{p})\, dp_x\, dp_y\, dp_z = \frac{1}{h^3}\, dp_x\, dp_y\, dp_z \tag{1-21}$$

This technique of counting continuum states has employed the periodic boundary condition. One can also calculate the density of energy states by considering particles confined to the inside of a box by an impenetrable wall.[1]

The result expresses only the spatial density of states. The particles may possess internal degrees of freedom that increase the total density of states. For instance, electrons are spin $s = \frac{1}{2}$ particle. The spatial density of states exists for each of the two projections of the spin along a chosen z axis. In the same vein, photons possess two additional degrees of freedom corresponding to the two possible modes of polarization. For both photons and electrons, therefore, the total density of states is twice as great as Eq. (1-21). The introduction of the box was, of course, a crutch, and its dimensions do not appear in the final answer. The density of states is independent of these dimensions.

The fact that there should be a volume h^3 associated with each distinguishable state in six-dimensional phase space can be approximately understood by the following argument. For states to be resolvable, they must be distinguishable in the face of the uncertainty principle, which demands that the product of the dynamically conjugate uncertainties be as great as Planck's constant, viz., $\Delta p_i\, \Delta x_i \approx h$. If we consider only particles of some specific momentum $\mathbf{p}$ in the range $dp_x\, dp_y\, dp_z$, the associated position uncertainties are

$$\Delta x \approx \frac{h}{dp_x} \qquad \Delta y \approx \frac{h}{dp_y} \qquad \Delta z \approx \frac{h}{dp_z}$$

The particles must be spread over a volume of uncertainty

$$V_{\text{par}} = \Delta x\, \Delta y\, \Delta z \approx \frac{h^3}{dp_x\, dp_y\, dp_z} \tag{1-22}$$

The demand for resolvability of particle states may be crudely interpreted as the demand that there be no more than one state within each volume of uncertainty. It follows that the number of resolvable states per unit volume is the reciprocal of the volume of each particle state. That is,

$$g(\mathbf{p})\, dp_x\, dp_y\, dp_z \approx \frac{1}{V_{\text{par}}} = \frac{1}{h^3}\, dp_x\, dp_y\, dp_z$$

which reexpresses Eq. (1-21). This simplified picture relates the density of continuum states to $1/h^3$ in a simple and intuitive (though not precisely correct) manner.

The statistical distributions of Eq. (1-21) apply to a gas in a state of thermodynamic equilibrium. In such a gas the particle momenta are isotropic. For that reason one usually needs to know the density of states with $|\mathbf{p}| = p$ in

[1] See Leighton, *op. cit.*, p. 159.

the range dp rather than the density of states with momentum $\mathbf{p}$ in the range $dp_x\, dp_y\, dp_z$. The volume of momentum space for which the radius vector $\mathbf{p}$ has constant magnitude p in the range dp is simply the volume of a spherical shell, $4\pi p^2\, dp$. By taking into account the doubling of states due to the two degrees of polarization for electrons and photons, the density of states may be written

$$g(p)\, dp = \frac{2}{h^3} 4\pi p^2\, dp \tag{1-23}$$

for photons and electrons.

Problem 1-6: Show that the density of free electrons of energy E in the range dE is, if the electrons are nonrelativistic,

$$n(E)\, dE = \frac{8\pi}{h^3} (2m_e{}^3)^{\frac{1}{2}} \frac{E^{\frac{1}{2}}\, dE}{\exp\,(\alpha + E/kT) + 1} \tag{1-24}$$

The size of the parameter α is determined from the condition that $\int n(E)\, dE = n_e$, the total density of free electrons.

At the low density of the outer layers of stars, the parameter α is such a large positive number that the unity term in the denominator of Eq. (1-24) becomes irrelevant. This situation reflects the unimportance of the Fermi-Dirac nature of the electrons at low density, where the density of electrons is given to high accuracy by Maxwell-Boltzmann statistics,

$$n_e(p)\, dp = Ae^{-p^2/2mkT} 4\pi p^2\, dp$$

where the normalization constant A is determined from the condition

$$\int n_e(p)\, dp = n_e$$

Problem 1-7: Show by performing the above integral that

$$n_e(p)\, dp = \frac{n_e 4\pi p^2\, dp}{(2\pi m_e kT)^{\frac{3}{2}}} \exp - \frac{p^2}{2m_e kT} \tag{1-25}$$

This equation expresses the Maxwell-Boltzmann distribution of electron momenta in thermal equilibrium.

Problem 1-8: Using the Einstein-Bose distribution for photons, the fact that $\alpha = 0$ for a photon gas, and the fact that the photon momentum is $p_\nu = h\nu/c$, show that the *energy density* of photons of frequency ν in the range $d\nu$ in thermal equilibrium is

$$u(\nu)\, d\nu = \frac{8\pi h\nu^3}{c^3} \frac{d\nu}{e^{h\nu/kT} - 1} \tag{1-26}$$

Problem 1-9: Show that the total energy density of the photon field in thermodynamic equilibrium is

$$u = \int u(\nu)\, d\nu = \frac{8\pi^5 k^4}{15c^3h^3} T^4 = aT^4 \tag{1-27}$$

The constant a is called the *Stefan-Boltzmann constant* and is equal to $a = 7.565 \times 10^{-15}$ erg cm^{-3} deg^{-4}.

Equation (1-27) shows that the energy density of photons in an equilibrium situation is a function only of the temperature, a result that will be demonstrated later from classical thermodynamic arguments. The immediate point of this entire discussion is that *the temperature of a gas is defined by matching the observed distribution of particle states to the appropriate equilibrium distribution for the same type of particles.* If the observed distribution functions do not match the equilibrium distributions, the gas is not in thermodynamic equilibrium, and a concept of temperature must be employed with caution. The single greatest simplification of the physics of stellar interiors results from the fact that the stellar interior is very nearly in the state of thermodynamic equilibrium, so close in fact that for most problems the assumption of thermodynamic equilibrium yields answers with nearly vanishing errors.

One method of assigning a temperature to a stellar photosphere is to match the energy spectrum of the photons leaving the star to those leaving a surface in thermodynamic equilibrium. Such a surface is called a blackbody, and the radiant flux leaving a blackbody is equal to that exiting from a hole cut in the surface of a container containing interior photons in thermal equilibrium. The energy flux leaving such a surface has the same spectrum shape as the internal energy density. The specific result of the calculation is that the power radiated per unit area per unit wavelength interval is

$$I_\lambda = \frac{2\pi c^2 h}{\lambda^5} \frac{1}{e^{ch/\lambda kT} - 1} \tag{1-28}$$

whereas the total power radiated per unit area is

$$\int_0^\infty I_\lambda \, d\lambda = \sigma T^4 \tag{1-29}$$

where the constant σ is equal to $ac/4$,

$$\sigma = 5.67 \times 10^{-5} \text{ erg cm}^{-2} \text{ sec}^{-1} \text{ deg}^{-4}$$

The spectrum represented by this formula is a smooth one, having a single maximum.

Problem 1-10: Show that Eq. (1-28) is the same spectral shape as Eq. (1-26). Note that Eq. (1-28) is displayed per wavelength interval.

Statistical mechanics provides astronomy with several techniques for ascribing a temperature to stellar photospheres. In each case the temperature is defined to be such that the value of some observed physical property most nearly resembles the value it would have in thermodynamic equilibrium. There are four simple concepts of stellar temperatures in common use: the effective temperature, the color temperature, the excitation temperature, and the ionization temperature. These temperatures are not observed to be equal to each other in stellar photospheres, the reason being that a stellar photosphere is not in thermodynamic equilibrium. The stellar photosphere is near the point where the pressure and density drop rapidly to near zero. Roughly speaking, there is approximately one mean free path of matter for an optical photon above the photosphere. But the combined facts of variation of optical depth with optical frequency, the temperature and density gradients, and the large net outward flux of radiation result in a disparity between the various surface temperatures in stars. This disparity is resolved by calculations of model atmospheres for stars which can yield simultaneously all the observed spectral features without recourse to definitions of special temperatures.

THE EFFECTIVE STELLAR TEMPERATURE

The effective temperature T_e of a star is defined as the temperature of a blackbody having the same radiated power per unit area. The stellar luminosity is related to the radius and effective temperature by

$$L = 4\pi R^2 \sigma T_e^4 \tag{1-30}$$

The direct determination of T_e is not generally possible because it demands knowledge of the surface area of a star, and stellar radii are not generally measurable because of the small angles they subtend at the earth. But for at least one important star, the sun, the effective surface temperature can be calculated from the surface brightness.

Problem 1-11: The radius of the sun is 16 min of arc as seen from the earth. The radiant flux at the top of the earth's atmosphere is 1.388×10^6 ergs cm^{-2} sec^{-1}. Calculate the effective temperature of the sun using only the data given. The quantity 1.388×10^6 ergs cm^{-2} sec^{-1} is called the *solar constant*. From its value and the distance of the sun the solar luminosity is calculated.

Because stellar radii are not measurable with high accuracy, it has been necessary to use some other techniques for determining effective temperatures. This technique is that of the model atmosphere, which relates the radiated power to other photospheric temperatures which are measurable. Equation (1-30) is then used to determine the stellar radius.

THE COLOR TEMPERATURE

This technique also matches the radiated photon power to a blackbody, but the matching is to the shape of the continuous spectrum rather than the integrated power; i.e., the observed continuous spectrum is matched to Eq. (1-28) rather than to Eq. (1-29).

Problem 1-12: Calculate the position of the maximum of the blackbody spectrum in the form $\lambda_{\max} T$ = const. The wavelength of maximum intensity in the solar spectrum is 5000 Å. Assuming the sun radiates as a blackbody, compute its surface temperature. Compare with the previous problem.

Ans: $\lambda_m T = 0.290$ cm °K.

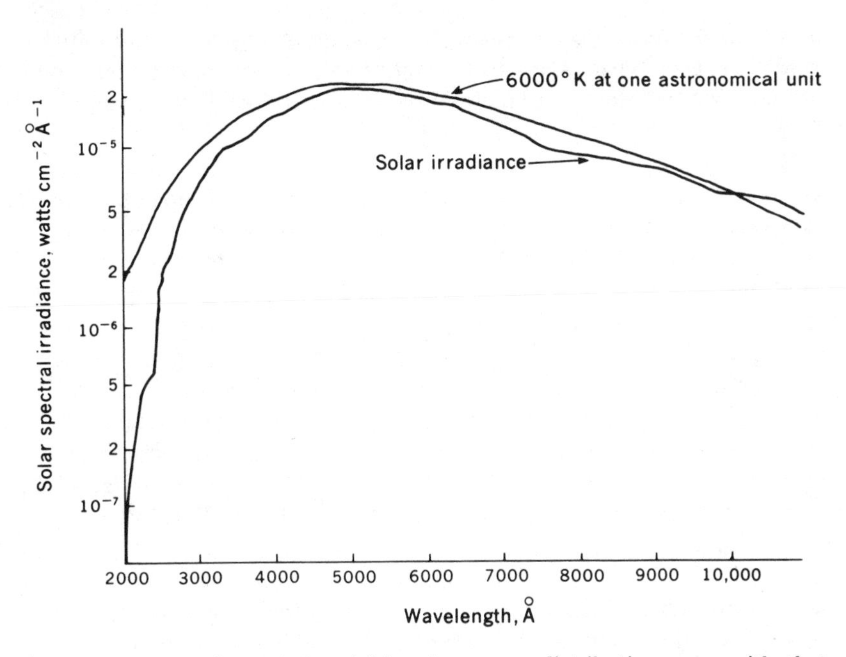

Fig. 1-5 Comparison of the visible solar energy-distribution curve with that from a blackbody at 6000°K. The overall resemblance is good, although the sun is quite deficient in the ultraviolet. [*D. P. Le Galley and A. Rosen (eds.), "Space Physics," p.* 111, *John Wiley & Sons, Inc., New York,* 1964.]

Superimposed upon the continuous spectrum are marked absorption features, prominent among which are the famous Fraunhofer lines for resonance absorption. Also observable are sizable discontinuities in the level of the continuous energy spectrum as the energy of the radiation crosses absorption thresholds. For instance, a drop in the intensity of the continuous spectrum occurs for frequencies higher than the Balmer limit of the hydrogen atom, since frequencies higher than this limit are capable of ionizing that fraction of hydrogen atoms lying in the first excited state. The composition of the stellar atmosphere can also influence the shape of the continuous spectrum of stars. For instance, ionized metals have numerous resonance lines in the ultraviolet. Thus, a star rich in metals may be expected to show less ultraviolet radiation than a metal-deficient star of equal temperature. These continuous absorption features distort the blackbody nature of the spectrum. However, with the aid of model atmospheres, it is still possible to calculate a surface temperature from the shape of the observed spectrum.[1] Figure 1-5 shows the observed continuous spectrum of the sun and a comparison blackbody spectrum for a surface temperature of 6000°K.

Absorption lines are not the only features which destroy the blackbody nature

[1] J. L. Greenstein (ed.), "Stellar Atmospheres," The University of Chicago Press, Chicago, 1960, contains several excellent review articles on the problems of stellar atmospheres.

of the continuum. The opacity of the stellar gases is a function of frequency; therefore, the depth of sight into a stellar surface depends upon the frequency of observation. Since the temperature of a star is a decreasing function of its radius, light of different frequencies will penetrate to different temperatures, and there will be an associated effect upon the shape of the continuous spectrum. An atmosphere is not in a state of thermodynamic equilibrium. Strictly speaking, thermodynamic equilibrium does not admit the possibility of temperature gradients. There may even be extremely hot, but extremely rare, gases above the surface of the star, such as the solar corona. The density of the corona is so low that it has almost no effect at all on either absorption or emission in the total continuous spectrum. Yet its temperature is some millions of degrees, a factor of several hundred higher than that of the surface of the star from which the continuous spectrum originates. The region of a star from which the bulk of the continuous spectrum originates is called its *photosphere*. That is where the approximation to the visible blackbody is. The mere existence of a solar corona, however, clearly indicates *some* violation of local thermodynamic equilibrium. These difficulties are, nonetheless, surmountable, and an average temperature may be calculated for the photosphere of a star from its observed continuous spectrum. Often it is not even necessary to measure a spectrum in detail. Sufficient data may be taken by broadband color photometry, using filters of various colors and recording the energy transmitted by the filter. Measurements of a star's apparent brightness in two or three colors are often ample to ascertain an approximate surface temperature of the star. The temperature of a true blackbody is completely determined by the intensity at two distinct wavelengths.

It is common, therefore, to speak of the color index of a star. This quantity is defined by measuring the luminosity of the star with a blue-sensitive photographic plate and with a yellow-sensitive plate and a yellow filter. The former quantity is called the *blue* or *photographic luminosity*, whereas the latter is called the *photovisual*, or simply *visual, luminosity*. The associated magnitudes are called B and V, the blue and the visual magnitude, respectively.[1] As stars become hotter, they get bluer; hence the magnitude measured in the blue compared with the magnitude in the visual spectrum gives an indication of the surface temperature of the star. The color index itself is defined as the difference in these magnitudes, $B - V$. In keeping with the inverse logarithmic relationship of magnitude to luminosity, the color index $B - V$ is greatest for cool stars and smallest (in fact, even negative) for very hot stars. By properly defining the way in which B and V are to be measured, astronomers are able to make a one-to-one correlation between the color index $B - V$ and the color temperature of the star. Figure 1-6 shows the sensitivity functions for the filters and detectors that define the B and V measurements.

[1] Technically, there is a slight difference between the older photographic and photovisual magnitudes M_{pg} and M_{pv} and the new blue and visual magnitudes B and V. These differences stem from different color classification systems for stars, and the reader is referred to Allen, *op. cit.*, p. 194, for technical definitions.

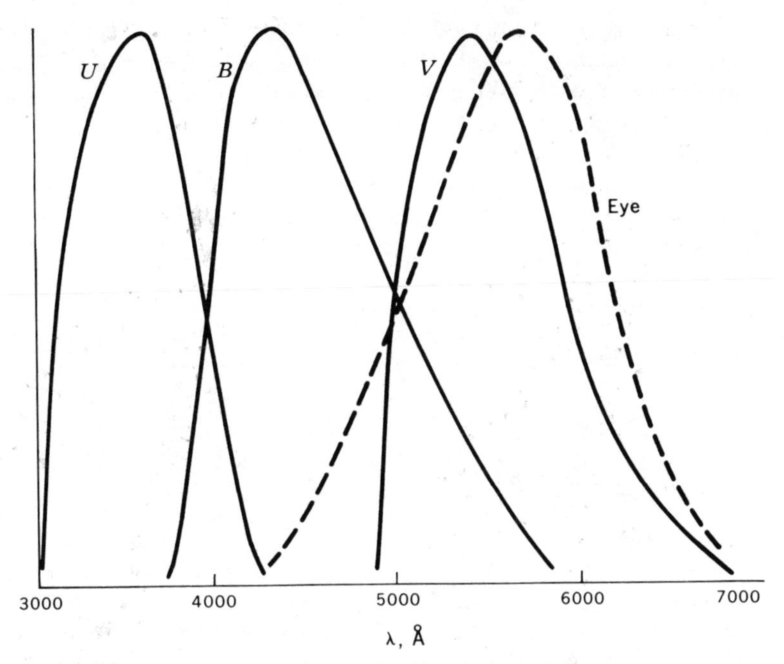

Fig. 1-6 A schematic comparison of the spectral responses of the U, B, and V detector systems used in multicolor photometry. The response of the human eye is shown for comparison.

THE EXCITATION TEMPERATURE

This method uses the fact that one given kind of atom or ion may have visible absorption lines originating from different initial states. Again, to take the simplest case, hydrogen certainly will absorb Lyman alpha (although that feature is not visible through the earth's atmosphere). There is also a certain fraction of hydrogen atoms which, as a result of thermodynamic equilibrium, are already in the first excited state of hydrogen. This fraction of the hydrogen atoms may absorb all the lines of the Balmer series of the spectrum of hydrogen. One may find, for instance, H_α in absorption. From the intensities of these absorption lines and from the intrinsic transition strengths of the atomic transitions involved, it is possible to calculate the ratios of the populations of the various excited states of atoms or ions. Let $\epsilon_{r,k}$ denote the excitation energy of the kth state of an r-times-ionized species of some nucleus; then Boltzmann's formula yields the population ratio of two states of the ion,

$$\frac{n_{r,k}}{n_{r,i}} = \frac{g_{r,k}}{g_{r,i}} \exp - \frac{\epsilon_{r,k} - \epsilon_{r,i}}{kT} \tag{1-31}$$

In this formula $g_{r,k}$ represents the statistical weight of the state (r,k). It is equal to the degeneracy of that particular level of the ion; i.e., it is the number of states with differing quantum numbers that have the same energy. When the Zeeman splitting of magnetic substates is neglected, all projections of the angular momentum are degenerate in energy, and we have

$$g_{r,k} = 2J_{r,k} + 1 \tag{1-32}$$

where $J_{r,k}$ is the angular momentum of the atomic state.[1]

Problem 1-13: It turns out by the nature of a pure $1/r$ potential that the hydrogen atom has additional degeneracy, such that states lie in degenerate groups characterized by a principal quantum number n.

$$E_n = -\frac{13.6}{n^2} \text{ ev} \qquad g_{H,n} = 2n^2$$

The visible spectrum of hydrogen is dominated by a red line called H_α, which corresponds to an atomic transition between the $n = 3$ and $n = 2$ levels. Calculate the frequency of H_α. Assuming that the temperature of the sun's surface is about 5700°K, what is the ratio of the population of the first excited state of hydrogen to the ground state of hydrogen?

Problem 1-14: Suppose that an ion has two effective electrons in the configuration $(3p)(4p)$. Give the spectroscopic designation of the states that can be formed from that configuration. Draw a schematic energy-level diagram showing the order you expect for those terms. If the configuration were $(3p)^2$ rather than $(3p)(4p)$, which of those terms would be forbidden by the exclusion principle?

Ans: See Leighton, *op. cit.*, p. 264.

[1] An appropriate introduction to the quantum mechanics of atomic states and their spectroscopic designations may be found in Leighton, *op. cit.*

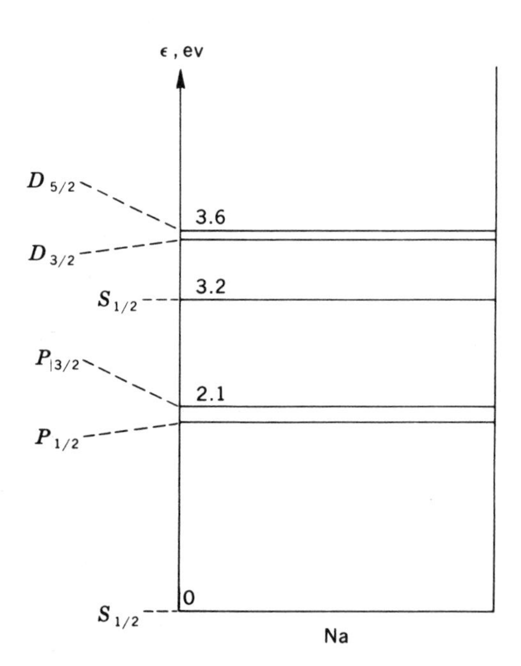

Fig. 1-7 An approximate term diagram for the electronic configuration of the element sodium. The excitation energy above the energy of the ground state is labeled by the quantum numbers of the configuration. The letter designates the orbital angular momentum of the electrons (in this case of a single-valence electron), and the subscript designates the total angular momentum of the states.

Problem 1-15: An approximate term diagram for sodium is shown in Fig. 1-7. (*a*) At what temperature are the combined populations of the two 3.6-ev states and the 3.2-ev state equal? (*b*) At that temperature, what fraction of all sodium excited states are in the $P_{\frac{1}{2}}$ state? (Ignore higher-lying states not shown.)

Astrophysicists make calculations of this type often. An excitation temperature of the stellar atmosphere may be determined in this way from the ratios of populations of two different excited states of the same atom. This procedure can often be followed for many different kinds of atoms as a check on inner consistency. Caution must be exercised in comparing excitation temperatures determined from differing species, however. Molecular transitions may indicate somewhat cooler temperatures than atomic-hydrogen transitions, for instance, if molecules exist only in relatively cool parts of the photosphere whereas appreciable population of the excited states of hydrogen occurs in deeper regions. To achieve complete consistency one must reproduce the complete stellar spectrum from a model atmosphere for the star.

The foregoing application of Boltzmann's formula is a simple example of the application of statistical mechanics to inverse reactions in thermal equilibrium. The population ratio of the two final states in an inverse pair of reactions is proportional to the ratio of the statistical weights of the two states times a Boltzmann factor favoring the state of lower energy. The Boltzmann factor is a negative exponential of the number of units of thermal energy kT that must be concentrated into one process to produce the final state of higher energy from the state of lower energy. In the case just considered, the inverse processes are just the photoexcitation of a state and its radiative deexcitation.

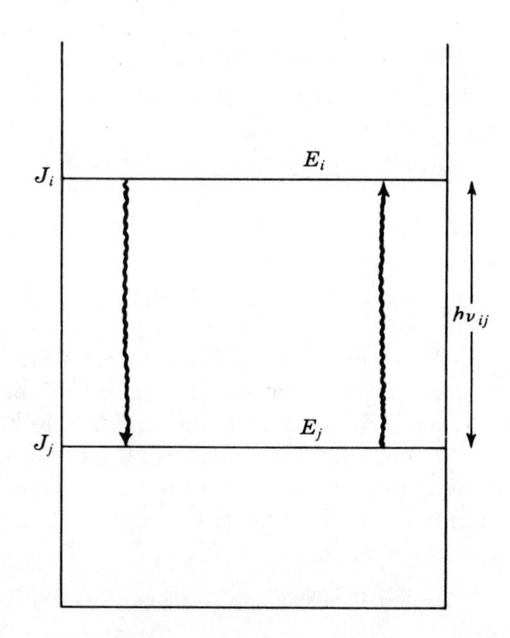

Fig. 1-8 Electromagnetic transitions between two states of an atom or ion may be accomplished by the emission or absorption of photons of frequency $h\nu_{ij} = E_i - E_j$. The inverse lifetimes of the states against the three types of transitions are given by the Einstein coefficients.

That the Boltzmann formula depends on the assumption of thermal equilibrium can be seen from the detailed balance of these inverse processes. Let ν_{ij} be the frequency of a photon emitted during a spontaneous transition between states i and j with $E_i - E_j = h\nu_{ij}$ (Fig. 1-8), and let the atoms be bathed by isotropic radiation with spectral energy density $u(\nu)$. The presence of the photon gas causes two other first-order transitions between the two states: (1) by absorbing a photon of energy $h\nu_{ij}$ an atom in state j may make an upward transition to state i, and (2) by interacting with a photon of energy $h\nu_{ij}$ an atom in state i may be *induced* to emit another photon of energy $h\nu_{ij}$ in the same direction as the stimulating photon. The rates of these photon-induced transitions must be proportional to $u(\nu_{ij})$. There are three transition probabilities, called the *Einstein coefficients*, which describe radiative transitions between states i and j: (1) A_{ij} is the probability per atomic state i per unit time of spontaneous radiative decay to state j; (2) $B_{ij}u(\nu_{ij})$ is the probability per atomic state i per unit time of induced radiative transitions from state i to state j; and (3) $B_{ji}u(\nu_{ij})$ is the probability per atomic state j per unit time of radiative absorption to state i. There is, furthermore, a definite quantum-mechanical relationship between these coefficients,[1] viz.,

$$B_{ij} = \frac{c^3}{8\pi h\nu_{ij}^3} A_{ij}$$

$$\frac{B_{ij}}{B_{ji}} = \frac{2J_j + 1}{2J_i + 1} = \frac{g_j}{g_i} \tag{1-33}$$

In thermodynamic equilibrium it is required that the number of transitions per unit time from i to j equal the number from j to i. Quantitatively, this requirement demands that

$$n_i[A_{ij} + B_{ij}u(\nu_{ij})] = n_j[B_{ji}u(\nu_{ij})] \tag{1-34}$$

Solving this expression for $u(\nu_{ij})$ yields

$$u(\nu_{ij}) = \frac{A_{ij}/B_{ij}}{N_jB_{ji}/N_iB_{ij} - 1} \tag{1-35}$$

Equation (1-35) can be matched with Eq. (1-26) only if

$$\frac{N_i}{N_j} = \frac{B_{ji}}{B_{ij}} \exp - \frac{h\nu_{ij}}{kT} = \frac{g_i}{g_j} \exp - \frac{h\nu_{ij}}{kT} \tag{1-36}$$

In this way the Boltzmann formula follows from the energy distribution of equilibrium radiation and the quantum-mechanical principles of inverse radiative reactions. Of course, the equation follows immediately from statistical thermodynamics and encompasses all pairs of inverse reactions; i.e., the excitation and deexcitation may involve inelastic-particle collisions as well as the radiation field, but the result remains the same as long as the system is in true thermodynamic equilibrium.

[1] Further discussion may be found in Sec. 3.3.

THE IONIZATION TEMPERATURE

A similar principle leads to a third method for determining the photospheric temperatures. In this method, the inverse reactions involved are the ionization of an atom or ion and the inverse recombination process. By comparing the strength of absorption lines of *two different states of ionization* of the same atom, it is often possible to determine the relative number densities of the two stages of ionization of that atom. For example, the H and K lines of singly ionized calcium often appear in the same spectrum with a line of neutral calcium at 4226 Å. The relative strength of these lines can be made to yield the relative numbers of singly ionized calcium atoms and neutral calcium atoms in a stellar photosphere. In such cases, that population ratio can also be used to determine a stellar temperature. The corresponding ionization equation was first derived by Saha,[1] and the equation bears his name today.

If we let n_{r+1} be the number density of the $(r + 1)$-times-ionized species of a certain atom, n_r be the number density of the r-times-ionized species of the same atom, and n_e be the number density of free electrons, the Saha equation demands that

$$\frac{n_{r+1}n_e}{n_r} = \frac{G_{r+1}g_e}{G_r}\frac{(2\pi m_e kT)^{\frac{3}{2}}}{h^3}e^{-\chi_r/kT} \tag{1-37}$$

In this equation G_{r+1}, G_r, and g_e are the so-called partition functions of the $(r + 1)$-times-ionized species, the r-times-ionized species, and the electron, respectively. The partition function of an atom is similar to, but slightly more complicated than, the statistical weight of a quantum-mechanical state. It is, in fact, the sum of the statistical weights of all bound states of the atom, each one weighted by the Boltzmann factor indicating the relative population of that level in the structure of that atom. Specifically, we have

$$G_r(T) = g_{r,0} + g_{r,1}e^{-\epsilon_{r,1}/kT} + g_{r,2}e^{-\epsilon_{r,2}/kT} + \cdots \tag{1-38}$$

where $g_{r,k}$ is again the statistical weight of the kth level of the r-times-ionized atom. The energies appearing in the Boltzmann exponentials are, of course, the energies of the excited states of the species r relative to the ground state of the species r.

The partition function possesses a physical interpretation related to the question: What fraction of the number density of an atom resides in a specific state of the atom? The ratio of the density of the species in that state to the density in any other *individual* state is given by the Boltzmann formula in Eq. (1-36). If the Boltzmann formula is summed to obtain the ratio of the density of ions in the state in question to the total density of the ion, it follows immediately that

$$\frac{n_{r,k}}{n_r} = \frac{g_{r,k}\exp(-\epsilon_{r,k}/kT)}{G_r} \tag{1-39}$$

[1] M. N. Saha, *Phil. Mag.*, **40**:472 (1920).

where G_r is the partition function of species r as given by Eq. (1-38). It will be obvious that $\sum_k \frac{n_{r,k}}{n_r} = 1$, a fact consistent with the definitions of $n_{r,k}$ and n_r.

A conceptual difficulty associated with the application of the partition function lies in the fact that it appears to be a sum of an infinite number of finite terms. All isolated atoms have an infinite number of excited states with much the same characteristics as the infinite number of excited states of the hydrogen atom. Strictly speaking, then, the partition function is infinite, and the fraction of atoms in any one excited state approaches zero. The solution to this apparent dilemma lies in the existence of a physical cutoff to the number of excited states that any one atom may have. Just as in the case of the hydrogen atom, the radius of the electron orbit for highly excited electrons is proportional to the principal quantum number of the state. In a real gas a point is reached at which the size of the atoms in highly excited states becomes comparable to the atomic separation. The electrons can then no longer be associated with any one atom but must pass into a band of continuous states, a situation somewhat analogous to metals. This physical limit to the number of bound states makes the partition function finite. In most applications, in fact, the numerical value of the partition function can be estimated by taking only a few terms (in many cases only one term) of the sum. In the stellar interior this phenomenon becomes quite important. Bound electrons greatly increase the opacity of matter to the thermal radiation, and it becomes necessary to consider the interactions in the gas in order to realistically calculate the number of bound electrons. This problem will be discussed further in Sec. 2-3.

It is therefore common to estimate partition functions by first estimating the temperature and the number of terms of the series that need to be kept to have a good approximation to the answer. It is also fortunate that the number ratios are, in general, considerably more sensitive to the temperature than to the ratio of the partition functions. For this reason, great accuracy in evaluation of the partition functions is often not necessary for astrophysical application. In the Saha equation one of the partition functions, viz., that of the electron, is particularly simple. Since the electron has no excited states, its partition function is just the statistical weight of the electron. Since the electron has a spin equal to $\frac{1}{2}$, its statistical weight is $g_e = 2$.

The other quantities appearing in the Saha equation have obvious meanings. They are the mass of the electron, the Boltzmann constant, the Planck constant, the temperature, and χ_r, which is the ionization energy of species r.

In many books on astrophysics the Saha equation is written in terms of the electron pressure instead of the electron density. In a perfect nondegenerate gas the electron pressure is given by $P_e = n_e kT$, so that an alternative expression for the Saha equation is

$$\frac{n_{r+1}}{n_r} p_e = \frac{G_{r+1}}{G_r} \frac{2(2\pi m_e)^{\frac{3}{2}}}{h^3} (kT)^{\frac{5}{2}} e^{-\chi_r/kT} \tag{1-40}$$

We shall, however, generally prefer to use Eq. (1-37). For computational purposes it is often desirable to take the logarithm of this equation and evaluate the constants:

$$\log \frac{n_{r+1}n_e}{n_r} = \log \frac{G_{r+1}}{G_r} + 15.6826 + \tfrac{3}{2} \log T - \frac{5039.95\chi_r}{T} \tag{1-41}$$

The temperature T is expressed in degrees Kelvin and the ionization potential χ_r in electron volts. The Saha equation is intrinsically more complicated than the Boltzmann formula because of three factors: (1) the explicit appearance of the electron density, (2) a more involved functional dependence upon the temperature, and (3) the cumbersome nature of partition functions as compared with statistical weights of atomic states. These difficulties are surmountable, however, and the equation is a valuable source of information concerning stellar temperatures.

The Saha equation can be derived from the statistics of photon and electron gases with the aid of the Einstein electromagnetic-transition coefficients. The calculation is quite similar to the one leading to the Boltzmann equation in that it is constructed from the detailed balance of the electrodynamic processes. The calculation will be easiest to follow in terms of single-bound-state models of the r- and $(r + 1)$-times-ionized species. Figure 1-9 shows these idealized atoms in an energy-level diagram indicating the ionization potential χ_r of species r. Now consider the three first-order electrodynamic processes involving combination and dissociation with an electron of momentum p: (1) spontaneous emission, i.e., radiative recombination; (2) induced emission, i.e., induced radiative recombination; and (3) absorption, i.e., photoionization. The rates of these three proc-

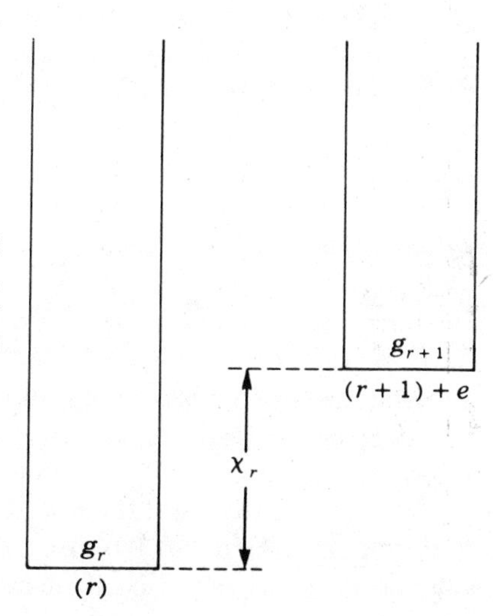

Fig. 1-9 An energy-level diagram for the ionization of idealized ions possessing only a single bound state. The r-times-ionized species must absorb energy greater than or equal to χ_r to make a transition to the $(r + 1)$-times-ionized species plus a free electron.

esses are as follows:

(*1*) *Spontaneous recombination:*

$$(r+1) + e\left(E = \frac{p^2}{2m}\right) \to r + \gamma\left(h\nu = \chi_r + \frac{p^2}{2m}\right)$$

Rate = (density of species $r+1$) × (density of electrons of momentum p) × (spontaneous Einstein emission coefficient)

$$= N_{r+1} n_e(p) A_{r+1,r} \tag{1-42}$$

(*2*) *Induced recombination:*

$$(r+1) + e\left(E = \frac{p^2}{2m}\right) + \gamma\left(h\nu = \chi_r + \frac{p^2}{2m}\right) \to r + 2\gamma\left(h\nu = \chi_r + \frac{p^2}{2m}\right)$$

Rate = (density of species $r+1$) × (density of electrons of momentum p) × $\left(\text{energy density of } h\nu = \chi_r + \frac{p^2}{2m} \text{ photons}\right)$ × (induced Einstein coefficient)

$$= N_{r+1} n_e(p) u(\nu_{r,r+1}) B_{r+1,r} \tag{1-43}$$

(*3*) *Photoionization:*

$$r + \gamma\left(h\nu = \chi_r + \frac{p^2}{2m}\right) \to (r+1) + e\left(E = \frac{p^2}{2m}\right)$$

Rate = (density of species r) × $\left(\text{energy density of } h\nu = \chi_r + \frac{p^2}{2m} \text{ photons}\right)$ × (Einstein absorption coefficient)

$$= N_r u(\nu_{r,r+1}) B_{r,r+1} \tag{1-44}$$

In the steady state characteristic of equilibrium the rate of ionizations equals the rate of recombinations; therefore

$$N_{r+1} n_e(p)[A_{r+1,r} + u(\nu_{r,r+1}) B_{r+1,r}] = N_r u(\nu_{r,r+1}) B_{r,r+1} \tag{1-45}$$

The equilibrium is given by

$$\frac{N_{r+1} n_e(p)}{N_r} = \frac{B_{r,r+1}/B_{r+1,r}}{1 + A_{r+1,r}/u(\nu_{r,r+1}) B_{r+1,r}} \tag{1-46}$$

In most cases of ionization equilibria the electron gas is nondegenerate, which is to say that the Maxwell-Boltzmann approximation to Fermi-Dirac statistics will apply. In that case $n_e(p)$ is given by Eq. (1-25). The energy density $u(\nu)$ is given by Eq. (1-26), and the ratios of Einstein coefficients are given by Eq. (1-33). The demonstration is, therefore, complete except for one important point involving the ratios of the B coefficients. That ratio is equal to the ratio of the statistical weight of the upper state to the statistical weight of the lower state. In the present problem, the upper state involves a free electron of momentum p, so that the statistical weight of that state must include the density of available momentum states. Thus

$$\frac{B_{r,r+1}}{B_{r+1,r}} = \frac{g_{r+1}g_e 4\pi p^2}{g_r} \tag{1-47}$$

Problem 1-16: Show from these assembled equations that

$$\frac{N_{r+1}n_e}{N_r} = \frac{g_{r+1}g_e}{g_r}\frac{(2\pi mkT)^{\frac{3}{2}}}{h^3} \exp - \frac{\chi_r}{kT} \tag{1-48}$$

Equation (1-48) is clearly the form taken by Eq. (1-37) for ions containing only a single state, but it is easy to generalize the result by considering real ions. In equilibrium Eq. (1-48) applies between *any* state r, k of species r and *any* state $r + 1$, j of species $r + 1$. In terms of the more realistic diagram shown in Fig.

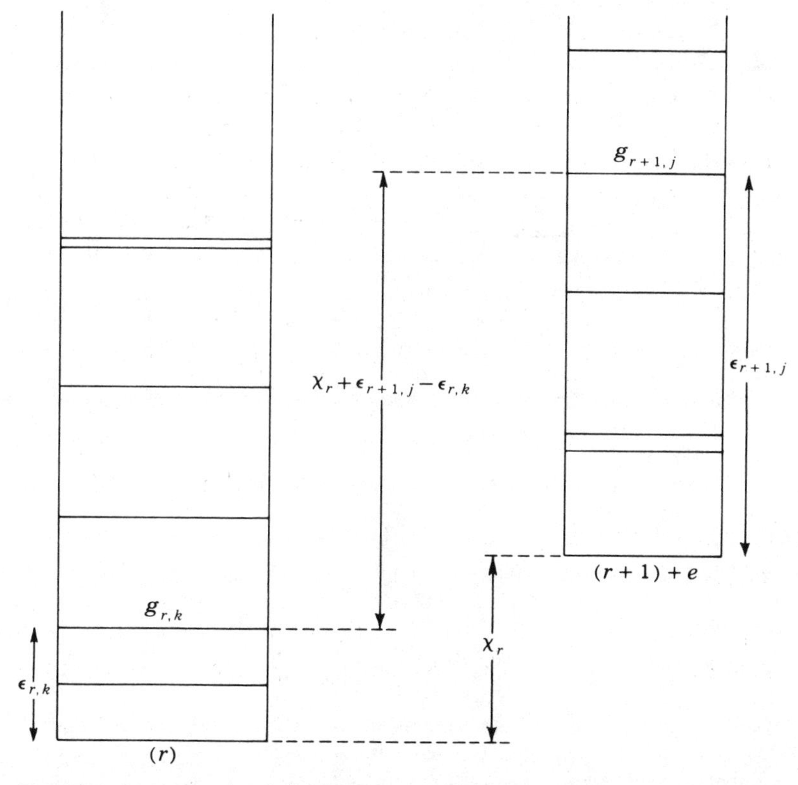

Fig. 1-10 An energy-level diagram for the ionization of ions possessing their full coterie of bound levels.

1-10, Eq. (1-48) should read

$$\frac{N_{r+1,j}n_e}{N_{r,k}} = \frac{g_{r+1,j}g_e}{g_{r,k}} \frac{(2\pi mkT)^{3/2}}{h^3} \exp - \frac{\chi_r + \epsilon_{r+1,j} - \epsilon_{r,k}}{kT} \tag{1-48'}$$

Problem 1-17: Show that the combination of Eqs. (1-48′) and (1-39) results in the Saha equation.

It will be clear from this simple derivation that the Saha ionization equation is actually correct in its stated form only for a nondegenerate electron gas. If the density were so high as to produce degeneracy, the number of states into which the ionized electron could be ejected would be reduced. Other complications, furthermore, such as *pressure ionization*,[1] accompany the high densities of degenerate configurations. This discussion is also inadequate in its neglect of collisional ionization and the inverse (three-body) recombination. In strict thermodynamic equilibrium, where the electron temperature equals the photon temperature, the same result is obtained. In those environments (e.g., solar corona, planetary nebulas) where deviations from local thermodynamic equilibrium are large, the Saha equation is not applicable. A grasp of the simple arguments leading to it will enable the reader to understand complicated situations more easily, however. The reader may also make a mental note that these same three radiative processes will be of importance in the discussion of the opacity of stellar matter. The ionization equation will, moreover, be important for other applications than the present context. It may be applied in any equilibrium process in which some species is dissociated into two other species in thermal equilibrium. Thus, the equation could, for instance, be applied to the dissociation of a molecule into its constituent atoms or even to the dissociation of a nucleus into two nuclei. We shall, in fact, use this same equation in many ways throughout other discussions.

Problem 1-18: Consider a gas cloud composed of hydrogen atoms (no H_2 molecules and no H^- ions). Relate the temperature at which one-half of the hydrogen atoms are ionized to the density in grams per cubic centimeter. In an interstellar gas cloud, the density may be 10^{-27} g/cm^3. How hot must that cloud be (if it is in thermal equilibrium) to be one-half ionized? In the outer layers of a star, the density may be of the order 10^{-4}. How hot must that stellar atmosphere be in order that the hydrogen be 50 percent ionized? You will notice that the required temperature changes by slightly more than 1 order of magnitude when the density changes by 23 orders of magnitude.

Problem 1-19: Show that in thermal equilibrium, the ratio of positive hydrogen ions to negative hydrogen ions at temperature T is given by

$$\frac{H^+}{H^-} = \frac{1}{n_e^2} \frac{8}{G_{H^-}} \frac{(2\pi m_e kT)^3}{h^6} \exp - \frac{\chi^- + \chi^+}{kT}$$

where χ^- and χ^+ are the binding energies of electrons to a neutral hydrogen atom and to a proton, respectively, and G_{H^-} is the partition function of the negative hydrogen ion. Equations similar to this one are very important for nuclear astrophysics.

[1] For a qualitative discussion of pressure ionization, see Chap. 2.

Problem 1-20: Actually, the interstellar medium is far from being in a state of thermodynamic equilibrium, because the spectral distribution of photons is a roughly planckian distribution characteristic of average starlight whereas the energy density of photons is comparable to that which a thermodynamic enclosure would have at a few degrees absolute. If the hydrogen ionization is predominantly due to starlight, and if the electrons have a maxwellian distribution corresponding to the same temperature T_e that characterizes the spectral shape of average starlight, but the photon energy density is WaT_e^4, argue that the level of hydrogen ionization is approximately

$$\frac{N(\mathrm{H}^+)}{N(\mathrm{H})} n_e = W \frac{(2\pi m_e kT)^{\frac{3}{2}}}{h^3} \exp - \frac{\chi_{\mathrm{H}}}{kT}$$

The factor W is called the *dilution factor* and is often used for approximate calculations.

Problem 1-21: Show that if hydrogen gas is in thermodynamic equilibrium, the ratio of the number of hydrogen atoms in the principal energy level n to the number of ionized hydrogen ions is

$$\frac{N_n(\mathrm{H})}{N(\mathrm{H}^+)} = \frac{n_e h^3 n^2}{(2\pi m_e kT)^{\frac{3}{2}}} \exp + \frac{\chi_{\mathrm{H}}}{n^2 kT}$$

Problem 1-22: A certain stellar atmosphere with a pressure 1,000 dynes/cm^2 consists entirely of hydrogen. By ignoring statistical weights, i.e., setting the ratio of partition functions equal to 1, find the temperature at which the H_2 molecules are 50 percent dissociated. The binding energy of the H_2 molecule is 4.48 ev. (The problem here is that the dissociated particles have comparable mass. How will that fact modify the ionization equation?)

These last two techniques for determining stellar temperatures may sometimes give slightly lower temperatures than the measurement of the blackbody continuum. The reason for this small difference is that many of the absorbing atoms which contribute to the absorption lines lie above the base of the photosphere and are therefore somewhat cooler. In a detailed theory of stellar atmospheres, a correction for this difference may be made.

The application of both the Boltzmann and the Saha equations depends upon the ability to convert the intensity of absorption lines into abundances of the absorbing species. The spectroscopic trace of an absorption line in a stellar spectrum reveals how much energy was absorbed by the specific transition from the continuous radiation. The conversion of the amount of energy absorbed by a spectral line into the abundances of the absorbing species involves the entire theory of line formation. This theory must be invoked to obtain the population ratios necessary for application of the Boltzmann and Saha equations. The same theory must also be invoked to calculate from stellar spectra the abundances of elements in stellar atmospheres. The theory of line formation is so specialized and so extensive that it will not be taken up in this book. The shape of an observed absorption line depends upon the physical conditions in the stellar atmosphere, i.e., its temperature, density, and composition, as well as upon the intrinsic strength of the atomic transition involved. There are several excellent treatises upon the theory of line formation. For our purposes, we shall simply note that there exists a way of converting the strength of absorption lines to abundances.

THE SPECTRAL TYPES

One of the earliest results of observational astronomy was the realization that there existed a correlation between properties of the stellar surface, such as its surface temperature, and the strength of specific absorption lines as seen in the spectrum of the star. In 1863 the Jesuit astronomer Angelo Secchi classified stars into four groups according to the prominent absorption lines in their spectra. An empirical classification scheme was subsequently developed in which stars were sorted into seven principal spectral types, each type being characterized by a certain range of surface temperature and the appearance of characteristic absorption lines. The detailed understanding of the correlation between surface temperatures and prominent absorption lines rests almost entirely in various applications of the Boltzmann and Saha formulas.

The principal classification groups of stars are traditionally labeled by the letters O, B, A, F, G, K, and M. The significances of the actual letters of the alphabet chosen to represent the spectral classes are mostly historical and are reminiscent of the analogous importance of the letters chosen to represent the various angular-momentum states in quantum mechanics. Each spectral class corresponds to a certain range in surface temperatures. Each of these major divisions or classes is further subdivided into 10 groups. For instance, the spectral type B is further subdivided into 10 subclasses labeled B0, B1, B2, . . . , B9, in order of decreasing temperature. A rough correlation of the surface temperatures of each spectral type with the prominent absorption lines appearing in the spectra of those stars is as follows:

Class O: Temperatures of 25,000°K and up. Lines of ionized helium are prominent. From the discussion of the Saha equation it is apparent that lines of ionized helium will appear only in such an extremely hot gas. Other atoms in high degrees of ionization are observed.

Class B: 25,000 to 11,000°K. The lines of hydrogen and neutral helium are conspicuous at class B0. Ionized oxygen and ionized carbon become strong at class B3. Neutral helium lines are strongest at class B5. Hydrogen lines become progressively stronger in the higher-numbered subdivisions of this class. By hydrogen lines we mean, of course, the Balmer series of hydrogen lines appearing in absorption. The intensity of such lines will, among other things, be proportional to the fraction of hydrogen atoms existing in the first excited state of hydrogen in thermal equilibrium. Thus, the strength of hydrogen lines is primarily determined by combined application of the Boltzmann and Saha formulas.

Class A: 11,000 to 7500°K. At class A0 hydrogen and ionized magnesium lines are strongest, whereas the helium and ionized oxygen lines have disappeared. Hydrogen lines weaken continuously in the higher subdivisions of this class, whereas ionized metals (Fe, Ti, Ca, etc.) strengthen. The hydrogen lines will continue to weaken as we progress toward cooler stars, because the temperature becomes less and less sufficient to maintain a significant fraction of hydrogen atoms in the first excited state. The lines of ionized metals are growing stronger because the relative number of metals in low degrees of ionization is increasing. In the hotter stars, the higher degree of ionization of the metals produces resonance lines that lie in the ultraviolet. The resonance lines of the slightly ionized metals, however, lie in the visible, and these grow stronger as the temperature cools.

Class F: 7500 to 6000°K. Class F0 is rich in lines of the ionized metals, the strongest being the *H* and *K* lines of singly ionized calcium. Metallic lines, particularly iron, strengthen and hydrogen lines weaken in the higher-numbered subdivisions of this class.

Class G: 6000 to 5000°K. In this class the lines of the neutral metals become strong, whereas the hydrogen lines continue to weaken. Lines of ionized calcium are very strong. Molecular bands of CN and CH appear. The sun belongs to the class G2.

Class K: 5000 to 3500°K. In general, molecular bands and lines of neutral metals become much stronger, whereas the lines of hydrogen and ionized metals continue to weaken. At K5 the lines of TiO are weakly visible.

Class M: 3500 to 2200°K. The characteristic feature of the spectrum of class M stars is the appearance of complex molecular oxide bands, of which TiO bands are strongest.

The majority of stars fall into one of these spectral classes. There are, however, some exceptions to this classification scheme: some stars have temperatures in a range parallel to one of the existing classes but show strikingly different spectral lines. Therefore, some additional spectral classes have been established which parallel the above classes in temperatures:

Class S: A low-temperature class parallel to class M. This class is still characterized by molecular oxide bands, but the most prominent feature is the ZrO bands. The elements Zr, Y, Ba, La, and Sr give strong atomic lines and oxide bands. Lines of neutral technetium are usually seen. It is believed that these abundances are enhanced because of nucleosynthesis within the interior which has been mixed to the surface.

Classes R and N (or Class C): Parallel in temperature to the ordinary classes K and M. The spectrum is characterized not by oxide bands but by molecular carbide bands, such as those of CN, C_2, and CH.

Class W: Extremely high-temperature type O objects, called *Wolf-Rayet stars*, with bright, broad emission lines of ionized helium and highly ionized carbon, oxygen, and nitrogen. Two sequences exist: (1) the WC stars have strong carbon lines and weak nitrogen lines, and (2) the WN stars have strong nitrogen lines and weak carbon lines. These stars are generally found to be emitting gas rapidly in space. The progression of spectral properties through the sequences

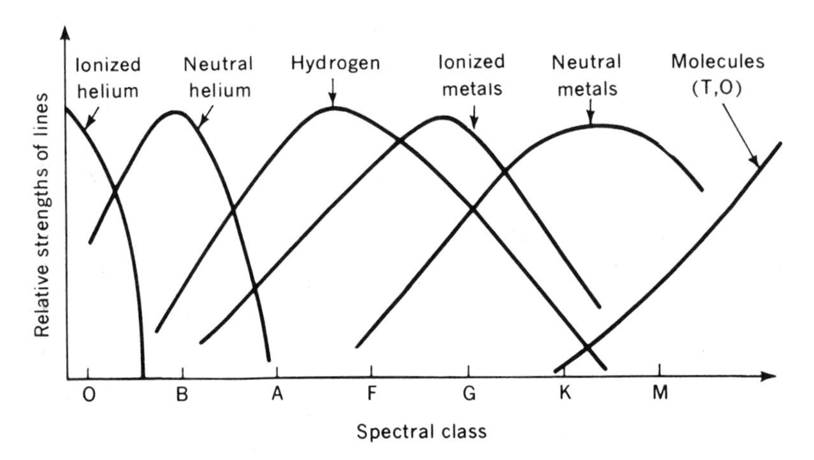

Fig. 1-11 The progression of selected spectral properties through the sequence of spectral classes. *(G. Abell, "Exploration of the Universe," Holt, Rinehart and Winston, Inc., New York,* 1964.)

of spectral classes and the correlation of color with surface temperature and spectral type are shown in Fig. 1-11.

Problem 1-23: What is the spectral type of a normal star having a maximum in its continuous spectrum at H_α?
Ans: K3.

Problem 1-24: What is the spectral type of a star for which the number of hydrogen atoms in the first excited state exceeds the number in the second excited state by the ratio of 4:1?

Problem 1-25: Absorption lines of singly ionized helium cannot be seen if the temperature is too low, because the helium is mostly neutral and what little ionized helium there is lies in the ground level and cannot make visible absorption lines. On the other hand, the lines cannot be seen if the temperature is too high, because all the helium becomes doubly ionized. The helium absorption lines arising from a state in level n of singly ionized helium will be strongest when the largest possible fraction of the helium lies in that level, i.e., when $\mathrm{He}_n{}^+/(\mathrm{He} + \mathrm{He}^+ + \mathrm{He}^{++})$, the fraction of He in state n of He^+, is a maximum. Show that this condition occurs when the quantity

$$G(\mathrm{He}^+) \exp\left[\left(1 - \frac{1}{n^2}\right)\frac{\chi^+}{kT}\right] + G(\mathrm{He}) \frac{n_e h^3}{2(2\pi mkT)^{\frac{3}{2}}} \exp\left\{\frac{\chi^0 + [(n^2-1)/n^2]\chi^+}{kT}\right\} + \frac{2(2\pi mkT)^{\frac{3}{2}}}{n_e h^3} \exp - \frac{\chi^+}{n^2 kT}$$

is a minimum. The ionization potential of neutral helium is $\chi^0 = 24.58$ ev, and χ^+ is the ionization potential of the He^+. Calculate for $n = 4$ and electron densities in the range $\log n_e \approx 19$ (the answer is not very sensitive to n_e) that temperature yielding the largest fraction of He^+ ions in the $n = 4$ level (assume $G = 1$). You may also notice that most of the helium is doubly ionized at this temperature. What is the spectral type?

One of the objects of this book will be to illustrate how the elements may be synthesized by nuclear reactions in stellar interiors. Normally, the atmosphere of a star shows the abundances of those elements which were present at the time the star formed. The newly formed elements in the stellar interiors do not usually migrate to the surface. However, examples of cases in which nuclear reactions in a stellar interior *have* changed the surface composition of the star into something abnormal may be seen in the classification list of stars already presented. For example, the S stars show strong enhancements of the spectral lines of those nuclei having exceptionally stable, or *magic*, neutron configurations, Zr, Y, Ba, La, and Sr. For some reason these particular elements are overabundant in the surfaces of class S stars. Apparently these species are produced in the interior of the star and mix to the surface of the star, an exception to the usual course of events. Further evidence of this fact may be seen in the appearance of the lines of Tc, an element that is unstable. Certainly the appearance of an unstable nuclear species must indicate the recent formation of that species. The nuclear processes going on in the interior of the type S stars apparently are such that the species with magic neutron configurations are produced in abundance. The details of this nuclear mechanism, called the *s* process, will be examined in Chap. 7.

Another example of an abnormal nuclear history may be seen in types R and N stars. In normal stars of that temperature, oxide bands are formed because oxygen, being normally more abundant than carbon, remain in excess after the formation of CO. The excess oxygen may then form less stable oxide bands. In R and N stars, such does not appear to be the case. Some nuclear process or vicissitude of evolutionary history has augmented carbon relative to oxygen; therefore, the CO exhausts the oxygen and leaves excess carbon for the formation of carbide bands. One of the most fascinating problems in stellar evolution and nucleosynthesis is that of separating abundance abnormalities into those contained in the star at birth and those produced by the star during its own lifetime.

1-3 MASS

The masses of stars can be measured only when they occur in a binary pair and when the orbital motion of the pair has been measured. In the best cases one can measure the separation of the pair of stars and their relative motion about their center of mass. The difference of the Doppler shift of lines from the two components can also be made to yield the orbital velocity. The masses are computed by a simple application of Kepler's third law. Let M_1 and M_2 represent the masses of the pair of stars; then

$$\frac{M_1 + M_2}{M_\odot + M_{\text{earth}}} = \frac{A^3}{P^2} \tag{1-49}$$

where A is the distance (semimajor axis) between the two stars expressed in astronomical units (1 AU = 1.496×10^{13} cm) and P is the period of the binary system in years. The ratio of the masses is determined in those additional cases where the center of mass of the pair can be found. It is then possible to calculate each mass individually.

Because of the restriction to observable binaries, the mass is certainly one of the most difficult quantities to determine for stars. The only star for which the mass is known with great precision is the sun:

$$M_\odot = 1.989 \pm 0.002 \times 10^{33} \text{ g}$$

That value is derived from the accurately known orbits of the planets.

Measured masses of stars are found to lie in the range of one-tenth of the solar mass to about 20 solar masses. Considerably larger masses probably have been observed, because whenever the inclination of the orbital plane cannot be determined, it is possible to calculate only a lower limit on the mass of a binary pair. There are theoretical indications that masses greater than 60 solar masses are unstable and that masses smaller than 0.08 solar masses do not become hot enough to cause nuclear reactions.

Interestingly enough, most stars with known masses display a striking relationship between their luminosity and their mass: the luminosity is proportional to a rather high power of the mass; that is,

$$L = \text{const} \times M^{\nu} \tag{1-50}$$

where ν lies in the range of 3.5 to 4.0. This relationship is found to be violated strongly only by the white dwarfs, which are much too faint for their observed masses. It is theoretically certain that the relationship is also violated for a class of stars called *giants*, whose masses have unfortunately not been measurable. Basically Eq. (1-50) may be thought of as an approximation applicable to a class called *main-sequence stars*. At any rate, the mass-luminosity law is important in that it roughly correlates the luminosity of main-sequence stars with their masses, a fact to be interpreted in terms of the theory of stellar structure.

1-4 RADIUS

The radii of stars, extremely important quantities for the theory of stellar evolution, have generally not been measurable. The recently developed technique of the intensity interferometer[1] has not yet yielded accurate results for a large number of stars. Only for a few dozen stars, mostly giants, do direct measurements exist. The values of known diameters range up to a thousand times that of the sun (Betelgeuse). These measurements are helpful, but for tests of stellar-structure theory one is generally forced to determine the radii from the Planck emission law

$$L = 4\pi R^2 \sigma T_e^{\,4} \tag{1-51}$$

[1] See *Sky and Telescope*, **28**:64 (1964) and *Science*, **153**:581 (1966) for a discussion of this technique.

Since the luminosity and effective surface temperatures of the stars are measurable by independent techniques, as we have seen, the radius of a star can be calculated. There are hidden problems, however, in Eq. (1-51), which logically stands as the definition of the effective temperature T_e. But stars are not blackbodies, so that the measured temperatures differ in some unknown way from the value of T_e that satisfies Eq. (1-51). What must be done is to construct model stellar atmospheres that are capable of reproducing the continuous spectrum and the line spectrum of various stars and to see how much energy is actually radiated by such an atmosphere. It turns out in practice that all surface temperatures are nearly equal for those stars radiating primarily in the visible portion of the spectrum, but Eq. (1-51) shows that an error in T_e is amplified in the determination of R. Because of inadequacies in the theory of convection, moreover, current models of stars are incapable of accurately predicting the radii of stars having surface convection zones. The problems of understanding stellar radii are vexing. The sun is still the only star whose radius is known with high precision: $R_\odot = 6.9598 \pm 0.0007 \times 10^{10}$ cm.

In summary, we observe that the science of stellar structure deals with four basic large-scale properties of a star which can be measured with varying degrees of accuracy. They are the luminosity, the surface temperature, the radius, and the mass. Also observable are certain spectral features from which the composition of stellar surfaces may be inferred. From this set of data, in itself a remarkable tribute to observational astronomy, the science of stellar evolution has been derived. The values of these macroscopic properties of stars lead immediately to the conclusion that stars are gaseous throughout. From their long lifetimes it may be presumed that the star must be a structure in hydrostatic equilibrium, the pressure at each point in the interior being just sufficient to support the weight of the layers above. The additional considerations of energy generation and transport make possible the construction of a theory of stellar evolution. An introduction to the physical principles of this theory constitutes a major objective of the subsequent chapters of this book.

1-5 ENERGETICS

The pressure, temperature, and density in the interior of a star are quantities related by the equation of state of the gas. The equation of state may be complicated, depending as it does upon specific physical effects that occur at various locations in the $T\rho$ plane, but the relationship $P = P(T,\rho)$ nonetheless exists. Even for a star in hydrostatic equilibrium, however, there remains considerable freedom of structure. A specified pressure may be produced by high density and a relatively low temperature or by a low density and a relatively high temperature. Finding the proper combination of temperature and density to provide a self-consistent stable structure for the star is the major problem of stellar structure, and the demand for hydrostatic equilibrium is in itself insufficient to generate a unique solution.

The rate at which energy is transported from the deep interior of the star to the surface to be radiated depends upon the temperature gradient of the stellar structure. Furthermore, the central temperature must have the proper value in order that the nuclear reactions which maintain the internal heat proceed at the rate required to counterbalance the luminosity. Because the rates of nuclear reactions are, in general, strongly temperature-dependent, the last condition imposes a severe restriction upon the allowable temperatures of stellar interiors. And it is the requirements regarding energy generation and transport that fix the structure of the star.

One associated problem is the question of how the energy is transported to the surface of the star from the interior. Interior heat is removed by four mechanisms, conduction, convection, radiation, and neutrinos. The process of heat conduction depends upon collisions between particles in which excess concentrations of thermal energy are transferred via elastic collisions to neighboring particles, a process that slowly spreads and distributes the excess energy. Although this process is extremely efficient in a metal, it is, except for degenerate electron gases, extremely inefficient in the gaseous state. The competition most often lies between radiation and convection as the means of transporting energy. If the hot region in a gaseous configuration releases its excess energy by radiating it to cooler parts of the gaseous configuration, the process is called *radiative transfer*. Because thermal emission is proportional to the fourth power of the temperature, a hot spot will radiate energy faster than it is receiving it from the cooler surroundings. If the opacity of the gas to radiation is not too large, the energy can be transported very efficiently by photons. Convection, too, is a relatively efficient way of transporting energy. A mass of gas in a hot region of a star may be moved bodily to a cooler region of the star where it may distribute its excess thermal energy. But by one (or a combination) of these mechanisms, the large thermal-energy store in the interiors of stars is transported to the surface and radiated. Whenever neutrinos are created in the stellar interior, however, they generally escape without interaction. Neutrino emission may be regarded as an instantaneous local heat sink.

Historically, it was difficult to determine the source of the large amounts of energy radiated by stars. Probably the best incomplete answer was provided by Helmholtz and Kelvin, who suggested that stars gain their energy from the work done by gravitation contraction. Because of the conservation of energy it is clear that a star cannot radiate more energy from this energy source than is liberated by its gravitational binding. The order of magnitude for the total gravitational binding of the sun is

$$\Omega = \frac{-GM^2}{R} = -\frac{(7 \times 10^{-8})(2 \times 10^{33})^2}{7 \times 10^{10}} = -4 \times 10^{48} \text{ ergs} \tag{1-52}$$

If the entire amount of this energy were radiated at the present luminosity of the sun, the age of the sun would be given by the formula

$$t_\odot L_\odot = 4 \times 10^{48} \tag{1-53}$$

Since $L_\odot = 3.9 \times 10^{33}$ ergs/sec, the age would be $t_\odot \approx 10^{15}$ sec $\approx$ 30 million years. This time is much too short for a maximum lifetime of the sun. It is known that the sun has existed over 100 times longer than this, because the age of the earth itself is about 4.6 billion years. Of course, one could argue that the earth is older than the sun. However, even with that unrealistic assumption the gravitation energy source would be insufficient. There exist fossilized algae whose ages are on the order of 1 billion years or more, a time which is again considerably longer than a gravitational contraction time. Thus the energy of contraction is not in itself sufficient for the sun.

It is now believed that stars provide their energy by conversion of rest mass into kinetic energy according to the Einstein mass-energy relationship

$$E = mc^2 \tag{1-54}$$

According to this interpretation, the sun is radiating energy at the expense of its own mass; i.e., the mass of the sun is slowly decreasing, being radiated away by some physical process that effects the conversion of matter into energy. The supply of energy is very large. The total rest-mass energy of the sun is given by

$$M_\odot c^2 = (2 \times 10^{33})(3 \times 10^{10})^2 = 2 \times 10^{54} \text{ ergs} \tag{1-55}$$

Since this supply of energy is almost a million times greater than the gravitational binding energy of the sun, it could in principle account for the sun's luminosity for a million times as long at the same rate of emission. Of course, there is no plausible way of converting all the mass into radiant energy. Only a certain fraction of the rest-mass energy of the sun can be converted. Even then it is necessary to find the detailed nuclear reactions by which this transmutation comes about. Fascinating astrophysical possibilities emerge from the details of these nuclear reactions.

Perhaps the strongest clue to the correct solution was the observation that stars are predominantly hydrogen. Approximately 90 percent of all nuclei in stars appear to be hydrogen nuclei. Because of the great binding energy of the helium nucleus, furthermore, the most abundant source of energy in stellar interiors appears to be the fusion of four hydrogen atoms into one helium atom. The amount of energy liberated by each such conversion is given by

$$(4M_{\mathrm{H}} - M_{\mathrm{He}})c^2 = 26.73 \text{ Mev} \tag{1-56}$$

The total rest-mass energy of 1 amu is equal to 931 Mev.[1] Since the rest mass of the system being considered is very nearly 4 amu, the fraction of the rest-mass

[1] In nuclear physics the Mev is a very convenient unit, one that will be employed in the nuclear discussions of this book. It is defined as the work done upon a unit atomic charge moved by an electric field through a potential difference of 10^6 volts, and in fact it was just such electrostatic particle accelerators that gave impetus to the experimental science of nuclear physics as we now know it. This energy unit is characteristic of the significant energies found in nuclear structure, moreover, in just the same way that the electron volt is a convenient unit for atomic physics.

energy converted into radiation in each of these transmutations is given by

$$\text{Fraction converted} = \frac{26.7}{4 \times 931} = 0.007 \tag{1-57}$$

i.e., the transmutation of hydrogen into helium will liberate 0.7 percent of the rest mass of the system in the form of energy. Even this energy source is sufficient to account for the lifetime of the sun, since the amount of available energy from this transmutation alone is then 0.007 of the total rest-mass energy of the sun, which is $(0.007)(2 \times 10^{54}) \approx 1.4 \times 10^{52}$ ergs. This available energy would sustain a total lifetime for the sun at its present rate of energy emission equal to

$$\text{Lifetime} = \frac{1.4 \times 10^{52}}{4 \times 10^{33}} = 3 \times 10^{18} \text{ sec} = 10^{11} \text{ years}$$

This very long time, 10^{11} years, is longer than the age of any known object, including our galaxy.

Problem 1-26: The mass of the hydrogen atom is equal to 1.007825 amu, and the mass of the helium atom is equal to 4.002603 amu. Calculate the energy released by nuclear binding when four hydrogen atoms combine to form one helium atom. How much energy is liberated when 1 g of hydrogen is converted to 1 g of helium?
Ans: 6.4×10^{18} ergs/g.

The energy given off by the conversion of hydrogen to helium comes very close to being the maximum amount of energy that can be liberated by nuclear reactions in stellar interiors. The release of the maximum available energy from nuclear reactions is accomplished by the conversion of hydrogen into that form of nuclear matter for which the binding energy per nucleon is the maximum. The nuclear species is an isotope of iron, Fe^{56}. However, the conversion of H into Fe^{56} liberates only 0.8 percent of the rest-mass energy. Apparently, therefore, the conversion of hydrogen to helium liberates seven-eighths of the total energy available from nuclear reactions.

The recognition that the source of energy in stars must be nuclear reactions, or even that the specific nuclear reaction likely to dominate is that which converts hydrogen into helium, is still not a sufficient explanation of the nuclear energy source in stars. We must also define the specific nuclear mechanism and reactions for bringing four protons together and somehow changing them into a helium atom, which is composed of two protons, two neutrons, and two electrons. Certain questions must be answered:[1] How many protons must come together at one point, and how and when do the two required beta decays occur? Historically, there was an even greater problem to be encountered in bringing two protons together, the coulomb repulsion between two positive charges. The work done in bringing two charges to a separation r is given by the potential energy

[1] In awarding the 1967 Nobel Prize in physics to Hans Bethe, the Swedish Academy of Science cited especially Bethe's discovery of the details of the fusion of hydrogen.

at the separation r,

$$V = \frac{q_1 q_2}{r} = \frac{1.44 Z_1 Z_2}{r(\text{fm})} \qquad \text{Mev} \tag{1-58}$$

where the first form is in cgs units and the second form is in units that will be most useful for nuclear reactions. In discussions of nuclear physics it is convenient to express energies in millions of electron volts and distances in units of 1 fm $= 10^{-13}$ cm. The fermi, or femtometer, is the unit of distance used in nuclear physics for two reasons. (1) The size of a nucleus is characteristically a few fermis. Since a nucleus does not have an unambiguous boundary, its size, or radius, is a somewhat uncertain quantity. However, the distribution of nucleons in the nucleus can be measured by observing the scattering of high-energy electrons from the nucleus. The results of such scattering experiments reveal that the nucleons are bound into a small volume with a radius of a few fermis. Additionally it is known from experimental nuclear reactions that the radius of a nucleus is a few fermis. (2) It is implicit in the present theory of the nuclear force that nucleons must approach to a separation of only a few fermis before the effects of the strong (but short-range) nuclear force can be felt. The nuclear force is believed to exist because of a *field* of pi mesons, with which nucleons are known to have very strong reactions, in much the same way that the electromagnetic force exists because there is a field of photons which interact strongly with charges. As such, the range of nuclear forces is determined theoretically by the Compton wavelength of the pi meson, $\hbar/m_\pi c = 1.41$ fm. For nuclear reactions to occur, two nucleons must come to a separation of approximately the range of the nuclear force plus the sum of their own radii, a separation of a few fermis.

The radius of a proton is measured to be approximately 0.8 fm, and the range of the nuclear force is about 1.4 fm. Accordingly two protons will interact only if they come to a separation of something like 2 to 3 fm. From Eq. (1-58) the potential energy for two protons at a separation of 2 fm is given by

$$V = \frac{1.44}{2} = 0.7 \text{ Mev}$$

This result indicates that two *classical* particles with a relative kinetic energy less than 700 kev could not come so close to each other as the distance 2 fm. The coulomb potential for like charges is somewhat like a hill between the two particles. Classically speaking, a particle either has sufficient energy to roll to the top of a hill, or it does not. If it does not, it turns around and rolls back down. On the other hand, the average thermal energy in a gas in thermal equilibrium is of the order kT, which is numerically equal to

$$kT = 0.862 \times 10^{-7} T \qquad \text{kev} \tag{1-59}$$

It is apparent that at the sun's center, where T is approximately 10^7 °K, the average thermal energy is only about 1 kev. This energy is considerably less than the 700 kev required to bring two protons to a separation of 2 fm, a distance characteristic of that necessary for them to interact via nuclear forces. We may view this dilemma in a slightly different way, asking instead to what separation protons of energy 1 kev can come.

$$\frac{1.44}{r_{\min}} = 10^{-3} \text{ Mev} \qquad r_{\min} \approx 10^3 \text{ fm}$$

Thus, the separation $r_{\min}$ at the *classical* turning point for two protons at 1 kev energy is approximately 1,000 times as great as the separation required for them to interact. Because of this difficulty it seemed for some time that the temperatures in stellar interiors were not adequate to cause nuclear reactions. Eventually it was shown by Gamow and his associates, who were considering the similar question of how an alpha particle can escape from the interior of a nucleus, that particles can penetrate barriers that they cannot surmount classically. That is, even though the coulomb potential is too high to be surmounted by a particle of a few kilovolts of energy, those particles may, with low probability, penetrate through the barrier. The introduction of this quantum-mechanical effect resolved the conflict by revealing that the fusion reactions could proceed in the interiors of stars at a rate sufficient to account for the energy liberated from the surface. This subject will be discussed in Chap. 4.

1-6 THE HERTZSPRUNG–RUSSELL DIAGRAM

A fundamental correlation between observable stellar properties was initiated by the Danish astronomer E. Hertzsprung in 1911, when he plotted the apparent magnitude of stars within a given star cluster against their colors. The American astronomer H. N. Russell made a similar investigation in 1913 of the absolute magnitudes of stars in the solar neighborhood. The *Hertzsprung-Russell diagram*, or simply H-R diagram, is the name given to the graph of a quantity that measures the luminosities of stars (luminosity, absolute bolometric magnitude, apparent visual magnitude, etc.) versus the effective surface temperature (or color index $B - V$ or spectral class). It is one of the most valuable correlations established in observational astronomy. For the purposes of this discussion we assume here that the fundamental quantities mentioned in the foregoing material are ascertainable for individual stars, i.e., the star's luminosity and effective surface temperature. Both the luminosity and effective temperature can be measured by related quantities, however. Thus, possible H-R diagrams might be (1) luminosity versus surface temperature, (2) luminosity versus color index $B - V$, (3) absolute bolometric magnitude versus spectral class, (4) absolute visual magnitude versus color, etc.

We shall mention here two observational problems associated with these diagrams. Often it is not possible to determine accurately the distance to faraway star clusters; therefore, the H-R diagrams of such objects are best plotted in terms of apparent magnitudes rather than absolute magnitudes. The relative magnitudes in such cases are much more accurate than the absolute magnitudes. In determining surface temperatures for distant stars it is often not feasible, because of time limitations, to use the many techniques available for nearby stars. The spectral lines are just too faint. The indicator of surface temperature that is easiest to measure for very faint objects is the color index, the blue magnitude minus the visual magnitude. For distant stellar clusters, therefore, it is common to display the H-R diagram in terms of the apparent magnitude plotted against the color index. The accuracy of a subsequent conversion to absolute bolometric magnitude and surface temperature depends upon the validity of the translating assumptions. Details of this very important problem may be found in books on observational astronomy. Remembering always that the problem does exist, we shall nonetheless think of the H-R diagram as the relationship of the absolute bolometric magnitude or luminosity to the surface temperature.

When such a diagram has been constructed for a large number of observable stars, it is clear that a very large percentage of the stars fall on a heavy diagonal curve, called the *main sequence*. This curve is such that the brightest objects in the sky are those with the highest surface temperatures and are blue in color.

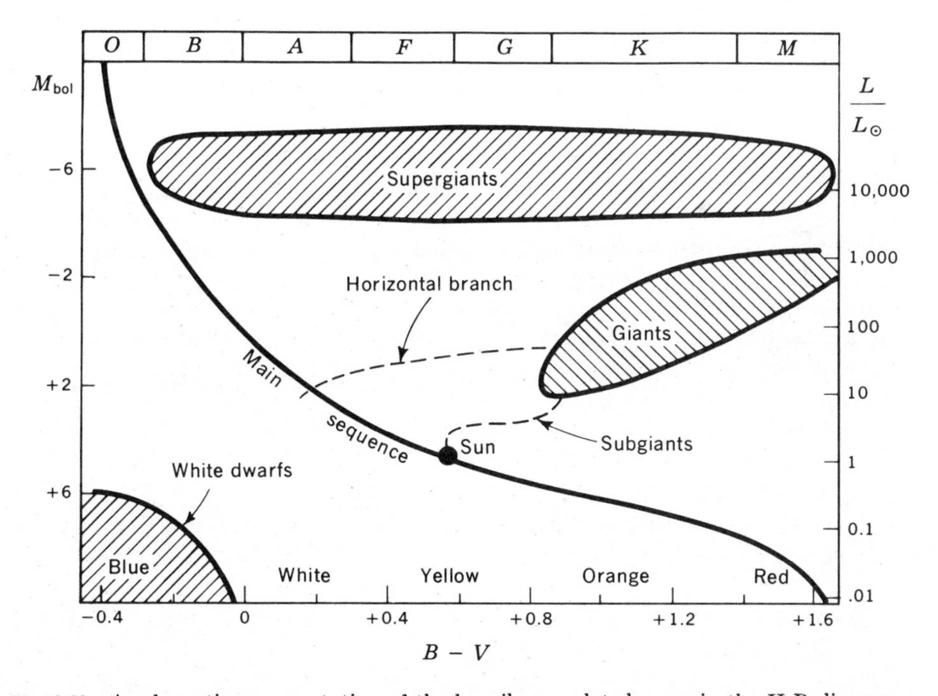

Fig. 1-12 A schematic representation of the heavily populated areas in the H-R diagram. A high percentage of stars lie near the main sequence. The next most populous groups are the white dwarfs and the giants. The subgiant and horizontal branches are conspicuous in those collections of stars having large numbers of giants, e.g., globular clusters.

Table 1-1 The main sequence†

Spectral type	*Absolute visual magnitude* M_v	*Color index* $B - V$	*Bolometric correction*	*Effective surface temperature* T_e, °K	*Color temperature* T_c, °K	*Absolute bolometric magnitude* M_{bol}	*Logarithm of luminosity* $\log \frac{L}{L_\odot}$
O5	−6.0	−0.45	4.6	35,000	70,000	−10.6	6.13
B0	−3.7	−0.31	3.0	21,000	38,000	−6.7	4.56
B5	−0.9	−0.17	1.6	13,500	23,000	−2.5	2.88
A0	0.7	0.00	0.68	9,700	15,400	0.0	1.88
A5	2.0	0.16	0.30	8,100	11,100	1.7	1.20
F0	2.8	0.30	0.10	7,200	9,000	2.7	0.80
F5	3.8	0.45	0.00	6,500	7,600	3.8	0.37
G0	4.6	0.57	0.03	6,000	6,700	4.6	0.05
G5	5.2	0.70	0.10	5,400	6,000	5.1	−0.15
K0	6.0	0.84	0.20	4,700	5,400	5.8	−0.43
K5	7.4	1.11	0.58	4,000	4,500	6.8	−0.83
M0	8.9	1.39	1.20	3,300	3,800	7.6	−1.15
M5	12.0	1.61	2.1	2,600	3,000	9.8	−2.03

† C. W. Allen, "Astrophysical Quantities," University of London Press, Ltd., London, 1963.

The dimmest objects in the sky are red in color and lie in the lower right-hand end of the main sequence. Figure 1-12 offers a schematic representation of the heavily populated areas in the H-R diagram. Our sun lies at a point about in the middle of the main sequence. The stars lying on the main sequence with luminosities less than that of the sun are often given the collective name *dwarfs*. The properties of the main sequence are listed in Table 1-1. The difference in the fourth and fifth columns reflects the difficulty of equating the several surface temperatures. In the lower main sequence, where the peak of the Planck spectrum is in the visible, the agreement is reasonably satisfactory, especially when the detailed features that distort the blackbody spectrum are taken into account. It is not surprising that disagreement grows for the hot main-sequence stars, considering that the visible continuum lies on the long-wavelength tail of the continuous spectrum.

Problem 1-27: If an O5 star were a 35,000°K blackbody, where would the peak in the continuous spectrum be? What fraction of the energy radiated by an O5 star is visible?
Ans: 830 Å, 1.4 percent.

The sixth column contains the absolute bolometric magnitude of the stars, which is the difference between its visual magnitude and the necessary bolometric correction. The final column lists the logarithm of the luminosity expressed in solar units, as in Eq. (1-14). This grouping of stars, called the main sequence, is remarkable in that it represents about 80 to 90 percent of observed stars. The main sequence consists primarily of those stars which are burning hydrogen in their stellar interiors as their source of energy. A few stars in advanced stages of evolution may also fall on or very near the main sequence, however.

Problem 1-28: Calculate the radii of B0, A5, and M0 main-sequence stars. Compare to the solar radius.

Several other important classes of stars appear as high concentrations on the H-R diagram, the most obvious being a large group of stars above and to the right of the main sequence. These stars are fairly luminous, having an absolute bolometric magnitude near zero, but they are very red in comparison with main-sequence stars of the same luminosity. This class of stars, whose detailed properties are given in Table 1-2, is called collectively the *red giants*. It will be obvious that the redness of these giants, as compared with their main-sequence counterparts of the same luminosity, is accounted for by their relatively large radii. The red giants are not actually very numerous as a class, representing only a few percent of the stars, but the class is easily discernible because of the large luminosities of these stars; i.e., the class is more prominent to the eye than in number.

A relatively smaller number of stars are found in old clusters to delineate a curve in the H-R diagram leading from the lower main sequence upward to the giant region. These stars, called *subgiants*, are believed to be stars whose envelopes are expanding while their helium cores contract to a point where the helium begins to produce energy by nuclear reactions. Stars are also found on a horizontal branch to the left of (bluer than) the giant region. These stars are believed to be in various phases of helium burning. The search for an adequate explanation of the relative numbers of stars at various positions on these two branches comprises a major objective of current research in stellar evolution.

There is also a class of very luminous stars ($L \approx 10^4 L_\odot$) of all colors that spreads a horizontal strip across the top of the H-R diagram. Collectively called

Table 1-2 Red giants†

Spectral type	*Absolute visual magnitude* M_v	*Color index* $B - V$	*Bolometric correction*	*Effective surface temperature* T_e, °K	*Color temperature* T_c, °K	*Absolute bolometric magnitude* M_{bol}	*Logarithm of luminosity* $\log \frac{L}{L_\odot}$
G0	1.8	0.65	0.1	5400	6000	1.7	1.21
G5	1.5	0.84	0.3	4700	5000	1.2	1.41
K0	0.8	1.06	0.6	4100	4400	0.2	1.81
K5	0.0	1.40	1.0	3500	3700	−1.0	2.29
M0	−0.3	1.65	1.7	2900	3400	−2.0	2.69
M5	−0.5	1.85	3.0		3000	−3.5	3.29

† C. W. Allen, "Astrophysical Quantities," University of London Press, Ltd., London, 1963.

supergiants, these stars are probably in advanced stages of stellar evolution and are perhaps approaching the end of their energy-generating lifetime.

Far below the main sequence in the lower left-hand part of the H-R diagram lies another important class of stars, the *white dwarfs*. These stars are very much smaller than the sun in radius, although many of them have masses comparable to the sun. Their densities must therefore be very high. With their very small surface area, they must have high surface temperatures, making them blue or white, in order to radiate their admittedly low luminosity. Numerically, the main sequence contains more stars than all other groups put together, but the white dwarfs are the next most numerous class, perhaps 10 percent or so of all stars. These stars constitute the "stellar graveyard" in that they represent the end products of stellar evolution. They consist of degenerate matter, having such a high density that the electrons have effectively filled all the available cells in momentum space. This situation results in large internal pressures, which are capable of supporting the structure. They have no internal energy sources left, so that their residual supply of thermal energy is being gradually radiated into space. They are presumably cooling off with no future expectation of active stellar life. Although there are few white dwarfs for which good parallax measurements exist, and although the appropriate relation between color indices and surface temperature is somewhat uncertain, as a result of peculiarities in the equation of state, the H-R diagram of the white dwarfs is still remarkable. It certainly indicates that these stars are very hot (when observable!) objects with very small radii. The average radius would appear to be approximately one-hundredth of the sun's radius.[1]

Problem 1-29: Calculate the radius of a white dwarf having a luminosity $L = 10^{-2}L_\odot$ and an effective temperature $T_e = 10^4$ °K.

The H-R diagrams of clusters of stars offer a number of important opportunities for the study of stellar evolution. Since the stars in such clusters are all essentially the same distance from the observer, their relative luminosities can be measured with high accuracy. The absolute luminosity of every star in the cluster can be determined as soon as the absolute luminosity of one star is found. The distance must often be determined by the apparent magnitude of a characteristic type of star whose absolute magnitude is believed known. Another advantage of clusters is that all stars in a cluster have essentially the same age; i.e., they probably formed at approximately the same time from the interstellar medium. Of course, all stars in a cluster cannot have formed at *exactly* the same time. It seems reasonable that formation times in a given cluster may differ by times of the order 10^8 years. Accordingly, the assumption of simultaneous formation is believed to be a good one for clusters having ages much greater than

[1] For a discussion of white dwarfs read W. J. Luyten, "Advances in Astronomy and Astrophysics," vol. 2, p. 199, Academic Press Inc., New York, 1963, and O. J. Eggen and J. L. Greenstein, *Astrophys. J.*, **141**:83 (1965).

10^8 years. It turns out that the H-R diagram of a star cluster is different from that of field stars in general, and, in addition, the clusters differ greatly from each other. Their very striking H-R diagrams provide important clues for determining the history of the galaxy.[1]

1-7 STELLAR POPULATIONS

Significant differences in average properties exist between the various types of star clusters, which may be sorted by various criteria into a sequence of associated characteristics, called *sequences of stellar populations.* This classification sequence apparently amounts to sorting the clusters according to their age and composition. The population concept has been a very useful one for the science of galactic dynamics and evolution. We shall consider briefly several criteria for making such a classification.

A *star cluster* is identified as a group of stars that have a much stronger gravitational attraction to each other than to general field stars (nonmembers). The number of stars in a cluster varies from about 10^5 for the richest clusters to loose associations of only a few stars. The richest clusters are massive spherical ones containing typically 10^5 stars called *globular clusters.* All are located very far from the sun, although a few are barely visible to the naked eye, appearing as a single fuzzy star. The distances to globular clusters are usually determined by the apparent magnitude of RR Lyrae variable stars characteristically found there. The diameter of the high-star-density region is typically tens of parsecs, and the central densities are as great as 10^3 pc^{-3}, a very high star density.

Problem 1-30: If the apparent magnitude of the $M_v = 0$ RR Lyrae variables in a given cluster is +15, what is the distance to the cluster?

The *open clusters* are more irregular groupings of a few to a few hundred stars arranged more or less at random and showing no concentration toward the cluster center. They are often called *galactic clusters* because they are found only in the disk of the galaxy, in contrast to the globulars, found far from the galactic plane. A special type of open cluster containing the most luminous main-sequence stars of type O and B is called an *association.* They are detectable by the fact that O stars are not randomly distributed in the sky but tend to be relatively near other O stars, usually as a group of several O, B, and often Wolf-Rayet stars, scattered over a characteristic dimension of 100 pc. A small open cluster is often found near the center of an association, and it is generally assumed that these stars share a common origin reflecting the peculiarities of star formation. Table 1-3 lists the gross characteristics of star clusters. The locations

[1] A rather complete discussion with extensive bibliography may be found in H. C. Arp, The Hertzsprung-Russell Diagram, in S. Flugge (ed.), "Handbuch der Physik," vol. 51, Springer-Verlag OHG, Berlin, 1958 (in English).

Table 1-3 Characteristics of star clusters†

	Number known	*Location in galaxy*	*Diameter, pc*	*Number of stars*	*Color of brightest stars*	*Density, stars pc³*	*Examples*
Globular	119	Corona and nucleus	50–100	10^4–10^5	Red	0.5–10^3	M 3
Open	867	Disk	10	50–10^3	Red or blue	0.1–10	Hyades, Pleiades
Associations	82	Spiral arms	30–200	10–100?	Blue	0.01	Orion

† G. Abell, "Exploration of the Universe," Holt, Rinehart and Winston, Inc., New York, 1964.

of these stars may be conceptually aided by the schematic drawing of the galaxy shown in Fig. 1-13.

The brightest stars in stellar systems generally show a striking correlation to the geometry of the system. In most galactic associations the brightest stars are blue giants of type O or B, whereas in globular clusters they are the luminous red giants. In open clusters the brightest stars are either upper-main-sequence stars or red giants, depending on the age of the cluster. A corresponding effect can be found in galaxies. Spiral galaxies contain types O and B stars as their brightest members, whereas elliptical galaxies contain luminous red giants as their brightest members. There are several reasons for believing these types of

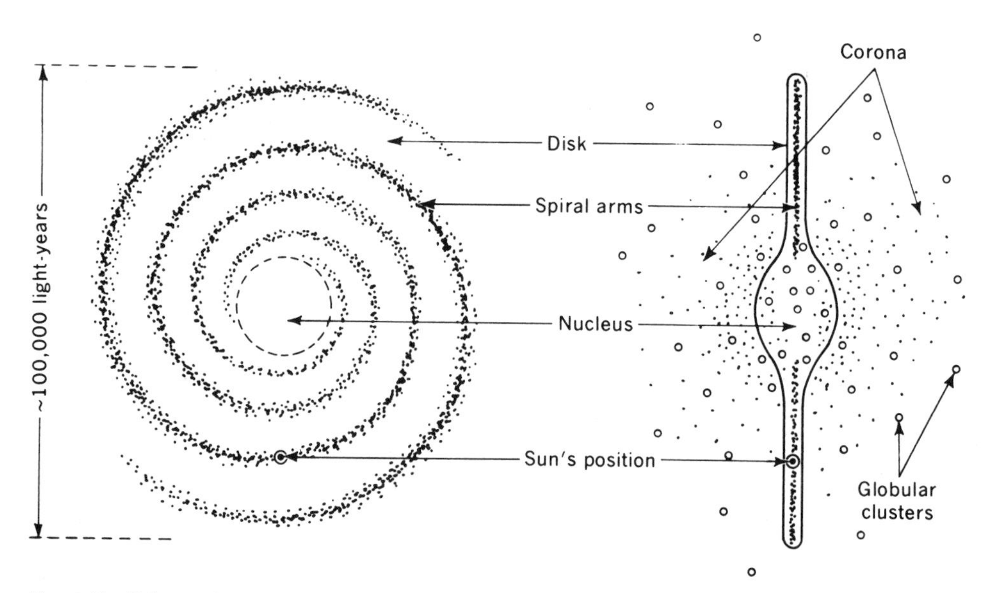

Fig. 1-13 Schematic representation of the galaxy. *(G. Abell, "Exploration of the Universe," Holt, Rinehart and Winston, Inc., New York,* 1964.)

clusters have different ages. (1) Highly luminous blue stars are found in or near clouds of interstellar gas and dust, out of which they have presumably just formed, not having had time enough to move away. The galactic clusters (or associations) in which blue stars are found often appear to be associated with large clouds of interstellar gas. (2) Some galactic associations with blue giants as their most luminous members are observed to be expanding from a common center. The rate of expansion and the present size of the cluster give an estimate of the age of the cluster. In some cases, this age is less than 10^8 years, and for certain associations only a few million years, a small age indeed in terms of stellar time. (3) For the globular clusters there is no direct observational estimate of their ages, but they contain no observable interstellar gas and are highly stable dynamical systems capable of long life. Age seems to account for the redness of the brightest stars in globular clusters. The theory of stellar evolution suggests that main-sequence stars turn red and luminous as they age. The quantitative correspondence with the H-R diagrams of the globular clusters indicates likely ages of 1 to 2×10^{10} years. It therefore appears that the globular clusters are old objects indeed. It is now conventional, following Baade, to divide stars into a relatively young class with blue giants as the most luminous members, called *population I*, and an old class with red giants as the most luminous members, called *population II*. Figures 1-14 and 1-15 show typical examples of two stellar clusters. The brightest star in the Pleiades, a galactic cluster of stars with a common origin, is blue. The cluster M 3, typical of globular clusters, consists of about 10^5 stars with spherical symmetry, and its brightest stars are red giants.

A related comparison of these two clusters stems from their H-R diagrams. It will be seen in Fig. 1-16 that the H-R diagram for the Pleiades falls along the main sequence. The H-R diagram of the globular cluster M 3 (Fig. 1-17), on the other hand, shows that the main sequence terminates at about $B - V = 0.4$. The concentration of stars then appears to veer off to the right, passing through the subgiant region into the more populated region of the red giants. The brightest stars in this cluster are red. The marked contrast of this M 3 diagram to that of the Pleiades, or almost any galactic cluster, is due to the evolution of stars from the upper tip of the remaining main sequence into the giant region. This evolution is typical for globular clusters and indicates an old age for them.

There also appears to be a correlation between the kinematical behavior of stars and their age. A useful measurement is that of a star's velocity relative to the orbital velocity about the galactic center. Stars in the solar neighborhood, for instance, may be arbitrarily divided into a high-velocity group and a low-velocity group. It has been found that the high-velocity group contains many red giants and has a collective H-R diagram similar to that of globular clusters. The clusters themselves have similar properties: galactic clusters have low relative velocities, often lingering near the interstellar gas out of which they have just formed, whereas globular clusters have high relative velocities, moving in large orbits about the galactic center. The average space velocity may also be

Fig. 1-14 The Pleiades, an open star cluster visible to the naked eye. Because the stars of this cluster are in close proximity to each other, they probably had a common origin. (*Dominion Astrophysical Observatory, J. A. Pearce.*)

correlated with the average distance from the galactic plane. Clearly, such a correlation should exist, because the stars with high kinetic energy may convert it into potential energy by rising to large distances from the galactic plane, whereas those with low kinetic energy cannot overcome the gravitational attraction presented by the mass of the galaxy. In keeping with this observation, the blue stars of types O and B are confined in a flat system to the galactic plane. The globular clusters, however, occupy a large spherical volume reaching to enormous distances from the galactic plane.

Fig. 1-15 The globular cluster M 3. This great spherical shower of about 10^5 stars had a common origin in the early epochs of the galaxy. (*Photograph from the Mount Wilson and Palomar Observatories.*)

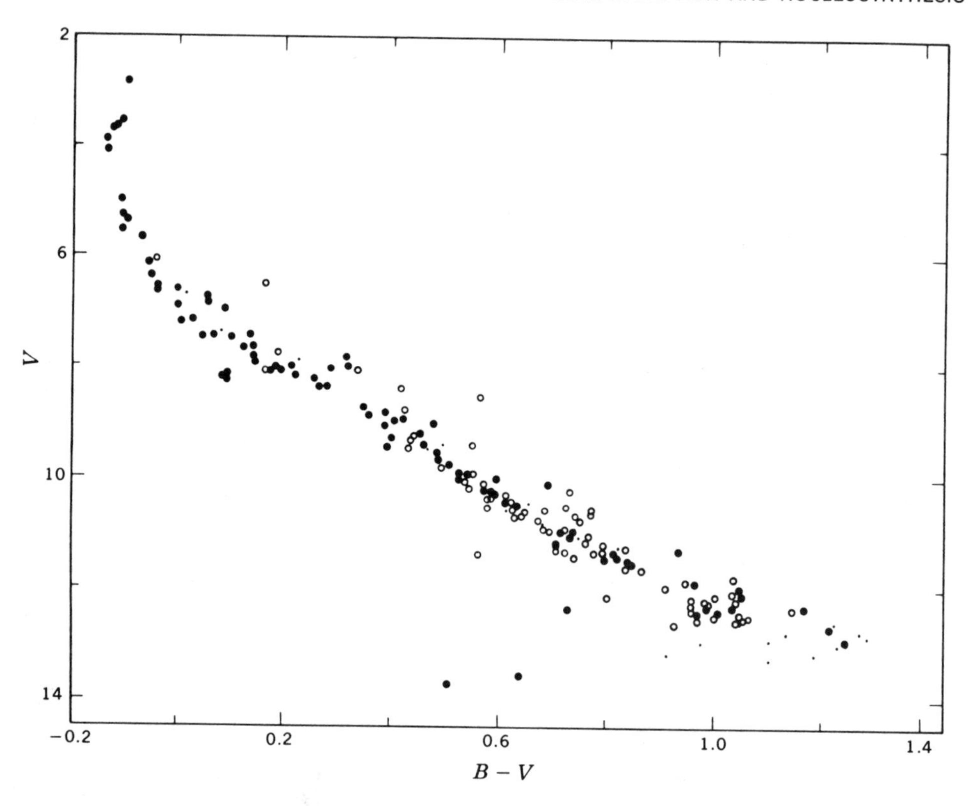

Fig. 1-16 Color-magnitude diagram of the Pleiades with reddening and absorption removed. This young galactic cluster falls very close to the zero-age main sequence, although the upper end appears to have moved rightward. [*After R. I. Mitchell and H. L. Johnson, Astrophys. J.*, **125**:418 (1957). *By permission of The University of Chicago Press. Copyright* 1957 *by The University of Chicago.*]

The correlation between age and average space velocity must certainly reflect the initial state of our galaxy. Suppose, for example, that our galaxy consisted of gases in a highly turbulent state at the time it formed and became an independent system. If the initial scale of the turbulence was quite large, massive quantities of gas may have moved coherently with high velocities relative to other large masses of gas. Any stars formed then will have inherited that same velocity. As time passed, however, the turbulent velocities of these large masses of gas must have dissipated into heat during viscous collisions. Eventually, the gas would have gravitationally contracted to the galactic disk, preserving only its angular momentum. Thus, late-forming stars and star clusters will have inherited low space velocities appropriate to the interstellar medium after the dissipation of the larger bulk velocities. If the initial state of the galaxy was as we have just imagined it, the observed correlation between ages and space velocities is not too difficult to accept. It must be kept clear, however, that these considerations of population classes apply to the statistics of groups of stars, and not to individual stars. For example, a few blue stars will individually have high relative velocities and may be found at large distances from the galactic plane. Some globular clusters may be found with low velocities. It is to the average properties of the objects under consideration that the generalizations of stellar populations may be applied.

Interesting photographic evidence of the spatial distribution of population classes in the spiral galaxy NGC 5194 can be seen in Figs. 1-18 and 1-19, which are, respectively, composites of a blue negative and a yellow positive, accentuating population I, and a blue positive and yellow negative, accentuating population II. By all odds the most dramatic feature of these photographs is the concentration of bright blue stars in the gas and dust lanes in the spiral arms. The older red component, on the other hand, occupies a much more nearly spherical volume. This may be an indication that spiral arms themselves are not static features of galaxies but are formed continuously by natural dynamic effects.

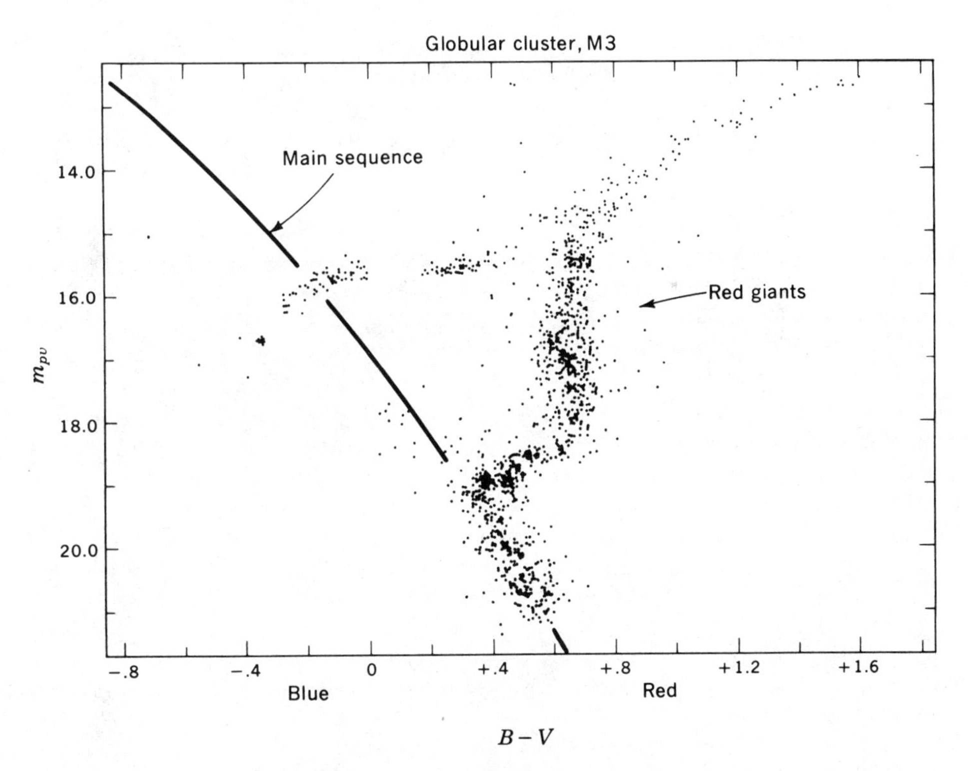

Fig. 1-17 Color-magnitude diagram of the old globular cluster M 3. The main sequence terminates near $B - V \approx 0.3$ and swerves upward along the subgiant branch into the giant region. [*After H. L. Johnson and A. Sandage, Astrophys. J.*, **124**:379 (1956). *By permission of The University of Chicago Press. Copyright* 1956 *by The University of Chicago.*]

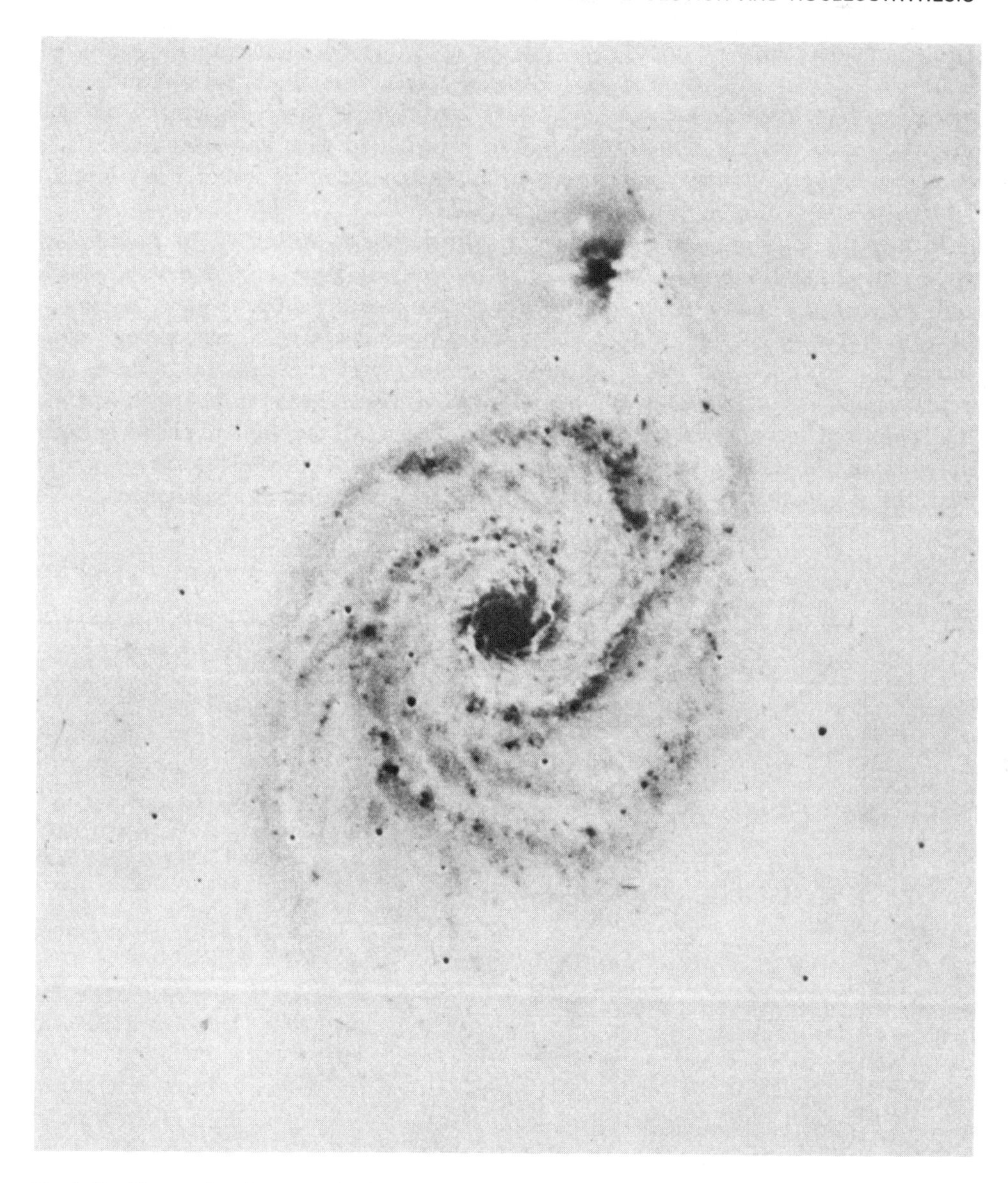

Fig. 1-18 Composite photograph of the galaxy NGC 5194 and its smaller companion NGC 5195 made by superposing a blue negative on a yellow positive. The resulting dark patches show the concentrations of blue stars in the dusty spiral arms where they are formed. (*Official U.S. Navy photograph.*)

Some further classification of population I stars has been made on the basis of spectroscopic differences. Recall that population I stars are relatively young groups of stars, their most frequent location being in or near the interstellar gas and dust that concentrates in the spiral arms of spiral galaxies. Since the sun has such a location, we may expect that the solar neighborhood is predominantly of population I. It is, therefore, only for the population I stars, or nearby stars, that extremely good statistics about spectral characteristics exist. A particularly useful study has been made of the strength of the absorption lines of metals in stellar atmospheres. Presumably, the strength of these absorption lines, for

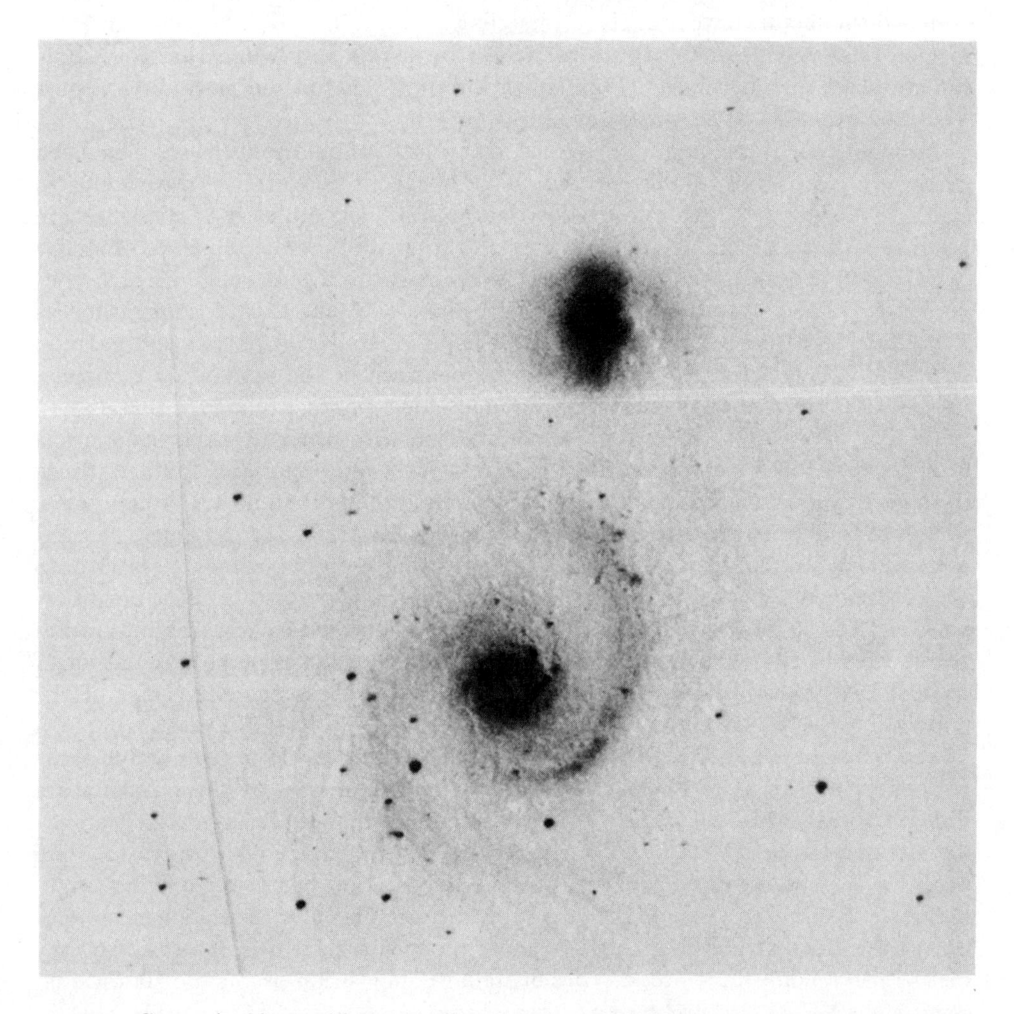

Fig. 1-19 Composite blue positive and yellow negative of NGC 5194/5. Comparison with Fig. 1-18 shows the yellow region to be much smoother and concentrated in the nucleus and inner arms. (*Official U.S. Navy photograph.*)

stars of the same spectral type, will be proportional to the abundance of the metallic elements in the stellar atmospheres. Because there are some small differences in the strength of these metallic lines in population I stars, these stars are further subdivided into two classes called *weak-line stars* and *strong-line stars*. The strong-line stars are found to have slightly smaller relative velocities than the weak-line stars. In keeping with the previous assumption about the meaning of relative velocities, the weak-line stars would be interpreted as the older of the population I class. Recalling that stellar atmospheres show in general the abundances of products that were there at the time the star formed, and noting that nucleosynthesis accompanying stellar evolution is believed to have gradually enriched the metal content of the interstellar medium, we would expect that stars formed relatively recently would be richest in metals and would therefore show the strongest metallic lines. Population I stars formed in the more distant past would be expected to be weak-line stars.

Although the foregoing division of stars into population classes has been extremely suggestive of the correlation with age, it is certainly misleading to think that stars fall into two or, for that matter, any number of classification groups. All the available evidence seems to indicate that the sequence of stellar populations is continuous and that divisions of it are a matter of research convenience. The correlation of population groups with age should be regarded as a working hypothesis for research in the fields of stellar evolution and galactic structure. It is definitely known that the youngest of the population I objects, the galactic associations containing luminous blue giants, are only a few million years old. By careful spectroscopic comparison with other G stars, the sun has been classified as a weak-line star, i.e., a relatively old population I star. Since from geophysical arguments the age of the sun is thought to be 4.6 billion years, an age of 5 billion years can be thought of as representative of the oldest of what astronomers call population I stars. At the other extreme, recent research into the evolution of stars has indicated ages for some globular clusters in the neighborhood of 10 to 20 billion years, an age roughly corroborated by research into radioactive ages of the elements. It may be, therefore, that extreme population II objects have an age as great as 20 billion years. These characteristics of the population classes are summarized in Table 1-4.

The differences in composition between the population classes are active areas for contemporary astronomical research. The abundances of the elements are difficult to ascertain accurately in any stellar atmosphere, let alone in the very distant population II stars. Yet the results of very ingenious observations are accruing at a rapid rate,[1] and the present information has some striking implications. Excluding those highly evolved and peculiar stars whose atmospheres reflect the element building actually going on within their own interior, we may distinguish among three broad types of stellar-composition problems: (1) the sun,

[1] See H. Hubenet (ed.), "Abundance Determinations in Stellar Spectra" (IAU Symposium No. 26), Academic Press Inc., London and New York, 1966.

Table 1-4 Sequence of stellar populations

Population class	*Typical members*	*Average velocity* V_z, *km/sec*	*Shape of subsystem*
Young population I	Blue giants, galactic clusters	8	Flat
Intermediate population I	Strong-line stars, solar neighborhood	10	Intermediate
Old population I	Weak-line stars	16	Intermediate
Mild population II	Majority of high-velocity stars	25	Intermediate
Extreme population II	Bright red giants, globular clusters, subdwarfs	75	Spherical

whose composition presumably represents the original composition of the solar system and is, as we have pointed out, an old population I star; (2) the very ancient stars, such as the members of globular clusters and the subdwarfs, or extreme population II stars, which show low metal-to-hydrogen ratios; and (3) the present interstellar medium and stars that are formed therefrom, the very young population I stars. In order to determine composition differences between stars from different population groups, it is advantageous to select stars which are otherwise as similar as possible in such large-scale properties as surface temperature and absolute magnitude. In this way one can hope to minimize the effects of differences in these two very important large-scale parameters. Both the class F dwarfs and the class G and class K giants are well represented among the brighter stars of various populations. As such, they have been the object of extensive abundance comparisons.

The results of these observations seem to indicate that nucleosynthesis in stars has slowly enriched the galaxy in heavy elements. This conclusion is quite firm if the initial composition of the gas now in the disk of the galaxy was the same as that of population II. There is still an unfortunate uncertainty regarding the helium abundance in extreme population II objects. It has been both tempting and common to think of the galaxy as being initially composed of hydrogen, but it is yet to be shown whether extreme population II objects are as deficient in helium as they are found to be in heavier elements. The difficulty with helium is that its lines can be excited only in hot stars, whereas extreme population II is almost entirely red. The most abundant elements other than hydrogen and helium, viz., carbon and oxygen, are smaller by at least a factor of 50 in extreme population II than they are in the sun. But their abundance in mild population II stars appears to be smaller by only a factor of 3 than it is in population I stars. It is not clear whether these abundances have increased noticeably with time in population I itself, however. This important question requires extensive accurate analyses of stellar atmospheres.

The situation is more dramatic for the heavy metals. Abundance analyses of certain K giants in globular clusters have shown heavy-metal concentrations that are less than 1 percent of the present-day metal concentrations in the interstellar medium. The elements Mn, Fe, and Ba are found in some instances to be rarer by a factor of 1,000 than in the present sun. At first glance, it would appear from these observations that some synthesis of carbon, nitrogen, and oxygen (and helium?) in our galaxy occurred more rapidly than the synthesis of the heavy metals did. That is not certain. But the understanding of these observations probably lies in the application of the principles of stellar evolution and nucleosynthesis, although uncertainties in the initial composition of the galaxy complicate the argument. There is no doubt that these abundance features are an essential part of the population classifications.

The scarcity of the metals in the population II groups led to the introduction of some extra classes of stars. One of the best examples is provided by a group of stars which falls just below the main sequence in ordinary color-versus-magnitude diagrams, a class of stars called *subdwarfs*. Spectra and color classifications were initially difficult to make in conventional systems based on normal compositions because of the weakness of the metallic lines in these stars. As previously mentioned, most ionized metals have their very numerous resonance lines lying in the ultraviolet. The result is a reduction in the amount of energy radiated in the ultraviolet by stars that are rich in metal content. For the subdwarfs the reverse is true. These high-velocity stars rather clearly belong to population II, having eccentric and far-reaching orbits about the center of the galaxy. When it was discovered that the low metal content allowed a much higher amount of energy to be radiated in the ultraviolet, it became apparent that a larger bolometric correction was required for low-metal stars than for high-metal stars. When the larger bolometric correction was made, it was found that these population II subdwarfs have essentially the same position on the H-R diagram as the population I dwarfs, their natural brothers.

In summation, it should be emphasized that there appear to be no sharp distinctions between population classes. Intermediate objects always seem to exist in any classification scheme. The ultimate observational aim in this regard seems to be the description of clusters in terms of their age, composition, and dynamic properties.[1]

1-8 STELLAR EVOLUTION

Science provides a description—a logical map that integrates countless individual observations into a prescription which neatly summarizes old facts and correctly predicts new ones. The science of stellar evolution describes how the observable properties of stars may sensibly change as time passes. Therein lies the peculi-

[1] See, particularly, O. J. Eggen, D. Lynden-Bell, and A. Sandage, *Astrophys. J.*, **136**:748 (1962); W. Baade, in C. Payne-Gaposhkin (ed.), "Evolution of Stars and Galaxies," Harvard University Press, Cambridge, Mass., 1963; A. Blaauw, The Concept of Stellar Populations, in L. H. Aller and D. B. McLaughlin (eds.), "Stars and Stellar Systems," The University of Chicago Press, Chicago, 1965.

arity of the science: man's lifetime and even recorded history are hopelessly short compared to most characteristic times over which we believe the appearance of a star may change. There are exceptions; very rarely stars explode, some stars are variable, and certain characteristics of the sun recur with an 11-year cycle, for instance; but basically the sun is thought to be as it was hundreds of millions of years ago. Indeed, for the vast majority of stars it seems that gross changes are hopelessly unobservable. Then how can such a science be given a firm observational basis?

The basic idea can most clearly be isolated by imagining the following situation. Suppose the billions of observable stars were identical at birth but had birthdays ranging continuously from today into the farthest reaches of history. A snapshot of the sky would then reveal the same sequence of stars as would be obtained by watching the history of a single star, with, however, one major difference: there would be no explicit clock labeling each star. By careful interplay of auxiliary observations with theoretical arguments based upon known physical principles, the stars of the snapshot could be ordered with respect to age. As clues were found relating one stellar age to another, a self-consistent description of the evolution of a single star could be developed. If one then adds the facts that stars are born with a spectrum of masses out of interstellar gas whose composition varies from place to place and time to time, one obtains the framework of the science of stellar evolution.

The history of stellar evolution as a quantitative physical science is dominated by four books. The first of these, "Gaskugeln," written by Emden in 1907, treated the theory of polytropic gas spheres (see Chap. 2). In this approximation the equation of state of the gas was combined with the condition of hydrostatic equilibrium to yield the properties of dynamically stable gas spheres.

Eddington, in his classic work "The Internal Constitution of the Stars" (1930), showed that the energy was commonly transported from the center to the surface by the process of radiative transfer. He developed the corresponding theory relating energy flux to temperature gradient and opacity in detail (see Chap. 3).

In a book[1] first published in 1939, Chandrasekhar reviewed and formalized the results of Emden and Eddington. Even more important, he applied the theory of quantum statistical mechanics to the degenerate gas and constructed therefrom a theory of the white dwarfs. Degenerate matter is now known to be of importance in many stages of stellar evolution, particularly of low-mass stars. Chandrasekhar's book was also the first to include a discussion of thermonuclear reactions as the internal source of stellar energy. The work of Bethe and von Weizsäcker showing that hydrogen could be fused into helium nuclei was occurring at just the time Chandrasekhar's monograph was being completed.

With the aid of the intensive laboratory work in nuclear physics conducted in the 1940's and 1950's it became possible to compute the rate at which energy is liberated in a hot gas. For the first time astronomers were able to compute

[1] S. Chandrasekhar, "An Introduction to the Study of Stellar Structure," Dover Publications, Inc., New York, 1957.

numerical models of stars in which full account is taken of the budget relating energy production and energy transfer in a static situation. Particularly influential was the work of Hoyle and coworkers in England, Schwarzschild and coworkers at Princeton, and Henyey and coworkers at Berkeley. These developments are detailed in the fourth milestone, "Structure and Evolution of the Stars," published by Schwarzschild in 1958.[1] This book has served as a point of departure for the large number of active workers in stellar evolution in the sixties. The electronic digital computer revolutionized the science, and the elegant hand integration techniques discussed in Schwarzschild's book are no longer required.

The chemical change brought about by the nuclear reactions provided the key to stellar evolution. It was found that homogeneous stars burning hydrogen lie on the main sequence and that stars with chemical inhomogeneities can lie in the red-giant region. It was, in fact, shown that the exhaustion of hydrogen over a sufficient central fraction of the star will cause it to move in the H-R diagram from the main-sequence to the giant region. Eddington had argued, however, that stellar rotation would provide circulational mixing of the stellar material (see Chap. 6). It remained for the work of Sweet and Mestel in the early 1950's to show that the mixing is much slower than Eddington had estimated and that the chemical inhomogeneities can develop. Present theories of stellar evolution are based almost entirely on models having concentric shells of differing composition.

The quantitative theory of stellar evolution confirms and strengthens the belief in the scale of ages outlined in Sec. 1-7. Calculations have been made of the time required for stars to age, and the results provide another relationship between the structure of the H-R diagram of star clusters and the division of the clusters into population classes. For example, in the schematic representation of the H-R diagram shown earlier (Fig. 1-12), a dashed line leads from the main sequence up toward the giant region and back again along a horizontal branch. In old star clusters the upper main sequence is not observed. Instead one finds that the locus of stars turns from the lower main sequence in just such a manner as outlined by this dashed curve. The H-R diagram of the globular cluster M 3 is a beautiful example of such an observation. To understand why that subgiant branch is observed instead of the upper main sequence it is necessary to use the results of the study of stellar evolution.

When stars form from gas in the interstellar medium, they are composed predominantly of hydrogen. They contract until the central temperature becomes sufficient to cause hydrogen thermonuclear reactions to begin, at which time they radiate energy at a rate equal to that liberated by the nuclear reactions in their interiors. They remain static structures for a lifetime that is determined by how long it takes them to consume the available hydrogen fuel in their centers. When the innermost 10 to 20 percent of the hydrogen in the core of the star has

[1] M. Schwarzschild, "Structure and Evolution of the Stars", Princeton University Press Princeton, N.J., 1958.

been exhausted, the star changes its observable characteristics in the sense that its outer regions expand and the inner core contracts. The net effect of this change in upper-main-sequence stars is to preserve roughly the luminosity of the star but to change its surface color toward the red. In this way stars are said to evolve off the main sequence into the giant region. The first stars to do so are the most massive members of the main sequence and also the most luminous. The mass-luminosity relationship is evidence of the fact that the luminosity is proportional to the third or fourth power of the mass; therefore, as the mass of stars increases, their rate of converting hydrogen into helium increases at an even faster rate. Thus a massive star will exhaust 10 percent of its hydrogen much more quickly than will a star of lower mass.

To see how this affects the H-R diagram of a cluster, let it be assumed that all stars within a given cluster form at the same time. Figure 1-20 then shows a schematic representation of the H-R diagram of that cluster of stars at three different points in time: (1) at the time of the cluster's formation, (2) after a relatively short time of perhaps some hundreds of millions of years, and (3) after a very long time, say, 10 billion years. It will therefore be expected that the appearance of the H-R diagram of a cluster of stars will be determined by the age of the cluster. The older the cluster is, the smaller the fraction of its upper main sequence that remains. This statement may be made semiquantitative since the conversion of hydrogen to helium releases 6.4×10^{18} ergs/g of hydrogen converted. Suppose then that a star of mass M must build up a helium core of mass fM before evolving from the main sequence. If X_{H} is taken to be the fraction of hydrogen by weight in the original uniform composition, the buildup of

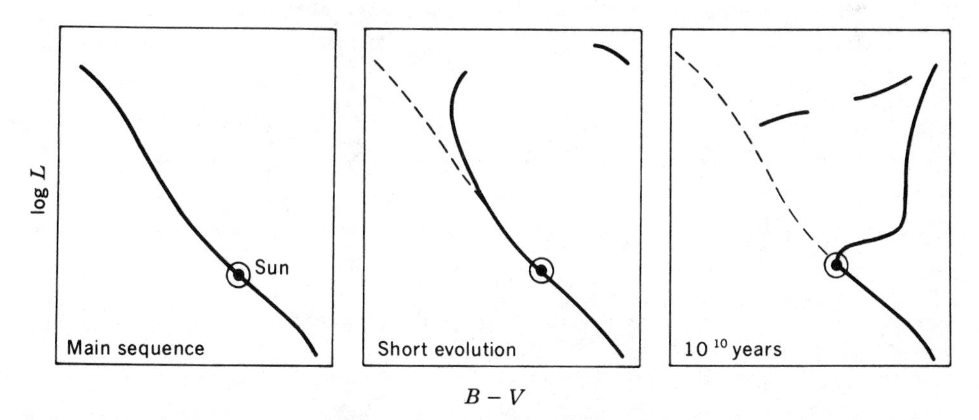

Fig. 1-20 Schematic representation of the H-R diagram of a cluster of stars at three different epochs in its history. After a short period of evolution, say about 10^8 years, the zero-age main sequence has become an evolved main sequence similar to the Pleiades. After a long period of evolution, say about 10^{10} years, the diagram resembles those of the globular clusters. Supergiants and white dwarfs have been omitted from this diagram because their participation is poorly understood.

the necessary hydrogen-exhausted core requires the conversion of $fX_{\rm H}M$ g of hydrogen into helium. During that time the total energy radiated from the star will be given by the mass of hydrogen converted times the energy released per gram converted:

$$\begin{aligned} E &= fX_{\rm H}M(6.4 \times 10^{18}) \\ &= 1.3 \times 10^{52} fX_{\rm H} \frac{M}{M_\odot} \qquad \text{ergs} \end{aligned} \tag{1-60}$$

The evolutionary lifetime T_E of the star on the main sequence may be estimated as the ratio of the total energy radiated during the main-sequence lifetime to the average luminosity of the star on the main sequence. Quantitatively, we have $T_E = E/L$, or

$$T_E = 1.1 \times 10^{11} fX_{\rm H} \frac{M/M_\odot}{L/L_\odot} \qquad \text{years} \tag{1-61}$$

Problem 1-31: Confirm Eq. (1-61).

The fraction of the stellar mass at the star's core that must be depleted of its hydrogen before the star expands rightward from the main sequence is near 15 percent for stars with a mass and composition near the solar values. Introducing this number yields

$$T_E = 12 \times 10^9 X_{\rm H} \frac{M/M_\odot}{L/L_\odot} \qquad \text{years} \tag{1-62}$$

An approximate statement of the mass-luminosity relationship may also be introduced:

$$\frac{M}{M_\odot} \approx \left(\frac{L}{L_\odot}\right)^{\frac{1}{4}} \tag{1-63}$$

If we take $X_{\rm H} = 0.6$ as characteristic of most stars, the lifetime becomes

$$T_E \approx 12 \left(\frac{L}{L_\odot}\right)^{-\frac{3}{4}} \times 10^9 \qquad \text{years} \tag{1-64}$$

Problem 1-32: Estimate the approximate main-sequence lifetime of a star with absolute visual magnitude $M_{\rm vis} = +3.0$; with $M_{\rm vis} = -2$.

By Eq. (1-64) we may roughly associate with each magnitude on the main sequence a time required for such a star to evolve from the main sequence. The most luminous stars in a stellar cluster still remaining on the main sequence can then be made to yield the age of the cluster. A reliable calculation of the age must proceed along a considerably more detailed route than the simple arguments leading to Eq. (1-64), which is by no means correct in detail. It is the

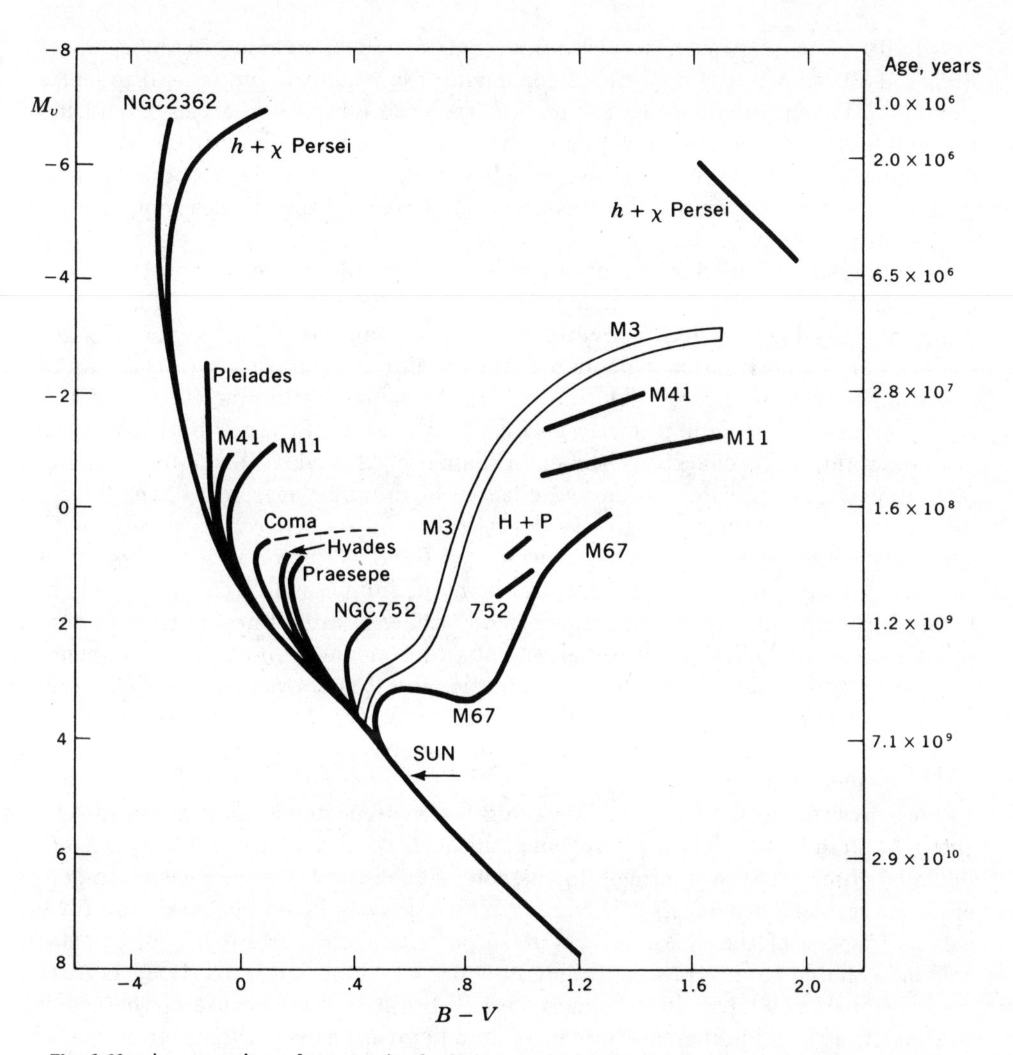

Fig. 1-21 A composite color-magnitude diagram of 10 galactic clusters and 1 globular cluster. Ages corresponding to the various main-sequence termination points are given along the right-hand ordinate. The zero-age main sequence is taken to be the blue envelope of the observed sets of main-sequence stars. Notice the rapidly evolved red giants in $h + \chi$ Persei, which are apparently no more than 2 million years old. Some white dwarfs are known in the Hyades, indicating that it is possible to form them in a few million years, either directly or as the end product of the evolution of upper-main-sequence stars. Curiously enough, the Hyades has no red giants. The oldest galactic cluster, M 67, is older than the sun and has scores of white dwarfs. Many fascinating problems are uncovered in the attempts to interpret the star densities in these diagrams quantitatively. [*After A. Sandage, Astrophys. J.*, **125**:435 (1957). *By permission of The University of Chicago Press. Copyright* 1957 *by The University of Chicago.*]

possibility of making such a calculation that should be evident to the reader.[1] Figure 1-21 shows a composite color-magnitude diagram for several galactic clusters. The approximate length of time required for stars of given magnitude to evolve off the main sequence is shown on the right. The crucial point for determination of the age of each cluster lies in determining the absolute magnitude of the turnoff point of the cluster, or equivalently the surface temperature of the turnoff point of the cluster.

The foregoing discussion has been a simplification in another respect; viz., the characteristics of a star also change somewhat during its lifetime on the main sequence. The conversion of hydrogen to helium at the stellar center changes the central composition continuously from the initial composition of the star to a hydrogen-exhausted core. During this time, the stellar luminosity and radius both increase, resulting in a nearly vertical small rise of the star's position in the H-R diagram. The concept of the main sequence, as used to this point, is therefore somewhat imprecise. Identical clusters of differing ages will have slightly different main sequences due to this effect: the main sequence rises slightly with age. As a more precise concept Johnson and Sandage introduced the notion of the *zero-age main sequence,* which is, as the name implies, the locus of hydrogen-burning stars of uniform composition (there having been no burning time to augment the central helium). For a given composition, the zero-age main sequence is a perfectly well defined concept. Operationally it is determined by close comparisons of clusters of differing ages. Unfortunately, the possible composition differences between clusters reintroduces ambiguity into the measured zero-age main sequence.[2]

The physical cause of the star's leaving the main sequence as it turns toward the right on its path to the giant region is the onset of the collapse of the hydrogen-depleted core. As the hydrogen in the core is exhausted, the core ceases to generate energy and becomes isothermal. When this condition becomes true for a sizable fraction of the mass, say about 15 percent, the central region of the star becomes unable to generate sufficient pressure to support the overlying layers.[3] As it begins to collapse, its temperature will begin to rise because of the gravitational work. This increase in central temperature causes the outer layers of the star to expand in order that the temperature gradient not be too great. The surface therefore reddens. The gravitational contraction will be stopped when energy-producing thermonuclear reactions start with helium nuclei.[4] The red giants are believed to gain their energy from just such a source. The star will remain somewhere in the giant region while helium is being converted into carbon

[1] Incorrect assumptions used in obtaining Eq. (1-64) are (1) an exact mass-luminosity relationship, (2) constancy of stellar mass, (3) constancy of main-sequence luminosity, (4) uniformity of original composition, and (5) uniformity of central mass fraction f.

[2] See, particularly, I. Iben, Jr., *Astrophys. J.*, **138**:452 (1963) for Hyades-Pleiades discussion.

[3] M. Schonberg and S. Chandrasekhar, *Astrophys. J*, **96**:161 (1942).

[4] Electron degeneracy halts the collapse in low-mass stars.

and oxygen in the deep interior of the star and while hydrogen is being depleted in a shell around the core of the star. Eventually, of course, a point will be reached where even the helium in the deep interior is also exhausted. At that time the star will undertake further gravitational contraction, and the by-products of helium burning begin to be heated toward the temperature at which they will interact. The details of stellar evolution regarding these points are by no means clear. It is obvious simply from the store of available energy, however, that after its time in the giant region the luminosity of the star must eventually begin to decrease unless the star ends its lifetime catastrophically. If it burns out slowly, its radius will shrink, and the surface temperature of the star will begin to increase again. Such a star will tend to move in some path that is not well determined from the giant region back down toward the lower left-hand corner of the H-R diagram, ending eventually in its probable fate as a white dwarf. In the trip from the red-giant region to the white-dwarf region of the H-R diagram, each star that is to remain visible must reduce its mass below a certain value known as the *Chandrasekhar limit*. Failing to achieve a sufficiently small mass, the star will encounter an instability leading to a supernova explosion. The details of mass loss of stars are also not at all well known. Stars may emit mass regularly, via a mechanism much like the solar wind, or spectacularly, as in a nova or even a supernova explosion. During the relatively rapid journey toward the white-dwarf region or toward the supernova, many exciting nuclear reactions occur in the stellar interior. These nuclear reactions probably account for the synthesis of most of the heavy elements that have been added to the interstellar medium since the formation of our galaxy. We shall attempt to trace the broad outline of the reactions that can occur and the possible effect they may have upon the evolution of the stars, but it will lie outside the scope of this book to discuss the details of the present status of the theory of stellar evolution.[1]

1-9 NUCLEOSYNTHESIS

The science of *nucleosynthesis* attempts to interpret the measured abundances of the nuclear species in terms of their nuclear properties and a set of environments in which nuclei can be synthesized by nuclear reactions. An older word, *nucleogenesis*, is now generally reserved for the question of the origin of matter itself, in whatever its primordial form. The connotations of the latter word reflect the primeval nature of that question. The science of nucleosynthesis is by common

[1] For such information the reader is referred to Aller and McLaughlin, *op. cit.*; R. F. Stein and A. G. W. Cameron (eds.), "Stellar Evolution," Plenum Press, New York, 1966; L. Gratton (ed.), "Star Evolution," Academic Press Inc., New York, 1963; and C. Hayashi, R. Hoshi, and D. Sugimoto, Evolution of the Stars, *Progr. Theoret. Phys. Kyoto Suppl.* 22, 1962. These volumes contain a host of modern articles with extensive bibliographies on special facets of stellar evolution and will often be referred to in this book. The chapters to follow will attempt only to introduce the basic physical principles common to the subject.

definition restricted to the attempt to trace in time and space the evolution of the chemical composition of the universe. The question of the origin and composition of primordial matter and the relationship of the primordial matter to the space-time structure of the universe is certainly one of the fundamental questions of natural philosophy. The present configuration of the universe can be factored into an initial configuration and its subsequent evolution, so that the cosmological questions of the initial composition may be postponed until the peculiarities of nucleosynthesis have been introduced. Indeed it may be said that one cannot fairly attack the cosmological problem of the genesis of nuclei until the science of evolution of nuclear composition by nucleosynthesis has been explored thoroughly.

The earliest attempts to construct a theory of the origin of the chemical elements relied upon the postulation of extreme conditions in a primordial state of the universe. Mayer and Teller[1] suggested that the abundances were determined by the fragmentation of polyneutron clumps from a large neutron ball. Gamow, Alpher, and Herman[1] also took neutrons as a starting point, assuming that in the subsequent expansion the decay of neutrons into protons would initiate a chain of neutron captures resulting in the observed abundances. The lack of stable nuclei at atomic weights 5 and 8 proved a difficult hurdle for this theory, and quantitative analysis shows many abundance features unaccountable by the theory. Klein, Beskow, and Treffenberg[1] suggested that nuclear statistical equilibrium in specific environments might account for the abundance features, but this suggestion also fails at key points.

Each of these early theories has attractive features that have influenced subsequent developments. In particular, the nonequilibrium theory of Gamow et al. first used *chains* of single-particle reactions, and reaction chains today are prominent in nucleosynthesis. Their specific suggestion of neutron-capture chains, moreover, has been widely employed to account for the abundances of elements heavier than iron. Although the equilibrium theory fails overall, the abundance peak near Fe^{56} may be intimately connected to the idea of nuclear statistical equilibrium introduced by Klein et al.

The major development of the last decade has been the placement of nucleosynthesis in the context of the evolution of stars rather than in the primordial state of the universe. Many different environments and compositions are encountered in the evolution of single stars, and, as a result, many types of naturally occurring nuclear reactions may be examined as potential sources for the elements. The basic working hypothesis, due primarily to von Weizsäcker, has been restated by Chandrasekhar as follows:

> *Apart from secondary effects of minor importance, the transmutation of elements is the entire cause of the presence of all elements in the stars; they are*

[1] A review of the development of these theories was written by R. A. Alpher and R. C. Herman, *Ann. Rev. Nucl. Sci.*, **2**:1 (1953).

all being synthesized continually in the stars which are assumed to have started as pure masses of hydrogen; further, transmutations are the only source of stellar energy.[1]

Nucleosynthesis in the primordial universe is still an active subject, however, especially since the discovery of a background of thermal radiation in the universe.[2] This thermal background, characterized by a 3°K spectrum, has been interpreted as the expanded residue of a primordial fireball. This primordial fireball may have produced many of the lightest elements,[3] but it now seems clear that most of those with nuclear charge $Z \geq 6$ are in fact the ashes of nuclear burning during stellar evolution. We shall adopt that viewpoint in this book, but no serious restriction of the relevance of the physical principles to be discussed is made thereby. The principles of nuclear astrophysics exist apart from specific assumptions regarding the relevant natural environments.

A necessary prerequisite for the development and testing of a theory is observational data. For nucleosynthesis the important data are the abundances of the nuclear species, now and as a function of time past. This information comes from the composition of the earth, of meteorites, and from the spectra of stars. Data from solar-system objects are regarded as a measure of the abundances in our region of the galaxy at the time of formation of the solar system. The composition of stellar surfaces is interpreted as a measure of the abundances at that point at the time that star formed unless it can be argued that nuclear products of that star have reached the surface. It was the development of abundance curves, along with their suggestive features to nuclear physicists, that exposed the many nuclear mechanisms now invoked in nuclear astrophysics.

Far and away the best abundance data in existence are those for the solar system. The pioneers in compiling abundance data were Goldschmidt, who performed geochemical analyses of terrestrial and meteoritic samples, and Russell, who first derived abundances from solar spectra. The 20 to 30 years following this work saw countless improvements culminating in influential reviews.[4] A modern estimate of these solar-system abundances is shown in Fig. 1-22. It is the features of this abundance curve that motivate most of the theories of nucleosynthesis. Many of these features will be discussed in subsequent chapters.

One of the most significant of the astronomical observations is that the heavy elements in extreme population II objects are less abundant compared to hydrogen than they are in population I objects by a factor of 100 or more. If the

[1] S. Chandrasekhar, "An Introduction to the Study of Stellar Structure," p. 469; reprinted from the Dover Publications edition, Copyright 1939 by The University of Chicago, as reprinted by permission of The University of Chicago.

[2] A. A. Penzias and R. W. Wilson, *Astrophys. J.*, **142**:419 (1965).

[3] P. J. E. Peebles, *Astrophys. J.*, **146**:542 (1966). See also R. V. Wagoner, W. A. Fowler, and F. Hoyle, *Astrophys. J.*, **148**:3 (1967).

[4] H. E. Suess and H. C. Urey, *Rev. Mod. Phys.*, **28**:53 (1956); and L. Goldberg, E. A. Muller, and L. H. Aller, *Astrophys. J. Suppl.*, **5**:1 (1960).

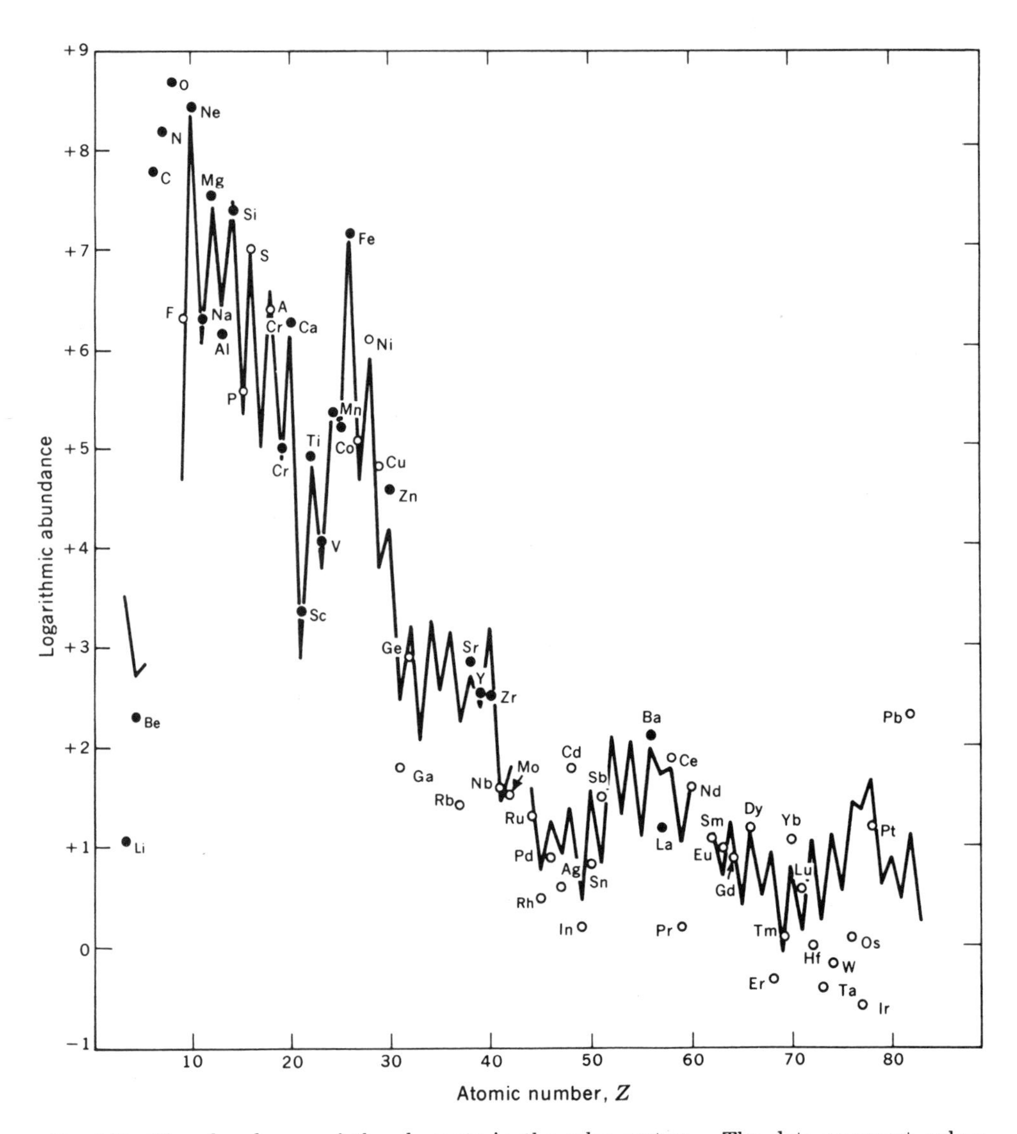

Fig. 1-22 The abundances of the elements in the solar system. The dots represent values obtained from the strengths of absorption lines in the spectrum of the sun, whereas the line represents the historic compilation of Suess and Urey, which was based mainly on chemical evidence from the earth and meteorites. Many of the estimates from both techniques have been improved since 1956, but the general features remain the same. It has been these abundance features which have inspired the nuclear physicists to seek the sets of thermonuclear circumstances that will reproduce this figure in a natural way.

initial chemical composition of the galaxy is taken to be uniform, and if no physical mechanisms capable of concentrating the heavy elements in the disk of the galaxy can be found, one is forced to the conclusion that the galaxy has synthesized at least 99 percent of its own heavy elements. This single fact is the primary motivation of the hypothesis that nucleosynthesis occurs in the natural evolution of stars.[1] As the stars evolve through their various phases of nuclear burning, the initial hydrogen (and to a lesser extent helium) are fused into heavier nuclei, which are assumed to be dispersed into the interstellar medium in the terminal phases of the stellar lifetime. This basic idea has been developed into an elaborate scheme for interpreting the abundances of the nuclear species. Of the large number of scientists who have contributed to this modern astrophysical science, the most influential have been A. G. W. Cameron, W. A. Fowler, F. Hoyle, E. E. Salpeter, and their associates, who have erected the nuclear superstructure, and E. M. Burbidge, G. R. Burbidge, J. L. Greenstein, and their associates, who have painstakingly analyzed the abundance features of stars in the attempt to correlate them with the appropriate nuclear mechanisms.

Although this basic idea has been found to work very well, there are several significant features unaccounted for by it. The major one is the question of the initial abundance of helium in the galaxy, a figure that is still uncertain because of the difficulty in detecting helium lines in the relatively cool population II objects. The stars on the blue end of the horizontal branch in population II have weak He lines compared to their population I counterparts. On the face of it this seems to indicate that helium is virtually absent in extreme population II.[2] If, on the other hand, the galaxy formed with a significant abundance of helium, it must be interpreted as the residual of some earlier history. Much the same situation occurs for rare light elements which are primarily bypassed in stellar nucleosynthesis, specifically D^2, He^3, Li, Be, and B. The initial galactic abundances of these species are unknown, and it may be that their abundances are primarily remnants of an early cosmological phase of the universe. Many alternative schemes have been suggested for the synthesis of the rare light nuclei, but the correct solution has not yet been demonstrated with certainty. The difficulty with these elements is that they are so rare in comparison with those species which are naturally synthesized as a by-product of the major nuclear burning stages that their small abundances may be accounted for by a variety of secondary processes of low efficiency. The theories adopt basically three different points of view regarding the rare light elements: (1) they are primordial in the galactic sense, in which case they carry cosmological information; (2) they are synthesized in nonthermal events at stellar surface, either spallation of heavier nuclei by energetic flare particles[3] or by shock phenomena in supernova envelopes;

[1] An excellent discussion has been presented by B. E. J. Pagel, Chemical Composition of Old Stars, in A. Beer (ed.), "Vistas in Astronomy," vol. 9, Pergamon Press, New York, 1968.

[2] L. W. Sargent and L. Searle, *Astrophys. J.*, **145**:652 (1966).

[3] W. A. Fowler, E. M. Burbidge, and G. R. Burbidge, *Astrophys. J. Suppl.*, **2**:167 (1955).

or (3) they were produced and/or modified in the early history of the solar system.[1] Although the solution of this problem has widespread implications, it will not be discussed further in this book, largely because the complicated arguments cannot be fully appreciated without an understanding of the *main line* of nucleosynthesis in stellar evolution, the principles of which constitute the burden of this book.

Not only does the astronomical evidence indicate a growth in time of the element abundances in the galaxy, but one finds considerable variations in abundance ratios from star to star as well. The extraordinary abundances observed in some stars seem to demand nucleosynthesis within the star itself. Since the surfaces of most stars retain their initial composition as the center evolves, the stars of peculiar composition seem to require either mixing of central material to the surface or large-scale mass loss of the outer layers. Probably the most dramatic evidence along these lines was the discovery by Merrill, in 1952, of spectroscopic lines in S-type stars of the element technetium, all of whose isotopes are radioactive, with half-lives shorter than a few million years, and not found terrestrially. The same stars are strongly overabundant in other elements, e.g., barium, that one now expects to be synthesized along with Tc. The combined evidence is a clear indication of nucleosynthesis within a given star. Much other evidence of this type has been observed, but we shall not attempt to chronicle this development here.[2]

The English physicist Lord Rutherford first produced another argument capable of demonstrating that the elements must have had some point of origin in time. The argument is based on the long-lived radioactive species naturally occurring in the solar system. Rutherford's argument was actually directed toward the determination of the solidification ages of rocks. By comparison of the density of helium atoms trapped following the alpha decay of uranium to the density of the parent uranium, he was able to establish ages greater than any known at that time. A slight extension of Rutherford's reasoning shows that the uranium itself cannot have an indefinitely great age. Because uranium decays to lead, one can argue that the abundance of the daughter lead would be infinitely greater than that of the parent uranium unless, in fact, uranium was created (synthesized) at times in the past not overwhelmingly greater than the uranium half-life. Today careful measurements of abundance ratios can be coupled with the details of the mechanisms of heavy-element nucleosynthesis to yield valuable information regarding the age of the elements.

The radioactive evidence for nucleosynthesis may be divided into two groups: (1) the naturally occurring long-lived radioactivities and (2) the extinct radioactivities. Because the earth is about 4.6 billion years old, the naturally occurring radioactivities (Th^{232}, U^{238}, U^{235}, Re^{187}, Rb^{87}, K^{40}) must necessarily have

[1] W. A. Fowler, J. L. Greenstein, and F. Hoyle, *Geophys. J. Roy. Astron. Soc.*, **6**:148 (1962).

[2] In their famous review paper E. M. Burbidge, G. R. Burbidge, W. A. Fowler, and F. Hoyle, *Rev. Mod. Phys.*, **29**:547 (1957) present much of the observational evidence.

very long half-lives. That feature allows them to be used to measure the time of galactic nucleosynthesis. The evidence is that heavy-element nucleosynthesis began long before the solar system formed but probably after the formation of the globular clusters. Although it is difficult to be precise because the various methods do not give concordant ages, it appears that the major early phase of galactic nucleosynthesis began somewhere between 8 and 15 billion years ago.

The extinct radioactivities, on the other hand, can be detected only by their daughter products. The significant ones have half-lives so short that they do not survive today yet sufficiently long that they could have been incorporated into the formation of solid bodies at the time the solar nebula withdrew from the interstellar gas. The most celebrated example is based on the discovery by John Reynolds that an isotope of xenon, Xe^{129}, is overabundant in certain meteorites. A majority of scientists believe that this overabundance is likely due to the trapping of I^{129}, with a half-life of 17 million years, in the meteorites at the time of their solidification. If that is so, the point to be established for the present discussion is this: no *single* event of galactic nucleosynthesis seems capable of producing the various radioactivities. The long-lived radioactivities integrate the rate of nucleosynthesis far into the past and indicate that nucleosynthesis occurred as far back as about 10 billion years ago, whereas the extinct radioactivities require nucleosynthesis shortly before the solar nebula condensed. *Both requirements may be satisfied by continuous nucleosynthesis;* so that one may observe that the abundances of radioactive parents and daughters in the solar system are both consistent with the notion that nucleosynthesis has occurred continuously in the galaxy as a by-product of the birth and death of stars.

Measurements of the relative abundances of the elements further strengthen the idea that nucleosynthesis occurs in stars. There is no doubt that hydrogen is the most abundant element in the galaxy. The next most abundant nuclear species is He^4, an unknown percentage of which may have been present initially in the galactic gas. But inasmuch as the primary nuclear burning stage in stars is that which fuses hydrogen into helium, one might also expect He^4 to be the major product of stellar nucleosynthesis. The next most abundant species after He^4 are those light nuclei with nuclear charge and mass numbers given by an integral multiple of the He^4 nucleus (C^{12}, O^{16}, Ne^{20}). It has seemed sensible to tentatively associate the synthesis of these nuclei (especially C^{12} and O^{16}) with a stellar burning stage in which helium nuclei are fused in a dense, hot helium gas.

The nuclear abundances generally decrease with increasing atomic weight until a large abundance peak is encountered, with Fe^{56} as its dominant member. That Fe^{56} is also a very special nucleus is revealed by the fact that the *nuclear binding energy per nucleon* has a maximum at Fe^{56}, which means that successive nuclear fusion reactions cease to liberate energy when all light nuclei have been fused into Fe^{56}: further fusion into heavier nuclei would *require* energy. Thus this major abundance feature may reflect the termination of the energy-generating stages of nuclear fusion. The phenomenon of the abundance peak may be viewed in a somewhat different way: under a wide range of conditions a nuclear gas assembles itself in statistical equilibrium into such a form that the binding energy per nucleon is maximized. This result of statistical mechanics (essentially a generalized Saha equation) economically characterizes the peak at iron, and the corresponding thermodynamic states are not unlike those anticipated in the terminal stages of stellar evolution.

It might at first seem that one would expect the synthesis of heavy elements to terminate with iron, at least if one envisions only equilibrium states of nuclear matter in which the total mass is converted to heavy nuclei. As an alternative, it was necessary to seek circumstances in which small traces of heavy nuclei could be processed to even heavier nuclei by nuclear reactions occurring as a *by-product* of the reactions among the dominant lighter nuclei. Capture of light charged nuclei (protons, for example) by heavy nuclei is inadequate because the coulomb barrier is much too large for barrier penetration at the temperatures characteristic of the burning of light nuclei. The solution appears to have been found by the demonstration that the very heavy nuclei can be formed efficiently by the capture of free neutrons liberated as a by-product of reactions between light charged particles. Such a strong correlation has been established between heavy nuclear abundances and their neutron-capture characteristics that the assumption must be considered as correct.

Thus by and large, one can say that the nuclear reactions in stars can produce the various nuclear species in abundances consistent with the anticipated reactions. The major features of the science are now so well demonstrated that few people doubt the existence of a correlation between nuclear abundances and nuclear properties. The major efforts are presently directed toward fitting the details of nuclear abundances to the various mechanisms and locating the sites of nucleosynthesis within the context of stellar evolution. It has become increasingly evident that the major problems lie not with synthesizing the elements in stars but with reinjecting them in a natural way into the interstellar medium. The question of the evolution of the chemical composition of the galaxy is inextricably entwined with the details of the rate of star formation throughout galactic history and the question of the composition of matter presently bound in stars.

Such considerations illustrate the nature of the problems in stellar evolution and nucleosynthesis. The remainder of this book is devoted to the basic physical principles from which the science may be constructed. In closing this introduction we note that the spirit of the inquiry was expressed by the poet:

I believe a leaf of grass is no less
than the journeywork of the stars.

WALT WHITMAN
"Leaves of Grass"

---

chapter 2

THERMODYNAMIC STATE OF THE STELLAR INTERIOR

The macroscopic properties of a star are intimately related to the microscopic phenomena occurring in the interior material. These phenomena and their rates depend upon the thermodynamic state of the material. One can calculate that in the interior environment the particles move very short distances compared to distances over which the temperature changes significantly before they collide with other particles. The rates of the fundamental atomic collision processes are, furthermore, very fast in comparison with rates of change of the local thermodynamic state. These facts enable one to assume a very important simplification in the description of the matter, viz., *local thermodynamic equilibrium*. In the state of thermodynamic equilibrium, all properties of matter are calculable in terms of the chemical composition, the density, and the temperature. In his pioneering book "The Internal Constitution of the Stars," Sir Arthur Eddington has given the following vivid description:

> *The inside of a star is a hurly-burly of atoms, electrons, and aether waves. We have to call to aid the most recent discoveries of atomic physics to follow the intricacies of the dance. We started to explore the inside of a star; we soon find ourselves exploring the inside of an atom. Try to picture the tumult! Dishevelled atoms tear along at 50 miles a second with only a few tatters left of the elaborate cloaks of electrons torn from them in the scrimmage. The lost electrons are speeding a hundred times faster to find new resting places. Look out!*

there is nearly a collision as an electron approaches an atomic nucleus; but putting on speed it sweeps round it in a sharp curve. A thousand narrow shaves happen to the electron in 10^{-10} of a second; sometimes there is a side-slip at the curve, but the electron still goes on with increased or decreased energy. Then comes a worse slip than usual; the electron is fairly caught and attached to the atom, and its career of freedom is at an end. But only for an instant. Barely has the atom arranged the new scalp on its girdle when a quantum of aether waves runs into it. With a great explosion the electron is off again for further adventures. Elsewhere two of the atoms are meeting full tilt and rebounding, with further disaster to their scanty remains of vesture.

As we watch the scene we ask ourselves, "Can this be the stately drama of stellar evolution?" . . .[1]

This chaotic situation is reduced to tractable proportions by application of the principles of statistical mechanics. Because thermodynamic equilibrium is quickly achieved on the atomic (but not nuclear) scale, the rates of all atomic (i.e., electromagnetic, but not nuclear) reactions equal those of their inverse reactions. The hurly-burly of the individual electron is replaced by a steady macroscopic state whose properties are embodied in the principles of statistical physics. The functions of state are determined by the chemical composition, density, and temperature. Foremost among these is the pressure $P = P(\rho,T)$, commonly called the *equation of state,* from which the star derives its structural support against gravity. The burden of this chapter will be the discussion of the equation of state and related phenomena.

Near the surface of the stars, the equation of state of the gas is extremely complicated. The atomic constituents of the outer layers are in varying degrees of ionization. Application of the Saha ionization equation reveals that the hydrogen constituent becomes almost completely ionized by the time the temperature has risen to about 10^4 °K, whereas the helium is almost completely ionized by the time the temperature has risen to 10^5 °K, at which temperature the heavier elements have also lost a sizable number of their electrons to the continuum and are in relatively high stages of ionization. For temperatures higher than 10^5 °K, it becomes increasingly more accurate, insofar as the pressure is concerned, to talk of a completely ionized gas. Other important properties of the gas, such as its internal energy and its opacity to radiation, are strongly dependent upon the degree of ionization. For the common stellar composition, in which hydrogen and helium comprise more than 95 percent of the mass, the pressure at temperatures greater than 10^5 °K can be calculated to high accuracy by assuming complete ionization. Significantly, a large fraction of the mass of most stars *does* lie at temperatures higher than 10^5 °K. The bulk of the structure of most stars is determined, therefore, by an equation of state appropriate to completely ionized matter. From an analytical point of view, it is extremely

[1] A. S. Eddington, "The Internal Constitution of the Stars," p. 19, Dover Publications, Inc., New York, 1959.

fortunate that this is so. A completely ionized gas behaves like a perfect gas to extremely high densities. Terrestrial matter reaches a density of only a few grams per cubic centimeter before it begins to resist compression, and the perfect-gas law begins to break down even before that density is reached. The rather large size of atoms and the interatomic forces between the electron clouds of the various atoms set a rather sudden limit to the density of un-ionized matter. The radii of nuclei, on the other hand, are only 10^{-5} of the radii of most atoms. A gas composed of nuclei and electrons, therefore, occupies only about 10^{-15} of the volume occupied by atoms. We may anticipate, therefore, that highly ionized matter can be compressed to extremely high densities before the perfect-gas law will break down as a result of the volume effect.

A *perfect gas* is defined as one in which there are no interactions between the particles of the gas. Although this criterion is never satisfied exactly in real gases, the approximation is physically sound if the average interaction energy between particles is much smaller than their thermal energies. This last condition may be satisfied by a weak interaction or by a sufficiently rarefied gas. In the ionized gas of a stellar interior the real interactions between particles are predominantly the coulomb interactions. It is fortunate that most physical circumstances in the stellar interior are such that the average coulomb energy of particles is much less than their characteristic kinetic energy, which is of the order kT for a nondegenerate gas. For this reason it will be adequate for most applications to use the equation of state of a perfect gas. We shall return later to the question of the real ionized gas and its applications.

2-1 MECHANICAL PRESSURE OF A PERFECT GAS

The microscopic source of pressure in a perfect gas is particle bombardment.[1] The reflection (or absorption) of these particles from a real (or imagined) surface in the gas results in a transfer of momentum to that surface. By Newton's second law ($F = dp/dt$), that momentum transfer exerts a force on the surface. The average force per unit area is called the *pressure*. It is the same mechanical quantity appearing in the statement that the quantity of work performed by the infinitesimal expansion of a contained gas is $dW = P\,dV$. In thermal equilibrium in stellar interiors, the angular distribution of particle momenta is isotropic; i.e., particles are moving with equal probabilities in all directions. When reflected from a surface, those moving normal to the surface will transfer larger amounts of momentum than those that glance off at grazing angles. Imagine that the surface in Fig. 2-1 is one of the surfaces of an evacuated can under particle bombardment. When particles are specularly reflected from that surface, the momentum transferred to the surface is $\Delta p_n = 2p \cos \theta$. Let $F(\theta,p)\,d\theta\,dp$ be the number of particles with momentum p in the range dp striking the surface per unit area per unit time from all directions inclined at angle θ to the normal in the range $d\theta$.

[1] In a nonperfect gas strong forces between the particles will represent an additional source or sink of energy for expansions and will therefore contribute to the pressure.

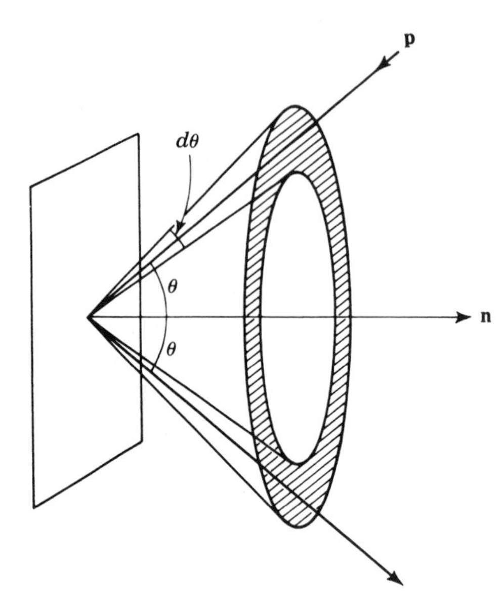

Fig. 2-1 A conical shell defining the set of directions having a spread $d\theta$ about the angle θ to the normal. The number of particles having $|\mathbf{p}| = p$ in the range dp that strike a unit area in unit time within this conical shell of directions is designated $F(\theta,p)\, d\theta\, dp$.

The contribution to the pressure from those particles is

$$dP = 2p \cos \theta\, F(\theta,p)\, d\theta\, dp \tag{2-1}$$

so that the total pressure is

$$P = \int_{\theta=0}^{\pi/2} \int_{p=0}^{\infty} 2p \cos \theta\, F(\theta,p)\, d\theta\, dp \tag{2-2}$$

In thermodynamic equilibrium, the angular distribution of momenta is isotropic, whereas the distribution of the magnitudes of the momenta is given by statistical mechanics. The flux $F(\theta,p)\, d\theta\, dp$ may be calculated as the product of the number density of particles with momentum p in the range dp moving in the cone of directions inclined at angle θ in the range $d\theta$ times the volume of such particles capable of passing through the unit surface in unit time. That volume is the volume of the parallelepiped shown in Fig. 2-2 and is equal to the product of v_p, the velocity associated with momentum p, and $\cos \theta$, the cross-sectional area of the column. That is,

$$F(\theta,p)\, d\theta\, dp = v_p \cos \theta\, n(\theta,p)\, d\theta\, dp \tag{2-3}$$

where $n(\theta,p)\, d\theta\, dp$ is the *number density* of particles moving in the prescribed cone. For isotropic radiation, furthermore, the *fraction* of the number of particles moving in the cone of directions at angle θ in the range $d\theta$ is just

$$\frac{n(\theta,p)\, d\theta\, dp}{n(p)\, dp} = \frac{2\pi \sin \theta\, d\theta}{4\pi} \tag{2-4}$$

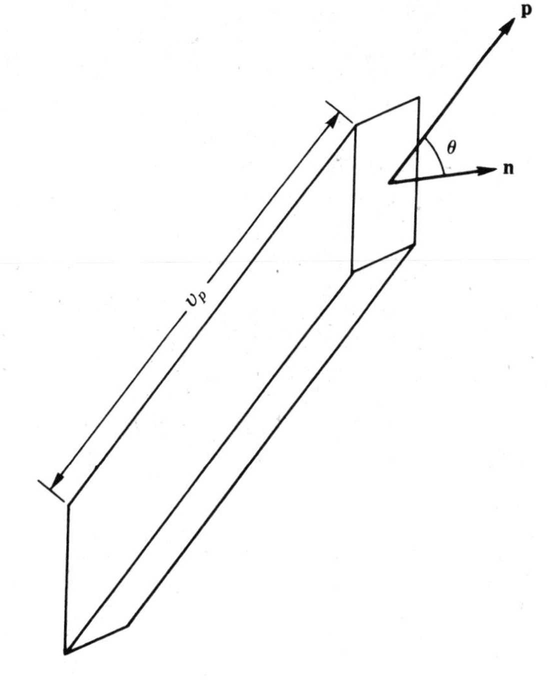

Fig. 2-2 The parallelepiped whose volume when multiplied by the density of particles about momentum **p** yields the number of particles per unit time of momentum **p** passing through the unit area **n**.

which is the fraction of the total spherical solid angle defined by the conical set of directions. The total number density of particles of momentum p in dp is $n(p)\,dp$. Evidently the gas pressure is

$$P = \int_0^{\pi/2} \int_0^\infty 2p \cos\theta \, v_p \cos\theta \, n(p) \, dp \, \tfrac{1}{2} \sin\theta \, d\theta \tag{2-5}$$

The explicit integration over angles is easily performed to yield

$$P = \frac{1}{3} \int_0^\infty p v_p n(p) \, dp \qquad \text{perfect gas} \tag{2-6}$$

This pressure integral, valid for a perfect isotropic gas, must be evaluated for several sets of relevant astrophysical circumstances. The relationship of v_p to p depends upon relativistic considerations, whereas the distribution $n(p)$ depends upon the type of particles and the quantum statistics. The simplest perfect gas is the monatomic nondegenerate nonrelativistic one considered in the next subsection, which will be followed by a discussion of the degenerate electron gas and then a discussion of radiation pressure.

THE PERFECT MONATOMIC NONDEGENERATE GAS

In the most common case for which the gas density is small enough to be nondegenerate and for which the thermal velocities are nonrelativistic, the pressure of a perfect gas is simply

$$P_g = NkT \tag{2-7}$$

where N is the number of free particles per unit volume.

Problem 2-1: From Chap. 1, the momentum distribution of a nondegenerate nonrelativistic gas in thermal equilibrium is maxwellian:

$$n(p)\,dp = \frac{N4\pi p^2\,dp}{(2\pi mkT)^{\frac{3}{2}}} \exp - \frac{p^2}{2mkT}$$

For a nonrelativistic gas, derive Eq. (2-7) from the pressure integral. The contribution from the several constituents of the gas are additive (Dalton's law of partial pressures). Is Eq. (2-7) also correct for relativistic velocities?

Let the mean molecular weight of the perfect gas be designated by μ. Then the density is

$$\rho = N\mu M_u \tag{2-8}$$

where M_u is the mass of 1 amu. The number of particles per unit volume can then be expressed in terms of the density and the mean molecular weight as

$$N = \frac{\rho}{\mu M_u} = \frac{N_0\rho}{\mu} \tag{2-9}$$

where $N_0 = 1/M_u$ is Avogadro's number and is equal to 6.0225×10^{23} mole^{-1}. Substitution into Eq. (2-7) gives the pressure of the gas in terms of the density and the mean molecular weight:

$$P_g = \frac{N_0 k}{\mu} \rho T \tag{2-10}$$

The mean molecular weight rather clearly will depend upon the composition of the gas. It is common to let X_Z represent the fraction of the gas *by weight* of element Z; that is, 1 g of gas contains X_Z g of the element of the atomic number Z. It follows that $\Sigma X_Z = 1$. Let us also suppose that each atom of element Z contributes n_Z free particles to the gas. For complete ionization, for instance, it will be true that $n_Z = Z + 1$, Z electrons plus the nucleus. Now let N_Z be the number density of atoms of element Z in the gas. From the definitions of all these quantities it is apparent that

$$N_Z = \frac{\rho_Z}{A_Z} N_0 = \rho \frac{X_Z}{A_Z} N_0 \tag{2-11}$$

Now the total number of free particles per cubic centimeter will be given by

$$N = \sum N_Z n_Z = \rho N_0 \sum \frac{X_Z n_Z}{A_Z} \tag{2-12}$$

where the sum is over all the elements Z. From a comparison of this last equation with Eq. (2-9), the mean molecular weight is given by

$$\frac{1}{\mu} = \sum \frac{X_Z n_Z}{A_Z} \tag{2-13}$$

It is conventional to use a slightly different terminology for the fraction by weight of the two most common elements in the stellar composition. In keeping with this convention, let X be the weight fraction of hydrogen, let Y be the weight fraction of helium, and let $1 - X - Y$ be the weight fraction[1] of all species heavier than helium. Then the mean molecular weight becomes

$$\mu = \left[\frac{X n_{\mathrm{H}}}{1.008} + \frac{Y n_{\mathrm{He}}}{4.004} + (1 - X - Y)\left\langle\frac{n_Z}{A_Z}\right\rangle\right]^{-1} \tag{2-14}$$

The quantity $\langle n_Z/A_Z\rangle$ is the average of n_Z/A_Z for $Z > 2$, each term being weighted proportional to X_Z.

Equation (2-14) can be further simplified for the case of complete ionization in the inner regions of stars. For complete ionization, the numbers of free particles contributed by the atoms of each element are $n_{\mathrm{H}} = 2$, $n_{\mathrm{He}} = 3$, and $n_Z = Z + 1$. When averaged over the species as they occur in nature, it is a convenient fact that the average atomic weight of element Z is approximately given by $A_Z = 2Z + 2$. The use of that approximation should be adequate in most cases where the fraction by weight of the species heavier than helium is small. With this approximation $\langle n_Z/A_Z\rangle$ in Eq. (2-14) becomes equal to $\frac{1}{2}$:

$$\mu \approx \frac{1}{2X + 3Y/4 + (1 - X - Y)/2} = \frac{2}{1 + 3X + 0.5Y} \tag{2-15}$$

It will also be convenient to have an auxiliary expression for the number density of electrons. Using exactly the same notation as above, we have

$$n_e = \sum N_Z(n_Z - 1) = \rho N_0 \sum \frac{X_Z}{A_Z}(n_Z - 1) \tag{2-16}$$

In the case of complete ionization $n_Z = Z + 1$, so that the number density of electrons becomes

$$n_e = \rho N_0 \sum \frac{X_Z Z}{A_Z} \qquad \text{complete ionization} \tag{2-17}$$

Insertion of the composition-by-weight parameters given above for hydrogen and helium yields

$$n_e = \rho N_0 \left[X + \frac{2Y}{4} + (1 - X - Y)\left\langle\frac{Z}{A_Z}\right\rangle\right] \tag{2-18}$$

where $\langle Z/A_Z\rangle$ is the average for $Z > 2$, the average being taken with respect to X_Z. If the fraction by weight of elements heavier than He is small, it is often

[1] It is common to denote this last weight fraction by Z. To avoid confusion with the nuclear charge in the present discussion, we forego that notation for the moment. We shall use the symbol Z later, however, where the context will make its meaning clear.

adequate to assume $\langle Z/A_Z \rangle \approx \frac{1}{2}$, in which case

$$n_e \approx \tfrac{1}{2}\rho N_0(1 + X) \tag{2-19}$$

It is also common to use a quantity called the *mean molecular weight per electron* μ_e, which is numerically equal to the average number of atomic mass units for each electron in the gas. From Eq. (2-17) it is evident that

$$\frac{1}{\mu_e} = \sum \frac{X_Z Z}{A_Z} \qquad n_e = \frac{\rho N_0}{\mu_e} \tag{2-20}$$

If the ionization is complete, and if $\langle Z/A_Z \rangle \approx \frac{1}{2}$ for $Z > 2$,

$$\mu_e = \frac{2}{1 + X} \tag{2-21}$$

It is advisable for the reader to pause long enough to gain familiarity with the composition parameters and to mentally evaluate the errors in the various approximations.

Problem 2-2: To be sure of understanding the mean molecular weight of the completely ionized gas, calculate and *interpret* the values of μ under the following circumstances: (*a*) all hydrogen, that is, $X = 1$, $Y = 0$; (*b*) all helium, that is, $X = 0$, $Y = 1$; (*c*) all heavy elements, that is, $X = 0$, $Y = 0$. Which of these three values is exactly given by the approximate equation (2-15)?

Problem 2-3: Calculate the mean molecular weight per electron μ_e for completely ionized conditions of all hydrogen ($X = 1$) and for all helium ($Y = 1$). Is Eq. (2-21) exact for $X = Y = 0.5$? Is it exact for $X = Z = 0.5$? What if the Z component is all C^{12} and O^{16}?

Problem 2-4: Show that for conditions under which Eq. (2-15) is valid, the rate of change of the mean molecular weight with respect to the heavy-element content Z, always holding the hydrogen fraction constant, is equal to $\mu^2/4$; that is,

$$\left(\frac{\partial \mu}{\partial Z}\right)_X = \frac{\mu^2}{4}$$

In calculations of stellar structure, and particularly of the structure of evolving stars, a large variety of compositions will be encountered. The statement was made in Chap. 1 that the average composition of the surfaces of population I stars and of the interstellar medium is more or less uniform. It is appropriate, therefore, at this time to present a simplified table of the abundances of the elements (Table 2-1), which are the best that can be inferred for population I objects. Most of the entries are derived from abundances of elements in the solar system, because those are the ones for which the most extensive data exist. The most important exceptions are He and Ne, which are observed only in objects hotter than the sun. It is common to think of the chemical composition of the solar system as a standard, against which other compositions are to be compared. This procedure is no more than a matter of convenience, however, and it must be remembered that the composition of our solar system has no special cosmo-

Table 2-1 Relative abundances of most common species in population I†

		Relative abundance	
Element	*Atomic weight*	*By number*	*By weight*
H	1	1,000	1,000
He‡	4	160	640
O	16	0.90	14
Ne‡	20	0.50	10
C	12	0.40	4.8
N	14	0.11	1.5
Si	28	0.032	0.9
Mg	24	0.025	0.6
S	32	0.022	0.7
A	40	0.008	0.3
Fe	56	0.004	0.2
Na	23	0.002	0.05
Cl	36	0.002	0.07
Al	27	0.002	0.05
Ca	40	0.002	0.08
F	19	0.001	0.02
Ni	59	0.001	0.06
>Ni	>60	$\sim 10^{-4}$	$\sim$0.01

† L. H. Aller, "The Abundance of the Elements," Interscience Publishers, Inc., New York, 1961.

‡ Because the sun is a G2 star, its helium abundance is not well known. The value in this table comes from the hotter B stars in the solar neighborhood, which are much younger than the sun. There are some indications that in the sun He/H $\approx$ 0.10 by number, which is about 60 percent of the amount of He found in B stars. A similar situation occurs for Ne, and it is more likely, but not certain, that in the sun Ne/O $\sim$ 0.1.

logical significance. A simple calculation reveals that the abundance parameters corresponding to Table 2-1 are

$$X = 0.60 \qquad Y = 0.38 \qquad Z = 0.02$$

These composition parameters may be thought of as characteristic of the majority of population I stars. It must be reemphasized, however, that it is in the deviations of composition from uniformity that some of the most intriguing problems of stellar evolution and nucleosynthesis are to be found.

Problem 2-5: The center of a certain star contains 60 percent hydrogen by weight and 35 percent helium by weight. Evaluate numerically the equation of state. What is the pressure at the center of the star if the density there is 50 g/cm^3 and the temperature is 15×10^6 °K?

Of course, some error has been introduced by simplifying assumptions made in obtaining the equation of state. Atoms are never completely ionized, and it is the Saha ionization equation that reveals the fraction of any given species that is ionized. In the relatively cool outer portions of a star, the number of free particles will depend upon the temperature. Elaborate techniques have been constructed for calculating a more realistic equation of state applicable to incomplete ionization. The reader who understands the ideas about it presented here, along with its restrictive assumptions, will have little trouble with a more sophisticated treatment of the equation of state.

Other than the lack of complete ionization in the cooler regions of the star, there are two extremely important physical circumstances that cause the equation of state for a perfect nondegenerate monatomic gas to be insufficient: (1) the pressure due to electromagnetic radiation in the interior of the star becomes comparable to the pressure due to particles, and (2) the electron gas becomes degenerate. We shall consider the second of these sets of circumstances, electron degeneracy, first.

ELECTRON DEGENERACY

Because electrons are particles with half-integral spin, the electron gas must obey Fermi-Dirac statistics. The density of electrons having momentum $|\mathbf{p}| = p$ in the range dp is accordingly

$$n_e(p)\, dp = \frac{2}{h^3} 4\pi p^2\, dp\, P(p) \tag{2-22}$$

where the occupation index for the Fermi gas is

$$P(p) = \left[\exp\left(\alpha + \frac{E}{kT}\right) + 1\right]^{-1} \tag{2-23}$$

That $P(p)$ has a maximum value of unity is an expression of the Pauli exclusion principle, to which electrons must adhere. When $P(p)$ is unity, all the available electronic states of the gas are occupied. It follows that the maximum density of electrons in phase space is

$$[n_e(p)]_{\max} = \frac{2}{h^3} 4\pi p^2 \tag{2-24}$$

It is this restriction upon the number density of electrons in momentum space which creates *degeneracy pressure*. If one continually increases the electron density, the electrons are forced into high-lying momentum states because the lower states are occupied. These high-momentum electrons will make a large contribution to the pressure integral.

For any given temperature and electron density n_e, the value of the parameter α is determined from the demand that

$$n_e = \int_0^\infty n_e(p)\, dp = n_e(\alpha,T) \tag{2-25}$$

This relationship will be explored quantitatively at a later time, but for the present we note from Eq. (2-23) that if α is a large positive number, $P(p)$ will be much less than unity for all energies. In this case the Fermi distribution reduces to the maxwellian distribution. As the electron density is increased at constant temperature, the parameter α becomes smaller, going to large negative values at high density.

In the limit of large negative α

$$P(p) = \begin{cases} 1 & \text{for } \dfrac{E}{kT} < |\alpha| \\ 0 & \text{for } \dfrac{E}{kT} > |\alpha| \end{cases} \qquad \text{complete degeneracy} \tag{2-26}$$

This transition occurs smoothly over a range of energy of several kT near the energy $E = |\alpha|kT$. If the energy $-\alpha kT$ is much larger than kT, the distribution function is nearly a step function. This limit is called *complete degeneracy*, and in this limit the quantity $|\alpha kT| = E_f$ is called the *Fermi energy*.

In the following discussion we shall calculate the pressure of a completely degenerate gas. The calculation will first be made for densities such that the energy E_f is nonrelativistic. It will then be repeated for densities high enough for E_f to correspond to relativistic electron velocities. Finally we shall calculate, in the nonrelativistic limit, the pressure of an electron gas for densities such that the distribution function is intermediate to the maxwellian and the completely degenerate distributions.

Complete degeneracy In a completely degenerate gas, the density is high enough so that all the available electron states having energies less than some maximum energy are filled. Since the total number density of electrons is to be finite, the density of states can be filled only up to some limiting value of the electron momentum

$$n_e(p)\,dp = \begin{cases} \dfrac{2}{h^3}\,4\pi p^2\,dp & p < p_0 \\ 0 & p > p_0 \end{cases} \tag{2-27}$$

It is clear that complete degeneracy is the state of minimum kinetic energy, the ground state, so to speak, of a degenerate perfect electron gas. The total number density of electrons in a completely degenerate electron gas is related to the maximum momentum by

$$n_e = \int_0^{p_0} n_e(p)\,dp = \frac{8\pi}{3h^3}\,p_0^3 \tag{2-28}$$

Inversion of this last equation shows that the maximum momentum of a completely degenerate gas is determined by the electron density:

$$p_0 = \left(\frac{3h^3}{8\pi}\,n_e\right)^{\frac{1}{3}} \tag{2-29}$$

The energy associated with the momentum p_0 is the Fermi energy.

The pressure of a completely degenerate perfect electron gas can be computed from the integral of Eq. (2-6) by inserting Eq. (2-27) for $n_e(p)$. Because it is also necessary to insert the velocity of a particle of given momentum, it is common to distinguish between a nonrelativistic and a relativistic degenerate electron gas.

Nonrelativistic complete degeneracy If the energy associated with p_0 is much less than $m_e c^2$, or 0.51 Mev, then $v_p = p/m$ for all momenta in the degenerate distribution, and the pressure integral is elementary:

$$P_{e,\mathrm{nr}} = \frac{8\pi}{15mh^3} p_0^5 \tag{2-30}$$

where nr signifies *nonrelativistic* electrons. Since the maximum momentum of the completely degenerate distribution is related to the electron density by Eq. (2-29), the electron pressure is determined by the electron density:

$$P_{e,\mathrm{nr}} = \frac{h^2}{20m}\left(\frac{3}{\pi}\right)^{\frac{2}{3}} n_e^{\frac{5}{3}} \tag{2-31}$$

The number density of the electrons may be written in terms of the mass density:

$$\begin{aligned} P_{e,\mathrm{nr}} &= \frac{h^2}{20m}\left(\frac{3}{\pi}\right)^{\frac{2}{3}} N_0^{\frac{5}{3}} \left(\frac{\rho}{\mu_e}\right)^{\frac{5}{3}} \\ &= 1.004 \times 10^{13} \left(\frac{\rho}{\mu_e}\right)^{\frac{5}{3}} \qquad \text{dynes/cm}^2 \end{aligned} \tag{2-32}$$

The value of μ_e is generally about 2 unless the gas contains considerable amounts of hydrogen. Inspection of this equation shows that the nonrelativistic-electron pressure varies as the $\frac{5}{3}$ power of the density. Since the pressure of a nondegenerate electron gas varies linearly with the density, it is clear that as the density increases, a point will be reached where the degenerate electron pressure becomes greater than the value that would be given by the formula for the pressure of a nondegenerate gas.

We may thereby define an approximate boundary line in the ρT plane, dividing it into regions of nondegenerate and degenerate gas, respectively, by the condition that the pressures given by the nondegenerate-gas equation and the completely degenerate electron-gas equation be equal. That is, when[1]

$$\frac{N_0 k}{\mu_e} \rho T = \frac{h^2}{20m}\left(\frac{3}{\pi}\right)^{\frac{2}{3}} (N_0)^{\frac{5}{3}} \left(\frac{\rho}{\mu_e}\right)^{\frac{5}{3}} \tag{2-33}$$

Numerical evaluation of this equation shows that the completely degenerate electron pressure exceeds the nondegenerate electron pressure when

$$\frac{\rho}{\mu_e} > 2.4 \times 10^{-8} T^{\frac{3}{2}} \qquad \text{g/cm}^3 \tag{2-34}$$

[1] It should perhaps be emphasized that Eq. (2-33) is never "true," since a gas cannot be simultaneously degenerate and nondegenerate. One might say that if ρ and T satisfy this equation, the state of the gas must be intermediate to nondegeneracy and complete degeneracy.

For densities greater than this value, the electron gas must be degenerate. Needless to say, the transition from nondegenerate to degenerate is not sudden and complete. The transition occurs smoothly for densities in the neighborhood of Eq. (2-34). The appropriate equation of state in the transition region will be discussed in the section on partial degeneracy.

It is instructive to apply Eq. (2-34) to two well-known astrophysical environments. At the center of the sun $\rho/\mu_e \approx 10^2$, and $T \approx 10^7$. For these values the inequality of Eq. (2-34) is strong in the opposite direction, so that one will anticipate using the nondegenerate electron pressure at the solar center. White-dwarf densities, on the other hand, are observationally known to be of order $\rho/\mu_e \approx 10^6$, whereas the interior temperatures are of order $T \approx 10^6$. For these values the inequality of Eq. (2-34) is strongly satisfied, and one must expect degeneracy pressure to dominate.

Relativistic complete degeneracy As the electron density is increased, the maximum momentum in a completely degenerate electron gas grows larger. Eventually a density is reached where the most energetic of the electrons in the degenerate distribution become relativistic. When that condition is reached, the substitution $v_p = p/m$ leading to Eq. (2-30) becomes incorrect. The velocity to be associated with the momentum p must be determined by relativistic kinematics.

Before calculating the pressure, let us estimate those densities for which it is necessary that some of the electrons be relativistic. For a relativistic particle, the total energy, which is the sum of the rest-mass energy plus the kinetic energy, forms a right triangle with the rest-mass energy and the momentum times the velocity of light, as illustrated in Fig. 2-3. The right-triangle relationship follows

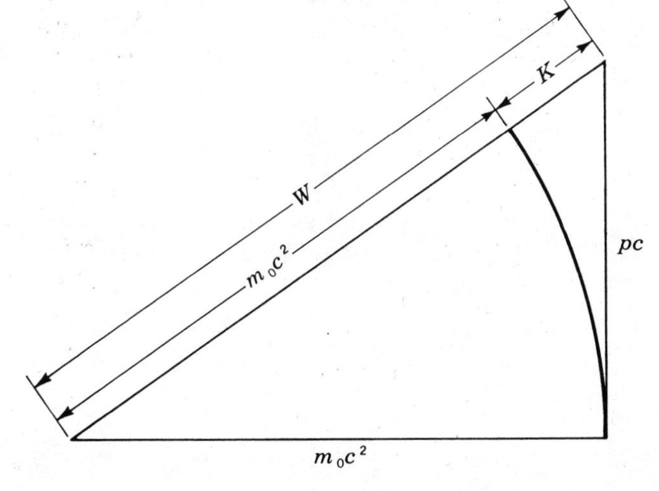

Fig. 2-3 The right triangle representing the relationship of the total energy of a particle to its momentum. The kinetic energy is the excess of the hypotenuse beyond the rest-mass energy.

from the relativistic expression for the total energy of a particle,

$$W^2 = p^2c^2 + m_0^2c^4 \tag{2-35}$$

where m_0 is the rest mass of the particle. The total energy is also given by the square of the velocity of light times the relativistic mass:

$$W = mc^2 = \frac{m_0c^2}{1 - (v/c)^2} \tag{2-36}$$

Equating W^2 in Eq. (2-35) to W^2 in Eq. (2-36) yields

$$pc = \frac{v}{c}W \tag{2-37}$$

What convenient order-of-magnitude criterion will ensure that particles are relativistic? One may say with adequate accuracy that particles become relativistic when v/c approaches unity and when the total energy W becomes appreciably greater than the rest-mass energy. As an order-of-magnitude criterion, it suffices to compute the density at which the electrons of maximum momentum have a total energy equal to, say, twice the rest-mass energy; from Eq. (2-37) the quantity pc will then be approximately $p_0c \sim 2m_0c^2$. On the other hand, the numerical value of p_0c is

$$p_0c = hc\left(\frac{3}{8\pi}n_e\right)^{\frac{1}{3}} = 6.12 \times 10^{-11}n_e^{\frac{1}{3}} \qquad \text{Mev} \tag{2-38}$$

In terms of the density and the mean molecular weight per electron, Eq. (2-38) may be expressed as

$$p_0c = 5.15 \times 10^{-3}\left(\frac{\rho}{\mu_e}\right)^{\frac{1}{3}} \qquad \text{Mev} \tag{2-39}$$

This last equation reveals that $p_0c \approx 2m_0c^2 \approx 1$ Mev when

$$\frac{\rho}{\mu_e} = 7.3 \times 10^6 \text{ g/cm}^3 \qquad \text{relativistic} \tag{2-40}$$

The natural conclusion is that as the density approaches this value, relativistic kinematics must be used in relating the velocity of an electron to its momentum. Densities this large are encountered in astrophysics, in white dwarfs, for instance.

The pressure integral for a completely degenerate gas may be evaluated without difficulty for relativistic particles. Since the momentum of a relativistic particle is given by Eqs. (2-36) and (2-37) as

$$p = \frac{m_0v}{[1 - (v/c)^2]^{\frac{1}{2}}} \tag{2-41}$$

one can determine by inversion that

$$v = \frac{p/m_0}{[1 + (p/m_0c)^2]^{\frac{1}{2}}} \tag{2-42}$$

Insertion of this value for v_p in the pressure integral yields

$$p_e = \frac{8\pi}{3mh^3} \int_0^{p_0} \frac{p^4\,dp}{[1 + (p/mc)^2]^{\frac{1}{2}}} \tag{2-43}$$

In Eq. (2-43) and those which follow, the electron rest mass is designated simply by m. For evaluation of this integral it is convenient to define a new parameter θ such that

$$\sinh\theta = \frac{p}{mc} \qquad dp = mc\cosh\theta\,d\theta$$

In terms of this new variable the pressure integral becomes

$$P_e = \frac{8\pi m^4c^5}{3h^3} \int_0^{\theta_0} \sinh^4\theta\,d\theta \tag{2-44}$$

which may be integrated to give

$$P_e = \frac{8\pi m^4c^5}{3h^3} \left(\frac{\sinh^3\theta_0\cosh\theta_0}{4} - \frac{3\sinh 2\theta_0}{16} + \frac{3\theta_0}{8} \right) \tag{2-45}$$

When written in terms of the Fermi momentum,

$$P_e = \frac{\pi m^4c^5}{3h^3} f(x) = 6.003 \times 10^{22} f(x) \qquad \text{dynes/cm}^2 \tag{2-46}$$

where

$$x = \frac{p_0}{mc} = \frac{h}{mc}\left(\frac{3}{8\pi}n_e\right)^{\frac{1}{3}}$$

$$f(x) = x(2x^2 - 3)(x^2 + 1)^{\frac{1}{2}} + 3\sinh^{-1} x \tag{2-47}$$

The numerical value of the dimensionless parameter x is

$$x = 1.195 \times 10^{-10} n_e^{\frac{1}{3}} = 1.009 \times 10^{-2} \left(\frac{\rho}{\mu_e}\right)^{\frac{1}{3}} \tag{2-48}$$

Problem 2-6: The limit of small x, that is, $p_0 \ll mc$, must correspond to nonrelativistic particles. Show that

$$f(x) \approx \tfrac{8}{5}x^5 - \tfrac{4}{7}x^7 + \cdots \qquad x \to 0$$

and confirm that the pressure obtained from this limiting value of $f(x)$ reduces to the completely degenerate nonrelativistic electron pressure determined previously in Eq. (2-30).

Problem 2-7: The limit of large x must correspond to highly relativistic degeneracy. Show that

$$f(x) \approx 2x^4 - 2x^2 + \cdots \qquad x \to \infty$$

Show that the pressure obtained by inserting this limiting value of $f(x)$ into Eq. (2-46) is identical to that obtained by letting $v_p = c$ in the integral for the pressure given in Eq. (2-6). Does that make sense? Evidently the pressure is proportional to $\rho^{\frac{4}{3}}$ at very high density.

Table 2-2 Pressure of a complete degenerate gas†

x	$f(x)$	x	$f(x)$
0	0	2.0	26.7
0.1	1.60×10^{-5}	2.1	32.9
0.2	5.05×10^{-4}	2.2	40.1
0.3	3.77×10^{-3}	2.3	48.4
0.4	1.55×10^{-2}	2.4	58.0
0.5	4.61×10^{-2}	2.5	68.9
0.6	0.111	2.6	81.2
0.7	0.232	2.7	95.2
0.8	0.436	2.8	110.8
0.9	0.756	2.9	128.3
1.0	1.23	3.0	1.48×10^2
1.1	1.90	3.5	2.80×10^2
1.2	2.82	4.0	4.85×10^2
1.3	4.05	4.5	7.85×10^2
1.4	5.63	5.0	1.21×10^3
1.5	7.64	5.5	1.78×10^3
1.6	10.1	6.0	2.53×10^3
1.7	13.2	6.5	3.49×10^3
1.8	16.9	7.0	4.71×10^3
1.9	21.4	8.0	8.07×10^3

† S. Chandrasekhar, "An Introduction to the Study of Stellar Structure," p. 392; reprinted from the Dover Publications edition, Copyright 1939 by The University of Chicago, as reprinted by permission of The University of Chicago.

Table 2-2 lists some numerical values of $f(x)$. From this table and Eq. (2-46) the electron pressure can be evaluated for cases of semirelativistic complete degeneracy. The quantity x is to be evaluated from Eq. (2-48). This result is correct only for a completely degenerate gas. Approximate relativistic expressions for a partially degenerate gas can be obtained if desired.[1] However, densities must exceed 10^6 g/cm^3 for a degenerate gas to be relativistic [Eq. (2-40)], for which the degeneracy will be essentially complete unless $T > 10^9$ °K [Eq. (2-34)]. Densities greater than 10^6 g/cm^3 at a temperature greater than 10^9 °K are probably found only in very late stages of stellar evolution. For all other classes of stars, degeneracy sets in at sufficiently low temperatures so that nonrelativistic kinematics should be adequate for the examination of partial degeneracy.

[1] See, for instance, S. Chandrasekhar, "An Introduction to the Study of Stellar Structure," p. 392, Dover Publications, Inc., New York, 1957, or D. H. Menzel, P. L. Bhatnagar, and H. K. Sen, "Stellar Interiors," p. 35, John Wiley & Sons, Inc., New York, 1963.

Probem 2-8: Show that the kinetic energy per unit volume of a completely degenerate gas is

$$\left(\frac{U}{V}\right)_{\text{kin}} = \frac{\pi m^4 c^5}{3h^3} g(x)$$

where $g(x) = 8x^3[(x^2+1)^{\frac{1}{2}} - 1] - f(x)$. Show also that $U \to \frac{3}{2}PV$ in the limit of small x and $U \to 3PV$ in the limit of large x.

Partial degeneracy The dividing line between degeneracy and nondegeneracy given in Eq. (2-34) defines only the region of the onset of degeneracy in the electron gas. That is, it indicates only the approximate condition under which electron degeneracy is becoming important in the equation of state. Actually, of course, there is a gradual transition from nondegeneracy toward complete degeneracy as the density rises. There is certainly no sharp transition between those extreme conditions. The electron occupation index gradually takes on the shape of a degenerate distribution with increase in density, as illustrated in Fig. 2-4.

The distribution of electron momenta is

$$n_e(p)\,dp = \frac{2}{h^3} \frac{4\pi p^2\,dp}{\exp(\alpha + E/kT) + 1} \tag{2-49}$$

where α is a number that depends upon the electron density and the temperature. That is, α is fixed by the requirement that the total number of electrons equal the electron density n_e:

$$n_e = \int_0^\infty \frac{2}{h^3} \frac{4\pi p^2\,dp}{\exp(\alpha + E/kT) + 1} = n_e(\alpha,T) \tag{2-50}$$

The integral for the pressure of the perfect electron gas becomes

$$P_e = \frac{8\pi}{3h^3} \int_0^\infty \frac{p^3 v_p\,dp}{\exp(\alpha + E/kT) + 1} \tag{2-51}$$

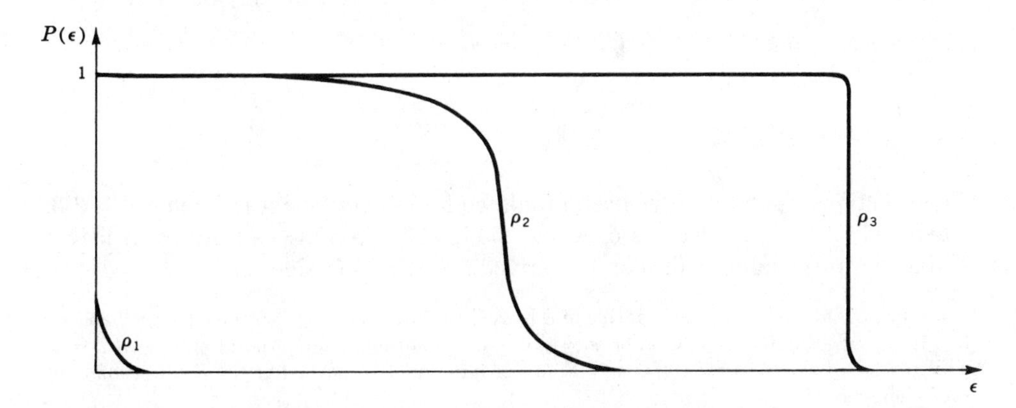

Fig. 2-4 Schematic illustration of the occupation index of an electron gas for three different degrees of degeneracy. In this particular case $\rho_3 > \rho_2 \gg \rho_1$ and $T_2 > T_3$.

As stated at the end of the last section, for temperatures of less than 10^9 °K nonrelativistic electron degeneracy sets in before relativistic degeneracy. Therefore, in considering the partially degenerate gas, we shall restrict ourselves to nonrelativistic kinematics, keeping in mind that the results will be somewhat in error for extremely high temperatures ($T > 10^9$). That is, we shall once again let $v_p = p/m$, whereupon

$$P_e = \frac{8\pi}{3h^3m} \int_0^\infty \frac{p^4\,dp}{\exp(\alpha + p^2/2mkT) + 1} \tag{2-52}$$

and

$$n_e = \frac{8\pi}{h^3} \int_0^\infty \frac{p^2\,dp}{\exp(\alpha + p^2/2mkT) + 1} \tag{2-53}$$

With the aid of a dimensionless energy $u = p^2/2mkT$, these equations may be written in the form

$$P_e = \frac{8\pi kT}{3h^3} (2mkT)^{\frac{3}{2}} \int_0^\infty \frac{u^{\frac{3}{2}}\,du}{\exp(\alpha + u) + 1} \tag{2-54}$$

$$n_e = \frac{4\pi}{h^3} (2mkT)^{\frac{3}{2}} \int_0^\infty \frac{u^{\frac{1}{2}}\,du}{\exp(\alpha + u) + 1} \tag{2-55}$$

These two equations constitute a parametric representation of the equation of state, the parameter being the quantity α. The parametric representation is made more explicit by conventionally defining two new functions,[1]

$$F_{\frac{1}{2}}(\alpha) = \int_0^\infty \frac{u^{\frac{1}{2}}\,du}{\exp(\alpha + u) + 1}$$
$$F_{\frac{3}{2}}(\alpha) = \int_0^\infty \frac{u^{\frac{3}{2}}\,du}{\exp(\alpha + u) + 1} \tag{2-56}$$

in which case the electron pressure and the electron density may be written as

$$P_e = \frac{8\pi kT}{3h^3} (2mkT)^{\frac{3}{2}} F_{\frac{3}{2}}(\alpha)$$
$$n_e = \frac{4\pi}{h^3} (2mkT)^{\frac{3}{2}} F_{\frac{1}{2}}(\alpha) \tag{2-57}$$

The functions $F_{\frac{1}{2}}$ and $F_{\frac{3}{2}}$ have been tabulated for selected values of α in Table 2-3. Their values for other values of α may be interpolated in the range of α listed, and asymptotic values will soon be derived for extreme values of α.

[1] In much of the literature the negative of α is used as the degeneracy parameter, in which case it is usually designated by η or Ψ; or $\Psi = \eta = -\alpha$. Another common notation is $-\alpha kT = \mu$, which is called the *chemical potential*. Many people prefer to normalize the F's in a different way, defining

$$U_n(\alpha) = \frac{1}{\Gamma(n+1)} F_n(\alpha)$$

Table 2-3 Table of Fermi-Dirac functions†

α	$\frac{2}{3}F_{\frac{3}{2}}$	$F_{\frac{1}{2}}$	α	$\frac{2}{3}F_{\frac{3}{2}}$	$F_{\frac{1}{2}}$
4.0	0.016 179	0.016 128	0.0	0.768 536	0.678 094
3.9	0.017 875	0.017 812	−0.1	0.839 082	0.733 403
3.8	0.019 748	0.019 670	−0.2	0.915 332	0.792 181
3.7	0.021 816	0.021 721	−0.3	0.997 637	0.854 521
3.6	0.024 099	0.023 984	−0.4	1.086 358	0.920 505
3.5	0.026 620	0.026 480	−0.5	1.181 862	0.990 209
3.4	0.029 404	0.029 233	−0.6	1.284 526	1.063 694
3.3	0.032 476	0.032 269	−0.7	1.394 729	1.141 015
3.2	0.035 868	0.035 615	−0.8	1.512 858	1.222 215
3.1	0.039 611	0.039 303	−0.9	1.639 302	1.307 327
3.0	0.043 741	0.043 366	−1.0	1.774 455	1.396 375
2.9	0.048 298	0.047 842	−1.1	1.918 709	1.489 372
2.8	0.053 324	0.052 770	−1.2	2.072 461	1.586 323
2.7	0.058 868	0.058 194	−1.3	2.236 106	1.687 226
2.6	0.064 981	0.064 161	−1.4	2.410 037	1.792 068
2.5	0.071 720	0.070 724	−1.5	2.594 650	1.900 833
2.4	0.079 148	0.077 938	−1.6	2.790 334	2.013 496
2.3	0.087 332	0.085 864	−1.7	2.997 478	2.130 027
2.2	0.096 347	0.094 566	−1.8	3.216 467	2.250 391
2.1	0.106 273	0.104 116	−1.9	3.447 683	2.374 548
2.0	0.117 200	0.114 588	−2.0	3.691 502	2.502 458
1.9	0.129 224	0.126 063	−2.1	3.948 298	2.634 072
1.8	0.142 449	0.138 627	−2.2	4.218 438	2.769 344
1.7	0.156 989	0.152 373	−2.3	4.502 287	2.908 224
1.6	0.172 967	0.167 397	−2.4	4.800 202	3.050 659
1.5	0.190 515	0.183 802	−2.5	5.112 536	3.196 598
1.4	0.209 777	0.201 696	−2.6	5.439 637	3.345 988
1.3	0.230 907	0.221 193	−2.7	5.781 847	3.498 775
1.2	0.254 073	0.242 410	−2.8	6.139 503	3.654 905
1.1	0.279 451	0.265 471	−2.9	6.512 937	3.814 326
1.0	0.307 232	0.290 501	−3.0	6.902 476	3.976 985
0.9	0.337 621	0.317 630	−3.1	7.308 441	4.142 831
0.8	0.370 833	0.346 989	−3.2	7.731 147	4.311 811
0.7	0.407 098	0.378 714	−3.3	8.170 906	4.483 876
0.6	0.446 659	0.412 937	−3.4	8.628 023	4.658 977
0.5	0.489 773	0.449 793	−3.5	9.102 801	4.837 066
0.4	0.536 710	0.489 414	−3.6	9.595 535	5.018 095
0.3	0.587 752	0.531 931	−3.7	10.106 516	5.202 020
0.2	0.643 197	0.577 470	−3.8	10.636 034	5.388 795
0.1	0.703 351	0.626 152	−3.9	11.184 369	5.578 378

Table 2-3 Table of Fermi-Dirac functions† (Continued)

α	$\frac{2}{3}F_{\frac{3}{2}}$	$F_{\frac{1}{2}}$	α	$\frac{2}{3}F_{\frac{3}{2}}$	$F_{\frac{1}{2}}$
−4.0	11.751 80	5.770 72	−8.0	52.901 73	15.380 48
−4.1	12.338 60	5.965 80	−8.1	54.453 85	15.662 24
−4.2	12.945 05	6.163 56	−8.2	56.034 24	15.945 80
−4.3	13.571 40	6.363 96	−8.3	57.643 07	16.231 14
−4.4	14.217 93	6.566 98	−8.4	59.280 52	16.518 26
−4.5	14.884 89	6.772 57	−8.5	60.946 78	16.807 14
−4.6	15.572 53	6.980 70	−8.6	62.642 01	17.097 76
−4.7	16.281 11	7.191 34	−8.7	64.366 39	17.390 13
−4.8	17.010 88	7.404 45	−8.8	66.120 09	17.684 23
−4.9	17.762 08	7.620 01	−8.9	67.903 29	17.980 04
−5.0	18.534 96	7.837 97	−9.0	69.716 16	18.277 56
−5.1	19.329 76	8.058 32	−9.1	71.558 86	18.576 77
−5.2	20.146 71	8.281 03	−9.2	73.431 57	18.877 68
−5.3	20.986 04	8.506 06	−9.3	75.334 45	19.180 26
−5.4	21.847 99	8.733 39	−9.4	77.267 68	19.484 51
−5.5	22.732 79	8.962 99	−9.5	79.231 41	19.790 41
−5.6	23.640 67	9.194 85	−9.6	81.225 82	20.097 96
−5.7	24.571 84	9.428 93	−9.7	83.251 06	20.407 15
−5.8	25.526 53	9.665 21	−9.8	85.307 30	20.717 97
−5.9	26.504 95	9.903 67	−9.9	87.394 71	21.030 42
−6.0	27.507 33	10.144 28	−10.0	89.513 44	21.344 47
−6.1	28.533 88	10.387 03	−10.1	91.663 65	21.660 13
−6.2	29.584 81	10.631 90	−10.2	93.845 52	21.977 38
−6.3	30.660 33	10.878 86	−10.3	96.059 18	22.296 22
−6.4	31.760 65	11.127 89	−10.4	98.304 81	22.616 64
−6.5	32.885 98	11.378 98	−10.5	100.582 56	22.938 62
−6.6	34.036 52	11.632 11	−10.6	102.892 59	23.262 17
−6.7	35.212 47	11.887 26	−10.7	105.235 05	23.587 28
−6.8	36.414 04	12.144 40	−10.8	107.610 10	23.913 93
−6.9	37.641 42	12.403 54	−10.9	110.017 89	24.242 12
−7.0	38.894 81	12.664 64	−11.0	112.458 57	24.571 84
−7.1	40.174 41	12.927 69	−11.1	114.932 31	24.903 09
−7.2	41.480 41	13.192 67	−11.2	117.439 24	25.235 86
−7.3	42.813 01	13.459 58	−11.3	119.979 53	25.570 13
−7.4	44.172 39	13.728 39	−11.4	122.553 32	25.905 91
−7.5	45.558 75	13.999 10	−11.5	125.160 76	26.243 19
−7.6	46.972 27	14.271 68	−11.6	127.802 01	26.581 95
−7.7	48.413 15	14.546 12	−11.7	130.477 20	26.922 20
−7.8	49.881 56	14.822 41	−11.8	133.186 50	27.263 93
−7.9	51.377 69	15.100 53	−11.9	135.930 04	27.607 12

Table 2-3 Table of Fermi-Dirac functions† (Continued)

α	$\frac{2}{3}F_{\frac{3}{2}}$	$F_{\frac{1}{2}}$	α	$\frac{2}{3}F_{\frac{3}{2}}$	$F_{\frac{1}{2}}$
−12.0	138.707 97	27.951 78	−14.0	201.709 50	35.142 97
−12.1	141.520 44	28.297 89	−14.1	205.242 49	35.517 00
−12.2	144.367 60	28.645 45	−14.2	208.812 95	35.892 38
−12.3	147.249 58	28.994 46	−14.3	212.421 01	36.269 08
−12.4	150.166 54	29.344 91	−14.4	216.066 81	36.647 12
−12.5	153.118 61	29.696 79	−14.5	219.750 48	37.026 49
−12.6	156.105 94	30.050 09	−14.6	223.472 15	37.407 18
−12.7	159.128 68	30.404 82	−14.7	227.231 96	37.789 18
−12.8	162.186 96	30.760 96	−14.8	231.030 03	38.172 50
−12.9	165.280 92	31.118 51	−14.9	234.866 50	38.557 12
−13.0	168.410 71	31.477 46	−15.0	238.741 50	38.943 04
−13.1	171.576 46	31.837 81	−15.1	242.655 15	39.330 27
−13.2	174.778 31	32.199 56	−15.2	246.607 59	39.718 79
−13.3	178.016 42	32.562 68	−15.3	250.598 95	40.108 59
−13.4	181.290 90	32.927 20	−15.4	254.629 36	40.499 69
−13.5	184.601 90	33.293 08	−15.5	258.698 93	40.892 06
−13.6	187.949 56	33.660 34	−15.6	262.807 81	41.285 71
−13.7	191.334 01	34.028 96	−15.7	266.956 12	41.680 64
−13.8	194.755 40	34.398 94	−15.8	271.143 98	42.076 83
−13.9	198.213 85	34.770 28	−15.9	275.371 53	42.474 29

† Taken from J. McDougall and E. C. Stoner, *Phil. Trans. Roy. Soc.*, **237**:67 (1938).

Problem 2-9: Show that in a perfect nonrelativistic electron gas

$$P_e = \frac{2}{3}\left(\frac{U}{V}\right)_{\text{kin}}$$

for *any degree of degeneracy.*

Problem 2-10: (*a*) Show that as $\alpha \to \infty$, $F_{\frac{3}{2}}/F_{\frac{1}{2}} \to \frac{3}{2}$, for which case $P_e \to n_e kT$, the pressure of a maxwellian electron gas. (*b*) Show that as $\alpha \to -\infty$, $F_{\frac{3}{2}}/F_{\frac{1}{2}} \to \frac{3}{5}u_0$, for which case $P_e \to (8\pi/15mh^3)p_0{}^5$, the pressure of a completely degenerate nonrelativistic electron gas.

From Eq. (2-57) it is apparent that

$$P_e = n_e kT\left(\frac{2}{3}\frac{F_{\frac{3}{2}}}{F_{\frac{1}{2}}}\right) \tag{2-58}$$

Thus, the factor $2F_{\frac{3}{2}}/3F_{\frac{1}{2}}$ measures the extent to which the electron pressure differs from that of a nondegenerate gas. This multiplication factor is plotted in Fig. 2-5 as a function of the parameter α. It can be seen that the gas pressure is essentially that of a nondegenerate gas for $\alpha > 2$.

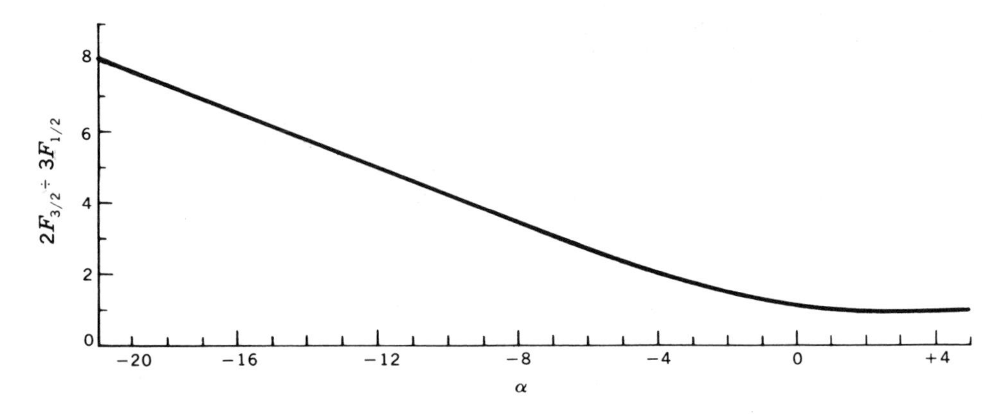

Fig. 2-5 The ratio $2F_{\frac{3}{2}}/3F_{\frac{1}{2}}$ as a function of the degeneracy parameter α. This ratio is equal to the ratio of the pressure of an electron gas to the pressure it would have if it were maxwellian at the same density.

On the other hand, Eq. (2-57) may be written in terms of the mass density,

$$\frac{\rho N_0}{\mu_e} = \frac{4\pi}{h^3} (2mkT)^{\frac{3}{2}} F_{\frac{1}{2}}(\alpha) \tag{2-59}$$

from which it follows that

$$\log\left(\frac{\rho}{\mu_e} T^{-\frac{3}{2}}\right) = \log F_{\frac{1}{2}}(\alpha) - 8.044 \tag{2-60}$$

This equation is plotted in Fig. 2-6, which relates $\log [(\rho/\mu_e)T^{-\frac{3}{2}}]$ to the degeneracy parameter α.

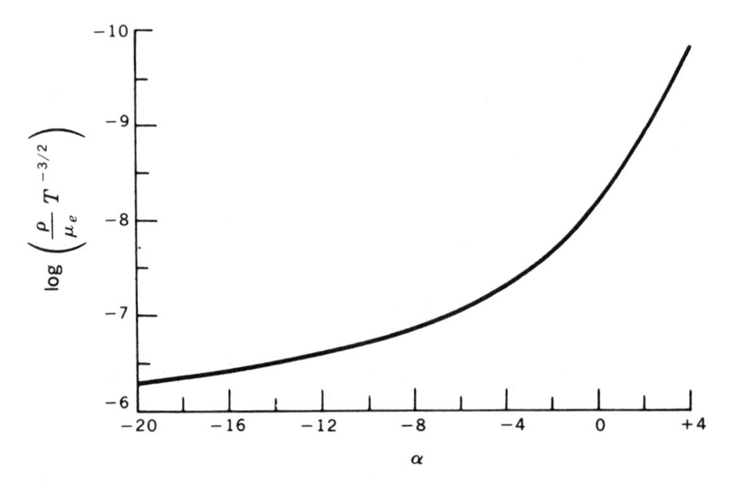

Fig. 2-6 The value of $(\rho/\mu_e)T^{-\frac{3}{2}}$ determines the degeneracy parameter α of an electron gas.

These equations describe the behavior of the equation of state of an electron gas in the partially degenerate region. For given ρ, T, Eq. (2-59) determines $F_{1/2}(\alpha)$, which in turn allows $F_{3/2}(\alpha)$ to be interpolated from Table 2-3. These calculations have used nonrelativistic kinematics because, in most stars, relativistic degeneracy is important only for such high densities that the degeneracy is essentially complete.

For many problems in nonrelativistic partial degeneracy, however, it is instructive to have appropriate expansions of the equation of state. Expansions that converge rapidly for weak degeneracy (nearly maxwellian) and for strong degeneracy (nearly complete) are easily obtained.

Weak nonrelativistic degeneracy For notational ease, define $\Lambda = \exp(-\alpha)$. Then for $\alpha > 0$, which is seen from Fig. 2-5 to correspond to weak degeneracy, the number Λ is less than unity. Then $F_{1/2}(\Lambda)$ may be expanded:

$$F_{1/2}(\Lambda) = \int_0^\infty \frac{u^{1/2}\,du}{(1/\Lambda)e^u + 1} = \int_0^\infty \Lambda e^{-u} u^{1/2} \frac{1}{1 + \Lambda e^{-u}}\,du$$

$$= \Lambda \int_0^\infty e^{-u} u^{1/2} [1 - \Lambda e^{-u} + (\Lambda e^{-u})^2 - (\Lambda e^{-u})^3 + \cdots]\,du \tag{2-61}$$

which may be integrated term by term to give

$$F_{1/2}(\Lambda) = -\frac{\sqrt{\pi}}{2} \sum_{n=1}^{\infty} \frac{(-1)^n \Lambda^n}{n^{3/2}} \qquad \Lambda < 1$$

or equivalently

$$F_{1/2}(\alpha) = -\frac{\sqrt{\pi}}{2} \sum_{n=1}^{\infty} \frac{(-1)^n e^{-n\alpha}}{n^{3/2}} \qquad \alpha > 0 \tag{2-62}$$

Then Eq. (2-57) becomes

$$n_e = \frac{2(2\pi mkT)^{3/2}}{h^3} \sum_{n=1}^{\infty} (-1)^{n+1} \frac{e^{-n\alpha}}{n^{3/2}} \qquad \alpha > 0 \tag{2-63}$$

Problem 2-11: Show by the same technique used in obtaining Eq. (2-63) that

$$P_e = \frac{2kT(2\pi mkT)^{3/2}}{h^3} \sum_{n=1}^{\infty} (-1)^{n+1} \frac{e^{-n\alpha}}{n^{5/2}} \qquad \alpha > 0 \tag{2-64}$$

Problem 2-12: For large α, the series may be approximated by one term. Show that Eqs. (2-63) and (2-64) then reduce to the maxwellian distribution.

Problem 2-13: Suppose that α is large enough for only the first two terms of the series to be important. Show, then, that

$$P_e \approx n_e kT \left[1 + \frac{n_e h^3}{2^{7/2}(2\pi mkT)^{3/2}} + \cdots\right]$$

$$= n_e kT(1 + 10^{-16.435} n_e T^{-3/2} + \cdots)$$

Strong nonrelativistic degeneracy The degeneracy becomes strong when α becomes a large negative number or, equivalently, when the parameter Λ becomes a large positive number. The expansion for large Λ employs a lemma due to Sommerfeld, which, as stated by Chandrasekhar,[1] is:

LEMMA: If $\phi(u)$ is a sufficiently regular function which vanishes for $u = 0$, then we have the asymptotic formula

$$\int_0^\infty \frac{du}{(1/\Lambda)e^u + 1} \frac{d\phi(u)}{du} = \phi(u_0) + 2\left[c_2\left(\frac{d^2\phi}{du^2}\right)_{u_0} + c_4\left(\frac{d^4\phi}{du^4}\right)_{u_0} + \cdots\right] \tag{2-65}$$

where $u_0 = \log \Lambda$ and $c_2, c_4, \ldots$ are numerical constants defined by

$$c_\nu = 1 - \frac{1}{2^\nu} + \frac{1}{3^\nu} - \frac{1}{4^\nu} + \cdots$$

The series for the constants c_ν can be summed.[2] For instance,

$$c_2 = \frac{\pi^2}{12} \qquad c_4 = \frac{7\pi^4}{720} \qquad c_6 = \frac{31\pi^6}{30{,}240}$$

Problem 2-14: By applying Sommerfeld's lemma to the integrals $F_{\frac{1}{2}}$ and $F_{\frac{3}{2}}$, show that

$$\begin{aligned} F_{\frac{1}{2}}(\alpha) &= \tfrac{2}{3}(-\alpha)^{\frac{3}{2}}\left(1 + \frac{\pi^2}{8\alpha^2} + \frac{7\pi^4}{640\alpha^4} + \cdots\right) \\ F_{\frac{3}{2}}(\alpha) &= \tfrac{2}{5}(-\alpha)^{\frac{5}{2}}\left(1 + \frac{5\pi^2}{8\alpha^2} - \frac{7\pi^4}{384\alpha^4} + \cdots\right) \end{aligned} \tag{2-66}$$

is a good expansion for $\alpha < -1$. These three-term expansions are accurate to three decimal places for $\alpha < -5.6$ and are quite useful for $\alpha < -3$.

Problem 2-15: Calculate $F_{\frac{1}{2}}(\alpha)$ and $\frac{2}{3}F_{\frac{3}{2}}(\alpha)$ for $\alpha = -3$ and compare the results with the values in Table 2-3.

Since

$$n_e = \frac{4\pi}{h^3}(2mkT)^{\frac{3}{2}}F_{\frac{1}{2}}(\alpha) \tag{2-67}$$

it is evident from Eq. (2-66) that the physical meaning of α in the limit of strong degeneracy is

$$-\alpha \approx \frac{1}{2mkT}\left(\frac{3h^3 n_e}{8\pi}\right)^{\frac{2}{3}} \tag{2-68}$$

[1] S. Chandrasekhar, "An Introduction to the Study of Stellar Structure," p. 389; reprinted from the Dover Publications edition, Copyright 1939 by The University of Chicago, as reprinted by permission of The University of Chicago.

[2] See, for instance, H. B. Dwight, "Tables of Integrals and Other Mathematical Data," eq. 47.2, p. 11, The Macmillan Company, New York, 1947.

which from Eq. (2-29) is

$$-\alpha \approx \frac{p_0^2}{2mkT} = \frac{E_f}{kT} \tag{2-69}$$

where E_f is the Fermi energy (the kinetic energy of an electron at the top of the Fermi sea). This result is the same one that was obtained from an inspection of the Fermi distribution function for large negative α. For incomplete degeneracy, however, the energies $|\alpha kT|$ and E_f have different definitions and physical meanings.

If the three-term expansion of $F_{\frac{1}{2}}(\alpha)$ is retained, Eq. (2-59) can be written as an approximate equation relating the value of α to the density and temperature:

$$(-\alpha)^{\frac{3}{2}}\left(1 + \frac{\pi^2}{8\alpha^2} + \frac{7\pi^4}{640\alpha^4} + \cdots\right) = 1.66 \times 10^8 \frac{\rho}{\mu_e} T^{-\frac{3}{2}} \qquad \text{for } \alpha < -3 \tag{2-70}$$

Problem 2-16: Show that the electron pressure is *twice* that of the maxwellian electron-gas formula when $\rho/\mu_e = 5.0 \times 10^{-8} T^{\frac{3}{2}}$. Compare this result with the approximate boundary of Eq. (2-34), which gave the density for which a *completely* degenerate gas formula yields the same pressure as the maxwellian gas formula.

The properties of the equation of state of the perfect electron gas are shown graphically in Fig. 2-7, where the ρT plane is divided into various zones according to the extent of the electron degeneracy. The diagonal line represents the approximate boundary between nondegenerate and degenerate electron gas as given by Eq. (2-34). In the neighborhood of this boundary the equation of state is to be evaluated from the parametric equations (2-57), which apply to partial degeneracy. For densities as high as indicated by Eq. (2-40), an electron gas becomes relativistic. This boundary is shown by the vertical line in Fig. 2-7. In the neighborhood of this line, the pressure of a completely degenerate gas can be evaluated from Eq. (2-46). For very high temperatures ($T > 10^9$) not considered in this discussion, the electron gas can be both relativistic and only partially degenerate. This situation presents a slightly more difficult form of the equation of state. We shall not consider it here. Suffice it to say that the Fermi statistics yield the same expression for the pressure as Eq. (2-53), the difference being that relativistic kinematics are to be used.

Several additional comments concerning a degenerate electron gas are appropriate at this time. With regard to the mechanical pressure which is to support a star, it is clear that the calculations presented here account only for the pressure due to the electrons. The contribution from the particle pressure of the nuclei in the gas must be added. Since nuclei are never degenerate in common stars, the pressure due to them is simply that of a maxwellian gas, whose equations have been developed previously. To calculate the partial pressure of this perfect nuclear gas one must, of course, use the appropriate value of the mean molecular weight. Since the electrons have in this case been accounted for independently, one must use only the mean molecular weight of the remaining ions

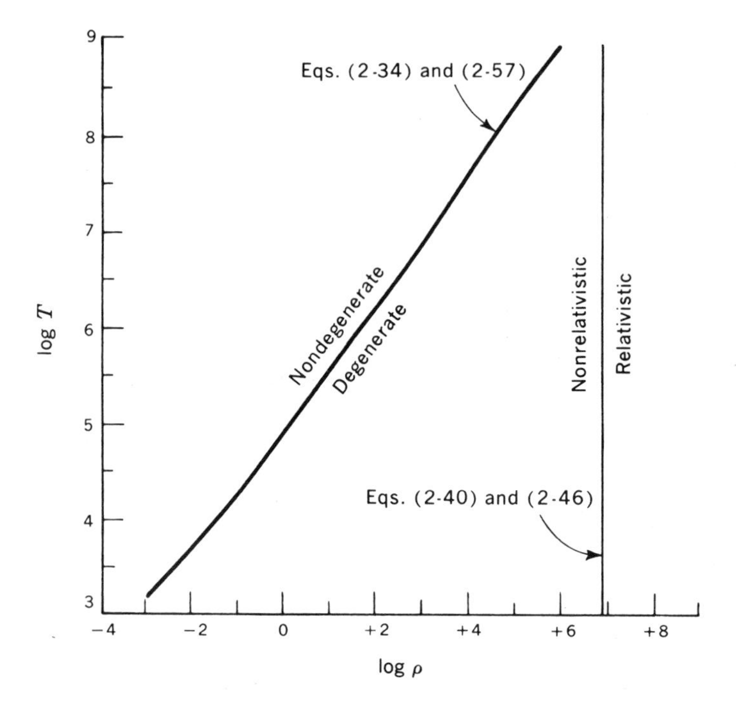

Fig. 2-7 Zones of the equation of state of an electron gas. The nonrelativistic transition region between nondegeneracy and extreme degeneracy is located according to Eq. (2-34), and the pressure is given by Eq. (2-57) in this region. As ρ approaches 10^7 g/cm^3, many of the electrons become relativistic, and the distribution becomes highly degenerate, in which case Eq. (2-46) adequately represents the pressure.

and nuclei. Let μ_i designate the mean molecular weight of the ions. The pressure due to particles is then the sum of the electron pressure and the nucleus pressure:

$$P_{\text{gas}} = P_e + \frac{N_0 k}{\mu_i} \rho T \tag{2-71}$$

In most practical cases where electron degeneracy does occur, the remaining nuclei are generally those of more advanced phases of stellar structure, consisting of helium nuclei, carbon nuclei, oxygen nuclei, or perhaps even heavier nuclei. In these circumstances the bulk of the pressure will be provided by the degenerate electron gas, the nuclei providing only a small additional term.

Problem 2-17: A gas composed of C^{12} and O^{16} has a density of 2.5×10^5 g/cm^3 at 10^8 °K. Is this gas in the degenerate or nondegenerate region of the equation of state? Assuming the degeneracy is complete, is it completely nonrelativistic, partially relativistic, or extremely relativistic? Calculate the electron pressure from Table 2-2. Assuming that the degeneracy is incomplete and nonrelativistic, calculate the electron pressure from Table 2-3. Why is the pressure calculated under assumptions of partial degeneracy greater than the pressure calculated for assumptions of complete degeneracy? Which numerical answer is more correct for the present problem? Why? What is the ratio of the electron pressure to the ion pressure?

Another interesting feature of the pressure of a completely degenerate gas is that it does not depend explicitly upon the temperature. Of course, at any finite temperature the electron gas is never completely degenerate, but in many cases the actual momentum distribution may be closely approximated by complete degeneracy. Whenever the energy associated with the momentum p_0 of the completely degenerate distribution greatly exceeds kT, the distribution of electron momenta will closely resemble that of complete degeneracy. It is in this case that the pressure is approximately independent of the temperature, being absolutely independent of the temperature for complete degeneracy.[1] This fact has the interesting consequence that a small rise in the temperature of an almost completely degenerate electron gas causes almost no change at all in the pressure. This last fact has far-reaching effects on stellar structure and on the evolution of stars. Those stages of stellar structure in which the electron gas is degenerate and is providing the main source of pressure for the gas must admit the possibility of abrupt rises in temperature with no corresponding increase in pressure. This situation actually occurs in certain stages of stellar evolution and leads to runaways in nuclear reaction rates (flash phenomena).

Problem 2-18: Show that the nonrelativistic electron pressure changes with temperature at constant volume according to

$$\left(\frac{\partial P_e}{\partial T}\right)_{n_e} = \frac{8\pi k}{3h^3}(2mkT)^{\frac{3}{2}}\left(\tfrac{5}{2}F_{\frac{3}{2}} - \tfrac{3}{2}F_{\frac{1}{2}}\frac{dF_{\frac{3}{2}}/d\alpha}{dF_{\frac{1}{2}}/d\alpha}\right)$$

$$= \frac{P_e}{T}\left(\frac{5}{2} - \frac{3}{2}\frac{F_{\frac{1}{2}}}{F_{\frac{3}{2}}}\frac{dF_{\frac{3}{2}}/d\alpha}{dF_{\frac{1}{2}}/d\alpha}\right)$$

The quantity in parentheses in the second expression is unity for a nondegenerate gas and zero for a completely degenerate gas. Confirm this by evaluating it with the aid of the appropriate expansions.

Another important feature of the degenerate electron distributions is related to the transport of heat energy in the interiors of stars. The normal processes of energy transport in stellar interiors are altered somewhat when the electron gas becomes degenerate. The most important fact is that heat conductivity, which normally plays a secondary role to radiative transport and to convective transport, becomes important. In the case of nondegeneracy, the mean free path of charged particles is so small that heat conduction is extremely inefficient. When an electron gas is degenerate, however, the mean free path of electrons becomes

[1] Mathematically one shows that $\partial P/\partial T$ is very small by making an expansion of the parametric equation of state and evaluating for noncomplete degeneracy. The reader is referred to Chandrasekhar, *op. cit.*, chap. 10.

quite long. In order for an energetic electron to lose energy, it must fall into a lower-lying cell in momentum space as well as impart a new energy and momentum to the particle from which it scatters. The filling up of the available states in momentum space below a certain level hinders this process and renders energetic electrons quite free to move about in even a partially degenerate electron gas. This very good conductivity will tend to make partially degenerate electron gases isothermal.

White-dwarf stars are, to good approximation, supported by a completely degenerate electron gas. As those stars radiate their thermal energy, becoming increasingly cooler, the nearly degenerate momentum distribution becomes increasingly rectangular. Eventually the thermal energy is radiated away, the temperature falls toward zero, the light goes out, and the object remains an inert mass supported by a dense sea of completely degenerate electrons, or so the story goes. This picture is in keeping with the observed properties of white dwarfs, which, from their observed masses and radii, are known to have densities as large as 10^6 g/cm^3.

Pioneers in stellar structure encountered a subtle paradox in contemplating the above picture, however. Faithful application of the hitherto successful ionization equation seemed to imply that ions and electrons recombine at low temperatures. Since the density of un-ionized matter is at most a few grams per cubic centimeter, it would appear necessary that white dwarf expand as it cools. Yet it could be shown that the thermal energy is, at all stages, insufficient to do the necessary gravitational work. Eddington expressed the paradox as follows:

> *I do not see how a star which has once got into this compressed condition is ever going to get out of it. So far as we know, the close packing of matter is only possible so long as the temperature is great enough to ionize the material. When the star cools down and regains the normal density ordinarily associated with solids, it must expand and do work against gravity.* The star will need energy to cool. *Sirius* comes *on solidifying will have to expand its radius at least tenfold, which means that 90 percent of its lost gravitational energy must be replaced. We can scarcely credit the star with sufficient foresight to retain more than 90 percent in reserve for the difficulty awaiting it. It would seem that the star will be in an awkward predicament when its supply of subatomic energy ultimately fails. Imagine a body continually losing heat but with insufficient energy to grow cold!*[1]

The physical basis for the resolution of this problem is the thermodynamic peculiarity of a degenerate gas: the temperature no longer corresponds to kinetic energy. The electrons in a zero-temperature degenerate gas must still have large kinetic energy if the density is great. The classical ionization equation showed that at high densities atoms become ionized as kT approaches the order of magnitude of the electron binding energy, which is when the kinetic energy of the free-

[1] *Op. cit.*, p. 172.

electron gas approaches the kinetic energy of the bound electrons. The same approximate result applies in degenerate circumstances. Atoms are in an ionized state when the kinetic energy of the electron gas exceeds the kinetic energy of a bound electron.

The approximate truth of this statement can be seen from the following considerations. In a completely degenerate gas, all available electron states with momentum less than p_0 are occupied. The exclusion principle thus forbids the presence of bound electrons unless they are bound so tightly that their momentum exceeds p_0, for otherwise there would be "too many" electrons in a momentum interval. Whereas a rigorous description of quantum statistics is considerably more complicated than this simple argument, the physical necessity of the result is evident.

The physical idea is also similar to that of the band structure of electronic states in solids. Ignore considerations of temperature completely for the moment. When the interatomic separations of atoms are large, the energy levels of electrons are just those associated with isolated atoms. Each energy level possesses a degeneracy equal to that of the atomic level times the total number of atoms. When the interatomic separation is decreased to the point where electronic levels of adjacent atoms overlap, however, a new feature is introduced by the exclusion principle. Since electrons are identical fermions, the mutual wave function of overlapping electrons must be antisymmetric in the electron coordinates. This antisymmetrization introduces a sharing of the indistinguishable electrons by all the atoms. In order that the electrons not be in exactly the same state, the many degenerate atomic energy levels of discrete energy regroup into a *continuous band* of energies for which each electron is shared by all atoms. The wave functions of those electrons in the band can be expressed by wave functions analogous to free electrons. This is what happens in a metal, for instance, for which the continuous band of *quasifree* electrons provides the source of electric conductivity. The same feature is carried to extremes at the densities of stellar interiors. Careful analysis shows that atoms are completely ionized by this mechanism for densities greater than about 10^3 g/cm^3 *independent of the temperature.* This physical effect has come to be called *pressure ionization,* and it resolves in a natural manner the paradox stated by Eddington.

This completes the introduction to the perfect electron gas. We have attempted to focus attention onto the physical principles rather than on the mathematical details. The serious student of stellar structure who has grasped these ideas may turn to more complete treatments for appropriate formulas applicable to the computation of physical problems.

THE PHOTON GAS

Particles are not the only source of mechanical pressure in a perfect gas. Pressure is also exerted by the radiation field in the interior of the star. By the radiation field we mean electromagnetic radiation, the omnipresent flux of photons inside a thermal enclosure. The pressure of the photon gas results from the fact that each quantum of electromagnetic energy $h\nu$ carries with it a momentum equal to $h\nu/c$. If we imagine these photons being specularly reflected from a mirror, it is clear that momentum will be transferred to the mirror by the photons.

If the environment is in thermodynamic equilibrium, the radiation flux is isotropic. In that case, the pressure integral for isotropic flux gives immediately the following interesting result for the radiation pressure P_r:

$$P_r = \frac{1}{3}\int_0^\infty \frac{h\nu}{c}\, cn(\nu)\, d\nu = \frac{1}{3}\int_0^\infty h\nu n(\nu)\, d\nu$$
$$= \tfrac{1}{3}u \qquad (2\text{-}72)$$

where u is the energy density of photons, which was shown in Chap. 1 to be given by $u = aT^4$.

It often happens in cases of physical interest that an enclosure is not strictly in thermodynamic equilibrium. In such objects as stars there exists a radiation field which is slightly anisotropic, resulting from the fact that there is a net excess of radiant energy flowing in one particular direction. In the case of a star, for instance, there is a net excess in the flux of electromagnetic radiation in the radial direction. At each point in the interior of such a star the situation corresponds nearly to one of thermodynamic equilibrium; i.e., the radiation field is nearly isotropic. It is convenient when considering such a slightly anisotropic radiation field to define the polar direction of a coordinate system as the direction of the net excess heat flow. In these cases of physical interest, azimuthal symmetry obtains about the direction of net flow. This symmetry corresponds to the assumption that there is no temperature gradient perpendicular to the direction of the net heat flow. In terms of this coordinate system we define a quantity called the *intensity of the radiation field* $I(\theta)$. Quantitatively, $I(\theta)\, d\Omega$ is the energy flux per square centimeter per second moving at a direction angle θ relative to the chosen axis inside a cone of directions defined by the solid angle $d\Omega$. Figure 2-8 shows the cross-sectional unit area inclined at angle θ to the chosen polar direction in the cone of directions corresponding to the solid angle $d\Omega$.

Let $u(\theta)\, d\Omega$ represent the *energy density* of radiation moving at angle θ in the set of directions $d\Omega$. It is clear from Fig. 2-8 that the flux $I(\theta)\, d\Omega$ passing through the unit area per second is given by the corresponding energy density $u(\theta)\, d\Omega$ times a unit column of length c, where c is the velocity of light:

$$I(\theta)\, d\Omega = cu(\theta)\, d\Omega \qquad (2\text{-}73)$$

Since the integral of the directed energy density over the total solid angle of 4π is just the total energy density, we have

$$u = \int^{4\pi} u(\theta)\, d\Omega = \frac{1}{c}\int^{4\pi} I(\theta)\, d\Omega \qquad (2\text{-}74)$$

Let H designate the net flux of energy transported per square centimeter per second in the polar direction. The flux per solid angle through a unit surface

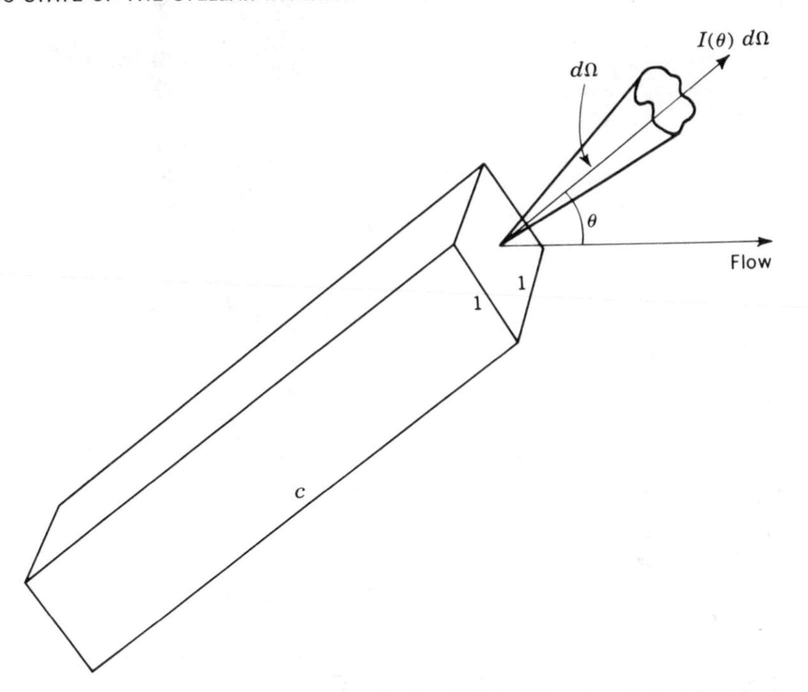

Fig. 2-8 The intensity $I(\theta)$ of the radiation field is defined such that $I(\theta)\ d\Omega$ is the energy flux moving in direction θ within the set of directions $d\Omega$.

normal to the polar direction is equal to $I(\theta) \cos \theta$, since the unit area has a projected area equal to $\cos \theta$ when viewed from the direction θ. It is apparent from Fig. 2-9 that the net flow of energy is given by

$$H = \int^{4\pi} I(\theta) \cos \theta \, d\Omega \tag{2-75a}$$

If, as is usually the case, the radiation field possesses azimuthal symmetry about the H axis,

$$H = 2\pi \int_0^{\pi} I(\theta) \cos \theta \sin \theta \, d\theta \tag{2-75b}$$

Evidently the heat flow vanishes for an isotropic radiation field [$I(\theta)$ = const].

From the relationship between the energy $E = h\nu$ and the momentum $p = h\nu/c$ of a single photon, it follows that the flux of radiant energy $I(\theta)$ corresponds to a momentum flux $I(\theta)/c$. The resulting radiation pressure may be visualized as the compression force on a spring separating two imaginary unit areas between which the radiation field is excluded, as shown in Fig. 2-10. If the radiation field $I(\theta)$ is to remain unaltered by this imaginary mechanical system, a photon absorbed in plate I must be emitted in the same direction from plate II. The

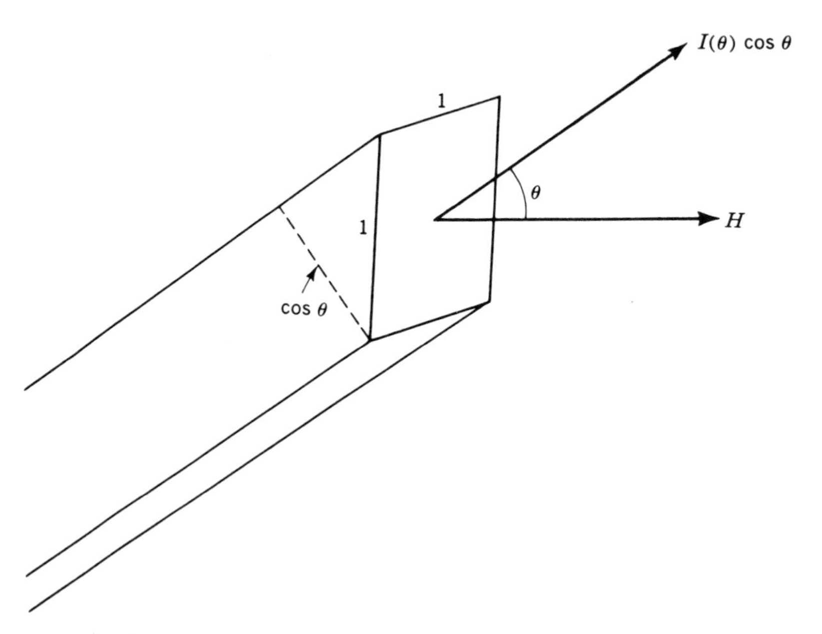

Fig. 2-9 The energy flux $I(\theta)\,d\Omega$ has associated with it an energy flow per unit area normal to the polar axis H equal to $I(\theta)\,d\Omega\cos\theta$.

resulting compressional force on the spring is, per unit area,

$$\begin{aligned} P_r &= \int^{4\pi} \frac{I(\theta)\cos\theta}{c} \cos\theta\, d\Omega \\ &= \frac{2\pi}{c}\int_0^\pi I(\theta)\cos^2\theta \sin\theta\, d\theta \end{aligned} \tag{2-76}$$

It is clear from these three simple calculations that the energy density u, the net flux of energy in the polar direction H, and the radiation pressure P_r are related to the three moments of the radiation field $I(\theta)$:

$$\begin{aligned} u &= \frac{1}{c}\int I(\theta)\, d\Omega = \frac{2\pi}{c}\int_0^\pi I(\theta)\sin\theta\, d\theta \\ H &= \int I(\theta)\cos\theta\, d\Omega = 2\pi\int_0^\pi I(\theta)\cos\theta\sin\theta\, d\theta \\ P_r &= \frac{1}{c}\int I(\theta)\cos^2\theta\, d\Omega = \frac{2\pi}{c}\int_0^\pi I(\theta)\cos^2\theta\sin\theta\, d\theta \end{aligned} \tag{2-77}$$

For cases appropriate to the interiors of stars, where near thermodynamic equilibrium obtains, the radiation field may be approximated by

$$I(\theta) = I_0 + I_1\cos\theta + \cdots \tag{2-78}$$

where I_0 represents the isotropic part of the radiation field and I_1 represents the anisotropy in the radiation field corresponding to the net flux in the polar direction. Equation (2-78) represents the first two terms in the Fourier cosine expansion of the general radiation field. The integrals corresponding to the above discussion are easily evaluated for this radiation field, yielding

$$u = \frac{2\pi}{c} \int (I_0 + I_1 \cos\theta) \sin\theta \, d\theta = \frac{4\pi}{c} I_0$$

$$H = 2\pi \int (I_0 + I_1 \cos\theta) \cos\theta \sin\theta \, d\theta = \frac{4\pi}{3} I_1 \tag{2-79}$$

$$P_r = \frac{2\pi}{c} \int (I_0 + I_1 \cos\theta) \cos^2\theta \sin\theta \, d\theta = \frac{4\pi}{3c} I_0$$

In this approximation the energy density in the radiation field and the radiation pressure are independent of the anisotropic term in the radiation field. On the other hand, the net heat flux carried by radiation flow is dependent upon the existence of the anisotropic term. The relationship between the energy density u and the radiation pressure P_r is the same as that for an isotropic radiation field:

$$P_r = \tfrac{1}{3}u = \tfrac{1}{3}aT^4 \tag{2-80}$$

where the constant $a = 7.565 \times 10^{15}$ ergs cm^{-3} deg^{-4}. For a more general form of the radiation field involving higher powers of $\cos\theta$ in the expansion, this relationship between the energy density and the radiation pressure is not strictly

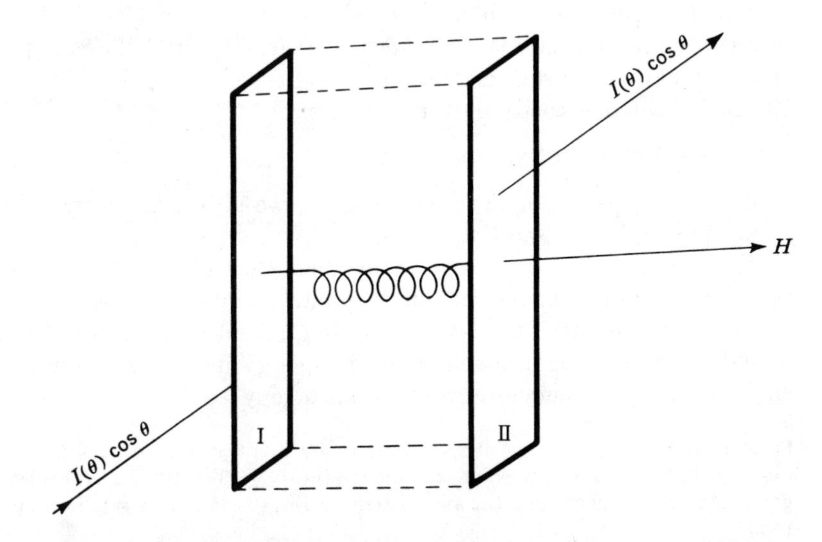

Fig. 2-10 Two imaginary plates separated by a spring. The radiation is excluded from the area between the plates, and each plate is required to emit the same radiation absorbed by its counterpart in order that the radiation field not be disturbed. The radiation pressure is the resulting compressional force on the spring.

correct, but in the cases of physical interest in the interiors of stars, the expression given by Eq. (2-78) is adequate. Judging from the relationships implied by Eq. (2-79), we may in fact rewrite Eq. (2-78) as

$$I(\theta) \approx \frac{c}{4\pi} u + \frac{3}{4\pi} H \cos \theta \tag{2-81}$$

We shall see shortly that the second term in Eq. (2-81) is numerically much smaller than the first in the interior of stars.

Problem 2-19: Suppose we have a radiation field in the form $I(\theta) = I_0 \exp(a \cos \theta)$. Note that the limit of small a corresponds to Eq. (2-78) with $a = I_1/I_0$. Calculate the relationship between u, H, and P_r. Does $P_r = u/3$? Do the expressions reduce to Eq. (2-79) to first order in a?

We may consider one slight variation of Eq. (2-75a) at this point in order to clear up a relationship used in Chap. 1. Instead of calculating the net energy flux inside an enclosure, as in Eq. (2-75a), we may ask for the total flux which would emerge from a hole cut in the surface of the container. This calculation will correspond to the energy emitted per unit area from the surface of a blackbody. In this case the appropriate limits of the integral in Eq. (2-75b) are only from 0 to $\pi/2$, resulting in an emission

$$J = 2\pi \int_0^{\pi/2} (I_0 + I_1 \cos \theta) \cos \theta \sin \theta \, d\theta = \pi I_0 + \frac{2\pi}{3} I_1 \tag{2-82}$$

For an isotropic radiation field (or invoking the fact that in all real cases of interest the second term is very much less than the first) the radiation per square centimeter from the surface of a blackbody is related to the internal energy density inside the blackbody by the relationship

$$J = \pi I_0 = \tfrac{1}{4} cu = \sigma T^4 \tag{2-83}$$

This is the source of the ratio between the two radiation constants introduced in Chap. 1; $\sigma = (c/4)a$ [see Eq. (1-29)].

We shall return to these three moments of the radiation field when we consider the question of the radiative transfer in stellar interiors. At that time the heat flow H will be related to the temperature gradient of the star. From Eq. (2-79) above, however, it is apparent that the energy density and the pressure in the enclosure are independent of a small anisotropy.

Problem 2-20: In a star like the sun, the entire luminosity originates from a region inside $r = 3 \times 10^{10}$ cm, at which point the temperature $T = 3 \times 10^6$ °K. Calculate from $L_\odot$ the energy flux H and show that the second term in Eq. (2-81) is very small compared to the first term.

The mechanical pressure of a perfect gas is to be computed as the sum of three terms:

$$P = P_{\text{ions}} + P_{\text{electrons}} + P_{\text{radiation}} \tag{2-84}$$

If the electron gas is nondegenerate, the sum of the first two terms is given by Eq. (2-10), where μ is the mean molecular weight of all free particles. If the electron gas is degenerate, the second term must be computed from one of the appropriate equations for degenerate electron pressure. In this case, the first term will be of the form of Eq. (2-10) except that μ will then represent the mean molecular weight of the ions only. Equation (2-84) will apply to all normal stellar interiors. It requires modification at extremely high temperatures ($T > 10^9$), when positron-electron pairs may be produced from energetic photons, and at high densities, where particle interactions may invalidate the perfect-gas approximation. If magnetic fields are present in low-density regions, Eq. (2-84) may require the addition of magnetic-field pressure. The difficult question of the coulomb interactions in the gas will be postponed to a later discussion.

Finally, we note that the pressure due to a perfect nondegenerate gas equals the pressure due to the radiation field when

$$\frac{N_0 k}{\mu} \rho T = \tfrac{1}{3} a T^4 \tag{2-85}$$

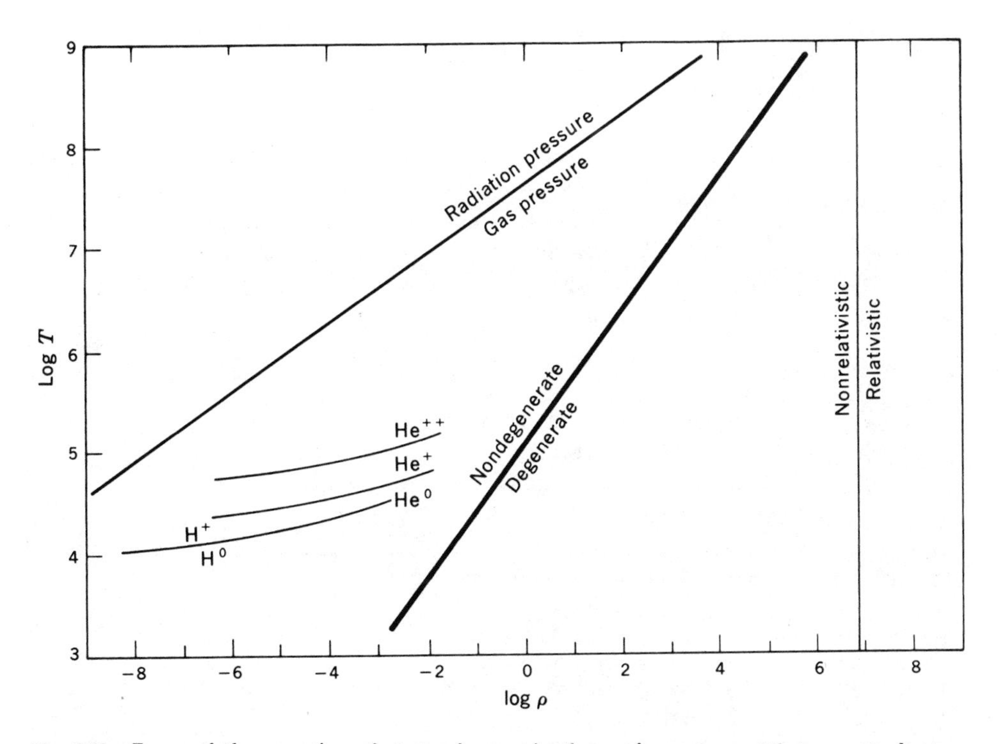

Fig. 2-11 Zones of the equation of state of a gas in thermodynamic equilibrium. Radiation pressure dominates the gas pressure in the upper left-hand corner. The remaining boundaries are similar to those in Fig. 2-7. Also included for comparison are the transition strips in a hydrogen-dominated gas between H^0 and H^+, between He^0 and He^+, and between He^+ and He^{++}.

or when

$$T = 3.20 \times 10^7 \left(\frac{\rho}{\mu}\right)^{\frac{1}{3}} \approx 3.6 \times 10^7 \rho^{\frac{1}{3}} \tag{2-86}$$

By this equation and the equations of the previous section the ρT plane may be roughly divided into three major regions in which:

(*1*) The pressure is dominated by the photon gas.
(*2*) The pressure is dominated by a nondegenerate gas.
(*3*) The pressure is due to a degenerate electron gas.

Figure 2-11 shows this rough division into the various zones of the equation of state, calculated for a composition of nearly all hydrogen, as is characteristic of most stellar interiors. We have included for comparison the various ionization zones of hydrogen and helium calculated from the Saha equation for a composition predominantly of hydrogen. These lines correspond to those values of temperature and density for which the ionization is 50 percent accomplished.

2-2 QUASISTATIC CHANGES OF STATE

A star is not a static thing. It undergoes large expansions and contractions in the course of its evolution. Individual mass elements in convection zones rise and fall along the radius of the star. The very stability of the stellar structure is determined by its response to small perturbations. In this section we introduce the physical principles, mainly thermodynamic, of slow expansions.

According to the first law of thermodynamics, the internal energy of a gas may be changed by adding or withdrawing a quantity of heat energy or by doing work *upon* the gas by expansion or contraction. Specifically

$$dU = dQ + dW \tag{2-87}$$

If the process of change is performed infinitely slowly, so that one can think of the state of the system at any moment as being one of equilibrium, the process is referred to as *quasistatic*. Since quasistatic processes can be conducted in a reverse sense, they are generally referred to as *reversible processes*. An infinitesimal change of the volume of an enclosure containing a gas requires an infinitesimal amount of mechanical work on the gas given by

$$dW = -P\,dV \tag{2-88}$$

We shall explicitly introduce this equation into the first law of thermodynamics, whereupon it may be written

$$dQ = dU + P\,dV \tag{2-89}$$

It is from considerations of the equation of state that the proper expression to be used for the pressure may be determined. The proper formula depends, as we have seen, upon the exact temperature and density of the gas in a star.

It is a fundamental proposition of thermodynamics that for quasistatic changes of state the heat increment may be written

$$dQ = T\,dS \tag{2-90}$$

where S is a function of state called the *entropy*. Because the entropy is a function of state, it may be computed for matter in thermodynamic equilibrium and depends only upon the equilibrium state of matter and not upon its past history. The change in entropy between two equilibrium states may be computed by the combination of Eqs. (2-90) and (2-89):

$$T\,dS = dU + P\,dV \tag{2-91}$$

wherein the change is evaluated along a sequence of equilibrium states reached by quasistatic changes. The internal energy is also a function of state, which means that for a given quantity of gas in equilibrium, U may be regarded as a function $U(V,T)$. For a quasistatic change of state of the gas, therefore,

$$dS = \frac{1}{T}\left[\left(\frac{\partial U}{\partial V}\right)_T + P\right] dV + \frac{1}{T}\left(\frac{\partial U}{\partial T}\right)_V dT \tag{2-92}$$

Because the entropy is a function $S(V,T)$ of the state of matter, dS may also be written

$$dS = \left(\frac{\partial S}{\partial V}\right)_T dV + \left(\frac{\partial S}{\partial T}\right)_V dT \tag{2-93}$$

Because the second partial derivatives are independent of the order in which they are taken, that is,

$$\frac{\partial}{\partial T}\frac{\partial S}{\partial V} = \frac{\partial}{\partial V}\frac{\partial S}{\partial T}$$

there exists an integrability condition for dS, viz.,

$$\frac{\partial}{\partial T}\left\{\frac{1}{T}\left[\left(\frac{\partial U}{\partial V}\right)_T + P\right]\right\} = \frac{\partial}{\partial V}\left[\frac{1}{T}\left(\frac{\partial U}{\partial T}\right)_V\right] \tag{2-94}$$

The application of these simple ideas may lead to quite useful results. As an example of this fact, we now show thermodynamically that the energy per unit volume of an equilibrium photon gas is proportional to the fourth power of the temperature.

THE STEFAN–BOLTZMANN LAW

In the interior of an evacuated container in thermal equilibrium, the energy density of photons can be a function only of the temperature: $u = u(T)$. If the contemplated box has volume V, the internal energy in the box is then

$$U = Vu(T) \tag{2-95}$$

whereupon the partial derivatives are

$$\left(\frac{\partial U}{\partial V}\right)_T = u(T) \qquad \left(\frac{\partial U}{\partial T}\right)_V = V\frac{du}{dT} \tag{2-96}$$

Independent of the photon energy density and spectrum, one knows that $P_r = u/3$ for an isotropic photon gas simply from the fact that photons are massless. (All massless particles move at the velocity of light and carry momentum $p = E/c$.) If these physical observations are inserted into Eq. (2-94), there results

$$\frac{\partial}{\partial T}\left(\frac{1}{T}\frac{4}{3}u\right) = \frac{\partial}{\partial V}\left(\frac{1}{T}V\frac{du}{dT}\right) \tag{2-97}$$

If the differentiation is carried out, one obtains

$$\frac{du}{u} = 4\frac{dT}{T} \tag{2-98}$$

which has the integral

$$u = aT^4 \tag{2-99}$$

This thermodynamic argument does not reveal the fact that the arbitrary constant of integration is the *Stefan-Boltzmann constant*, but it does show that the elementary thermodynamic properties of a photon gas are consistent with the quantum-statistical properties discussed in Chap. 1.

SPECIFIC HEATS AND ADIABATIC CHANGES OF A PERFECT NONDEGENERATE GAS

Because a nondegenerate gas represents the simplest case, and because many interesting concepts were historically introduced from considerations of such a gas, we shall first discuss the concepts of specific heats and adiabatic changes within that context.

The equation of state of the nondegenerate gas may be written in a thermodynamically more convenient form by introducing the *specific volume*, defined to be the volume of one gram of gas, in place of its reciprocal quantity, the density. Then Eq. (2-10) becomes

$$PV = \frac{R}{\mu}T \tag{2-100}$$

where R is the molar gas constant with a value $R = N_0k = 8.314 \times 10^7$ ergs mole^{-1} deg^{-1}. The ratio R/μ may be thought of as the gas constant for 1 g of gas, and its value depends upon the composition of the gas. Equation (2-100) combined with $U = U(T)$ defines the properties of a perfect nondegenerate gas. The first law of thermodynamics,

$$dQ = dU + P\,dV \tag{2-101}$$

has units of ergs per gram when V is defined to be the specific volume. For some other purposes one might want to discuss the thermodynamics in terms of 1 mole of gas, in which case V would be the volume of 1 mole, the factor μ would not appear in Eq. (2-100), and Eq. (2-101) would then have units of ergs per mole. We shall normally choose to work with 1 g of gas, however.

In keeping with the assumption $U = U(T)$, Eq. (2-101) can be written

$$dQ = \frac{dU}{dT}\,dT + P\,dV \tag{2-102}$$

The specific heats of the gas are defined in the following way. Let α be a function of the physical variables. Then the rate of heat addition per unit rise in temperature, all the time keeping the function α constant, is called the specific heat at constant α, and is designated by c_α:

$$c_\alpha = \left(\frac{dQ}{dT}\right)_\alpha \tag{2-103}$$

This rate is to be determined from Eq. (2-102) in such a way that α remains constant. This demand is met in principle by changing the physical variables from (T,V), as in Eq. (2-102), to the pair (T,α). Then there exists an alternative differential expression

$$dQ = F_1\,dT + F_2\,d\alpha \tag{2-104}$$

where F_1 and F_2 represent two functions of the physical variables. When this is accomplished, the value $c_\alpha = F_1$ can be read off.

For instance, the specific heat at constant volume can be read immediately from Eq. (2-102):

$$c_V = \left(\frac{dQ}{dT}\right)_V = \frac{dU}{dT} \tag{2-105}$$

In the special case where the gas particles possess no excited states, e.g., an ionized gas, the internal energy is simply the kinetic energy of translation, and c_V is a constant. To calculate the specific heat at constant pressure c_p, the first law must be changed to a form of Eq. (2-104) where α is to be equal to P. This may be easily done with the help of the equation of state, Eq. (2-100), whose differential is

$$P\,dV + V\,dP = \frac{R}{\mu}\,dT \tag{2-106}$$

This result allows Eq. (2-102) to be transformed to

$$dQ = \left(\frac{dU}{dT} + \frac{R}{\mu}\right)dT - V\,dP \tag{2-107}$$

which has the desired form. In conjunction with Eq. (2-105) it is evident that for a perfect nondegenerate gas

$$c_p = c_V + \frac{R}{\mu} \tag{2-108}$$

Problem 2-21: Show for a nonperfect gas $[U = U(V,T)]$ that

$$c_P - c_V = \left[\left(\frac{\partial U}{\partial V}\right)_T + P\right]\left(\frac{\partial V}{\partial T}\right)_P$$

Demonstrate that the right-hand side reduces to R/μ for a perfect gas. This result will be useful in regions of partial ionization.

For many thermodynamic applications, the ratio of the specific heats, c_P/c_V, is an important quantity, hereafter designated by γ. From the point of view of the classical kinetic theory of gases, γ depends upon the number of degrees of freedom associated with the molecules of the gas. Quantum mechanics has explained the fact that γ also depends upon the temperature. When kT becomes smaller than rotational or vibrational quantum of energy, those degrees of freedom "freeze out."[1] Specifically,

$$\gamma = 1 + \frac{2}{f} \tag{2-109}$$

where f is the number of "unfrozen" degrees of freedom (translational, rotational, vibrational) of a molecule. In stellar interiors the gases are ionized particles possessing only the three translational degrees of freedom. Hence, for applications in the interiors of stars

$$\gamma = \gamma_{\text{perfect monatonic gas}} = \tfrac{5}{3} \tag{2-110}$$

Problem 2-22: For 1 mole of a perfect monatomic nondegenerate gas

$$U = \tfrac{3}{2}N_0kT \qquad \text{and} \qquad PV = N_0kT$$

Show that the amounts of heat required to raise the temperature by an amount ΔT at constant volume and constant pressure, respectively, are

$$\Delta Q_V = \tfrac{3}{2}N_0k\,\Delta T \qquad \Delta Q_P = \tfrac{5}{2}N_0k\,\Delta T$$

Compare with Eq. (2-110).

Problem 2-23: Show that the ratio of the isothermal to isentropic compressibilities is always equal to the ratio of the specific heats c_P/c_V, just as it is for a perfect gas. In other words, show that

$$\frac{\chi_T}{\chi_S} = \frac{c_P}{c_V} \qquad \chi_T \equiv -\frac{1}{V}\left(\frac{\partial V}{\partial P}\right)_T \qquad \chi_S \equiv -\frac{1}{V}\left(\frac{\partial V}{\partial P}\right)_S$$

The notation constant S means constant entropy, or $dQ = 0$.

By employing the equation of state (2-100) and Eq. (2-105) the first law may be written for an ideal gas:

$$dQ = c_V\,dT + \frac{RT}{\mu V}\,dV \tag{2-111}$$

[1] See, for instance, "F. K. Richtmyer, E. H. Kennard, and T. Lauritsen, "Introduction to Modern Physics," 5th ed., p. 405, McGraw-Hill Book Company, New York, 1955.

An adiabatic change is defined as a quasistatic change of state during which no heat is added, that is, $dQ = 0$. Setting dQ equal to zero and substituting $R/\mu = c_P - c_V$ from Eq. (2-108) yields

$$c_V \frac{dT}{T} + (c_P - c_V) \frac{dV}{V} = 0 \qquad \text{ideal gas} \tag{2-112}$$

For an ideal gas c_P and c_V are constants, in which case Eq. (2-112) may be integrated:

$$TV^{\gamma-1} = \text{const} \tag{2-113}$$

Problem 2-24: With the aid of the equation of state, show that the track of the quasistatic adiabatic change in Eq. (2-113) has these three equivalent forms:

$$TV^{\gamma-1} = \text{const} \qquad PV^{\gamma} = \text{const} \qquad P^{1-\gamma}T^{\gamma} = \text{const} \tag{2-114}$$

Problem 2-25: Show that the differential adiabatic change of the perfect nondegenerate gas is

$$\frac{dT}{T} + (\gamma - 1)\frac{dV}{V} = 0 \qquad \frac{dP}{P} + \frac{\gamma}{1-\gamma}\frac{dT}{T} = 0 \qquad \frac{dP}{P} + \gamma\frac{dV}{V} = 0 \tag{2-115}$$

When the effects of radiation are considered in the next section, we shall see that the values of γ are not the same in each of the three equations (2-115), nor are they constant along the adiabatic track, nor are they equal to the ratio of specific heats.

QUASISTATIC CHANGES IN AN ENCLOSURE CONTAINING MATTER AND RADIATION

To good approximation the matter in a stellar interior is in thermodynamic equilibrium. Thus the particle gas at each interior point is accompanied by a photon gas characteristic of the local temperature. The presence of the photons introduces two important effects into quasistatic expansions, radiation pressure and ionization change.

Effects of radiation pressure From Eq. (2-86) it can be concluded that radiation pressure will be important only for temperatures high enough so that matter is essentially completely ionized. It will also be true in practically all common cases that radiation pressure will be relatively unimportant except for densities low enough for the electron gas to be nondegenerate. Thus with a small sacrifice in generality we can evaluate the effect of radiation pressure by considering adiabatic expansions of an ideal nondegenerate monatomic gas plus radiation pressure. Then the pressure becomes

$$P = P_g + P_r = \frac{N_0 k}{\mu} \rho T + \tfrac{1}{3} a T^4 \tag{2-116}$$

Since the particle internal energy is, for monatomic particles, just the kinetic energy, the internal energy per gram is

$$U = aT^4V + \frac{N_0}{\mu}\left(\frac{3}{2}kT\right) \tag{2-117}$$

where V is the volume of 1 g of gas. Now for a quasistatic change, the first law of thermodynamics requires

$$dQ = \left(\frac{\partial U}{\partial T}\right)_V dT + \left(\frac{\partial U}{\partial V}\right)_T dV + P\,dV \tag{2-118}$$

and the partial derivatives are evaluated as

$$\left(\frac{\partial U}{\partial V}\right)_T = aT^4 \qquad \left(\frac{\partial U}{\partial T}\right)_V = 4aT^3V + \frac{3}{2}\frac{N_0k}{\mu} \tag{2-119}$$

Then Eq. (2-118) can be written at once as

$$dQ = \left(4aT^3V + \frac{3}{2}\frac{N_0k}{\mu}\right)dT + \left(\tfrac{4}{3}aT^4 + \frac{N_0k}{\mu}\frac{T}{V}\right)dV \tag{2-120}$$

The differential expression for an adiabatic change is obtained by setting $dQ = 0$. For many purposes it is useful to rearrange that equation into a form resembling that for the adiabatic changes of a particle gas. Following Chandrasekhar, it is useful to define the *adiabatic exponents* Γ_1, Γ_2, and Γ_3 by the equations

$$\frac{dP}{P} + \Gamma_1\frac{dV}{V} = 0 \tag{2-121a}$$

$$\frac{dP}{P} + \frac{\Gamma_2}{1-\Gamma_2}\frac{dT}{T} = 0 \tag{2-121b}$$

$$\frac{dT}{T} + (\Gamma_3 - 1)\frac{dV}{V} = 0 \tag{2-121c}$$

where the changes of state involved are adiabatic.

The definitions of the adiabatic exponents are made in this form to retain the analogy to the corresponding equations (2-115) for a perfect nondegenerate particle gas, for which all three adiabatic exponents are equal to γ. For a perfect gas with constant γ it is possible to integrate the equations immediately, of course, as in Eq. (2-114). The present equations are not immediately integrable because, as we shall see, the adiabatic exponents are functions of the thermodynamic state. Rather clearly, these definitions demand that $\Gamma_3 - 1 = (\Gamma_2 - 1)\Gamma_1/\Gamma_2$.

Now from Eq. (2-116) it follows that

$$\begin{aligned} dP &= \left(\tfrac{4}{3}aT^4 + \frac{N_0k}{\mu}\frac{T}{V}\right)\frac{dT}{T} - \frac{N_0k}{\mu}\frac{T}{V}\frac{dV}{V} \\ &= (4P_r + P_g)\frac{dT}{T} - P_g\frac{dV}{V} \end{aligned} \tag{2-122}$$

The substitution of dP/P into Eq. (2-121a) yields

$$(4P_r + P_g)\frac{dT}{T} + [\Gamma_1(P_r + P_g) - P_g]\frac{dV}{V} = 0 \tag{2-123}$$

Equation (2-120), on the other hand, can be written for $dQ = 0$ as

$$(12P_r + \tfrac{3}{2}P_g)\frac{dT}{T} + (4P_r + P_g)\frac{dV}{V} = 0 \tag{2-124}$$

Comparison of Eqs. (2-123) and (2-124) defines Γ_1 in terms of the partial pressures

$$\frac{\Gamma_1(P_r + P_g) - P_g}{4P_r + P_g} = \frac{4P_r + P_g}{12P_r + \tfrac{3}{2}P_g} \tag{2-125}$$

It is conventional to designate by β the fraction of the total pressure contributed by the particle pressure

$$\beta P = P_g \qquad (1 - \beta)P = P_r \tag{2-126}$$

Problem 2-26: By substituting Eq. (2-126) into (2-125), show that

$$\Gamma_1 = \frac{32 - 24\beta - 3\beta^2}{24 - 21\beta} \tag{2-127}$$

For a particle gas ($\beta = 1$), Γ_1 reduces to $\frac{5}{3}$, the value of γ for a monatomic gas. For a photon gas ($\beta = 0$), Γ_1 reduces to $\frac{4}{3}$, the value of γ for a photon gas.

The second adiabatic exponent is found by similar algebraic steps. Substituting dP/P from Eq. (2-122) into Eq. (2-121b) and comparing with Eq. (2-124) gives

$$\frac{12(1 - \beta) + \tfrac{3}{2}\beta}{4 - 3\beta + \Gamma_2/(1 - \Gamma_2)} = -\frac{4 - 3\beta}{\beta} \tag{2-128}$$

Problem 2-27: Solve for Γ_2 and show that it has the same limits as Γ_1 for $\beta \to 0, 1$.

$$\Gamma_2 = \frac{32 - 24\beta - 3\beta^2}{24 - 18\beta - 3\beta^2} \tag{2-129}$$

Problem 2-28: Show that

$$\Gamma_3 = \frac{32 - 27\beta}{24 - 21\beta} \tag{2-130}$$

Each adiabatic exponent decreases monotonically from a value of $\frac{5}{3}$ for $\beta = 1$ to a value of $\frac{4}{3}$ for $\beta = 0$. Thus the logarithmic derivatives in Eqs. (2-121a) to (2-121c) are seen to be related by coefficients (Γ_1, Γ_2, Γ_3) whose values depend on the relative importance of gas pressure in the total pressure.

Later discussions will indicate instabilities in stellar structure associated with values of Γ less than $\frac{4}{3}$. A careful analysis shows that stellar instability occurs if Γ_1, when appropriately averaged over a star, becomes smaller than $\frac{4}{3}$. The present discussion shows that radiation pressure cannot (in the absence of positron-electron-pair production or ionization) reduce Γ below $\frac{4}{3}$. However, it is quite clear that stars dominated internally by radiation pressure will have adiabatic exponents close to $\frac{4}{3}$. Their binding energy is relatively small compared to stars for which particles provide the pressure.[1] The adiabatic exponents will also be relevant in the model of convection to be discussed later and to the theory of pulsation.

From the equations developed, it is possible to calculate the specific heats at constant volume and at constant pressure for an enclosure containing an ideal monatomic gas and radiation. From Eq. (2-120) we see that

$$C_V = \left(\frac{dQ}{dT}\right)_V = 4aT^3V + \frac{3}{2}\frac{N_0k}{\mu} = \frac{3}{2}\frac{N_0k}{\mu}\left(1 + \frac{8aT^4/3}{N_0k/\mu V}\right)$$

$$C_V = c_V\left(1 + \frac{8P_r}{P_g}\right) = c_V\left[1 + \frac{8(1-\beta)}{\beta}\right] = c_V\frac{8-7\beta}{\beta} \tag{2-131}$$

where $c_V = 3N_0k/2\mu$ is the specific heat of the particle gas alone.

Problem 2-29: Show that

$$C_P = c_V\frac{\frac{32}{3} - 8\beta - \beta^2}{\beta^2} \tag{2-132}$$

Problem 2-30: It was proved in Prob. 2-23 that $c_p/c_V = \chi_T/\chi_S$ for any gas. Evaluate these compressibilities for the present case of a monatomic gas plus radiation and prove thereby that

$$\frac{C_P}{C_V} = \frac{\Gamma_1}{\beta} \tag{2-133}$$

Check this result by Eqs. (2-131) and (2-132). What is the meaning of the infinity as $\beta \to 0$?

Equations (2-121) are not integrable as they stand. It is a simple matter, however, to rearrange variables until an integrable form of the adiabatic track is obtained. From the second law of thermodynamics, dQ/T must be the differential of the entropy S. From Eq. (2-120) we see that for an ideal monatomic nondegenerate gas

$$dS = \left(4aT^2V + \frac{3N_0k}{2\mu T}\right)dT + \left(\tfrac{4}{3}aT^3 + \frac{N_0k}{uV}\right)dV \tag{2-134}$$

Problem 2-31: Confirm that S satisfies the integrability condition

$$\frac{\partial^2 S}{\partial T\,\partial V} = \frac{\partial^2 S}{\partial V\,\partial T}$$

[1] See the discussion of the virial theorem later in this section for a simplified demonstration.

Show also that the elimination of the variable V in favor of a new variable $W = T^3V$ allows dS to be written as

$$dS = -\frac{3}{2}\frac{N_0k}{\mu}\frac{dT}{T} + \frac{4a}{3}\,dW + \frac{N_0k}{\mu}\frac{dW}{W} \tag{2-135}$$

Equation (2-135) can be integrated by inspection.

$$S = \text{const} + \frac{N_0k}{\mu}\ln\frac{T^{3/2}}{\rho} + \frac{4a}{3}\frac{T^3}{\rho} \tag{2-136}$$

The second term is just the entropy per gram of an ideal nondegenerate monatomic gas, whereas the third term measures the entropy per gram of the photon component. The entropies are additive. Evidently the last term can be written in terms of the partial-pressure ratio, so that alternate forms are

$$S = \text{const} + \frac{N_0k}{\mu}\left(\ln\frac{T^{3/2}}{\rho} + 4\frac{P_r}{P_g}\right) \tag{2-137a}$$

and

$$S = \text{const} + \frac{N_0k}{\mu}\left(\ln\frac{T^{3/2}}{\rho} + 4\frac{1-\beta}{\beta}\right) \tag{2-137b}$$

The increase in entropy of the final state f over that of the initial state i is

$$\Delta S = \frac{N_0k}{\mu}\left\{\ln\left[\left(\frac{T_f}{T_i}\right)^{3/2}\frac{\rho_i}{\rho_f}\right] + 4\left(\frac{1-\beta_f}{\beta_f} - \frac{1-\beta_i}{\beta_i}\right)\right\} \tag{2-138}$$

If a portion of a stellar interior is allowed to expand (or contract) reversibly without exchanging heat with its surroundings, then $\Delta S = 0$, and the change is an adiabatic one. Two examples of such changes *might be* (1) the gravitational contraction of a stellar core following the exhaustion of a nuclear fuel supply and (2) the expansion of a rising convective mass of gas. In no case is it obvious that such a change will be adiabatic. That no heat be exchanged demands that the expansion occur in times short enough so that only an insignificant fraction of the photon energy can diffuse into the surroundings during the expansion. For any physical expansion, therefore, it will be necessary to compute the heat exchange by radiative transfer during the expansion. If the entropy increase during the expansion,

$$\Delta S = \int_i^f \frac{dQ}{T}$$

can be computed, Eq. (2-138) still provides one relationship between the initial and final states. Of course, these equations are valid only for temperatures high enough for ionization to be complete and for densities low enough for electrons to be nondegenerate. More elaborate formulas can be derived to apply to more general circumstances.

Problem 2-32: A gas composed of equal numbers of C^{12} and O^{16} is initially at a density of 100 g/cm³ and a temperature of 10^8 °K. What is its density after an adiabatic compression to a temperature of 10^9 °K?

The foregoing treatment was based on the assumption of a nondegenerate gas. Because the entropies are additive, however, the results can be generalized to a partially degenerate gas by separately computing the entropy for the Fermi electron gas. Using the fact that for any degree of degeneracy the internal energy per gram of a nonrelativistic gas is $U = \frac{3}{2}PV$, where V is the specific volume, the first law can be expressed as

$$dQ = dU + P\,dV = \tfrac{5}{2}P\,dV + \tfrac{3}{2}V\,dP \tag{2-139}$$

With the aid of Eq. (2-57) for the electron pressure, dQ becomes

$$dQ = \frac{5}{2}\left[\frac{8\pi kT}{3h^3}(2mkT)^{3/2}F_{3/2}(\alpha)\right]dV + \tfrac{3}{2}V\left[\frac{5}{2}\frac{8\pi k}{3h^3}(2mkT)^{3/2}F_{3/2}(\alpha)\right]dT$$
$$+ \tfrac{3}{2}V\left[\frac{8\pi kT}{3h^3}(2mkT)^{3/2}\frac{dF_{3/2}}{d\alpha}\right]d\alpha \tag{2-140}$$

Although this expression is in terms of three increments, dV, dT, and $d\alpha$, it will be realized that these increments are not independent. From the fact that

$$n_e = \frac{\rho N_0}{\mu_e} = \frac{N_0}{\mu_e}\frac{1}{V} = \frac{4\pi}{h^3}(2mkT)^{3/2}F_{1/2}(\alpha) \tag{2-141}$$

it follows that dV can be expressed in terms of dT and $d\alpha$. This operation yields

$$dV = -\frac{\mu_e V^2}{N_0}\,dn_e = -\frac{\mu_e V^2}{N_0}\left[\frac{4\pi}{h^3}(2mkT)^{3/2}\left(\frac{dF_{1/2}}{d\alpha}\,d\alpha + \tfrac{3}{2}F_{1/2}\frac{dT}{T}\right)\right]$$
$$= -\frac{N_0}{\mu_e n_e}\left(\frac{dF_{1/2}/d\alpha}{F_{1/2}}\,d\alpha + \frac{3}{2}\frac{dT}{T}\right) \tag{2-142}$$

When this expression for dV is inserted into Eq. (2-140), the result for $dS = dQ/T$ is

$$dS = \frac{N_0 k}{\mu_e}\left(-\frac{5}{3}\frac{F_{3/2}}{F_{1/2}^2}\frac{dF_{1/2}}{d\alpha} + \frac{dF_{3/2}/d\alpha}{F_{1/2}}\right)d\alpha \tag{2-143}$$

It will be noticed that dS is a function of α times the increment $d\alpha$. It follows that the electron entropy is a function only of α.

Problem 2-33: It can be shown that $dF_{3/2}/d\alpha = -3F_{1/2}/2$. Confirm this result from the expansions appropriate for weak degeneracy, Eq. (2-62), and from the expansions appropriate for strong degeneracy, Eq. (2-66). Then show that the function

$$S(\alpha) = \frac{N_0 k}{\mu_e}\left(\frac{5}{2}\frac{F_{3/2}}{F_{1/2}} + \alpha\right) \tag{2-144}$$

reproduces dS.

From the additive properties of the entropy per gram, the generalized form of the entropy becomes

$$S = \text{const} + \frac{N_0 k}{\mu_i} \ln \frac{T^{\frac{3}{2}}}{\rho} + \frac{N_0 k}{\mu_e} \left[\frac{5}{3} \frac{F_{\frac{3}{2}}(\alpha)}{F_{\frac{1}{2}}(\alpha)} + \alpha \right] + \frac{4a}{3} \frac{T^3}{\rho} \tag{2-145}$$

The three terms are the entropies of the ions, the electrons, and the radiation, respectively. The result is correct only if the electrons are nonrelativistic, however.

Problem 2-34: Can you derive an expression for the entropy of a partially degenerate relativistic electron gas?

Effects of ionization If the ionization of matter is incomplete, quasistatic changes will be accompanied by changes in the degree of ionization. From the Saha equation it follows that even a slight rise in the temperature of a partially ionized gas may considerably increase the ionization, which in turn may require a large amount of energy. A large energy requirement for a small temperature rise corresponds to a large value of the specific heat. It will therefore be expected that the thermodynamics of a partially ionized gas will differ considerably from that of a completely ionized (or completely neutral) gas. The differences occur because the number of free particles per gram is not constant and because energy is required to increase the number of free particles.

As the simplest of examples we shall compute c_V for a partially ionized gas of pure hydrogen. Now by definition $c_V = (dQ/dT)_V = (\partial U/\partial T)_V$, so that the first requirement is an expression for the internal energy of the partially ionized gas. Actually only the change in internal energy must be correctly represented. This end may be accomplished by noting that each free particle possesses translational energy equal to $3kT/2$, that an amount of energy approximately equal to $\chi_{\rm H}$ is required for each hydrogen ionization, and that the internal energy of atoms represented by their population of excited states will change with temperature. For hydrogen the situation is simplified by the fact that the fractional population of excited states is very small, so that the internal energy per neutral atom can be neglected, and the average ionization potential can be taken to be that for the ground state. Then if N, H, and H$^+$ represent, respectively, the numbers per unit volume of free particles, of neutral hydrogens, and ionized hydrogens, the internal energy per gram is very nearly

$$U(T,V) = \tfrac{3}{2} NkTV + \chi_{\rm H} {\rm H}^+ V \tag{2-146}$$

where V is the specific volume. Then

$$dU = \left[\tfrac{3}{2} NkV + \tfrac{3}{2} kTV \left(\frac{\partial N}{\partial T} \right)_V + \chi_{\rm H} V \left(\frac{\partial {\rm H}^+}{\partial T} \right)_V \right] dT + \left[\tfrac{3}{2} NkT + \tfrac{3}{2} kTV \left(\frac{\partial N}{\partial V} \right)_T + \chi_{\rm H} {\rm H}^+ + \chi_{\rm H} V \left(\frac{\partial {\rm H}^+}{\partial V} \right)_T \right] dV \tag{2-147}$$

and

$$C_V = \left(\frac{\partial U}{\partial T}\right)_V = \tfrac{3}{2}NkV + \tfrac{3}{2}kTV\left(\frac{\partial N}{\partial T}\right)_V + \chi_{\rm H}V\left(\frac{\partial {\rm H}^+}{\partial T}\right)_V \tag{2-148}$$

From the condition of charge neutrality in pure hydrogen,

$$n_e = {\rm H}^+ \tag{2-149}$$

and the density of free particles is

$$N = {\rm H} + {\rm H}^+ + n_e = {\rm H} + 2{\rm H}^+ \tag{2-150}$$

Because the electron mass is negligible in comparison with the proton mass, the specific volume is related to the number densities by

$${\rm H}^+ + {\rm H} = \frac{N_0}{V} \tag{2-151}$$

where N_0 is Avogadro's number. From these relationships between number densities it follows that

$$\left(\frac{\partial N}{\partial T}\right)_V = \left(\frac{\partial {\rm H}^+}{\partial T}\right)_V = -\left(\frac{\partial {\rm H}}{\partial T}\right)_V \tag{2-152}$$

Thus

$$c_V = \tfrac{3}{2}NkV\left[1 - \frac{2T}{3N}\left(\frac{3}{2} + \frac{\chi_{\rm H}}{kT}\right)\left(\frac{\partial {\rm H}}{\partial T}\right)_V\right] \tag{2-153}$$

Inasmuch as

$$\tfrac{3}{2}NkV = c_V{}^{(0)} \tag{2-154}$$

is the specific heat for a constant number of particles, the second term in Eq. (2-153) may be regarded as a correction term. The partial derivative in the second term must be evaluated from the Saha equation:

$$\frac{{\rm H}^+ n_e}{\rm H} = \frac{({\rm H}^+)^2}{\rm H} = \frac{(2\pi mkT)^{\frac{3}{2}}}{h^3}\exp -\frac{\chi_{\rm H}}{kT} = g(T) \tag{2-155}$$

By eliminating $\rm H^+$ in favor of H with Eq. (2-151) we obtain

$$\frac{(N_0/V - {\rm H})^2}{\rm H} = g(T) \tag{2-156}$$

so that

$$-\left[\frac{(N_0/V)^2}{{\rm H}^2} + 1\right]\left(\frac{\partial {\rm H}}{\partial T}\right)_V = \left(\frac{\partial g}{\partial T}\right)_V \tag{2-157}$$

Problem 2-35: Show by further manipulation that

$$\left(\frac{\partial H}{\partial T}\right)_V = -\frac{1}{T}\left(\frac{3}{2} + \frac{\chi_{\rm H}}{kT}\right)\frac{{\rm H}^+{\rm H}}{{\rm H}^+ + 2{\rm H}} \tag{2-158}$$

Equation (2-153) now yields

$$c_V = c_V^{(0)} \left[1 + \frac{2}{3} \left(\frac{3}{2} + \frac{\chi_H}{kT} \right)^2 \frac{H^+H}{(H + 2H^+)(2H^+ + H)} \right] \tag{2-159}$$

where $c_V^{(0)} = 3NkV/2$. Because the sum of H^+ and H is constant, the correction factor has a maximum at $H^+ = H$, the stage of 50 percent ionization.

Problem 2-36: It has already been pointed out that for most common densities hydrogen ionizes near 10^4 °K. Using that approximation for T, estimate the maximum enhancement factor for $c_V^{(0)}$.

Problem 2-37: Using the equation of state $PV = NkT$, show that

$$c_p = c_V^{(0)} \left[\frac{5}{3} + \frac{1}{3} \left(\frac{5}{2} + \frac{\chi_H}{kT} \right)^2 \frac{H^+H}{(H^+ + H)^2} \right] \tag{2-160}$$

The specific heats for the pure hydrogen gas are shown in Fig. 2-12 as a function of the percent of hydrogen ionized. It can be seen that both c_V and c_P have maximum values about 30 times as great as their normal values. In examination of this figure it should be noted that $NV = N_0$ for neutral hydrogen and $NV = 2N_0$ for completely ionized hydrogen. Thus the specific heats per gram are twice as great at 100 percent ionization as for neutrality, and there is a slight asymmetry in the curves toward high ionization.

It is also evident that the ratio c_P/c_V drops below its normal value of $\frac{5}{3}$, leading one to wonder about the values of the adiabatic exponents in the regions of partial ionization. The first thing to notice is that the adiabatic exponents differ from each other in an ionization zone and they all differ from the ratio c_P/c_V. The adiabatic exponents are again defined by

$$\begin{aligned} &\frac{dP}{P} + \Gamma_1 \frac{dV}{V} = 0 \\ &\frac{dP}{P} + \frac{\Gamma_2}{1 - \Gamma_2} \frac{dT}{T} = 0 \\ &\frac{dT}{T} + (\Gamma_3 - 1) \frac{dV}{V} = 0 \end{aligned} \tag{2-161}$$

for adiabatic changes. Each function can easily be derived in terms of the specific heats and certain other partial derivatives. We first note the following relationship between the specific heats whenever the internal energy $U(V,T)$ is not independent of the volume, as is the case in regions of partial ionization. From the first law,

$$dQ = \left(\frac{\partial U}{\partial T} \right)_V dT + \left[\left(\frac{\partial U}{\partial V} \right)_T + P \right] dV \tag{2-162}$$

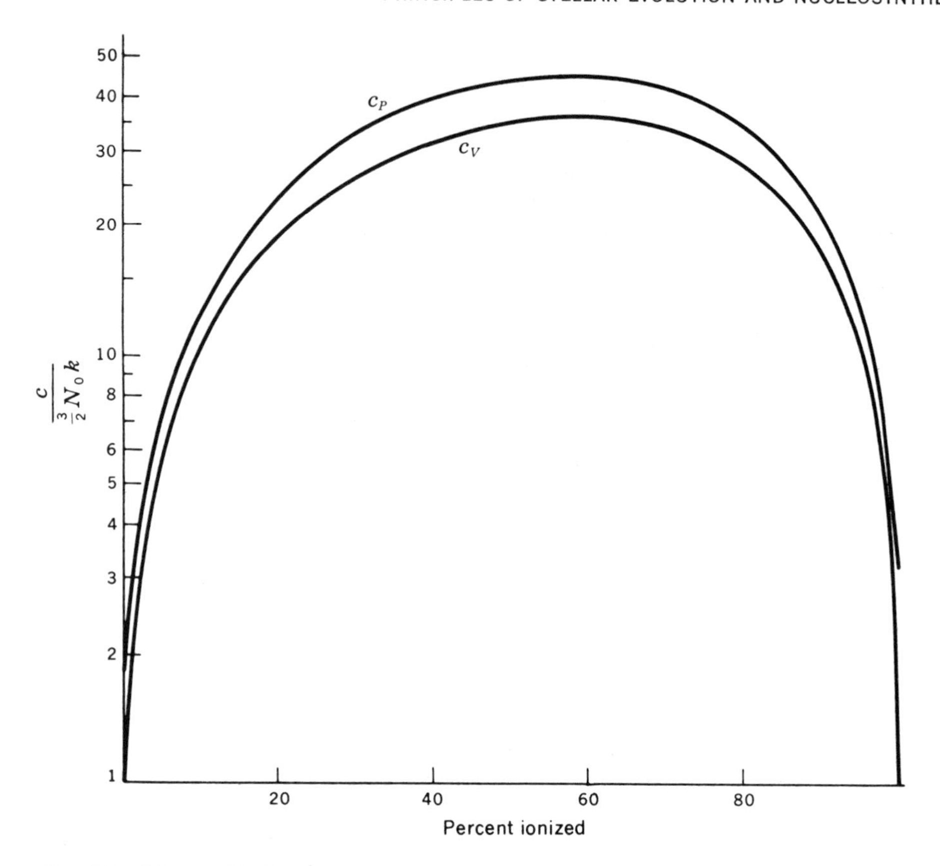

Fig. 2-12 The specific heats per gram c_P and c_V for a pure hydrogen gas as a function of its degree of ionization. In regions of partial ionization both specific heats are increased by large factors as a result of the large energy requirements for ionization changes. The specific heats at full ionization are twice as great as those of the neutral gas because the number of particles per gram is twice as great. Numerical values are given in Table 2-4.

and from the definitions of specific heats it follows that

$$\begin{aligned} c_P = \left(\frac{dQ}{dT}\right)_P &= \left(\frac{\partial U}{\partial T}\right)_V + \left[\left(\frac{\partial U}{\partial V}\right)_T + P\right]\left(\frac{\partial V}{\partial T}\right)_P \\ &= c_V + \left[\left(\frac{\partial U}{\partial V}\right)_T + P\right]\left(\frac{\partial V}{\partial T}\right)_P \end{aligned} \tag{2-163}$$

By rearrangement we have

$$\left(\frac{\partial U}{\partial V}\right)_T + P = (c_P - c_V)\left(\frac{\partial T}{\partial V}\right)_P \tag{2-164}$$

which yields in the first law

$$dQ = c_V\,dT + (c_P - c_V)\left(\frac{\partial T}{\partial V}\right)_P dV \tag{2-165}$$

For an adiabatic change

$$\frac{dT}{T} + \frac{(c_P - c_V)}{c_V T} \left(\frac{\partial T}{\partial V}\right)_P dV = 0 \qquad \text{adiabatic} \tag{2-166}$$

Comparison with Eq. (2-161) yields

$$\Gamma_3 - 1 = \frac{V}{T} \frac{c_P - c_V}{c_V} \left(\frac{\partial T}{\partial V}\right)_P \tag{2-167}$$

To obtain Γ_1 it is necessary only to express dQ as a differential in dP and dV. To do so we express dT as

$$dT = \left(\frac{\partial T}{\partial P}\right)_V dP + \left(\frac{\partial T}{\partial V}\right)_P dV \tag{2-168}$$

so that

$$dQ = \left(\frac{\partial U}{\partial T}\right)_V \left(\frac{\partial T}{\partial P}\right)_V dP + \left[\left(\frac{\partial U}{\partial T}\right)_V + P + \left(\frac{\partial U}{\partial T}\right)_V \left(\frac{\partial T}{\partial V}\right)_P\right] dV \tag{2-169}$$

With the aid of Eq. (2-164) this expression reduces to

$$dQ = c_V \left(\frac{\partial T}{\partial P}\right)_V dP + c_P \left(\frac{\partial T}{\partial V}\right)_P dV \tag{2-170}$$

Setting $dQ = 0$ and dividing by $c_V(\partial T/\partial P)_V P$ yields

$$\frac{dP}{P} + \frac{c_P(\partial T/\partial V)_P}{P c_V (\partial T/\partial P)_V} dV = 0 \qquad \text{adiabatic} \tag{2-171}$$

By use of the cyclic relation,

$$\left(\frac{\partial T}{\partial V}\right)_P = -\left(\frac{\partial T}{\partial P}\right)_V \left(\frac{\partial P}{\partial V}\right)_T \tag{2-172}$$

We obtain for an adiabatic change

$$\frac{dP}{P} - \frac{V}{P} \frac{c_P}{c_V} \left(\frac{\partial P}{\partial V}\right)_T \frac{dV}{V} = 0 \qquad \text{adiabatic} \tag{2-173}$$

Comparison with Eq. (2-161) finally gives Γ_1:

$$\Gamma_1 = -\frac{c_P}{c_V} \frac{V}{P} \left(\frac{\partial P}{\partial V}\right)_T \tag{2-174}$$

Problem 2-38: Show that

$$\frac{\Gamma_2}{1 - \Gamma_2} = -\frac{c_P T}{(c_P - c_V)(\partial T/\partial P)_V P} \tag{2-175}$$

Equations (2-167), (2-174), and (2-175) define the adiabatic exponents. Their evaluation in an ionization zone requires use of the formulas for c_P and c_V as well as evaluation of the partial derivatives appearing in the expressions. These quantities are relatively simple in the example of a pure hydrogen gas, as illustrated by the next problem.

Table 2-4 Properties of a hydrogen gas near $T = 10^4$ °K

Fraction ionized	Γ_1	Γ_2	Γ_3	$\frac{c_V}{\frac{3}{2}N_0k}$	$\frac{c_P}{\frac{3}{2}N_0k}$
0.00	1.6666	1.6666	1.6666	1.0000	1.6666
0.05	1.2097	1.1593	1.1662	5.8974	7.3037
0.10	1.1688	1.1160	1.1214	10.5264	12.8572
0.15	1.1537	1.1000	1.1049	14.8650	18.2437
0.20	1.1460	1.0920	1.0965	18.8891	23.3797
0.25	1.1415	1.0872	1.0915	22.5716	28.1816
0.30	1.1386	1.0842	1.0884	25.8826	32.5659
0.35	1.1368	1.0822	1.0864	28.7882	36.4492
0.40	1.1356	1.0810	1.0851	31.2503	39.7478
0.45	1.1349	1.0803	1.0844	33.2262	42.3783
0.50	1.1347	1.0801	1.0842	34.6670	44.2572
0.55	1.1349	1.0803	1.0844	35.5176	45.3010
0.60	1.1356	1.0810	1.0851	35.7147	45.4261
0.65	1.1368	1.0822	1.0864	35.1856	44.5490
0.70	1.1386	1.0842	1.0884	33.8465	42.5862
0.75	1.1415	1.0872	1.0915	31.6003	39.4542
0.80	1.1460	1.0920	1.0965	28.3336	35.0695
0.85	1.1537	1.1000	1.1049	23.9133	29.3486
0.90	1.1688	1.1160	1.1214	18.1820	22.2079
0.95	1.2097	1.1593	1.1662	10.9524	13.5640
1.00	1.6666	1.6666	1.6666	2.0000	3.3333

Problem 2-39: Show for the pure hydrogen gas that

$$\Gamma_1 = \frac{10(\mathrm{H}^+ + \mathrm{H})^2 + 2(\frac{5}{2} + \chi_\mathrm{H}/kT)^2\mathrm{H}^+\mathrm{H}}{3(\mathrm{H}^+ + 2\mathrm{H})(2\mathrm{H}^+ + \mathrm{H}) + 2(\frac{3}{2} + \chi_\mathrm{H}/kT)^2\mathrm{H}^+\mathrm{H}} \tag{2-176}$$

Analogous expressions for Γ_2 and Γ_3 can be derived with the necessary manipulation.

The values of the adiabatic exponents for the hydrogen gas at 10^4 °K are shown in Table 2-4 along with the values of c_V and c_P in units of $\frac{3}{2}N_0k$. The striking feature is that all three adiabatic exponents fall to values near unity for partial ionization. Table 2-4 was computed by simply assuming $T = 10^4$ °K, but the values will not be much different if, for each density, one actually computes the temperature corresponding to the desired degree of ionization. These adiabatic exponents are displayed on an expanded scale in Fig. 2-13.

The situation in a star will be further complicated by the mixture in composition, because each element will undergo stages of ionization at appropriate temperatures.[1] Since hydrogen will probably be the dominant element in the outer portions of the star, the adiabatic exponents will be similar to those of the table

[1] See P. Ledoux, Stellar Stability and Stellar Evolution, in L. Gratton (ed.), "Star Evolution," Academic Press Inc., New York, 1963; also a series of papers by C. Rouse, *Astrophys. J.*, **134**:435 (1961); **135**:599 (1962); **137**:1286 (1963); **139**:339 (1963).

in the hydrogen ionization zone. After hydrogen ionization the helium will ionize at a higher temperature, but because of its smaller abundance the helium ionization will not usually cause such a large drop in the adiabatic exponents. Heavier elements will be of even less importance, but it will be prudent to remember that the adiabatic exponents may be significantly smaller than $\frac{5}{3}$ even up to temperatures of around 10^6 °K.

Several important effects in stellar evolution are related to the small adiabatic exponents in ionization zones. Because a star with $\bar{\Gamma} < \frac{4}{3}$ is unstable, hydrostatic equilibrium cannot be achieved in star formation until the hydrogen, and perhaps the helium, has been ionized in the interior. Even in hydrostatic equilibrium many stars pulsate because of peculiarities traceable to the ionization zones. The pulsation phenomenon is discussed in Chap. 6.

Finally we would mention that in convection zones the temperature gradient is usually assumed to be the adiabatic temperature gradient. The presence of par-

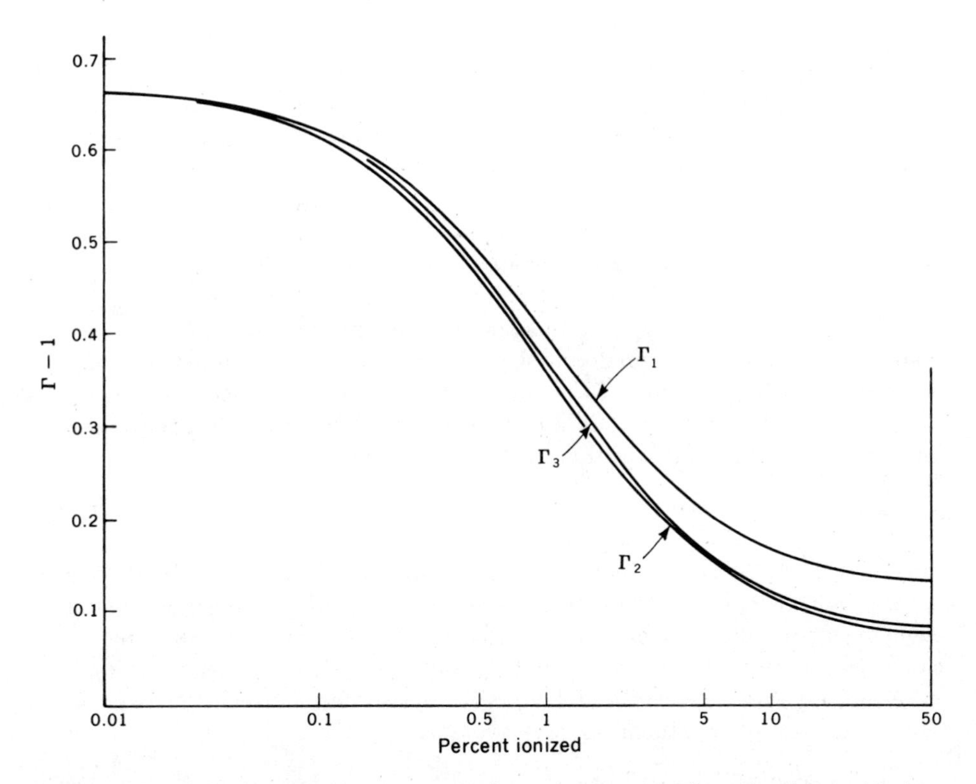

Fig. 2-13 The adiabatic exponents of a pure hydrogen gas as a function of its degree of ionization. Only the initial 50 percent is shown because the second 50 percent is its mirror image. The exponents change rapidly in the regions 0 to 1 or 99 to 100 percent ionization. Between 5 and 95 percent ionization, the values are considerably less than $\frac{4}{3}$ and therefore have a destabilizing influence on the structure. Numerical values are given in Table 2-4.

tial ionization will modify that temperature gradient from the ones computed under the $\gamma = \frac{5}{3}$ assumption.

Radiation has another interesting effect upon the adiabatic exponents at high temperatures. In the presence of the coulomb field of a nucleus a photon may create a positron-electron pair if $h\nu > 2m_ec^2$:

$$\gamma + Z \rightarrow Z + e^+ + e^-$$

This process is *thermodynamically similar to ionization* in that increased temperature is accompanied by an increase in the number of free particles at great expense to the energy of the photon gas. It is physically clear that the specific heats are increased thereby. Interestingly enough, it turns out that Γ_1 drops below $\frac{4}{3}$ for $T > 1 \times 10^9$ °K if $\rho = 10^3$ g/cm³ and for $T > 5 \times 10^9$ °K if $\rho = 10^6$ g/cm³. Adiabatic exponents are somewhat academic quantities in those conditions, however, because the associated annihilation of the electron pairs into neutrino pairs produces such a large heat loss that adiabatic changes are not really possible.[1]

HYDROSTATIC EQUILIBRIUM AND UNIFORM CONTRACTION

The pressure plays a primary role in the structure of stars because it provides resistance against gravitational collapse. Consider a static massive sphere held together by gravity. If a volume element of gas is to be held mechanically at a certain position in a star, neither being expelled outward by pressure nor falling to the center of gravitational attraction, then it will be necessary for the pressure and gravity forces on a volume element to sum to zero. This condition is called *hydrostatic equilibrium*. The balance in the case of a spherical gas cloud like the one illustrated in Fig. 2-14 leads to a simple differential equation for the pressure gradient. Consider the small cylindrical volume element with axis of length dr parallel to the radius vector at the point r and having a cross-sectional area equal to dA. Let dP be the pressure increment associated with dr. Then the radial force on this volume element due to the pressure differential is

$$F_P = P\,dA - (P + dP)\,dA = -dP\,dA \tag{2-177}$$

Since the pressure will actually decrease in the radial direction, the differential dP will be negative, and the pressure force will then be positive. If the volume element is not to be accelerated upward by F_P, it is necessary that it be exactly cancelled by the central gravitational force on the volume element. If dm is the mass of the volume element and $M(r)$ is the mass interior to the spherical surface at radius r, the gravitational force is

$$F_G = -G\frac{M(r)\,dm}{r^2} \tag{2-178}$$

[1] For a thorough discussion see W. A. Fowler and F. Hoyle, *Astrophys. J. Suppl.*, **9**:201 (1964), app. B.

where

$$M(r) = \int_0^r \rho(r)4\pi r^2 \, dr \tag{2-179}$$

The gravitational constant $G = 6.670 \times 10^{-8}$ dyne cm² g⁻². In writing these equations, we have clearly considered only the spherically symmetric problem. The important perturbation of rotation is discussed in Chap. 6. Because the mass of the small volume element is $\rho \, dA \, dr = dm$, we have

$$0 = F_P + F_G = -\frac{dP}{dr} - G\frac{\rho M}{r^2} \tag{2-180}$$

This force balance, together with the definition of Eq. (2-179), constitutes one of the most important conditions to be satisfied by static stellar structures. It applies equally well to the quiescent stages of contraction and expansion that normally occur during a stellar lifetime. These slow changes in stellar structures require times of the order of 10^6 years for their accomplishment, whereas any violations of Eq. (2-180) would cause sizable changes of structure to occur in a matter of hours. Only in highly dynamic situations is this condition violated, the most common examples being pulsating stars and exploding stars. If

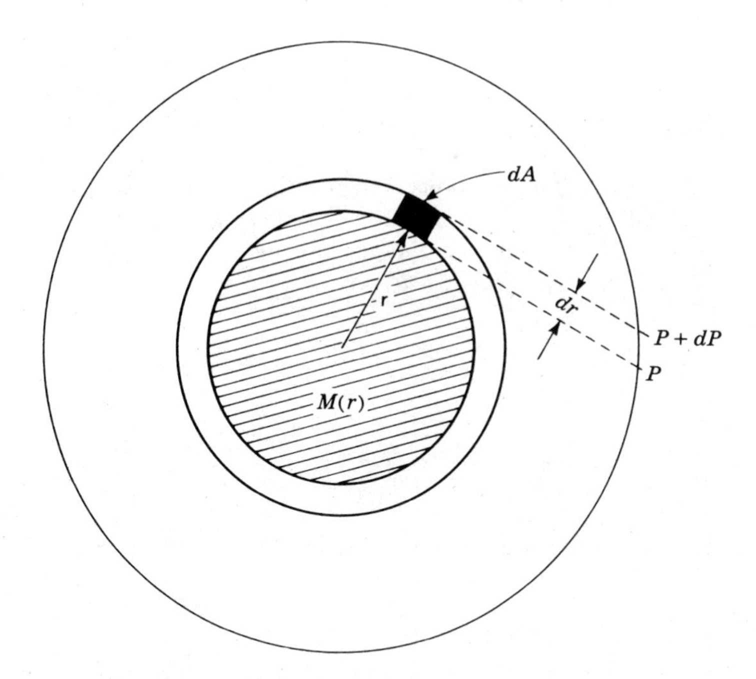

Fig. 2-14 In the hydrostatic equilibrium of a spherical body under self-gravitation, the weight of a small mass element is counterbalanced by the excess pressure on its lower surface.

very rapid adjustments of the structure are to be made, the inertial force resulting from the acceleration term must be added to the force balance. The result is rather clearly

$$\frac{dP}{dr} = -\rho \frac{GM}{r^2} - \rho \frac{d^2r}{dt^2} \tag{2-181}$$

Since GM/r^2 is the local acceleration of gravity $g(r)$ at the interior point, it is clear that the last term will be negligible unless the mass acceleration d^2r/dt^2 is a significant fraction of g. Such a situation would be tantamount to free fall or free explosion; in the vast majority of the less dramatic cases, Eq. (2-180) holds with high accuracy.

Even a casual examination of the condition of hydrostatic equilibrium reveals that very high central pressures will be demanded to support objects as massive as characteristic stars. In fact, we may estimate the order of magnitude of the sun's central pressure simply from the observed number $\mathfrak{M}_\odot = 2 \times 10^{33}$ g and $R_\odot = 7 \times 10^{10}$ cm. We can consider a point at half the solar radius and assume that (1) the pressure gradient is roughly the central pressure divided by the radius, (2) the density is roughly the average density of the sun, and (3) the interior mass is roughly one-half a solar mass. Making these insertions, we find that the central pressure is to order of magnitude given by

$$P_{c\odot} \approx \frac{6}{4\pi} \frac{G\mathfrak{M}_\odot{}^2}{R_\odot{}^4} = 5 \times 10^{15} \text{ dynes/cm}^2 = 5 \times 10^9 \text{ atm} \tag{2-182}$$

Problem 2-40: Perform the steps leading to Eq. (2-182) and make the numerical substitutions.

The rough substitutions made for the properties of the median point of the sun may not appeal to some readers; however, the following inequality, due to Milne, places a lower limit on the central pressure of a star. Equations (2-180) and (2-179) may be combined to

$$\frac{dP}{dr} = -\frac{GM}{4\pi r^4} \frac{dM}{dr} \tag{2-183}$$

Since

$$\frac{d}{dr}\left(P + \frac{GM^2}{8\pi r^4}\right) = \frac{dP}{dr} + \frac{GM}{4\pi r^4} \frac{dM}{dr} - \frac{GM^2}{2\pi r^5} \tag{2-184}$$

the cancellation by Eq. (2-183) of the first two terms on the right-hand side shows that

$$\frac{d}{dr}\left(P + \frac{GM^2}{8\pi r^4}\right) < 0 \tag{2-185}$$

Since this function in parentheses decreases with increasing r, it must be greater at the center, where its value is P_c, than it is at the radius of the star, where its value is $G\mathfrak{M}^2/8\pi R^4$, where $\mathfrak{M}$ is the total mass. Thus

$$P_c > \frac{G\mathfrak{M}^2}{8\pi R^4} = 4.4 \times 10^{14} \left(\frac{\mathfrak{M}}{\mathfrak{M}_\odot}\right)^2 \left(\frac{R_\odot}{R}\right)^4 \qquad \text{dynes/cm}^2 \tag{2-186}$$

This lower limit is a full order of magnitude less than the more realistic estimate in Eq. (2-182), but the lower limit has the advantage of demonstrating rigorously that very high central pressures in terms of terrestrial standards are required. It is also physically interesting to note that the pressure estimated in Eq. (2-182) can be generated only if the temperature is around 10^7 °K for a gas with an average density near that of the sun. The point of this discussion is that *the elementary demand for hydrostatic equilibrium sets at once the order of magnitude of the physical variables in a stellar interior.*

At various stages in their lifetimes stars undergo expansions and contractions. For instance, the condition of hydrostatic equilibrium sets in long before a newly formed star has shrunk to its eventual main-sequence radius. It must undergo a rather extended period of slow contraction before settling into its static configuration. Or, as another example, when the necessary amount of hydrogen has been exhausted from a main-sequence core, it once again contracts gravitationally whereas the outer regions expand. It is of interest, therefore, to examine the thermodynamics of the simplest type of contraction, that of a *uniform contraction.* Chandrasekhar has stated the definition of such a process in the following words: "An expansion or contraction of a spherical distribution of matter is said to be uniform if the distance between any two points is altered in the same way as the radius of the configuration."[1] That is, if the subscripts f and i are associated with the final and initial configurations, whose *corresponding points* are related by

$$\mathbf{r}_f = y\mathbf{r}_i \qquad dV_f = y^3\,dV_i \tag{2-187}$$

then y is to be for a uniform contraction a constant scale factor that represents the ratio of the size of the final sphere to the size of the initial sphere. The final density is related to the initial density by

$$\rho_f = y^{-3}\rho_i \tag{2-188}$$

The condition of hydrostatic equilibrium is maintained throughout, so that

$$\frac{dP_f}{dr_f} = -\frac{GM_f}{r_f^2}\rho_f \qquad \frac{dP_i}{dr_i} = -\frac{GM_i}{r_i^2}\rho_i \tag{2-189}$$

If these equations refer to *corresponding points,* as defined in Eq. (2-187), the interior masses will be equal: $M_f = M_i$. Then

$$dP_f = -\frac{GM_f}{r_f^2}\rho_f\,dr_f = -\frac{GM_i}{(yr_i)^2}y^{-3}\rho_i y\,dr_i = y^{-4}\frac{GM_i}{r_i^2}\rho_i\,dr_i$$

$$= y^{-4}\,dP_i \tag{2-190}$$

[1] *Op. cit.,* p. 45.

Since Eq. (2-190) applies to the pressure difference between corresponding points, it applies equally well to the pressures themselves, so that

$$P_f = y^{-4} P_i \tag{2-191}$$

Problem 2-41: Show that a uniform contraction of a perfect nondegenerate gas results in the following ratios for the physical variables:

$$\frac{P_f}{P_i} = \left(\frac{R_i}{R_f}\right)^4 \qquad \frac{\rho_f}{\rho_i} = \left(\frac{R_i}{R_f}\right)^3 \qquad \frac{T_f}{T_i} = \frac{R_i}{R_f} \tag{2-192}$$

To the extent that contractions of real stars resemble uniform contractions of a perfect gas, the pressure rises much more rapidly than the temperature.

THE VIRIAL THEOREM

The virial theorem is a statistical statement about mutually interacting particles. Consider a general system of mass points m_i with positions $\mathbf{r}_i$ under influence of the force $\mathbf{F}_i$. Then the equations of motion are

$$\frac{d}{dt}\mathbf{p}_i = \mathbf{F}_i \tag{2-193}$$

Consider the quantity $\sum_i \mathbf{p}_i \cdot \mathbf{r}_i$, where the sum is over all the particles of the system. The time derivative of the sum is

$$\frac{d}{dt}\sum_i \mathbf{p}_i \cdot \mathbf{r}_i = \sum_i \frac{d\mathbf{p}_i}{dt} \cdot \mathbf{r}_i + \sum_i \mathbf{p}_i \cdot \frac{d\mathbf{r}_i}{dt} \tag{2-194}$$

The second term in a nonrelativistic case is just $\Sigma m_i v_i^2 = 2K$, where K is the total kinetic energy of the assembly of particles. The left-hand side may be rewritten as

$$\frac{d}{dt}\sum m_i \frac{d\mathbf{r}_i}{dt} \cdot \mathbf{r}_i = \frac{d}{dt}\sum \frac{1}{2}\frac{d}{dt}(m_i r_i^2) = \frac{1}{2}\frac{d^2 I}{dt^2} \tag{2-195}$$

where $I = \sum_i m_i r_i^2$ is the *spherical moment of inertia.* By taking these features into account, Eq. (2-194) may be written as

$$\frac{1}{2}\frac{d^2 I}{dt^2} = 2K + \sum_i \mathbf{F}_i \cdot \mathbf{r}_i \tag{2-196}$$

The sum in Eq. (2-196) is called the *virial of Clausius.* It includes only the long-range field forces and the external forces, for all *collisions at a point* contribute two terms whose sum is zero. For static configurations, furthermore, the moment of inertia is constant, giving

$$K = -\frac{1}{2}\sum_i \mathbf{F}_i \cdot \mathbf{r}_i \qquad \text{static} \tag{2-197}$$

As a first application of this important theorem, consider 1 g of gas contained under pressure in a box of volume V. The forces in the virial fall into two categories, the external pressure force at the surface of the container and the interparticle forces. For the external pressure force we have

$$\sum_{\text{pressure}} \mathbf{F}_i \cdot \mathbf{r}_i = \int_{\text{surface}} (-P)\, d\mathbf{S} \cdot \mathbf{r} = -P \int_S \mathbf{r} \cdot \mathbf{n}\, dS \tag{2-198}$$

where $\mathbf{n}$ is a unit vector normal to the surface. By Gauss' theorem this integral becomes

$$-P \int_V \nabla \cdot \mathbf{r}\, dV = -3PV \tag{2-199}$$

The interparticle forces on the other hand may be thought of as occurring in pairs:

$$\sum_i \mathbf{F}_i \cdot \mathbf{r}_i = \sum_{\text{pairs}} \mathbf{F}_{ij} \cdot \mathbf{r}_i + \mathbf{F}_{ji} \cdot \mathbf{r}_j \tag{2-200}$$

where $\mathbf{F}_{ij} = -\mathbf{F}_{ji}$ is defined as the force on particle i due to particle j. Reassembling the virial equation, we have

$$K = \tfrac{3}{2}PV - \frac{1}{2} \sum_{\text{pairs}} \mathbf{F}_{ij} \cdot (\mathbf{r}_i - \mathbf{r}_j) \tag{2-201}$$

Consider first the case of the perfect gas, where the interparticle forces are zero by definition. The kinetic energy K is in the nondegenerate case the product of $\frac{3}{2}kT$ times the number of particles. Thus with N representing the number of free particles per unit volume, we have

$$P = NkT \tag{2-202}$$

in keeping with Eq. (2-7).

In the case of a star, the situation is different. The gas experiences two forces. The virial contains electric forces from coulomb collisions between the charged particles of the ionized gas as well as gravitational forces. Although electric forces are intrinsically much stronger than gravitational forces, their contribution to the virial is generally quite small. The strong electric forces maintain charge neutrality in the ionized gas, so that there is no net electric force from the bulk of the star. Only in close scattering events do the particles experience unbalanced electric forces, and in those cases the contribution to the virial consists of two nearly equal and opposite terms. There will exist a weak radial electric field that prohibits the positive nuclei from falling to the center and leaving the electrons behind, but this field exerts equal and opposite forces on the electrons and nuclei within any small volume element dV and hence does not contribute to the virial. So let us temporarily ignore the coulomb force, a subject to which we shall return for closer analysis in Sec. 2-3. For a perfect-gas star, the only forces on the particles are gravitational, and explicitly

$$\mathbf{F}_{ij} = -\frac{Gm_i m_j}{(r_{ij})^3} (\mathbf{r}_i - \mathbf{r}_j) \tag{2-203}$$

The static virial theorem becomes in this case

$$K = -\frac{1}{2} \sum_{\text{pairs}} \mathbf{F}_{ij} \cdot (\mathbf{r}_i - \mathbf{r}_j) = \frac{1}{2} \sum_{\text{pairs}} \frac{Gm_i m_j}{r_{ij}} \tag{2-204}$$

Each term in the sum equals the negative of the potential energy due to the interaction of m_i with m_j. When summed over all pairs,

$$K = -\frac{\Omega}{2} \tag{2-205}$$

where Ω is the total potential energy of the star. This well-known result of the inverse-square force is itself often referred to as the *virial theorem*. Because the kinetic energy of nondegenerate particles in a star is measured by the average temperature, Eq. (2-205) has the very important consequence that the temperature of a stellar interior rises as the star contracts gravitationally.

Problem 2-42: Consider the Bohr model of the hydrogen atom with an electron circling a proton. Show that Eq. (2-205) is satisfied.

Notice that K was defined to be the translational kinetic energy of particles and does not include the energy of internal degrees of freedom (rotational, vibrational, or excitation energy) or the kinetic energy of trapped photons. For the sake of illustrating an important principle in a simple way, consider a perfect gas characterized by a uniform value of the adiabatic exponent γ. It follows that for adiabatic expansions

$$dU = -P\,dV$$

$$\frac{dT}{T} + (\gamma - 1)\frac{dV}{V} = 0 \tag{2-206}$$

the second condition coming from Eq. (2-115). By elimination of dV we have

$$\begin{aligned} dU &= \frac{PV}{T}\frac{dT}{\gamma - 1} \\ &= \frac{N_0 k}{\mu}\frac{dT}{\gamma - 1} = \frac{2}{3}\frac{dK}{\gamma - 1} \end{aligned} \tag{2-207}$$

The assumption of uniform γ allows Eq. (2-207) to be integrated over the entire star:

$$K = \tfrac{3}{2}(\gamma - 1)U \tag{2-208}$$

Let us define an energy E as the sum of the gravitational energy Ω and the internal energy U. Then by the virial theorem

$$\begin{aligned} E &= U + \Omega = U - 2K \\ &= -(3\gamma - 4)U = \frac{3\gamma - 4}{3(\gamma - 1)}\Omega \end{aligned} \tag{2-209}$$

The energy E contains all the particle energy but does not include photons. Some straightforward implications of Eq. (2-209) follow.

(*1*) For the perfect monatomic gas $\gamma = \frac{5}{3}$ and $E = -U = \Omega/2$. The total energy (excluding that in photons) is negative and equal to half of the gravitational binding energy. But the total amount of energy must be conserved. Thus, as a star shrinks, one half of the binding energy appears as thermal motion whereas the other half must go into the production of radiation, most of which is usually lost into space. This reasoning lay behind the discussion in Chap. 1 of the inadequacy of gravity as a source of energy for the full lifetimes of stars.

This result has great significance for stellar evolution, however. Static stellar structures are those in which the luminosity is replenished by energy-liberating nuclear reactions in the interior, but whenever a given nuclear fuel is exhausted in the interior, a gravitational contraction ensues to replace the energy radiated from the star. This contraction increases the magnitude of Ω, one-half of which appears as additional kinetic energy; i.e., *the temperature rises.* This rise continues until the temperature becomes sufficient to ignite the next source of nuclear fuel, which usually halts the contraction. Thus it is that gravitation provides the driving source which compels the stars to evolve through a sequence of nuclear burning phases.

(*2*) The energy E changes from negative to positive as γ decreases through a value $\gamma = \frac{4}{3}$. If the sum of internal and gravitational energies is positive, the mass can fly apart. This result would be contradictory to the initial assumption of the virial theorem that the mass is static $\left(\frac{1}{2}\frac{d^2I}{dt^2} = 0\right)$, and hence that assumption must itself be incorrect for $\gamma \leq \frac{4}{3}$. Thus dynamic radial instabilities occur when γ reaches a value as low as $\frac{4}{3}$. In real stars, the constant γ must be generalized to include effects of radiation pressure and ionization. When that generalized γ has an average value over the star that is less than $\frac{4}{3}$, radial instabilities ensue.[1] Since $\gamma = \frac{4}{3}$ for a photon gas, we note that a star internally dominated by radiation pressure will have a relatively "loose" structure. Effects of ionization can cause γ to drop well below $\frac{4}{3}$ in the regions of ionization.

Problem 2-43: The gravitational binding of the sun is of the order

$$\Omega \approx -\frac{GM^2}{R} = -4 \times 10^{48} \text{ ergs}$$

Since the sun is predominately a perfect gas, the thermal kinetic energy is about $K \approx 2 \times 10^{48}$ ergs by the virial theorem. If the internal temperature of the sun were $10^7(1 - r/R)^2$ °K,

[1] An introduction to stellar stability is contained in the discussion of pulsation in Chap. 6. The principles of stability are complicated in detail, however, and a full discussion cannot be included in this book. Fortunately there are monographs that expertly summarize the current state of knowledge. See particularly P. Ledoux, Stellar Stability, in L. H. Aller and D. B. McLaughlin (eds.), "Stellar Structure," The University of Chicago Press, Chicago, 1965, and other papers cited in that review.

what is the total trapped energy in the radiation field? Apparently most of the radiant energy escaped during the contraction.

The virial theorem can be approached in a different way that employs only macroscopic quantities. The assumption that the star is static, which is necessary to draw any useful information from the virial theorem, is expressible by the condition of hydrostatic equilibrium:

$$\frac{dP}{dr} = -\rho \frac{GM(r)}{r^2} \tag{2-210}$$

Multiply both sides of this equation by

$$V(r)\, dr = \tfrac{4}{3}\pi r^3\, dr \tag{2-211}$$

Then one obtains

$$\begin{aligned} V(r)\, dP &= -\tfrac{1}{3} 4\pi r^2 \rho\, dr \frac{GM(r)}{r} \\ &= -\frac{1}{3}\frac{GM}{r}\, dM \end{aligned} \tag{2-212}$$

When this equation is integrated over the star, the left-hand side becomes

$$\int V\, dP = PV \Big|_0^R - \int P\, dV$$

In an idealized star, the pressure goes to zero near the surface $r = R$, whereas the volume vanishes near the center. The right-hand side of Eq. (2-212), on the other hand, is just one-third the gravitational potential energy. Thus we have

$$-3\int P\, dV = \Omega \tag{2-213}$$

The result is somewhat more general than the previous treatment because no assumption has been made about the nature of the interior particles. However the earlier results are easily recovered as special cases. For the perfect gas

$$P = \begin{cases} \dfrac{2}{3}\left(\dfrac{K}{V}\right) & \text{nonrelativistic} \\ \dfrac{1}{3}\left(\dfrac{K}{V}\right) & \text{relativistic} \end{cases} \tag{2-214}$$

which gives immediately the earlier result

$$\Omega = \begin{cases} -2K & \text{nonrelativistic} \\ -K & \text{relativistic} \end{cases} \tag{2-215}$$

whereas the total energy of a monatomic gas

$$E = \begin{cases} \Omega + K = \dfrac{\Omega}{2} & \text{nonrelativistic} \\ 0 & \text{relativistic} \end{cases} \tag{2-216}$$

As a final application of the virial theorem we shall consider what effect the coulomb interactions of the ionized gas are likely to have on the value of the pressure. To do so, return to Eq. (2-201) describing the nonrelativistic gas in a container. In this case we ignore the gravitational interaction in the sum over pairs. Because the coulomb force between pairs is an inverse-square force, the sum is easily related to the coulomb potential energy.

Problem 2-44: Show that for the ionized nonrelativistic gas in a box

$$K = \tfrac{3}{2}PV - \tfrac{1}{2}U_c \tag{2-217}$$

where U_c is the coulomb potential energy.

From Eq. (2-217) we see that the pressure on the walls is

$$P = \frac{2}{3}\frac{K}{V} + \frac{1}{3}\frac{U_c}{V} \tag{2-218}$$

In the nondegenerate case

$$P = NkT + \frac{1}{3}\frac{U_c}{V} \tag{2-219}$$

where N is the number of free particles per unit volume. The interesting point is that there is an additional *coulomb pressure* equal to one-third of the coulomb energy density. The reason that this term has been neglected in the discussion of the virial theorem is that it was argued to be small. If the positive charges are, on the average, equidistant from both positive and negative charges, the net coulomb energy is zero. The gravitational energy dominates in the star because all the individual interactions occur with the same sign, whereas the individually much greater coulomb interactions cancel out because of charge neutrality. The coulomb interactions do play a significant role in some phenomena, however, because the plasma polarizes; i.e., the electrons cluster near the ions to some extent. A discussion of this phenomenon follows.

2-3 THE IONIZED REAL GAS

Gases having interactions between constituent particles are called *real gases.* The name reflects the fact that nature always provides some interaction, and for the dynamics of the gas the relevant question is the relative magnitude of the average interaction and the average kinetic energy. In an ionized gas the strongest forces are the coulomb forces between the charges. Much less important are the interactions between magnetic moments and the nuclear-force interactions. Although the latter is quite strong, the particles are seldom within the range of the nuclear force except at nuclear density ($\rho > 10^{14}$ g/cm^3).

For the stellar interior, the important problem is to ascertain which properties of the real gas may introduce observable consequences into astrophysics. The first of these is that the particle pressure deviates from that of an ideal gas, the size of the effect depending upon the coulomb energy density. A second important consequence is that the energetics of atomic and nuclear reactions is modified, the most important application being to the ionization equilibrium. And third, the transport phenomena in the ionized gas are influences by collective phenomena of the plasma.

In the presence of forces, the internal energy of a monatomic gas must include the potential energy of the interactions

$$U = \sum \frac{p^2}{2m} + \Phi \tag{2-220}$$

where Φ is the potential energy. Because Φ depends upon the average interparticle distance, it is a density-dependent quantity. The pressure is given by the change in internal energy associated with adiabatic compression,

$$dU_{\text{ad}} = -P\,dV \tag{2-221}$$

If the internal energy is density-dependent because of the interactions, a corresponding pressure results.

All the important effects to be discussed occur because the charge density is made nonuniform by the coulomb interactions. To have a concise notation for the composition of the ionized gas, let $\bar{n}_Z$ represent the *average* number density of each species of charge Z, with $Z = -1$ corresponding to the electron density. Because the gas is macroscopically neutral,

$$\sum_Z Z\bar{n}_Z = 0 \tag{2-222}$$

To calculate the electrostatic energy of the gas one must know the electrostatic potential ϕ_Z at each charge Z *due to all the other particles of the gas.* Then the coulomb energy per unit volume is

$$\left(\frac{U}{V}\right)_c = \frac{1}{2}\sum_Z eZ\bar{n}_Z\phi_Z \tag{2-223}$$

In the material to follow we shall estimate the important effects for two limiting cases of physical interest, the nearly perfect gas and the zero-temperature gas.

THE NEARLY PERFECT GAS AT LOW DENSITY

The gas may be called *nearly perfect* if the coulomb energy between particles is much smaller than their thermal energy. If the ions have charge Z, the coulomb energy between neighbors is of order $(Ze)^2/r$, where $r \approx n_Z^{-\frac{1}{3}}$ is the average separation of the ions. For this coulomb energy to be small compared to kT requires that

$$n_Z \ll \left(\frac{kT}{Z^2e^2}\right)^3 \tag{2-224}$$

For an ionized gas satisfying this low-density requirement, the coulomb interactions may be estimated in a manner devised by Debye and Hückel.[1] Each charge will polarize the neighborhood to some extent. On the average, a spherically symmetric but inhomogeneously charged ion cloud surrounds each ion. The potential around each ion of charge Z_i *due to all charges* will be a shielded coulomb potential $V_i(r)$, where r is the distance from the ion, as illustrated in Fig. 2-15. The density of each type of ion is described by some function $n_Z(r)$. The potential energy of each charge Z in the field $V_i(r)$ of the ion sphere will be $ZeV_i(r)$. Hence from Boltzmann's formula the average density will be perturbed to

$$n_Z = \bar{n}_Z \exp - \frac{ZeV_i}{kT} \tag{2-225}$$

The coefficient is set equal to $\bar{n}_Z$ because V_i falls rapidly to zero as one moves away from the shielded charge Z_i, where the density of n_Z must fall to its average value.

The potential in the ion cloud V_i is related to the charge density by Poisson's equation

$$\nabla^2 V_i = -4\pi\rho_q = -4\pi e \Sigma Z n_Z \tag{2-226}$$

[1] This treatment follows L. D. Landau and E. M. Lifschitz, "Statistical Physics," chap. 7, Addison-Wesley Publishing Company, Inc., Reading, Mass., 1958.

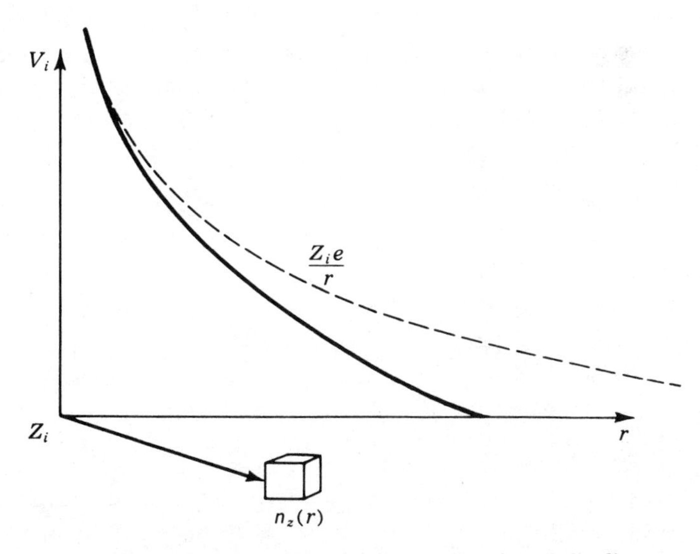

Fig. 2-15 The electrostatic potential V_i as a function of the distance from the ion of charge Z_i. This potential drops more rapidly than the coulomb potential Z_ie/r because of electron clustering near the ion.

Consistent with the assumption that the coulomb energy is small compared to kT is the assumption that each density may be expanded:

$$n_Z = \bar{n}_Z \exp - \frac{ZeV_i}{kT} \approx \bar{n}_Z \left(1 - \frac{ZeV_i}{kT} + \cdots\right) \tag{2-227}$$

Substitution of this approximation in Poisson's equation gives

$$\nabla^2 V_i \approx -4\pi e \sum Z\bar{n}_Z \left(1 - \frac{ZeV_i}{kT}\right) = + \frac{4\pi e^2}{kT} \sum Z^2 \bar{n}_Z V_i \tag{2-228}$$

the last step resulting from the charge-neutrality condition.

The equation for the potential in the ion sphere may now be written:

$$\nabla^2 V_i - \kappa^2 V_i = 0 \tag{2-229}$$

where κ is an inverse length defined by

$$\kappa^2 = \frac{4\pi e^2}{kT} \sum Z^2 \bar{n}_Z \tag{2-230}$$

Problem 2-45: Show that the spherically symmetric solution of Eq. (2-229) is

$$V_i = \text{const } e^{-\kappa r}/r \tag{2-231}$$

From the result of the preceding problem and the fact that $V_i \to eZ_i/r$ as $r \to 0$, the potential in the ion sphere must be

$$V_i = \frac{eZ_i e^{-\kappa r}}{r} \tag{2-232}$$

Thus the field drops off rapidly for distances $r > 1/\kappa$. The length $1/\kappa = R_D$ is called the *Debye-Hückel radius* and is a measure of the size of the ion cloud. The condition of weak interaction energy (compared to kT) can easily be seen to be equivalent to the assumption that $1/\kappa$ is much greater than the average distance between neighboring particles.

The potential at Z_i due to all other charges in its ion cloud may be obtained by expanding Eq. (2-232) very near to Z_i:

$$V_i(r) = \frac{eZ_i}{r} - eZ_i\kappa + \cdots \tag{2-233}$$

where the higher-order terms in the expansion vanish as $r \to 0$. The first term is the coulomb field of the ion itself, whereas the second must be the potential at the ion due to the other charges. It is perhaps instructive to notice that the potential at the ion (but not elsewhere) is the same as if the charge Z_i were surrounded by a charge $-Z_i$ spread over a spherical shell at the Debye-Hückel radius. The potential at any charge Z due to all other charges is then

$$\phi_Z \approx -eZ\kappa = -\frac{eZ}{R_D} \tag{2-234}$$

where

$$R_D = \left(\frac{kT}{4\pi e^2 \Sigma Z^2 \bar{n}_Z}\right)^{\frac{1}{2}} \tag{2-235}$$

Problem 2-46: In the sum $\Sigma Z^2 \bar{n}_Z$ the electrons are represented by $Z = -1$. Show with aid of the charge-neutrality condition that the sum may be expressed as a sum *over positive ions only* as

$$\sum_Z Z^2 \bar{n}_Z = \sum_{+Z} (Z^2 + Z)\bar{n}_Z = \sum_{+Z} (Z^2 + Z) \frac{\rho X_Z}{A_Z} N_0 \tag{2-236}$$

It will frequently be convenient to define a quantity dependent only upon the composition by

$$\zeta \equiv \sum_{+Z} (Z^2 + Z) \frac{X_Z}{A_Z} \tag{2-237}$$

whereupon

$$R_D = \left(\frac{kT}{4\pi e^2 \rho N_0 \zeta}\right)^{\frac{1}{2}} \tag{2-238}$$

The coulomb potential energy per unit volume is immediately obtainable by inserting Eq. (2-234) into Eq. (2-223):

$$\left(\frac{U}{V}\right)_c = -e^3 \left(\frac{\pi}{kT}\right)^{\frac{1}{2}} \left(\sum \bar{n}_Z Z^2\right)^{\frac{3}{2}} = -e^3 \left(\frac{\pi}{kT}\right)^{\frac{1}{2}} (\rho N_0 \zeta)^{\frac{3}{2}} \tag{2-239}$$

With the aid of these results it becomes possible to discuss two rather complex features of the nearly perfect ionized gas, the pressure due to the coulomb energy and the effective ionization potential. Although the phenomena themselves are general, the correctness of the subsequent results is limited to the temperature-density domain for which the Debye-Hückel treatment is valid. Since the kinetic energy per unit volume is just $\frac{3}{2}kT$ times the number of particles per unit volume, the Debye-Hückel treatment is valid only if $(\rho N_0/\mu)\frac{3}{2}kT$ is much greater than Eq. (2-239).

Coulomb pressure In analyzing thermodynamic changes of state one considers a specific quantity of gas. If that quantity is chosen to be 1 g, its internal energy U is the internal energy density times the volume of 1 g ($V = 1/\rho$). The coulomb energy per gram is then

$$U_c = -e^3 \left(\frac{\pi}{kTV}\right)^{\frac{1}{2}} (N_0 \zeta)^{\frac{3}{2}} = \frac{B}{(VT)^{\frac{1}{2}}} \tag{2-240}$$

By so defining a constant B the dependence $U_c \propto (VT)^{-\frac{1}{2}} = (\rho/T)^{\frac{1}{2}}$ is simple and explicit. Furthermore, let the number of particles in 1 g of gas be represented simply by N. Then the internal energy and pressure can be simply expressed as the corresponding quantities for the ideal gas plus the corrections necessitated by the coulomb imperfection:

$$U = \tfrac{3}{2}NkT + U_c \qquad P = \frac{N}{V}kT + P_c \tag{2-241}$$

The differential of the entropy is

$$dS = \frac{1}{T}\left[\left(\frac{\partial U}{\partial V}\right)_T + P\right] dV + \frac{1}{T}\left(\frac{\partial U}{\partial T}\right)_V dT \tag{2-242}$$

where the partial derivatives are

$$\begin{aligned} \left(\frac{\partial U}{\partial V}\right)_T &= \left(\frac{\partial U_c}{\partial V}\right)_T = -\frac{1}{2}\frac{B}{V^{\frac{3}{2}}T^{\frac{1}{2}}} \\ \left(\frac{\partial U}{\partial T}\right)_V &= \tfrac{3}{2}Nk - \frac{1}{2}\frac{B}{V^{\frac{1}{2}}T^{\frac{3}{2}}} \end{aligned} \tag{2-243}$$

Insertion of these values gives

$$dS = \frac{1}{T}\left(-\frac{1}{2}\frac{B}{V^{\frac{3}{2}}T^{\frac{1}{2}}} + \frac{N}{V}kT + P_c\right) dV + \frac{1}{T}\left(\tfrac{3}{2}Nk - \frac{1}{2}\frac{B}{V^{\frac{1}{2}}T^{\frac{3}{2}}}\right) dT \tag{2-244}$$

Problem 2-47: Show that the integrability condition on dS demands that

$$\frac{\partial}{\partial T}\frac{P_c}{T} = -\frac{1}{2}\frac{B}{V^{\frac{3}{2}}T^{\frac{5}{2}}} \tag{2-245}$$

The integral of Eq. (2-245) is

$$\frac{P_c}{T} = \frac{1}{3}\frac{B}{V^{\frac{3}{2}}T^{\frac{3}{2}}} + f(V) \tag{2-246}$$

where $f(V)$ is some arbitrary function of V. But since the coulomb pressure must vanish as $T \to \infty$, the function $f(V) = 0$. By reincorporating the definition of B it follows that

$$P_c = \frac{1}{3}\left(\frac{U}{V}\right)_c = -\frac{e^3}{3}\left(\frac{\pi}{kT}\right)^{\frac{1}{2}} (\rho N_0 \zeta)^{\frac{3}{2}} \tag{2-247}$$

The coulomb pressure is negative, corresponding to the fact that compression of the charges performs work rather than requires work. This calculation has confirmed Eq. (2-219), which was derived from the virial theorem.

The total pressure is then

$$P = \frac{N_0 k}{\mu}\rho T - \frac{e^3}{3}\left(\frac{\pi}{kT}\right)^{\frac{1}{2}} (\rho N_0 \zeta)^{\frac{3}{2}} \tag{2-248}$$

This result can be valid only if the second term is a small correction to the perfect-gas pressure, as can easily be seen by the following consideration. The coulomb pressure is one-third of the coulomb-energy density, whereas the perfect-gas pressure is two-thirds of the kinetic-energy density; but the Debye-Hückel treatment is valid only if the coulomb-energy density is much less than the kinetic-energy density. Quantitatively the criterion demands

$$U_c \ll U_{\text{kin}}$$

$$e^3 \left(\frac{\pi}{kT}\right)^{\frac{1}{2}} (\rho N_0 \zeta)^{\frac{3}{2}} \ll \frac{3}{2} \frac{N_0 k}{\mu} \rho T \tag{2-249}$$

$$T \gg 10^5 \rho^{\frac{1}{3}} \zeta$$

This condition will not be satisfied in major portions of many stars, so that Eq. (2-247) for the coulomb pressure should be used with caution.

Problem 2-48: Compute the coulomb pressure and the perfect-gas pressure for a helium gas at $T = 10^6$ °K and $\rho = 10^{-2}$ g/cm³.
Ans: $P_c = -4.8 \times 10^9$ dynes/cm², $P_g = 6.3 \times 10^{11}$ dynes/cm².

Depression of the continuum and effective ionization potential Interactions between gas particles alter the energetics of the gas. The ionization of an ion in a perfect gas is no different from the ionization in a vacuum, but in a real ionized gas the potential energy of both the bound and free electrons is altered by the coulomb energy. In the present section these effects will be discussed within the framework of the Debye-Hückel model.

From Eq. (2-233) the average potential around each electron is

$$V_e(r) = -\frac{e}{r} + e\kappa - \cdots \tag{2-250}$$

so that the potential energy of each free electron due to interactions with other charges is

$$(\text{PE})_e = -e^2\kappa = -\frac{e^2}{R_D} \tag{2-251}$$

This negative energy reflects the fact that the free electron is actually bound to the plasma as a whole, even though it is free to navigate through the plasma. The ionization of an atom requires sufficient energy to liberate the electron to the continuum of states with zero kinetic energy. But the energy of a zero-kinetic-energy electron is $E = -e^2/R_D$, and one commonly says that the continuum of states (as opposed to discrete bound states) *has been depressed* by an amount $E_0 = e^2/R_D$, and the energy required to ionize a bound electron is accordingly reduced.

The energy required for ionization is further reduced by the fact that the energy of the bound electron state is also changed and in the direction such that the binding energy is reduced. This effect comes about because the electron moves in a shielded potential rather than a pure $1/r$ potential. Consider a single bound electron. Since it is sensible to assume that the radius of the ground-state orbital is much less than R_D, the electron bound to charge Z moves in a potential that is approximated by

$$V(r) \approx \frac{Ze}{r} - \frac{(Z-1)e}{R_D} \tag{2-252}$$

The first term is just the $1/r$ potential from the charge Z that provides the normal binding energy for the atom, whereas the second term is the potential due to the Debye sphere surrounding the ion of charge $Z - 1$. The potential energy of the

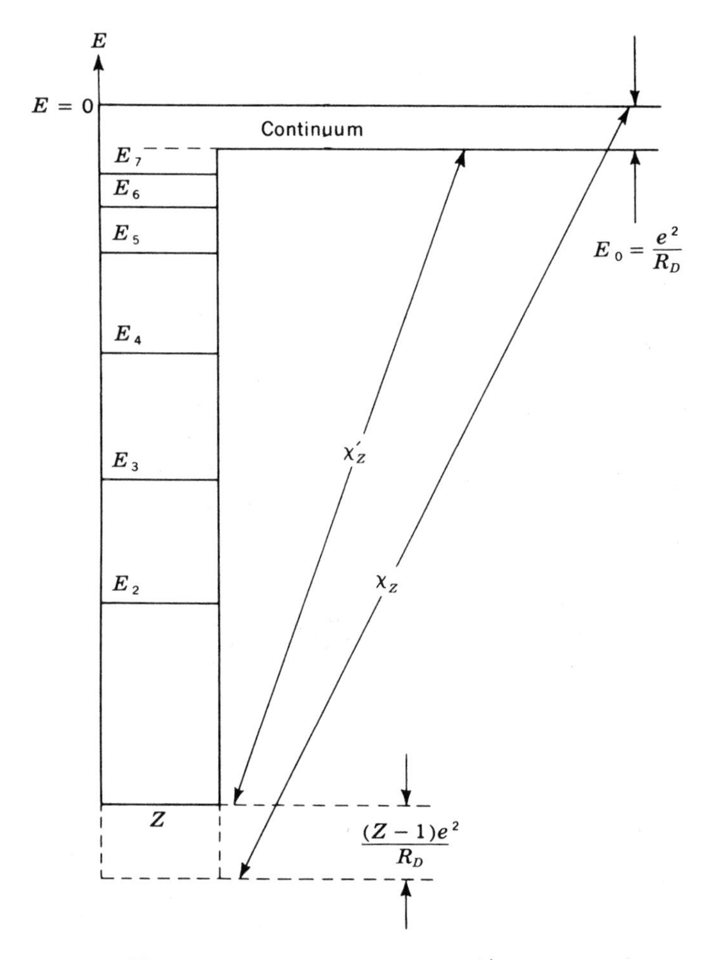

Fig. 2-16 The effective ionization potential χ'_Z. The laboratory ionization potential is reduced by coulomb interactions in the dense plasma. The energy of continuum electrons is lowered by e^2/R_D, whereas the energy of a tightly bound electron with Bohr radius $a \ll R_D$ is raised by $(Z - 1)e^2/R_D$. Thus the ionization potential is effectively reduced by the amount Ze^2/R_D.

bound electron is

$$(\mathrm{PE})_e \approx -\frac{Ze^2}{a} + \frac{(Z-1)e^2}{R_D} \tag{2-253}$$

where a is the radius of the orbital. (For a hydrogenlike ion $a = a_0/Z$.)

From Eq. (2-253) it can be seen that, to first approximation, the ground-state energy is elevated by an amount $(Z-1)e^2/R_D$. Figure 2-16 shows the energy-level diagram of an ion, where χ_Z is the laboratory ionization potential and χ'_Z is the effective ionization potential in the plasma. In the calculation of the ionization equilibrium, therefore, the laboratory ionization potential must be replaced by an *effective ionization potential* whose value is

$$\chi'_Z = \chi_Z - \frac{Ze^2}{R_D} \tag{2-254}$$

The excited bound states are also shifted upward in energy, so that some of the states that are bound in the laboratory move up into the continuum of states in a dense plasma. This phenomenon resolves in a natural way the paradox mentioned in Chap. 1, viz., that the partition functions appear to be infinite. In the crudest approximation, one would simply regard all those states bound by less than Ze^2/R_D in the laboratory as being unbound in the star; and even though Ze^2/R_D might be a much smaller energy than χ_Z, that energy range is the one which contains the infinite number of bound states, so that its elimination not only makes the partition functions finite but often means that they can be estimated by the sum of only a few terms. In a more exact calculation, the excited-state energies will be increased by an amount different from that of the ground state because there are smaller energy shifts because of the differences in radii of the excited states. The present simplified discussion should be regarded only as an illustration of the dominant terms of a problem that is extremely complicated in its details.[1] Correction of the ionization potential is an important part of the calculation of the radiative opacity of the gas.

Problem 2-49: Estimate the partition function for a five-times-ionized carbon atom (hydrogen-like) if it is embedded in a hydrogen gas at $T = 5 \times 10^5$ °K and $\rho = 5 \times 10^{-3}$ g/cm³. In the simple approximation of the text, how much is the binding energy of that electron reduced from its normal value of 36 Ry?

Ans: $G_c = 2.6$, $\Delta\chi \approx -1.0$ Ry.

THE ZERO-TEMPERATURE IONIZED GAS

As the density of the gas increases, the coulomb energy increases, and one says that the gas becomes less perfect. At densities high enough for the electron gas to be degenerate, however, that rule of thumb becomes incorrect. A degenerate gas becomes more perfect as the density rises. This peculiarity comes about

[1] J. C. Stewart and K. D. Pyatt, *Astrophys. J.*, **144**:1203 (1966), and C. A. Rouse, *Phys. Rev.*, **159**:41 (1967).

because the relevant energy for the gas becomes the Fermi energy E_f rather than the thermal energy kT. The perfect-gas condition $E_c \ll kT$ must be replaced by $E_c \ll E_f$, but because the Fermi energy increases as a higher power of the density than the coulomb energy, a nearly perfect degenerate gas is more nearly perfect at higher density. Quantitatively the argument goes as follows. If $\langle r \rangle$ is an appropriate average separation of electrons from the ions of charge Z, then the condition that the gas be perfect becomes

$$\frac{Ze^2}{\langle r \rangle} \ll E_f \tag{2-255}$$

or

$$Ze^2 n_e^{\frac{1}{3}} \ll \frac{\hbar^2}{2m} (3\pi^2 n_e)^{\frac{2}{3}} \tag{2-256}$$

This equation clearly becomes true for sufficiently large values of the electron density, in fact, for $n_e > 10^{23} Z^3$, a condition that seems always satisfied in astrophysical degenerate electron gases. The consequence of this fact is that the electron gas is nearly uniform; i.e., the electronic charge density is, in first approximation, constant.

As was discussed earlier, as long as $E_f \gg kT$, the state of the electron gas is almost independent of T. The same cannot be said of the ions, because their energy is predominantly thermal. For arbitrary temperature, the configuration of the gas may be quite complicated, but it is not too difficult to see what must happen as the temperature is reduced to very low values. Assume for this discussion that the density is great enough so that the matter must remain *pressure-ionized* even at zero temperature. As $T \to 0$ the thermal energy of the ions vanishes, so that they will assume a configuration that minimizes the potential energy. This stable configuration is such that the ions maximize the inter-ion separation and is more like a lattice than a gas. For small displacements of an ion from its equilibrium point in the lattice, there will be a restoring force resulting from an increase in the coulomb potential energy. The result is that each ion is situated in a harmonic-oscillator potential. Now although a classical oscillator can have zero energy, a quantum oscillator cannot. It must oscillate with a zero-point energy equal to $\frac{3}{2}\hbar\omega$, where ω is the frequency of the oscillator ($\omega = \sqrt{k/m}$, where k is the force constant of the equivalent oscillator). Salpeter[1] and others have discussed this phenomenon and showed that the zero-point energy will be less than the coulomb energy holding the ions in the lattice structure and hence the lattice structure is correct at zero temperature. Although most interesting astrophysical environments are not strictly at zero temperature, it may turn out at high density that the oscillator energy dominates kT, in which case a zero-temperature approximation is sensible. Because the major coulomb effects are easy to estimate in this limit, we now turn our attention to those effects in the zero-temperature ionized gas.

[1] E. E. Salpeter, *Astrophys. J.*, **134**:669 (1961).

Draw an imaginary sphere of radius R_Z around each ionized nucleus of charge Z such that the sphere contains Z free electrons which neutralize the entire sphere. That is, the gas is imagined as being divided into neutral spheres about each nucleus Z which contain the Z electrons closest to that nucleus. Assume, furthermore, that the Z electrons in each ion sphere are spread uniformly over the volume of the sphere and that the average electron density is unperturbed by the nuclear charges. (These assumptions are incorrect in detail, but as a simple approximation they allow an easy numerical estimate of the physical effects being considered. The nuclear charge will actually polarize the electron density to a degree dependent upon density and temperature.)

Since the ion spheres are spherically symmetric and neutral, the electrostatic potential energy of electrons in a given ion sphere can be calculated from the geometry of that sphere alone. And the first point to notice is that the free electrons do possess an average potential energy *which is negative.* Consider the total electrostatic energy of one sphere. It may be calculated as the sum of the potential energies from electron-electron interactions and of electron-nucleus interactions.

(*1*) *The electron-electron interaction:* To assemble the uniform sphere with total charge Z_e requires electrostatic energy equal to

$$U_{ee} = + \int_0^{R_Z} \frac{q_r\,dq}{r} \tag{2-257}$$

as illustrated in Fig. 2-17, where

$$q_r = \left(\frac{r}{R_Z}\right)^3 Ze \qquad dq = \frac{4\pi r^2\,dr}{\frac{4}{3}\pi R_Z{}^3} Ze \tag{2-258}$$

The integration gives

$$U_{ee} = +(Ze)^2 \frac{3}{R_Z{}^6} \int_0^{R_Z} r^4\,dr = \frac{3}{5}\frac{(Ze)^2}{R_Z} \tag{2-259}$$

(*2*) *The electron-nucleus interaction:* To assemble the uniform sphere of charge Ze about the nucleus Z requires electrostatic energy

$$U_{eZ} = -Ze \int_0^{R_Z} \frac{dq}{r} = -\frac{3}{2}\frac{(Ze)^2}{R_Z} \tag{2-260}$$

Thus the total electrostatic energy of the ion sphere is

$$U = U_{ee} + U_{eZ} = \left(\frac{3}{5} - \frac{3}{2}\right)\frac{(Ze)^2}{R_Z} = -\frac{9}{10}\frac{(Ze)^2}{R_Z} \tag{2-261}$$

Since this electrostatic energy is shared by Z electrons, an electron in a given ion

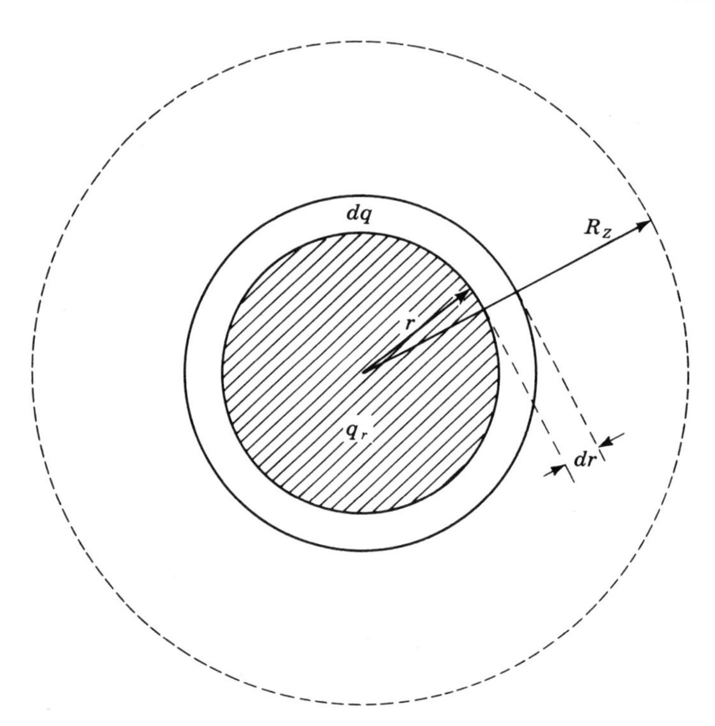

Fig. 2-17 The ion sphere in a degenerate zero-temperature plasma. The electron gas is uniform in these circumstances, and each ion Z is surrounded by a sphere of radius R_Z which contains Z free electrons. In this zero-temperature limit other ions are excluded from the sphere so that $\frac{4}{3}\pi R_Z{}^3 n_e = Z$.

sphere has an average potential energy equal to

$$\langle -eV \rangle_Z = -\frac{9}{10}\frac{Ze^2}{R_Z} \tag{2-262}$$

This negative energy physically reflects the fact that the electrons are actually bound to the ionized gas. This is again the idea behind the depression of the continuum, viz., that continuum electrons moving with kinetic energy $p^2/2m$ have total energy equal to $p^2/2m + E_0$, where E_0 is the electron potential energy averaged over all the ion spheres of the system.

The magnitude of the continuum depression can be calculated with this model. The number of ion spheres per unit volume of type Z is

$$n_Z = \frac{\rho X_Z}{A_Z} N_0 \tag{2-263}$$

and each one contains Z electrons with potential energy $\langle -eV \rangle_Z$, so that the potential energy per unit volume from ion spheres of charge Z is

$$\langle -eV\rangle_Z Z n_Z = \langle -eV\rangle_Z \frac{ZX_Z}{A_Z}\rho N_0 \tag{2-264}$$

and the total electron potential energy per unit volume is just the sum over all ion spheres:

$$\left(\frac{U}{V}\right)_e = \rho N_0 \sum_Z \langle -eV\rangle_Z \frac{ZX_Z}{A_Z} \tag{2-265}$$

Since the total number of electrons per unit volume is $n_e = \rho N_0/\mu_e$, the average potential energy per electron is

$$E_0 = \frac{(U/V)_e}{n_e} = \mu_e \sum_Z \langle -eV\rangle_Z \frac{ZX_Z}{A_Z} \tag{2-266}$$

To evaluate E_0 numerically requires evaluation of $\langle -eV\rangle_Z$ for each ion sphere. It is evident that for this purpose the radii of the ion spheres are required. Since the electrons are assumed to be uniformly distributed, the volume of each ion sphere must be such that

$$n_e(\tfrac{4}{3}\pi R_Z^3) = Z \tag{2-267}$$

or

$$R_Z = \left(\frac{3}{4\pi}\frac{Z}{n_e}\right)^{\frac{1}{3}} \tag{2-268}$$

Substituting into Eq. (2-262) give

$$E_0 = -\mu_e \sum_Z \frac{9}{10}\frac{Ze^2}{\left(\frac{3}{4\pi}\frac{Z}{n_e}\right)^{\frac{1}{3}}}\frac{ZX_Z}{A_Z} = -\frac{9}{10}\left(\frac{4\pi}{3}\right)^{\frac{1}{3}}\mu_e n_e^{\frac{1}{3}} e^2 \sum_Z \frac{Z^{\frac{5}{3}}X_Z}{A_Z} \tag{2-269}$$

or

$$E_0 = -\frac{9}{10}\left(\frac{4\pi}{3}\mu_e^2 \rho N_0\right)^{\frac{1}{3}} e^2 \left(X + 2^{-\frac{1}{3}}Y + \sum_{Z>2}\frac{Z^{\frac{5}{3}}X_Z}{A_Z}\right) \tag{2-270}$$

The most useful way to evaluate this energy is to multiply numerator and denominator by the Bohr radius $a_0 = 0.528 \times 10^{-8}$ cm and to note that $e^2/a_0 = 2\mathcal{R}$, where $\mathcal{R}$ is the Rydberg constant and numerically equals the binding energy of the hydrogen ground state. Then the continuum depression in Rydbergs is

$$E_0 = -1.30(\mu_e^2\rho)^{\frac{1}{3}}\left(X + 0.79Y + \sum_{Z>2}\frac{Z^{\frac{5}{3}}X_Z}{A_Z}\right) \quad \text{Ry} \tag{2-271}$$

It is evident that, in this approximation, the continuum depression depends upon the density as $\rho^{\frac{1}{3}}$ and upon the composition. (In a more complete treatment, E_0 also depends upon the temperature, a dependence that results from the spatial nonuniformity of the electron density and the breakup of the ion lattice.) The continuum depression is also seen to increase with the average Z of the ions, being smallest for essentially pure hydrogen. For population I composition ($X = 0.60$, $Y = 0.38$, $Z = 0.02$)

$$E_0 \approx -1.44\rho^{\frac{1}{3}}\mathcal{R} \tag{2-272}$$

All free electrons in this approximation are considered as having total energy $E = p^2/2m + E_0$. This fact makes a slight change in the previous expression for the number density of electrons. Now we have

$$\begin{aligned} n_e(p)\,dp &= \frac{2}{h^3}\frac{4\pi p^2\,dp}{\exp\,(\alpha + E/kT) + 1} \\ &= \frac{2}{h^3}\frac{4\pi p^2\,dp}{\exp\,(\alpha + E_0/kT + p^2/2mkT) + 1} \end{aligned} \tag{2-273}$$

Thus

$$n_e = \int_0^\infty n_e(p)\,dp = \frac{4\pi}{h^3}\,(2mkT)^{\frac{3}{2}}F_{\frac{1}{2}}\left(\alpha + \frac{E_0}{kT}\right) \tag{2-274}$$

Exactly the same thing occurs in the pressure integral for a perfect gas, so that the equation of state is represented, to first approximation, by the same parametric representation as before, where α is replaced by $\alpha + E_0/kT$. The coulomb energy does contribute a correction to the pressure which must be computed separately, however.

The fact that the coulomb potential energy is negative and increases with density means that it must reduce the pressure from the value appropriate to a noninteracting gas. Although a compression must do work against the free-particle momenta, it does negative work against the coulomb field.

Problem 2-50: Show from the first law of thermodynamics that the coulomb pressure from the homogeneous-ion-sphere model is

$$P_c = -\frac{3}{10}\left(\frac{4\pi}{3}\right)^{\frac{1}{3}}\frac{(\rho N_0)^{\frac{4}{3}}}{\mu_e^{\frac{1}{3}}}\,e^2\left(X + 2^{-\frac{1}{3}}Y + \sum_{Z>2}\frac{Z^{\frac{5}{3}}X_Z}{A_Z}\right) \tag{2-275}$$

It will be noticed from Eq. (2-275) that

$$P_c = \frac{1}{3}\left(\frac{U}{V}\right)_c \tag{2-276}$$

where $(U/V)_c$ is the energy density in the coulomb field. It is remarkable that the relationship of the pressure to the coulomb-energy density is the same as that obtained for the photon gas and for the Debye-Hückel gas, in both cases confirming the result of the virial theorem. Of course the calculation is inexact. In the present case, for example, the next largest correction (much smaller than this one) comes from distortion of the electron density. This calculation proceeds from a Thomas-Fermi model of ion sphere.[1] The coulomb corrections to the pressure are not always negligible and should, for example, be included in models of white dwarfs.[2]

Problem 2-51: Show that the numerical value of the coulomb pressure is

$$P_c = -5.68 \times 10^{12} \mu_e^{-\frac{4}{3}} \rho^{\frac{4}{3}} \left(X + 0.79Y + \sum_{Z>2} \frac{Z^{\frac{2}{3}} X_Z}{A_Z} \right) \quad \text{dyne/cm}^2 \tag{2-277}$$

Problem 2-52: Calculate in this approximation the density for which completely (pressure) ionized iron has zero pressure.
Ans: $\rho = 260 \text{ g/cm}^3$.

Just as in the Debye-Hückel case, the coulomb interactions reduce the ionization potential of bound electrons. Again to simplify the matter, assume that the only bound electrons are in orbitals having a Bohr radius much smaller than the ion sphere:

$$a_{Zn} = \frac{n^2}{Z} a_0 < R_Z \tag{2-278}$$

Then to the normal energy

$$E_{Zn} = -\frac{1}{2} \frac{Ze^2}{a_{Zn}} \tag{2-279}$$

must be added the interaction of an electron at the center of the ion sphere to the other electrons of the sphere:

$$E'_{Zn} = E_{Zn} + \frac{3}{2} \frac{Ze^2}{R_Z} \tag{2-280}$$

Problem 2-53: Show that in Rydbergs

$$E'_{Zn} = -\frac{Z^2}{n^2} + 2.16 \left(\frac{\rho Z^2}{\mu_e} \right)^{\frac{1}{3}} \quad \text{Ry} \tag{2-281}$$

Equation (2-281) is only a rough approximation because it has assumed (1) a homogeneous electron density, (2) that the nuclear charge Z is unshielded by free electrons, and (3) that the bound-electron wave function is unperturbed by the free electrons. A more satisfactory treatment would actually involve the solution for the bound-electron wave function in a shielded rather than a pure coulomb potential. If, however, the bound-electron radius is much less than the ion-sphere radius, this expression will give a reasonable approximation to the energy of the bound electron.

[1] The reader is referred to Salpeter, *op. cit.*, for this treatment.

[2] For a thorough discussion of coulomb effects on white dwarfs see H. Van Horn, *Astrophys. J.*, **151**:227 (1968).

Now the *effective ionization potential* is the difference in energy between the lowest energy (zero kinetic energy) of the continuum electron states and the energy (negative) of the bound electron; that is,

$$\begin{aligned}\chi_{\text{eff}} &= E_0 - E'_{Zn} = E_0 + \frac{1}{2}\frac{Ze^2}{a_{Zn}} - \frac{3}{2}\frac{Ze^2}{R_Z} \\ &= \frac{Z^2}{n^2} - 2.16\left(\frac{\rho Z^2}{\mu_e}\right)^{\frac{1}{3}} - 1.30(\mu_e{}^2\rho)^{\frac{1}{3}}\left(X + 0.79Y + \sum_{Z>2}\frac{Z^{\frac{5}{3}}X_Z}{A_Z}\right) \quad \text{Ry} \qquad (2\text{-}282)\end{aligned}$$

The first term is the normal low-density ionization potential; $\chi = Z^2/n^2$ Ry. The remaining terms represent the reduction in ionization brought about by the coulomb interactions. These interactions reduce χ by a constant amount, the continuum depression E_0, plus another term which varies from state to state. This reduction will rapidly cut off the partition functions of ions by restricting their allowable states to the more tightly bound ones. When the energy of a bound electron is less negative than E_0, the electrons are to be considered as free. We also note here another view of the phenomenon of pressure ionization, because there exists, for each Z, a sufficiently high density that $\chi_{\text{eff}} < 0$.

Problem 2-54: Estimate the density at which iron must be completely pressure-ionized.
Ans: $\rho \approx 2 \times 10^4$.

A summary is in order at this point. The coulomb interactions render the ionized gas imperfect. The two most important effects are a negative coulomb pressure and a reduction of the ionization potentials. These effects have been quantitatively investigated for two extreme environments: (1) at sufficiently low density and high temperature, kT dominates coulomb energies, and the nondegenerate gas may be treated approximately by the Debye-Hückel method; (2) at high density and low temperature, the electron gas is degenerate and near perfect, but the ions are forced into a lattice because their coulomb energies exceed kT. The treatments given were only the simplest approximations possible in a very difficult subject, so that the formulas derived should be regarded with caution inasmuch as the assumptions made may not be satisfied in most astrophysical environments.

As a final note, the reader may be interested to know that the same type of problem occurs in the equation of state of nuclear matter. The properties of neutron stars, if they exist, depend upon the corrections to the independent-particle-model equation of state necessitated by interactions in the gas at nuclear density. In this case, of course, the perturbing interactions stem from the nuclear force rather than the coulomb force.[1]

Another application of the coulomb interactions will be postponed until Chap. 4, where it will be seen that the energetics of nuclear reactions is shifted by a

[1] J. N. Bahcall and R. A. Wolf, *Phys. Rev.*, **140**:1445 (1965).

sufficient amount to alter the reaction rates. This phenomenon is generally called *electron screening* in nuclear reactions.

2-4 POLYTROPES

The static configuration of a gaseous sphere held together by self-gravitation must satisfy the condition of hydrostatic equilibrium. The essence of the equation of hydrostatic equilibrium is that the pressure at each point in a stellar interior is sufficient to just balance the weight of the overlying layers of the star. Furthermore, the pressure itself is determined by the equation of state applicable to the local conditions in the stellar interior. These considerations do not in themselves determine the structure of a star. Any specified pressure that may be required to support the overlying layers is obtainable from an unlimited number of combinations of density and temperature at that point. What are clearly needed are more conditions on the density and temperature in a stellar interior that relate to other physical processes that go on there.

It was about the beginning of the twentieth century when several notable physicists, viz., Lane, Ritter, Kelvin, Emden, and Fowler, considered the question of what limitations could be placed on the structure of a star just from the condition of hydrostatic equilibrium alone. They quickly concluded that some other condition relating the physical variables in the stellar interior is necessary in order to be able to specify the structure. The necessary relationships are to be found in the production and transport of thermal energy, subjects to be discussed in subsequent chapters. One explicit auxiliary condition that has been found to correspond to certain idealized physical situations, however, is a pressure expressible in terms of some power of the density only. For historical reasons[1] the assumed pressure-density relationship is written as

$$P = K\rho^{(n+1)/n} \tag{2-283}$$

where the number n is called the polytropic index. Gaseous spheres in hydrostatic equilibrium in which the pressure and density are related by Eq. (2-283) at each point along the radius are called *polytropes*. The constant K depends upon the nature of the polytrope. It was shown by Lane and Emden that if a polytropic pressure-density relation is assumed, the properties of the structure can be computed.

Since, of course, *any* explicit relationship between the pressure and the density would make possible the solution for the structure of a self-gravitating gaseous sphere in hydrostatic equilibrium, one might ask why a relationship of the form

[1] The nomenclature is patterned after quasistatic changes of state of an ideal gas for which a generalized specific heat is held constant. It was found by early workers in kinetic theory that for such changes of state, called *polytropic changes* by R. Emden in his classical treatise "Gaskugeln," B. G. Teubner, Leipzig, 1907, the variables change along a track $P = K\rho^{\gamma'}$, where γ' is determined by C_P, C_V, and C, the specific heat characterizing the process. These matters of historical interest are elegantly summarized in Chandrasekhar, *op. cit.*, chaps. 2 and 4.

of Eq. (2-283) should be chosen for study. The reason lies in the fact that some idealized physical circumstances for a star would lead naturally to equations of that form. To clarify this point, we shall consider examples of stars that can be represented by polytropes.

As a first example, we may follow Kelvin in considering a star that is "boiling" in a state that he described as *adiabatic convective equilibrium*. If the whole interior of a star is completely convective, mass elements are both rising and falling in the interior of a star. A star is said to be in convective equilibrium if any mass element after rising and falling from its initial temperature and density to a new temperature and density finds itself at the same temperature and density as the surroundings. The convective equilibrium is adiabatic if the convective cells move without heat exchange. It will be evident that if some mechanism continuously stirs and mixes the entire interior of a star, it must soon come to a condition of convective equilibrium, for any differences in temperature and density of the surroundings in a star from those of an element that has risen from some lower portion of a star will quickly be eliminated. If radiation pressure is an unimportant determinant in the structure, adiabatic changes are of the form

$$P = K\rho^\gamma \tag{2-284}$$

where $\gamma = \frac{5}{3}$ for an ideal monatomic gas. If such a rising or falling element is, furthermore, at the same conditions of temperature, density, and pressure as the surrounding matter at all times, it follows that the run of pressure and density in the star is such that

$$P(r) = K\rho(r)^{\frac{5}{3}} \tag{2-285}$$

It seems, therefore, that a star in convective equilibrium in which radiation pressure is not important is a polytrope of exponent $\gamma = \frac{5}{3}$, which is also a polytrope of index $n = 1.5$. In such a way Kelvin was led quite naturally to at least one physical possibility that would correspond to the structure of a polytrope.

As a second example, consider a star in which radiation pressure is not unimportant. We have defined the quantity β such that

$$P_g = \frac{N_0 k}{\mu} \rho T = \beta P \qquad P_r = \tfrac{1}{3} a T^4 = (1 - \beta) P \tag{2-286}$$

for a nondegenerate gas.

Equating the values of the pressure from these two equations gives immediately *at each point*

$$T = \left(\frac{N_0 k}{\mu} \frac{3}{a} \frac{1 - \beta}{\beta} \right)^{\frac{1}{3}} \rho^{\frac{1}{3}} \tag{2-287}$$

Since we also have

$$P = \frac{N_0 k}{\mu} \frac{\rho T}{\beta} \tag{2-288}$$

we see that

$$P = \left[\left(\frac{N_0k}{\mu}\right)^4 \frac{3}{a} \frac{1-\beta}{\beta^4}\right]^{\frac{1}{3}} \rho^{\frac{4}{3}} \tag{2-289}$$

This equation is true at each point in the interior of the star we are considering. The ratio of gas pressure to total pressure β does, in general, depend upon the distance from the center of the star. *If*, however, one had a special configuration in which the quantity β was a constant, i.e., such that the gas pressure was a constant fraction of the total pressure throughout the star, then the expression in brackets in Eq. (2-289) reduces to a constant, and one has an equation of the form

$$P = K\rho^{\frac{4}{3}} \tag{2-290}$$

This model star would correspond to a polytrope of polytropic exponent $\frac{4}{3}$ or of index 3. We can see later that this particular polytrope corresponds more closely to stars in radiative equilibrium, i.e., stars for which the energy is transported by radiative transfer rather than by convection. It will, in fact, be shown that the polytrope of index 3 corresponds to that star in radiative equilibrium such that at each distance r from the center of the star, the product of the energy liberated per unit mass from all the material interior to r times the opacity of the gas at the point r is a constant. The properties of a nondegenerate polytrope of index 3 have also been highly developed in the analytical study of gaseous configurations, especially by Eddington. This model star is frequently called the *standard model*. We shall use the standard model often in an attempt to get a first approximation to the runs of temperatures and densities in the interiors of stars.

A third example may be provided by stars supported by the pressure of a completely degenerate electron gas (white dwarfs?). That pressure has been shown to vary as $\rho^{\frac{5}{3}}$ or $\rho^{\frac{4}{3}}$, according to whether the electron velocities are nonrelativistic or relativistic. The corresponding polytropes can provide useful insights into their structure.

These examples give some indication of the physical reasons that lie behind considering the structure of gaseous spheres in hydrostatic equilibrium for which the pressure and density are related by an equation of the form of Eq. (2-283). Motivated by the fact that the density is proportional to T^n in a nondegenerate polytrope of index n, a convenient definition is

$$\rho \equiv \lambda\phi^n \tag{2-291}$$

where λ is a scaling parameter whose value depends upon the definition of the quantity ϕ. This representation for the run of density throughout the star will be convenient for the study of polytropes, where we shall identify the parameter λ with the central density of the star, thereby normalizing the function ϕ to unity at the center. For this representation, the pressure is

$$P = K\rho^{(n+1)/n} = K\lambda^{(n+1)/n}\phi^{n+1} \tag{2-292}$$

The solution for the structure of a polytrope, then, depends upon the coupling of Eq. (2-283) to the condition of hydrostatic equilibrium

$$\frac{dP}{dr} = -\rho \frac{GM_r}{r^2} \frac{dM_r}{dr} = 4\pi r^2 \rho \tag{2-293}$$

from which it follows that

$$\frac{1}{r^2}\frac{d}{dr}\left(\frac{r^2}{\rho}\frac{dP}{dr}\right) = \frac{1}{r^2}\frac{d}{dr}(-GM_r) = -4\pi G\rho \tag{2-294}$$

Substitution of the values of pressure and density for a polytrope of index n into this last equation yields

$$(n+1)K\lambda^{1/n}\frac{1}{r^2}\frac{d}{dr}\left(r^2\frac{d\phi}{dr}\right) = -4\pi G\lambda\phi^n \tag{2-295}$$

This equation can be made more attractive by defining a unit of length

$$a \equiv \left[\frac{(n+1)K\lambda^{(1-n)/n}}{4\pi G}\right]^{\frac{1}{2}} \tag{2-296}$$

and by defining a dimensionless distance variable $\xi = r/a$, whereupon Eq. (2-295) reduces to

$$\frac{1}{\xi^2}\frac{d}{d\xi}\left(\xi^2\frac{d\phi}{d\xi}\right) = -\phi^n \tag{2-297}$$

This equation is called the *Lane-Emden equation* for the structure of a polytrope for index n. The solution for ϕ as a function of ξ completely determines the structure of the polytrope except for the choice of the central density. By setting λ equal to the central density, it is easy to see that the temperaturelike variable ϕ must obey certain boundary conditions at the center of the star, i.e., at $\xi = 0$; viz.,

$$\phi = 1 \qquad \frac{d\phi}{d\xi} = 0 \qquad \text{at } \xi = 0 \tag{2-298}$$

Problem 2-55: Show that $(d\phi/d\xi)_0 = 0$. *Hint:* Expand the equation of hydrostatic equilibrium near the center.

The solution ϕ which satisfies the Lane-Emden equation of index n under these boundary conditions is called the *Lane-Emden function of index n.*

Explicit solutions of the Lane-Emden equation for general values of n apparently do not exist. For values of n other than $n = 0$, 1, and 5, numerical techniques must be employed for the determination of the Lane-Emden function. We first note that if $\phi(\xi)$ is a solution of the equation, then $\phi(-\xi)$ is also a solution. This observation implies that if ϕ is expressed as a power series in ξ, only even powers of ξ appear; that is,

$$\phi(\xi) = C_0 + C_2\xi^2 + C_4\xi^4 + \cdots \tag{2-299}$$

Problem 2-56: By substituting Eq. (2-299) into the Lane-Emden equation, show that the first three terms of the series for ϕ are

$$\phi = 1 - \tfrac{1}{6}\xi^2 + \frac{n}{120}\xi^4 - \cdots \tag{2-300}$$

By taking a sufficient number of terms in this alternating series for ϕ, one can calculate the solution to any desired accuracy for values of $\xi < 1$. For values of $\xi > 1$, this solution can be continued from the differential equation by standard numerical methods. The solutions are found to decrease monotonically from the center and for values of n less than 5 have a zero for some finite value of ξ, say, $\xi = \xi_1$. At $\xi = \xi_1$ it is clear that ϕ's being equal to zero makes the pressure vanish, and the configuration may be said to have a physical boundary at that point. Table 2-5 lists the value of ξ_1 and the derivative of ϕ at $\xi = \xi_1$ for the various Lane-Emden functions of index n. We shall shortly see that the value of ξ_1 and the slope of ϕ at $\xi = \xi_1$ are important for determining large-scale properties of the various gaseous configurations.

It is evident that the structure of each polytrope is specified in terms of the dimensionless length ξ. An inspection of the length a used in forming this dimensionless length shows that its value is determined by two numbers; the first is the constant K, which occurs in Eq. (2-283) relating the pressure to the density, and the second is the parameter λ, which we have taken to be the central density for the solution of this problem. It is apparent, therefore, that each Lane-Emden function ϕ_n represents, for a specified value of the constant K, a one-parameter family of solutions, the parameter being the central density λ. As an example, we may turn to the polytropes that were considered in introducing this whole discussion. In the case of the completely convective polytrope of index 1.5, it is easy to see that the value of K is determined by the particular adiabat of the gas. In adiabatic convective equilibrium the entropy per gram of material is constant, and it specifies the value of K.

Problem 2-57: Ignore the entropy of the radiation and show from Eq. (2-136) that the entropy per gram can be written

$$S = \frac{3}{2}\frac{N_0 k}{\mu} \ln \frac{P}{\rho^{5/3}} + \text{const} \tag{2-301}$$

that is, $S = S(K)$.

In the case of the standard model, or polytrope of index 3, we see from Eq. (2-289) that the value of the constant K is given by

$$K = \left[\left(\frac{N_0 k}{\mu}\right)^4 \frac{3}{a}\frac{1-\beta}{\beta^4}\right]^{1/3} \tag{2-302}$$

where β represents the ratio of the gas pressure to the total pressure and is a constant throughout the standard model. The selection of a value of K in either of these two instances still allows a complete run of corresponding solutions as determined by the dentral density λ. It is quite clear that considerable leeway still exists for the actual structure of the various polytropes being considered. To understand the way in which these various factors come into play, we need to consider several more large-scale properties of the configuration that can be derived from the material presented so far.

(*1*) *Radius:* The radius of the configuration is by definition determined by the first zero of the Lane-Emden function of order n. Thus

$$R = a\xi_1 = \left[\frac{(n+1)K}{4\pi G}\right]^{\frac{1}{2}} \lambda^{(1-n)/2n}\xi_1 \tag{2-303}$$

(*2*) *Mass:* The mass M interior to the normalized radius ξ is given by

$$M(\xi) = \int_0^{a\xi} 4\pi r^2 \rho \, dr = 4\pi a^3 \int_0^{\xi} \lambda \phi^n \xi^2 \, d\xi \tag{2-304}$$

By using the Lane-Emden equation itself, the integral in Eq. (2-304) can be transformed to

$$\begin{aligned} M(\xi) &= -4\pi a^3 \lambda \int_0^{\xi} \frac{d}{d\xi} \xi^2 \frac{d\phi}{d\xi} \, d\xi \\ &= -4\pi a^3 \lambda \xi^2 \frac{d\phi}{d\xi} \end{aligned} \tag{2-305}$$

Substituting for the unit of length a and evaluating the above expression at $\xi = \xi_1$

Table 2-5 Constants of the Lane-Emden functions†

n	ξ_1	$-\xi_1^2 \left(\frac{d\phi}{d\xi}\right)_{\xi=\xi_1}$	$\frac{\rho_c}{\bar{\rho}}$
0	2.4494	4.8988	1.0000
0.5	2.7528	3.7871	1.8361
1.0	3.14159	3.14159	3.28987
1.5	3.65375	2.71406	5.99071
2.0	4.35287	2.41105	11.40254
2.5	5.35528	2.18720	23.40646
3.0	6.89685	2.01824	54.1825
3.25	8.01894	1.94980	88.153
3.5	9.53581	1.89056	152.884
4.0	14.97155	1.79723	622.408
4.5	31.83646	1.73780	6,189.47
4.9	169.47	1.7355	934,800
5.0	∞	1.73205	∞

† S. Chandrasekhar, "An Introduction to the Study of Stellar Structure," p. 96; reprinted from the Dover Publications edition, Copyright 1939 by The University of Chicago, as reprinted by permission of The University of Chicago.

gives the total mass of the star:

$$\mathfrak{M} = -4\pi \left[\frac{(n+1)K}{4\pi G} \right]^{\frac{3}{2}} \lambda^{(3-n)/2n} \left(\xi^2 \frac{d\phi}{d\xi} \right)_{\xi_1} \tag{2-306}$$

We note that in the case $n = 3$, the mass depends only upon K and is independent of λ. The product of ξ_1^2 times the slope of ϕ evaluated at the first zero, ξ_1, is one of the quantities listed in Table 2-5.

Problem 2-58: Show that the mass of the standard-model polytrope of index 3 is given numerically by

$$\mathfrak{M} = 18.0 \frac{\sqrt{1-\beta}}{\mu^2\beta^2} \mathfrak{M}_\odot \tag{2-307}$$

For a given composition μ, the mass determines the value of β.

Problem 2-59: Imagine that a white dwarf is a body supported by the pressure of completely degenerate electrons. As the mass of the structure is increased, the central density becomes so high that the degeneracy becomes relativistic at the center, such that (confirm this)

$$P_c \rightarrow 1.244 \times 10^{15} \left(\frac{\rho}{\mu_e} \right)^{\frac{4}{3}} \quad \text{dynes/cm}^2 \tag{2-308}$$

and falls off to nonrelativistic degeneracy in the outer portions of the star. As the mass is continually increased, the star shrinks to ever higher densities and ever smaller radius, until the electrons become highly relativistic everywhere. Then Eq. (2-308) is applicable throughout the star. Show that at this point the mass is

$$\mathfrak{M} = \frac{5.80}{\mu_e^2} \mathfrak{M}_\odot \tag{2-309}$$

This mass is called the *Chandrasekhar limit,* since Chandrasekhar showed that this was the maximum mass that could be supported by electron degeneracy. (Other physical effects such as rotation and inverse beta decay have been ignored.) It seems clear that this value must represent a limiting mass because the electrons can be relativistic throughout only if the mass is sufficient to squeeze the volume to a point. Inasmuch as white dwarfs are observed to have nonzero radii, their masses must be less than Eq. (2-309). The question of what happens if the mass exceeds this limit is a difficult one and will not be considered here.

(*3*) *Ratio of mean density to central density:* The mean density of the configuration is given by the total mass of the configuration divided by the volume of the configuration, whereas the central density is equal to λ. Thus, the ratio of the mean density to the central density can be determined from Eqs. (2-303) and (2-306) to be

$$\frac{\bar{\rho}}{\rho_c} = -\frac{3}{\xi_1} \left(\frac{d\phi}{d\xi} \right)_{\xi_1} \tag{2-310}$$

It is evident that the ratio of mean density to central density depends only upon the index of the polytrope. In fact, we may take this to be the main feature of the polytropic index; viz., the extent to which the matter is concentrated toward the center of the star. Although we have not shown it here, the ratio of the central density to the mean density varies between the limits of unity (a star of uniform density) for a polytrope of index zero to value of infinity (a star infinitely concentrated toward the center) for a polytrope of index 5, passing through all intermediate values as n increases from zero to 5.

Problem 2-60: Show that the central density in the standard model exceeds the mean density by the factor $\rho_c/\bar{\rho} = 54.2$.

(*4*) *The central pressure:* Since ϕ is normalized to unity at $\xi = 0$ by the interpretation of λ, the central pressure may be written, from Eq. (2-292), as

$$P_c = K\lambda^{(n+1)/n} \tag{2-311}$$

To express the central pressure in terms of macroscopic properties, we note that Eq. (2-303) can be written in the form

$$R = \left[\frac{(n+1)}{4\pi G}\,\xi_1{}^2\right]^{\frac{1}{2}} [K\lambda^{(1-n)/n}]^{\frac{1}{2}} \tag{2-312}$$

from which

$$K\lambda^{(1-n)/n} = \frac{4\pi R^2 G}{(n+1)\xi_1{}^2} \tag{2-313}$$

Hence, the central pressure is given by

$$\begin{aligned} P_c &= (K\lambda^{(1-n)/n})\lambda^2 = K\lambda^{(1-n)/n}\rho_c{}^2 \\ &= \frac{4\pi R^2 G}{(n+1)\xi_1{}^2}\left[\frac{\xi_1}{3}\,\frac{1}{(d\phi/d\xi)_{\xi_1}}\right]^2 \bar{\rho}^2 \\ &= \frac{1}{4\pi(n+1)(d\phi/d\xi)_{\xi_1}{}^2}\,\frac{G\mathfrak{M}^2}{R^4} \end{aligned} \tag{2-314}$$

Problem 2-61: Show that for the standard model the central pressure is given numerically by

$$P_c = 1.24 \times 10^{17} \left(\frac{\mathfrak{M}}{\mathfrak{M}_\odot}\right)^2 \left(\frac{R_\odot}{R}\right)^4 \quad \text{dynes/cm}^2 \tag{2-315}$$

This result gives a larger and more realistic estimate of the central pressure of the sun than the earlier rough arguments because it allows for the central condensation.

(*5*) *The central temperature:* The central temperature may be computed from the central pressure and the central density by use of the appropriate equation of state. For the case of the ideal ionized nondegenerate gas the relevant relationships are

$$P_g = \frac{N_0 k}{\mu}\,\rho_c T_c = \beta_c P_c \tag{2-316}$$

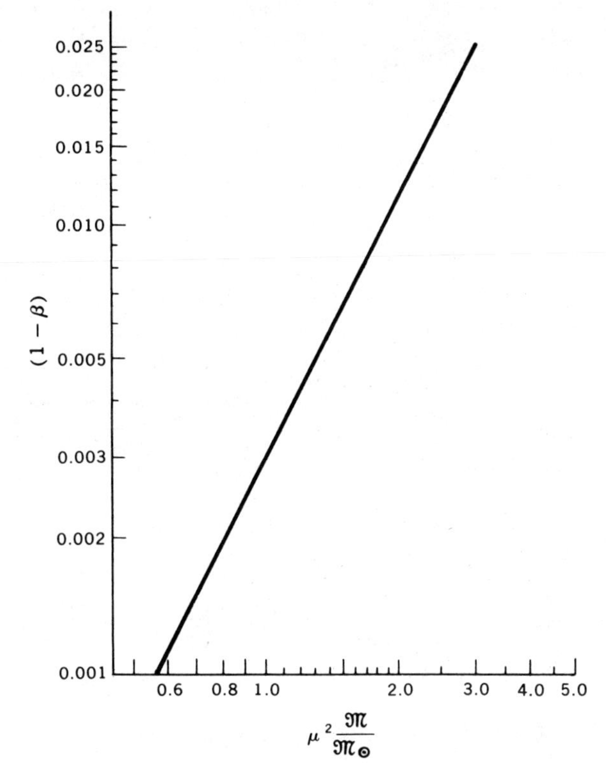

Fig. 2-18 The value of $1 - \beta$ for the standard model of low-mass stars.

where P_c equals the total pressure (gas plus radiation) at the center of the star. Thus the central temperature is given by

$$T_c = \frac{\mu}{N_0 k} \frac{\beta_c P_c}{\rho_c} \tag{2-317}$$

Problem 2-62: Show that for the standard model the central temperature is

$$T_c = 4.6 \times 10^6 \mu\beta \left(\frac{\mathfrak{M}}{\mathfrak{M}_\odot}\right)^{\frac{2}{3}} \rho_c^{\frac{1}{3}} \tag{2-318}$$

We may note that the equation above for the central temperature of the standard model contains quantities that are not independent of each other. In Prob. 2-58 it was demonstrated that the mass of the standard model is related to the ratio of gas pressure to total pressure; therefore, for a fixed value of μ it is apparent that β is a function of the mass (or vice versa), although the solution cannot be written explicitly. Figures 2-18 and 2-19 show graphically the dependence of β upon the quantity $\mu^2(\mathfrak{M}/\mathfrak{M}_\odot)$. These figures indicate the growing importance of radiation pressure with increasing mass. Since the mean molecular

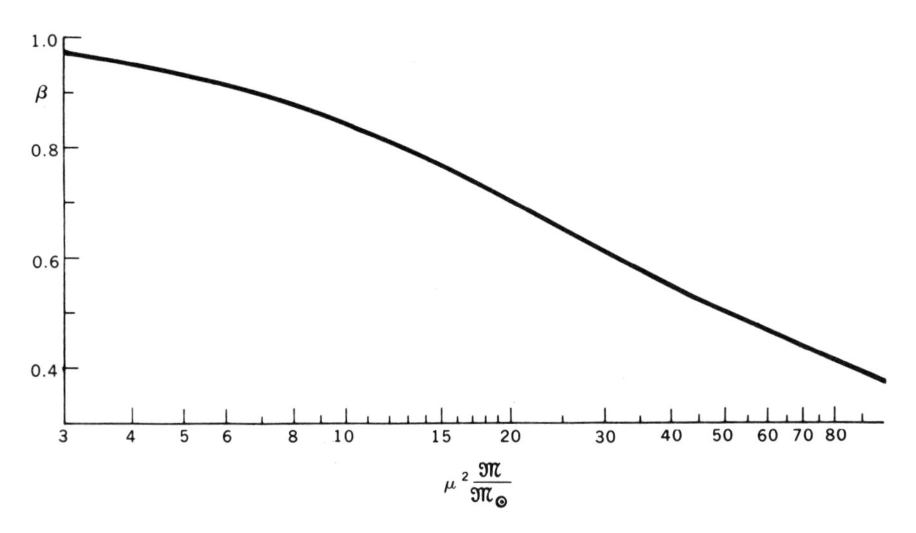

Fig. 2-19 The value of $1 - \beta$ for the standard model of high-mass stars.

weight μ lies between $\frac{1}{2}$ and 2, the quantity $\mu^2(\mathfrak{M}/\mathfrak{M}_\odot)$ is not greatly different from the mass expressed in solar masses. At any rate, this additional constraint must be taken into account when using Eq. (2-318) for the central temperature of the standard model.

We may in fact use Eq. (2-318) at this point to make an estimate of the central temperatures of main-sequence stars. Main-sequence stars certainly are not polytropes of index 3, but we may nonetheless expect to see the trend of central temperatures by representing all main-sequence stars by the standard model. It is clear that the central density is an unknown in Eq. (2-318); however, an earlier problem demonstrated that $\rho_c = 54.2\bar{\rho}$ for the standard model, so that Eq. (2-318) may be altered to read

$$T_c = 17.4 \times 10^6 \mu\beta \left(\frac{\mathfrak{M}}{\mathfrak{M}_\odot}\right)^{\frac{2}{3}} \rho^{\frac{1}{3}} \tag{2-319}$$

Problem 2-63: A certain type O star has $\mathfrak{M} = 30\mathfrak{M}_\odot$, $R = 6.6R_\odot$, $X = 0.70$, and $Y = 0.30$. Estimate the importance of radiation pressure and the central temperature by approximating the star by the standard model. A much better calculation[1] on electronic computers yields $\beta_c = 0.77$, $T_c = 3.7 \times 10^7$.

The mean density $\bar{\rho}$ is just the mass divided by the volume, so that we can obtain these properties of main-sequence stars from Table 1-1. In fact, when the properties of Table 1-1 are coupled with the mass-luminosity relationship and with the fact that $\bar{\rho}_\odot = 1.4 \text{ g/cm}^3$, it appears that an approximation to the

[1] R. Stothers, *Astrophys. J.*, **138**:1074 (1963).

average density of main-sequence stars is

$$\bar{\rho} \approx \frac{1.4}{\mathfrak{M}/\mathfrak{M}_\odot} \qquad \text{g/cm}^3 \tag{2-320}$$

Problem 2-64: Using the data of Table 1-1, show that a main-sequence star of $6\mathfrak{M}_\odot$ has an average density approximately equal to 0.28 g/cm³. What is the percentage difference between this value and that inferred from Eq. (2-320)? How much of a percentage error in the central temperature would be introduced by that percentage error in the average density?

It is of some interest that the density of main-sequence stars decreases with increasing mass. This fact is a consequence of the virial theorem, which demands higher temperatures for higher values of the potential energy of self-gravitation. These higher temperatures become sufficient to support the star in a more distended configuration of lower density.

It is also clear from Fig. 2-18 that $\beta \approx 1$ for stars of main-sequence mass. The mean molecular weight has a value near 0.7 for the centers of stars that have partially depleted their hydrogen (say, $X \approx 0.5$, $Y \approx 0.5$). By combining these approximations, Eqs. (2-319) and (2-320) yield

$$T_c \approx 14 \times 10^6 \left(\frac{\mathfrak{M}}{\mathfrak{M}_\odot}\right)^{\frac{1}{3}} \tag{2-321}$$

as anticipated central temperatures for main-sequence stars. When the question of thermonuclear reactions is considered, Eq. (2-321) will prove a helpful guide.

Many other interesting physical quantities can be calculated for these model stars called polytropes. An extensive discussion of the mathematical considerations related to this well-developed subject will be found in the monograph on stellar structure by Chandrasekhar. Furthermore, we have considered only uniform polytropes of homogeneous composition, whereas it is possible to divide stars into polytropic shells or mixed polytropes. A great deal of intuitive appreciation for the complexities of the physics of stellar structure may be obtained by an extensive analysis of the structure of polytropes. We have employed only the simplest features in this section as an introduction to the subject. Modern research has shown that the usefulness of polytropes is for the most part limited to this introductory acquaintance. Accurate and detailed models of the structure of real stars may be obtained only from detailed computer calculation. The additional physics needed to make these detailed calculations will constitute the burden of the following chapter.

chapter 3

ENERGY TRANSPORT IN THE STELLAR INTERIOR

What are the fundamental principles governing the structure of a star? The criterion of hydrostatic equilibrium is a good one as far as it goes, but the pressure required at a given point can be supplied by infinitely varied combinations of density and temperature. The assumption of a polytropic pressure-density relationship is convenient in that it does allow the stellar structure to be determined by the condition of hydrostatic equilibrium, but except for a few cases, one has little assurance that the resulting polytropes correspond to real stars. For most stars it is to be expected that the pressure-density relationship will vary throughout the star. From the considerations to be presented in this chapter it will be apparent that the luminosity of a static star is determined from two independent conditions: (1) the rate of energy flow, and hence the luminosity $\mathfrak{L}$, is determined by the temperature gradient and the details of energy transport; and (2) the luminosity of a *static* star must equal the rate at which energy is being liberated by nuclear reactions in the interior of the star. For a correct model of a star, the two luminosity criteria must yield the same answer, a situation seldom true for polytropes. The conditions that have allowed accurate models of stars to be constructed are those provided by the details of energy generation and transport in the stellar interior. The latter subject constitutes the topic of this chapter, and thermonuclear energy generation will be introduced in Chap. 4.

The transport of energy is caused by a temperature gradient. The common ways by which a star transfers heat from its hotter parts to its cooler parts are by the mechanisms of radiative transfer, convection, or conduction. The fourth major mechanism, neutrino emission, emits energy directly from the interior into space without interactive transport through the stellar matter. Since the equilibrium energy density in the radiation field is $u = aT^4$, the hot portions of the star contain more photon energy than the cool portions do. Since no walls exist to contain the photons, they may diffuse toward cooler regions, hindered only by their atomic interactions with the stellar matter. The associated mechanism of heat transport is called *radiative transfer* and is proportional, among other things, to the temperature gradient. If the energy is carried by the particles themselves, rather than by photons, the heat transport is called *conduction*. It is also proportional to the temperature gradient but is, except in a degenerate electron gas, much less efficient than radiative transfer. There is a limit to the rate at which energy can be transported by these two mechanisms, however, for if the temperature gradient becomes too large, the gas develops convective instability. Then rising bubbles of gas are hotter than their surroundings and continue to rise until they dissipate their excess heat content by equilibrating with the lower-temperature environment. We shall see that this extremely efficient mechanism results in stellar interiors in a temperature gradient that is calculable from hydrostatic equilibrium, the so-called *adiabatic temperature gradient*. And lastly, the mechanisms of neutrino emission will be outlined.

3-1 ENERGY BALANCE

One of the principles that must be satisfied in all of science is that of energy conservation. The stars radiate large amounts of energy from their surface, and if the interior energy is not to disappear, the radiated energy must be replaced in the interior. It was pointed out in connection with the virial theorem that self-gravitating gaseous spheres will be provided with gravitational energy during a contraction, but the amount of energy liberated in this way is not sufficient to maintain a stellar luminosity for a stellar lifetime. Most stellar structures are essentially static; i.e., the distribution of temperature, density, and pressure in the star hardly changes with time. If that is to be the case, the power radiated must be supplied at the same rate in the stellar interior. As is now commonly known, this energy is provided by exothermic nuclear reactions that proceed near the center of the star.

The rate at which energy is liberated by thermonuclear reactions in the interior of a star will be a function of position, being determined by the local values of the density ρ, the temperature T, and the set of composition parameters $\{X_Z\}$. We shall designate the power liberated per gram of stellar matter by nuclear reactions by the symbol ϵ:

$$\epsilon(\text{ergs g}^{-1}\text{ sec}^{-1}) = \epsilon(\rho,T,\{X_Z\})$$

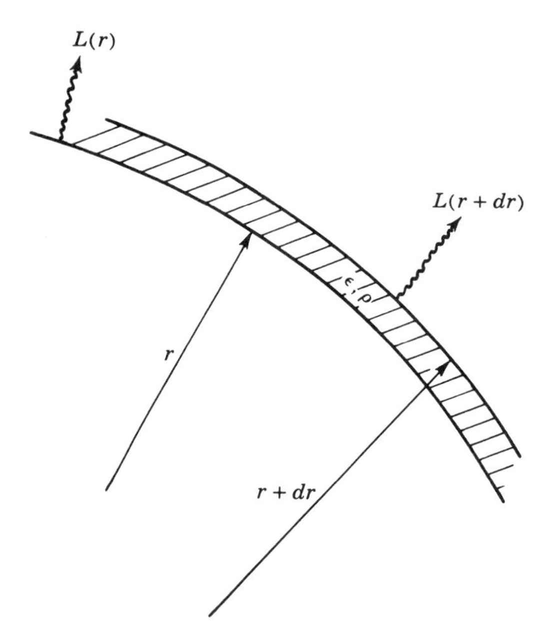

Fig. 3-1 The energy balance in a static spherical shell of thickness dr. The net rate of radial energy flow through a sphere of radius r is designated by $L(r)$. The value of $L(r + dr)$ is greater than $L(r)$ in a static spherical body by virtue of the thermonuclear power generated within the shell. In a nonstatic spherical star the rate of change of the entropy of the gas within the shell must also be included in the balance.

From the definition of ϵ it follows that the rate at which energy is liberated per cubic centimeter of stellar matter is given by the product $\rho\epsilon$. Then for a *perfectly static* stellar structure, the balance of energy can be written

$$\mathcal{L} = \int_V \epsilon\rho \, dV = \int_0^R \epsilon\rho 4\pi r^2 \, dr \tag{3-1}$$

where $\mathcal{L}$ is the total luminosity of the star. We might emphasize at this point that no stellar structure is exactly static. For even when the energy liberated is being replaced by nuclear reactions, these same nuclear reactions are slowly changing the composition of the stellar interior in which they are occurring, leading thereby to very slow changes in the overall structure of the star. This type of change is called *stellar evolution*. Changes of this type usually occur so slowly that as a good approximation one can think of evolution as *a sequence of static structures*.

Equation (3-1) can be expressed in a differential form. The balance of energy that is represented by that equation must also hold in a shell located at each value of the radius r. If we denote the energy flowing outward through a spherical surface at radius r by $L(r)$ and the energy flowing outward through a spherical surface at $r + dr$ by the quantity $L(r + dr)$, as in Fig. 3-1, then clearly

$$L(r + dr) - L(r) = \epsilon(r)\rho(r)4\pi r^2 \, dr$$

or

$$\frac{dL(r)}{dr} = \epsilon(r)\rho(r)4\pi r^2 \tag{3-2}$$

Equation (3-2) will represent one of the major relationships that must be satisfied in a *static* stellar structure. It will often be expressed as a mass gradient instead of radial gradient:

$$\frac{dL(r)}{dM(r)} = \epsilon(r)$$

$$M(r) = \int_0^r 4\pi r^2 \rho \, dr \tag{3-3}$$

One might ask whether this equality must be maintained at all times. The answer is clearly *no*. If the nuclear energy sources were at any time turned off, the ensuing energy imbalance would cause diminution of the internal energy of the star. That energy loss would start a slow contraction of the stellar structure. However, the virial theorem indicates that a contraction of the star may liberate in the form of radiant energy one-half of the increase in the gravitational energy. This energy source could keep the star burning for a rather long period of time. The point is that no cataclysmic consequences would occur from the sudden absence of an energy source in a star. On the contrary, the star's settling into configurations of smaller and smaller radius would be quite slow.

The analog of Eq. (3-3) when the star is slowly contracting can easily be written. It is a simple application of the first law of thermodynamics. The amount of heat that must be added per gram per second to stellar matter is equal to the rate of increase in internal energy plus the rate at which the mass of gas does expansion work:

$$\frac{dQ}{dt} = \frac{dU}{dt} + P\frac{dV}{dt} \tag{3-4}$$

This equation has units (cgs) of ergs per gram per second if V is the specific volume ($V = 1/\rho$). The rate of heat addition during quasistatic changes is given in terms of thermodynamic functions by $dQ/dt = T\,dS/dt$. Reconsideration of the spherical shell of Fig. 3-1 shows that

$$\frac{dQ}{dt} = \epsilon(r) - \frac{L(r+dr) - L(r)}{\rho 4\pi r^2 \, dr} \tag{3-5}$$

or

$$\frac{dL(r)}{dM(r)} = \epsilon - T\frac{dS}{dt} \tag{3-6}$$

The special case of the static structure is easily recovered by setting the time derivative equal to zero. On the other hand, if the star is expanding quasistatically, Eq. (3-6) gives the amount by which ϵ must exceed dL/dM. It will obviously be an important equation in the computation of stellar models during evolutionary change.

For example, if the gas is a simple monatomic gas without significant radiation pressure, Eq. (2-137) shows

$$\begin{aligned} S &= \frac{N_0 k}{\mu} \ln \frac{T^{3/2}}{\rho} + \text{const} \\ &= \frac{3}{2} \frac{N_0 k}{\mu} \ln \frac{P}{\rho^{5/3}} + \text{const} \end{aligned} \tag{3-7}$$

Then

$$\begin{aligned} T \frac{dS}{dt} &= T \frac{3}{2} \frac{N_0 k}{\mu} \frac{\rho^{5/3}}{P} \frac{d}{dt} \frac{P}{\rho^{5/3}} \\ &= \tfrac{3}{2} \rho^{2/3} \frac{d}{dt} \frac{P}{\rho^{5/3}} \end{aligned} \tag{3-8}$$

For such a gas, then, we have

$$\frac{dL}{dM} = \epsilon - \tfrac{3}{2} \rho^{2/3} \frac{d}{dt} \frac{P}{\rho^{5/3}} \tag{3-9}$$

Problem 3-1: Show that if radiation pressure is significant but the gas is nondegenerate,

$$\frac{dL}{dM} = \epsilon - \frac{3}{2} \frac{N_0 k T}{\mu} \frac{d}{dt} \ln \frac{e^{8y/3} y^{2/3}}{T} \tag{3-10}$$

where $y = (1 - \beta)/\beta$.

It must be emphasized that Eq. (3-6) does not determine the luminosity of a star. It simply shows what the luminosity must be for the rate of energy generation by nuclear reactions (coupled where necessary by the energy liberated from gravitational work) to balance the loss by radiation. The actual rate of flow of energy is determined by the mechanisms of the energy transport. Whether this energy transport is due to radiative transfer, convection, or even conduction, its magnitude is determined in some manner by the temperature gradient of the star. If, on the other hand, neutrinos are emitted by stellar matter, the power per gram of emitted neutrinos is entered in Eq. (3-6) as a negative ϵ. Thus we shall find that a star of prescribed structure automatically has a certain outflow of energy, regardless of whether that energy is simultaneously being replaced by nuclear reactions or not. In fact, since the nuclear reactions and neutrino emission are themselves dependent upon temperature, density, and composition, we can see that quite a delicate balance must be struck between the structure of the star, i.e., the temperature gradient, and the temperature and composition. It is just the establishment of this balance that is the determining factor of stellar structure.

3-2 RADIATIVE TRANSFER

It has been noted that there are basically three means by which the energy liberated in stellar interiors can be transferred to the surface to be radiated, viz., conduction, convection, and radiation. Of the three, conduction turns out to be the least important, at least in terms of the frequency of environments in which it is important. The bulk of all energy transport in stars occurs by the mechanisms of radiative transfer and convection. Let us now consider the first of these two mechanisms. The idea of radiative transfer is simply that photons emitted thermally in hot regions of a star and absorbed in cooler regions transfer energy from the hotter regions to the cooler regions. The effectiveness of the transport will be a function, among other things, of the temperature gradient and of the ability of photons to travel freely from one region of the star to another. It happens that photons in stars are able to travel distances of only about 1 cm or less before having some interaction with matter, so that it is only because of the small temperature differences that may exist over distances of the order of 1 cm that radiative transfer can occur at all. Since the thermal emission of matter is proportional to the temperature to the fourth power, we shall expect that, in general, the rate of energy transport will be proportional to the gradient of the fourth power of the temperature. Furthermore, the effectiveness of the transport will depend upon how opaque the gas is to the characteristic photons, i.e., how far they can travel before interacting with matter. The opacity of the stellar gas will be the subject of Sec. 3-3. It should be physically clear that the existence of a temperature gradient will be a necessary condition for radiative transfer of a thermal type, for if no temperature gradient exists, and if matter is in thermodynamic equilibrium, the density of photons of all frequencies is isotropic, and no net flux of radiation in any direction can occur.

The physical basis of the theory of radiative transfer in stars can easily be seen in the following simplified problem. Consider two plane semi-infinite blackbodies with a small temperature difference dT, as shown in Fig. 3-2. What is the heat transfer per unit area from the hotter plane to the cooler plane? The solution is quite simple. The emission per unit area from the surface of the blackbody is

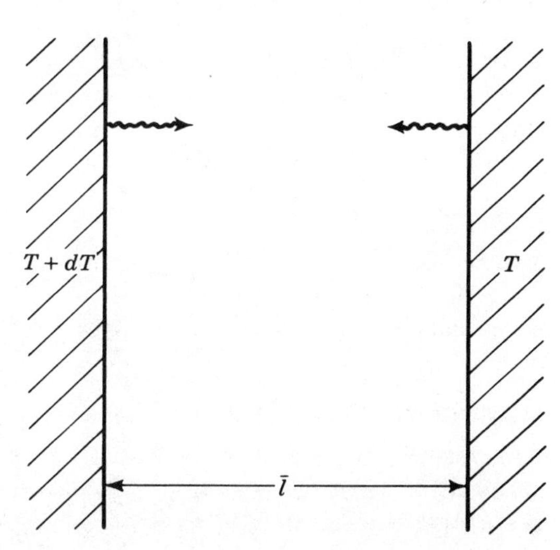

Fig. 3-2 A simplified model of radiative transfer in a plane stratified atmosphere is given by two plane blackbodies separated by the average distance a photon travels before absorption.

just equal to σT^4, whereas a blackbody absorbs all radiation that falls upon it. Thus, the net difference in the exiting flux over the entering flux per unit area is given by

$$H = \sigma[(T + dT)^4 - T^4] \approx 4\sigma T^3 \, dT \tag{3-11}$$

This simple expression determines the heat flow in a vacuum between two parallel blackbodies of temperature difference dT. The problem may seem oversimplified with regard to anything really occurring in the world, but a star can be thought of as just such stratified temperature layers with the temperature increasing toward the center of the star. The temperature difference between two such stratified blackbody layers will be given by the product of the temperature gradient and the characteristic distance photons travel before absorption; that is, $dT = (dT/dx)\bar{l}$, where $\bar{l}$ is something like an average mean free path of photons in stellar matter. Thus, the energy flux per unit area may be written as

$$H = -4\bar{l}\sigma T^3 \frac{dT}{dx} \tag{3-12}$$

where the minus sign indicates that the heat flux is opposite to the temperature gradient.

It is clear from the foregoing equation that the heat flux in a stellar interior will be determined by the temperature structure of the star, the only unknown remaining in Eq. (3-12) being the average distance $\bar{l}$ traveled by photons before their absorption by the stellar matter. This length is determined by the various ways in which photons interact with matter. The reduction in intensity of a beam of photons as it passes through matter of density ρ is conventionally written as

$$\frac{dI}{dx} = -\bar{\kappa}\rho I \tag{3-13}$$

where $\bar{\kappa}$ is the mass absorption coefficient and is a function of the photon frequency. The bar over κ indicates the necessity of some type of average over the photon-frequency spectrum. Over a limited range of distances for which $\bar{\kappa}$ and ρ may be a constant, Eq. (3-13) has an integral form:

$$I(x) \propto \exp -\bar{\kappa}\rho x \tag{3-14}$$

From this integral form it is apparent that $(\bar{\kappa}\rho)^{-1}$ is the approximate distance $\bar{l}$ that photons travel before absorption. Noting also that the emission constant for a blackbody is $\sigma = ca/4$, Eq. (3-12) becomes approximately

$$H \approx -\frac{ac}{\bar{\kappa}\rho} T^3 \frac{dT}{dx} \tag{3-15}$$

This result, obtained by a simple argument, is correct to factors of order unity (there is an error of a factor $\frac{4}{3}$). It also does not show how the average mass absorption coefficient is to be determined. We shall return to these problems during a detailed discussion of the sources of the opacity to photons in a stellar interior. What this simple example does show is that in the presence of a temperature gradient in a gas that is approximately in thermal equilibrium, the flux of energy is proportional to the gradient T^4 and to $(\bar{\kappa}\rho)^{-1}$.

In Chap. 2 we introduced the intensity of the radiation field I, the net radiation flux per unit area H, and the radiation pressure P_r. The radiation field intensity may be decomposed into an integral over the specific intensity in the frequency interval $d\nu$ such that

$$I(\theta) = \int_0^\infty I_\nu(\theta)\, d\nu \tag{3-16}$$

It will be necessary to consider each frequency separately, because the number density of the photons and the energy of each photon as well as the absorption coefficient of stellar matter for those photons are all quantities that depend upon the photon frequency. We may then define the mass absorption coefficient for frequency ν in the following way. Think of a monochromatic pencil of radiation[1] moving in some specified direction. The specific intensity in this small solid angle will be altered by absorption from its value I_ν to a value $I_\nu + dI_\nu$ upon passing through a thickness ds of matter of density ρ:

$$dI_\nu \equiv -\kappa_\nu \rho I_\nu\, ds \tag{3-17}$$

The quantity κ_ν is called the *mass absorption coefficient of frequency* ν and is defined by this equation. The coefficient κ_ν will, of course, be a function of the frequency and of the thermodynamic state and chemical composition of the material. Equation (3-17) represents the total energy flux removed from the pencil by interaction with matter in moving through the thickness ds. Accordingly, it includes energy lost both by processes of true absorption and by processes of scattering. It is clear, however, that absorption and scattering are physically different processes, and we shall have to differentiate between them throughout the following discussion. By *true absorption* we shall mean interactions of photons with matter whereby the energy of the photon is converted into some other form of energy, e.g., the ionization of an atom. By *scattering* we mean simply those processes by which the direction of the photon is changed. In the deep interiors of stars, where matter is almost completely ionized, the bulk of scattering occurs from Thompson scattering by free electrons, whereas in cool stellar atmospheres the bulk of the scattering may be due to scattering from molecules. Therefore, when thinking of this reduction of the specific intensity we must not assume that the decrease in photons corresponds to photons that have disappeared. Quite the contrary; it will often happen that the energy lost from the pencil will reappear as scattered radiation moving in other directions. The true-absorption process corresponds to incident photons that disappear from the radiation field.

[1] *Pencil of radiation* is a term used to designate a set of directions contained in a differential of solid angle $d\Omega$ centered about a specific direction, in this case to be defined by the angle θ to the radial direction.

We shall differentiate between the true absorption and the scattering absorption by designating the corresponding opacities with different subscripts. Specifically, we shall write for true absorption

$$dI_\nu = -\kappa_{\nu a}\rho I_\nu \, ds \tag{3-18}$$

and for scattering absorption

$$dI_\nu = -\kappa_{\nu s}\rho I_\nu \, ds \int_{\Omega'} p(\cos\theta') \frac{d\Omega'}{4\pi} \tag{3-19}$$

In this last equation we have introduced the *scattering phase function* $p(\cos\theta')$ which gives the angular distribution of the scattered energy removed from the beam, the angle θ' being defined as the angle of the scattered radiation relative to the direction of the incident pencil. Furthermore, the phase function is normalized such that

$$\int p(\cos\theta') \frac{d\Omega'}{4\pi} = 1 \tag{3-20}$$

Thus, the total reduction in the pencil is the sum of the reductions due to true absorption and due to scattering. Accordingly,

$$dI_\nu = -(\kappa_{\nu a} + \kappa_{\nu s})\rho I_\nu \, ds \tag{3-21}$$

The scattering phase function appears at first glance to be somewhat of a needless complexity. Such is not the case, and once its very simple meaning is appreciated, no great difficulties will be met in applying it to most of the problems dealing with stellar interiors. For instance, if the scattering is isotropic, $p(\cos\theta') = 1$. In that case the energy removed from the beam by scattering is equally redistributed to all solid angles $d\Omega'$. In fact, for the *removal* of specific intensity from the beam, the scattering phase function *is* a needless complexity, since the integral of the phase function over all solid angles into which scattering may occur is defined to be unity. It is for the computation of the amount of energy scattered *into* the beam from incident pencils in other directions that the scattering phase function will be needed. For scattering from most atomic particles and from free electrons in stellar interiors, the angular distribution is that of Rayleigh scattering, which is proportional to $1 + \cos^2\theta'$.

Problem 3-2: Show that the normalization of the phase function for Rayleigh scattering is

$$p(\cos\theta) = \tfrac{3}{4}(1 + \cos^2\theta) \tag{3-22}$$

Of course, radiation is not only absorbed in stellar interiors; it is also emitted by the material in question. The equation that describes radiative transfer computes the balance between true absorption, scattering, and true emission. The sources of emission are divided into true emission and scattered emission, in analogy to the division of absorption into true absorption and scattering. We define $j_\nu(\theta)$ to be the *true emission* of radiation of frequency ν per unit frequency interval in the direction θ per unit solid angle from each gram of stellar matter. Thus, the specific intensity in the pencil of radiation at angle θ will be increased by true emission in moving through a distance ds by the amount

$$dI_\nu(\theta) = +j_\nu(\theta)\rho \, ds \tag{3-23}$$

In the general theory of radiative transfer, this emission term may be quite troublesome, but in stellar interiors it is greatly simplified by the assumption of local thermodynamic equilibrium. This assumption is that matter radiates spontaneously in stellar interiors exactly as it would in thermodynamic equilibrium; i.e., the stellar matter is approximately described as having a local temperature T, and the assumption of local thermodynamic equilibrium is that matter at temperature T radiates exactly as it would if it were in surroundings in thermodynamic equilibrium at the same temperature. This assumption should be accurate enough for our purpose. In Prob. 2-20 it was demonstrated that the deviation from local thermodynamic equilibrium for the radiation field in stellar interiors is of the order 10^{-10}; that is, the second term in Eq. (2-81), which represents the anisotropy in the radiation field, is of the order 10^{-10} of the first term. Interior stellar matter is therefore extremely close to the state of thermodynamic equilibrium. The temperature gradient is very small, furthermore, so the fact that photons travel only about 1 cm before interaction implies that they are absorbed at essentially the same temperature at which they were emitted. In thermodynamic equilibrium the emission per gram is very simple, for what is then required is that the material radiate exactly the same power that it absorbs and with the same spectrum of frequencies, viz., the spectrum appropriate to thermodynamic equilibrium. We can easily see what this requirement implies for the true-emission coefficient j_ν.

Consider a thin slab of stellar matter having unit cross-sectional area and thickness dx, as illustrated in Fig. 3-3. What is the rate of absorption of energy

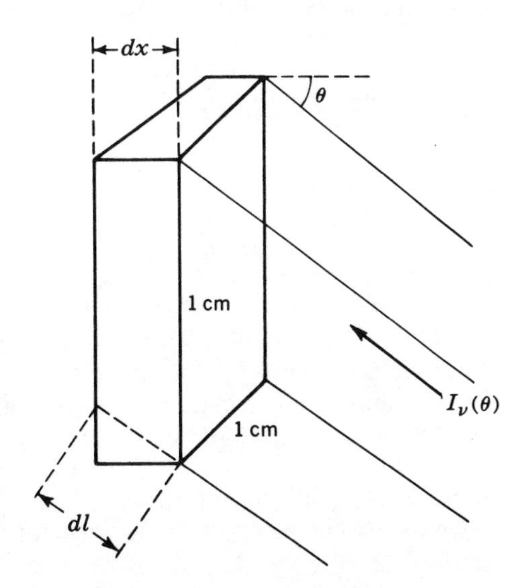

Fig. 3-3 A thin slab of unit area and thickness dx absorbs energy from the radiation field $I_\nu(\theta)$. In thermodynamic equilibrium the slab must reradiate the same power.

from the radiation field by this mass element? From Eq. (3-18) we may write that the energy absorbed in 1 sec from the radiation field moving in a direction θ to the normal to the unit slab per unit frequency interval about ν is given by

$$dE_\nu(\theta) = -\kappa_{\nu a}\rho \, dl \, I_\nu(\theta) \cos\theta \tag{3-24}$$

The factor $I_\nu(\theta) \cos\theta$ is equal to the specific intensity impinging upon a unit area at an inclination angle θ. The absorbing path length in the slab is $dl = dx/(\cos\theta)$. Then the true absorption of energy at frequency ν is given by the integral over all solid angles:[1]

$$dE_\nu = -\kappa_{\nu a}\rho \, dx \int I_\nu(\theta) \, d\Omega = \kappa_{\nu a}\rho \, dx \, cu_\nu \tag{3-25}$$

Since the mass contained in the slab of unit area is just $\rho \, dx$, the power truly absorbed per unit frequency interval per unit mass is just equal to

$$\frac{dE_\nu}{dm} = -\kappa_{\nu a} c u_\nu \tag{3-26}$$

The emission from the slab must have exactly the same value, or deviations from local thermodynamic equilibrium can be shown to occur; i.e., the material will heat up, cool off, or change the spectrum in such a way that some cycle can be constructed for doing work from the radiation field. This balance may be represented by demanding that

$$4\pi j_\nu = \kappa_{\nu a} c u_\nu \tag{3-27}$$

where the factor of 4π takes account of the fact that the true-emission coefficient j_ν is defined per unit solid angle and that in local thermodynamic equilibrium the true emission is isotropic. The combination of quantities $cu_\nu/4\pi$ *in thermodynamic equilibrium* is given the special symbol $B_\nu(T)$ to distinguish it from the quantity $cu_\nu/4\pi$ for those cases when the real energy density is *not* that of local thermodynamic equilibrium:

$$j_\nu = \kappa_{\nu a} \frac{cu_\nu}{4\pi} = \kappa_{\nu a} B_\nu(T) \tag{3-28}$$

where

$$B_\nu(T) = \frac{2h\nu^3}{c^2} \frac{1}{\exp(h\nu/kT) - 1} \tag{3-29}$$

$B_\nu(T)$ is called the *source function* for true emission in local thermodynamic equilibrium. This balance between absorption and emission in local thermodynamic equilibrium is called *Kirchhoff's law*.

One must be cautious to avoid making a mistake in extending properties of radiation computed for conditions of local thermodynamic equilibrium to the

[1] The meaning of u_ν in this equation is that of the energy density of radiation of frequency ν per unit frequency interval. We have used Eq. (2-74) for the specific frequency ν; that is, $\int I_\nu(\theta) \, d\Omega = cu_\nu$.

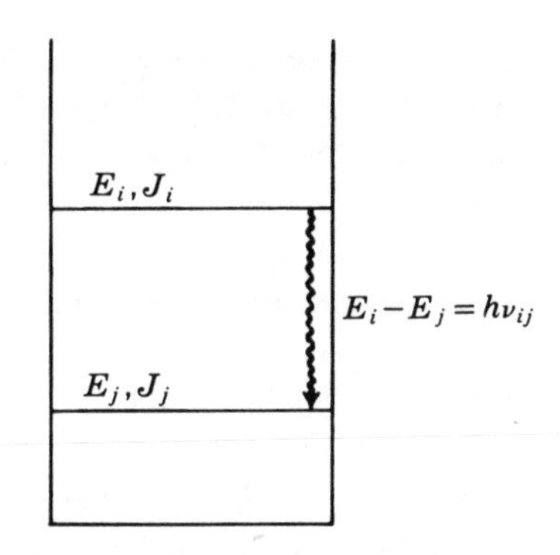

Fig. 3-4 Radiative transitions between two states of an atom are of three types, spontaneous and induced decays from the upper state to the lower state and absorption by the lower state. The transition matrix element is the same for each transition, so that the ratios of their rates are determined entirely by statistical weights and by the density of photons at energy ν_{ij}.

slightly anisotropic radiation field appropriate in stellar interiors, where, strictly speaking, thermodynamic equilibrium does not exist. Part of the emission contained in the equilibrium statement of Kirchhoff's law is spontaneous emission resulting solely from material temperature, and part of the emission is induced emission. The induced emission comes from atomic transitions *caused by* the radiation field. These transitions are exactly those which were considered in the discussion of the Boltzmann equation in Chap. 1, where it was pointed out that there exists a probability measured by the Einstein coefficient B_{ij} for the radiation field to cause downward transitions in atomic structure. Induced emission is a type of *sympathetic reaction,* and it produces radiation of the same frequency moving in the same pencil of direction as the incident radiation. It is correct to think that the incident radiation causes the state to make a downward transition faster than it would have if it were left alone.

To see how induced emission should be treated, return to the two states of an atom shown in Fig. 3-4. The emission resulting from radiative transitions of state i to state j in thermodynamic equilibrium has two terms. The first is a spontaneous transition rate, which is, per unit solid angle, equal to $(1/4\pi)N_iA_{ij}$. The second is the induced decay rate caused by encounters of atoms in state i with photons of frequency ν_{ij}, which can cause a second photon to be emitted in the same direction, the rate of which per unit solid angle is equal to $(1/4\pi)N_iB_{ij}u_{\nu ij}$. We can compute from these two expressions the *fraction* of the total emission that is spontaneous:

$$\frac{\text{Spontaneous}}{\text{Total}} = \frac{N_iA_{ij}}{N_iA_{ij} + N_iB_{ij}u_{\nu_{ij}}} = \frac{1}{1 + (B_{ij}/A_{ij})u_{\nu_{ij}}} \tag{3-30}$$

From the ratio of Einstein coefficients it follows immediately that

$$\frac{\text{Spontaneous emission}}{\text{Total emission}} = 1 - e^{-h\nu/kT} \tag{3-31}$$

The quantum theory of radiation requires that this ratio of spontaneous to total emission in local thermodynamic equilibrium apply to any mechanism of true absorption. There are mechanisms other than transitions between discrete atomic states that represent true emission and absorption. For instance, ionization is an example of true absorption, whereas its inverse process, radiative recombination, is an example of true emission. Quantum-mechanically it can be shown that the radiative recombination of an ion and an electron with the emission of a photon of frequency ν_{ij} can be induced to occur by the presence of a second photon of frequency ν_{ij}. In this manner one reaches a ratio of spontaneous emission for radiative recombination to total emission by radiative recombination that is the same as Eq. (3-31).

If deviations from strict thermodynamic equilibrium occur, the emission must be separated into two terms. The spontaneous emission is still determined by the temperature of the matter and the source function $B_\nu(T)$, whereas the induced emission is proportional to the actual specific intensity of radiation $I_\nu(\theta)$, instead of to the isotropic function $B_\nu(T)$. Quantitatively, this statement can be expressed by

$$j_\nu(\theta) = \kappa_{\nu a}(1 - e^{-h\nu/kT})B_\nu(T) + \kappa_{\nu a}e^{-h\nu/kT}I_\nu(\theta) \tag{3-32}$$

It is easy to see that in thermodynamic equilibrium this expression for the emission reduces to Kirchhoff's law, since in that case the specific intensity $I_\nu(\theta)$ is equal to $B_\nu(T)$.

There is yet another term that must appear in the emitted power, the energy introduced into the pencil by scattered photons. From the definition of the scattering phase function, one can immediately write down that the energy scattered per unit solid angle into the pencil moving in the direction θ, ϕ from a pencil moving in the direction θ', ϕ' is given by

$$j_{\nu,\mathrm{scattered}} = \kappa_{\nu s}\frac{1}{4\pi}\int^{\pi}\int^{2\pi} p(\theta,\phi;\theta',\phi')I_\nu(\theta',\phi') \sin\theta'\, d\theta'\, d\phi' \tag{3-33}$$

where $p(\theta,\phi;\theta',\phi')$ is the scattering phase function corresponding to the angle between the direction of the pencil under consideration (θ,ϕ) and any other pencil (θ',ϕ'). The geometry is shown in Fig. 3-5.

At this point the energy balance that determines the equation of transfer can be established. Consider a small pencil of directions contained in a solid angle $d\Omega$ about an angle θ relative to the direction of the temperature gradient (direction of net heat flow). Further consider a small circular cylinder having unit cross section and length dl coaxial with the pencil of radiation under consideration, as shown in Fig. 3-6. It is now a simple matter to write a statement of the conservation of energy for the radiation flowing in the chosen pencil of directions. With regard to the cylinder constructed in Fig. 3-6, we demand that the radiation leaving the top of the cylinder equal the sum of the radiation that enters the bottom of the cylinder minus the absorption in the cylinder plus the emission in the cylinder, all within the chosen pencil of directions.

The power per unit area exiting from the top of the cylinder is given by

$$\left(\frac{dQ}{dt}\right)_{\mathrm{top}} = -I_\nu(r + dr, \theta)\, d\Omega \tag{3-34}$$

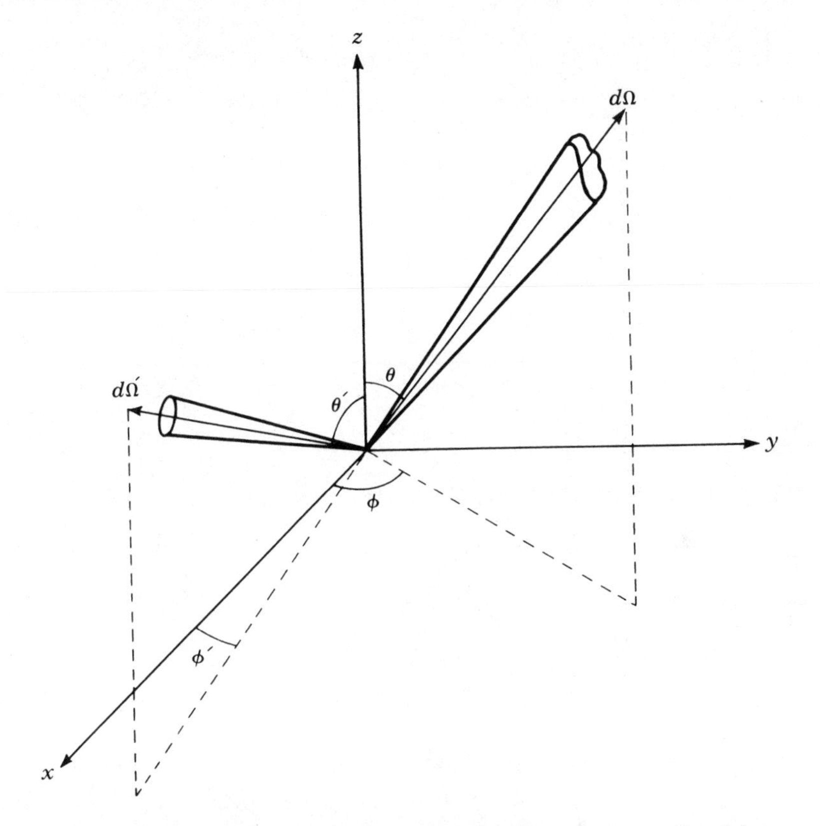

Fig. 3-5 The geometry of photon scattering. Photons are scattered into the solid angle $d\Omega$ from all other pencils designated by $d\Omega'$. The z axis is regarded as being the direction of the temperature gradient so that the radiation field has azimuthal symmetry.

whereas the power per unit area entering the bottom is given by

$$\left(\frac{dQ}{dt}\right)_{\text{bot}} = +I_\nu(r,\theta)\,d\Omega \tag{3-35}$$

The power per unit area absorbed during transit of the cylinder is given by Eq. (3-21) as

$$\left(\frac{dQ}{dt}\right)_{\text{abs}} = -(\kappa_{\nu a} + \kappa_{\nu s})\rho I_\nu(r,\theta)\,dl\,d\Omega \tag{3-36}$$

The power per unit area emitted into the pencil of direction inside the cylinder is a sum of three terms, the spontaneous emission, the induced emission, and the scattered emission. From our earlier discussion, it is evident that the total emis-

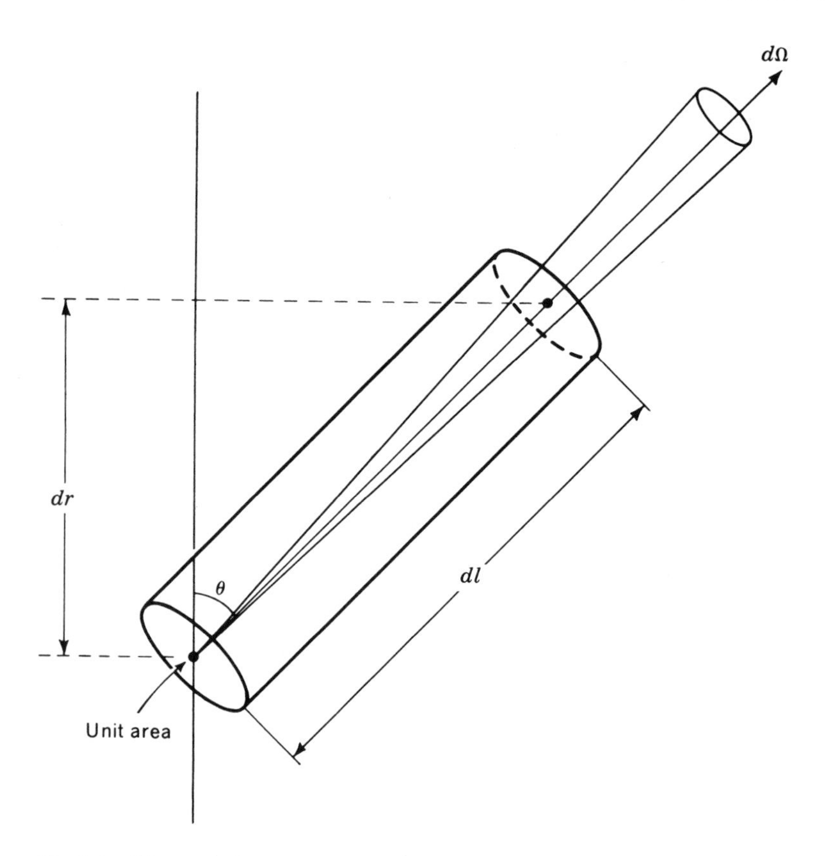

Fig. 3-6 The equation of radiative transfer is obtained by applying the conservation of energy to photons absorbed and emitted within this cylinder and within the pencil of directions indicated by $d\Omega$.

sion is just the sum of Eqs. (3-32) and (3-33):

$$\left(\frac{dQ}{dt}\right)_{\text{em}} = +\kappa_{\nu a}\rho\, dl\,[(1 - e^{-h\nu/kT})B_\nu(T) + e^{-h\nu/kT}I_\nu(r,\theta)]\,d\Omega$$
$$+ \kappa_{\nu s}\rho\, dl\,\frac{d\Omega}{4\pi}\int^{\pi}\int^{2\pi} p(\theta,\phi;\theta',\phi')I_\nu(r,\theta',\phi')\sin\theta'\,d\theta'\,d\phi \quad (3\text{-}37)$$

By summing these various contributions with the proper algebraic signs, the statement of conservation of energy may be written

$$I_\nu(r + dr,\,\theta) - I_\nu(r,\theta) = \frac{\partial I_\nu}{\partial r}\,dr$$
$$= -(\kappa_{\nu a} + \kappa_{\nu s})I_\nu(r,\theta)\rho\, dl + \kappa_{\nu a}(1 - e^{-h\nu/kT})B_\nu(T)\rho\, dl$$
$$+ \kappa_{\nu a}e^{-h\nu/kT}I_\nu(r,\theta)\rho\, dl$$
$$+ \kappa_{\nu s}\frac{\rho\, dl}{4\pi}\int_{\Omega'} p(\theta\phi;\theta'\phi')I_\nu(r,\theta',\phi')\,d\Omega' \quad (3\text{-}38)$$

Dividing Eq. (3-38) through by the total mass $\rho\, dl$ of the cylinder and regrouping terms gives

$$\begin{aligned}\frac{1}{\rho\, dl}\frac{\partial I_\nu}{\partial r}\,dr &= \frac{1}{\rho}\frac{\partial I_\nu}{\partial r}\cos\theta \\ &= -(\kappa_{\nu a}^* + \kappa_{\nu s})I_\nu(r,\theta) + \kappa_{\nu a}^* B_\nu(T) \\ &\qquad + \kappa_{\nu s}\frac{1}{4\pi}\int_{\Omega'} p(\theta\phi;\theta'\phi')I_\nu(r,\theta',\phi')\,d\Omega' \end{aligned} \tag{3-39}$$

where we have used the fact that $dr/dl = \cos\theta$, and where we have defined a *reduced absorption coefficient* to be

$$\kappa_{\nu a}^* = \kappa_{\nu a}(1 - e^{-h\nu/kT}) \tag{3-40}$$

Equation (3-39) is the desired equation of radiative transfer for a plane parallel atmosphere under conditions of local thermodynamic equilibrium. What is desired for the computation of the structure of the star, however, is an equation of transfer that gives the energy flow per unit area directly in terms of an *average* opacity and the temperature gradient. This result may be achieved by multiplying the equation of transfer by $\cos\theta$ and integrating over all solid angles $d\Omega$. The various terms in the integral become

$$(i)\quad \int \frac{1}{\rho}\frac{\partial I_\nu}{\partial r}\cos^2\theta\,d\Omega = \frac{1}{\rho}\frac{\partial}{\partial r}\int I_\nu\cos^2\theta\,d\Omega = \frac{c}{\rho}\frac{\partial P_\nu}{\partial r} \tag{3-41}$$

where we have used the spectral form of Eq. (2-77),

$$(ii)\quad \int(\kappa_{\nu a}^* + \kappa_{\nu s})I_\nu\cos\theta\,d\Omega = -(\kappa_{\nu a}^* + \kappa_{\nu s})H_\nu \tag{3-42}$$

Also from Eq. (2-77),

$$(iii)\quad \int\kappa_{\nu a}^* B_\nu(T)\cos\theta\,d\Omega = 0 \tag{3-43}$$

since $B_\nu(T)$ is isotropic,

$$(iv)\quad \kappa_{\nu s}\frac{1}{4\pi}\int_\Omega\int_{\Omega'}\cos\theta\, p(\theta\phi;\theta'\phi')I_\nu(r,\theta',\phi')\,d\Omega\,d\Omega' \tag{3-44}$$

The value of this integral depends upon the form of the scattering phase function. However, it is not difficult to see that the integral vanishes if p contains only *even powers* of the cosine of the angle between the scattering beams. Ignore for a moment the integration over $d\Omega'$ and consider first the integration over $d\Omega$. Pick any incident pencil defined by θ', ϕ' and hold that pencil fixed. Then the integral over $d\Omega$ may be thought of as sums of pairs of pencils taken in the direction of θ and in the direction opposite to θ. For these two pencils, p has the same value if it involves only even powers of $\cos\theta$, whereas the factor $\cos\theta$ itself occurring in the integrand has equal but opposite values for the two pencils. Hence the sum of the two terms vanishes, as does therefore the integral. The remaining discussion will be limited to scattering phase functions having that property.

The result of these operations on the equation of transfer leaves then a very simple equation, which is

$$\frac{c}{\rho}\frac{\partial P_\nu}{\partial r} = -(\kappa_{\nu a}^* + \kappa_{\nu s})H_\nu(r) \tag{3-45}$$

It was shown in Chap. 2 that even in the case of a slightly anisotropic radiation field the radiation pressure $P_\nu = \frac{1}{3}u_\nu$ and that the energy density u_ν is still that of thermodynamic equilibrium. What is desired then is simply the total heat flux per unit area, which is

$$H = \int_0^\infty H_\nu \, d\nu = -\frac{c}{3\rho}\int_0^\infty \frac{1}{\kappa_{\nu a}^* + \kappa_{\nu s}}\frac{du_\nu}{dr}\, d\nu \tag{3-46}$$

Since the energy density u_ν is a function only of the temperature in the approximation of local thermodynamic equilibrium,

$$\frac{du_\nu}{dr} = \frac{du_\nu}{dT}\frac{dT}{dr} \tag{3-47}$$

Hence,

$$H = -\frac{c}{3\rho}\int_0^\infty \frac{1}{\kappa_{\nu a}^* + \kappa_{\nu s}}\frac{du_\nu}{dT}\frac{dT}{dr}\, d\nu \tag{3-48}$$

The temperature gradient can, of course, be withdrawn from the integral over frequencies. The equation will also have a nicer symmetry if it is multiplied and divided by the integral

$$\int_0^\infty \frac{du_\nu}{dT}\, d\nu = \frac{d}{dT}\int u_\nu \, d\nu = \frac{du}{dT} = 4aT^3 \tag{3-49}$$

for then the equation for the heat flow becomes

$$H = -\frac{4ac}{3\rho}T^3\frac{dT}{dr}\frac{\displaystyle\int_0^\infty \frac{1}{\kappa_{\nu a}^* + \kappa_{\nu s}}\frac{du_\nu}{dT}\, d\nu}{\displaystyle\int_0^\infty \frac{du_\nu}{dT}\, d\nu} \tag{3-50}$$

Since the function u_ν in thermodynamic equilibrium differs from the Planck source function $B_\nu(T)$ only by a constant, it is conventional to replace the ratio involving u_ν in Eq. (3-50) by a similar ratio involving the function $B_\nu(T)$.

It is apparent that the ratio of integrals in Eq. (3-50) is simply a normalized average of the inverse opacity, the average to be taken in a specified way over the radiation spectrum. Following Rosseland, we define the *Rosseland mean opacity* κ by

$$\frac{1}{\kappa} = \frac{\displaystyle\int_0^\infty \frac{1}{\kappa_{\nu a}[1 - \exp(-h\nu/kT)] + \kappa_{\nu s}}\frac{dB_\nu}{dT}\, d\nu}{\displaystyle\int_0^\infty \frac{dB_\nu}{dT}\, d\nu} \tag{3-51}$$

With this definition of κ, the radiative heat flux becomes

$$H = \frac{4ac}{3\kappa\rho} T^3 \frac{dT}{dr} \tag{3-52}$$

which is just what we argued it should be before we started. The lengthy calculation has discovered the factor $\frac{4}{3}$, which was not obvious in the beginning, and has also demonstrated how the average opacity is to be determined from the spectral opacity.

The Rosseland mean opacity has several interesting features. It is an average over frequency

$$\frac{1}{\kappa} = \left\langle \frac{1}{\kappa_{\nu a}[1 - \exp(-h\nu/kT)] + \kappa_{\nu s}} \right\rangle \tag{3-53}$$

the average being taken with respect to the weighting factor

$$\frac{dB_\nu}{dT} = \frac{2h^2\nu^4}{c^2kT^2} \frac{\exp(h\nu/kT)}{[\exp(h\nu/kT) - 1]^2} \tag{3-54}$$

From Eq. (3-53) it is apparent that the sources of radiative opacity are essentially additive except for the fact that the true-absorption coefficients are reduced by the factor $1 - \exp(-h\nu/kT)$ to correct for induced emission. This correction factor is clearly such as to devalue low-energy ($h\nu < kT$) absorption. The weighting function dB_ν/dT, on the other hand, has the following physical significance: the photon frequencies most important for radiative transfer are those for which the difference in the product of photon number density times photon energy between two points of slightly different temperature is maximal. (This statement ignores, however, the possible frequency dependence of the opacity.) The factor dB_ν/dT devalues the opacity at very low and at very high frequencies. Thus the Rosseland mean is a specific compromise between those frequencies for which the opacity is the greatest and those frequencies for which the greatest number of effective photons exist.

Problem 3-3: Show that the weighting factor dB_ν/dT is such as to place the greatest weight on the opacity near $\nu = 4kT/h$.

Equation (3-52) gives the heat flow per unit area by radiative transfer. To obtain the net outflow of energy through a shell of radius r, we need only multiply by $4\pi r^2$. Thus

$$L(r) = -4\pi r^2 \frac{4ac}{3\kappa\rho} T^3 \frac{dT}{dr} \tag{3-55}$$

This last equation, along with the definition of the mean opacity, represents one of the basic equations of stellar structure (whenever radiative transfer is the dominant mode of energy transport).

Several interesting integral properties of stars in radiative equilibrium can be demonstrated. Equation (3-55) for the radiative energy flow can be rearranged as an expression for the gradient of the radiation pressure:

$$\frac{dP_r}{dr} = -\frac{\kappa\rho}{4\pi c r^2} L(r) \tag{3-56}$$

If the star is, in addition, static, then

$$\frac{dL(r)}{dr} = 4\pi r^2 \rho \epsilon \tag{3-57}$$

and

$$\frac{dP}{dr} = -\frac{GM(r)}{r^2}\rho \tag{3-58}$$

Division of Eq. (3-56) by Eq. (3-58) yields

$$\frac{dP_r}{dP} = \frac{\kappa L(r)}{4\pi c G M(r)} \tag{3-59}$$

Define the quantity $\eta(r)$ to be a ratio of the average rate of nuclear energy generation interior to point r to the corresponding average for the whole star:

$$\eta(r) \equiv \frac{\bar{\epsilon}_r}{\epsilon} = \frac{L/M}{\mathfrak{L}/\mathfrak{M}} \tag{3-60}$$

where $\mathfrak{L}$ and $\mathfrak{M}$ are the luminosity and mass of the star. In terms of η, Eq. (3-59) becomes

$$\frac{dP_r}{dP} = \frac{\mathfrak{L}}{4\pi c G \mathfrak{M}} \kappa\eta \tag{3.61}$$

The integral of this equation from an interior point r to the surface is

$$P_r(r) = \frac{\mathfrak{L}}{4\pi c G \mathfrak{M}} \int_0^{P(r)} \kappa(r)\eta(r)\, dP = \frac{\mathfrak{L}}{4\pi c G \mathfrak{M}} P(r)\overline{\kappa\eta} \tag{3-62}$$

where $\overline{\kappa\eta}$ is the *pressure average* of $\kappa\eta$ between $r = r$ and $r = R$.

Equation (3-62) is the proof of a theorem due to Strömgren: *the ratio of the radiation pressure to the total pressure at a point inside a star in radiative equilibrium is proportional to the average value of $\kappa\eta$ for the regions exterior to the point r, the average being taken with respect to dP, where P is the total pressure.* A particular case obtained from evaluation at the center is

$$\mathfrak{L} = \frac{4\pi c G \mathfrak{M}(1 - \beta_c)}{\overline{\kappa\eta}} \tag{3-63}$$

where $\overline{\kappa\eta}$ is pressure averaged over the entire star and β_c is the ratio of gas pressure to total pressure at the stellar center. Chandrasekhar has termed Eq. (3-63) the *luminosity formula* for stars in radiative equilibrium.

A second particular case of interest occurs if $\kappa\eta$ is a constant, in which case $\overline{\kappa\eta(r)}$ is also a constant, whereupon, from Eq. (3-62), $\beta(r)$ must then be constant. In the discussion of polytropes in the last chapter, we pointed out that constant β

yields the nondegenerate polytrope of index 3, the standard model. Thus the constancy of β, in stars in radiative equilibrium, depends on the constancy of $\kappa\eta$. From the discussions that will follow later in this chapter it will be apparent that κ increases by several orders of magnitude from the center to the surface, whereas η decreases by several orders of magnitude from its central value $\eta_c = \epsilon_c/\bar{\epsilon}$ to its value at the surface $\eta(R) = 1$. This compensating feature accounts for the fortuitously good resemblance of the standard model to real stars.[1]

Problem 3-4: What is the luminosity of a standard model of $1M_\odot$ in which $\kappa\eta = 1$ (cgs) and $\mu = \frac{1}{2}$. Recall that $\beta = \beta(M)$.

The application of crude scaling arguments to Eq. (3-55) can also provide some insight into the mass-luminosity relationship for main-sequence stars. Assume for the sake of argument that the values of stellar opacity are relatively insensitive to the mass of the star. In Chap. 2 the central temperature of the standard model was found to scale with mass as

$$T_c \propto \mathfrak{M}^{\frac{2}{3}}\bar{\rho}^{\frac{1}{3}}$$

whereas $\bar{\rho} \propto \mathfrak{M}/R^3$. Thus the characteristic internal temperatures scale as $T_c \propto \mathfrak{M}/R$. Assuming also that the characteristic temperature gradients scale as $dT/dr \propto T_c/R$, the luminosity would scale according to Eq. (3-55) as

$$\mathfrak{L} \propto R^2 \frac{(\mathfrak{M}/R)^3}{\mathfrak{M}/R^3} \frac{\mathfrak{M}/R}{R} = \mathfrak{M}^3 \tag{3-64}$$

This result is not unlike the observed facts for main-sequence stars. The conclusion that the luminosity should be proportional to the third (or so) power of the mass will soon be indicated by another point of view, the temperature dependence of thermonuclear reaction rates.

3-3 OPACITY OF STELLAR MATTER

Evaluation of the rate of energy transport by radiative transfer requires calculation of the Rosseland mean opacity, but before the mean opacity can be calculated, one needs detailed knowledge of the atomic absorption cross sections as a function of the photon frequency. All the processes which impede the free motion of a photon must be calculated and added together. We shall find that the opacity of the gas depends rather strongly on both the composition of the gas and its thermodynamic state. Thus the Rosseland mean opacity is a function of state, both by virtue of the average over the Planck spectrum and by virtue of its dependence upon the state of matter, and one can write for each composition a function $\kappa = \kappa(\rho,T)$.

Photons interact with matter in a large number of ways. Most of these inter-

[1] This and other theorems on radiative equilibrium may be found in S. Chandrasekhar, "An Introduction to the Study of Stellar Structure," chap. 6, Dover Publications, Inc., New York, 1957.

actions have been exhaustively studied, both in the laboratory and by the theoretical techniques of quantum electrodynamics, and the agreement between experiment and theory is essentially perfect, being limited only by experimental uncertainty or by the approximations necessary in making some calculations. Fortunately it turns out that from regions just beneath the photospheres of stars all the way to their centers the radiative opacity is almost entirely due to four basic types of event. With very little loss of generality, therefore, we may restrict ourselves to an introductory discussion of those four processes, which may be described as follows:

(*1*) *Bound-bound absorption:* This is the absorption of a photon by an atom during which a bound electron makes a transition to a bound state of higher energy. It is a true-absorption process, and its inverse is the normal emission of light by atoms accompanying downward transitions.
(*2*) *Bound-free absorption:* This is the absorption of a photon by an atom during which a bound electron makes a transition to a continuum state, also called *photoionization*. It is a true-absorption process, and its inverse is radiative recombination.
(*3*) *Free-free absorption:* This is the absorption of a photon by a continuum electron as it passes an ion and makes a transition to another continuum state of higher energy. It is a true-absorption process, and its inverse is called *bremsstrahlung*.
(*4*) *Scattering from free electrons:* This is the scattering of photons by individual free electrons in the gas, commonly called *Compton scattering*, or in the nonrelativistic approximation usually applicable in stars, *Thomson scattering*. It is not true absorption, inasmuch as the scattered photon energy equals the incident energy.

Calculation of the cross section for these processes involves the quantum mechanics of atomic transitions. For most astrophysical energies the electromagnetic field can be treated as a classical perturbation on a quantum-mechanical atomic system, although in a complete theoretical treatment the electromagnetic field is also quantized. The technique employed to calculate the transition rates is called *time-dependent perturbation theory*.

The quantum theory of the opacity of matter depends upon the interactions of charged particles and electromagnetic waves such that energy is either absorbed from the electromagnetic wave or emitted as an electromagnetic wave. In the absorption or emission process the charged particles must change their state. We shall introduce the physical ideas within the framework of the quantum mechanics of atomic transitions, an outline of which follows.[1]

[1] The treatment is standard. As a guide we follow the discussion in E. Merzbacher, "Quantum Mechanics," chaps. 19 and 20, John Wiley & Sons, Inc., New York, 1961. The point of the following material is not so much to provide a discourse on quantum mechanics as it is to illustrate how the opacity problem fits within that framework.

When the hamiltonian is independent of time, the development in time of the particle wave function is relatively straightforward. The Schrödinger equation

$$i\hbar \frac{d\psi(t)}{dt} = H\psi(t) \tag{3-65}$$

is first solved for the orthonormal energy eigenstates given by

$$H\psi_n = E_n\psi_n \tag{3-66}$$

The wave function is then expressible (by completeness) as a linear superposition of energy eigenstates

$$\psi(t) = \sum_n c_n \exp\left(-\frac{i}{\hbar} E_n t\right) \psi_n \tag{3-67}$$

which clearly satisfies Eq. (3-65). This wave function can be matched to an initial state $\psi(t_0)$ by choosing the coefficients as

$$c_n = \langle\psi_n|\psi(t_0)\rangle \exp\left(\frac{i}{\hbar} E_n t_0\right) \tag{3-68}$$

where the bracket represents the overlap integral of the initial state with the energy eigenfunction. It can be seen that if the particles are initially in one of the energy eigenstates, all other values of c_n are zero for all time: the state is stationary.

In characteristic opacity problems, the stationary states of charged particles are altered by a momentary perturbation, a passing electromagnetic pulse, for instance, or the time-varying electric field produced by the near passage of a charged particle. In such cases, the total hamiltonian for the particles being considered is not constant in time, and there are no stationary states. But if the perturbation is limited in space and time, the system may be regarded as being in a stationary state before the perturbation appears and in a stationary state after the perturbation has vanished. If the final state differs from the initial state, it is said that the perturbation has produced a *transition*. This sequence of events suggests that the total hamiltonian be considered as a sum,

$$H = H_0 + V(t) \tag{3-69}$$

where H_0 is the time-independent operator and $V(t)$ is the time-dependent perturbation. (The treatment is still correct, and very useful, even if V is independent of the time and represents only a small perturbation to an easily solvable hamiltonian.) Both before and after the transition the system is regarded as being approximately in a stationary state of H_0. If the perturbation V were in fact absent, the eigenfunctions would be given by the equation

$$H_0\psi_n^{(0)} = E_n^{(0)}\psi_n^{(0)} \tag{3-70}$$

where the zero superscripts designate the quantities as being the zero-order approximations relevant to the unperturbed hamiltonian. In the absence of perturbation, the development of the wave function in time would again be as in Eq. (3-67). The presence of the perturbation renders that treatment incorrect, but it will be legitimate to expand $\psi(t)$ in terms of the unperturbed eigenfunctions if the coefficients are now regarded as functions of time:

$$\psi(t) = \sum_n c_n(t) \exp\left(-\frac{i}{\hbar} E_n^{(0)} t\right) \psi_n^{(0)} \tag{3-71}$$

The coefficients

$$c_n(t) = \langle \psi_n^{(0)} | \psi(t) \rangle \exp\left(\frac{i}{\hbar} E_n^{(0)} t\right) \tag{3-72}$$

are the probability amplitudes for finding the system in the nth unperturbed state.

Problem 3-5: By substituting Eq. (3-71) into the equation of motion and using the orthonormality of eigenstates, show that the coefficients c_n change in time according to

$$i\hbar \frac{dc_k}{dt} = \sum V_{kn} c_n e^{i\omega_{kn} t} \tag{3-73}$$

where $\hbar\omega_{kn} \equiv E_k^{(0)} - E_n^{(0)}$ and $V_{kn} \equiv \langle \psi_k^{(0)} | V | \psi_n^{(0)} \rangle$. V_{kn} is called the *matrix element* of the perturbation between unperturbed eigenstates k and n. The notation used here is a standard shorthand:

$$\langle \psi_k | V | \psi_s \rangle \equiv \int_{\text{space}} \psi_k^*(x,y,z) V(x,y,z,t) \psi_s(x,y,z)\, dx\, dy\, dz$$

Solution of Eq. (3-73) will delineate the way in which the state of the system changes. As each c_n changes, the probability that the system is in the unperturbed eigenstate $\psi_n^{(0)}$ changes as $|c_n(t)|^2$.

Suppose for simplicity that the system at the initial time $t_0 = -\infty$ is an eigenstate s of the unperturbed hamiltonian. Suppose furthermore that H_0 possesses only discrete energy levels. The treatment can easily be extended to continuous eigenstates later. Then the initial conditions are described by

$$c_s(-\infty) = 1 \qquad c_k(-\infty) = 0 \qquad \text{for } k \neq s \tag{3-74}$$

If the perturbation is weak enough and acts for a time short enough for the system to have only a small probability of admixing other states k into the initial state s, then the approximation

$$c_s(t) \approx 1 \gg c_k(t) \tag{3-75}$$

can be made. Then from Eq. (3-73) the coefficient $c_k(t)$ is given approximately by

$$i\hbar \frac{dc_k}{dt} = V_{ks} e^{i\omega_{ks} t} \tag{3-76}$$

If the perturbation is of short duration and sufficiently small in magnitude, each amplitude c_k will remain small throughout the passage of the pulse, and this fact has been used in writing Eq. (3-76). Because $V_{ks}(t)$ is a transient, the integral converges to a definite value:

$$c_k(+\infty) = -\frac{i}{\hbar}\int_{-\infty}^{+\infty} V_{ks}e^{i\omega_{ks}t}\,dt \tag{3-77}$$

The probability that a transition to state k has occurred is equal to $|c_k(+\infty)|^2$. It is proportional to the absolute square of the Fourier component of the perturbation matrix element V_{ks} evaluated at the transition frequency ω_{ks}.

As an example of these formulas and their application to the opacity problem, consider an atom in a radiation field. To simplify the problem, electron spin will be neglected, and it will be assumed that only one electron participates in the interaction with the radiation. In the absence of radiation, the hamiltonian of the electron is

$$H_0 = \frac{p^2}{2m} + V_c \tag{3-78}$$

where V_c is the coulomb interaction of the electron with the ion to which it is bound. The perturbation of the atom is due to an external electromagnetic field, which can be described by a scalar potential ϕ plus a vector potential $\mathbf{A}$. When this perturbation is added, the total hamiltonian becomes[1]

$$H = \frac{[\mathbf{p} + (e/c)\mathbf{A}]^2}{2m} + V_c - e\phi \tag{3-79}$$

Because the electromagnetic field has no sources near the atom, the field can be described in terms of $\mathbf{A}$ alone:

$$\phi = 0 \qquad \nabla \cdot \mathbf{A} = 0 \qquad \nabla^2\mathbf{A} - \frac{1}{c^2}\frac{\partial^2\mathbf{A}}{\partial t^2} = 0 \tag{3-80}$$

Problem 3-6: Using the fact that in coordinate representation the electron's momentum operator is $\mathbf{p} = (h/i)\nabla$, show that the hamiltonian is

$$H = H_0 + \frac{e}{mc}\mathbf{A}\cdot\mathbf{p} + \frac{e^2}{2mc^2}A^2 \tag{3-81}$$

In first-order interactions, the final term gives a much smaller effect than the second term, and so the perturbing potential for atoms is $V = (e/mc)\mathbf{A}\cdot\mathbf{p}$. It turns out that the term in A^2 is important for scattering of electromagnetic waves from free electrons, however, so that one should not forget its presence in case the $\mathbf{A}\cdot\mathbf{p}$ term gives no first-order transition.

An arbitrary pulse of radiation can be written as a Fourier superposition of harmonic plane waves

$$\mathbf{A}(\mathbf{r},t) = \int_{-\infty}^{\infty} \mathbf{A}(\omega)\exp\left[-i\omega\left(t - \frac{\mathbf{n}\cdot\mathbf{r}}{c}\right)\right]d\omega \tag{3-82}$$

[1] This expression is also the classical one. See H. Goldstein, "Classical Mechanics," p. 222, Addison-Wesley Publishing Company, Inc., Reading, Mass., 1953. The electron charge is $q_e \equiv -e$.

where $\mathbf{n}$ is the unit vector in the propagation direction and Fourier component $\mathbf{A}(\omega)$ must satisfy $\mathbf{A}^*(\omega) = \mathbf{A}(-\omega)$ in order that $\mathbf{A}$ be real. The condition $\nabla \cdot \mathbf{A} = 0$ further requires the wave to be transverse: $\mathbf{n} \cdot \mathbf{A}(\omega) = 0$.

The perturbation V is then written

$$V = \frac{e}{mc} \int_{-\infty}^{\infty} \exp \left[-i\omega \left(t - \frac{\mathbf{n} \cdot \mathbf{r}}{c} \right) \right] \mathbf{A}(\omega) \cdot \mathbf{p} \, d\omega \tag{3-83}$$

The matrix element V_{ks} between two states whose wave functions are symbolized by the bra $\langle k|$ and the ket $|s\rangle$ is

$$V_{ks} = \frac{e}{mc} \int_{-\infty}^{\infty} \langle k| \exp \left(i \frac{\omega}{c} \mathbf{n} \cdot \mathbf{r} \right) \mathbf{p} |s\rangle \cdot e^{-i\omega t} \mathbf{A}(\omega) \, d\omega \tag{3-84}$$

giving the transition amplitude

$$c_k(+\infty) = -\frac{ie}{\hbar mc} \iint_{-\infty}^{\infty} \langle k| \exp \left(i \frac{\omega}{c} \mathbf{n} \cdot \mathbf{r} \right) \mathbf{p} |s\rangle \cdot \mathbf{A}(\omega) e^{i(\omega_{ks} - \omega)t} \, dt \, d\omega \tag{3-85}$$

The integral $\int_{-\infty}^{\infty} e^{i(\omega_{ks} - \omega)t} \, dt$ is equivalent to $2\pi\delta(\omega_{ks} - \omega)$, where $\delta(\omega_{ks} - \omega)$ is the so-called *delta function*,[1] defined such that

$$\int_{-\infty}^{\infty} f(\omega) \delta(\omega_{ks} - \omega) \, d\omega = f(\omega_{ks}) \tag{3-86}$$

It follows that

$$c_k(+\infty) = -\frac{2\pi ie}{\hbar mc} \langle k| \exp \left(i \frac{\omega_{ks}}{c} \mathbf{n} \cdot \mathbf{r} \right) \mathbf{p} |s\rangle \cdot \mathbf{A}(\omega_{ks}) \tag{3-87}$$

The result shows that only the radiation field at frequency ω_{ks} contributes to the absorption. This is the frequency condition originally postulated by Bohr, viz., that the photons absorbed in making an atomic excitation between states k and s must have frequency $\hbar\omega_{ks} = E_k - E_s$.

This treatment has corresponded to the absorption of photons of frequency ω_{ks} in the excitation process from s to k, but the same pulse can also cause downward transitions from k to s provided atoms exist in the state k. This process corresponds to stimulated emission. Because the energy of the transition is then $\omega_{sk} = -\omega_{ks}$, and because the perturbation hamiltonian is a *hermitian operator*,[2] which means that $\int \psi_k^* V \psi_s = \int (V\psi_k)^* \psi_s$, it is immediately demonstrable that the transition amplitude from k to s is

$$c_{k \to s} = c_{s \to k}^*$$

which means that the two transition probabilities are equal. This property is known as *detailed balancing*. The excess energy $\hbar\omega_{sk}$ appears as an added photon of energy $\hbar\omega_{sk}$ in the radiation field. When the electron spin and its coupling to form the total angular momentum of the atomic states are taken into account,

[1] See Merzbacher, *op. cit.*, p. 80.

[2] See *ibid.*, p. 141.

along with the different number of substates in the initial and final levels of an atom, one obtains the ratio of two of the Einstein coefficients described in Eq. (1-33):

$$\frac{B_{ks}}{B_{sk}} = \frac{2J_s + 1}{2J_k + 1} \tag{3-88}$$

If we let the electromagnetic pulse be plane polarized with the direction of polarization $\mathbf{e}$, then $\mathbf{A}(\omega) = A(\omega)\mathbf{e}$, and the transition probability is

$$|c_k(+\infty)|^2 = \frac{4\pi^2 e^2}{\hbar^2 m^2 c^2} |A(\omega_{ks})|^2 \left| \langle k| \exp\left(i\frac{\omega_{ks}}{c}\mathbf{n}\cdot\mathbf{r}\right)\mathbf{p}\cdot\mathbf{e}|s\rangle \right|^2 \tag{3-89}$$

The stellar opacity problem deals with the concept of the absorption cross section, which enters into the mass absorption coefficient. From the photon point of view, the cross section is given by the ratio of the number of photons in the small frequency interval $d\omega$ absorbed per atom to the total number of photons in the interval $d\omega$ per unit area to which the atoms were exposed. The discussion of the radiation field has been presented in a classical way, however, so that the equivalent operation is to compare the average energy absorbed per atom from the pulse to the total energy of the pulse, again within a small frequency interval. From the classical field point of view, the energy flux in the pulse is given by Poynting's vector, which is in gaussian cgs units,

$$\mathbf{S} = \frac{c}{4\pi}\mathbf{E}\times\mathbf{H} \tag{3-90}$$

The field vectors $\mathbf{E}$ and $\mathbf{H}$ for the electromagnetic waves being discussed are derivable from the vector potential according to

$$\mathbf{E} = -\frac{1}{c}\frac{\partial\mathbf{A}}{\partial t} \qquad \mathbf{H} = \nabla\times\mathbf{A} \tag{3-91}$$

By employing these relationships along with the representation of the radiation field used in Eq. (3-82), the energy flux is

$$\begin{aligned}\mathbf{S} &= \frac{c}{4\pi}\mathbf{E}\times\mathbf{H} = \frac{c}{4\pi}\left(-\frac{1}{c}\frac{\partial\mathbf{A}}{\partial t}\right)\times(\nabla\times\mathbf{A}) \\ &= -\frac{1}{4\pi}\left\{\int_{-\infty}^{\infty} -i\omega A(\omega)\mathbf{e}\exp\left[-i\omega\left(t-\frac{\mathbf{n}\cdot\mathbf{r}}{c}\right)\right]d\omega\right\} \\ &\qquad \left\{\frac{1}{c}\int_{-\infty}^{\infty} i\omega' A(\omega')(\mathbf{n}\times\mathbf{e})\exp\left[-i\omega'\left(t-\frac{\mathbf{n}\cdot\mathbf{r}}{c}\right)\right]d\omega'\right\}\end{aligned} \tag{3-92}$$

Because the wave is transverse ($\mathbf{e}\cdot\mathbf{n} = 0$), the triple product is $\mathbf{e}\times(\mathbf{n}\times\mathbf{e}) = \mathbf{n}$, giving

$$\mathbf{S} = -\frac{\mathbf{n}}{4\pi c}\iint_{-\infty}^{\infty} \omega\omega' A(\omega)A(\omega'')\exp\left[-i(\omega+\omega')\left(t-\frac{\mathbf{n}\cdot\mathbf{r}}{c}\right)\right]d\omega\,d\omega' \tag{3-93}$$

The energy carried by the pulse through unit area is given by the time integral of the flux as the pulse passes by

$$E = \int_{-\infty}^{\infty} \mathbf{S}(t) \cdot \mathbf{n}\, dt \tag{3-94}$$

This time integral again yields the delta function $2\pi\delta(\omega + \omega')$, and since $A^*(\omega) = A(-\omega)$, it follows that

$$E = \frac{1}{2c}\int_{-\infty}^{\infty} \omega^2|A(\omega)|^2\, d\omega = \frac{1}{c}\int_0^{\infty} \omega^2|A(\omega)|^2\, d\omega = \int_0^{\infty} E(\omega)\, d\omega \tag{3-95}$$

Thus the energy per unit frequency interval in the pulse is $E(\omega) = (\omega^2/c)|A(\omega)|^2$, and that value can be inserted in Eq. (3-89) when forming the energy absorbed from the pulse:

$$\hbar\omega_{ks}|c_k(+\infty)|^2 = \frac{4\pi^2 e^2 E(\omega_{ks})}{\hbar m^2 c\omega_{ks}} \left| \langle k| \exp\left(i\frac{\omega_{ks}}{c}\mathbf{n}\cdot\mathbf{r}\right)\mathbf{p}\cdot\mathbf{e}|s\rangle \right|^2 \tag{3-96}$$

The absorption cross section at frequency ω is defined as

$$\sigma(\omega) = \frac{\text{energy absorbed/initial state } s \text{ at frequency } \omega}{\text{incident energy/unit area at frequency } \omega} \tag{3-97}$$

where both numerator and denominator are measured in the same infinitesimal frequency interval. Apparently Eq. (3-96) is now in that form. The absorption events are basically of three types: (1) the bound-bound absorption, characterized by line transitions between discrete states; (2) the bound-free absorption events, where the final electron state lies in the continuum; and (3) the free-free absorption events, where both the final and initial electron states are in the continuum. The type of process determines the wave functions used for the states represented by $\langle k|$ and $|s\rangle$.

(1) BOUND-BOUND ABSORPTION

If the two states are discrete, the absorbed energy is at the unique frequency ω_{ks}. From Eq. (3-96) the cross section must be (designating $e^2/\hbar c \equiv \alpha = 1/137.04$)

$$\sigma(\omega) = \frac{4\pi^2\alpha}{m^2\omega_{ks}} \left| \langle k| \exp\left(i\frac{\omega_{ks}}{c}\mathbf{n}\cdot\mathbf{r}\right)\mathbf{p}\cdot\mathbf{e}|s\rangle \right|^2 \delta(\omega - \omega_{ks}) \tag{3-98}$$

such that the integral $\int E(\omega)\sigma(\omega)\, d\omega$ gives the energy absorbed from the pulse. The form of this cross section is such that it is infinite for an infinitesimally narrow range of frequencies at ω_{ks}, as represented by the delta function. Actual atomic lines are not infinitely sharp in energy because the initial and final electron states are not. All excited states have finite lifetimes against decay. Because time and energy are conjugate variables, a localization in time must introduce an uncertainty in energy in the spirit of the uncertainty principle. Since the lifetime τ represents the uncertainty in time with which the state can be localized, the uncertainty in energy Γ must satisfy $\tau\Gamma = \hbar$. When account is taken of the finite width, the analog of Eq. (3-98) must be

$$\int_{\Delta\omega} \sigma(\omega)\, d\omega = \frac{4\pi^2\alpha}{m^2\omega_{ks}} \left| \langle k| \exp\left(i \frac{\omega_{ks}}{c} \mathbf{n} \cdot \mathbf{r}\right) \mathbf{p} \cdot \mathbf{e} |s\rangle \right|^2 \tag{3-99}$$

where $\Delta\omega$ is a small frequency band that contains the line shape, that is, $\hbar\,\Delta\omega$ is of the order Γ, and it is assumed that the spectral energy $E(\omega)$ remains roughly constant at $E(\omega_{ks})$ over the width of the line. Equation (3-99) measures the area under the cross-sectional curve. The shape of the cross section itself is not given, although it has a peak at $\omega = \omega_{ks}$ and a width of order Γ. The idealized line shape can be seen from the following argument, however.

When placed in a quasistationary state $\psi_k^{(0)}$, the wave function will not only change its phase in time, as for a stationary state, but will also decay exponentially with its mean life τ, giving a wave function of the form

$$\psi_k(t) \approx \exp\left(-\frac{t}{2\tau}\right) \psi_k^{(0)} \exp\left(-\frac{i}{\hbar} E_k^{(0)} t\right) \qquad \text{for } t > 0 \tag{3-100}$$

such that $\int |\psi_k|^2\, dV = \exp(-t/\tau)$. But because the energy operator is

$$E_{\text{op}} = -\frac{\hbar}{i} \frac{\partial}{\partial t}$$

the probability that the state has energy E is given by a Fourier decomposition into energy eigenfunctions:

$$\psi(t) = \int_{-\infty}^{\infty} \phi(E) \exp\left(-\frac{i}{\hbar} Et\right) dE \tag{3-101}$$

where $|\phi(E)|^2$ measures the probability that the state has energy E. By the Fourier integral theory $\phi(E)$ must then be proportional to

$$\phi(E) \propto \int_0^{\infty} \exp\left[\frac{i}{\hbar}(E - E_k^{(0)})t - \frac{t}{2\tau}\right] dt \tag{3-102}$$

The probability that the state has energy E must be unity when summed over all energies.

Problem 3-7: Show that the normalized probability distribution for the state energy is

$$P(E)\, dE = \frac{\hbar}{2\pi\tau} \frac{dE}{(E - E_k^{(0)})^2 + (\hbar/2\tau)^2} \tag{3-103}$$

It is evident that the probability has fallen to one-half maximum at the energy $E_{\frac{1}{2}} = E_k^{(0)} \pm \hbar/2\tau$, giving a full width at half maximum equal to $\Gamma = \hbar/\tau$, as inferred from the uncertainty principle. Although it is off the present subject, we note that the same state profile will be important for nuclear states when considering the thermonuclear reaction rates (Chap. 4). For present purposes we note from this result that the line-absorption cross section takes the form

$$\sigma(\omega) = \frac{4\pi^2\alpha}{m^2\omega_{ks}} \left| \langle k| \exp\left(i \frac{\omega_{ks}}{c} \mathbf{n} \cdot \mathbf{r}\right) \mathbf{p} \cdot \mathbf{e} |s\rangle \right|^2 \frac{\Gamma/2\pi\hbar}{(\omega - \omega_{ks})^2 + (\Gamma/2\hbar)^2} \tag{3-104}$$

This cross section displays the characteristic Lorentz resonance shape observed in absorption lines. In much of the astrophysical literature the cross sections are discussed in terms of so-called *oscillator strengths* of the transition. For a linear harmonic oscillator with direction of motion parallel to the polarization **e**, the integrated cross section is equal to[1]

$$\int_{\Delta\omega} \sigma(\omega)\, d\omega = \frac{2\pi^2 e^2}{mc} = \frac{2\pi^2 \hbar \alpha}{m} \tag{3-105}$$

Actual lines are measured against this standard by multiplying the oscillator cross section by a factor f, the oscillator strength, which represents how strongly the line absorbs. Specifically

$$\sigma(\omega) \equiv \frac{2\pi^2 e^2}{mc} f_{ks} \frac{\Gamma/2\pi\hbar}{(\omega - \omega_{ks})^2 + (\Gamma/2\hbar)^2} \tag{3-106}$$

Evidently the oscillator strength is contained in the matrix element of Eq. (3-104):

$$f_{ks} = \frac{2}{\hbar m \omega_{ks}} \left| \langle k| \exp\left(i \frac{\omega_{ks}}{c} \mathbf{n} \cdot \mathbf{r}\right) \mathbf{p} \cdot \mathbf{e} |s\rangle \right|^2 \tag{3-107}$$

The evaluation of the importance of atomic lines in the radiative transfer–opacity problem depends upon the extent to which lines block out the spectrum. This in turn depends upon the oscillator strengths of the lines, the density of the lines in frequency, and the widths of the lines. With regard to the last point it will be noted only that the lines are much broader than their laboratory counterparts, because all atomic states are broadened by the state of the plasma—Doppler broadening, Stark broadening, and collisional broadening. Other perturbations caused by the charged-particle gas shorten the lifetimes of states and therefore increase their widths. Several different attempts have been made to estimate the line widths in the ionized gas. The subject is difficult but fortunately does not appear to be extremely important to the interior opacity problem, where the number of absorbing lines is usually great enough for the overall opacity to be relatively insensitive to the uncertainties in line widths and shapes.[2]

Quantitative application of these results depends upon the simplification and evaluation of the matrix element. The commonest and most important simplification results from the observation that the wavelength of the incident light is usually long compared to the dimensions of the absorber. That is,

$$\lambda = \frac{2\pi c}{\omega_{ks}} \gg \bar{r} \tag{3-108}$$

where $\bar{r}$ represents some average dimension over which the interaction occurs. In that case the exponential operator in the matrix element may advantageously

[1] See *ibid.*, p. 454, for the quantum treatment, or W. Panofsky and M. Phillips, "Classical Electricity and Magnetism," eq. 21-59, Addison-Wesley Publishing Company, Inc., Reading, Mass., 1955, for the classical treatment, which gives the same result.

[2] A. N. Cox, J. N. Stewart, and D. D. Eilers, *Astrophys. J. Suppl.*, **11**:1 (1965).

be expanded,

$$\exp\left(i\frac{\omega_{ks}}{c}\mathbf{n}\cdot\mathbf{r}\right) \approx 1 + i\frac{\omega_{ks}}{c}\mathbf{n}\cdot\mathbf{r} + \cdots \tag{3-109}$$

and because each term is successively smaller, the matrix element may be approximated by the first nonvanishing term in the corresponding series of matrix elements

$$\langle k|\mathbf{p}\cdot\mathbf{e}|s\rangle + \langle k|i\frac{\omega_{ks}}{c}(\mathbf{n}\cdot\mathbf{r})\mathbf{p}\cdot\mathbf{e}|s\rangle + \cdots \tag{3-110}$$

The rapidity of this convergence can be seen by noting from hydrogenlike atoms that $\hbar\omega_{ks}$ must be of the order of the binding Z^2e^2/a_0, whereas $\bar{r}$ must be of the order a_0/Z. Thus

$$\frac{\omega_{ks}\bar{r}}{c} = \frac{Z^2e^2}{\hbar c a_0}\frac{a_0}{Z} = Z\frac{e^2}{\hbar c} = \alpha Z = \frac{Z}{137} \tag{3-111}$$

is a measure of the ratio of the intrinsic size of succeeding terms in the expansion. The first term of the expansion gives, as we shall see, the electric-dipole matrix element (commonly called *allowed* since these are the strongest lines except when the matrix element vanishes), whereas the second term gives the electric-quadrupole and magnetic-dipole matrix elements (commonly called *forbidden* since they are generally important only if the electric-dipole transition is forbidden by the vanishing of its matrix element). Only the electric-dipole transitions need be considered in the stellar opacity problem, although forbidden transitions have very important astrophysical applications in low-density gases such as gaseous nebulas. Higher-moment transitions are strong in *nuclear transitions* because characteristic nuclear energies are 10^6 times as great as atomic energies, and although the nuclear size is about 10^{-4} of the atomic size, the ratio $\omega\bar{r}/c$ is no longer extremely small. In fact, the adjective *forbidden* is not even used in discussion of nuclear gamma rays.

What must be evaluated in the electric-dipole transitions is then the component of the matrix element $\langle k|\mathbf{p}|s\rangle$ in the direction of the photon polarization $\mathbf{e}$. Although this analysis can be carried through with aid of the momentum operator itself, the common practice is to further simplify the problem with the aid of a formal quantum-mechanical result. Suppose

$$H_0 = \frac{p^2}{2m} + V(x,y,z) = -\frac{\hbar^2}{2m}\left(\frac{\partial^2}{dx^2} + \frac{\partial^2}{\partial y^2} + \frac{\partial^2}{\partial z^2}\right) + V(x,y,z) \tag{3-112}$$

as in the case of an atom. Then the *commutator* of any position coordinate with the unperturbed hamiltonian is

$$H_0x - xH_0 = -\frac{\hbar^2}{2m}\left(\frac{\partial^2}{\partial x^2}x - x\frac{\partial^2}{\partial x^2}\right) = -2\frac{\hbar^2}{2m}\frac{\partial}{\partial x} \tag{3-113}$$

Problem 3-8: Confirm Eq. (3-113). Keep in mind that this commutator is an operator on some other function of position.

Evidently

$$H_0\mathbf{r} - \mathbf{r}H_0 = -\frac{i\hbar}{m}\mathbf{p} \tag{3-114}$$

With this operator result, the matrix element can be transformed to

$$\langle k|\mathbf{p}|s\rangle = \frac{im}{\hbar}\langle k|H_0\mathbf{r} - \mathbf{r}H_0|s\rangle \tag{3-115}$$

By the hermitian property of H_0 the first term yields the energy eigenvalue of state $\langle k|$, and the second yields the energy eigenvalue of state $|s\rangle$:

$$\begin{aligned}\langle k|\mathbf{p}|s\rangle &= \frac{im}{\hbar}(E_k^{(0)} - E_s^{(0)})\langle k|\mathbf{r}|s\rangle \\ &= im\omega_{ks}\langle k|\mathbf{r}|s\rangle\end{aligned} \tag{3-116}$$

With the aid of this result, the integrated cross section in Eq. (3-99) becomes

$$\int_{\Delta\omega}\sigma(\omega)\,d\omega = 4\pi^2\alpha\omega_{ks}|\langle k|\mathbf{r}\cdot\mathbf{e}|s\rangle|^2 \tag{3-117}$$

and the oscillator strength becomes

$$f_{ks} = \frac{2m\omega_{ks}}{\hbar}|\langle k|\mathbf{r}\cdot\mathbf{e}|s\rangle|^2 \tag{3-118}$$

In the interior of a star the directions and polarizations of the photons are isotropic. It is accordingly appropriate to average the product $\mathbf{r}\cdot\mathbf{e}$ over the directions:

$$\overline{|\langle k|\mathbf{r}\cdot\mathbf{e}|s\rangle|^2} = \tfrac{1}{3}|\langle k|\mathbf{r}|s\rangle|^2 \tag{3-119}$$

For the problem of bound-bound absorption the states k and s are spatially represented by wave functions. If electron spin is ignored, as is justified for electric-dipole absorption, and if only the electron making the transition is considered, then the functions

$$\psi_k = R_{klm}(r)Y_l^m(\theta,\phi) \tag{3-120}$$

are the eigenfunctions of the bound-state Schrödinger equation[1]

$$-\frac{\hbar^2}{2m}\nabla^2\psi_k + V(r)\psi_k = E_k\psi_k \tag{3-121}$$

The spherical harmonics[2] for $m \geq 0$,

$$Y_l^m(\theta,\phi) = \left[\frac{2l+1}{4\pi}\frac{(l-m)!}{(l+m)!}\right]^{\frac{1}{2}}(-1)^m e^{im\varphi}P_l^m(\theta,\phi) \tag{3-122}$$

are orthonormal in that

$$\int_0^{2\pi}\int_0^{\pi} Y_l^{m*}(\theta,\phi)Y_{l'}^{m'}(\theta,\phi)\sin\theta\,d\theta\,d\phi = \delta_{l,l'}\delta_{m,m'} \tag{3-123}$$

[1] See Merzbacher, *op. cit.*, chap. 9.

[2] For $m < 0$ we have $Y_l^m = (-1)^m Y_l^{-m*}$.

and the radial functions are normalized such that

$$\int_0^\infty |R_{klm}(r)|^2 r^2\,dr = 1 \tag{3-124}$$

For the special case of a hydrogenlike atom, $V = -Ze^2/r$, and the functions R_{klm} become the well-known Laguerre polynomials of the hydrogen atom. For more complex ions the effective potential must be found by an approximate self-consistent technique, from which the function R can be calculated by numeric techniques and tabulated. For the moment we may ignore the form of $R(r)$ and obtain the *selection rules* for electric-dipole radiation.

Problem 3-9: Show that the matrix element may be written

$$\begin{aligned}|\langle k|\mathbf{r}|s\rangle|^2 &= \tfrac{1}{2}|\langle k|x+iy|s\rangle|^2 + \tfrac{1}{2}|\langle k|x-iy|s\rangle|^2 + |\langle k|z|s\rangle|^2 \\ &= \tfrac{1}{2}|\langle k|r\sin\theta\, e^{i\phi}|s\rangle|^2 + \tfrac{1}{2}|\langle k|r\sin\theta\, e^{-i\phi}|s\rangle|^2 + |\langle k|r\cos\theta|s\rangle|^2\end{aligned}$$

With the results of the preceding problem, the three matrix elements are

$$\begin{aligned}\langle n'l'm'|x \pm iy|nlm\rangle &= R_{nl}^{n'l'} \int Y_{l'}^{m'*} \sin\theta\, e^{\pm i\phi} Y_l^m\, d\Omega \\ \langle n'l'm'|z|nlm\rangle &= R_{nl}^{n'l'} \int Y_l^{m'*} \cos\theta\, Y_l^m\, d\Omega\end{aligned} \tag{3-125}$$

where

$$R_{nl}^{n'l'} \equiv \int_0^\infty R_{n'l'}(r) R_{nl}(r) r^3\,dr \tag{3-126}$$

The spherical harmonics have the property that the angular integrations vanish and no electric-dipole transitions occur unless

$$l - l' = \pm 1 \qquad m - m' = 0, \pm 1 \tag{3-127}$$

These are the familiar angular-momentum selection rules for electric-dipole transitions. If $l' = l + 1$,

$$\begin{aligned}|\langle n'l', m \pm 1|\mathbf{r}|nlm\rangle|^2 &= \frac{1}{2}\frac{(l \pm m + 1)(l \pm m + 2)}{(2l+1)(2l+3)} (R_{nl}^{n'l'})^2 \\ |\langle n'l'm|\mathbf{r}|nlm\rangle|^2 &= \frac{(l + m + 1)(l - m + 1)}{(2l+1)(2l+3)} (R_{nl}^{n'l'})^2\end{aligned} \tag{3-128a}$$

and if $l' = l - 1$,

$$\begin{aligned}|\langle n'l', m \pm 1|\mathbf{r}|nlm\rangle|^2 &= \frac{1}{2}\frac{(l \mp m)(l \mp m - 1)}{(2l+1)(2l-1)} (R_{nl}^{n'l'})^2 \\ |\langle n'l', m|\mathbf{r}|nlm\rangle|^2 &= \frac{(l + m)(l - m)}{(2l+1)(2l-1)}\end{aligned} \tag{3-128b}$$

The total transition rate from a given state nlm to all the substates of the level $n'l'$ is obtained by summing over the allowed values of m'. The result for

$l' = l + 1$ is

$$\sum_{m'} |\langle n'l'm'|\mathbf{r}|nlm\rangle|^2 = \frac{l+1}{2l+1} (R_{nl}^{n'l'})^2 \tag{3-129}$$

and for $l' = l - 1$

$$\sum_{m'} |\langle n'l'm'|\mathbf{r}|nlm\rangle|^2 = \frac{l}{2l+1} (R_{nl}^{n'l'})^2 \tag{3-130}$$

These matrix elements yield the oscillator strengths and the cross sections. It is worth noting that the transition rate is independent of the m quantum number of the particular initial substate.

Problem 3-10: Let the upper of two atomic states have angular momentum l' and the lower have angular momentum l. Show that the ratio of the absorption rate per atomic state l to the induced-emission rate per atomic state l' is given by

$$\frac{B_{ll'}}{B_{l'l}} = \frac{2l'+1}{2l+1}$$

As an example, consider the Lyα absorption cross section, i.e., from the 1s to the 2p level of hydrogen. We have in this case

$$f_{ks} = \frac{2m\omega_{ks}}{\hbar} |\langle k|\mathbf{r}\cdot\mathbf{e}|s\rangle|^2 = \frac{2m\omega_{ks}}{3\hbar} |\langle k|\mathbf{r}|s\rangle|^2$$

$$= \frac{2m\omega_{ks}}{3\hbar} (R_{10}^{21})^2 \tag{3-131}$$

inasmuch as $l' = l + 1$ and $l = 0$ in Eq. (3-129). For this transition

$$\hbar\omega_{ks} = 1 - \frac{1}{2^2}\text{ Ry} = \frac{3}{4}\frac{e^2}{2a_0}$$

where a_0 is the Bohr radius ($a_0 = \hbar^2/me^2$).

Problem 3-11: Show that the oscillator strength is a dimensionless number.

Problem 3-12: The normalized radial wave functions for the 1s and 2p levels of hydrogen are

$$R_{1s} = 2a_0^{-\frac{3}{2}}e^{-r/a_0}$$

$$R_{2p} = \frac{1}{2\sqrt{6}} a_0^{-\frac{5}{2}} r e^{-r/2a_0}$$

Calculate the square of the radial matrix element[1] $(R_{10}^{21})^2$.
Ans: $1.66a_0^2$.

[1] The radial matrix elements for hydrogen may be found in E. U. Condon and G. H. Shortley, "The Theory of Atomic Spectra," p. 133, Cambridge University Press, New York, in a slightly different notation.

Problem 3-13: Calculate the oscillator strength for Lyα.
Ans: $f = 0.42$.

Problem 3-14: The lifetime of the $2p$ state is given by $\hbar/\Gamma = 0.16 \times 10^{-8}$ sec. Calculate the *peak cross section* at the Lyα resonance.
Ans: 7.2×10^{-11} cm^2.

Although this example is the simplest of all atomic absorption cross sections, it has illustrated the major features of the physics.

(2) BOUND-FREE ABSORPTION

In the ionization process the absorption of a photon by an ion leads to the emission of an electron into a continuum state. The wave functions ψ_k and ψ_s must accordingly correspond to those for a continuum state and a bound state, respectively. The conservation of energy requires the photon energy to be equal to the sum of the electron binding energy plus its kinetic energy upon ejection:

$$\hbar\omega_{ks} = \chi + \frac{p^2}{2m} \tag{3-132}$$

These features alone introduce several differences from the bound-bound absorption case. First, there exists a threshold in frequency for each ionization process such that $\sigma = 0$ if $\hbar\omega < \chi$. Second, we expect a smoothly varying cross section for $\hbar\omega > \chi$ inasmuch as suitable continuum states of all energies are available.

A brief simplified review of the laboratory interaction of x-rays with matter will provide a good starting point for the discussion. It is found that when a beam of monochromatic x-rays of intensity I_0 is passed through a thickness x of absorbing material, the intensity decreases exponentially with absorbing path length. One may write $I = I_0 \exp(-\mu x)$, where μ is called the *linear absorption coefficient.* By changing the density of the absorbing material, it is found that the attenuation is proportional, for a given absorber, not to the path length but rather to the number of absorbing atoms in a column of unit cross-sectional area. (For instance, a gas target could easily be compressed for comparisons at differing density.) The fact that the reduction in intensity is exponential in the number of absorbing atoms per cubic centimeter led directly to an interpretation based on the *probability per atom* of absorbing photons. That probability is measured by the atomic absorption cross section, which is defined by imagining one atom in a flux of photons:

$$\sigma = \frac{\text{probability of photon absorption/unit time}}{\text{flux of photons}} \qquad \frac{\text{sec}^{-1}}{\text{cm}^{-2}\,\text{sec}^{-1}}$$

It then follows that the reduction of intensity in traversal of thickness dx of absorber having $N_0\rho/A$ absorbing atoms per cubic centimeter is given by

$$dI = -\sigma \frac{N_0\rho}{A} I\, dx \tag{3-133}$$

According to Eq. (3-13), then, the mass absorption coefficient κ is

$$\kappa = \frac{N_0}{A}\sigma \tag{3-134}$$

Experiments reveal that for a target of given atomic species, the mass absorption coefficient varies with wavelength in the manner illustrated in Fig. 3-7. The opacity is found to rise monotonically until it reaches a critical wavelength, whereupon it drops discontinuously, to be followed by another smooth rise, etc. The opacity is due to the absorption of a photon, its energy being converted into that necessary to ionize an electron plus the kinetic energy of the liberated electron. The discontinuous drops in absorption occur when the wavelength becomes so long that the photon energy $\hbar\nu = \hbar c/\lambda$ becomes insufficient to ionize a given type of electron. For instance, the first abrupt drop in Fig. 3-7 occurs for wavelengths too long to ionize the most tightly bound atomic electrons (the K electrons; $n = 1$). The next abrupt drop occurs for wavelengths too long to ionize the $n = 2$ (L shell) electrons (there are actually three different classes L_{I}, L_{II}, and L_{III} of $n = 2$ electrons with approximately equal binding energy). Between the shell edges, it is found that the absorption increases as λ^3 and as the fourth power of the charge of the nucleus of the atom, so that the atomic cross section can be represented by the formula

$$\sigma \approx \begin{cases} C_K Z^4 \lambda^3 + b_K & \lambda < \lambda_K \\ C_{L_{\text{I}}} Z^4 \lambda^3 + b_{L_{\text{I}}} & \lambda_K < \lambda < \lambda_{L_{\text{I}}} \end{cases} \tag{3-135}$$

where the constants b are small and the characteristic wavelengths are

$$\frac{hc}{\lambda_K} = \chi_K \qquad K \text{ binding energy}$$

$$\frac{hc}{\lambda_{L_{\text{I}}}} = \chi_{L_{\text{I}}} \qquad L_{\text{I}} \text{ binding energy}$$

and so on. The first two constants are measured to be $C_K = 2.25 \times 10^{-2}\ \text{cm}^{-1}$ and $C_{L_{\text{I}}} = 0.33 \times 10^{-2}\ \text{cm}^{-1}$.

From the shape of the observed cross sections one can see that $\sigma(\omega)$ is largest near the threshold and falls off smoothly at higher frequencies. In the stellar opacity problem there can clearly be no absorption unless there are bound electrons. From the Saha equation it is clear that at moderate density there will be no bound levels except those for which χ is comparable to, or greater than, kT. Since the most important frequencies in the Rosseland mean are of order of a few times kT, we shall expect absorption near the thresholds to dominate in the stellar opacity problem. This fact introduces considerable computational complexity into the calculation for this reason: for photons with energy slightly in excess of the threshold the ejected electrons have small kinetic energy, which in turn means that the continuum wave function ψ_k will be appreciably perturbed by the coulomb potential of the ion from which it is ejected. For accurate calcula-

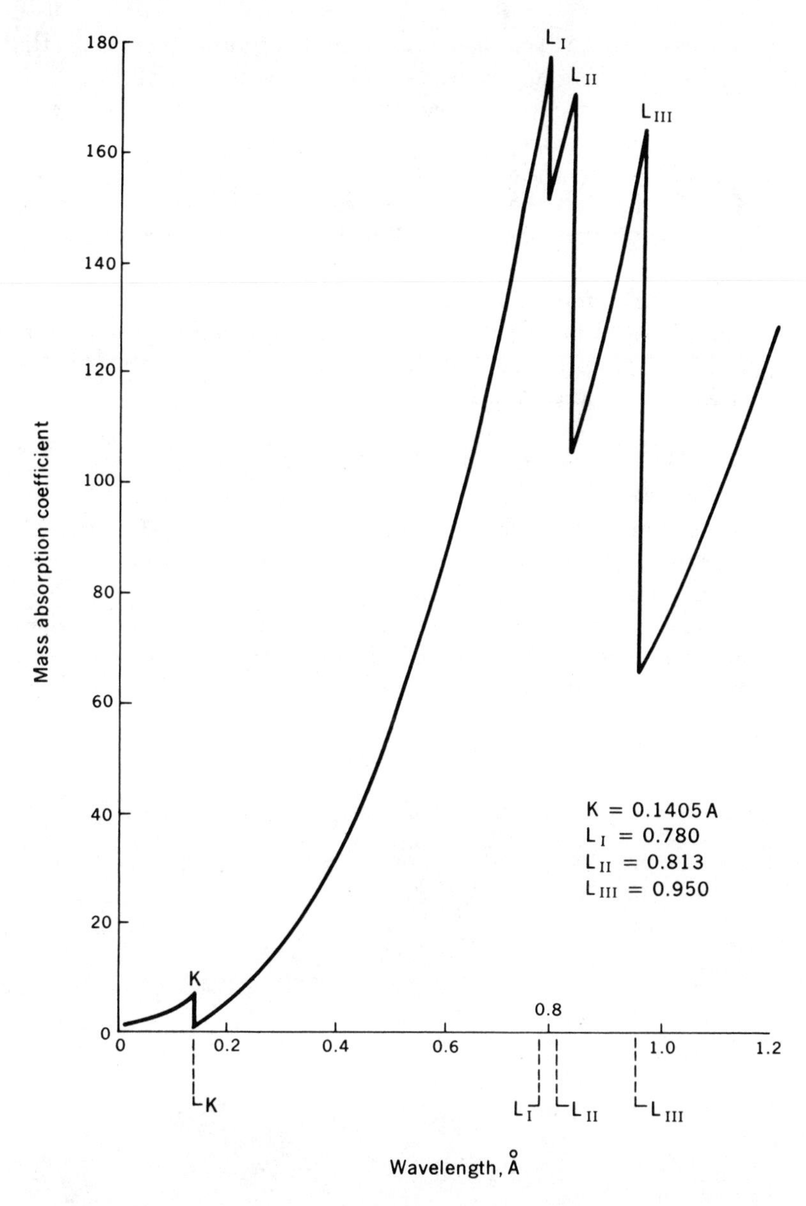

Fig. 3-7 The dependence of the mass absorption coefficient of neutral atoms of lead on the wavelength λ. Photons having wavelengths shorter than the K absorption edge have sufficient energy to photoeject a K electron from the atom, and the coefficient is found to decrease as λ^3. Larger absorption edges beginning at longer wavelengths correspond to the less tightly bound electrons in the L shells. *(From F. K. Richtmyer, E. H. Kennard, and T. Lauritsen, "Introduction to Modern Physics," 5th ed. Copyright* 1955. *McGraw-Hill Book Company. Used by permission.)*

tions, therefore, it is necessary to use coulomb functions for ψ_k, that is, free wave functions describing a particle moving in a coulomb field.

As a simpler example of the process, however, we shall ignore the coulomb interaction and assume that the ejected particles may be represented by plane waves. This assumption can be correct only for photon energies much greater than χ in order that the ejected electron not be too slow. We shall also assume that the energies are small enough to be nonrelativistic for the electron. It will also be assumed that the bound-state wave function is hydrogenlike.

The fact that the final state lies in the continuum means that there is a continuum of transition probabilities $|C_k(+\infty)|^2$, one for each final electron state. Thus if $d\sigma(\omega)$ represents the *differential cross section* for photoejection into a differential solid angle $d\Omega$, its value is

$$d\sigma(\omega) = \frac{4\pi^2\alpha}{m^2\omega}\left|\langle k|\exp\left(i\frac{\omega}{c}\mathbf{n}\cdot\mathbf{r}\right)\mathbf{p}\cdot\mathbf{e}|s\rangle\right|^2\frac{\Delta n}{\Delta\omega} \tag{3-136}$$

where Δn is the number of continuum electron eigenstates in the solid angle $d\Omega$ and in an energy band corresponding to the frequency interval $\Delta\omega$ about the average energy $E_k = \hbar\omega - \chi = \hbar\omega + E_s$.

To evaluate this expression a useful crutch is to imagine the problem to be confined to a very large cube of dimension L containing the atom. Then the normalized electron plane waves are

$$\psi_k = L^{-\frac{3}{2}}\exp(i\mathbf{k}\cdot\mathbf{r}) \qquad E_k = \frac{\hbar^2k^2}{2m} \tag{3-137}$$

From the fact that the number of continuum eigenstates with momentum $\mathbf{p}$ in the intervals $\Delta p_x\,\Delta p_y\,\Delta p_z$ is (see Chap. 1)

$$\Delta n = \frac{L^3\,\Delta p_x\,\Delta p_y\,\Delta p_z}{h^3} = \frac{L^3 4\pi p^2\,dp}{h^3} \tag{3-138}$$

it is easily shown that

$$\frac{\Delta n}{\Delta E} = \frac{m^{\frac{3}{2}}\sqrt{E}\,L^3}{\sqrt{2}\,\pi^2\hbar^3} \tag{3-139}$$

Problem 3-15: Derive Eq. (3-139).

It follows that within a differential solid angle $d\Omega$

$$\frac{\Delta n}{\Delta\omega} = \frac{d\Omega}{4\pi}\frac{m^{\frac{3}{2}}\sqrt{E}\,L^3}{\sqrt{2}\,\pi^2\hbar^2} \tag{3-140}$$

If it is assumed that the photoejection occurs from the K shell, the initial wave function is

$$\psi_s = R_{1s}Y_0{}^0 = \frac{1}{\sqrt{\pi}}\left(\frac{Z}{a}\right)^{\frac{3}{2}}\exp -\frac{Zr}{a} \tag{3-141}$$

Assemblage of these results in Eq. (3-136) yields

$$\frac{d\sigma(\omega)}{d\Omega} = \frac{\alpha k}{2\pi^2 m\hbar\omega}\left(\frac{Z}{a}\right)^3 \left|\int \exp\,(-i\mathbf{k}\cdot\mathbf{r})\,\exp\left(i\frac{\omega}{c}\mathbf{n}\cdot\mathbf{r}\right)\mathbf{e}\cdot\frac{\hbar}{i}\nabla\,\exp\left(-\frac{Zr}{a}\right)dV\right|^2 \tag{3-142}$$

Evaluation of the integral can be simplified by using the fact that the momentum operator is hermitian and that $\mathbf{e}\cdot\mathbf{n} = 0$; thus

$$\begin{aligned}\int \exp\left[i\left(-\mathbf{k}\cdot\mathbf{r} + \frac{\omega}{c}\mathbf{n}\cdot\mathbf{r}\right)\right]\mathbf{e}\cdot\frac{\hbar}{i}\nabla\,\exp\left(-\frac{Zr}{a}\right)dV \\ = \mathbf{e}\cdot\int\left\{\frac{\hbar}{i}\nabla\,\exp\left[i\left(\mathbf{k}-\frac{\omega}{c}\mathbf{n}\right)\cdot\mathbf{r}\right]\right\}^*\exp\left(-\frac{Zr}{a}\right)dV \\ = \hbar\mathbf{e}\cdot\mathbf{k}\int\exp\left[i\left(\frac{\omega}{c}\mathbf{n}-\mathbf{k}\right)\cdot\mathbf{r} - \frac{Zr}{a}\right]r^2\,dr\,d\Omega\end{aligned} \tag{3-143}$$

Now define the vector

$$\mathbf{q} \equiv \mathbf{k} - \frac{\omega}{c}\mathbf{n} \tag{3-144}$$

Inasmuch as $\hbar\mathbf{k}$ is the electron momentum and $\hbar(\omega/c)\mathbf{n}$ is the photon momentum, $\hbar\mathbf{q}$ must represent the momentum transfer.

Problem 3-16: The angular part of the integral is contained in the factor $\exp\,(-i\mathbf{q}\cdot\mathbf{r})$. Show that

$$\int \exp\,(-i\mathbf{q}\cdot\mathbf{r})\,d\Omega = 4\pi\frac{\sin qr}{qr}$$

Then evaluate the radial integral using the trick

$$\int r\sin qr\,\exp\left(-\frac{Zr}{a}\right)dr = -\frac{d}{dq}\int\cos qr\,\exp\left(-\frac{Zr}{a}\right)dr$$

Gathering all the factors yields

$$\frac{d\sigma(\omega)}{d\Omega} = \frac{32\alpha\hbar k(\mathbf{e}\cdot\mathbf{k})^2}{m\omega}\left(\frac{Z}{a}\right)^5\left(\frac{Z^2}{a^2}+q^2\right)^{-4} \tag{3-145}$$

This formula, which is in good agreement with observations for $\hbar\omega \gg \chi$, can easily be converted into a total cross section for the same energy region. Let the angle between the photon direction and that of the ejected electron be designated by θ, and let the angle between the photon polarization and the plane of the photon and electron momenta be designated by ϕ, as illustrated in Fig. 3-8. Then from $\mathbf{q} = \mathbf{k} - (\omega/c)\mathbf{n}$ it follows that

$$\left(\frac{Z}{a}\right)^2 + q^2 = \left(\frac{Z}{a}\right)^2 + k^2 + \left(\frac{\omega}{c}\right)^2 - 2k\frac{\omega}{c}\cos\theta \tag{3-146}$$

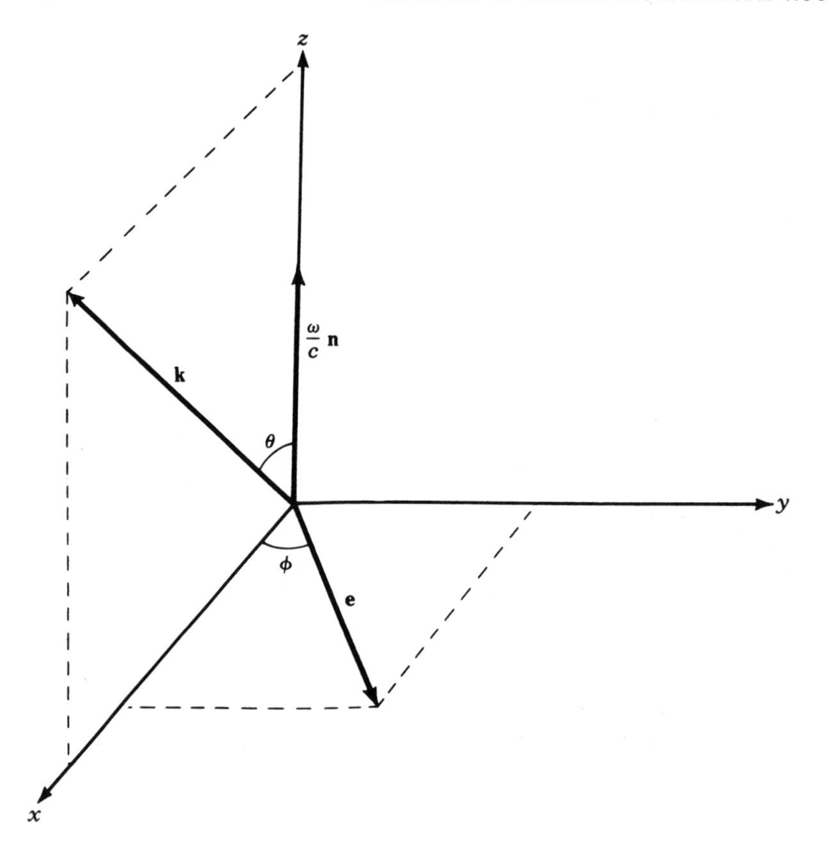

Fig. 3-8 Kinematic vectors for the photoionization process. The direction of photon propagation is **n**, and its polarization **e** makes an angle ϕ with the plane defined by **n** and the wave number **k** of the ejected electron. For the opacity problem one needs the total cross section obtained by averaging over ϕ and integrating the differential cross section over θ.

From the conservation of energy, furthermore,

$$\frac{\hbar^2 k^2}{2m} = \hbar\omega - \chi = \hbar\omega - \frac{Z^2 e^2}{2a} \tag{3-147}$$

where the Bohr radius $a = \hbar^2/me^2$.

Problem 3-17: Show that

$$\frac{Z^2}{a^2} + q^2 = 2\frac{m\omega}{\hbar}\left(1 - \frac{\hbar k}{mc}\cos\theta + \frac{\hbar\omega}{2mc^2}\right)$$

Consistent with the assumption that the electron velocity is nonrelativistic we may ignore the photon energy in comparison with the electron rest-mass energy and replace $\hbar k/mc = v/c = \beta$. Then the differential cross section becomes

$$\begin{aligned}\frac{d\sigma(\omega)}{d\Omega} &= 2\alpha k(\mathbf{e}\cdot\mathbf{k})^2\left(\frac{\hbar}{m\omega}\right)^5\left(\frac{Z}{a}\right)^5(1-\beta\cos\theta)^{-4} \\ &= 2\alpha k^3\left(\frac{\hbar}{m\omega}\right)^5\left(\frac{Z}{a}\right)^5\frac{\sin^2\theta\cos^2\phi}{(1-\beta\cos\theta)^4}\end{aligned} \tag{3-148}$$

Problem 3-18: The total cross section may be obtained by integrating over angles. Realizing that the factor β may be ignored in the nonrelativistic domain, show that

$$\sigma(\omega) = \frac{8\pi\alpha}{3}\left(\frac{Z}{a}\right)^5\left(\frac{\hbar}{m\omega}\right)^5 k^3 \tag{3-149}$$

Both the electron momentum and the photon frequency appear in Eq. (3-149). From Eq. (3-148) it can be seen, however, that for $\hbar\omega \gg \chi$ we have $k^2 \approx 2m\omega/\hbar$. It follows that at high frequency

$$\sigma(\hbar\omega \gg \chi) \approx \frac{2}{3}\frac{\alpha^{\frac{9}{2}}}{\pi^{\frac{5}{2}}a^{\frac{3}{2}}}Z^5\lambda^{\frac{7}{2}} \tag{3-150}$$

This formula is not quite the same form as the empirical results, but near the absorption edge, where the cross section is largest, one has approximately $\lambda^{\frac{7}{2}} \approx \lambda^3\lambda_k^{\frac{1}{2}}$, where $hc/\lambda_k = Z^2e^2/2a$. Thus near the edge $Z^5\lambda^{\frac{7}{2}} \propto Z^4\lambda^3$. Of course, this entire treatment is not valid near the absorption edge, but the $Z^4\lambda^3$ dependence is approximately correct.

The approximation of replacing the free-electron wave function by a plane wave leading to this result is called the *Born approximation*. It is inadequate near the threshold, where the electron wave function is strongly perturbed by the coulomb interaction. Because these are the most important frequencies for the stellar opacity problem, it is necessary to use exact coulomb waves for ψ_k. This calculation will not be detailed here because of its complexity, but the results are that the cross section for ionization of a hydrogenic electron in a state with principal quantum number n may conveniently[1] be written

$$\begin{aligned}\sigma_{\text{b-f}} &= \frac{64\pi^4 me^{10}}{3\sqrt{3}\,ch^6}\frac{Z^4}{n^5}\frac{g(\nu,n,l,Z)}{\nu^3} \\ &= 2.82\times10^{29}\frac{Z^4}{n^5\nu^3}g(\nu,n,l,Z) \qquad \text{cm}^2\end{aligned} \tag{3-151}$$

where ν is the photon frequency. This result, except for the factor g, was first derived by Kramers[2] in a semiclassical calculation. The factor g is commonly

[1] For a detailed discussion of this procedure see W. J. Karzas and R. Latter, *Astrophys. J. Suppl.*, **6**:167 (1961), and H. Y. Chiu, "Stellar Physics," Blaisdell Publishing Company, Waltham, Mass., 1968.

[2] *Phil. Mag.*, **46**:836 (1923).

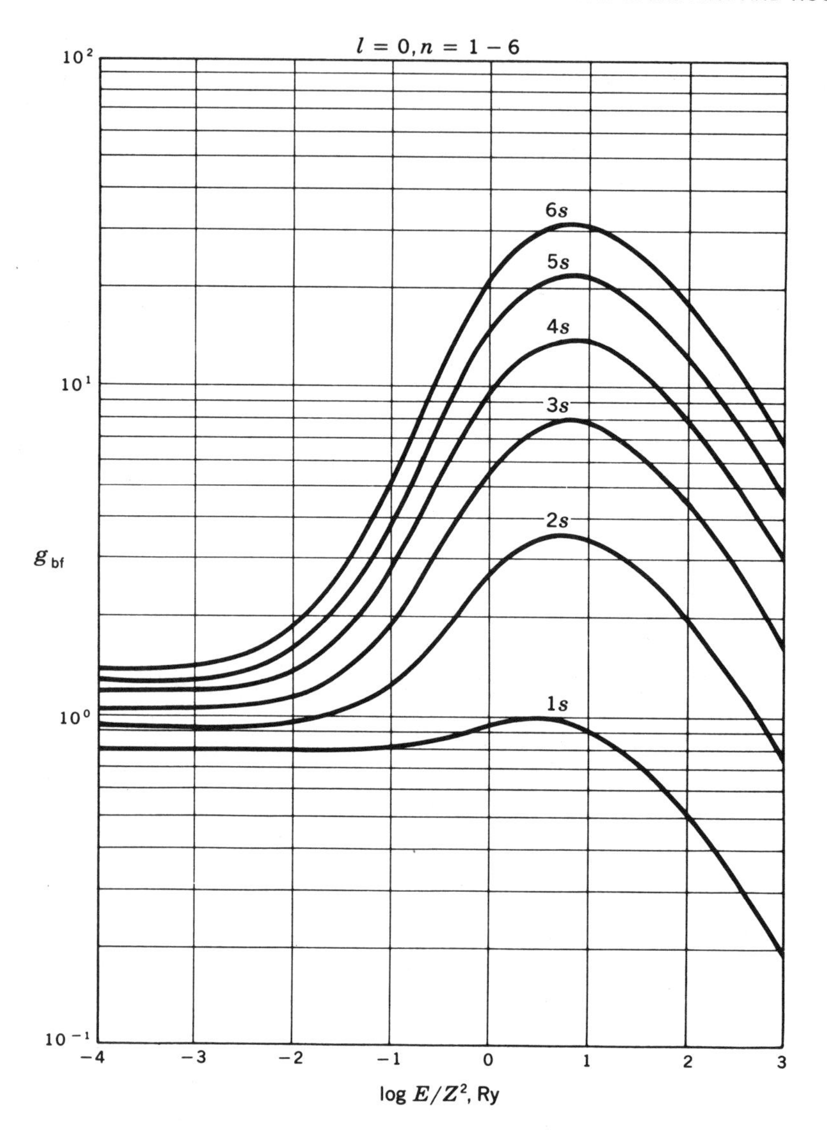

Fig. 3-9 The bound-free Gaunt factors for the photoionization of $l = 0$ electrons from hydrogenlike atoms. The abscissa is the energy of the liberated electron in units of Z^2 Rydbergs. Near the absorption edge ($\log E = -\infty$), where the cross section is the greatest, the Gaunt factors are near unity. [*W. J. Karzas and R. Latter, Astrophys. J. Suppl.,* **6**:167 (1961). *By permission of The University of Chicago Press. Copyright* 1961 *by The University of Chicago.*]

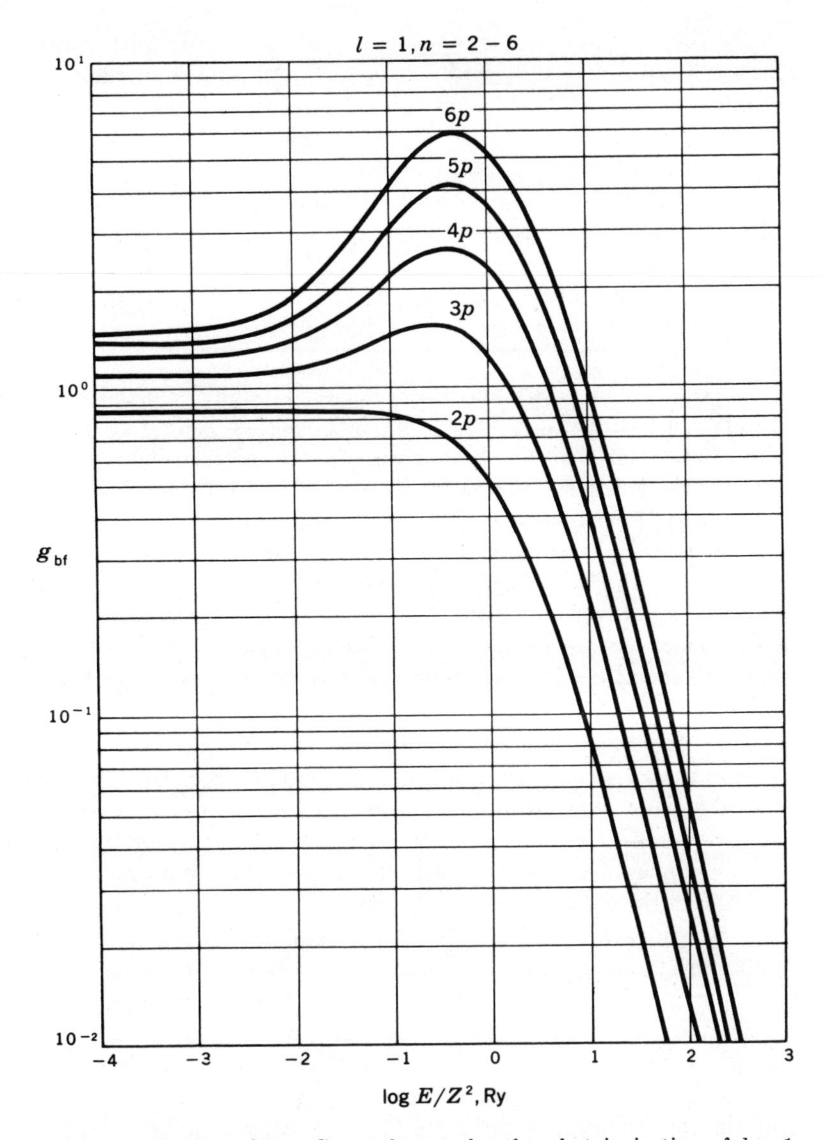

Fig. 3-10 The bound-free Gaunt factors for the photoionization of $l = 1$ electrons from hydrogenlike atoms. [*W. J. Karzas and R. Latter, Astrophys. J. Suppl.*, **6**:167 (1961). *By permission of The University of Chicago Press. Copyright* 1961 *by The University of Chicago.*]

called the *Gaunt factor*. Its value depends upon the initial-state quantum numbers and is a slowly varying function of the photon frequency. For hydrogenlike atoms, the Gaunt factors are almost all within 20 percent of unity near the absorption edges, where the cross sections are the largest. Recent accurate calculations of hydrogenlike Gaunt factors can be found in Karzas and Latter.[1] Their values for $l = 0$ and $l = 1$ are shown in Figs. 3-9 and 3-10, where they are plotted against the free-electron energy expressed in Rydbergs per Z^2. It can be seen that they are nearly constant and near unity near the threshold. This feature has the advantage that in the calculation of the Rosseland mean an appropriate average g may be used, and the frequency dependence can be regarded as simply ν^{-3}.

In the interior of a star, where matter is highly ionized, there will be a large number of hydrogenlike ions from the elements of high atomic number. (When an ion has more than one bound electron, the nuclear charge Z is usually replaced by an effective nuclear charge Z^* that takes into account the shielding of the nucleus by other bound electrons—no mean calculation in itself!) Each element in the mixture contributes its share to the total opacity. In analogy with Eq. (3-134) the bound-free opacity contributed by each bound-electron state is the product of the ionization cross section from that state and the number of such bound electrons per gram. One must sum at each photon frequency the absorptions due to all processes. The result is that photons can be absorbed by liberating bound electrons from all those shells of all atomic species for which the binding energy is less than the photon energy. Although the absorption cross section is largest near the edge, one must sum over possible absorptions far from the edge, since a large number of bound electrons per unit volume in some shell may dominate the absorption far from the edge. For increasingly higher frequencies, one must sum over an increasing number of initial bound states.

Apparently the bound-free opacity will depend upon the numbers of electrons that are bound in the various atomic states. When the temperature becomes so high that the ionization is complete, the bound-free opacity falls to zero. But the average abundance of iron is so high that its bound-free absorption by the K-shell electrons holds the bound-free opacity up even for temperatures of several million degrees. What must be done is to calculate, at each temperature and density and chemical composition, the complete ionization equilibrium to obtain the occupation numbers for each atomic state. This is a very difficult calculation in itself, for it introduces not only the uncertainty inherent in the structure of the many-electron atom but also other collective features such as the coulomb interactions with the plasma.

Problem 3-19: Show that *in the limit of high ionization* the average number of bound electrons of principal quantum number n is

$$n_n = n_e n^2 \frac{h^3}{2(2\pi mkT)^{\frac{3}{2}}} e^{+\chi_n/kT}$$

[1] *Loc. cit.*

(3) FREE-FREE ABSORPTION

The free-free absorption process is imagined as follows: a free electron of momentum $\mathbf{p}_s$ approaches an ion of charge Z, and in the process of scattering it also interacts with the radiation field, absorbing a photon of energy $\hbar\omega$, and exits with a new momentum $\mathbf{p}_k$. The conservation of energy applies to the overall process, so that the relation

$$\frac{p_k^2}{2m} = \frac{p_s^2}{2m} + \hbar\omega \tag{3-152}$$

shows that energy was truly absorbed from the radiation field and converted into kinetic energy of particles. The inverse of this process is called bremsstrahlung and is the corresponding process of true emission.

That the ion of charge Z plays a fundamental role in this process can be easily seen by trying to ignore it. Suppose we naïvely try to calculate the matrix element

$$V_{ks} = \langle k| \exp\left(i\frac{\omega}{c}\mathbf{n}\cdot\mathbf{r}\right)\mathbf{p}\cdot\mathbf{e}|s\rangle$$

in the Born approximation of representing ψ_k and ψ_s by plane waves. Then the cross section would be proportional to the absolute square of

$$\begin{aligned}\int_V \exp(-i\mathbf{k}_k\cdot\mathbf{r})\exp\left(i\frac{\omega}{c}\mathbf{n}\cdot\mathbf{r}\right)\mathbf{e}\cdot\frac{\hbar}{i}\nabla\exp(i\mathbf{k}_s\cdot\mathbf{r})\,dV \\ = \hbar\mathbf{e}\cdot\mathbf{k}_s\int\exp(-i\mathbf{q}\cdot\mathbf{r})r^2\,dr\,d\Omega \\ = \mathbf{e}\cdot\mathbf{p}_s\frac{4\pi}{q}\int_0^\infty \sin qr\,r\,dr\end{aligned} \tag{3-153}$$

where

$$\mathbf{q} = \mathbf{k}_k - \frac{\omega}{c}\mathbf{n} - \mathbf{k}_s$$

Although the radial integral does not appear to be well defined, it can be thought of as the limit

$$\int \sin qr\,r\,dr = \lim_{\alpha\to 0}\int \sin qr\,e^{-\alpha r}r\,dr = \lim_{\alpha\to 0}\frac{2\alpha}{(\alpha^2+q^2)^2} \tag{3-154}$$

This limit clearly equals zero unless $q = 0$; that is, the matrix element vanishes unless $\mathbf{k}_k = \mathbf{k}_s + (\omega/c)\mathbf{n}$, which can be seen to be a statement of the conservation of momentum in the interaction of photon and electron. But this requirement is incompatible with the conservation-of-energy requirement in Eq. (3-152).

Problem 3-20: Confirm the incompatibility. Evidently a free electron cannot truly absorb a photon.

The presence of the ion solves this problem by itself absorbing some of the momentum. (Because of its large mass that momentum transfer absorbs negligible kinetic energy, however.)

Proper treatment of this process requires that the interaction of the electron with the coulomb field of the ion be included. The interaction hamiltonian must be expressed as a sum of two terms,

$$H_{\text{tot}} = H_{\text{free}} + V_c + H_{\text{int}} \tag{3-155}$$

where $V_c = -Ze^2/r$ is the coulomb interaction and H_{int} represents the interaction of the charges with the radiation field.

At this point one faces a choice. If the unperturbed hamiltonian is taken to be $H_0 = H_{\text{free}} + V_c$, the zeroth-order wave functions $\psi_k^{(0)}$ are coulomb wave functions representing the scattering of the electron by the ion. Then the perturbation theory can be carried out in first order in the interaction with the electromagnetic field, although the formal manipulations are rather messy. Alternatively, one may regard H_0 as being H_{free}, in which case the zeroth-order wave functions are plane waves but the perturbing hamiltonian is the sum of $V_c + H_{\text{int}}$. This is the Born approximation again, but because neither piece of the perturbation hamiltonian acting alone can cause the transition in question, the simultaneous treatment of these two perturbations requires a more complicated form of perturbation theory than that outlined so far. Whereas the transition probability is proportional to $|V_{ks}|^2$ in first order, when that matrix element vanishes, as it does in this case, a higher order of the theory must be invoked. In second order the matrix element $\langle k|V|s\rangle$ must be replaced by[1]

$$V_{ks} \rightarrow \sum_m \frac{V_{km}V_{ms}}{E_s - E_m}$$

This effective transition matrix element describes the transition as a two-step process in which a *virtual* transition from the initial state s to an *intermediate state* m is followed by a second virtual transition from the state m to the final state k, summing over all intermediate states. An interesting feature is that energy need not be conserved in the intermediate state m inasmuch as it exists for only a very short time, during which energy conservation may be violated (the uncertainty principle). In the overall transition from s to k energy will be conserved, of course, as in Eq. (3-152). The factor involving the interaction of the electron with the radiation field conserves momentum, as we have seen, but the interaction with the ion changes the momentum by any amount, the ion absorbing the recoil. Thus there are two time sequences for the intermediate state, as illustrated schematically in Fig. 3-11. (1) The electron absorbs the photon by the H_{int} term, yielding electron momentum $\mathbf{k}_m = \mathbf{k}_s + (\omega/c)\mathbf{n}$. Then the second matrix element due to Ze^2/r causes a scattering transition from $\mathbf{k}_m$ to $\mathbf{k}_k$. (2) The coulomb matrix element first causes a transition from $\mathbf{k}_s$ to $\mathbf{k}_m$, with $\mathbf{k}_m$ chosen such

[1] See, for instance, Merzbacher, *op. cit.*, chap. 20.

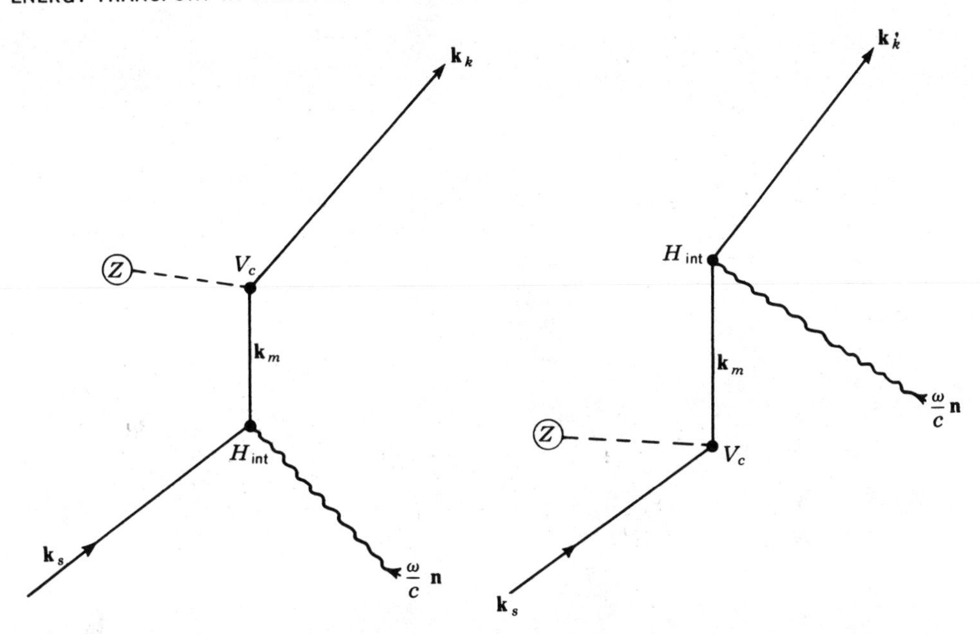

Fig. 3-11 Feynman diagrams illustrating the calculation of free-free absorption in second-order perturbation theory. The incoming plane-wave electron is perturbed by the coulomb potential V_c and by the interaction H_{int}, with the electromagnetic wave being absorbed. The perturbations act in either order, and energy need not be conserved in the *virtual* intermediate state. Alternatively, the calculation can be made in first-order perturbation theory if coulomb waves are used for the unperturbed wave functions.

that H_{int} has a momentum-conserving matrix element for the absorption of the photon; $\mathbf{k}_k = \mathbf{k}_m + (\omega/c)\mathbf{n}$. The sums must be performed for both sequences of intermediate states and added together. Then by the "golden rule" of time-dependent perturbation theory, the transition probability per unit time (or the transition rate) is

$$r = \frac{2\pi}{\hbar}\left|\sum_m \frac{V_{km}V_{ms}}{E_s - E_m}\right|^2 \rho(E_k) \tag{3-156}$$

where $\rho(E_k)$ is the density of final states.

Because of the complexities of this calculation it will not be repeated here. Calculations of the inverse process (bremsstrahlung) may be found in the literature.[1] The relationship between the bremsstrahlung cross section and the free-free absorption cross section is discussed by Karzas and Latter.[2] The result is that in the nonrelativistic region the cross section for an ion of charge Z to absorb

[1] See, for instance, W. Heitler, "Quantum Theory of Radiation," 3d ed., p. 242, Oxford University Press, Fair Lawn, N.J., 1954; H. Bethe and E. E. Salpeter, "Quantum Mechanics of One- and Two-electron Atoms," Springer-Verlag OHG, Berlin, 1957.

[2] *Loc. cit.*

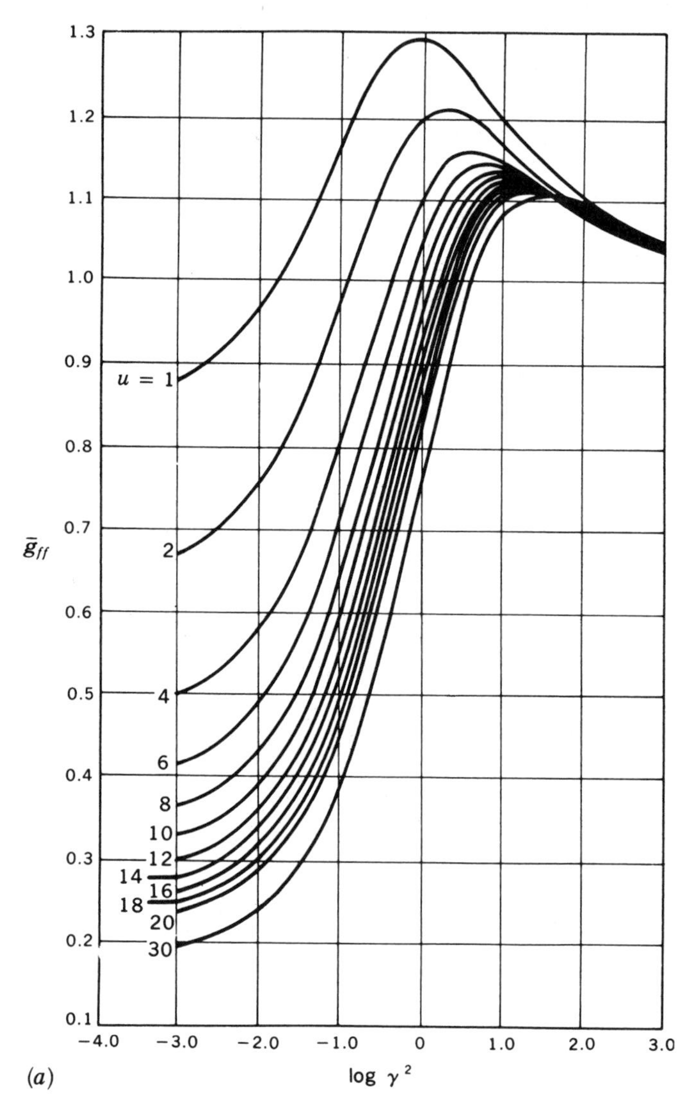

Fig. 3-12 The free-free Gaunt factors averaged over the maxwellian energy spectrum of the free electrons. (*a*) The abscissa is $\gamma^2 = Z^2/kT$, with kT expressed in Rydbergs. Each curve is labeled by the photon energy $u = h\nu/kT$; (*b*) the roles of abscissa and curve parameter are reversed. [*W. J. Karzas and R. Latter, Astrophys. J. Suppl.*, **6**:167 (1961). *By permission of The University of Chicago Press. Copyright* 1961 *by The University of Chicago.*]

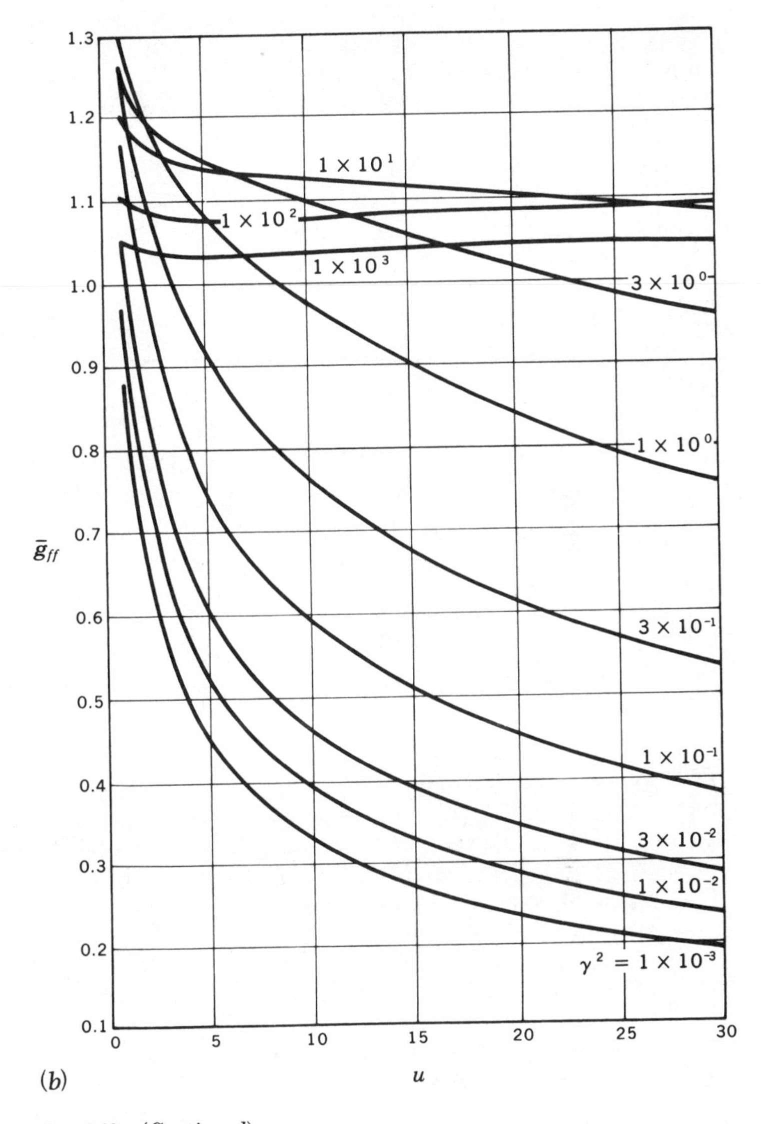

Fig. 3-12 *(Continued)*

a photon of frequency ν by a free-free transition can be written

$$d\sigma_{\text{f-f}}(Z,\nu,v) = \frac{4\pi Z^2 e^6 g_{\text{f-f}}(v,\nu)}{3\sqrt{3}\, hcm^2 v\nu^3}\, n_e(v)\, dv \tag{3-157}$$

where v is the electron velocity, $n_e(v)\,dv$ is the density of electrons having velocity v in the range dv, and $g_{\text{f-f}}(v,\nu)$ is called the *Gaunt factor for free-free transitions.* The Gaunt factor depends upon the electron velocity and the energy of the absorbed photon, and is defined in this conventional way as a correction to a semiclassical calculation of Kramers.[1]

If it is assumed that the electron gas is nondegenerate, as is most often the case when the free-free opacity is of astrophysical importance, the electrons possess a maxwellian velocity distribution:

$$n_e(v)\,dv = 4\pi n_e \left(\frac{m}{2\pi kT}\right)^{3/2} \exp\left(-\frac{mv^2}{2kT}\right) v^2\,dv$$

Then the average free-free cross section for photons of frequency ν is

$$\bar{\sigma}_{f\text{-}f}(Z,\nu) = \int_{v=0}^{\infty} d\sigma_{f\text{-}f}(Z,\nu,v) \tag{3-158}$$

Problem 3-21: Carry out this integral to show

$$\bar{\sigma}_{f\text{-}f} = \frac{16\pi^2 Z^2 e^6 n_e}{3\sqrt{3}\,(2\pi m)^{3/2}(kT)^{1/2}} \frac{1}{\nu^3} \int_0^{\infty} e^{-x} g_{f\text{-}f}(x)\,dx \tag{3-159}$$

where $x = mv^2/2kT$.

The integral in Eq. (3-159) is called the *temperature-averaged Gaunt factor:*

$$\bar{g}_{f\text{-}f}(\nu,Z,T) = \int_0^{\infty} e^{-x} g_{f\text{-}f}(x,\nu,Z,T)\,dx \tag{3-160}$$

The values of average free-free Gaunt factors are shown in Fig. 3-12. Each curve is labeled by the photon frequency $u = h\nu/kT$. Then $\bar{g}_{f\text{-}f}$ is plotted against the ratio of Z^2 to the temperature in Rydbergs. A little study will show that $\bar{g}_{f\text{-}f}$ will be expected to be near unity in almost all astrophysical circumstances.

Numerically Eq. (3-159) becomes

$$\bar{\sigma}_{f\text{-}f}(Z,\nu,T) = 3.69 \times 10^8 \frac{Z^2 n_e \bar{g}_{f\text{-}f}}{T^{1/2}\nu^3} \quad \text{cm}^2 \tag{3-161}$$

The free-free opacity is obtained from multiplication by the number of ions of type Z per gram and summing over Z:

$$\kappa_{f\text{-}f}(\nu) = \sum \frac{X_z N_0}{A_z} \bar{\sigma}_{f\text{-}f}(Z,\nu) \tag{3-162}$$

This contribution to the opacity must be added to other sources before performing the average over frequencies required for the determination of the Rosseland mean, but because the free-free opacity dominates the total opacity in some important areas of the ρT plane, it is instructive to examine the Rosseland mean of the free-free opacity alone. With the substitution $u = h\nu/kT$, Eq. (3-51) can

[1] Its detailed value is given in *ibid.*, eq. (16) together with several graphs of its characteristic values.

easily be reduced to

$$\frac{1}{\kappa} = \frac{2\pi k^4}{ac^3h^3}\int_0^\infty \frac{u^4 \exp 2u\, du}{\kappa(u)[\exp (u) - 1]^3} \tag{3-163}$$

which is applicable to all types of true absorption. For the specific case of free-free absorption, however, the opacity can be written approximately as

$$\kappa_{\text{f-f}}(\nu) \approx K\nu^{-3} = K\left(\frac{h}{kT}\right)^3 u^{-3} \tag{3-164}$$

where K is a constant dependent only upon the composition if we temporarily ignore the variation of $g_{\text{f-f}}(\nu)$ with frequency. The resulting expression is

$$\frac{1}{\kappa} = \frac{2\pi k^4}{ac^3h^3}\left(\frac{kT}{h}\right)^3 \frac{1}{K}\int_0^\infty \frac{u^7 \exp 2u\, du}{[\exp (u) - 1]^3} \tag{3-165}$$

Inasmuch as the integrand has a maximum at $u = 7$, the average Gaunt factor to be used in calculating K is approximately $\langle g_{\text{f-f}}\rangle \approx \bar{g}_{\text{f-f}}(u = 7)$. Evidently Eq. (3-165) may be thought of as being of the form

$$\frac{1}{\kappa} = \frac{1}{K\langle\nu^{-3}\rangle} \tag{3-166}$$

where $\langle\nu^{-3}\rangle$ is the average of ν^{-3} introduced by the Rosseland mean.

Problem 3-22: Show by evaluating (approximately) the integral in Eq. (3-165) that

$$\langle\nu^{-3}\rangle = \left[197\left(\frac{kT}{h}\right)^3\right]^{-1} \tag{3-167}$$

The results of this problem show that the act of performing the Rosseland mean for free-free opacity is approximately equivalent to replacing ν^3 by $197(kT/h)^3$ in the formula for $\kappa_{\text{f-f}}(\nu)$. Note that kT/h is the frequency of a *thermal photon*, and so the effective photon energy for free-free absorption is

$$(h\nu)_{\text{eff}} \approx 5.82kT \qquad \text{free-free} \tag{3-168}$$

With the aid of this result, Eq. (3-161) for the Rosseland mean cross section for free-free absorption is

$$\begin{aligned}\langle\sigma_{\text{f-f}}\rangle &= 2.07 \times 10^{-25}\frac{Z^2 n_e}{T^{3.5}}\bar{g}_{\text{f-f}}(u = 7) \qquad \text{cm}^2 \\ &= 1.25 \times 10^{-1}\frac{Z^2\bar{g}_{\text{f-f}}}{\mu_e}\frac{\rho}{T^{3.5}} \qquad \text{cm}^2 \\ &\approx 6.25 \times 10^{-2}(1 + X)Z^2\bar{g}_{\text{f-f}}\rho T^{-3.5} \qquad \text{cm}^2\end{aligned} \tag{3-169}$$

From Eq. (3-162) the Rosseland mean opacity becomes

$$\langle \kappa_{\text{f-f}} \rangle = 7.53 \times 10^{22} \frac{\rho}{\mu_e T^{3.5}} \sum_Z \frac{Z^2 \bar{g}_{\text{f-f}}(Z, u = 7) X_z}{A_z} \qquad \text{cm}^2/\text{g} \tag{3-170}$$

Kramers first made this demonstration, and for that reason any opacity of the form $\rho T^{-3.5}$ is called a *Kramers opacity*. Approximate formulas with that density and temperature dependence have frequently been used in investigations of stellar structure because it turns out the bound-free opacity follows approximately the same form over a limited region. For detailed computations of stellar structure on digital computers, however, it is preferable to interpolate opacities from published tables.

If the electron gas is partially degenerate, the averaging process is somewhat more complex in that the Fermi-Dirac distribution must be used for $n_e(v)$, and a factor should be included to contain the probability that the final electron state is not already occupied. In the higher range of electron density, moreover, shielding (collective interactions) may play a role in the absorption cross section. The cross sections discussed here are those for a pure coulomb potential and are slightly larger than those obtained with a Debye-Hückel potential, say.

(4) SCATTERING FROM ELECTRONS

The fourth major source of radiative opacity in the interior of a star results from the scattering of photons by free electrons in the gas. This scattering opacity is always present when there are free electrons, but because the cross section is so small, it is dominated by the bound-free opacity until the ionization is essentially complete and by the free-free opacity until the temperature is sufficiently great. In the high-temperature range free electrons represent the major impediment to the propagation of a photon.

In one way the quantum mechanics of scattering is more complicated than the true-absorption processes discussed previously; viz., two photons are involved rather than one. In a full theory of quantum electrodynamics the vector potential $\mathbf{A}$ is expanded linearly in terms of *creation and destruction operators for single photons;* therefore, in the other three processes mentioned the interaction term $\mathbf{p} \cdot \mathbf{A}$ gives rise to the absorption of a single photon. It is this quantization of the radiation field that theoretically describes the fact that energy is generally emitted as a single photon with transition energy $\hbar\omega$ rather than as several photons. But the matrix element for scattering must involve the destruction of the initial photon and the creation of the final (scattered) photon, as indicated schematically in Fig. 3-13, and must accordingly be quadratic in the vector potential $\mathbf{A}$. If

$$H_{\text{int}} = \frac{e}{mc} \mathbf{p} \cdot \mathbf{A} + \frac{e^2}{2mc^2} \mathbf{A} \cdot \mathbf{A} \equiv H_{\text{int}}^{(1)} + H_{\text{int}}^{(2)} \tag{3-171}$$

then the matrix elements quadratic in $\mathbf{A}$ involve the second-order perturbation of

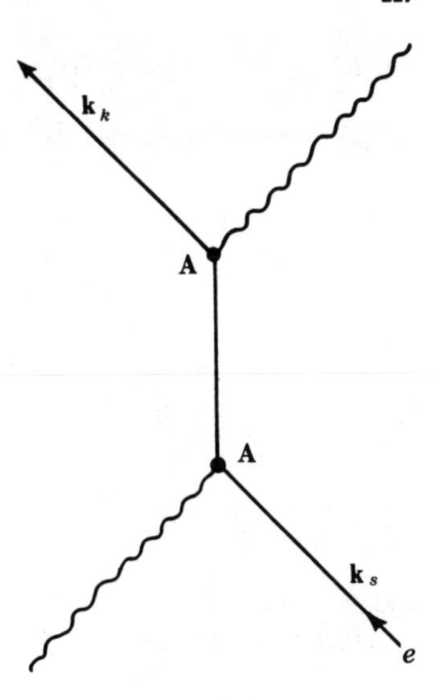

Fig. 3-13 Feynman diagram for the scattering of a photon from a free electron. The interaction with the electromagnetic field must occur twice in this process, with the result that the cross section is proportional to the square of the length e^2/mc^2.

$H^{(1)}_{\text{int}}$ and the hitherto neglected $H^{(2)}_{\text{int}}$:

$$\langle k|H_{\text{int}}|s\rangle = \sum_m \frac{\langle k|H^{(1)}|m\rangle\langle m|H^{(1)}|s\rangle}{E_s - E_m} + \langle k|H^{(2)}|s\rangle \tag{3-172}$$

It is immediately obvious that this matrix element is proportional to e^2 (instead of e), giving a cross section proportional to e^4 (instead of e^2). The actual calculation of the cross section will not be described here.[1] It turns out that for photon energies much less than mc^2 the answer can be obtained simply from classical electromagnetism.

When an electromagnetic wave is incident on a free particle of charge $-e$ and mass m, the particle will be accelerated by the electric field of the wave. This acceleration will cause radiation to be emitted in directions other than that of the incident plane wave, viz., scattering. For photon energies much less than mc^2 the scattered radiation has the same frequency as the incident radiation in the stellar rest frame. A common result of classical electrodynamics is that the power radiated into solid angle $d\Omega$ at angle ψ to the direction of acceleration $\mathbf{a}$ is, for nonrelativistic particles,

$$dP = \frac{e^2}{4\pi c^3} a^2 \sin^2 \psi \, d\Omega \tag{3-173}$$

The scattered wave is polarized in the plane containing $\mathbf{a}$ and the direction of viewing. If the plane wave is moving initially in the Z direction with propagation number $k = 2\pi/\lambda$ and unit polarization vector $\boldsymbol{\varepsilon}$ (in xy plane), then the

[1] See Heitler, *op. cit.*, p. 211 or J. M. Jauch and F. Rorlich, "Theory of Photons and Electrons," chap. 11, Addison-Wesley Publishing Company, Inc., Reading, Mass., 1955.

electric field can be written

$$\mathbf{E}(z,t) = \boldsymbol{\varepsilon} E_0 \cos(kz - \omega t) \tag{3-174}$$

Then from Newton's second law of motion

$$\mathbf{a} = -\boldsymbol{\varepsilon} \frac{e}{m} E_0 \cos(kz - \omega t) \tag{3-175}$$

The radiated power is a function of time since a itself is a function of time, but because we are interested only in the average power radiated (scattered), we content ourselves with the average value of a^2, which is

$$\overline{a^2} = \frac{1}{2} \frac{e^2}{m^2} E_0^2 \tag{3-176}$$

Substitution of Eq. (3-176) into Eq. (3-173) yields the following average for the scattered power:

$$\begin{aligned} \frac{dP}{d\Omega} &= \frac{e^2}{4\pi c^3} \frac{1}{2} \frac{e^2}{m^2} E_0^2 \sin^2 \psi \\ &= \frac{c}{8\pi} E_0^2 \left(\frac{e^2}{mc^2} \right)^2 \sin^2 \psi \end{aligned} \tag{3-177}$$

The differential scattering cross section may be defined in this classical calculation as the ratio of the power scattered into unit solid angle to the incident power per unit area:

$$\frac{d\sigma}{d\Omega} = \frac{\text{power radiated/unit solid angle}}{\text{incident power/unit area}} \tag{3-178}$$

But the flux of energy per unit area in an electromagnetic wave is just given by Poynting's vector for a plane wave:

$$S = \frac{\text{power}}{\text{area}} = \frac{c}{4\pi} \overline{E^2} = \frac{c}{8\pi} E_0^2 \tag{3-179}$$

Substitution yields the differential scattering cross section,

$$\frac{d\sigma}{d\Omega} = \left(\frac{e^2}{mc^2} \right)^2 \sin^2 \psi \tag{3-180}$$

The reader may note with some interest that the constants appearing here are those which come from the square of the term $H_{\text{int}}^{(2)}$ in the interaction hamiltonian. The number $r_0 = e^2/mc^2 = 2.818 \times 10^{-13}$ cm is called the *classical radius of the electron* because it represents the radius of a shell of charge e having potential energy equal to the rest-mass energy of the electron. Apparently it also gives the effective area of an electron to a photon.

Equation (3-180) is still not in the most useful form for our needs. The angle ψ is the angle between the direction of observation (at angle θ to the z axis) and the acceleration of the charge (the polarization direction $\boldsymbol{\varepsilon}$, which we now take to define

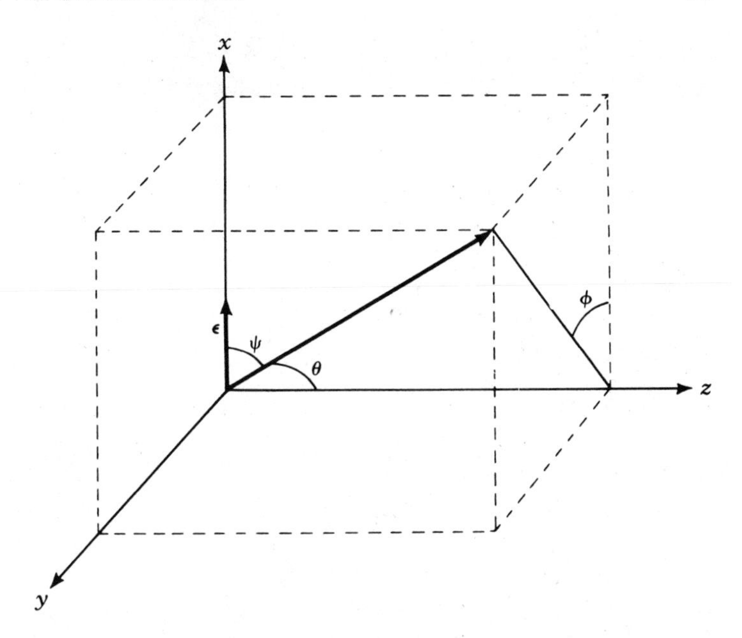

Fig. 3-14 Kinematic vectors of the scattering of a photon from an electron. (*a*) The direction of photon propagation is taken as the z axis; (*b*) $\boldsymbol{\varepsilon}$ is the polarization of the photon; (*c*) the photon is scattered by angle θ and makes an angle ψ with $\boldsymbol{\varepsilon}$.

the x axis). The situation is illustrated in Fig. 3-14, from which it is evident that

$$\sin^2 \psi = 1 - \sin^2 \theta \cos^2 \phi \tag{3-181}$$

If the initial wave is unpolarized, the average value of $\sin^2 \psi$ viewed at any one direction θ to the propagation vector of the initial wave is

$$\overline{\sin^2 \psi} = 1 - \sin^2 \theta \, \overline{\cos^2 \phi} = 1 - \tfrac{1}{2} \sin^2 \theta \tag{3-182}$$

This average gives the *Thomson formula* for the differential scattering cross section of unpolarized photons:

$$\frac{d\sigma}{d\Omega} = \left(\frac{e^2}{mc^2}\right)^2 \frac{1}{2} (1 + \cos^2 \theta) \tag{3-183}$$

Several comments should be made about the Thomson scattering cross section. It is not valid for relativistic particles or for photon energies comparable to mc^2. Since these restrictions are violated only for very high temperatures ($T > 10^9$ °K), it appears that Eq. (3-183) will be satisfactory to explain the scattering of photons from free particles in ordinary stellar interiors. The value of the cross section is seen to be proportional to $(1/m)^2$. Free-electron scattering is, therefore, much more important than scattering from nuclei, which can henceforth be neglected.

From an earlier problem we see that in terms of the normalized scattering phase function the differential cross section may be expressed as

$$d\sigma = \frac{8\pi}{3}\left(\frac{e^2}{mc^2}\right)^2 p(\cos\theta)\,\frac{d\Omega}{4\pi}$$

$$= \sigma_T \left[p(\cos\theta)\,\frac{d\Omega}{4\pi}\right] \tag{3-184}$$

where σ_T is total cross section for scattering of photons into all angles by a free electron, and is called the *Thomson cross section:*

$$\sigma_T \equiv \frac{8\pi}{3}\left(\frac{e^2}{mc^2}\right)^2 = 0.665 \times 10^{-24}\ \text{cm}^2 \tag{3-185}$$

The Thomson cross section is independent of frequency; thus the term $\kappa_{\nu s}$ in the Rosseland mean will simply be a constant if the entire scattering opacity is due to free electrons. Other types of scattering do occur in stellar structure. For example, photons scatter from ions and molecules in the cooler outer regions of a star, where ionization is incomplete. The cross section for scatterings of this type (essentially from harmonically bound electrons) is frequency-dependent and is called *Rayleigh scattering.* Inclusion of this type of scattering is important in the construction of model atmospheres for stars, but for these problems the entire method of radiative transfer is different. It is sometimes unsatisfactory to use the assumption of local thermodynamic equilibrium in stellar atmospheres, which assumption was instrumental in leading to the definition of the Rosseland mean opacity for radiative transfer. The complications of atmospheric transfer are great and have led to the special science of model atmospheres for stars (and planets). Our major concern in this book, however, is with the problems of the stellar interior. For temperatures high enough for the scattering opacity to be significant in comparison with true absorption, it is almost entirely from the Thomson electron scattering cross section.

Equation (3-185) gives the energy removed from a pencil *per electron.* Since the opacity is defined in terms of the energy removed per gram, its value may be obtained simply by the product of σ_T and the number of free electrons per gram.

Problem 3-23: Show that in the limit of complete ionization

$$\kappa_{es} = \frac{0.400}{\mu_e} \approx 0.200(1 + X_H) \tag{3-186}$$

where X_H is the hydrogen mass fraction.

Equation (3-186) has been found adequate for most of the problems of stellar interiors. It certainly does not apply when the temperature drops to the point where the majority of the electrons are not free. In those cases other types of scattering also occur. The problem is academic, however, since other sources of opacity (bound-free, free-free transitions) become many orders of magnitude greater than κ_{es} in those circumstances, and scattering can be neglected entirely in the total opacity.

Problem 3-24: Show that in an ionized gas composed of hydrogen and helium the electron-scattering opacity exceeds the free-free opacity only if

$$T > 4.5 \times 10^6 \rho^{0.286} \qquad \kappa_{es} > \kappa_{f\text{-}f} \tag{3-187}$$

The Thomson cross section is a particularly simple and convenient one. Because it is frequency-independent, there is no Rosseland mean to perform (if other opacity sources are negligible). Because the scattering phase function is proportional to $1 + \cos^2 \theta$, the cancellation made in the diffusion theory of radiative transfer is also satisfied. Unfortunately this happy situation is not valid at the highest temperature ranges of importance in stellar structure. When the photon energies become significant fractions of mc^2, recourse must be made to the quantum-mechanical calculation. The differential cross section is then given by the Klein-Nishina formula

$$d\sigma = \frac{d\Omega}{4\pi} \frac{\sigma_T \frac{3}{4}(1 + \cos^2 \theta)}{(1 + 2\epsilon \sin^2 \frac{1}{2}\theta)^2} \left[1 + \frac{4\epsilon^2 \sin^4 \frac{1}{2}\theta}{(1 + \cos^2 \theta)(1 + 2\epsilon \sin^2 \frac{1}{2}\theta)} \right] \tag{3-188}$$

where $\epsilon = h\nu/mc^2$. The integrated cross section is

$$\sigma = \tfrac{3}{4}\sigma_T \left\{ \frac{1 + \epsilon}{\epsilon^2} \left[\frac{2 + 2\epsilon}{1 + 2\epsilon} - \frac{\ln (1 + 2\epsilon)}{\epsilon} \right] + \frac{\ln (1 + 2\epsilon)}{2\epsilon} - \frac{1 + 3\epsilon}{(1 + 2\epsilon)^2} \right\} \tag{3-189}$$

Problem 3-25: Show that $\lim_{\epsilon \to 0} \sigma = \sigma_T$.

At high temperatures these exact formulas introduce significant corrections to the problem of radiative transfer. Not only does the scattering cross section decrease below the Thomson value, but it can be seen that the scattering phase function no longer contains only even powers of $\cos \theta$. Thus the scattering of photons into and out of a pencil of radiation no longer cancels.[1] The energy of the scattered photon, furthermore, is significantly reduced by the scattering process at high energy because the fractional shift in frequency $\Delta\nu/\nu$ is of order $h\nu/mc^2$. The effective result is that the opacity is even smaller than that indicated by the Klein-Nishina formula.

Problem 3-26: Show that for moderate photon energies the angular factor $1 + \cos^2 \theta$ in the Thomson differential cross section becomes approximately

$$(1 + \cos^2 \theta) \left[1 + \frac{h\nu}{mc^2} (1 - \cos \theta) \right]^{-2}$$

[1] D. H. Sampson, *Astrophys. J.*, **129**:734 (1959). For the inclusion of electron degeneracy, see also C. Chin, *Astrophys. J.*, **142**:1481 (1966), for the effects of degeneracy.

Finally, it should be noted that at high energy the photons may create a positron-electron pair in the field of a nucleus; $\gamma + Z \rightarrow Z + e^+ + e$ if $h\nu > 2mc^2$. The cross section is of order

$$\sigma_{\text{pair}} \approx Z^2 \times 10^{-3}\sigma_T \qquad h\nu > 2mc^2$$

and so this cross section will not be an important source of opacity. The created positrons and electrons may be important, however. They increase the number of scatterers for the scattering opacity, and their annihilation may result in an important source of neutrinos.

(5) THE TOTAL RADIATIVE OPACITY

In the stellar interior the four sources of opacity discussed previously are the only ones of major importance. Which form of opacity dominates depends upon the thermodynamic state, but the general trend is as follows. At low temperature, when a significant number of nuclei are only partially ionized, the opacity is dominated by bound-bound and bound-free absorption by the bound electrons. As the ionization nears completion, the opacity due to free-free transitions becomes dominant, but because the Rosseland mean of $\kappa_{\text{f-f}}$ decreases with increasing temperature, a temperature will ultimately be reached where the scattering by free electrons dominates. Of course, all forms of opacity contribute simultaneously, and what must be done is first to sum the frequency-dependent opacities, correcting the true absorption for induced emission,

$$[\kappa_{\text{b-b}}(\nu) + \kappa_{\text{b-f}}(\nu) + \kappa_{\text{f-f}}(\nu)](1 - e^{-h\nu/kT}) + \kappa_s$$

and then form the Rosseland mean. It should also be emphasized that in general the sum of the Rosseland mean opacities of each component is not equal to the Rosseland mean of the sum. That is, if $\kappa_1(\nu)$ and $\kappa_2(\nu)$ represent two different sources of opacity with Rosseland means

$$\frac{1}{\langle\kappa_1\rangle} = \int \frac{f(\nu)}{\kappa_1(\nu)}\,d\nu \qquad \frac{1}{\langle\kappa_2\rangle} = \int \frac{f(\nu)}{\kappa_2(\nu)}\,d\nu$$

then $\langle\kappa_1\rangle + \langle\kappa_2\rangle \neq \langle\kappa_1 + \kappa_2\rangle$.

Problem 3-27: Show that the Rosseland means $\langle k_1\rangle$ and $\langle k_2\rangle$ are additive if they are both true-absorption opacities and if they have the same frequency dependence, e.g., free-free absorption from two different ions Z_1 and Z_2.

The Rosseland mean of an individual component therefore has little meaning in radiative transfer unless that opacity is the dominant one. It will be instructive, however, to compare the means of specific components of the opacity to the mean of the total opacity as a nonlinear measure of the importance of the component. The following simplified problem of a pure hydrogen gas illustrates some of these features.

Problem 3-28: Calculate the following details of the opacity of a pure hydrogen gas at a density $\rho = 10^{-3}$ g/cm³. (*a*) At what temperatures are the Rosseland means of the free-free opacity and bound-free opacity equal? (Use the Debye-Hückel model to estimate the degree of ionization.) (*b*) What are the values at that temperature of $\langle \kappa_{\text{b-f}} \rangle = \langle \kappa_{\text{f-f}} \rangle$, κ_{es}, and $\langle \kappa_{\text{tot}} \rangle$? (*c*) At what temperature does $\langle \kappa_{\text{f-f}} \rangle = \kappa_{\text{es}}$? (*d*) Is the total opacity in this last case equal to $\langle \kappa_{\text{f-f}} \rangle + \kappa_{\text{es}}$? *Note:* The lengthy calculation of part (*a*) contains several subtleties for the serious student. The temperature is found by self-consistent iteration.

The trends in opacity problems can best be illustrated with a few figures. Figure 3-15 shows a rough division of the ρT plane into the dominant sources of opacity for an average population I mixture. In the high-temperature region the electron-scattering opacity $\kappa_{\text{es}} = 0.20(1 + X_{\text{H}})$ dominates. As the temperature decreases, $\kappa_{\text{b-f}}$ overtakes κ_{es} at low densities and $\kappa_{\text{f-f}}$ overtakes κ_{es} at inter-

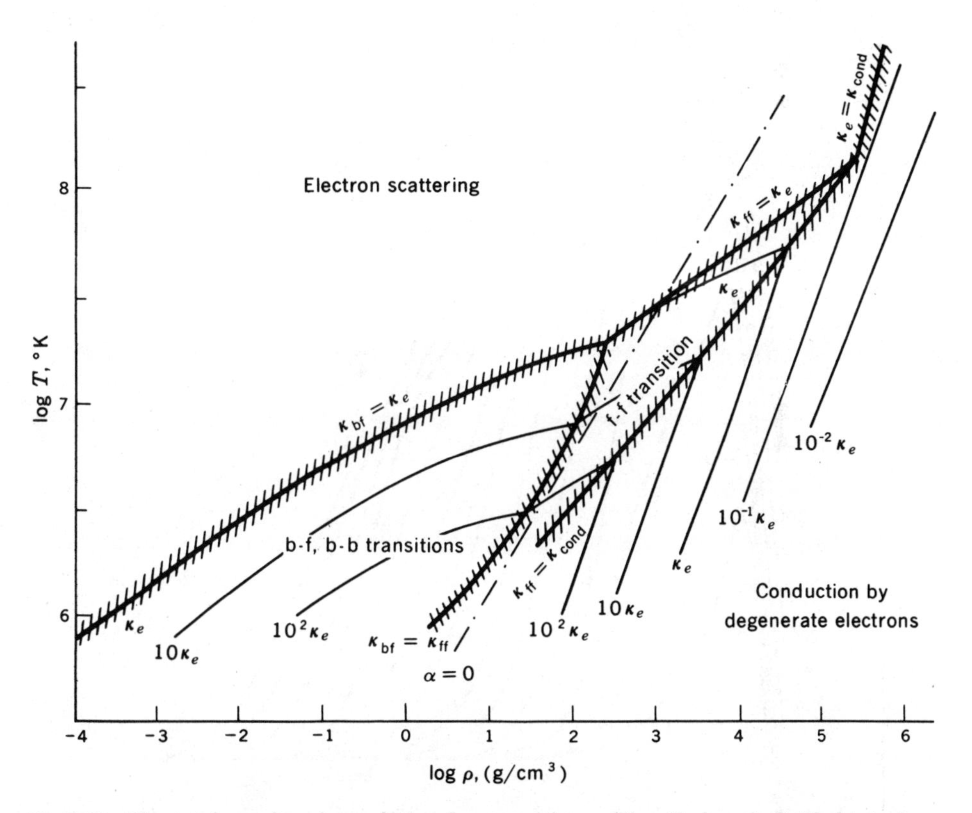

Fig. 3-15 The total opacity of population I composition. The ρT plane is divided into four domains according to which opacity source is the most important for energy transport, electron scattering, bound-free transitions, free-free transitions, and conduction by degenerate electrons (to be discussed in Sec. 3-4). The lines designating these boundaries are cross-hatched. Contours of constant opacity are labeled by the value of κ in terms of the opacity κ_e due to electrons. A dashed line shows where the degeneracy parameter $\alpha = 0$. (*After C. Hayashi, R. Hoshi, and D. Sugimoto, Progr. Theoret. Phys. Kyoto, Suppl.* 22, 1962.)

mediate densities. At a given value of temperature (say 3×10^6 °K) the dominant mechanisms are electron scattering at very low density, bound absorption at low density, free-free absorption at intermediate density, and electron conduction (to be discussed later) at high density. (Electron conduction is not a radiative opacity. It is a physically different mechanism of heat transport, which becomes more efficient than radiative transport at very high density.) Also shown as fine-lined curves are the loci of constant opacity, each curve being labeled by a factor indicating how much greater the opacity is than the electron-scattering opacity.

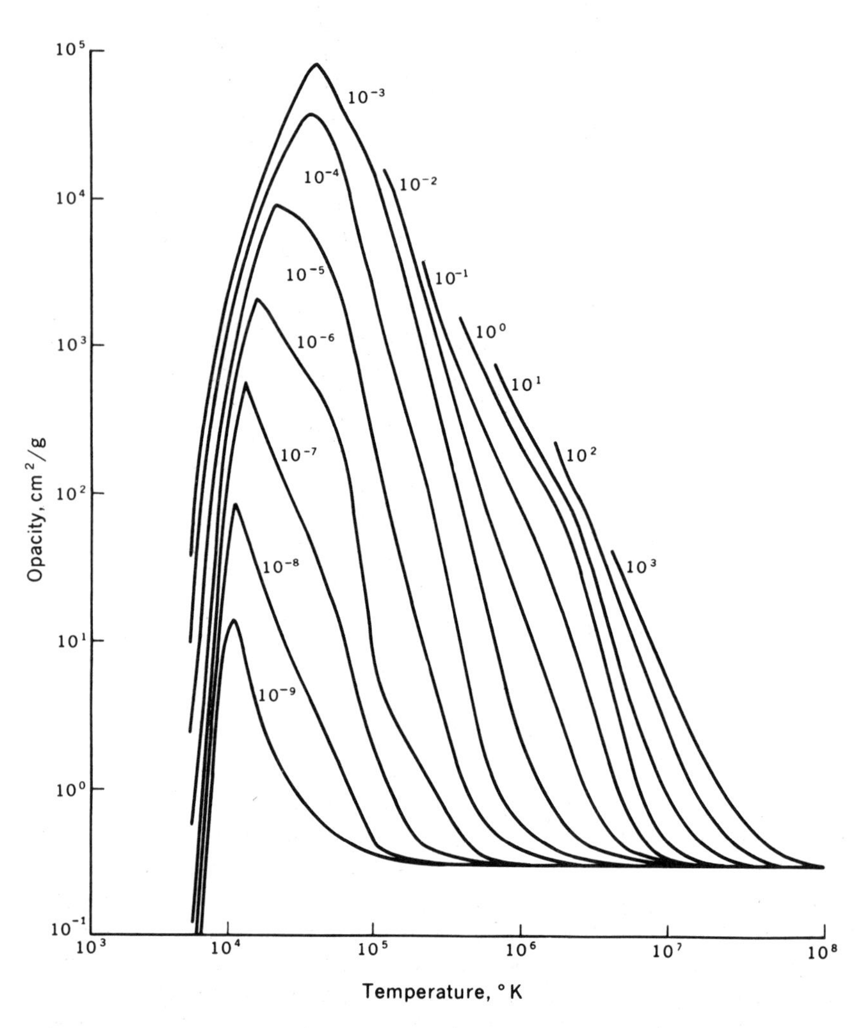

Fig. 3-16 The total opacity of material of solar composition as a function of temperature. Each curve is labeled by the value of the density. The range of values shown was chosen to illustrate the characteristic values of the opacity within the sun. [*After D. Ezer and A. G. W. Cameron, Icarus,* **1**:422 (1963).]

Arthur Cox and his coworkers[1] at Los Alamos have computed opacities for 11 different composition mixtures. Many of their results for the composition $X = 0.596$, $Y = 0.384$, $Z = 0.020$ serve as outstanding illustrations of the results obtained from detailed computations. Figure 3-16 shows the value of the total opacity as a function of temperature for various values of the density. Two outstanding features of this figure deserve explicit comments. The large peak between 10^4 and 10^5 °K is due to the ionization of the principal constituents, hydrogen and helium. At lower temperatures there are relatively few photons with sufficient energy for these ionization processes, so that the Rosseland mean of the cross section is small, whereas at higher temperatures the ionization greatly reduces the number of bound electrons per gram, with the result that the bound absorption must again be small. The important result is that the largest radiative opacities are found in the hydrogen and helium ionization zones in stars. As a result the temperature gradients required to radiatively transport energy through these ionization zones are so large that the zones are almost always unstable against convection. Second, we note that the opacity approaches the electron-scattering opacity at high temperature.

One of the outstanding contributions of the Los Alamos group has been their demonstration of the importance of the hitherto neglected bound-bound absorption. Figure 3-17 shows their calculation of the ratio of the total opacity including bound-bound absorption to the opacity neglecting that contribution. Apparently the bound-bound absorption can increase the opacity by as much as a factor of 3 in special regions of the ρT plane. A diagonal line shows the present conditions in the sun, from which it can be seen that the contribution of the lines is very relevant in astrophysics. This result is for a given composition, but much the same thing is to be expected for other compositions. Of course, the bound absorptions and the free-free absorptions are both strongly dependent upon the fraction by weight Z of the heavy elements. Thus in first approximation the true absorption is proportional to Z. For a given heavy-element concentration Z, on the other hand, it is not difficult to see how the opacity depends upon the abundance of hydrogen relative to helium. The opacity is approximately proportional to the number of free electrons per gram for the following reasons: (1) from the Saha equation it follows that the number of bound electrons in a highly ionized gas is proportional to the free-electron density, so that for a given temperature and mass density the number of bound electrons is proportional to μ_e^{-1}; (2) the free-free absorption and the electron scattering are explicitly dependent upon the number of free electrons per gram, i.e., to μ_e^{-1}. Since $\mu_e^{-1} \approx (1 + X)/2$, the opacities for fixed density, temperature, and heavy-element concentration are nearly proportional to $1 + X$.

Although the Rosseland means are nonadditive, the comparison of the mean of a constituent opacity to the mean of the total opacity is nonetheless a useful indication of the importance of the constituent. Several curves of this type will

[1] For a review see A. Cox, Stellar Coefficients and Opacities, in L. H. Aller and D. B. McLaughlin (eds.), "Stellar Structure," The University of Chicago Press, Chicago, 1965; also Cox, Stewart, and Eilers, *loc. cit.*

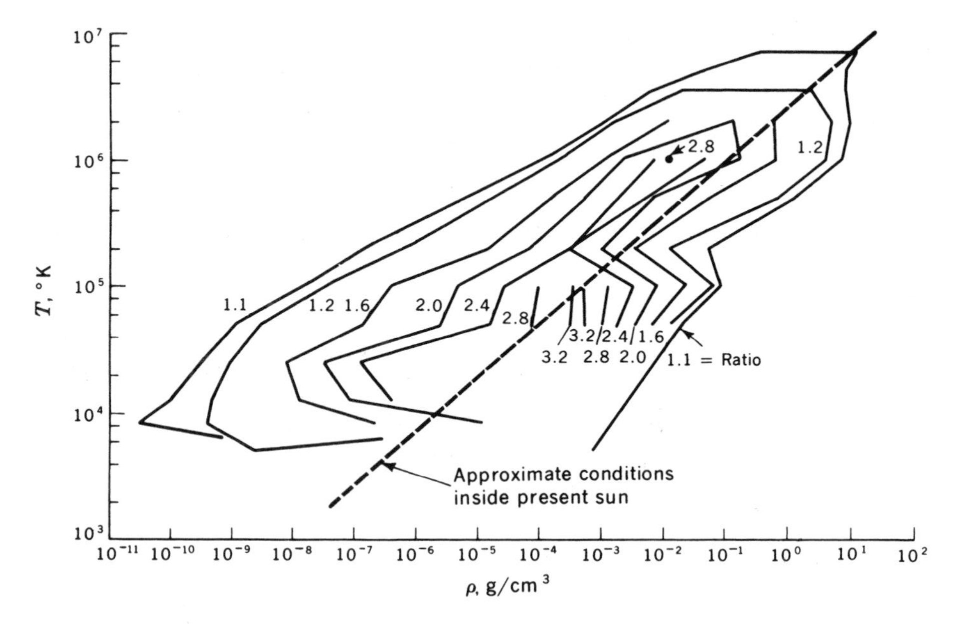

Fig. 3-17 The ratio of the total radiative opacity with discrete atomic lines included to its value without the discrete lines. The conditions within the sun are shown as a point of reference. It can be seen that atomic transitions increase the value of the opacity by a significant factor in the sun. [*Reprinted from A. Cox, Stellar Coefficients and Opacities, in L. H. Aller and D. B. McLaughlin (eds.), "Stellar Structure," The University of Chicago Press, Chicago,* 1965. *By permission of The University of Chicago Press. Copyright* 1965 *by The University of Chicago.*]

be presented. They are also based upon the work of Cox et al.[1] on a mixture $X = 0.596$, $Y = 0.384$, $Z = 0.020$. The *number fractions* are:

H	0.859
He	0.139
C	0.000342
N	0.000096
O	0.000766
Ne	0.000431
Na	0.000002
Mg	0.000022
Al	0.000001
Si	0.000027
Ar	0.000029
Fe	0.000004

[1] Cox, Stewart, and Eilers, *loc. cit.*

Figure 3-18 indicates how much of the opacity is due to hydrogen by presenting the ratio of the Rosseland mean of the bound-bound and bound-free opacity from hydrogen to the mean of the total opacity. It is evident that hydrogen contributes most of the opacity in the ionization zone near 10^4 °K. At temperatures higher than 10^5 °K the density of neutral hydrogen becomes insignificant.

Figure 3-19 shows, in a higher temperature-density domain, the importance of free-free opacity from H and He nuclei only. The decrease at high temperature reveals the $T^{-3.5}$ dependence of the free-free opacity, electron scattering taking over at high temperature. It is equally apparent that free-free opacity is a much more important source at high density than at low density. In fact, we have seen that $\kappa_{f\text{-}f}$ is proportional to the density whereas κ_{es} is constant.

Figure 3-20 shows the fraction of the total opacity due to C, N, O, and Ne, the most abundant of the minor constituents of the gas. In the lower-temperature regions of the graph, these elements contribute significantly, especially at

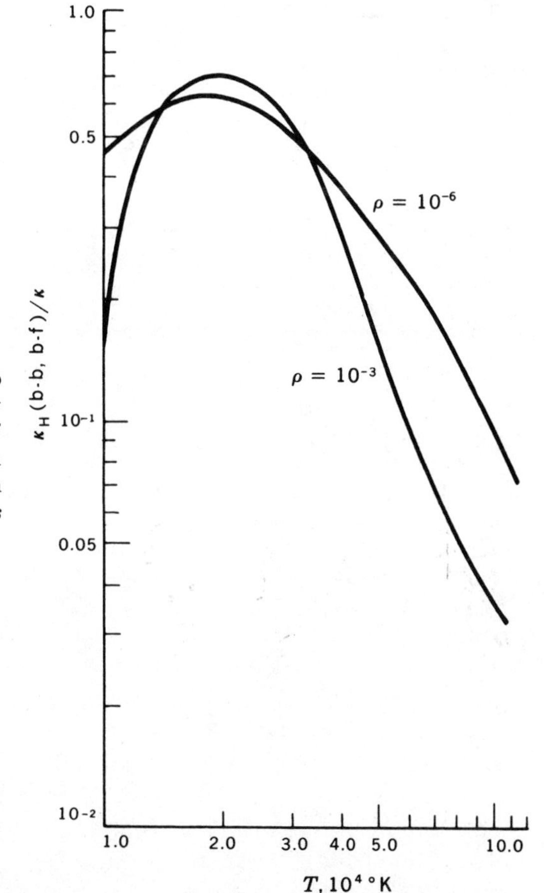

Fig. 3-18 Ratio of the opacity due to hydrogen bound states to the total opacity in a population I gas. This source of opacity is very important in the hydrogen ionization zone, but it is negligible deep in the interior. (*Prepared from calculations provided by Arthur N. Cox.*)

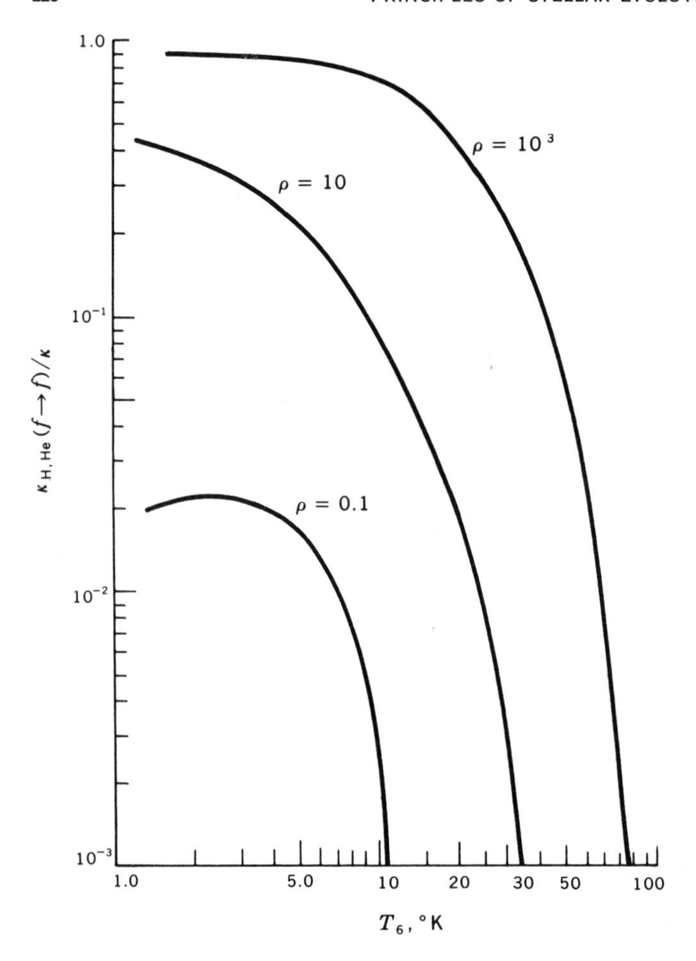

Fig. 3-19 Ratio of the opacity due to free-free transitions in the field of H and He nuclei to the total opacity in a population I gas. The transitions are most important at relatively high densities and at moderate temperatures. Their importance decreases rapidly with increasing temperature for $T_6 > 10$. (*Prepared from calculations provided by Arthur N. Cox.*)

lower densities. This contribution is largely due to the bound-absorption processes. It can be seen that at densities as great as $\rho = 10^3$, on the other hand, these elements never contribute more than 10 percent of the total opacity, the reason being that free-free absorption from the more abundant H and He is very strong at densities that large, as was seen in Fig. 3-19. But at a density $\rho = 0.1$ and temperature $T = 10^6$, electrons bound to these nuclei account for about half of the total opacity of the gas.

And finally Fig. 3-21 shows the ratio of the electron-scattering opacity to the total opacity. Electron scattering becomes the dominant opacity near 10^8 °K for the three densities shown. It takes over more rapidly at low density, where the competition from free-free absorption is less strong than at high density. Below 10^6 °K electron scattering is negligible for all densities of interest.

Much more interesting information is contained in the details of the Los Alamos results. It should be mentioned that in numerical construction of stellar models, analytic approximations to opacities are generally discarded in favor of opacity tables, which may be stored in a computer memory as a grid between whose points the opacity can be interpolated for any T, ρ, and composition. Outstand-

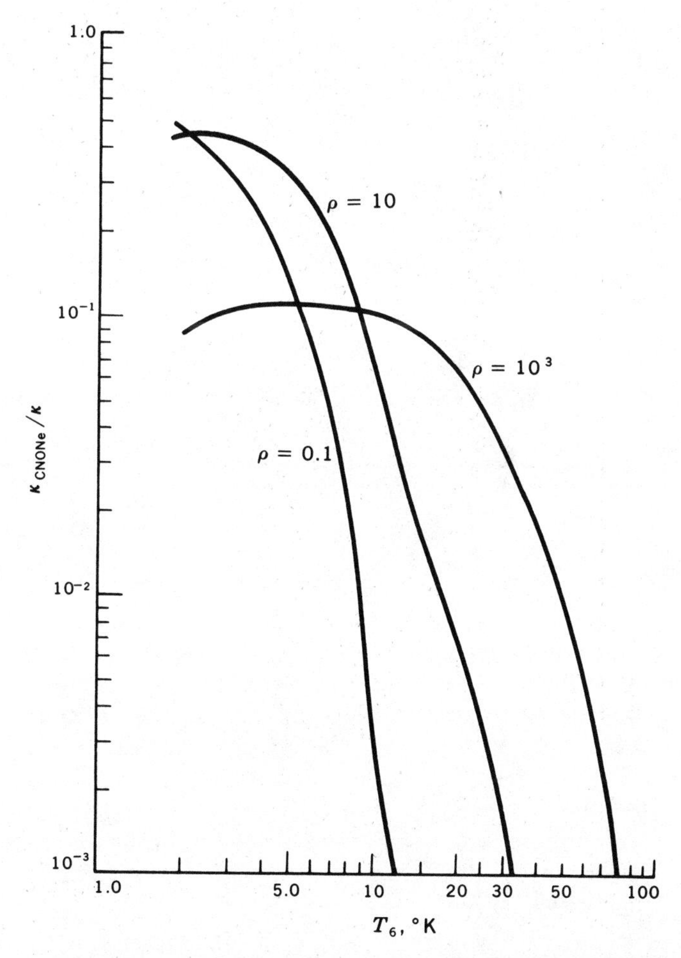

Fig. 3-20 Ratio of the opacity due to C, N, O, and Ne to the total opacity in a population I gas. (*Prepared from calculations provided by Arthur N. Cox.*)

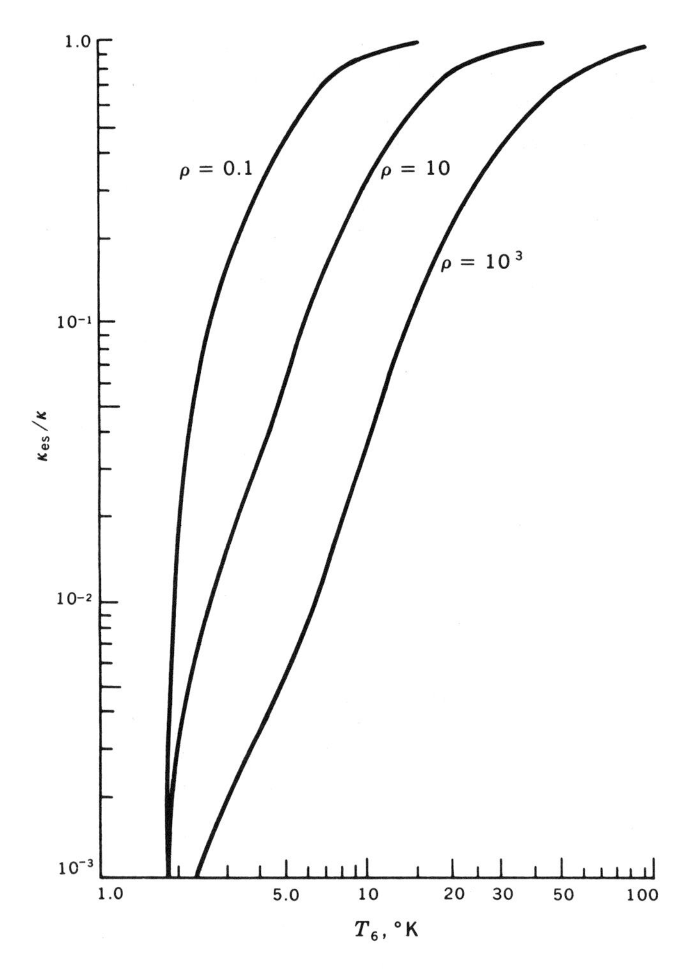

Fig. 3-21 Ratio of the opacity due to scattering from electrons to the total opacity in a population I gas. At high temperatures the opacity is dominated by electron scattering. The ratio decreases markedly with increasing density as a result of an increase in the importance of the other opacity sources. (*Prepared from calculations provided by Arthur N. Cox.*)

ing among these tables are those of Cox et al.,[1] who evaluate opacities for 11 different astrophysical compositions. Of course, interpolation within a grid is only a particularly complex form of an analytic opacity, and it is sometimes found by comparison against computed tables that a relatively simple formula may adequately represent the opacity over the range of thermodynamic states anticipated for a particular calculation. The advantage of an analytic approximation, say the Kramers opacity $\kappa = \kappa_0 \rho T^{-3.5}$, is that it often allows interesting theorems to be analytically established from the differential equations. The outstanding account of analytic investigations of stellar structure is to be found in Chandrasekhar.[2]

[1] *Ibid.*

[2] S. Chandrasekhar, "An Introduction to the Study of Stellar Structure," Dover Publications, Inc., New York, 1957.

In particular, he derives (following Eddington) the luminosity of the standard model with Kramers opacity. From the luminosity formula, Eq. (3-63), we have

$$\mathfrak{L} = \frac{4\pi c G \mathfrak{M}(1 - \beta_c)}{\overline{\kappa\eta}} \tag{3-190}$$

Both the factors $1 - \beta$ and $\kappa\eta$ are constant within the standard model. If the opacity is taken to be

$$\kappa = \kappa_0 \rho T^{-3.5} \tag{3-191}$$

then

$$\overline{\kappa\eta} = \kappa_0 \eta_c \rho_c T_c^{-3.5}$$

which is, for a nondegenerate gas, equal to

$$\overline{\kappa\eta} = \kappa_0 \eta_c \frac{a}{3} \frac{\mu}{N_0 k} \frac{\beta_c}{1 - \beta_c} T_c^{-\frac{1}{2}} \tag{3-192}$$

Problem 3-29: Use the foregoing formulas along with the others appropriate to the standard model, Eq. (2-318) for T_c and Eq. (2-307) for the relationship of $\mathfrak{M}$ to $1 - \beta_c$, to show that for the standard model

$$\frac{\mathfrak{L}}{\mathfrak{L}_\odot} = 1.8 \times 10^{25} \frac{1}{\kappa_0 \eta_c} \left(\frac{\mathfrak{M}}{\mathfrak{M}_\odot}\right)^{5.5} \left(\frac{R_\odot}{R}\right)^{0.5} (\mu\beta_c)^{7.5} \tag{3-193}$$

Although this formula results in a somewhat steeper mass-luminosity relationship than the observed $\mathfrak{L}/\mathfrak{L}_\odot \approx (\mathfrak{M}/\mathfrak{M}_\odot)^4$, it does show that the mass enters as a fairly high power. Equation (3-193) also shows that for a given mass and energy source, the luminosity depends upon composition roughly as

$$\mathfrak{L} \propto \frac{\mu^{7.5}}{\kappa_0} \tag{3-194}$$

This rule of thumb will aid in understanding the main sequence.

Chandrasekhar also demonstrates[1] that the outer layers of stars in radiative equilibrium having Kramers opacity and low radiation pressure are very similar to the standard model. Furthermore, he demonstrates[2] that the interior structure of such stars is a polytrope of index $n = 3.25$ if the energy generation ϵ is constant, again much like the standard model. If the energy is all liberated at the center, on the other hand, the polytropic index varies continuously from the value $n = 3.25$ at the radiative surface to the value $n = 1.5$ at the convective center. Such conclusions are very illuminating, and although they will not be repeated in this book, their study is recommended to the serious student.

The energy-transport problem is probably the most difficult one of the many that enter into the theory of the structure of stars, and one of the most involved

[1] *Ibid.*, chap. 8.

[2] *Ibid.*, chap. 9.

aspects of that problem is the radiative opacity. Much has been accomplished, but much more remains to be done.

3-4 CONDUCTION

Consider the motion of charged particles in an ionized gas. If the gas is nondegenerate, each particle has average energy $E_T = \frac{3}{2}kT$. That energy corresponds to a velocity $V_T = (2E_T/m)^{\frac{1}{2}} = (3kT/m)^{\frac{1}{2}}$. It is apparent that electron velocities are greater than ion velocities by a factor like

$$\frac{V_e}{V_{\text{ion}}} = \left(\frac{m_{\text{ion}}}{m_e}\right)^{\frac{1}{2}} \approx 43\sqrt{A_{\text{ion}}} \tag{3-195}$$

If the electron gas is degenerate, the ratio is even larger, since the electrons are forced into higher momentum states whereas the ions retain the same thermal velocity. Thus it is physically correct to think of the electron gas as being composed of fast-moving particles relative to a nearly stationary background of positive ions.

If the electron gas has a temperature gradient, it seems perfectly clear that the high-velocity, high-density, high-temperature electrons will migrate more rapidly toward lower temperatures than the converse. If that is so, a net flux of electrons will move down the temperature gradient! Such a particle flux in a star obviously contradicts the usual assumption of constant particle density in a static star. It would also destroy charge neutrality. But after the charge neutrality is thrown slightly out of balance, the resultant electric field will retard the further efflux of electrons. Thus a somewhat more refined picture of the gas in a star will include a small radial electric field capable of canceling the tendency of the heavy ions to settle toward the center and the tendency of the fast-moving electrons to migrate outward. Under those circumstances, there will be no net motion of electrons in any direction, and the configuration may be regarded as static.

There may, on the other hand, be a net transfer of energy without any net transfer of electrons. Figure 3-22 shows a unit area normal to the direction of the temperature gradient (chosen to be the x direction). The number of electrons per unit time passing through unit area in the $\pm x$ direction is designated by $j_\pm$. Particle conservation demands that $j_+ = j_-$. On the other hand, the average energy $\bar{E}_+$ of the particles passing through in the $+x$ direction need not equal the average energy $\bar{E}_-$ of those passing through in the $-x$ direction, especially if the electrons on the left (say) are hotter than those on the right. The net energy flux $Q_x = j_+\bar{E}_+ - j_-\bar{E}_-$ need not vanish, and, in fact, it will not vanish in a stellar interior. This simple discussion shows that the conduction heat flux will be proportional to the product of the electron current $j_\pm$ and the excess energy per particle moving in the direction of the heat flow. The value of j will depend only on the density and velocity of the electrons. It is the calculation of the excess energy $\bar{E}_+ - \bar{E}_-$ that is the difficult problem in conductive transport.

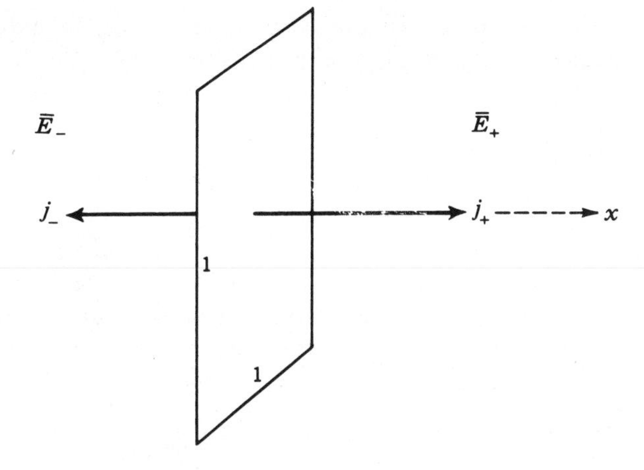

Fig. 3-22 The electron current per unit area in the $+x$ direction is j_+, and the average energy of those electrons is $\bar{E}_+$. The current j_- in the $-x$ direction must equal j_+ in the steady state, but there is nonetheless a net energy transport if $\bar{E}_- \neq \bar{E}_+$.

One can see that this excess energy will be roughly proportional to the product of how steeply the average energy changes with position times the average distance traveled by electrons between (major) collisions. The hotter electrons and ions will tend to lose energy to their cooler neighbors immediately down the temperature gradient. The many collisions will thereby transfer energy in a diffusion process not completely unlike that of radiative transfer. On the other hand, this mechanism of transporting energy is usually less efficient than radiative transfer, a fact that essentially comes about because photons travel much farther between collisions than electrons do; i.e., photons can "see a bigger temperature drop" than particles can. The following problem demonstrates this fact for a characteristic stellar environment.

Problem 3-30: Consider the problem of an ionized carbon gas at 10^6 °K and a density $\rho = 1$ g/cm^3. The radiative opacity is dominated by free-free transitions, and the Rosseland free-free absorption coefficient is given by Eq. (3-169):

$$\langle \sigma_{f\text{-}f} \rangle = 2.07 \times 10^{-25} \frac{6^2(6.02 \times 10^{23})/2}{(10^6)^{3.5}} = 2.3 \times 10^{-21} \text{ cm}^2$$

The $1/e$ length for the characteristic photon is then 4.5×10^{20} ions per square centimeter.

The same ions will scatter nondegenerate electrons according to the Rutherford differential cross section

$$\frac{d\sigma}{d\Omega} = \left(\frac{Ze^2}{mv^2}\right)^2 \frac{1}{(1 - \cos\theta)^2} = 5.2 \times 10^{-27} \left[\frac{Z}{E(\text{Mev})}\right]^2 \frac{1}{(1 - \cos\theta)^2} \qquad \text{cm}^2/\text{sterad}$$

Compute the number of ions per square centimeter in a $1/e$ length for electrons of average thermal energy to backscatter ($\theta > \pi/2$) in a single scattering. (It turns out that many small-angle scatterings accomplish the same thing in an even shorter distance.)

Ans: 3.1×10^{16} ions per square centimeter.

Now describe at each point $\mathbf{r}$ the distribution of electron velocities by a function $f(\mathbf{r},\mathbf{v})$. The notation $f(\mathbf{r},\mathbf{v})$ is used as shorthand for $f(x,y,z,v_x,v_y,v_z)$; that is, it is a scalar function of position and of the magnitude and direction of the electron velocity. Let it physically represent the fraction of electron momentum states occupied. In thermodynamic equilibrium (no temperature gradient) the distribution function is isotropic and is equal to (for electrons)

$$f_0 = \frac{1}{\exp(\alpha + E/kT) + 1} \tag{3-196}$$

The subscript zero is attached to designate f_0 as the equilibrium distribution. Let the discussion be restricted to nonrelativistic electrons.[1] The energy E is just the kinetic energy minus the continuum depression E_0 (regarded here as a positive number), which accounts, as we have seen, for the fact that the electrons are not electrostatically free. Then, defining

$$\xi = -\alpha kT + E_0 \qquad K = E + E_0 = \tfrac{1}{2}mv^2 \tag{3-197}$$

we have

$$f_0 = \frac{1}{\exp[(K - \xi)/kT] + 1} \tag{3-198}$$

This distribution function is isotropic; its value depends only upon the magnitude of $\mathbf{v}$. The *steady-state* distribution function in a temperature gradient will differ somewhat from f_0, however, and in a nonisotropic manner.

The current of particles and current of kinetic energy along any axis can be expressed in terms of the distribution function:

$$j_x = \frac{2m^3}{h^3} \int v_x f(\mathbf{v})\, d^3v \tag{3-199}$$

$$Q_x = \frac{2m^3}{h^3} \int K(v) v_x f(\mathbf{v})\, d^3v \tag{3-200}$$

Problem 3-31: Deduce Eqs. (3-199) and (3-200).

The task is to compute the energy flow Q_x subject to the steady-state demand that there be no particle current; that is, $j_x = 0$. (Here and in what follows we shall think of the x axis as being in the same direction as the temperature gradient.)

For small temperature gradient it is useful to linearize the problem in the first-order perturbations of the distribution function; i.e., it will be useful to write

$$f(\mathbf{v}) = f_0(v) + g(\mathbf{v}) \tag{3-201}$$

if $g(\mathbf{v})$, the perturbation of f_0, is very small compared to f_0. This perturbation is the explicit result of three physical features: (1) the temperature gradient, (2) the electric field, and (3) collisions. Since the distribution function is to be constant

[1] Conduction will be important only in a degenerate gas. The restriction to nonrelativistic electrons therefore restricts the calculation to densities less than a few times 10^6 g/cm^3. The relativistic problem may be treated in a similar way, as for instance in Chiu, *op. cit.*

in time in spite of the fact that it is not constant in space or direction, its time rate of change must be zero:

$$\frac{df}{dt} = \left(\frac{\partial f}{\partial t}\right)_{\text{temp grad}} + \left(\frac{\partial f}{\partial t}\right)_{\text{elec field}} + \left(\frac{\partial f}{\partial t}\right)_{\text{coll}} = 0 \tag{3-202}$$

Each of these three circumstances would cause the distribution function to change in a calculable way, but since f is unchanging, it must be of such a form that the physically distinct sources of change cancel each other. The derivatives in Eq. (3-202) are not partial derivatives in the mathematical sense (time nowhere appears explicitly) but are total derivatives with respect to specific partial causes. The student will be well advised to clarify this distinction in his own mind. The first two terms will be rather easy to evaluate, whereas the third will prove difficult.

Let us first ask how it is that a temperature gradient will lead to change of distribution function. Consider an infinitesimal pencil of particle velocities about the velocity $\mathbf{v}$. The number of such particles in a small volume will, after a short time, be replaced by new particles that have entered the volume from some other starting point. Because the points differ in temperature, the fraction of states occupied at the two points differs. The situation is illustrated in Fig. 3-23. If $f(\mathbf{r},\mathbf{v})$ is the distribution function at $\mathbf{r}$ at time $t = 0$, after a short time t it will have been replaced by particles that were at $t = 0$ at the position $\mathbf{r}' = \mathbf{r} - \mathbf{v}t$. The rate of change is

$$\left(\frac{\partial f}{\partial t}\right)_{\nabla T} = \lim_{t\to 0} \frac{f(\mathbf{r} - \mathbf{v}t, \mathbf{v}) - f(\mathbf{r},\mathbf{v})}{t} = -v_x \frac{\partial f}{\partial x} - v_y \frac{\partial f}{\partial y} - v_z \frac{\partial f}{\partial z} \tag{3-203}$$

Since the temperature gradient is taken to be in the x direction, the factors $\partial f/\partial y$ and $\partial f/\partial z$ vanish. Thus

$$\left(\frac{\partial f}{\partial t}\right)_{\nabla T} = -v_x \frac{df}{dx} = -v_x \frac{df}{dT}\frac{dT}{dx} \tag{3-204}$$

Since the gradient dT/dx is assumed small, the retention of first-order terms in its effect requires only the use of df_0/dT in place of df/dT. That is, the change of the distribution function with respect to temperature will be given in zeroth order by the change of the equilibrium function.

Problem 3-32: Show that

$$\frac{df_0}{dx} = -\frac{\partial f_0}{\partial K}\left(\frac{d\xi}{dT} + \frac{K - \xi}{T}\right)\frac{dT}{dx} \tag{3-205}$$

Do you understand that the term $d\xi/dT$ is required because all the factors that entered the definition of ξ are functions of the local thermodynamic state?

The change in f due to the electric field can be computed in a similar manner. In this case, however, the acceleration changes the *velocity* at all points in space. That is, the change in the distribution function results from the motion of parti-

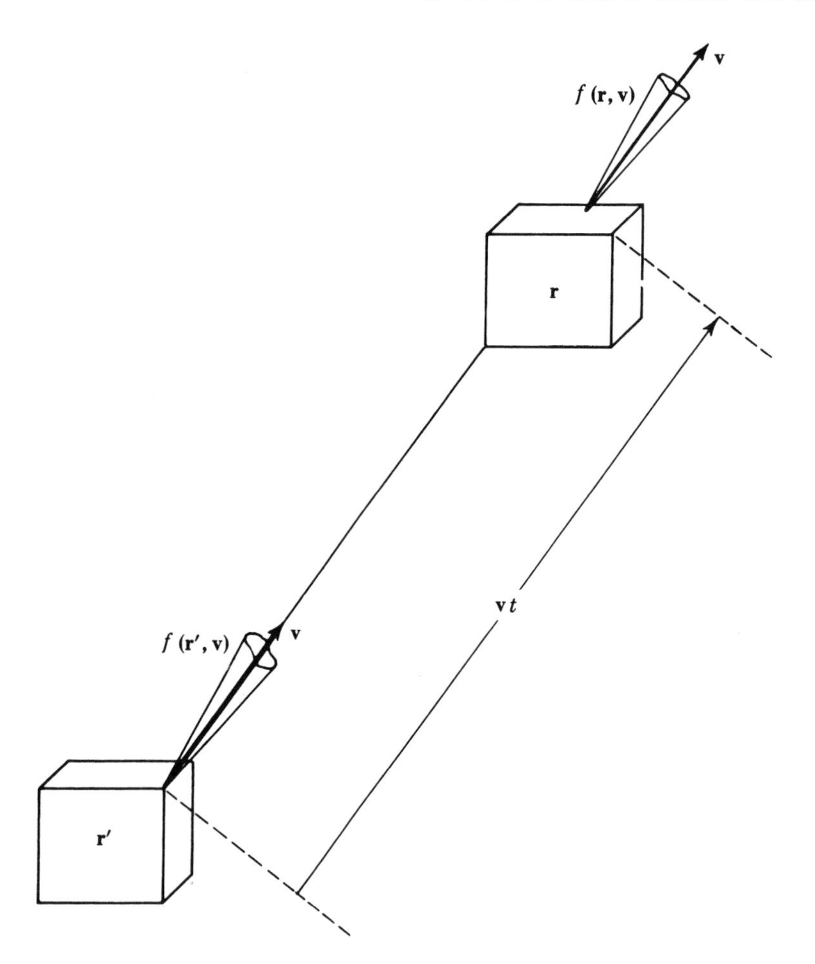

Fig. 3-23 The rate of change of the distribution function due to the existence of a temperature gradient. In a short time interval t the particles at $\mathbf{r}'$ having velocity $\mathbf{v}$ move to the point $\mathbf{r}$, but $f(\mathbf{r}',\mathbf{v}) \neq f(\mathbf{r},\mathbf{v})$ if there is a temperature gradient, so that this spatial migration tends to alter the distribution function.

cles in velocity space rather than in coordinate space. In a short time Δt, the electric field $\mathbf{E}$ would change velocity to a velocity $\mathbf{v} - (e/m)\mathbf{E}\,\Delta t$, thereby changing $f(\mathbf{r},\mathbf{v})$ to $f(\mathbf{r}, \mathbf{v} - (e/m)\mathbf{E}\,\Delta t)$. From this we conclude that

$$\left(\frac{\partial f}{\partial t}\right)_{\text{elec field}} = -\frac{e}{m} E_x \frac{\partial f}{\partial v_x} \tag{3-206}$$

where the electric field is, like ∇T, in the x direction. To first order in the perturbations f may again be approximated by f_0, since $E_x \rightarrow 0$ in the equilibrium state; i.e., the linear dependence on E_x already makes this expression correct to first order in the perturbations.

From Eq. (3-202) it must follow that

$$\left(\frac{\partial f}{\partial t}\right)_{\text{coll}} = v_x \frac{df}{dT}\frac{dT}{dx} + \frac{e}{m} E_x \frac{\partial f}{\partial v_x} \qquad (3\text{-}207)$$

Equation (3-207) is called the *Boltzmann transport equation.* When, as in this case, the perturbation to f will be small, the linearized Boltzmann transport equation results:

$$\left(\frac{\partial f}{\partial t}\right)_{\text{coll}} = v_x \frac{df_0}{dT}\frac{dT}{dx} + \frac{e}{m} E_x \frac{\partial f_0}{\partial v_x} \qquad (3\text{-}208)$$

Problem 3-33: Show by manipulating the right-hand side of Eq. (3-208) that

$$\left(\frac{\partial f}{\partial t}\right)_{\text{coll}} = v_x \frac{\partial f_0}{\partial K}\left[eE_x - \left(\frac{d\xi}{dT} + \frac{K-\xi}{T}\right)\frac{dT}{dx}\right] \qquad (3\text{-}209)$$

The calculation of conduction usually proceeds along the following lines.[1] One next tries to calculate from the details of the scattering processes the rate at which the distribution function would be changed by collisions. That is, given the electron $f(\mathbf{v})$, one calculates explicitly the rates of electron scatterings into and out of specific volumes in phase space. In fact, for $f(\mathbf{v}) = f_0(v) + g(\mathbf{v})$, we try to calculate the term in the form

$$\left(\frac{\partial f}{\partial t}\right)_{\text{coll}} = -\frac{g(\mathbf{v})}{\tau(v)} \qquad (3\text{-}210)$$

where $\tau(v)$ is the *relaxation time* for the perturbation of the distribution function. That is, if the temperature gradient and electric field were suddenly turned off, we see that the distribution function would relax to its equilibrium value due to collisions according to

$$f(\mathbf{v},t) = f_0(v) + g(\mathbf{v},0)e^{-t/\tau(v)} \qquad (3\text{-}211)$$

It remains to be shown, of course, that Eq. (3-210) can conveniently be obtained in such a form. If it can, the idea will be to insert it in Eq. (3-209) and to calculate $g(\mathbf{v})$ in terms of E_x and dT/dx. That $g(\mathbf{v})$ can then be used to calculate j_x and Q_x. Finally, the demand that $j_x = 0$ will allow E_x to be eliminated, thereby giving Q_x in terms of dT/dx.

The calculation of the scattering rates in a very dense plasma is a difficult problem, however. The coulomb force actually yields an infinite total cross section for an isolated particle, but in the plasma collective effects become very

[1] The treatment is that of L. Mestel, *Proc. Cambridge Phil. Soc.*, **46**:331 (1950). It is a relatively simple argument that gives the correct answer for the most common circumstance, but see the later discussion. Mestel's treatment was adapted from that of R. E. Marshak, *Ann. New York Acad. Sci.*, **1941**:49, who was apparently the first to adapt the theory of metallic conductivity to the stellar problem.

important in determining the scattering. The energy is of course carried by the electrons. Because conduction is important only in a degenerate electron gas, the electrons are faster than the ions by at least a factor of 100. Thus the ions can be considered as static, but their average position relative to each other does turn out to be important. Since differing problems are met in considering the scattering from electrons and the scattering from the ions, we shall briefly consider them both.

The scattering of electrons by electrons is usually ignored except in a special sense to be mentioned shortly. For one thing the scattering from electrons must be smaller than from ions because the scattering cross section is proportional to Z^2. But even more compelling reasons are to be found in the degeneracy of the electron gas. For reasons mentioned in Chap. 2 the electron gas becomes more perfect as the density increases. Thus the best zero-order approximation to a degenerate electron gas is a uniform sea of negative charge. And to the extent that the charge density is uniform, the electrons do not scatter from it at all. In the scattering of an electron from an electron, furthermore, the momenta of both electrons are charged; but the rate of any process involving the change of momentum of an electron in a degenerate gas must be multiplied by the probability that the final electron state is unoccupied. This feature significantly reduces the scattering of electrons from ions, and it will suppress electron-electron scattering even more because *both* electrons must have an empty final state. For all these reasons electron-electron scattering is ignored in the conductivity calculation. The exception occurs to the extent that the ions are shielded by ion-induced perturbations of the uniform electron density.

The scattering from the ions, then, constitutes the major scattering with which we are concerned. Here, too, the collective effects may be very important. One effect is that the ions also shield each other in the following sense. One can think of an electron as scattering from a given ion if it comes so close to the ion that the interaction is dominated by that ion. As one imagines larger and larger impact parameters, which lead to more and more forward scattering, a point is eventually reached where the impact parameter is halfway to the next ion.[1] In such cases the electron cannot be thought of as scattering from an individual ion. This cutoff has the happy feature of removing the infinity in the coulomb-scattering cross section, an infinity that results from extreme forward scattering due to collisions at very large impact parameters.

This cutoff due to the collective action of neighboring ions suggests another difficult feature of the calculation: the scattering depends upon the average relative position of the ions. The value of the conductivity depends upon whether the positions of the ions are correlated (an extreme example being a lattice) or uncorrelated (an extreme example being a classical gas). The amount of scatter-

[1] The situation is actually much more complicated. See, for instance, R. S. Cohen, L. Spitzer, and P. McR. Routly, *Phys. Rev.*, **80**:230 (1950); L. Spitzer and R. Harm, *Phys. Rev.*, **89**:977 (1953), and references contained therein, particularly to the work of Chandrasekhar.

ing depends upon this correlation. If the ions are arranged in a perfect lattice, the potential is exactly periodic, and the electrons move as if they were free. For example, in many solid-state problems where the lattice is nearly perfect the scattering is due solely to imperfections or impurities in the lattice structure. If the distribution of scattering centers is random, on the other hand, the scattering is due to an incoherent sum of the scattering from individual ions. An example is the gaseous state. Thus in a high-density degenerate gas the conductivity depends upon the state of the ion component. The ions are, in fact, most often in a state analogous to a gas, but at very high density and low temperature, the coulomb interactions of the gas become greater than kT, and the ions arrange themselves into a solidlike lattice. For intermediate conditions the ions behave more like a liquid. The division of the ρT plane according to the state of the ion component is shown in Fig. 3-24. The upshot is that a general calculation of the conductivity should proceed by a fundamental treatment capable of incorporating the positional correlation of the ion component. In the treatment to follow, however, which follows Mestel's analysis, the ions are regarded as being independent and uncorrelated. In defense of this approximation it may only be surmised that for those astrophysical calculations for which the exact value of the conductivity is important, the ions are usually in the gaseous state. For such high density that the ions are liquidlike or solidlike the conductivity is so large that the medium will be very nearly isothermal. And inasmuch as the importance of conductivity in stellar structure lies in its role in the determination of the temperature gradient required to transport a given heat flux, the greatest accuracy is required in the lower range of conductivities. We shall therefore use Mestel's treatment of uncorrelated ions, although it seems fair to say that much more work could usefully be performed on the conductivity problem.[1]

Let $R_+(\mathbf{r},\mathbf{v})\, dx\, dy\, dz\, dv_x\, dv_y\, dv_z$ be the rate at which electrons in the volume $dx\, dy\, dz$ at $\mathbf{r}$ are scattered into the velocity-space volume $dv_x\, dv_y\, dv_z$ about velocity $\mathbf{v}$, and let $R_-(\mathbf{r},\mathbf{v})$ be the corresponding differential rate for scattering out of that volume. R_+ and R_- can be computed from the distribution function and the differential-scattering cross section for electrons on ions. The differential-scattering cross section $d\sigma/d\Omega\ (\theta,v)$ has the following meaning: if an ion is bombarded by a flux F ($\text{cm}^{-2}\ \text{sec}^{-1}$) electrons, the number of electrons scattered per unit time by an angle θ into a solid angle $d\Omega$ is $F(d\sigma/d\Omega)\, d\Omega$. The scattering occurs because of the coulomb force, and the differential cross section is a function of the charge of the scattering ion, the velocity of the electron, and the scattering angle θ. Let us therefore picture a pencil of electrons of velocity $\mathbf{v}'$ in the range d^3v' and calculate the rate of scattering into and out of that pencil. The distribution function has been so defined that the number density of electrons moving in the pencil is the product of the number of continuum states in

[1] Considerable clarification has been provided by W. B. Hubbard, *Astrophys. J.*, **146**:858 (1966), whose approach by way of the Kubo relations avoids dependence on the Boltzmann equation.

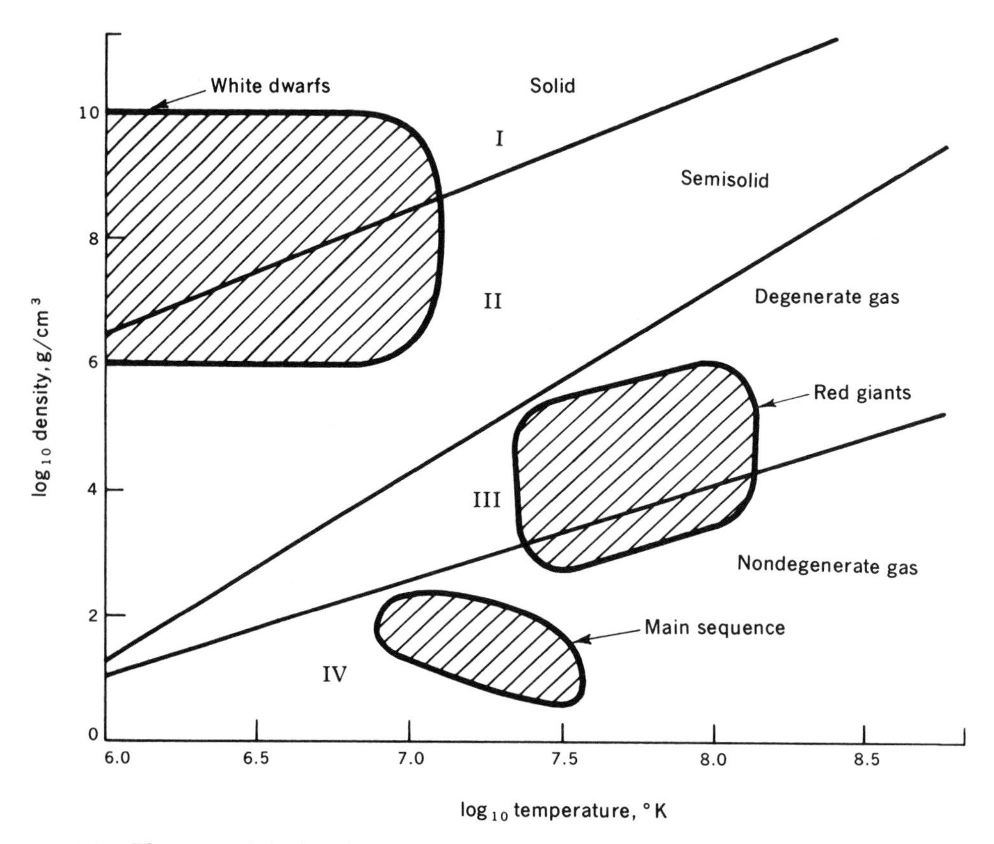

Fig. 3-24 The state of the ions in an ionized medium depends upon temperature and density. Those two quantities determine the relative importance of the coulomb interaction energy per ion and kT. When U_c/kT is greater than unity, the ions are forced into a lattice structure in order to minimize the potential energy, and the state may be described as solid. In region II, $U_c/kT \sim 1$, the ion spheres interpenetrate to a considerable degree, and the state may be described as semisolid. At the somewhat smaller values of U_c/kT in region III, the electron gas remains strongly degenerate, but the ions now move as a nondegenerate gas. In region IV both ions and electrons are nondegenerate gases. The conductivity depends not only upon the state of the electron gas but on the state of the ion gas as well, by virtue of the degree of correlation among ionic scattering centers. [*R. A. Wolf, Phys. Rev.*, **137**:B1634 (1965).]

the pencil times the distribution function:

$$n_e(\mathbf{v}')\, d^3v' = \frac{2m^3}{h^3}\, d^3v'\, f(\mathbf{v}') \tag{3-212}$$

With this definition the distribution function $f(\mathbf{v}')$ *is the probability that an available electron state is occupied.* [Note that Eq. (3-212), as well as all that follows, is limited to nonrelativistic momenta, since the number of continuum states $(2/h^3)\, d^3p'$ has been replaced by $(2m^3/h^3)\, d^3v'$. It should also be clear that the symbol d^3v' is being used to designate the differential volume element $dv'_x\, dv'_y\, dv'_z$ in velocity space.] The flux of electrons in the pencil is the product of the number density of electrons in the pencil times their velocity:

$$\mathbf{F}(\mathbf{v}')\, d^3v' = \mathbf{v}' \frac{2m^3}{h^3} f(\mathbf{v}')\, d^3v' \tag{3-213}$$

It will be more convenient to represent the flux of particles in terms of the cone of directions $d\Omega'$ in which they move. Just as a differential volume in space $dx\, dy\, dz$ can be replaced by $r^2\, dr\, d\Omega$, so the flux of particles of velocity v' in the interval dv' moving in the set of directions $d\Omega'$ can be written

$$\mathbf{F}(\mathbf{v}')v'^2\, dv'\, d\Omega' = \mathbf{v}' \frac{2m^3}{h^3} f(\mathbf{v}')v'^2\, dv'\, d\Omega' \tag{3-214}$$

Now consider the fluxes contained in two cones of particles as shown in Fig. 3-25. We next compute the rate of change of the flux of particles in the $(\mathbf{v},d\Omega)$ pencil by scattering into and out of that pencil. Only two other considerations must first be mentioned. The Rutherford scattering of an electron from an ion does not (to high accuracy) change the speed of the electron, only its direction. Thus for the scattering into and out of the phase-space pencils, $|\mathbf{v}| = |\mathbf{v}'|$. Second, $d\sigma/d\Omega$ is usually taken to mean the cross section computed without considerations of the exclusion principle, but conduction will eventually prove most important in degenerate gases, where the exclusion principle *is* important. Thus the differential cross section for scattering into the pencil at $\mathbf{v}$ must be multiplied by $1 - f(\mathbf{v})$. Since $f(\mathbf{v})$ is, as we have seen, the probability that an electron state

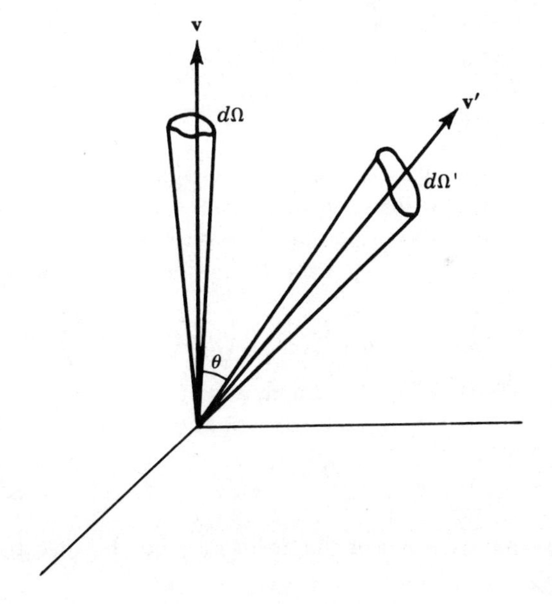

Fig. 3-25 Scattering of electrons from the pencil of directions $d\Omega'$ into the pencil $d\Omega$ alters the distribution function because $f(\mathbf{r},\mathbf{v}') \neq f(\mathbf{r},\mathbf{v})$.

is occupied, the factor $1 - f(\mathbf{v})$ represents the probability that it is empty, i.e., that the scattered electron has an unoccupied final state to go into.

The rate of increase in the electron flux in the pencil $(\mathbf{v},dv,d\Omega)$ due to scatterings of electrons from other pencils $(\mathbf{v}',dv',d\Omega')$ is

$$R_+(\mathbf{v})v^2\,dv\,d\Omega = N_i \int_{\Omega'} v' \frac{2m^3}{h^3} f(\mathbf{v}')v'^2\,dv'\,d\Omega' \frac{d\sigma(\theta)}{d\Omega} [1 - f(\mathbf{v})]\,d\Omega \tag{3-215}$$

where by elasticity we have $v = v'$ and $dv = dv'$, and where θ is the angle between $\mathbf{v}$ and $\mathbf{v}'$ and N_i is the number density of ions. Likewise the loss of electron density from the pencil is

$$R_-(\mathbf{v})v^2\,dv\,d\Omega = N_i \int_{\Omega'} v \frac{2m^3}{h^3} f(\mathbf{v})v^2\,dv\,d\Omega \frac{d\sigma(\theta)}{d\Omega} [1 - f(\mathbf{v}')]\,d\Omega \tag{3-216}$$

Several extraneous factors can now be eliminated. The difference between R_+ and R_- must represent the rate of change due to collisions of the specific electron flux in the pencil:

$$[R_+(\mathbf{v}) - R_-(\mathbf{v})]v^2\,dv\,d\Omega = \frac{2m^3}{h^3}\left[\frac{\partial f(\mathbf{v})}{\partial t}\right]_{\text{coll}} v^2\,dv\,d\Omega \tag{3-217}$$

Taking the indicated difference and recalling that the *speeds* are equal, one obtains

$$\left[\frac{\partial f(\mathbf{v})}{\partial t}\right]_{\text{coll}} = vN_i \int_{\Omega'} \frac{d\sigma(\theta)}{d\Omega} \{f(\mathbf{v}')[1 - f(\mathbf{v})] - f(\mathbf{v})[1 - f(\mathbf{v}')]\}\,d\Omega \tag{3-218}$$

which is the desired result. For the present problem, this result takes an even simpler form when it is recalled that we are assuming that the distribution function is only slightly perturbed from its equilibrium value by the small temperature gradient. That is, $f(\mathbf{v}) = f_0(v) + g(\mathbf{v})$. Expanding the products of distribution functions yields

$$\left[\frac{\partial f(\mathbf{v})}{\partial t}\right]_{\text{coll}} = vN_i \int_{\Omega'} \frac{d\sigma(\theta)}{d\Omega} [g(\mathbf{v}') - g(\mathbf{v})]\,d\Omega' \tag{3-219}$$

This equation emphasizes the fact that it is the nonisotropic nature of the perturbed distribution function that causes the change. (Remember that $|v| = |v'|$.)

Two expressions, one from the Boltzmann transport equation and one from the last explicit calculation, have now been derived for $(\partial f/\partial t)_{\text{coll}}$. For convenience Eq. (3-209) from the Boltzmann equation will be written

$$\left(\frac{\partial f}{\partial t}\right)_{\text{coll}} = v_x \frac{\partial f_0}{\partial K}\left[eE_x - \left(\frac{d\xi}{dT} + \frac{K - \xi}{T}\right)\frac{dT}{dx}\right] \equiv v_x \Pi(v) \tag{3-220}$$

where $\Pi(v)$, so defined, is a function of the magnitude of v but not its direction. From Eqs. (3-219) and (3-220) we seek a relaxation solution of the form

$$\left(\frac{\partial f}{\partial t}\right)_{\text{coll}} = -\frac{g(\mathbf{v})}{\tau(v)} \tag{3-221}$$

An equation of this form may be obtained by the following manipulations. Put the desired form, $g(\mathbf{v}) = -\tau(v)(\partial f/\partial t)$, into Eq. (3-219):

$$\left(\frac{\partial f}{\partial t}\right)_{\text{coll}} = +vN_i\tau(v) \int \frac{d\sigma}{d\Omega} \left\{ \left[\frac{\partial f(\mathbf{v})}{\partial t}\right]_{\text{coll}} - \left[\frac{\partial f(\mathbf{v}')}{\partial t}\right]_{\text{coll}} \right\} d\Omega' \qquad (3\text{-}222)$$

Substitution of Eq. (3-220) from the Boltzmann equation into the integrand of Eq. (3-222) produces

$$v_x\Pi(v) = \left(\frac{\partial f}{\partial t}\right)_{\text{coll}} = vN_i\tau(v)\Pi(v) \int \frac{d\sigma}{d\Omega}\,(v_x - v_x')\,d\Omega' \qquad (3\text{-}223)$$

From this last equation it is apparent that $\tau(v)$ can be found if the integral yields something proportional to v_x. If it does, v_x cancels out, and a τ that is a function only of the magnitude of v results.

The next step in the calculation thus becomes the evaluation of the integral in Eq. (3-223). Careful definition of a coordinate system permits successful analytic evaluation of the integral over Ω', but the result can be seen directly from Fig. 3-26. The cross section $d\sigma/d\Omega$ is a function of the angle between $\mathbf{v}$ and $\mathbf{v}'$. Thus the integral over Ω' can be thought of as an azimuthal rotation of $\mathbf{v}'$ about $\mathbf{v}$, for which the cross section is constant, followed by a polar integral from the $\mathbf{v}$ axis. In the azimuthal integral of $\mathbf{v}'$ about $\mathbf{v}$, the average value of $\mathbf{v}'$ along $\mathbf{v}$ is $v \cos \theta$. Thus the average of v_x' over the azimuthal integral (for which $d\sigma/d\Omega$ is a

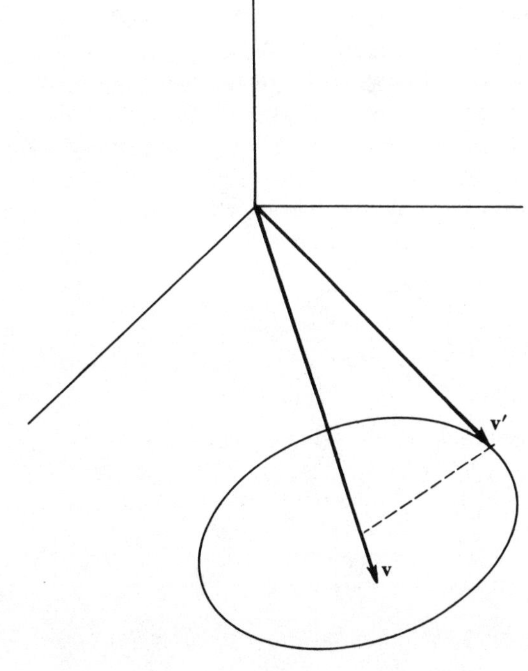

Fig. 3-26 The integral over Ω' can be thought of as a rotation of $\mathbf{v}'$ about $\mathbf{v}$ followed by a polar integral from the $\mathbf{v}$ axis.

constant) is $v'_x = v_x \cos \theta$. Then the remainder of the integral over Ω' is just the polar integral. Thus Eq. (3-223) becomes

$$v_x = N_i v \tau(v) \int_\theta \frac{d\sigma(\theta)}{d\Omega} (v_x - v_x \cos \theta) 2\pi \sin \theta \, d\theta \tag{3-224}$$

The velocity v_x divides out, and inversion yields

$$\frac{1}{\tau(v)} = 2\pi v N_i \int_0^\pi \frac{d\sigma(\theta)}{d\Omega} (1 - \cos \theta) \sin \theta \, d\theta \tag{3-225}$$

as a collision relaxation time in the desired form.

The Rutherford cross section for scattering from an isolated charge is

$$\frac{d\sigma}{d\Omega} = \left(\frac{Ze^2}{mv^2}\right)^2 \frac{1}{(1 - \cos \theta)^2} \tag{3-226}$$

If this cross section is integrated in Eq. (3-225), an infinite result obtains. It is a well-known fact that the total coulomb-scattering cross section of charged particles is infinite because the long-range coulomb force scatters particles with indefinitely large impact parameters into nonzero angles. The divergence of the Rutherford cross section comes from the forward scattering.

In the dense plasma of a stellar interior, however, such a picture is unrealistic because of the collective effects. Reconsider the homogeneous-ion-sphere model of an ionized ion in a stellar interior, as described in Sec. 2-3. The nuclear charge Z is shielded by an ion sphere of radius R_Z such that $\frac{4}{3}\pi R_Z{}^3 n_e = Z$. Thus for impact parameters greater than R_Z, a classical picture of coulomb scattering would give a vanishing cross section (instead of an infinite one!) because the nuclear charge is completely shielded for larger impact parameters. This radius R_Z is also the average half-distance to the next ion, if one wants to think of it that way. And in fact, there will be some shielding even for smaller impact parameters. A quantum-mechanical calculation of scattering from a shielded ion results in a cross section that tends toward some constant value at small angles. As an approximation, let the cross section be described by

$$\frac{d\sigma}{d\Omega} = \begin{cases} \left(\dfrac{Ze^2}{mv^2}\right)^2 \dfrac{1}{(1 - \cos \theta)^2} & \theta > \theta_0 \\ \text{const} = \dfrac{d\sigma(\theta_0)}{d\Omega} & \theta < \theta_0 \end{cases} \tag{3-227}$$

where θ_0 is some small angle that characterizes this cutoff due to shielding.

Problem 3-34: Show that with this approximation to the coulomb scattering the relaxation time for $\theta_0 \ll 1$ is approximately

$$\frac{1}{\tau} \approx 2\pi v N_i \left(\frac{Ze^2}{mv^2}\right)^2 \left(\frac{1}{2} + \ln \frac{2}{1 - \cos \theta_0}\right)$$

The first term is evidently quite negligible for $\theta_0 \ll 1$.

It is the usual treatment to take

$$\frac{1}{\tau} = 2\pi v N_i \left(\frac{Ze^2}{mv^2}\right)^2 \ln \frac{2}{1 - \cos \theta_0} \tag{3-228}$$

and to perform some detailed calculations to see what the appropriate value of θ_0 is. When one calculates the scattering from a shielded potential by some standard technique of quantum mechanics, e.g., the Born approximation, the result is

$$\theta_0 \approx \frac{\hbar}{mvR_z} \tag{3-229}$$

The essential correctness of this simple result can be motivated from an application of the indeterminacy principle. The product of the uncertainties of two conjugate dynamical variables (in this case angle and angular momentum) must exceed $\hbar$. The uncertainty of the angular momentum must be less than mvR_z to ensure that the electron strikes the ion sphere; hence the angle of the beam must be uncertain by $\Delta\theta \geq \hbar/mvR_z$. This angular uncertainty must then correspond (at least roughly) to the minimum scattering angle of the problem. In using this approximation to θ_0 it will be necessary to decide what to use for the average electron momentum. For the time being, however, we shall be content to note from Eq. (3-228) that $\tau(v)$ is inversely proportional to the function

$$\Theta(v) \equiv \tfrac{1}{2} \ln \frac{2}{1 - \cos \theta_0(v)} \tag{3-230}$$

Having obtained a relaxation time that is a function only of electron speed (as opposed to its direction), let us return to the larger problem of relating $\tau(v)$ to the conductivity. The forms we have found valid for the distribution function are

$$f(\mathbf{v}) = f_0 + g(\mathbf{v}) = f_0 - v_x\tau(v)\Pi(v) \tag{3-231}$$

The whole idea is to insert the distribution function into the equations for particle flux and heat flux. The unperturbed distribution function yields no fluxes since $\int v_x \, d\Omega = 0$. Insertion of $g(\mathbf{v})$ into the fluxes yields

$$j_x = 0 = \int 2\frac{m^3}{h^3} d^3v \, v_x{}^2\tau(v) \frac{df_0}{dK}\left[eE_x - \left(\frac{d\xi}{dT} + \frac{K - \xi}{T}\right)\frac{dT}{dx}\right] \tag{3-232}$$

and

$$Q_x = -\int 2\frac{m^3}{h^3} d^3v \, v_x{}^2K\tau(v) \frac{df_0}{dK}\left[eE_x - \left(\frac{d\xi}{dT} + \frac{K - \xi}{T}\right)\frac{dT}{dx}\right] \tag{3-233}$$

The first of these equations will allow the elimination of eE_x from the second, which will give Q_x in terms of dT/dx. To see how this can be done, define a modified electric force eE_x' by

$$eE_x' \equiv eE_x - \left(\frac{d\xi}{dT} - \frac{\xi}{T}\right)\frac{dT}{dx} \tag{3-234}$$

Since E_x and ξ are functions of the local thermodynamic state but not of particle velocity, the combination eE'_x can be pulled out of the integral over velocity space. Equations (3-232) and (3-233) can then be written

$$eE'_x \int d^3v\, v_x^2\tau(v) \frac{df_0}{dK} = \frac{1}{T}\frac{dT}{dx} \int d^3v\, v_x^2 K\tau(v) \frac{df_0}{dK} \tag{3-235}$$

and

$$\frac{h^3}{2m^3} Q_x = -eE'_x \int d^3v\, v_x^2\tau(v) K \frac{df_0}{dK} + \frac{1}{T}\frac{dT}{dx} \int d^3v\, v_x^2 K^2\tau(v) \frac{df_0}{dK} \tag{3-236}$$

Problem 3-35: Reduce the above integrals by transforming the integral over d^3v to a spherical integral about the x axis, that is, $v_x = v \cos\theta$, $d^3v = 2\pi \sin\theta\, d\theta\, v^2\, dv$. Then by changing the integral over v to an integral over K, show that Eqs. (3-235) and (3-236) become

$$eE'_x \int_0^\infty \frac{K^3}{\Theta}\frac{df_0}{dK}\, dK = \frac{1}{T}\frac{dT}{dx} \int_0^\infty \frac{K^4}{\Theta}\frac{df_0}{dK}\, dK \tag{3-237}$$

and

$$Q_x = -\frac{16m}{3h^3Z^2e^4N_i}\left(eE'_x \int^\infty \frac{K^4}{\Theta}\frac{df_0}{dK}\, dK - \frac{1}{T}\frac{dT}{dx}\int^\infty \frac{K^5}{\Theta}\frac{df_0}{dK}\, dK\right) \tag{3-238}$$

The results of the previous problem reduce the calculation of λ_c to a matter of algebra. Recalling that the definition of λ_c is $Q_x = -\lambda_c(dT/dx)$, it follows that

$$\lambda_c = \frac{16m}{3h^3Z^2e^4TN_i}\left[\frac{\left(\int_0^\infty \frac{K^4}{\Theta}\frac{df_0}{dK}\, dK\right)^2}{\int_0^\infty \frac{K^3}{\Theta}\frac{df_0}{dK}\, dK} - \int_0^\infty \frac{K^5}{\Theta}\frac{df_0}{dK}\, dK\right] \tag{3-239}$$

The four integrals may, in principle, be evaluated numerically for each thermodynamic environment. What is usually done, however, is to withdraw Θ from the integrals by replacing it by an appropriate average. It is, in fact, Eq. (3-239) that shows explicitly how Θ is to be averaged. Inasmuch as the entire treatment is an approximate one, reasonably reliable estimates of the importance of conductivity in energy transfer can be obtained from the following simple argument. The function Θ is a slowly varying function of velocity:

$$\Theta = \tfrac{1}{2} \ln \frac{2}{1 - \cos(\hbar/mvR_z)} \tag{3-240}$$

whereas the remainder of the integrand has strong zeros for very low and very high values of the kinetic energy. Therefore a large error in Θ will not be made if it is simply evaluated for an important intermediate electron velocity.

Problem 3-36: Assume that for a nondegenerate gas the most important velocities are those near the peak in the distribution ($K \approx \frac{3}{2}kT$) and that the shielding radius R_Z is approximately that of a sphere containing Z electrons. (Perhaps one should use the Debye radius instead.)

Show then that

$$\theta_0 \approx 0.058 \left[\frac{\rho(1+X)}{T_7^{\frac{3}{2}} Z} \right]^{\frac{1}{3}} \qquad \text{nondegenerate} \tag{3-241}$$

Assume for a degenerate gas that the most important electron velocities are those at the top of the Fermi sea since they are the ones that can most easily find vacant final states into which to scatter. Show for that case that

$$\theta_0 = \left(\frac{4}{9\pi Z} \right)^{\frac{1}{3}} = 0.53 Z^{-\frac{1}{3}} \qquad \text{degenerate} \tag{3-242}$$

Examination of the dependence of Θ on θ_0 will reveal that only small errors in the conductivity result even if these characteristic values of θ_0 are in error by a factor of 2. Show that an error of a factor of 2 in θ_0 can make *at most* an error of a factor of 1.74 in λ_c.

Problem 3-37: If the values of Θ in the integrals of Eq. (3-239) are all replaced by a constant average value, a further simplification of the conductivity integrals can be made. Show by integrating by parts that

$$\int_0^\infty K^n \frac{df_0}{dK}\, dK = -n(kT)^n F_{n-1}\left(-\frac{\xi}{kT}\right)$$

where

$$F_{n-1}\left(-\frac{\xi}{kT}\right) = \int_0^\infty \frac{x^{n-1}\, dx}{\exp\,(x - \xi/kT) + 1} \tag{3-243}$$

is a Fermi-Dirac integral of the same type as defined in Chap. 2.

In terms of the Fermi-Dirac integrals, Eq. (3-239) becomes

$$\lambda_c = \frac{16mk^5T^4}{3h^3Z^2e^4\Theta N_Z} \frac{15F_2F_4 - 16F_3^2}{3F_2} \tag{3-244}$$

The argument of the Fermi-Dirac functions is $-\xi/kT = \alpha - E_0/kT$, which is just the degeneracy parameter reduced by the continuum depression. In stellar problems involving Fermi statistics, the potential energy of the electron is always eliminated by absorbing it into a redefinition of the degeneracy parameter. To save space the continuum depression will not be explicitly written. Rather the argument of the Fermi functions will, as in Chap. 2, be called α.

Equation (3-244) is valid for a single type of ion. If there are several types of ions present, the factor $N_Z Z^2$ is replaced by the sum $\Sigma N_Z Z^2$. It is obvious from the derivation that this result is correct only for nonrelativistic electrons. Relativistic expressions may be found in the literature.[1]

The immediate question of astrophysical interest is how efficient conduction is as a means of heat transport. To really place this question in perspective, it is necessary to compare conduction to radiative transfer, the other mechanism of diffusive energy flow. For conduction we have computed a conductivity, whereas

[1] E. Schatzman, in S. Flugge (ed.), "Handbuch der Physik," vol. 51, Springer-Verlag OHG, Berlin, 1958, or Chiu, *op. cit.*

for radiation we have computed an opacity. It is common to compare them in the following way. The conductive heat flow Q_x can be written

$$Q_x = -\lambda_c \frac{dT}{dx} \equiv -\frac{4ac}{3\rho\kappa_c} T^3 \frac{dT}{dx} \tag{3-245}$$

where the second equality is the definition of a *conductive opacity*

$$\kappa_c = \frac{4acT^3}{3\rho\lambda_c} \tag{3-246}$$

The usefulness of such a definition is that it places the heat-conduction equation in the same form as that of radiative transfer. The heat flux by radiative transfer is proportional to $1/\kappa$, whereas the heat flux by conduction is proportional to $1/\kappa_c$, the other factors being identical. Thus it is clear that conduction will be more important than radiative transfer only if $\kappa_c < \kappa_r$. Both processes clearly occur simultaneously, so that the total rate of heat flow is just the sum of the rates from each mechanism. This means that the radiative-transfer equation can be made to contain both effects if a generalized opacity is defined by

$$\frac{1}{\kappa} = \frac{1}{\kappa_r} + \frac{1}{\kappa_c} \tag{3-247}$$

If $\kappa_c \ll \kappa_r$, then $\kappa \approx \kappa_c$; conversely, if $\kappa_c \gg \kappa_r$, then $\kappa \approx \kappa_r$. The opacities represent a "resistance" to heat flow and have an electrical analogy in the resistance to current, inasmuch as the well-known resistance of two parallel resistors R_1 and R_2 is

$$\frac{1}{R} = \frac{1}{R_1} + \frac{1}{R_2}$$

The two methods of diffusive transfer are parallel, and hence their resistances to heat flow (opacity) add like parallel resistors. Note that this situation is quite unlike that of the various sources of opacity to photons ($\kappa_{\text{b-f}}$, $\kappa_{\text{f-f}}$, κ_e, etc.), for which the resistances add in series (but in a complicated way because of the essential difference between scattering and absorption). As soon as a conductivity has been calculated, it can be converted into an opacity by Eq. (3-246) for direct comparison with the radiative (Rosseland mean) opacity at the same point.

WEAK DEGENERACY—THE MAXWELLIAN GAS

In Chap. 2 the Fermi integrals were evaluated in the limit of large positive α to be

$$\begin{aligned} F_s(\alpha) &= \int_0^\infty e^{-(u+\alpha)} u^s (1 - e^{-(u+\alpha)} + e^{-2(u+\alpha)} - \cdots) \\ &= s!e^{-\alpha}\left(1 - \frac{e^{-\alpha}}{2^{s+1}} + \cdots\right) \end{aligned} \tag{3-248}$$

Using this result, it is a simple matter to evaluate the combination of Fermi integrals in the limit of large α:

$$\frac{15F_2F_4 - 16F_3{}^2}{3F_2} \approx 24e^{-\alpha} \tag{3-249}$$

It was also shown in Chap. 2 that the limit of large α corresponds to a maxwellian gas and that in that limit

$$e^{-\alpha} \approx \frac{n_e h^3}{2(2\pi mkT)^{\frac{3}{2}}} \tag{3-250}$$

Putting these results into Eq. (3-244) gives

$$\lambda_c = \frac{16\sqrt{2}}{\pi^{\frac{3}{2}}} \frac{k^{\frac{7}{2}}T^{\frac{5}{2}}}{m^{\frac{1}{2}}e^4 Z\Theta} \tag{3-251}$$

or in cgs units

$$\lambda_c = 7.82 \times 10^{-5} \frac{T^{\frac{5}{2}}}{Z\Theta} \tag{3-252}$$

Problem 3-38: Show that the retention of the next term in the expansion gives

$$\lambda_c \approx \frac{128 m_e k^5 T^4}{h^3 e^4 Z^2 N_z \Theta} e^{-\alpha}(1 - \tfrac{13}{32}e^{-\alpha}) \tag{3-253}$$

where

$$\frac{n_e h^3}{2(2\pi mkT)^{\frac{3}{2}}} \approx e^{-\alpha}\left(1 - \frac{e^{-\alpha}}{2^{\frac{3}{2}}}\right)$$

Problem 3-39: Using the approximations outlined thus far, calculate the conductive opacity of a helium gas at a temperature of 10^8 °K and a density of 10^3 g/cm^3. How does κ_c compare with the radiative opacity?
Ans: $\kappa_c \gg \kappa_r$.

STRONG DEGENERACY

In the limit of strong degeneracy (large negative α) the Fermi integrals can be evaluated explicitly with the aid of the Sommerfeld lemma, as outlined in Chap. 2.

Problem 3-40: By using the Sommerfeld lemma, show that the leading term in the conductivity for large negative α is

$$\lambda_c \approx \frac{16\pi^2 mk^5T^4}{9h^3e^4Z^2N_z\Theta}(-\alpha)^3 \tag{3-254}$$

Retention of terms of order α results in a multiplicative correction factor $1 + 6\pi^2/5\alpha^2 \cdots$.

From Eq. (2-68) it is apparent that for values of alpha large enough to justify retaining only the first term of the expansion,

$$-\alpha \approx \frac{1}{2mkT}\left(\frac{3h^3n_e}{8\pi}\right)^{\frac{2}{3}} \tag{3-255}$$

Insertion of this value in Eq. (3-254) gives for the degenerate (nonrelativistic) conductivity

$$\begin{aligned}\lambda_c &= \frac{k^2h^3Tn_e^2}{32e^4m^2Z^2N_z\Theta} \\ &= \frac{1}{32}\frac{1}{\mu_e}\frac{k^2h^3N_0\rho T}{e^4m^2Z\Theta} \\ &= 2.36 \times 10^3 \frac{1}{\mu_e}\frac{\rho T}{Z\Theta}\end{aligned} \tag{3-256}$$

Problem 3-41: Calculate the conductive opacity of a helium gas at 10^8 °K and a density $\rho = 4 \times 10^5$ g/cm^3. Show also that near those conditions the opacity depends upon temperature and density as $\kappa_c \propto (T/\rho)^2$.

From the results of the previous problem it will be apparent that κ_c becomes very small at high density. The physical reason for this fact is that in the limit of high degeneracy, it becomes very difficult for an electron to scatter, because most of the states are occupied. Thus the mean free path of an electron becomes quite long, enabling it to traverse a significant temperature drop even for small temperature gradient.

It is not an easy matter, however, to compare the conductive opacity under such conditions to the radiative opacity. The formulas we have developed for κ_r are not applicable to a degenerate gas. Even though $\kappa_c < \kappa_{es}$ in circumstances like the previous problem, it is not clear that the conductive opacity is less than the true radiative opacity. Filling up available states will also inhibit the scattering of a photon from an electron, since the electron momentum must change according to the photon recoil. It does turn out, however, to be true that the opacity is essentially conductive in highly degenerate matter.

INTERMEDIATE DEGENERACY

For the conductivity in the important intermediate region of partial degeneracy one must resort to tabulated values of the combination $(15F_2F_4 - 16F_3^2)/3F_2$. Cox and his coworkers at Los Alamos have tabulated the quantity

$$\frac{\lambda_c Z^2 N_z \Theta}{N_0 T^4 \exp(-\alpha)} = \frac{16mk^5e^\alpha}{3h^3e^4N_0}\frac{15F_2F_4 - 16F_3^2}{3F_2} \tag{3-257}$$

which is a function only of α. Their results are reproduced in Table 3-1. Included in Table 3-1 is the value of $\theta_0 Z^{\frac{1}{3}}$, which they find also to be a function only of α. The values of θ_0 in the table are somewhat different from the crude approximations used in this text because those authors used more elaborate techniques for averaging Θ. The results listed in Table 3-1 may be used to estimate λ_c for conditions of intermediate degeneracy. The comparison with

Table 3-1 Conductivity integrals†

α	$\theta_0 Z^{\frac{1}{3}}$	$\lambda_c \dfrac{\rho Z^2 \Theta}{T^4 \mu_{\text{ion}}} e^{\alpha}$, *erg moles sec*$^{-1}$ *cm*$^{-4}$ *deg*$^{-6}$ $\times 10^{-13}$
4.0	0.1548	6.252
3.0	0.2146	6.218
2.0	0.2946	6.142
1.0	0.3952	5.948
0.2	0.4842	5.634
−0.4	0.5494	5.268
−1.0	0.6084	4.779
−1.6	0.6582	4.188
−2.2	0.6982	3.543
−2.8	0.7292	2.895
−3.4	0.7531	2.290
−4.0	0.7712	1.759
−5.0	0.7926	1.070
−6.0	0.8066	0.6133
−7.0	0.8161	0.3346
−8.0	0.8228	0.1752

† Reprinted from A. Cox, Stellar Absorption Coefficients and Opacities, in L. H. Aller and D. B. McLaughlin (eds.), "Stellar Structure." By permission of the University of Chicago Press. Copyright 1965 by The University of Chicago.

radiative opacity in this case is a difficult problem because the weak degeneracy also lowers the electron-scattering opacity and the free-free absorption. The conductive opacity of a helium gas as calculated from Table 3-1 is shown in Fig. 3-27.

It will be apparent that the physics of diffusive energy transport is decidedly complicated. We have presented only the simplest considerations in this chapter. Much more important detailed work remains to be done on the associated problems. Opacity is an extremely important determining factor in stellar structure, and confidence in computed opacities presently limits the confidence with which stellar structure can be fixed with computer models. It is to be hoped that the next few years will bring increasingly accurate opacity calculations along the lines of the efforts of Cox and his coworkers at Los Alamos. The treatment of conductivity also needs refinement, especially in extremely dense matter, where the large coulombic energy tends to correlate the ion positions. Ion correlation has been neglected completely in this treatment of conduction. Its effect will be to increase the conductivity, and the increase is probably large when the ions have been forced into a lattice.

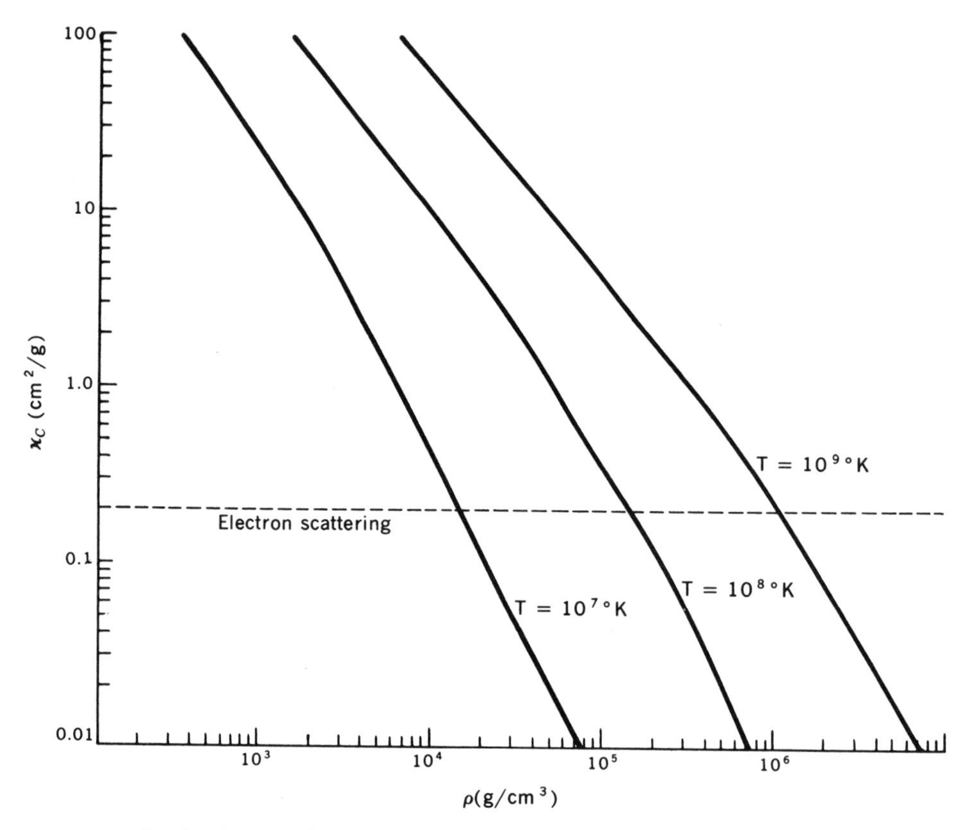

Fig. 3-27 Conductive opacity of a helium gas. The conductivities were computed from Table 3-1. The density at which the conductive opacity equals the radiative opacity due to scattering from the electrons is seen to increase by roughly one order of magnitude for each order-of-magnitude increase in the temperature.

3-5 CONVECTIVE INSTABILITY OF THE TEMPERATURE GRADIENT

In the diffusive mechanisms of energy transport the heat flux is proportional to the temperature gradient and inversely proportional to the total opacity:

$$H = \frac{4ac}{3\kappa\rho} T^3 \frac{dT}{dr} \tag{3-258}$$

To the extent that the stellar structure resembles the standard-model polytrope of index 3 the ratio T^3/ρ is constant. In that approximation the flux is proportional to $\frac{1}{\kappa}\frac{dT}{dr}$. In a static stellar model the heat flux must be sufficiently great to carry out all the energy liberated within a given sphere. Evidently this demand establishes the temperature gradient. Inasmuch as a rather large temperature gradient may be required to carry large fluxes or to carry a moderate flux through a region of very high opacity, it is sensible to question whether the temperature gradient can increase without bound. The answer is that it cannot, as demonstrated by a simple argument presented in 1906 by Karl Schwarzschild.[1] What he showed is that an instability to convective gas motion occurs if the temperature gradient is too great.

A system is stable if it presents a restoring force to arbitrary imagined displacements. In convective motion matter moves in a coherent fluid way rather than as individual particles, and one commonly visualizes macroscopic mass elements as rising and falling in the interior. The way to test for stability against this kind of motion is to give a mass element a small radial displacement and see whether it keeps going or falls back to its original position. It may be helpful in considering such displacement to imagine the mass element as being contained in a perfectly elastic balloon whose only function is that of maintaining the identity of the mass element.

Suppose now that the mass element is displaced a small distance dr without exchanging heat energy with the environment, as illustrated in Fig. 3-28. The pressure forces will immediately expand the balloon until it has the same pressure as the environment. But in establishing this pressure balance, the density ρ^* inside the balloon will not in general equal the surrounding density $\rho(r + dr)$. Because the perturbation has been made adiabatically, the density ρ^* after the displacement will be related to the pressure change in the environment by the

[1] K. Schwarzschild, *Göttingen Nachr.*, **1906**:41.

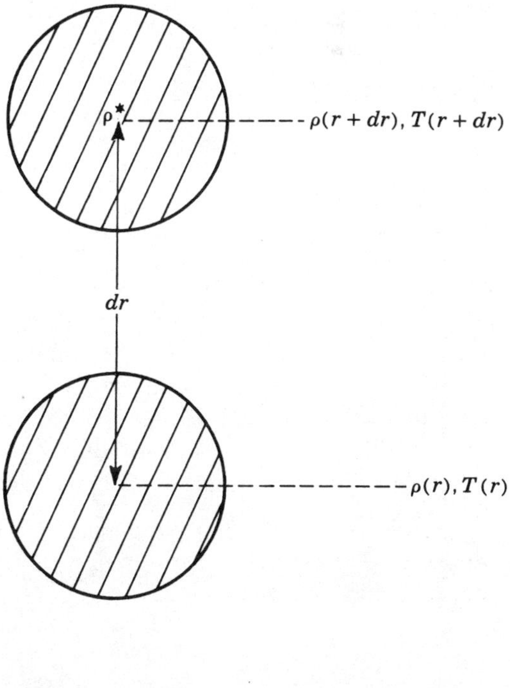

ρ^*

$\rho(r + dr), T(r + dr)$

dr

$\rho(r), T(r)$

Fig. 3-28 Convective instability occurs when the density ρ^* of a gas element which has been adiabatically elevated by a distance dr is less than the surrounding density.

first adiabatic exponent:

$$\frac{dP}{P} = -\Gamma_1 \frac{dV}{V} = \Gamma_1 \frac{d\rho}{\rho} \tag{3-259}$$

where $dP = P(r + dr) - P(r) = (dP/dr)\,dr$. Thus the expanded density will be

$$\rho^* = \rho(r) + (d\rho)_S = \rho(r) + \frac{1}{\Gamma_1}\frac{\rho}{P}\left(\frac{dP}{dr}\right) dr \tag{3-260}$$

Now, if $\rho^* > \rho(r + dr)$, the displaced element will be denser than the surroundings and will settle back under the influence of gravity. If, on the other hand, $\rho^* < \rho(r + dr)$, the displaced element will experience a net bouyant force that will cause it to continue upward. Thus the stability condition becomes

$$\text{Stability condition} = \begin{cases} \text{stable} & \text{if } \rho^* > \rho(r + dr) \\ \text{unstable} & \text{if } \rho^* < \rho(r + dr) \end{cases} \tag{3-261}$$

Because $\rho(r + dr) = \rho(r) + (d\rho/dr)\,dr$, we have stability if

$$\frac{1}{\Gamma_1}\frac{\rho}{P}\frac{dP}{dr} > \frac{d\rho}{dr} \tag{3-262}$$

The stability condition can easily be expressed in terms of the temperature gradient. Physically it is clear that if $\rho^* > \rho(r + dr)$, we must have $T^* < T(r + dr)$ in order for the balloon to have the same pressure as the environment. Because the temperature decreases radially, it is also clear that the stability condition demands that the temperature *decrement* for a radial adiabatic displacement be greater than the temperature decrement of the environment. Thus the layer is stable if

$$\left|\left(\frac{dT}{dr}\right)_{\text{star}}\right| < \left|\left(\frac{dT}{dr}\right)_{\text{ad}}\right| \tag{3-263}$$

where $(dT/dr)_{\text{ad}}$ is the so-called *adiabatic temperature gradient*. It is defined by the second adiabatic exponent. Inasmuch as both gradients are negative, the *algebraic* condition for stability is

$$\left(\frac{dT}{dr}\right)_{\text{star}} > \left(1 - \frac{1}{\Gamma_2}\right)\frac{T}{P}\left(\frac{dP}{dr}\right)_{\text{star}} \tag{3-264}$$

Many readers will find it easier to remember the stability condition in terms of the absolute magnitudes, as in Eq. (3-263). *If the temperature changes too rapidly with distance, instability against convection exists.*

The occurrence of this instability limits the energy flux that can be carried diffusively outward in a star. If the macroscopic-transfer equation is solved for the temperature gradient

$$\frac{dT}{dr} = -\frac{3}{4ac}\frac{\kappa\rho}{T^3}\frac{L(r)}{4\pi r^2} \tag{3-265}$$

the stability condition becomes

$$-\frac{3}{4ac}\frac{\kappa\rho}{T^3}\frac{L(r)}{4\pi r^2} > \left(1 - \frac{1}{\Gamma_2}\right)\frac{T}{P}\frac{dP}{dr} \tag{3-266}$$

The pressure gradient may be eliminated with the aid of the condition of hydrostatic equilibrium, resulting in the condition

$$L(r) \leq \frac{16\pi acG}{3\kappa}\left(1 - \frac{1}{\Gamma_2}\right)\frac{T^4}{P}M(r) \tag{3-267}$$

If the luminosity required to maintain energy balance exceeds this amount, the energy will have to be carried by convective transport.

Problem 3-42: Show that if the equation of state is that for an ideal nondegenerate gas, the luminosity is limited in zones of radiative equilibrium to (in cgs units)

$$L(r) \leq 1.22 \times 10^{-18}\frac{\mu T^3}{\kappa\rho}M(r)$$

A physically simple motivation for the fact that convection frequently occurs in the hydrogen and helium ionization zones may be found in the results of the preceding problem. If a star is approximated by the standard model, T^3/ρ is a constant, and $L(r)$ is limited to a number proportional to $M(r)/\kappa$. In the outer layers $M(r)$ is also nearly constant, so that the maximum radiative luminosity is proportional to κ^{-1}. Because the opacity is so large in the ionization zones (see Fig. 3-16), the luminosity there often exceeds the upper limit for radiative equilibrium.

Convection is being discussed as a mechanism of energy transport, but we have not yet demonstrated that the convective instability results in a transport of energy. The demonstration is very simple. Suppose that the medium is unstable, in which case an adiabatically rising element is less dense and hotter than the environment. Because the balloon temperature exceeds that of the surroundings, heat will leak from the balloon to the surroundings. The net effect is the transport of heat to material at a lower temperature.

Problem 3-43: Make an adiabatic displacement downward in the unstable situation. Show that the element continues to fall and that the tendency toward thermalization will cool the underlying gas.

Presumably, then, convection is characterized by macroscopic fluid motions, which must be directed both upward and downward to conserve mass, and which have the effect of heating the outer regions and cooling the inner regions. The heat transfer results from the fact that the magnitude of the temperature gradient exceeds the magnitude of the adiabatic temperature gradient. The moving element would remain at the surrounding temperature if the two gradients were equal, and there would be no thermal imbalance to be equalized. It is evident that the heating of outer layers and the cooling of underlying layers has a tendency to reduce the magnitude of the temperature gradient. Convection will in fact reduce the gradient until it is just sufficient to carry the excess thermal energy at the appropriate rate. Such a state is called *convective equilibrium.*

For the problem of stellar structure one needs to know how great the temperature gradient must be to carry the luminosity of the star. We have argued only that $|dT/dr|$ must exceed $|dT/dr|_{\text{ad}}$ in order that energy be transported by convection, but what is clearly needed is some indication of the extent to which the adiabatic gradient must be exceeded to carry a given amount of energy. The solution of this problem is extremely difficult, in fact no general solution has yet been found, but fortunately it can be argued that this uncertainty has no large effect upon the *internal* structure of stars. This conclusion results from the great efficiency of convection as a means of energy transport. By a great efficiency we mean that a very slight excess in the temperature gradient is ample to transport the luminosity convectively. The necessary excess in the temperature gradient is so small compared to the temperature gradient itself that it becomes a good approximation to replace the actual temperature gradient by the adiabatic temperature gradient. The argument runs as follows.

One must first devise a physical model for convective energy transport. The simplest and most popular is the *mixing-length model.* Each mass element is envisioned as rising or falling adiabatically for a distance l, called the *mixing length*, which is characteristic of the environment in some way. After traveling that distance the mass element thermalizes with the local environment. After adiabatically rising a distance l, a mass element will be hotter than the surroundings by an amount approximately given by

$$\Delta T = \left(\left|\frac{dT}{dr}\right| - \left|\frac{dT}{dr}\right|_{\text{ad}}\right) l = l\,\Delta\nabla T \tag{3-268}$$

where $\Delta\nabla T$ is a symbol defined as the excess of the absolute magnitude of the temperature gradient over the absolute magnitude of the adiabatic temperature gradient. The mass element then thermalizes at constant pressure, releasing an amount of heat per unit mass given by

$$\Delta Q = c_p\,\Delta T = c_p l\,\Delta\nabla T \tag{3-269}$$

If the average velocity of the adiabatic cells as they pass any given level is designated by $\bar{V}$, the average excess heat flux is

$$H = \rho\bar{V}\,\Delta Q = \rho\bar{V}c_p l\,\Delta\nabla T \tag{3-270}$$

Evaluation of this heat flux requires an estimate of the average velocity $\bar{V}$ of the convective cells. This estimate can be made by assuming that the mass element is accelerated for the distance l by a bouyant force given by Archimedes' principle. Inasmuch as $\Delta\rho = 0$ at the outset of a cell's motion and is

$$\Delta\rho = l\left(\left|\frac{d\rho}{dr}\right| - \left|\frac{d\rho}{dr}\right|_{\text{ad}}\right) = l\,\Delta\nabla\rho$$

after rising a distance l, the average density deficiency in the rising element may be taken to be

$$\overline{\Delta\rho} = \tfrac{1}{2}l\,\Delta\nabla\rho \tag{3-271}$$

The average bouyant force per unit volume is then

$$\bar{F} = g\,\overline{\Delta\rho} = \tfrac{1}{2}gl\,\Delta\nabla\rho \tag{3-272}$$

where $g = GM/r^2$ is the local gravity. This average force per unit volume causes a radial acceleration

$$a = \frac{\bar{F}}{\rho} = \frac{gl}{2\rho}\,\Delta\nabla\rho \tag{3-273}$$

Acting through a distance l, this acceleration produces a terminal velocity $V = \sqrt{2al}$, which may be taken to be twice the average velocity. Thus we obtain

$$\bar{V} = \frac{1}{2}\left(\frac{gl^2}{\rho}\,\Delta\nabla\rho\right)^{\frac{1}{2}} = \frac{1}{2}\left(\frac{GM}{\rho r^2}\,\Delta\nabla\rho\right)^{\frac{1}{2}} l \tag{3-274}$$

Problem 3-44: Show that for the ideal monatomic nondegenerate gas

$$\Delta\nabla\rho = \frac{\rho}{T}\,\Delta\nabla T$$

This result may be used because convection occurs only in nondegenerate regions.

Assembling these approximate results yields

$$H = c_p\rho\left(\frac{GM}{Tr^2}\right)^{\frac{1}{2}}(\Delta\nabla T)^{\frac{3}{2}}\frac{l^2}{2} \tag{3-275}$$

According to this model of convection, the heat flux is proportional to the excess of the temperature gradient to the $\frac{3}{2}$ power and to the square of the average mixing length. From experience with convection in fluids one estimates that the cells will not dissipate until they have moved a sufficiently great distance such that the pressure and density will have changed by a significant fraction of their initial values, i.e., until they have moved a distance of the order of a scale height, which is usually at least 10^9 cm in a star. With such values for the mixing length it is a simple matter to demonstrate that very small values of $\Delta\nabla T$ are required for normal fluxes.

Problem 3-45: Show that if the entire solar luminosity were carried by convection through the midway point of the sun with a mixing length of 10^9 cm, the required excess of the temperature gradient would be only about 10^{-6} of the temperature gradient itself.

From arguments similar to that in the previous problem it is believed that *whenever convection occurs in the interior of a star, the temperature gradient is very nearly the adiabatic temperature gradient.* That is, $\Delta\nabla T$ is very important as far as the heat flux is concerned but is negligibly small as far as the structure of a star is concerned. For a prescription for stellar models, one is forced to compare the temperature gradient that would be necessary to transport the flux radiatively with the adiabatic temperature gradient at each point; if the adiabatic gradient is greater in magnitude, the layer is stable, and the flux is carried radiatively; whereas if the required radiative gradient is the greater, the flux is carried convectively along a temperature gradient given to high accuracy by

$$\frac{dT}{dr} = \left(1 - \frac{1}{\Gamma_2}\right)\frac{T}{P}\frac{dP}{dr} \tag{3-276}$$

Although the approximation of this prescription appears to be a good one in the interior of a star, it runs into difficulty in the important outer layers of stars, where convection so often occurs. The difficulty arises because the pressure and density change so rapidly there compared to their small values. The mixing length accordingly becomes very short, but because the heat flux is proportional to

$$H \propto \rho l^2(\Delta\nabla T)^{\frac{3}{2}}$$

the low density and small mixing length together demand a large value for $\Delta\nabla T$, so large, in fact, that it is no longer adequate to use the adiabatic temperature gradient in the computation of the structure of the outer layers. One can also see that the exact temperature gradient is needed to determine the radius (and hence T_e) of a stellar model. Suppose the bottom of a surface convection zone is imagined to be fixed in temperature and position. The additional radial distance required for that temperature to fall to its near-zero photospheric value clearly depends upon the size of the temperature gradient. Accordingly we find that one of the most vexing problems in stellar structure today is the inability to calculate stellar radii with precision for those stars having surface convection zones. In general one can say that stars will possess a surface convection zone when the photospheric temperature is low enough for the hydrogen to be largely neutral, for in that case there must exist an interior region of very large opacity in the hydrogen ionization zone. If, on the other hand, the photosphere is hot enough for the hydrogen to be ionized there, the underlying layers will probably be in radiative equilibrium. Accordingly we note that the radii of the upper-main-sequence stars can be computed with greater precision than those of the lower-main-sequence stars.

What is badly needed is a theory of convection that compellingly relates the temperature gradient to the heat flux. Several authors have expressed this need as being the need for a theory of the *effective* mixing length, even though the idealized mixing-length model is inappropriate. Examination of characteristic numbers of fluid mechanics shows that the convection in a star will be turbulent rather than orderly. It is accompanied, furthermore, by continuous exchange of energy via photons, microturbulence, hydromagnetic waves, and viscous interactions of one mass element with another. The theories of turbulent convection are mathematically complicated,[1] however, and in keeping with the introductory nature of this book we shall drop the subject at this point.

As a final point on convection we add that the macroscopic velocities involved are negligible in comparison with the thermal velocities of the particles, so that a negligible kinetic energy is contained in the collective motions. The turbulent pressure, on the other hand, may need to be included in the outer layers. Since the turbulent mixing is fast in comparison with evolution times, one generally regards each convective region as being of homogeneous composition. Material of different composition injected at any point into a convection zone becomes distributed uniformly over that zone in a time small compared to evolutionary time scales.

In spite of the fact that detailed knowledge of convection is probably not required for adequate understanding of the structure of a stellar interior, much work remains to be done to develop a theory adequate for the outer layers. It is particularly discouraging to have uncertainties in surface convection undermine the confidence with which model stars can be placed on the H-R diagram. It is that placement which provides the basic test of our understanding of the stellar interior, and so the deficiency in knowledge jeopardizes the whole theory to some degree.

3-6 NEUTRINO EMISSION

The power carried by the neutrino flux has not traditionally been associated with the problem of energy transport in stars, but it is now evident that neutrino power is the dominant consideration in the late evolutionary stages of many stars. As a final subject in this chapter on energy transport in the stellar interior we shall outline the role of neutrino emission in stellar evolution and the basic neutrino-emitting mechanisms of astrophysical importance. The point to be emphasized at once is that neutrino emission is dissimilar to the other mechanisms of energy transport in one major way; viz., neutrinos interact so weakly with matter that they usually emerge directly from the stellar interior without any interaction with stellar matter. The other mechanisms require the transport of internal energy to the surface, from which it can be radiated, and as a consequence the rate of energy outflow is related to the temperature gradient of the star. Because neutrinos emerge directly from their point of origin, the energy outflow is given directly by the rate at which neutrinos are produced. Insofar as stellar structure is concerned, the creation of neutrinos in stellar matter is equivalent to a local refrigeration.

The existence of a massless, spin-$\frac{1}{2}$ particle called the neutrino was hypothesized by Pauli in 1933. The hypothesis was made as the simplest way of simul-

[1] Some comments on the theory of turbulent convection along with references to major works in the field will be found in P. Ledoux, Stellar Stability, in Aller and McLaughlin, *op. cit.* An interesting variation involving magnetic tangles has been discussed by A. Finzi and R. A. Wolf, *Astrophys. J.* (1968).

taneously saving fundamental conservation laws of physics that were apparently being violated in nuclear beta decay. In the simplest decay, which suffices to illustrate the utility of the hypothesis, the neutron decays into a proton, electron, and antineutrino:

$$n \rightarrow p + e^- + \bar{\nu}$$

Before the postulation of the neutrino the apparent decay $n \rightarrow p + e^-$ seemed to violate three conservation laws: (1) momentum—the combined momenta of p and e^- did not sum to the momentum of the neutron; (2) energy—the combined kinetic energy of p and e^- varied from decay to decay, although it was observed to have an upper limit; (3) spin angular momentum—there is no way to couple $s = \frac{1}{2}$ for both p and e^- to yield $s = \frac{1}{2}$ for the neutron. The emission of a neutral particle of half-integral spin and small (probably zero) mass in the decay allowed each conservation law to be satisfied. In the following year, 1934, Enrico Fermi proposed a theory of nuclear beta decay that not only saved the conservation laws by the use of Pauli's neutrino idea but also explained such features as the electron energy spectrum and the dependence of beta-decay rates on the total energy liberated by the decay. Fermi's theory has survived to this day in an expanded and enriched form that has been indicated by countless ingenious laboratory experiments revealing the nature of the interaction involved and by beautiful theoretical formulations, mostly notably by Feynman and Gell-Mann.

Of particular importance were the experiments made by Cowan, Reines, and their collaborators to detect the neutrino.[1] Not only was the neutrino shown to exist, but the neutrino emitted in electron beta decay was found to differ from that emitted in positron beta decay or in electron capture. This fact fits beautifully into the field theories of half-integral-spin particles, which seemed to demand that fermions be created in particle-antiparticle pairs. New conservation laws requiring the conservation of the number of nucleons and the number of leptons were found to be consistent with both field theory and experimental fact. *Lepton* is a Greek word meaning "light thing" and is used to stand for the class of light particles involved in the beta-decay processes. This conservation of numbers is achieved by demanding that the number of particles minus the number of antiparticles in each class be constant. For example, in the decay

$$n \rightarrow p + e^- + \bar{\nu}$$

the number of nucleons is unity, and the numbers of leptons is zero before the decay. The same numbers remain after the decay if the electron is taken to be "the particle" of the positron-electron pair in the lepton class and the neutrino emitted in this decay is taken to be "the antineutrino" of the neutrino-antineutrino pair in the lepton class. The field theory presents a logical equivalence between the creation of a particle and the destruction of its antiparticle. Thus instead of creating an antineutrino in the above decay, a neutrino may be

[1] For a review see F. Reines, *Ann. Rev. Nucl. Sci.*, **10**:1 (1960).

destroyed:

$$\nu + n \rightarrow p + e^-$$

The neutron in this reaction has decayed by absorbing a neutrino rather than by emitting an antineutrino. The same operation is legitimate for the electrons; viz., the emission of the electron can logically be replaced by the destruction of a positron:

$$n + \beta^+ \rightarrow p + \bar{\nu}$$

This reaction also may go the other way in time,

$$p + \bar{\nu} \rightarrow n + \beta^+$$

and in fact this very reaction was used by Cowan and Reines to measure the cross section for the interaction of antineutrinos with hydrogen, and the value they found was $\sigma = (11 \pm 4) \times 10^{-44}$ cm^2. Numbers of this size are characteristic for the interaction of neutrinos with matter. These are very small cross sections. By contrast the interaction of light with atoms is of order 10^{-16} cm^2, and the interaction of nuclei with each other by the nuclear force is of order 10^{-24} cm^2. Because of its weakness this interaction is called the *weak interaction*. Its weakness is also measured by the relative slowness of beta decay when compared with electromagnetic transitions of the same energy. Dimensionless measures of the relative strengths of the known interactions are (1) nuclear π-meson field = 10, (2) electromagnetic field = $\frac{1}{137}$, (3) weak interaction = 10^{-23}, and (4) gravity = 10^{-45}. One might note that gravity is the weakest interaction by far, and its effect would be undetected to this day were it not for the possibility of assembling huge masses; but it is absurd to contemplate the measurement of gravity by the scattering of one elementary particle from another, as one does for the other forces.

When the variety of interactions of neutrinos with matter is examined, one finds that an average cross section for neutrinos of energy E_ν to interact with other particles is of order

$$\sigma_\nu \approx 10^{-44} \left(\frac{E_\nu}{m_e c^2}\right)^2 \qquad \text{cm}^2$$

In most forms of neutrino emission from stars the factor $E_\nu/m_e c^2$ is within a factor 10 of unity, so that 10^{-44} cm^2 represents a crude estimate of neutrino cross sections. Consider matter of unit density, which is about the average density of most stars. Then each cubic centimeter contains about 10^{24} particles. It follows that the mean free path of characteristic stellar neutrinos is about 10^{20} cm, which is equal to 10^9 solar radii! No wonder then that most neutrinos shoot right out of the star with the velocity of light, at which they all travel.

There is an important exception to the transparency of stellar matter to neutrinos, the supernova. In the dynamic collapse of a supernova core to near nuclear densities it is found that the large thermal energy of the emitted neutrinos, coupled with the considerably greater amount of mass per column of unit area, causes the imploded core to be partially opaque. For this problem, which cannot be adequately covered in a book of this type, one must develop a theory of neutrino transfer. Only under very extreme conditions of condensation, such as an imploded supernova core, can a significant number of neutrinos actually interact with stellar matter, and so in most cases one considers neutrino energy as being lost to the internal energy of the star.

The weak interaction is further complicated by the fact that other fermions participate. The muon is very much like a "heavy electron" in its interactions. For instance, electron capture by a proton,

$$p + e^- \rightarrow n + \nu$$

has an analogous muon capture,

$$p + \mu^- \rightarrow n + \nu'$$

where the prime superscript on the last neutrino emphasizes the fact that the neutrinos involved in muon weak interactions differ from those in electron weak interactions. The ν' in the second reaction cannot be absorbed by a neutron and change to $p + e^-$ by the inverse of the first reaction because $\nu' \neq \nu$. Thus the particles with which one deals in weak interactions are proton p, neutron n, electron e^-, electron neutrino ν, negative muon μ^-, mu neutrino ν', and the antiparticle for each of them. The question is how one can construct a theory which not only summarizes which of the host of possible reactions actually occur in nature but also correctly predicts their rates. The detailed theory involves far too many complications for a book at this level, but the general structure of the theory will be illuminating in light of the large number of weak reactions of importance in astrophysics. A sketch of this general structure follows.

In each of the example reactions one sees the involvement of four fermions. The weak interaction is envisioned as one which occurs when four fermions (but not any four) come together at a point (or at least a very small region) in space-time. The four interacting fermions are always comprised of any two of the following pairs:

(*1*) Antineutron proton ($\bar{n}p$)
(*2*) Antiproton neutron ($\bar{p}n$)
(*3*) Positron neutrino ($\bar{e}\nu$)
(*4*) Antineutrino electron ($\bar{\nu}e$)
(*5*) Positive muon mu neutrino ($\bar{\mu}\nu'$)
(*6*) Mu antineutrino muon ($\bar{\nu}'\mu$)

The interaction is constructed in terms of these six pairs of particles.

The weak reactions also change the particle population of the universe. Returning to the initial example of the decay of the neutron, for instance, one finds that the reaction destroys a neutron and creates a proton, an electron, and an antineutrino. Thus the interaction must involve operators that create and destroy fermions but in combinations that conserve the lepton number and the baryon number. These operators are just the six pairs of fermions listed above, and in writing the interaction, the particle symbol may conveniently be taken to represent an operator capable either of creating the particle or of destroying its antiparticle. For example, the pair $(\bar{n}p)$ is taken to mean a product of two operators with the following possible interpretations:

(*1*) Creation of antineutron and proton
(*2*) Destruction of neutron and creation of proton
(*3*) Destruction of neutron and destruction of antiproton
(*4*) Creation of antineutron and destruction of antiproton

Because the fermions are grouped only in pairs of nucleon-antinucleon or lepton-antilepton operators, the restriction of the interaction to these pairs will have the effect of conserving both baryons and leptons.

The rate of weak reactions is calculated from the "golden rule" of time-dependent perturbation theory,

$$\text{Rate} = \frac{2\pi}{\hbar} |\langle k|H_{\text{int}}|s\rangle|^2 \rho(E) \tag{3-277}$$

where k and s represent the final-state and initial-state wave functions, H_{int} is the weak-interaction hamiltonian capable of matrix elements between k and s, and $\rho(E)$ is the density of states for the final particles. Perhaps the greatest clarification of Fermi's original theory lay in his demonstration that beta-decay rates were energy-dependent in a way largely accounted for by the phase space $\rho(E)$ available to the ejected electron-neutrino pair, in quite the same way that the electric-dipole photon-emission rates in atomic transitions are dominated by the phase space available to the emitted photon. In order for the weak-interaction hamiltonian to have matrix elements between states k and s having differing populations of fermions, it must contain as factors the appropriate creation and destruction operators for the fermions involved. In terms of the Feynman–Gell-Mann theory now believed applicable to weak interactions, the hamiltonian is written as the square of a Fermi interaction current,

$$H_{\text{int}} = \sqrt{2}\, g J_\mu^\dagger J_\mu \tag{3-278}$$

where g is a universal coupling constant measuring the strength of the weak interaction and the interaction current J_μ is a space-time four-vector:[1]

$$J_\mu = (\bar{n}\gamma_\mu a p) + (\bar{e}\gamma_\mu a \nu) + (\bar{\mu}\gamma_\mu a \nu') \tag{3-279}$$

[1] We shall write only the *strangeness-conserving* terms of the interaction current because they seem to be the only ones of importance to astrophysics, but the weak decays of strange particles involve a strangeness-changing current that is analogous to the one discussed here.

The γ_μ are the vector Dirac operators γ_1, γ_2, γ_3, and γ_4, and

$$a = \tfrac{1}{2}(1 - i\gamma_1\gamma_2\gamma_3\gamma_4)$$

is another Dirac operator giving the proper mixture of vector and axial-vector currents in the so-called V − A theory suggested by experimental results. The vector $J_\mu^\dagger$ is the hermitian conjugate of J_μ. For the purposes of this outline the nature of the Dirac operators need not concern us, and if the components of the interaction current are suppressed, we may write in a schematic way

$$J = (\bar{n}p) + (\bar{e}\nu) + (\bar{\mu}\nu')$$
$$J^\dagger = (\bar{p}n) + (\bar{\nu}e) + (\bar{\nu}'\mu)$$

In the universal weak-interaction theory the pairs of fermion operators are weighted equally in the current and share the universal strength g of the weak interaction.

In the matrix elements of H_{int}, therefore, the four fermions occurring in any one term are those obtained by the product of any pair in J_μ with any pair in $J_\mu^\dagger$. It is this restriction of the fermion groupings that may occur which gives rise to the selection rules determining which reactions are possible. The corresponding experimental situation is interesting in that the cross-product terms in the interaction are all well studied in the laboratory and share the universal strength of the interaction, whereas the squared terms in the interaction are difficult to observe and have not been well verified. This situation is of particular importance to astrophysics, inasmuch as the squared terms give rise to important reactions capable of causing significant neutrino emission in stars.

THE CROSS-PRODUCT TERMS

The interaction term arising from the product of any pair in J_μ with a different pair in $J_\mu^\dagger$ gives rise to interactions in which the four fermions are different types of particles. The corresponding reactions have been well studied in the laboratory, where it has been found that each of the three cross-product terms exists in nature with equal[1] intrinsic strength. It will be instructive to see what reactions come from these products.

The interaction $(\bar{n}p)(\bar{\nu}e)$ **and** $(\bar{p}n)(\bar{e}\nu)$ As described earlier, these products of operators may give rise to any of the following events:

$$n \rightarrow p + \bar{\nu} + e^-$$
$$p \rightarrow n + \nu + e^+$$
$$\nu + n \rightarrow p + e^-$$
$$e^+ + n \rightarrow p + \bar{\nu}$$

[1] Experimental uncertainty combined with uncertainty in the magnitude of some required theoretical corrections results in an uncertainty of a few percent.

as well as those obtained by changing each particle to its antiparticle and/or reversing the arrows of the reaction. What these terms produce is the known varieties of nuclear beta decay. If that nucleus consisting of Z protons and N neutrons with atomic weight $A = Z + N$ is designated by the convenient symbol (Z,A), the major nuclear beta decays[1] of importance can be written as follows.

(*1*) *Negative beta decay:* An internal nuclear neutron is converted to a proton,

$$(Z - 1, A) \rightarrow (Z,A) + e^- + \bar{\nu}$$

Since the electron and antineutrino must be emitted with nonzero kinetic energy, this reaction is energetically possible only if

$$M_n(Z - 1, A) > M_n(Z,A) + m_e$$

where $M_n(Z,A)$ represents the mass of the *nucleus* (Z,A). By adding $(Z - 1)m_e$ to both sides of the equation the inequality can be expressed in terms of the masses $M_A(Z,A)$ of the *atom* (Z,A):

$$M_A(Z - 1, A) > M_A(Z,A) \tag{3-280}$$

The two conditions actually differ slightly in that there is a very small (almost always negligible) difference in the binding energy of the atomic electrons in the two atoms. Because the atomic binding energy actually contributes to the energy of the decay, the second form is more nearly correct, but for our purposes the two equations can be thought of as synonymous.

(*2*) *Positive beta decay:* An internal nuclear proton is converted to a neutron,

$$(Z + 1, A) \rightarrow (Z,A) + e^+ + \nu$$

This reaction is energetically possible only if

$$M_n(Z + 1, A) > M_n(Z,A) + m_e$$

Problem 3-46: Show that in terms of atomic masses the energy requirement for positron decay becomes

$$M_A(Z + 1, A) > M_A(Z,A) + 2m_e \tag{3-281}$$

(*3*) *Electron capture:* An internal proton is converted to a neutron,

$$e^- + (Z + 1, A) \rightarrow (Z,A) + \nu$$

This reaction is energetically possible only if

$$m_e + M_n(Z + 1, A) > M_n(Z,A)$$

or alternatively if

$$M_A(Z + 1, A) > M_A(Z,A) \tag{3-282}$$

[1] A thorough discussion of nuclear beta decay may be found in M. A. Preston, "Physics of the Nucleus," chap. 15, Addison-Wesley Publishing Company, Inc., Reading, Mass., 1962.

In the laboratory this reaction usually proceeds by the capture of a K-shell ($1s$) electron because those electrons have the greatest probability of being found at the nucleus, where the reaction must occur. In a star the situation will usually differ in that the nuclei are generally ionized. In that case the capture will be of free electrons in the gas, and the rate will accordingly depend upon the electron density.[1] In principle the electron kinetic energy should also be included in the energy balance, but it too is usually negligible in comparison with the other energies. A major exception occurs at extremely high density, where the Fermi energy of the degenerate electron gas may become sufficiently great to cause otherwise stable nuclei to capture the energetic electrons from the top of the Fermi distribution. For example, a hydrogen gas at high density may be forced to capture free electrons and change to a neutron gas. This mechanism may be responsible for triggering a special type of supernova explosion.
(*4*) *Free-positron capture:*[2] An internal neutron is converted to a proton. This must also be considered if a significant density of free positrons exists in the gas. It has not been demonstrated that this last alternative is ever of astrophysical importance, however, and the first three reactions may be thought of as the dominant nuclear decays in astrophysics.

One other general comment must be made about the energetics of all nuclear decays. In the laboratory nuclei exist only in their ground states, and the nuclear and atomic masses are those of the ground states of the species. In the hot thermal environment of a stellar interior, however, the nuclei may have a small but significant admixture of excited states. Because the intrinsic decay rate of an excited state may be much greater than that of the ground state, which may even be stable, the beta-decay rates in stars may be strongly temperature-dependent, a feature first exploited by Cameron.[3] Figure 3-29 shows a two-level nucleus (usually all that need be considered) with ground-state spin J_1 and decay rate $\lambda_\beta(1)$ and an excited state of excitation E^*, spin J_2, and decay rate $\lambda_\beta(2)$. The nuclear partition function for $(Z - 1, A)$ is in this case

$$G(Z - 1, A) = (2J_1 + 1) + (2J_2 + 1) \exp - \frac{E^*}{kT} \tag{3-283}$$

The fraction of nuclei $(Z - 1, A)$ in the ground state is then

$$P_1(Z - 1, A) = \frac{2J + 1}{G(Z - 1, A)} \tag{3-284}$$

and the fraction in the excited state is

$$P_2(Z - 1, A) = \frac{2J_2 + 1}{G(Z - 1, A)} \exp - \frac{E^*}{kT} \tag{3-285}$$

[1] J. N. Bahcall, *Astrophys. J.*, **139**:318 (1964).

[2] H. Reeves and P. Stewart, *Astrophys. J.*, **141**:1432 (1965).

[3] A. G. W. Cameron, *Astrophys. J.*, **130**:452 (1959).

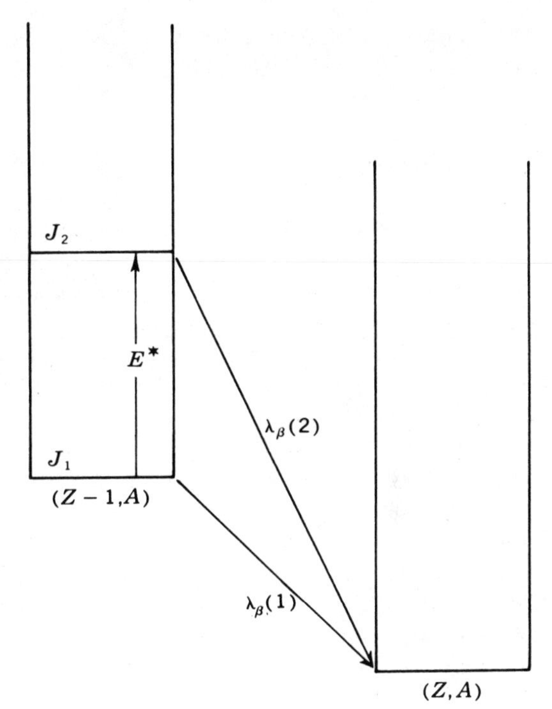

Fig. 3-29 Excited-state beta decay occurs when thermally populated excited states undergo beta decay. The effective decay rate may be very temperature-dependent if the beta-decay rate $\lambda_\beta(2)$ of the excited state is much greater than that of the ground state.

The total beta-decay rate of the species is then given by the weighted sum of the two rates:

$$\lambda_\beta(Z - 1, A) = P_1(Z - 1, A)\lambda_\beta(1) + P_2(Z - 1, A)\lambda_\beta(2) \tag{3-286}$$

These formulas can easily be generalized to an arbitrary number of excited states. Because $\lambda_\beta(2)$ may be many orders of magnitude greater than $\lambda_\beta(1)$ in some cases, the total decay rate may be strongly temperature-dependent. If the ground state of $(Z - 1, A)$ were stable, the excited-state beta decay could provide a means by which stable species undergo beta decay in a star.[1]

Problem 3-47: The nucleus In^{115} is nearly stable in the laboratory, having a half-life of 5×10^{14} years for beta decay to Sn^{115}, but an excited state of In^{115} at $E^* = 0.335$ Mev decays to Sn^{115} with a half-life of 4.5 hr. Compute the total half-life of In^{115} in thermal equilibrium at $T = 7.5 \times 10^8$ °K.
Ans: $\tau_{\frac{1}{2}} = 0.83$ year.

The excited states may be involved in positron decay or in free-electron capture, moreover, so that one must be alert to the possible role of excited states in

[1] More esoteric reactions can have the same effect. P. B. Shaw, D. D. Clayton, and F. C. Michel, *Phys. Rev.*, **140**:B1433 (1965) showed that a photon may be absorbed in a virtual beta decay, thereby contributing its energy to the energy of the decay. This mechanism seems to be less important than excited-state beta decay, however.

all types of nuclear beta decay in stars. The energy criteria previously written down must be modified in such cases to include the excited-state energy. The simplest way is to replace $M_n(Z,A)$ by $M_n(Z,A) + E^*/c^2$, where E^* is the excitation energy and $M_n(Z,A)$ is the mass of the nuclear ground state.

A certain amount of nuclear beta decay accompanies the cycles of nuclear reactions which provide the internal heat source for the stars. For example, there exist sets of reactions capable of fusing four hydrogen atoms into a helium atom (or said another way, four protons into an alpha particle). But because the alpha particle consists of two protons and two neutrons, two positive beta decays must occur somewhere in the cycle. Without knowledge of the details of the reaction chain we may nonetheless write

$$4p \rightarrow \alpha + 2e^+ + 2\nu$$

As noted in Chap. 1, this accomplishment liberates 6.4×10^{18} ergs/g of hydrogen converted. The energy created as neutrinos, however, must be subtracted from the total energy release insofar as heat input to the interior is concerned. This procedure is typical of the handling of the neutrino production that accompanies the energy-generating processes; viz., *the effects of neutrino emission in energy generation are deployed at their source by subtracting the average neutrino energy from the total energy liberated.* These corrections are usually only a few percent.

There does exist one important nuclear decay cycle, however, that has nothing to do with energy generation but serves only as a producer of neutrinos. The process of alternate electron capture and beta decay of the same nucleus was invented by Gamow and Schönberg and is commonly called the *Urca process.*[1] Specifically

$$e^- + (Z,A) \rightarrow (Z - 1, A) + \nu \tag{3-287}$$

is followed by

$$(Z - 1, A) \rightarrow (Z,A) + e^- + \bar{\nu} \tag{3-288}$$

The sum of the two reactions yields

$$e^- + (Z,A) \rightarrow e^- + (Z,A) + \nu + \bar{\nu} \tag{3-289}$$

A neutrino-antineutrino pair has been produced with no change of composition, but the neutrinos run off with energy. It is not immediately clear where that energy came from except that it must somehow have been provided by the thermal environment. From the energy conditions one sees that *both* the electron-capture reaction and the beta decay cannot happen spontaneously. Thus energy must be provided to one or both reactions to allow them to occur. Energy may be provided to the first in the form of kinetic energy of the captured electron or in the form of excited-state energy of the nucleus (Z,A). The beta decay may be provided energy by occurring from an excited state of $(Z - 1, A)$. Thus thermal energy in the form of electron kinetic energy or excitation energy of an

[1] Named after a Rio de Janeiro casino where the customer lost little by little.

excited state, or both, is lost to two neutrinos by the full cycle. The calculation of neutrino loss rates by the Urca process is a detailed problem in applied nuclear physics. Not only does it depend in obvious ways on temperature and density, but the detailed nuclear properties of the catalyst nuclei (Z,A) determine whether the process can happen efficiently. Its rate is therefore strongly dependent upon the nuclear composition.

The actual rate of energy loss by neutrinos accompanying nuclear beta decay is small in most astrophysical applications. The neutrinos emitted during nuclear energy-generation cycles provide a small correction to the energy-generation rate, and the Urca process becomes important only at very high density and temperature. At sufficiently large temperature and density the nuclear matter usually achieves the state of nuclear statistical equilibrium, to be described later, which is characterized by heavy-element abundances usually in the range $50 < A < 70$. Tsuruta[1] has calculated the details of the Urca neutrino-loss rates from nuclear matter in statistical equilibrium. Her results are shown in Fig. 3-30, where each curve is labeled by the Fermi energy E_f of the electron gas. It will be noted from Chap. 2 that Fermi energies of a few Mev for a relativistic degenerate electron gas[2] correspond to large densities:

$$E_f(\text{Mev}) \approx 0.5 \times 10^{-2}\rho^{\frac{1}{3}} \qquad \begin{array}{l}\text{relativistic} \\ \rho > 10^7\end{array} \tag{3-290}$$

At the lower densities found in the common stars other *nonnuclear* mechanisms of neutrino emission have more astrophysical importance than the Urca process. They will be discussed later.

The interaction $(\bar{\mu}\nu')(\bar{\nu}e)$ **and** $(\bar{e}\nu)(\bar{\nu}'\mu)$ These operator products also define a variety of reactions. We need list only the commonest example, muon decay,

$$\mu^- \rightarrow e^- + \nu' + \bar{\nu}$$

The rates observed for this cross product of interaction-current pairs are consistent with its sharing the same strength g which characterizes nuclear beta decay. Inasmuch as muons are probably not of importance in stellar evolution, however, they will not appear in the remainder of the book. This reaction and the next have been included only to display the logical structure of the particle operators in the weak-interaction current.

The interaction $(\bar{\mu}\nu')(\bar{p}n)$ **or** $(\bar{n}p)(\bar{\nu}'\mu)$ This product describes, among other things, the carefully observed capture of muons by hydrogen:

$$\mu^- + p \rightarrow n + \nu'$$

[1] S. Tsuruta, Ph.D. thesis, Columbia University, New York, 1964. A good treatment is H. Y. Chiu, *Ann. Phys.*, **15**:1 (1961). The most thorough calculations and tabulations are those of C. J. Hansen, Ph.D. thesis, Yale University, New Haven, Conn., 1967.

[2] At zero temperature the total Fermi energy, including rest mass, is

$$W_F = (0.511\ \text{Mev})[1 + (\rho/\mu_e)^{\frac{2}{3}}(9.82 \times 10^5\ \text{g/cm}^3)^{-\frac{2}{3}}]^{\frac{1}{2}}$$

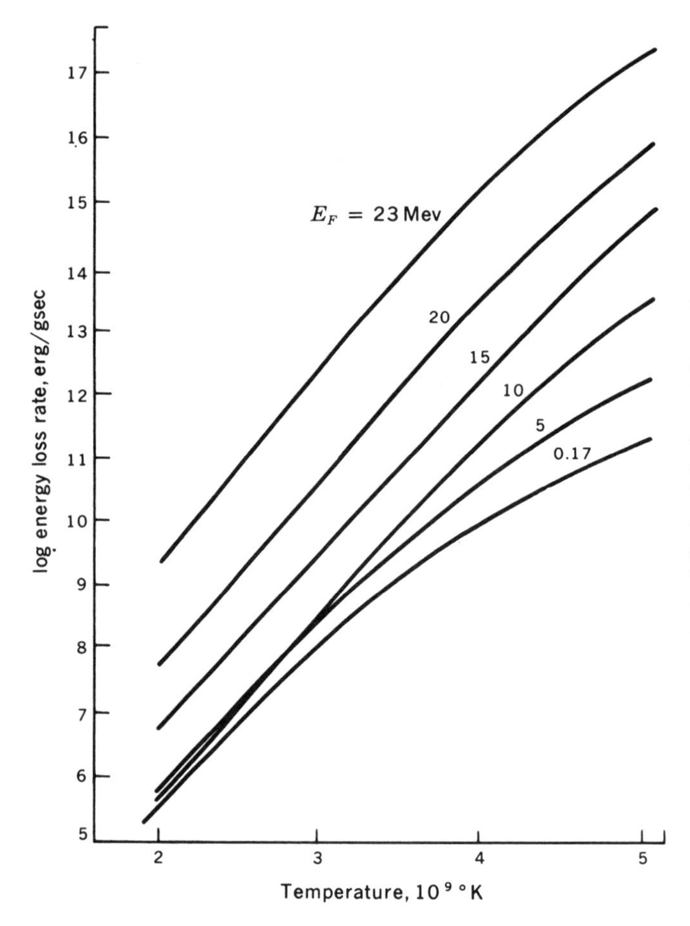

Fig. 3-30 The rate of liberation of neutrino energy by the nuclear Urca process from matter in nuclear statistical equilibrium. The power is strongly temperature-dependent. The density dependence is reflected by the Fermi kinetic energy of the degenerate electron gas, the values of which label the curves. [*S. Tsuruta and A. G. W. Cameron, Can. J. Phys.,* **43**:2056 (1965).]

This reaction shares the strength g of the other two. It is the equality of these cross-product terms that constitutes the major evidence for the universality of the weak interaction. The *squared* or *self-interaction* terms in the hamiltonian, which are the ones that give rise to the major neutrino-emitting reactions in astrophysics, have either not been observed at all or have only been inferred.

Problem 3-48: Assuming the hamiltonian to be correct, do the reactions

$n + \nu \rightarrow \mu^+ + e^-$ and $p + \bar{n} \rightarrow e^+ + \nu$

occur in nature?

THE SELF-INTERACTION TERMS

One could have constructed a weak-interaction theory that arbitrarily yields zero for the squared terms in the current-current interaction, thereby leaving only the cross products observed so far. From a theoretical point of view it has been more attractive to consider each product of pairs as occurring with equal strength. Laboratory evidence for one of the self-interaction terms has been claimed, furthermore, though probably not conclusively. The question of whether these self-interaction terms do in fact occur is an important one for astrophysics, because they are capable of producing more dramatic energy loss than nuclear beta decay does. Let us consider briefly the three self-interaction products.

The interaction $(\bar{n}p)(\bar{p}n)$ This interaction leads to the reaction

$$n + p \rightarrow n + p$$

which represents the scattering due to the weak interaction of a proton from a neutron. The scattering amplitude (cross section) for this process is much too small to be observed directly. Its effects are masked by the stronger nuclear force, which gives rise to the same event. Because the weak interaction violates the symmetry of parity (equivalence of rates to those in a space-inverted coordinate system), whereas the nuclear interaction does not, the existence of this weak-scattering process will admix into every nuclear state a small amount of the opposite parity,[1] and this admixture may lead to observable effects. Although the parity-nonconserving amplitudes for electromagnetic transitions are small compared to the parity-conserving ones, the interference of the two amplitudes, which is linear rather than quadratic in the small amplitude, should be measurable by modern techniques. Michel[2] has made a detailed investigation of the nuclear transitions capable of demonstrating the existence of this weak force. The quest was undertaken experimentally by Boehm and coworkers, who successfully detected a small parity admixture in a gamma decay in Ta^{181} in an amount consistent with this current-current hypothesis of the weak interaction.[3] At the present time it appears that one can say that experimental evidence for the nucleon self-interaction terms exists. If this is correct, the assumption that all the self-interaction terms exist seems even more plausible than before.

The interaction $(\bar{\mu}\nu')(\bar{\nu}'\mu)$ This product leads to the scattering of mu neutrinos from muons. It seems completely undetectable and will be ignored, inasmuch as muons are not generally believed to be important in stellar evolution, although they may play a role in the gravitationally collapsed supernova core.

The interaction $(\bar{e}\nu)(\bar{\nu}e)$ These operator products may be viewed as the destruction of an electron and a neutrino accompanied by the creation of an electron and a neutrino,

$$e^- + \nu \rightarrow e^- + \nu$$

which once again describes a weak scattering, this time of neutrinos (or antineutrinos) from electrons (or positrons). The cross section is expected to be

[1] Parity in nuclear states will be discussed further in Chap. 4.

[2] F. C. Michel, *Phys. Rev.*, **133**:B329 (1964).

[3] F. Boehm and E. Kankeleit, *Phys. Rev. Letters*, **14**:312 (1964).

of order 10^{-44} cm^2, so that even near a reactor flux of 10^{14} ν/cm^{-2} sec^{-1} about 10^6 moles is required to produce one scattering event per second! In spite of such enormous difficulty, elaborate experiments are being contemplated in an attempt to detect this very important interaction. It is not primarily the scattering of neutrinos from electrons that is of interest astrophysically but rather the several other reactions that will be possible if the scattering interaction exists. It is, in fact, from this one single term that all the important neutrino-emitting reactions (with the exception of nuclear beta decay) arise. It should be noted that this is a nonnuclear process. It involves only an interaction between neutrinos and electrons. It follows from quantum electrodynamics that an interaction with an electron automatically implies interaction with photons, inasmuch as a photon is electrodynamically a virtual positron-electron pair, and interactions with virtual particles are the same as with the free particles. (For instance, photons may scatter from other photons by the same principle.) Thus this self-interaction term of the hamiltonian, if it exists, links neutrinos to the most abundant constituents of the gas, photons and electrons. There are, moreover, a variety of ways in which this interaction will be expected to appear. It will lie outside the scope of this book to calculate the rates involved, but an enumeration of the major processes follows.[1]

(1) $e^+ + e^- \rightarrow \nu + \bar{\nu}$: This reaction is obtainable from neutrino-electron scattering simply by transposing an electron and a neutrino by switching them to antiparticles, an operation consistent with the field-theoretic interpretation of the particle operators. The annihilation of a positron-electron pair into two photons is well known. It comes about via the electromagnetic interaction. Because the weak rates are so small in comparison with electromagnetic rates, the branching ratio is calculated as

$$\frac{e^+ + e^- \rightarrow \nu + \bar{\nu}}{e^+ + e \rightarrow \gamma + \gamma} \approx 10^{-19} \tag{3-291}$$

Any terrestrial attempt to observe this very small fractional competition would be hopeless, but if this self-interaction term in the weak current-current hamiltonian exists, as assumed, this reaction may be the most prolific source of neutrino emission in nature.

Another interesting electrodynamic feature arises from the interpretation of a photon as a virtual positron-electron pair. If the weak interaction is capable of converting the electron pair into a neutrino pair, as assumed, the possibility arises that photons may be replaced by a neutrino pair if the process is allowed by conservation laws.

Problem 3-49: It is well known that a photon, even if it has sufficient energy, cannot spontaneously transform into a real electron pair without violating conservation of momentum or of energy. This results from the finite electron mass. The neutrino, however, is also massless.

[1] Probably the most complete account available is in Chiu, "Stellar Physics," *loc. cit.* A very extensive bibliography of work prior to 1964 is contained in W. A. Fowler and F. Hoyle, *Astrophys. J. Suppl.*, **9**:201 (1964). The reader should consult these works for more details.

It must therefore travel with the velocity of light and satisfy the relation $E = pc$, just as a photon does. Spin-$\frac{1}{2}$ particles traveling at the velocity of light are in a state of definite *helicity*, however, which is to say that the spin is either parallel or antiparallel to the momentum. The current interpretation is that the neutrino is left-handed and the antineutrino is right-handed. Show, then, that the reaction $\gamma \rightarrow \nu + \bar{\nu}$ is not prohibited by conservation of energy and linear momentum but that it is prohibited by those principles if the spins of the neutrino pair are required to couple to unity to match the spin of the photon.

It turns out that the photon can be replaced by a neutrino pair when it interacts with a third body, in much the same way that the high-energy photon becomes capable of conversion to an electron pair during interaction with a third body.

One is not used to thinking, perhaps, of the presence of positrons in a star, because they annihilate quickly into two photons with the dense electron gas. But at high temperature ($T > 10^9$ °K) quite a large density of photons have energies in excess of the threshold for pair production in the field of an ion:

$$h\nu > 2m_ec^2 \qquad \text{for pair production}$$

At sufficiently high temperatures, therefore, pairs are also created at a large rate. Thus if n_+ designates the density of positrons, its value will be given by the differential equation

$$\frac{dn_+}{dt} = \text{pair production rate} - \frac{n_+}{\tau_{\text{ann}}} \tag{3-292}$$

where τ_{ann} is the mean lifetime of free positrons against annihilation with the free electrons. Because of the short lifetimes involved, the positron density will quickly achieve an equilibrium value such that $dn_+/dt = 0$. In that case the equilibrium density of free positrons is numerically equal to the number produced per unit volume within the annihilation lifetime:

$$(n_+)_{\text{eq}} = \text{pair production rate} \times \tau_{\text{ann}} \tag{3-293}$$

At temperatures of several billion degrees the pair production rate becomes so large that the positron density becomes a significant fraction of the electron density. Of course the positron density also depends upon the electron density, because the annihilation lifetime is inversely proportional to the electron density.

Since equilibrium is achieved so quickly, one may assume that the positron density is always in equilibrium. In that case the details of the creation and annihilation reactions can be ignored, and one can calculate the positron density from statistical mechanics alone. One sets the number densities equal to the usual integral over the Fermi-Dirac momentum spectrum and uses the auxiliary condition relating the *chemical potential*[1] of the positrons to that of the electrons.

[1] See L. D. Landau and E. M. Lifschitz, "Statistical Physics," Addison-Wesley Publishing Company, Inc., Reading, Mass., 1958, for the meaning of chemical potential and *ibid.*, chap. 11, for the positron density. Fowler and Hoyle, *loc. cit*, present a thorough discussion of this neutrino mechanism.

The result is that [compare Eq. (2-53)]

$$n_\pm = \frac{1}{\pi^2}\left(\frac{m_e c}{\hbar}\right)^3 \int_0^\infty \frac{\chi^2\,d\chi}{\exp\,[\theta(\chi^2+1)^{\frac{1}{2}} \pm \phi] + 1} \tag{3-294}$$

$$n_\pm = \frac{1}{\pi^2}\left(\frac{m_e c}{\hbar}\right)^3 \int_1^\infty \frac{\omega(\omega^2-1)^{\frac{1}{2}}\,d\omega}{\exp\,(\theta\omega \pm \phi) + 1} \tag{3-295}$$

where $\theta = m_e c^2/kT = 5.930/T_9$, $\chi = p/m_e c$, ω is the total energy of electron or positron in units of $m_e c^2$, and ϕ is the chemical potential divided by kT. For any temperature these equations determine n_+ and n_- parametrically, the parameter being ϕ. The parameter ϕ can be determined from the condition

$$n_- - n_+ = n_e^{(0)} \tag{3-296}$$

where $n_e^{(0)}$ is the electron density required to neutralize the nuclei ($n_e^{(0)}$ would equal the electron density were it not increased somewhat by production of pairs). If the gas is nondegenerate (as expected at high temperatures), and if $kT < m_e c^2$, the integrals may conveniently be expanded to give approximately

$$n_\pm \approx \mp \frac{n_e^{(0)}}{2} + \left[\left(\frac{n_e^{(0)}}{2}\right)^2 + n_1^2\right]^{\frac{1}{2}} \tag{3-297}$$

where

$$n_1 \approx \frac{1}{\sqrt{2}}\left(\frac{m_e kT}{\pi\hbar^2}\right)^{\frac{3}{2}} \exp - \frac{m_e c^2}{kT}$$

$$\approx 1.52 \times 10^{29} T_9^{\frac{3}{2}} \exp - \frac{5.93}{T_9} \qquad \text{cm}^{-3} \tag{3-298}$$

Problem 3-50: Calculate the positron number density at 10^9 °K in an oxygen gas having a nuclear density of 10^3 g/cm^3.

Once the number densities n_+ and n_- are determined, the rate of neutrino emission can be calculated from

$$\frac{d\nu}{dt} = 2n_+n_-\langle\sigma v\rangle \tag{3-299}$$

where σ is the cross section for annihilation into a neutrino pair and $\langle\sigma v\rangle$ is the average value of the cross section times the relative velocity of positron and electron, the average being taken with respect to the velocity distribution of the particles. To calculate energy loss during stellar evolution, however, one needs not the number of neutrinos emitted per second but the power emitted in the form of neutrinos. In the annihilation the total energy W, including the rest masses of the electron pair, is converted into neutrino energy. Thus the neutrino power per unit volume is

$$\frac{du_\nu}{dt} = n_+n_-\langle\sigma v W\rangle \tag{3-300}$$

The annihilation cross section turns out to be (for the moderate energies found in stars)[1]

$$\sigma = 1.42 \times 10^{-45} \frac{c}{v} (\omega^2 - 1) \qquad \text{cm}^2 \tag{3-301}$$

where v is the relative velocity of the pair and ω is the center-of-momentum total energy (including rest mass) in units of $m_e c^2$. This cross section has the characteristic smallness of all neutrino cross sections because it must be proportioned by g^2. Performing the integral over the spectrum is somewhat tedious, the answer being expressible in terms of modified Hankel function.[2] Reasonably good simple approximations exist for nondegenerate electrons in the nonrelativistic limit (low temperature) and in the relativistic limit (very high temperature):

$$\frac{du_\nu}{dt} \approx 4.9 \times 10^{18} T_9{}^3 \exp \frac{-11.86}{T_9} \qquad T_9 < 1 \tag{3-302}$$

$$\frac{du_\nu}{dt} \approx 4.6 \times 10^{15} T_9{}^9 \qquad T_9 > 3 \tag{3-303}$$

The temperature T_9 is in units of billions of degrees, and the power emitted is, in cgs units, ergs per cubic centimeter per second. The importance of these results can be seen by noting that at high temperature the neutrino luminosity may radiate the thermal energy in a short time. It will also be noted that the neutrino power is independent of the density. This surprising result occurs because in a nondegenerate gas the product $n_+ n_-$ is independent of the density, as may be easily confirmed from Eq. (3-297). Because the positrons are destroyed by electrons, their concentration tends to be inversely proportional to the electron density, and the product of the concentration is essentially constant.

Problem 3-51: In a gas at a temperature 3×10^9 °K, what is the ratio of the energy density in the thermal photon gas to the neutrino power? Discuss the implications.
Ans: About 2 hr.

The results of numerical calculations are conveniently displayed in Fig. 3-31. The upper solid curve is the neutrino luminosity (measured by the left-hand ordinate), which rises asymptotically as T^9. Also shown as dashed curves are the high- and low-temperature approximations to the neutrino luminosity as given in Eqs. (3-302) and (3-303). The middle solid curve shows the product $n_+ n_-$ (as measured by the right-hand ordinate). That product increases like the sixth power of the temperature. The bottom solid curve shows the averaged product $\langle \sigma v W \rangle$, and is measured by the right-hand ordinate under the hiatus. It rises roughly like T^3.

(2) $\gamma_{\text{plasmon}} \rightarrow \bar{\nu} + \nu$: Even though a photon is virtually a positron-electron pair, it cannot decay into a neutrino pair because the neutrinos, with their pecul-

[1] At ultrahigh energy the reaction may proceed via an intermediate Bose particle.

[2] See Fowler and Hoyle, *op. cit.*, p. 213.

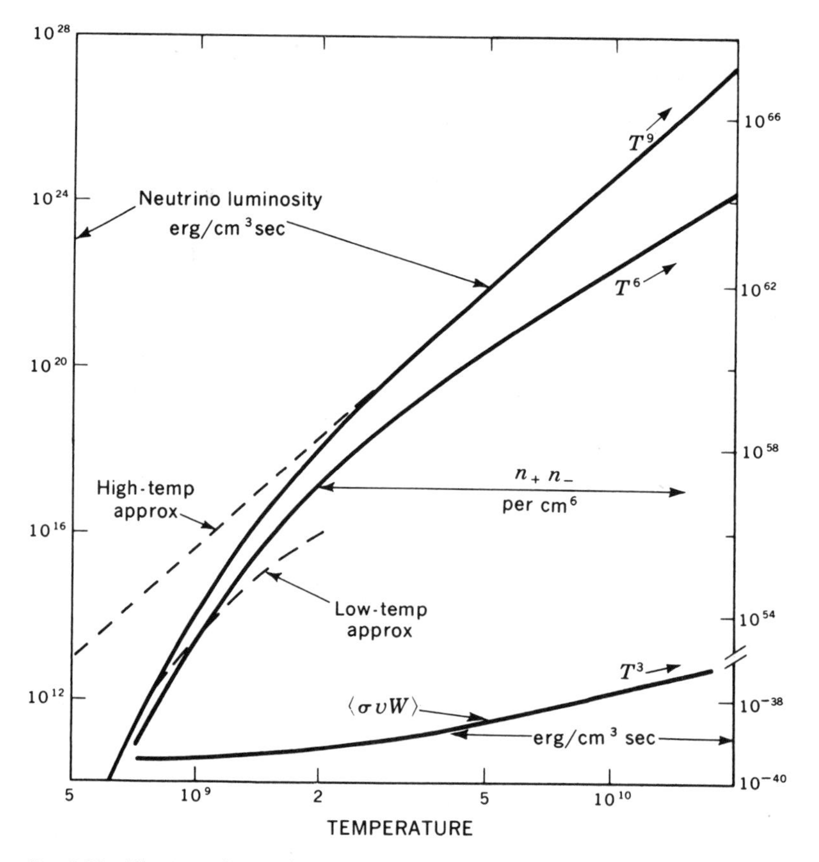

Fig. 3-31 The neutrino emissivity from the annihilation of positron-electron pairs, the product n_+n_- of the pair densities, and the average $\langle\sigma v W\rangle$ for their reaction are shown as functions of the temperature. Note that the latter two ordinates are located on the right-hand margin. This neutrino source depends upon the universality of the weak interaction. [*W. A. Fowler and F. Hoyle, Astrophys. J. Suppl.,* **9**:201 (1964). *By permission of The University of Chicago Press. Copyright* 1964 *by The University of Chicago.*]

iar helicity properties, cannot couple to unit spin unless they move in opposite directions. If the neutrinos must move in opposite directions, the decay cannot conserve energy and momentum, inasmuch as all three particles obey the energy-momentum relationship $E = pc$. Although this situation prevents the decay in a vacuum, the situation is more complicated in a star. The stellar plasma is a dialectric for photon propagation, such that the *dispersion relation* for angular frequency and wave number becomes

$$\omega^2 = k^2c^2 + \omega_0^2 \tag{3-304}$$

where ω_0 is called the *plasma frequency*. In a nondegenerate plasma, the plasma frequency is given by the well-known expression[1]

$$\omega_0^2 = \frac{4\pi n_e e^2}{m_e} \qquad \text{nondegenerate} \tag{3-305}$$

or, in cgs units,

$$\omega_0 = 5.6 \times 10^4 n_e^{\frac{1}{2}}$$

In nondegenerate stellar interiors this frequency is so small in comparison with the frequency of a *thermal photon*, $\hbar\omega_{\text{th}} \approx kT$, that no significant modification of photon processes results. However, $\hbar\omega_0$ becomes comparable to kT in many high-density astrophysical situations. The electron gas in these cases is degenerate, a feature which modifies the plasma frequency somewhat:

$$\omega_0^2 = \frac{4\pi e^2 n_e}{m_e}\left[1 + \left(\frac{\hbar}{m_e c}\right)^2 (3\pi^2 n_e)^{\frac{2}{3}}\right]^{-\frac{1}{2}} \qquad \text{degenerate} \tag{3-306}$$

Returning now to the photon moving through the dialectric of the dense plasma, we observe that the dispersion relation

$$\omega^2 = k^2c^2 + \omega_0^2$$

is kinematically equivalent to the energy-momentum relation for a massive particle:

$$E^2 = p^2c^2 + m^2c^4 \tag{3-307}$$

Such an electromagnetic wave, when quantized, has been called a *plasmon*, and it behaves like a relativistic Bose particle with a rest mass.

Problem 3-52: Show that the equivalent rest mass for the plasmon is

$$m_{\text{plasmon}} = \frac{\hbar\omega_0}{c^2}$$

The dispersion relation describes the fact that the electromagnetic wave in the plasma has, for a given momentum, an excess energy. The energy excess allows the wave to decay into a neutrino pair in which the neutrino and antineutrino move in opposite directions. In a sense then, the plasmon is energetically unstable against the neutrino-decay mode, whereas the free photon is not.

A corollary is that the *minimum energy* of a propagating photon is $\hbar\omega_0$. The medium is completely absorbing, and electromagnetic disturbances are damped out rather than propagated for $\omega < \omega_0$. In the theory of radiative transfer this feature is taken into account by setting the opacity equal to infinity for $\omega < \omega_0$. This usually has very little effect on the Rosseland mean, because in situations where radiative transfer is important in stars one usually finds that $\omega_0 \ll kT/\hbar$, and the values of the opacity at very low frequencies do not influence the energy transfer. In the degenerate gas, where $\hbar\omega_0$ becomes greater than kT, however,

[1] T. Stix, "The Theory of Plasma Waves," McGraw-Hill Book Company, New York, 1962.

the spectrum of electromagnetic waves is considerably altered by the frequency threshold imposed by the plasma. The plasmons still obey Bose-Einstein statistics, but since they must be thermally excited, the electromagnetic waves can be "frozen out" if $\hbar\omega_0 > kT$. At high densities one finds that only the waves far out on the exponential tail of the Planck spectrum can propagate. Under those circumstances the energy density in the radiation field may fall considerably below the value $u = aT^4$ for a blackbody spectrum at the same temperature.[1] This feature is in itself probably of no importance for stellar structure, inasmuch as the equation of state of the gas is dominated by the degenerate electron gas, and the major importance of plasmons probably lies in their decay into neutrinos.

Because of these thermal properties of plasmons the associated neutrino power will depend upon the energy ratio,

$$x \equiv \frac{\hbar\omega_0}{kT} = \frac{3.345 \times 10^{-4}(\rho/\mu_e)^{\frac{1}{2}}}{T_9[1 + 1.0177 \times 10^{-4}(\rho/\mu_e)^{\frac{2}{3}}]^{\frac{1}{4}}} \tag{3-308}$$

It has been shown that[2] in ergs per cubic centimeter per second

$$\frac{du_\nu}{dt} = 3.07 \times 10^{21} \left(\frac{kT}{m_e c^2}\right)^9 x^9 F(x) \tag{3-309}$$

where

$$F(x) = \sum_{n=1}^{\infty} \frac{K_2(nx)}{nx} \tag{3-310}$$

and the function K_2 is a modified Bessel function of the second kind.

For large values of x, a situation implying that the plasma quantum is comparable in energy to kT, the series for $F(x)$ can be approximated by the first term,

$$F(x) \approx \left(\frac{\pi}{2}\right)^{\frac{1}{2}} x^{-\frac{3}{2}} e^{-x} \tag{3-311}$$

which leads to

$$\frac{du_\nu}{dt} \approx 3.85 \times 10^{21} \left(\frac{\hbar\omega_0}{m_e c^2}\right)^{7.5} \left(\frac{kT}{m_e c^2}\right)^{1.5} e^{-x} \tag{3-312}$$

For small values, $x < 0.5$, the series for $F(x)$ converges very slowly but is well approximated by

$$F(x) \approx \frac{2.404}{x^3} + \frac{1}{x}\left[\left(\frac{1}{2} \ln x\right) - 0.5966\right] + \frac{x}{96}[(\ln x) - 2.851] \tag{3-313}$$

[1] C. L. Inman, *Astrophys. J.*, **142**:201 (1965).

[2] C. L. Inman and M. Ruderman, *Astrophys. J.*, **140**:1025 (1964). These published rates have been decreased by a factor of 4 as a result of an observation of Zaidi's.

For very small x the leading term dominates, giving

$$\frac{du_\nu}{dt} \approx 7.4 \times 10^{21} \left(\frac{kT}{m_e c^2}\right)^3 x^6 \tag{3-314}$$

This limit corresponds to a high-temperature limit, where emission by the annihilation of pairs usually dominates the plasmon emission. Note that each neutrino power has been expressed in ergs per cubic centimeter per second. If one wishes to define a neutrino term in the energy generation, then

$$\epsilon_\nu \equiv -\frac{1}{\rho}\frac{du_\nu}{dt} \tag{3-315}$$

Figure 3-32 shows the neutrino power per gram radiated from plasmon decay as a function of density for several different temperatures. As the density increases for a given temperature, the neutrino power initially increases, because the increasing plasmon mass liberates increasing energy in each decay. But the plasmon modes require energy for their excitation, which must be derived from the thermal environment, so that when $\hbar\omega_0$ becomes greater than kT, the number of plasmons decreases as $\exp(-\hbar\omega_0/kT)$; that is, the plasmons are frozen out at a very high density, and the neutrino power, after passing through a maxi-

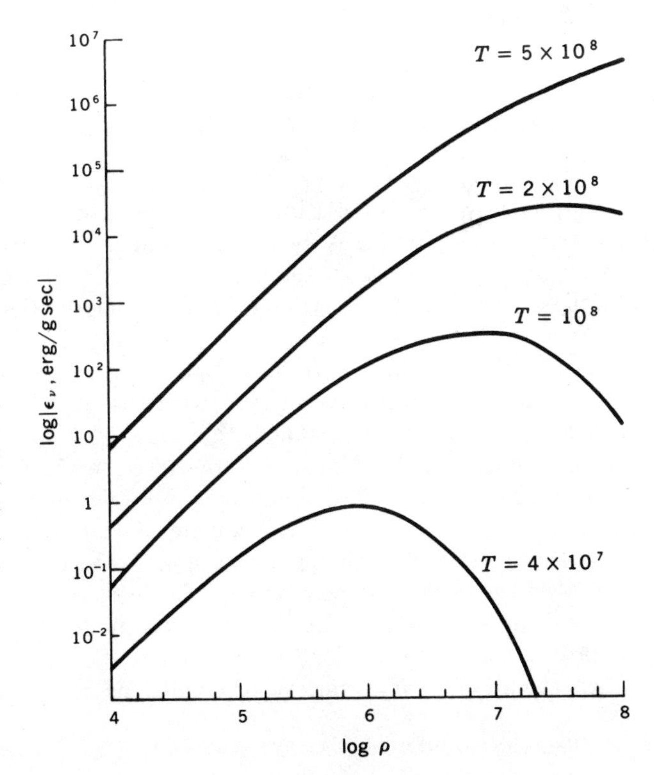

Fig. 3-32 The neutrino emissivity resulting from the decay of plasmon modes into neutrino pairs. At each value of the temperature the emissivity increases with density until the plasma frequency becomes so large that its quantum modes cannot be thermally excited. This source depends upon the universality of the weak interaction. [*Constructed from numerical tables in C. L. Inman and M. Ruderman, Astrophys. J.*, **140**:1025 (1964).]

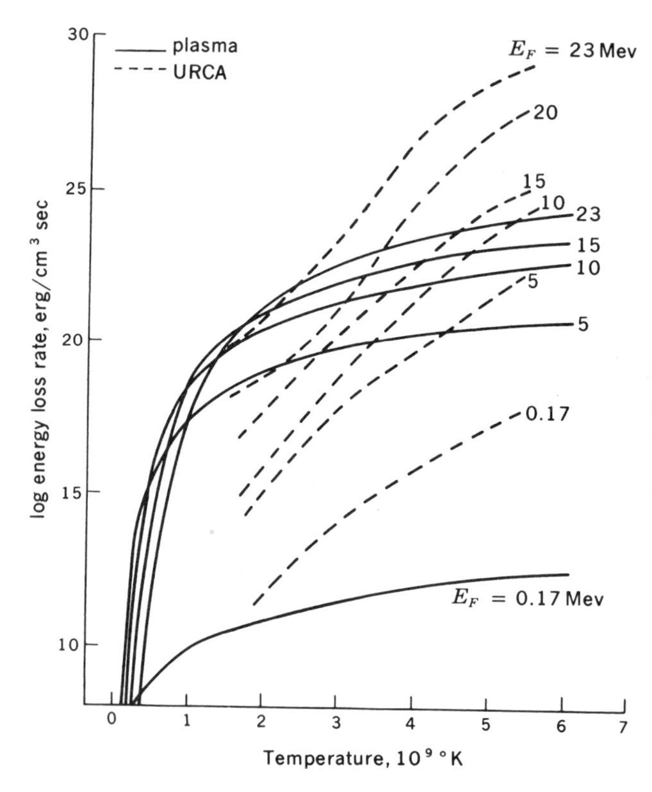

Fig. 3-33 The competition between the Urca neutrino emissivity from matter in nuclear statistical equilibrium and the plasma emissivity at high temperature and density. The former does not depend upon the universality of the weak interaction, whereas the latter does. [*S. Tsuruta and A. G. W. Cameron, Can. J. Phys.*, **43**:2056 (1965).]

mum, falls off with further increases in density. For the same reason increasing the temperature increases the neutrino power by increasing the density of excited plasmons. At very high temperatures, however, the annihilation of positron-electron pairs becomes a stronger neutrino source than plasmon decay.

At a combination of extremely high density and temperature, such as may be found in the imploded core of a supernova or a neutron star, the plasma neutrinos may be competitive instead with the nuclear Urca process. Figure 3-33 shows this rather special competition, again labeled by the Fermi energy of the electron gas. At the densities encountered in the calmer stages of stellar evolution, the plasma neutrinos give way to annihilation neutrinos at high temperature.
(*3*) $\gamma + e \rightarrow e + \nu + \bar{\nu}$: The analog of Compton scattering, with the exception that a neutrino pair replaces the outgoing photon, this process has come to be called the *photoneutrino process*. The electron modifies the momentum-energy balance in such a way that the emerging photon can appear as a neutrino pair. One Feynman diagram for the reaction is shown in Fig. 3-34. The rates have been calculated[1] for a variety of conditions in the electron gas. We shall only

[1] V. Petrosian, G. Beaudet, and E. Salpeter, *Phys. Rev.*, **154**:1445 (1967).

summarize approximate nonrelativistic expressions for the neutrino power in cgs units, ergs per cubic centimeter per second:

(*i*) Nonrelativistic nondegenerate

$$\frac{du_\nu}{dt} \approx 0.98 \times 10^8 \frac{\rho}{\mu_e} T_9{}^8 \tag{3-316}$$

(*ii*) Nonrelativistic degenerate

$$\frac{du_\nu}{dt} \approx 4.8 \times 10^{11} \left(\frac{\rho}{\mu_e}\right)^{\frac{1}{3}} T_9{}^9 \tag{3-317}$$

The photoneutrino process competes with pair annihilation only at temperatures sufficiently low so that electron pairs cannot be created, and it competes with the plasma neutrino process only at densities so low that the plasmon quantum $\hbar\omega_0$ becomes trivially small. Figure 3-35 shows the domains in the temperature-density plane which are dominated by these three neutrino sources arising from the $(\bar{e}\nu)(\bar{\nu}e)$ interaction term (the nuclear Urca neutrinos are not included, although they do compete at high temperature and high density, as was shown in Fig. 3-33). There exist many other mechanisms for neutrino emission in stars,[1] but since they appear to be less efficient than those discussed so far, they are omitted here.

[1] See Fowler and Hoyle, *loc. cit.*, or Chiu, "Stellar Physics."

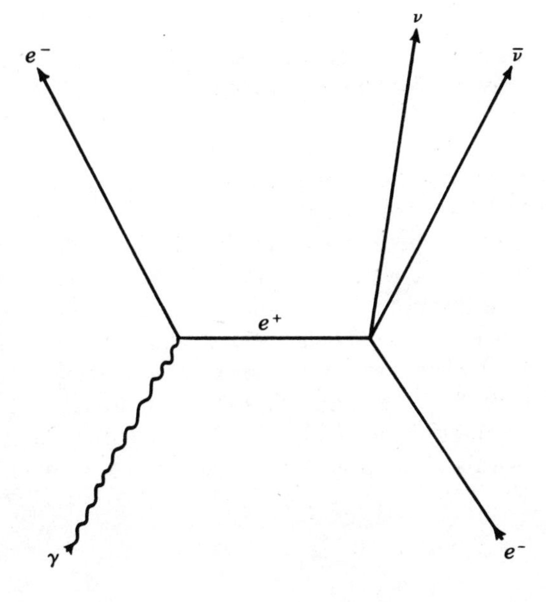

Fig. 3-34 A Feynman diagram for the photoneutrino process. It can be seen that one vertex involves the electromagnetic annihilation of a positron-electron pair, which is well known, whereas the other vertex involves the unchecked annihilation into neutrino pairs. The rate for the neutrino emissivity depends therefore upon the same weak-interaction physics as the annihilation of real pairs.

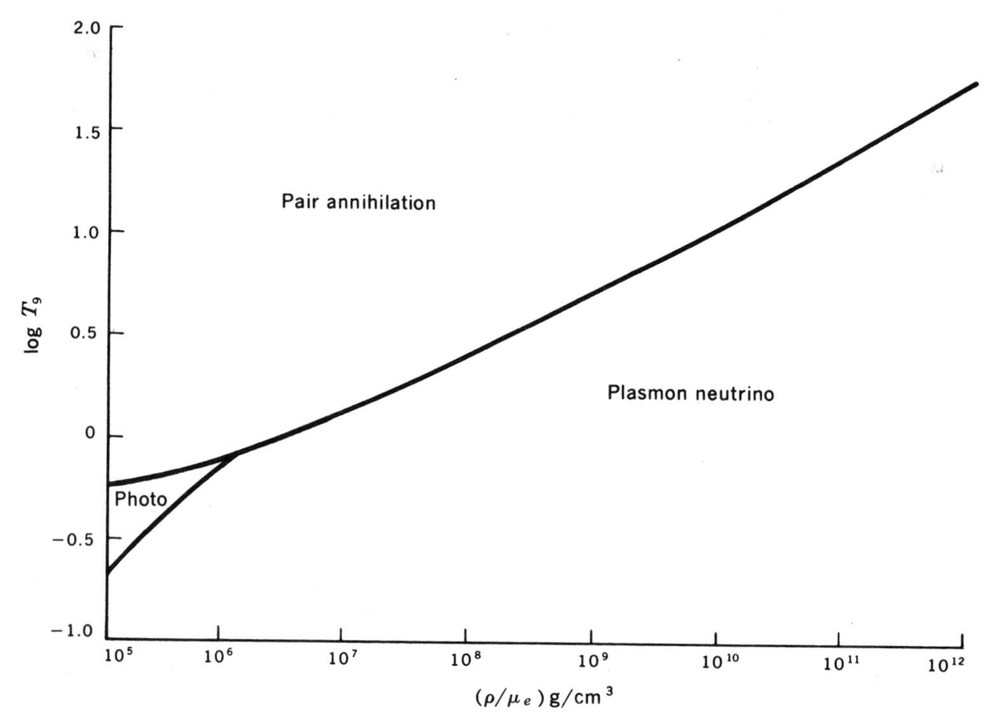

Fig. 3-35 The temperature-density plane is divided into three regions showing the dominant neutrino-emitting mechanisms arising from the $(\bar{e}\nu)(\bar{\nu}e)$ interaction. If this interaction shares the strength of a universal weak interaction, these neutrino-emitting processes are very important in stellar evolution.

Problem 3-53: Calculate the temperature at $\rho/\mu_e = 10^5$ g/cm^3 for which the photoneutrino emissivity equals the plasma-neutrino emissivity.

In summary, it will be noted that, except for the corrections to the energy-generating nuclear reactions, the importance of neutrino emission in stellar evolution seems to be limited to the late evolutionary stages. The most important of these, the electron-pair annihilation into a neutrino pair, the plasmon decay into a neutrino pair, and the photoneutrino pair, stem from an interaction that has not yet been observed in the physics laboratory, the $(\bar{e}\nu)(\bar{\nu}e)$ term in the current-current interaction. The neutrino emission becomes potentially important whenever the temperature exceeds 5×10^8 °K and/or whenever the density exceeds 10^5 g/cm^3. It seems fair to say that a cloud of uncertainty hangs over calculations of the terminal stages of stellar evolution until the proof or disproof of this weak-interaction term has been demonstrated convincingly.

chapter 4

THERMONUCLEAR REACTION RATES

The heart of stellar evolution and nucleosynthesis is the thermonuclear reaction. In the foregoing discussions, the rate of energy liberation per gram of stellar material has been designated by the symbol ϵ. It is the fusion of light nuclei into heavier nuclei that liberates kinetic energy (at the expense of mass) and serves as the interior source of the energy radiated from the surface. The condition that the power liberated internally balance the power radiated from the surface determines a steady state in the structure of the star. That situation cannot be a truly static one, however, because the very reactions that liberate energy necessarily change the chemical composition of the stellar interior. It is the slow change of chemical composition that causes the structure of the star to evolve. If, after a finite lifetime, a star ejects all or part of its mass into space, the chemical composition of the interstellar medium will have been altered by the thermonuclear debris. Stated most simply, it is the working hypothesis of the stellar nucleosynthesist that all or part of the heavy elements found in our galaxy have been synthesized in the interiors of stars by these same fusion reactions. For these reasons the subject of thermonuclear reaction rates is a focal point of this book.

A complete science of thermonuclear reaction rates is formidable. It involves complicated details of nuclear physics, many of which are still unsolved. The mechanism of each reaction must be scrutinized to achieve assurance of the proper prescription for the stellar reaction rate. Still there are a few basic physical principles that are common to the computation of all thermonuclear reaction rates, and it is to these general principles that we address ourselves in this chapter.

4-1 KINEMATICS AND ENERGETICS

A nuclear reaction in which a particle a strikes a nucleus X producing a nucleus Y and a new particle b is symbolized by

$$a + X \rightarrow Y + b$$

For example, a reaction in which a deuteron strikes a C^{12} nucleus producing a C^{13} nucleus and a proton is written

$$d + C^{12} \rightarrow C^{13} + p$$

An alternative notation in common usage is

$$X(a,b)Y \qquad C^{12}(d,p)C^{13}$$

The incoming or outgoing particle may often be a photon, as in the reaction

$$p + N^{14} \rightarrow O^{15} + \gamma \qquad N^{14}(p,\gamma)O^{15}$$

In all such nuclear reactions, the total energy, momentum, and angular momentum are conserved quantities.

The necessity for conserving the total linear momentum in a nuclear reaction suggests that the kinematical description be in terms of the motion of the center of mass (or strictly speaking the center of momentum) of the nuclear system plus the motion of the particles relative to their center of mass. Just as in classical mechanics, the total energy or the total momentum of the system may be expressed as the sum of the energy or momentum in the center-of-mass system plus the energy or momentum of the motion of the center of mass itself. The conservation of momentum demands, among other things, that the motion of the center of mass be unaltered by the reaction.

For two particles of masses m_1 and m_2 and nonrelativistic velocities $\mathbf{v}_1$ and $\mathbf{v}_2$,

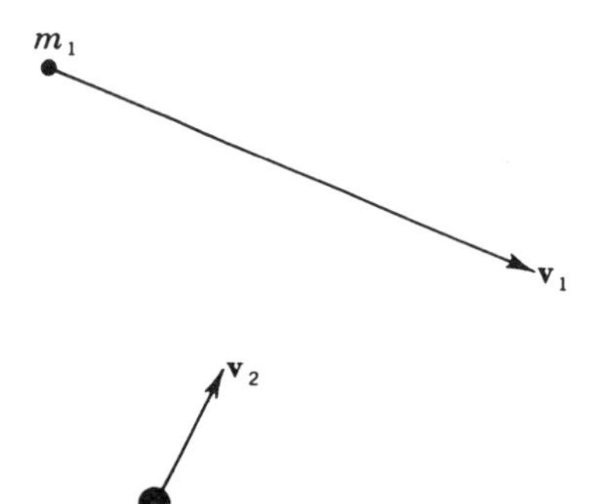

Fig. 4-1 A particle of mass m_1 and velocity $\mathbf{v}_1$ in collision with a particle of mass m_2 and velocity $\mathbf{v}_2$.

the velocity $\mathbf{V}$ of the center of mass is given by the value of the momentum:

$$m_1\mathbf{v}_1 + m_2\mathbf{v}_2 = (m_1 + m_2)\mathbf{V}$$

or

$$\mathbf{V} = \frac{m_1\mathbf{v}_1 + m_2\mathbf{v}_2}{m_1 + m_2} \tag{4-1}$$

This discussion will be entirely restricted to the nonrelativistic kinematics appropriate to the low kinetic energy in stellar interiors.

The momentum of particle 1 relative to the center of mass is

$$m_1(\mathbf{v}_1 - \mathbf{V}) = \frac{m_1 m_2}{m_1 + m_2}(\mathbf{v}_1 - \mathbf{v}_2) = \mu\mathbf{v} \tag{4-2}$$

where μ is the *reduced mass*

$$\mu = \frac{m_1 m_2}{m_1 + m_2} \tag{4-3}$$

and $\mathbf{v}$ is the *relative velocity* of m_1 and m_2

$$\mathbf{v} = \mathbf{v}_1 - \mathbf{v}_2 \tag{4-4}$$

Problem 4-1: Confirm Eq. (4-2). Write an analogous expression that is relativistically correct.

In the same manner the momentum of m_2 relative to the center of mass is

$$m_2(\mathbf{v}_2 - \mathbf{V}) = -\mu\mathbf{v} \tag{4-5}$$

Thus in the center of mass the particles approach each other with equal and opposite momenta; the total momentum is zero in the center-of-mass system. We can think of the whole center-of-mass picture as sliding along with the velocity $\mathbf{V}$, as in Fig. 4-2. The conservation of momentum will be satisfied by demanding that the velocity of the center of mass be unchanged by the collision and that the total momentum in the center-of-mass system be zero after the collision just as before. The kinetic energy before the collision is

$$(\mathrm{KE})_i = \tfrac{1}{2}m_1 v_1^2 + \tfrac{1}{2}m_2 v_2^2 \tag{4-6}$$

Problem 4-2: Use Eqs. (4-2) and (4-5) to show that

$$(\mathrm{KE})_i = \tfrac{1}{2}(m_1 + m_2)V^2 + \tfrac{1}{2}\mu v^2 \tag{4-7}$$

Equation (4-7) indicates that the kinetic energy of the two particles can be thought of as the sum of those associated with a mass $m_1 + m_2$ moving with the velocity of the center of mass plus a mass μ moving with the relative velocity $\mathbf{v}$. The first term is the kinetic energy of the center of mass itself, which must be

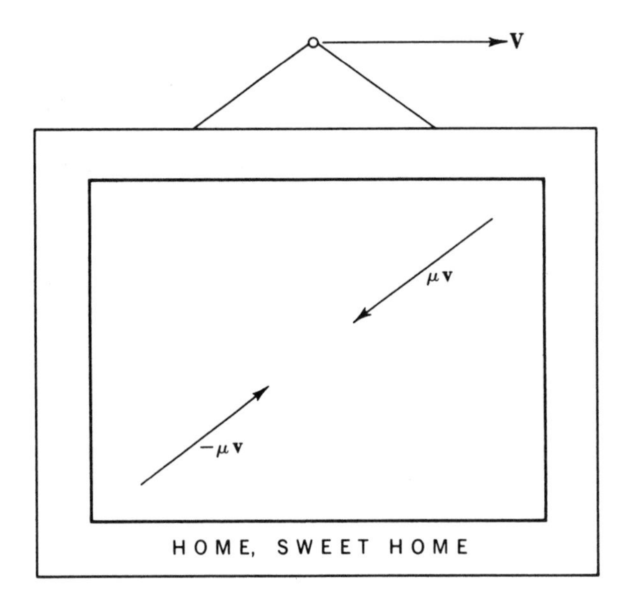

Fig. 4-2 The collision of Fig. 4-1 in the center-of-mass coordinates. In the frame of reference moving with the velocity **V** of the center of mass of the two particles, the particles collide with equal and opposite momenta $\pm\mu\mathbf{v}$.

the same after the collision as before it.[1] The second term, $\mu v^2/2$, represents that portion of the kinetic energy available for doing work against any force separating the two particles. It is commonly called the *kinetic energy in the center-of-mass system.*

Problem 4-3: If it is imagined that the particles are held apart by a repulsive spring, it is clear that the extent to which the spring may be compressed by relative motion is most simply obtained using the center-of-mass system in Fig. 4-2, because only the equal and opposite components of the particle momenta can be expended in doing work against the spring without violating the law of momentum conservation. Show that this energy in the center-of-mass system is

$$(\mathrm{KE})_{\mathrm{cm},i} = \frac{(\mu v)^2}{2m_1} + \frac{(\mu v)^2}{2m_2} = \frac{1}{2}\mu v^2 \tag{4-8}$$

These nonrelativistic formulas are applicable to nuclear reactions only if the combined mass of the final particles equals the combined mass of the initial particles. But the source of new kinetic energy comes from a *reduction* of mass according to the Einstein relationship

$$\Delta\mathrm{KE} = -\Delta M c^2$$

In low-energy nuclear reactions, however, $\Delta M/M \approx 10^{-3} \rightarrow 10^{-4}$, so that the assumption of constant mass is accurate to better than 0.1 percent. For our

[1] This statement once again assumes that the total mass of the particles will change by a kinematically negligible amount in the reaction. The assumption is consistent with the assumption that the particles may be treated nonrelativistically.

purposes it is adequate to consider that to be equality. Since the kinetic energy of the center of mass is accordingly unchanged by the reaction, the kinetic energy in the center-of-mass system must be increased or decreased according to whether the final mass is less than or greater than the initial mass.

Thus if we return to the reaction

$$a + X \rightarrow Y + b$$

the conservation-of-energy principle demands the equality

$$E_{aX} + (M_a + M_X)c^2 = E_{bY} + (M_b + M_Y)c^2 \tag{4-9}$$

where E_{aX} is the center-of-mass kinetic energy of a and X and E_{bY} is the kinetic energy in the center of mass of the bY system. The second terms on each side of the equation represent the fact that the sums of the rest masses before and after the reaction are not necessarily exactly equal and that kinetic energy may be either liberated or absorbed by that inequality. This is the well-known Einstein mass-energy relationship. The masses involved are the masses of the nuclei a, X, b, and Y. For instance, in the example $C^{12}(d,p)C^{13}$

$$\begin{aligned} E_{d,C^{12}} &+ c^2[M(d \text{ nucleus}) + M(C^{12} \text{ nucleus})] \\ &= E_{p,C^{13}} + c^2[M(\text{proton}) + M(C^{13} \text{ nucleus})] \end{aligned} \tag{4-10}$$

Notice, however, that the net amount of electric charge is conserved in normal nuclear reactions. It follows that the number of electrons in the neutral atoms is equal on both sides of the reaction equation. Thus we may, if we choose, replace nuclear masses by atomic masses, since the same number of electron rest masses is added thereby to both sides of the equation. The replacement of nuclear masses by atomic masses does introduce a small error due to the difference in the binding energies of the electrons on the two sides of the equation; however, the difference in total binding of atomic electrons on the two sides of the equation is very small compared to the difference in nuclear masses. We shall, therefore, use atomic masses in this book, always keeping in mind that errors in the mass-energy relationship on the order of a few electron volts are introduced. Thus we shall write, instead of Eq. (4-10),

$$E_{d,C^{12}} + c^2[M(D^2) + M(C^{12})] = E_{p,C^{13}} + c^2[M(H^1) + M(C^{13})] \tag{4-11}$$

where the masses are those of the neutral atoms. The great utility in using atomic masses is that these are the quantities that are traditionally measured in atomic-mass measuring experiments, i.e., a mass spectrograph.

Another quantity that is clearly equal on both sides of the nuclear-reaction equation is the total number of nucleons (*nucleon* is the generic name of a neutron or a proton). Hence the atomic weight, which is defined as the *integer* nearest in value to the exact mass expressed in atomic mass units, remains the same on both sides of the mass-energy equation. The energy balance itself is not disturbed, therefore, by subtracting the atomic weight times the rest-mass energy of 1 amu from both sides of the equation. The masses then become the *excesses of mass* over the integral number of atomic mass units. We define the *atomic mass excess in units of energy* by the quantity

$$\begin{aligned}\Delta M_{AZ} &= (M_{AZ} - AM_u)c^2 \\ &= [M_{AZ}(\text{amu}) - A]c^2 M_u \end{aligned} \qquad (4\text{-}12)$$

where M_u is the mass of 1 atomic mass unit (amu), defined as one-twelfth of the mass of the neutral C^{12} atom.[1] In the convenient numerical units of Mev, Eq. (4-12) becomes

$$\Delta M_{AZ} = 931.478(M_{AZ} - A) \qquad \text{Mev}$$

where 931.478 is the rest-mass energy of 1 amu in Mev and M_{AZ} is the mass of species (A,Z) in atomic mass units. With this definition the energy-balance equation becomes

$$E_{aX} + (\Delta M_a + \Delta M_X) = E_{bY} + (\Delta M_b + \Delta M_Y) \qquad (4\text{-}13)$$

where ΔM are expressed in energy units, generally in Mev, as defined above. Table 4-1 shows the values of the atomic mass excesses for various atoms expressed in units of Mev. Notice that the atomic mass excess of C^{12} is zero by definition. A numerical example of the use of Table 4-1 may be useful at this point. For the reaction $C^{12}(d,p)C^{13}$, for instance, we see from the mass table that

$$E_{d,C^{12}} + 13.1359 + 0 = E_{p,C^{13}} + 3.1246 + 7.2890$$

or

$$E_{p,C^{13}} = E_{d,C^{12}} + 2.7223 \text{ Mev} \qquad (4\text{-}14)$$

There is an increase of kinetic energy equal to 2.722 Mev for each such reaction.

Problem 4-4: Calculate the energy generated by the reaction $He^3(He^3,2p)He^4$.
Ans: 12.860 Mev.

4-2 CROSS SECTION AND REACTION RATE

The energy-balance equation yields the energy liberated by each nuclear reaction. From that energy and the number of reactions per unit volume per second the energy liberated per unit volume per second can be calculated as a simple product. Thus the second half of the calculation of energy-liberation rates involves the use of the concept of the cross section for a reaction. The cross section is a measure of the probability per pair of particles of occurrence of a reaction. Consider once more the reaction $a + X \rightarrow Y + b$. Envision some nuclei of type X being bom-

[1] Before 1960 the atomic mass unit was defined as one-sixteenth of the O^{16} atom.

Table 4-1 Atomic mass excesses†

Z	*Element*	*A*	*M* − *A*, *Mev*	*Z*	*Element*	*A*	*M* − *A*, *Mev*
0	*n*	1	8.07144			19	3.33270
1	H	1	7.28899			20	3.79900
	D	2	13.13591	9	F	16	10.90400
	T	3	14.94995			17	1.95190
	H	4	28.22000			18	0.87240
		5	31.09000			19	−1.48600
2	He	3	14.93134			20	−0.01190
		4	2.42475			21	−0.04600
		5	11.45400	10	Ne	18	5.31930
		6	17.59820			19	1.75200
		7	26.03000			20	−7.04150
		8	32.00000			21	−5.72990
3	Li	5	11.67900			22	−8.02490
		6	14.08840			23	−5.14830
		7	14.90730			24	−5.94900
		8	20.94620	11	Na	20	8.28000
		9	24.96500			21	−2.18500
4	Be	6	18.37560			22	−5.18220
		7	15.76890			23	−9.52830
		8	4.94420			24	−8.41840
		9	11.35050			25	−9.35600
		10	12.60700			26	−7.69000
		11	20.18100	12	Mg	22	−0.14000
5	B	7	27.99000			23	−5.47240
		8	22.92310			24	−13.93330
		9	12.41860			25	−13.19070
		10	12.05220			26	−16.21420
		11	8.66768			27	−14.58260
		12	13.37020			28	−15.02000
		13	16.56160	13	Al	24	0.1000
6	C	9	28.99000			25	−8.9310
		10	15.65800			26	−12.2108
		11	10.64840			27	−17.1961
		12	0			28	−16.8554
		13	3.12460			29	−18.2180
		14	3.01982			30	−17.1500
		15	9.87320	14	Si	26	−7.1320
7	N	12	17.36400			27	−12.3860
		13	5.34520			28	−21.4899
		14	2.86373			29	−21.8936
		15	0.10040			30	−24.4394
		16	5.68510			31	−22.9620
		17	7.87100			32	−24.2000
8	O	14	8.00800	15	P	28	−7.6600
		15	2.85990			29	−16.9450
		16	−4.73655			30	−20.1970
		17	−0.80770			31	−24.4376
		18	−0.78243			32	−24.3027

Table 4-1 Atomic masses excesses† (Continued)

Z	*Element*	*A*	*M − A, Mev*
15	P	33	−26.3346
		34	−24.8300
16	S	30	−14.0900
		31	−18.9920
		32	−26.0127
		33	−26.5826
		34	−29.9335
		35	−28.8471
		36	−30.6550
		37	−27.0000
		38	−26.8000
17	Cl	32	−12.8100
		33	−21.0140
		34	−24.4510
		35	−29.0145
		36	−29.5196
		37	−31.7648
		38	−29.8030
		39	−29.8000
		40	−27.5000
18	Ar	34	−18.3940
		35	−23.0510
		36	−30.2316
		37	−30.9509
		38	−34.7182
		39	−33.2380
		40	−35.0383
		41	−33.0674
		42	−34.4200
19	K	36	−16.7300
		37	−24.8100
		38	−28.7860
		39	−33.8033
		40	−33.5333
		41	−35.5524
		42	−35.0180
		43	−36.5790
		44	−35.3600
		45	−36.6300
		46	−35.3400
		47	−36.2500
20	Ca	38	−21.6900
		39	−27.3000
		40	−34.8476
		41	−35.1400
		42	−38.5397
		43	−38.3959
		44	−41.4596
		45	−40.8085
		46	−43.1380
		47	−42.3470
		48	−44.2160
		49	−41.2880
21	Sc	40	−20.9000
		41	−28.6450
		42	−32.1410
		43	−36.1740
		44	−37.8130
		45	−41.0606
		46	−41.7557
		47	−44.3263
		48	−44.5050
		49	−46.5490
		50	−44.9600
22	Ti	42	−25.1230
		43	−29.3400
		44	−37.6580
		45	−39.0020
		46	−44.1226
		47	−44.9266
		48	−48.4831
		49	−48.5577
		50	−51.4307
		51	−49.7380
		52	−49.5400
23	V	46	−37.0600
		47	−42.0100
		48	−44.4700
		49	−47.9502
		50	−49.2158
		51	−52.1989
		52	−51.4360
		53	−52.1800
		54	−49.6300
24	Cr	48	−42.8130
		49	−45.3900
		50	−50.2490
		51	−51.4472
		52	−55.4107
		53	−55.2807
		54	−56.9305
		55	−55.1130
		56	−55.2900
25	Mn	50	−42.6480
		51	−48.2600
		52	−50.7020

Table 4-1 Atomic mass excesses† (Continued)

Z	*Element*	*A*	*M − A, Mev*
25	Mn	53	−54.6820
		54	−55.5520
		55	−57.7048
		56	−56.9038
		57	−57.4800
		58	−55.6500
26	Fe	52	−48.3280
		53	−50.6930
		54	−56.2455
		55	−57.4735
		56	−60.6054
		57	−60.1755
		58	−62.1465
		59	−60.6599
		60	−61.5110
		61	−59.1300
27	Co	54	−47.9940
		55	−54.0140
		56	−56.0310
		57	−59.3389
		58	−59.8380
		59	−62.2327
		60	−61.6513
		61	−62.9300
		62	−61.5280
		63	−61.9200
28	Ni	56	−53.8990
		57	−56.1040
		58	−60.2280
		59	−61.1587
		60	−64.4707
		61	−64.2200
		62	−66.7480
		63	−65.5160
		64	−67.1060
		65	−65.1370
		66	−66.0550
29	Cu	58	−51.6590
		59	−56.3590
		60	−58.3460
		61	−61.9840
		62	−62.8130
		63	−65.5831
		64	−65.4276
		65	−67.2660
		66	−66.2550
		67	−67.2910
		68	−65.4100
30	Zn	60	−54.1860
		61	−56.5800
		62	−61.1230
		63	−62.2170
		64	−66.0003
		65	−65.9170
		66	−68.8810
		67	−67.8630
		68	−69.9940
		69	−68.4250
		70	−69.5500
		71	−67.5200
		72	−68.1440
31	Ga	63	−56.7200
		64	−58.9280
		65	−62.6580
		66	−63.7060
		67	−66.8650
		68	−67.0740
		69	−69.3262
		70	−68.8970

† Based on the scale $C^{12} \equiv 0$; 1 amu = 931.478 Mev. This table of masses, prepared by T. Lauritsen, is largely adapted from the comprehensive review by J. H. E. Mattauch, W. Thiele, and A. H. Wapstra, *Nucl. Phys.*, **67**:1 (1965). Terminal zeros are generally not significant digits.

barded by a uniform flux of particles of type *a*, as illustrated in Fig. 4-3. The cross section for the reaction under consideration is defined under such circumstances as

$$\sigma(\text{cm}^2) = \frac{\text{number of reactions/nucleus } X\text{/unit time}}{\text{number of incident particles/cm}^2\text{/unit time}} \tag{4-15}$$

The name cross section arose because of the units of area and because the formula

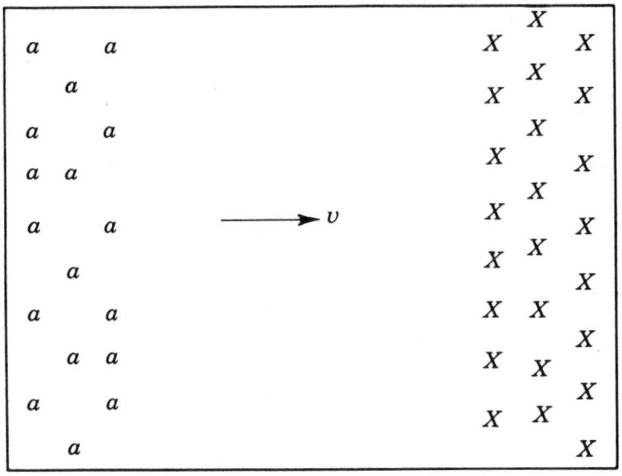

Fig. 4-3 Stationary particles of type X are bombarded with a uniform flux of particles of type a. This situation corresponds to that of the nuclear experiment in the laboratory.

for the number of reactions per unit time can also be computed by assuming that each nucleus X has a cross-sectional area σ and that a reaction occurs each time an a particle strikes in that area. Although such a picture is not physically correct, it is sometimes a helpful mnemonic. Notice that this definition of the cross section is actually symmetric in the two types of particles, since the relative velocity is the same viewed from either particle. This definition actually defines $\sigma(v)$, since it is assumed that the flux of particles has relative velocity $\mathbf{v}$.

Now suppose that the target nuclei X are considered to be in the form of a gas of uniform density N_X. Then the reaction rate per unit volume will be given by the product of σN_X and the flux of particles of type a. Suppose further that the flux of particles of type a is due to the uniform translation with velocity $\mathbf{v}$ of a uniform gas of type a particles having number density N_a. Then the flux of a particles is given by the product $\mathbf{v}N_a$, so that finally we can write the expression for the reaction rate,

$$r = \sigma(v)vN_aN_X \tag{4-16}$$

If both a and X particles are moving, v is the magnitude of their relative velocity.

In a mixture of gases in the state of thermodynamic equilibrium, there exists some *spectrum of relative velocities* between particles of type a and type X, just as a and X *individually* have well-defined velocity spectra. Call this relative velocity spectrum $\phi(v)$, and let it be defined such that $\int \phi(v)\, dv = 1$. In that case $\phi(v)\, dv$ is to be the probability that the relative velocity of the pair of particles has magnitude v in the range dv. Then, generalizing Eq. (4-16), the total reaction rate per unit volume is

$$r_{aX} = N_aN_X \int_0^\infty v\sigma(v)\phi(v)\, dv = N_aN_X\langle\sigma v\rangle \tag{4-17}$$

where the bracketed quantity $\langle\sigma v\rangle$ is a common notation for the average value of the product of relative velocity times cross section. The problem of computing thermonuclear reaction rates reduces to the evaluation of $\langle \sigma v \rangle$ for the reactions that occur in stellar interiors.

As one last point here, we note a small but important correction to Eq. (4-17), which is correct only insofar as particle a is not identical to particle X. The product $N_a N_X$ is equal to the total number of unique pairs of particles (a,X) per unit volume. However, if the reaction occurs between identical particles of type a, the total number of pairs of particles per unit volume is not $N_a{}^2$ but rather $\frac{1}{2}N_a{}^2$. The factor $\frac{1}{2}$ must, therefore, be introduced into the reaction-rate equation if the two types of interacting particles are identical, in order to avoid counting each pair of particles twice. This factor can be formally introduced into the reaction-rate formulas as

$$r_{aX} = (1 + \delta_{aX})^{-1} N_a N_X \langle \sigma v \rangle \tag{4-18}$$

where δ_{aX} is the Kronecker delta, defined as unity if $a = X$ and as zero otherwise. We shall also use the notation

$$\lambda \equiv \langle \sigma v \rangle \tag{4-19}$$

which will be called the *reaction rate per pair of particles*. Then

$$r_{aX} = \lambda_{aX}(1 + \delta_{aX})^{-1} N_a N_X \tag{4-20}$$

Problem 4-5: Make the following thought experiment. Suppose $a = X$ and you think of the density N_a as being divided into two components, half of which are stationary and half of which are moving with translational velocity $\mathbf{v}_0$. Then the target density is $\frac{1}{2}N_a$, and the flux is $\frac{1}{2}N_a \mathbf{v}_0$, giving a reaction rate

$$r = \tfrac{1}{4}N_a{}^2 v_0 \sigma(v_0)$$

which, on the face of it, differs from Eq. (4-18) by a factor of 2. Resolve this discrepancy. *Hint:* What is $\phi(\mathbf{v})$?
Ans: $\phi(\mathbf{v}) = \frac{1}{2}\delta(\mathbf{v} - \mathbf{v}_0) + \frac{1}{2}\delta(\mathbf{v})$.

In discussions of time scales in nuclear astrophysics it is often useful to compute the mean lifetimes of nuclear species in a given environment. Explicitly one defines the lifetime $\tau_a(X)$ of species X against reactions with species a, such that the rate of change of the abundance of X due to reactions with a satisfies the equation

$$\left(\frac{\partial N_X}{\partial t}\right)_a = -\frac{N_X}{\tau_a(X)} \tag{4-21}$$

The partial derivative in Eq. (4-21) does not have the usual mathematical meaning; in this case it is a mnemonic indicating the rate of change of N_X due to a partial cause. Since $(\partial N_X/\partial t)_a$ is also equal to $-r_{aX}$, we have

$$\tau_a(X) = (1 + \delta_{aX})^{-1} \frac{N_X}{r_{aX}} = (\lambda_{aX} N_a)^{-1} \tag{4-22}$$

A little thought reveals that the factor $1 + \delta_{aX}$ appearing in Eq. (4-20) does not appear in the lifetime calculation; i.e., if $a \neq X$, the factor is unity, so that its inclusion is superfluous, and if $a = X$, the rate r_{aX} is divided by 2, but each reaction destroys two particles, in which case $(\partial N_X/\partial t)_a = -2r_{aa}$, and the product again yields Eq. (4-22).

Problem 4-6: Show that if species X can be destroyed by several reactions, the total lifetime $\tau(X)$ is given by

$$\frac{1}{\tau(X)} = \sum \frac{1}{\tau_i(X)}$$

What is the form of the function $\phi(v)$ which enters the calculation of the reaction rate? The nuclei in stellar interiors are, with the exception of neutron stars, always nondegenerate. In a state of thermodynamic equilibrium, the differing types of nuclei will separately be described by Maxwell-Boltzmann distributions of velocities. It is, perhaps, not so obvious what the distribution of *relative* velocities between two different sets of particles will be. A little thought will show that it is also maxwellian.

From the statistics of Chap. 1 it is apparent that particles of type 1 have a distribution of velocities $\mathbf{v}_1$ given by

$$N_1(\mathbf{v}_1)\, dv_{1x}\, dv_{1y}\, dv_{1z} = N_1 \left(\frac{m_1}{2\pi kT}\right)^{\frac{3}{2}} \exp\left(-\frac{m_1 v_1^2}{2kT}\right) dv_{1x}\, dv_{1y}\, dv_{1z} \tag{4-23}$$

It is then clear that the reaction rate will involve a double integral over

$$N_1(\mathbf{v}_1)\, dv_{1x}\, dv_{1y}\, dv_{1z} N_2(\mathbf{v}_2)\, dv_{2x}\, dv_{2y}\, dv_{2z} = N_1 N_2 \frac{(m_1 m_2)^{\frac{3}{2}}}{(2\pi kT)^3} \exp\left(-\frac{m_1 v_1^2 + m_2 v_2^2}{2kT}\right) d^3v_1\, d^3v_2 \tag{4-24}$$

which physically represents the product of the probability that particle 1 has velocity $\mathbf{v}_1$ in the range d^3v_1 times the probability that particle 2 has velocity $\mathbf{v}_2$ in the range d^3v_2.

In terms of the relative velocity and the velocity of the center of mass, the individual velocities become

$$\begin{aligned} \mathbf{v}_1 &= \mathbf{V} + \frac{m_2}{m_1 + m_2}\mathbf{v} \\ \mathbf{v}_2 &= \mathbf{V} - \frac{m_1}{m_1 + m_2}\mathbf{v} \end{aligned} \tag{4-25}$$

We have already pointed out that the total kinetic energy appearing in the exponential of Eq. (4-24) may also be written as the sum of the kinetic energy of the center of mass and the kinetic energy of relative motion in the center-of-mass coordinates. The probability product may thus be expressed as

$$N_1(\mathbf{v}_1)\, d^3v_1\, N_2(\mathbf{v}_2)\, d^3v_2 = N_1 N_2 \frac{(m_1 m_2)^{\frac{3}{2}}}{(2\pi kT)^3} \exp\left[-\frac{(m_1 + m_2)V^2}{2kT} - \frac{\mu v^2}{2kT}\right] d^3v_1\, d^3v_2 \tag{4-26}$$

Because the cross section appearing in the reaction-rate integral is a function only of the relative velocity, the integral over the velocity of the center of mass can be done at once if the integral over $d^3v_1\, d^3v_2$ can be related simply to an integral over $d^3V\, d^3v$. This transformation can easily be done with the aid of the theory of jacobian determinants, which, in its simplest form, states that given two functions $f(x,y)$ and $g(x,y)$ of two variables x and y, an integral over $dx\, dy$ may be replaced by an integral over $df\, dg$, but the ratios of the two differential areas are given by the absolute magnitude of the determinant of partial derivatives:

$$\frac{dA_{f,g}}{dA_{x,y}} = \text{magnitude of} \begin{vmatrix} \dfrac{\partial f}{\partial x} & \dfrac{\partial f}{\partial y} \\ \dfrac{\partial g}{\partial x} & \dfrac{\partial g}{\partial y} \end{vmatrix} \tag{4-27}$$

The jacobian determinant in the case at hand has unit magnitude, as can easily be seen by considering one component of the transformation. The ratio of the differential area $dv_{1x}\, dv_{2x}$ to the differential area $dV_x\, dv_x$ is given by the magnitude of the determinant

$$\begin{vmatrix} \dfrac{\partial v_{1x}}{\partial V_x} & \dfrac{\partial v_{1x}}{\partial v_x} \\ \dfrac{\partial v_{2x}}{\partial V_x} & \dfrac{\partial v_{2x}}{\partial v_x} \end{vmatrix} = \begin{vmatrix} 1 & \dfrac{m_2}{m_1 + m_2} \\ 1 & \dfrac{-m_1}{m_1 + m_2} \end{vmatrix} = -1 \tag{4-28}$$

Using this fact and the definition of the reduced mass, one sees that the probability product in Eq. (4-26) may be written as the product of two factors,

$$\left\{\left(\frac{m_1 + m_2}{2\pi kT}\right)^{3/2} \exp\left[-\frac{(m_1 + m_2)V^2}{2kT}\right] d^3V\right\} \left[\left(\frac{\mu}{2\pi kT}\right)^{3/2} \exp\left(-\frac{\mu v^2}{2kT}\right) d^3v\right] \tag{4-29}$$

the first of which represents a Maxwell-Boltzmann velocity distribution of the velocity of the center of mass and the second of which represents a Maxwell-Boltzmann distribution of relative velocity. Because these distributions are normalized, the integral over d^3V can be done at once and yields unity. Thus the reaction-rate integral has reduced to

$$\begin{aligned} r &= \int N_1(\mathbf{v}_1)N_2(\mathbf{v}_2)v\sigma(v)\, d^3v_1\, d^3v_2 \\ &= N_1N_2 \int v\sigma(v) \left(\frac{\mu}{2\pi kT}\right)^{3/2} \exp\left(-\frac{\mu v^2}{2kT}\right) d^3v \end{aligned} \tag{4-30}$$

which is to correspond to Eq. (4-17). The correspondence is made by noting that d^3v may be replaced by $4\pi v^2\, dv$, giving a probability that the relative velocity has magnitude v in the interval dv equal to that obtained from a maxwellian based upon the reduced mass:

$$\phi(v)\, dv = \left(\frac{\mu}{2\pi kT}\right)^{3/2} \exp\left(-\frac{\mu v^2}{2kT}\right) 4\pi v^2\, dv \tag{4-31}$$

The corresponding reaction rate is

$$r_{12} = (1 + \delta_{12})^{-1} N_1 N_2 \langle \sigma v \rangle$$
$$= (1 + \delta_{12})^{-1} N_1 N_2 4\pi \left(\frac{\mu}{2\pi kT} \right)^{\frac{3}{2}} \int_0^\infty v^3 \sigma(v) \exp\left(- \frac{\mu v^2}{2kT} \right) dv \qquad (4\text{-}32)$$

The calculation of $\langle \sigma v \rangle$, which is required for the reaction rates, reduces to performing the integral

$$\lambda = \langle \sigma v \rangle = 4\pi \left(\frac{\mu}{2\pi kT} \right)^{\frac{3}{2}} \int_0^\infty v^3 \sigma(v) \exp\left(- \frac{\mu v^2}{2kT} \right) dv \qquad (4\text{-}33)$$

To calculate the thermonuclear reaction rates in stars, additional expressions giving the details of $\sigma(v)$ for the important reactions will be required.

4-3 NONRESONANT REACTION RATES

The integral required for the calculation of the reaction rate per pair of particles has been established in the preceding section. The essential feature still missing in the calculation is the value of the nuclear cross section itself. It was pointed out in Chap. 1 that nuclear reactions in the centers of the stars can proceed only because the reacting particles penetrate the repulsive coulomb barrier that separates them. It was noted there that the coulomb barrier between two particles is given by

$$V = \frac{Z_1 Z_2 e^2}{R} = \frac{1.44 Z_1 Z_2}{R(\text{fm})} \qquad \text{Mev}$$

whereas the kinetic energy of the interacting particles is determined by a Maxwell-Boltzmann distribution of velocities corresponding to a thermal energy

$$kT = 8.62 \times 10^{-8} T \qquad \text{kev}$$

It is obvious from a comparison of these two numbers that for temperatures on the order of tens to hundreds of millions of degrees, the average kinetic energies of interacting particles are many orders of magnitude smaller than the coulomb barriers which separate them. The particles with the best chance of penetrating this coulomb barrier are those having the largest energies in the Maxwell-Boltzmann distribution; however, the expression for $\phi(v)$ given in Eq. (4-31) shows that the number of pairs of particles with center-of-mass energies much greater than kT decreases rapidly with energy. The upshot is that some compromise must be struck between the demand for the most energetic particles in the distribution and the rapidly decreasing number of particles of higher energy.

Gamow first showed, in connection with the problem of alpha decay, that the probability for two particles of charge Z_1 and Z_2 moving with relative velocity v

to penetrate their electrostatic repulsion is proportional to the factor

$$\text{Penetration} \propto \exp - \frac{2\pi Z_1 Z_2 e^2}{\hbar v} \tag{4-34}$$

It follows that the cross sections for nuclear reactions will also tend to be proportional to such a factor, since reactions can hardly occur unless particles penetrate this repulsion. We shall consider the reasoning leading to this functional dependence at a later time, when we discuss the penetration probabilities for a pair of particles. At the present time, we shall simply use this general fact as a guide to the construction of a *suitable representation* of the nuclear cross section at low energy. It will also become apparent later that the quantum-mechanical interaction between two particles is always proportional to a geometrical factor, $\pi\lambda^2$, where λ is the de Broglie wavelength:

$$\pi\lambda^2 \propto \left(\frac{1}{p}\right)^2 \propto \frac{1}{E} \tag{4-35}$$

At low energy both Eqs. (4-34) and (4-35) are rapidly varying functions of the energy. These considerations provide the motivation for choosing to define the cross section at low energy as a product of three separate energy-dependent factors:

$$\sigma(E) \equiv \frac{S(E)}{E} \exp - \frac{2\pi Z_1 Z_1 e^2}{\hbar v} \tag{4-36}$$

This equation is to be thought of as defining the factor $S(E)$. That is,

$$S(E) \equiv \sigma(E) E \exp \frac{2\pi Z_1 Z_2 e^2}{\hbar v} \tag{4-37}$$

The advantage of writing the cross section in this way is that two of the strongly energy-dependent factors appearing in the nuclear cross sections are factored explicitly, leaving a residual function of energy, $S(E)$, which may, in favorable circumstances, be extremely simple itself. This factor, $S(E)$, represents the intrinsically nuclear parts of the probability for the occurrence of a nuclear reaction, whereas the other two explicit factors represent well-known energy dependences that are nonnuclear in nature. The fortunate upshot is that when the interaction energy of the pair of particles is not nearly equal to an energy at which the two particles resonate in a quasistationary state (a discussion of which will appear later), the factor $S(E)$ is often found to be constant, or at least a slowly varying function of energy over a limited energy range.

These features can be illustrated with the experimental facts regarding the nuclear reaction $C^{12}(p,\gamma)N^{13}$. Figure 4-4 shows the measured cross section for this reaction as a function of the laboratory energy of protons striking a C^{12} target. Since nuclear cross sections represent very small areas, it is traditional to use as the unit of cross-sectional area the *barn*, which is defined as 1 barn $= 10^{-24}$ cm^2. The convenience of this numerical unit results from the natural size of the de Broglie geometrical cross section for kev nucleons, viz., $\pi\bar{\lambda}^2 = 657E^{-1}$ barn, where E is in kev and the reduced mass is 1 amu. That

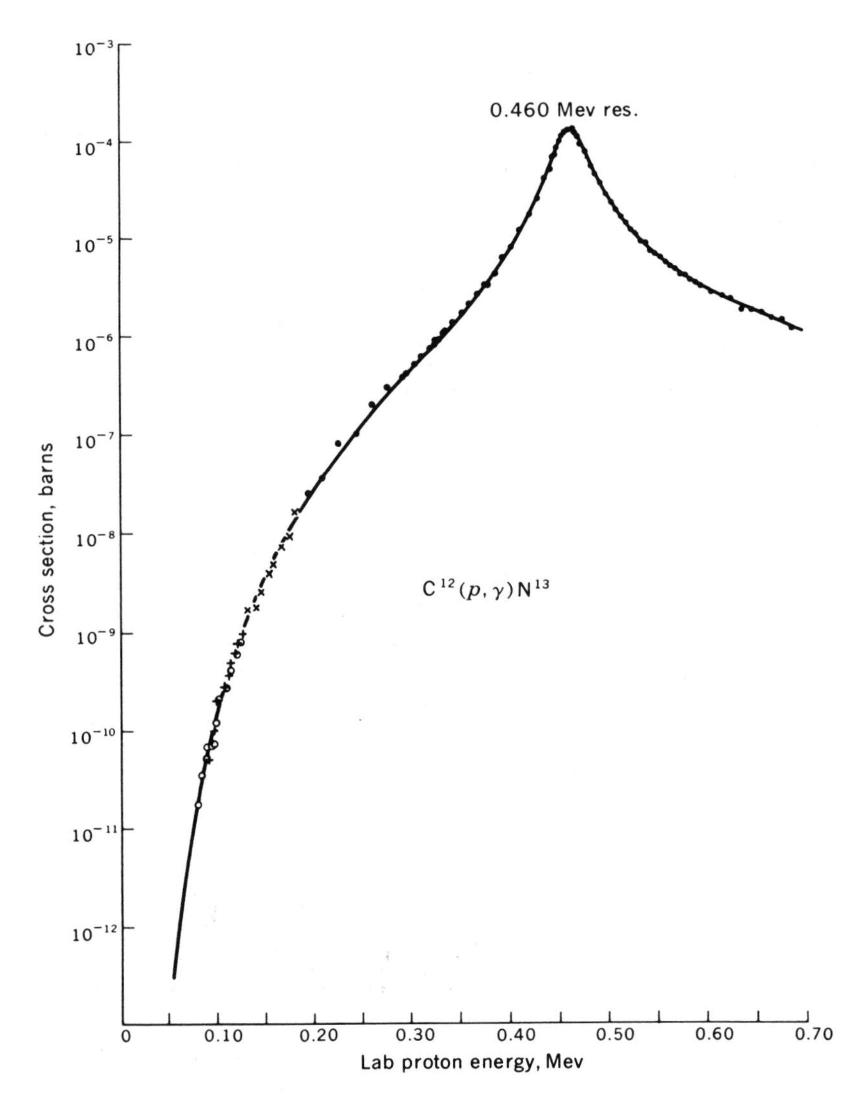

Fig. 4-4 The measured cross section for the reaction $C^{12}(p,\gamma)N^{13}$ as a function of laboratory proton energy. A four-parameter theoretical curve has been fitted to the experimental points. An extrapolation to $E_p = 0.025$ Mev, which is an interesting energy for this reaction in astrophysics, appears treacherous. *(Courtesy of W. A. Fowler and J. L. Vogl.)*

unit of cross section is used in Fig. 4-4. The energy abscissa is seen to be the laboratory proton energy, $\frac{1}{2}m_pv^2$, whereas the energy to be used in Eq. (4-37) is the energy of the pair of particles in the center of mass, which is

$$E = \frac{1}{2}\frac{m_1m_2}{m_1+m_2}v^2 = \tfrac{1}{2}\mu v^2 \tag{4-38}$$

It is evident that the energy of a pair of particles in their center of mass is related to the laboratory energy of particle 1 by the relationship

$$E = \frac{m_2}{m_1 + m_2} \mathrm{E}_{1,\mathrm{lab}} \tag{4-39}$$

Several interesting features are immediately obvious from Fig. 4-4. The cross section has a maximum of about 10^{-4} barn near the energy $E_{\mathrm{lab}} = 460$ kev and falls by seven orders of magnitude as the energy falls from 500 to 100 kev. It is further apparent that near 100 kev, the cross section is changing by about one order of magnitude per 25 kev. In other words, the nuclear cross sections for the interactions of charged particles vary extremely rapidly with energy at low energies. The maximum in this cross section at 460 kev is due to a resonance in the compound N^{13} system. Such resonances will be discussed later.

The point to be made at this time is that the rapidly falling cross section at low energies is due almost entirely to the effects of the exponential factor in the cross section. This exponential, sometimes called the *Gamow velocity factor*, is proportional to the probability of penetration through the coulomb repulsion. Quantitative definitions of the penetration factors will be described later. As factual evidence for the foregoing statements, the nuclear cross-section factor $S(E)$, as defined in Eq. (4-37), is shown in Fig. 4-5, along with the experimental data, which are plotted as points. The interesting fact is that the cross-section factor $S(E)$ is seen to be almost independent of energy, changing by less than a factor of 2 between 0 and 100 kev. Whereas the cross section itself changed by an order of magnitude in 25 kev near 100 kev, the cross-section factor changes by not more than 10 percent or so in 25 kev near 100 kev. These facts corroborate the statement that the cross-section factor is quite often a slowly varying function of energy that can be represented over a limited energy range as either a constant or a slowly increasing linear function of energy. We shall return to these two instructive figures often in the material that follows.

Problem 4-7: Show that if the cross section is written

$$\sigma(E) = \frac{S(E)}{E} \exp - bE^{-\frac{1}{2}} \tag{4-40}$$

the value of the parameter b is

$$b = 31.28 Z_1 Z_2 A^{\frac{1}{2}} \qquad \mathrm{kev}^{\frac{1}{2}} \tag{4-41}$$

where A is the reduced atomic weight, defined to be

$$A = \frac{A_1 A_2}{A_1 + A_2} = \frac{\mu}{M_u} \tag{4-42}$$

and M_u is, as before, the mass of 1 amu. That is, if the center-of-mass energy E is expressed in units of kev, Eq. (4-41) may be used for numerical convenience. It is conventional to use these energy units in preference to cgs units in nuclear astrophysics.

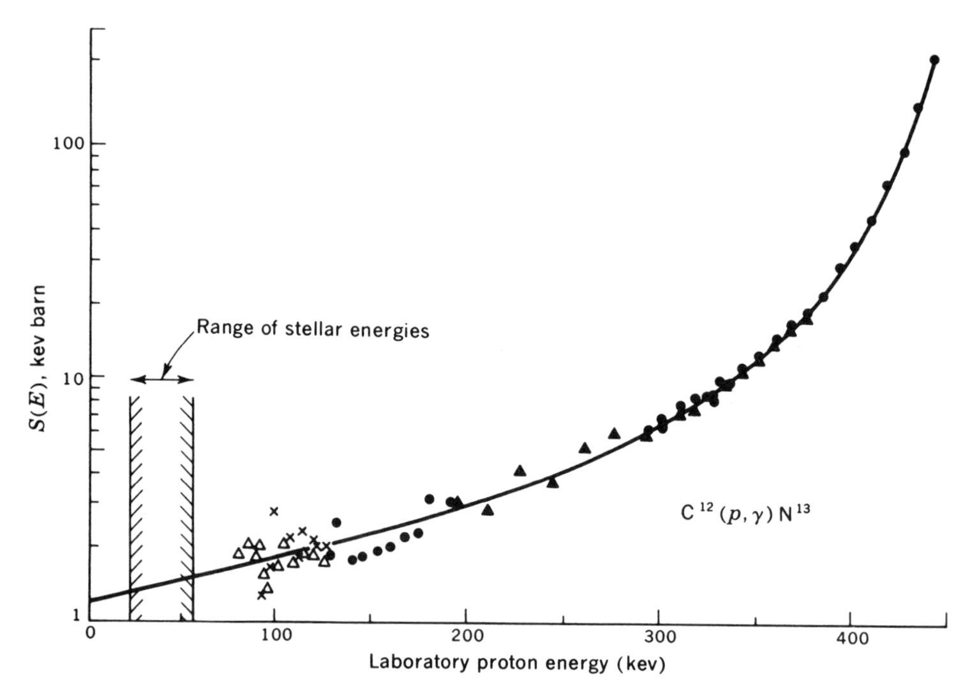

Fig. 4-5 The cross-section factor $S(E)$ for the radiative capture of protons by C^{12}. The differing types of data points represent five different experiments performed at different times and laboratories by the workers indicated. Detailed references and discussion may be found in D. F. Hebbard and J. L. Vogl, *Nucl. Phys.*, **21**:652 (1960). This curve is more readily extrapolated than the one in Fig. 4-4.

The velocity distribution may be written as the following normalized energy distribution:

$$\psi(E)\, dE = \phi(v)\, dv = -\frac{2}{\sqrt{\pi}}\frac{E}{kT}\exp\left(-\frac{E}{kT}\right)\frac{dE}{(kTE)^{\frac{1}{2}}} \tag{4-43}$$

In the nonresonant-reaction case, the cross-section factor $S(E)$ is slowly varying over the range of energies that are important in stellar interiors, and so in that case Eq. (4-37) may be a useful substitution for $\sigma(E)$ in the calculation of the reaction rate per pair of particles:

$$\begin{aligned}
\lambda = \langle\sigma v\rangle &= \int_0^\infty \sigma(E)v(E)\psi(E)\, dE \\
&= \int_0^\infty \frac{S(E)}{E}\exp\left(-bE^{-\frac{1}{2}}\right)\sqrt{\frac{2E}{\mu}}\frac{2}{\sqrt{\pi}}\frac{E}{kT}\exp\left(-\frac{E}{kT}\right)\frac{dE}{(kTE)^{\frac{1}{2}}} \\
&= \left(\frac{8}{\mu\pi}\right)^{\frac{1}{2}}\frac{1}{(kT)^{\frac{3}{2}}}\int_0^\infty S(E)\exp\left(-\frac{E}{kT} - bE^{-\frac{1}{2}}\right) dE
\end{aligned} \tag{4-44}$$

The behavior of the integrand is largely determined by the exponential factor, since it is a rapidly varying function of energy. Notice that since exp $(-E/kT)$ goes rapidly to zero for large E whereas exp $(-bE^{-\frac{1}{2}})$ goes rapidly to zero for small E, the major contribution to the integral will come from values of the energy that are such that the exponential factor is near its maximum. It will soon be apparent that most stellar reactions occur in a fairly narrow band of stellar energies, so narrow that the factor $S(E)$ will have a nearly constant value over the band of energies. This effective range of stellar energies was schematically indicated in Fig. 4-5 for the $C^{12}(p,\gamma)N^{13}$ reaction. A good approximation to Eq. (4-44) will be obtained by replacing $S(E)$ by its (nearly constant) value at the energy for which the exponential factor is maximal. Let S_0 represent that constant value [strictly speaking the *average* value of $S(E)$, the average being taken with respect to the exponential factor]. There results

$$\lambda = \left(\frac{8}{\mu\pi}\right)^{\frac{1}{2}} \frac{S_0}{(kT)^{\frac{3}{2}}} \int_0^\infty \exp\left(-\frac{E}{kT} - \frac{b}{\sqrt{E}}\right) dE \tag{4-45}$$

which can be evaluated by approximating the integrand by an appropriate gaussian.

Such a procedure is the simplest example of a method of doing a certain class of integrals, called the *method of steepest descent.* The method is applicable to integrals of the form

$$\int g(x)e^{-f(x)}\, dx$$

where $g(x)$ is a slowly varying function of x and the function $f(x)$ has a value much larger than unity and a single sharp minimum at x_0. In those circumstances, the integral may be approximated by expanding $f(x)$:

$$\begin{aligned} f(x) &= f(x_0) + f'(x_0)(x - x_0) + f''(x_0)\frac{(x - x_0)^2}{2} + \cdots \\ &\approx f(x_0) + f''(x_0)\frac{(x - x_0)^2}{2} \end{aligned}$$

since the first derivative vanishes at the minimum, and higher terms are discarded as being important only for those relatively large values of $x - x_0$ for which $f(x) \gg f(x_0)$, a fact necessitating that there be little contribution to the integral. Then a good estimate for the value of the integral becomes

$$g(x_0)e^{-f(x_0)} \int_{-\infty}^\infty \exp\left[-f''(x_0)\frac{(x - x_0)^2}{2}\right] dx$$

which has an elementary value.

Problem 4-8: Show that the integral is approximately

$$\int g(x)e^{-f(x)}\, dx \approx \sqrt{\frac{2\pi}{f''(x_0)}}\, g(x_0)\, e^{-f(x_0)}$$

Of course this approximation is useless unless $f(x)$ has the properties prescribed for it.

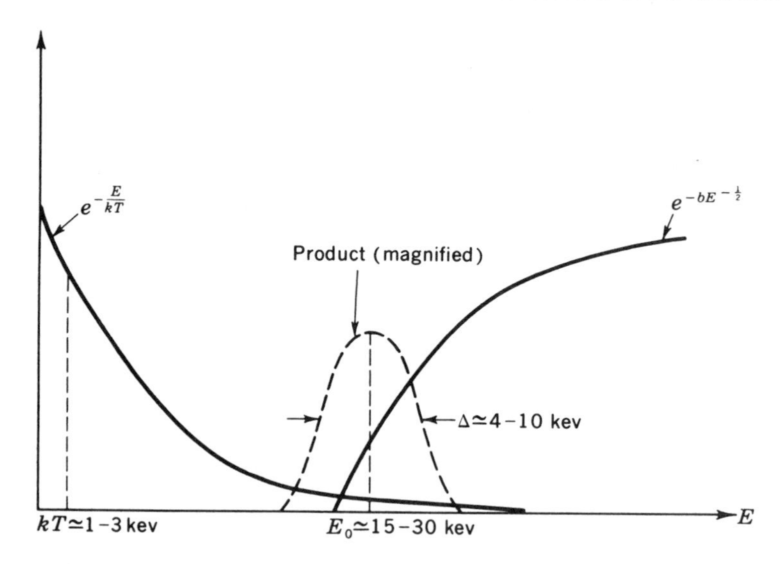

Fig. 4-6 The dominant energy-dependent factors in thermonuclear reactions. Most of the reactions occur in the high-energy tail of the maxwellian energy distribution, which introduces the rapidly falling factor exp $(-E/kT)$. Penetration through the coulomb barrier introduces the factor exp $(-bE^{-\frac{1}{2}})$, which vanishes strongly at low energy. Their product is a fairly sharp peak near an energy designated by E_0, which is generally much larger than kT. The peak is pushed out to this energy by the penetration factor, and it is therefore commonly called the *Gamow peak* in honor of the physicist who first studied the penetration through the coulomb barrier.

Standard techniques exist for determining the extent of the error made in the method of steepest descents. It will be more instructive to apply such analysis to the problem at hand, however, than to concern ourselves further with the general technique.

The integrand in Eq. (4-44) is a sharply peaked function, being the product of an exponential that vanishes at large energy, exp $(-E/kT)$, and an exponential that vanishes at low energy, exp $(-bE^{-\frac{1}{2}})$, as illustrated schematically in Fig. 4-6. All other things being equal, the particles that are most effective in causing nuclear reactions are those pairs having energies near E_0. The value of E_0 is determined from the location of the maximum of the integrand:

$$\frac{d}{dE}\left(\frac{E}{kT} + bE^{-\frac{1}{2}}\right)_{E=E_0} = \frac{1}{kT} - \frac{1}{2}bE_0^{-\frac{3}{2}} = 0 \quad .$$

or

$$E_0 = \left(\frac{bkT}{2}\right)^{\frac{2}{3}} \tag{4-46}$$

Problem 4-9: Show that

$$E_0 = 1.220(Z_1{}^2Z_2{}^2AT_6{}^2)^{\frac{1}{3}} \qquad \text{kev} \tag{4-47}$$

where T_6 is the temperature in millions of degrees. This energy is frequently called the *most effective energy for thermonuclear reactions.*

Evaluation of Eq. (4-47) shows that for normal light nuclei and temperatures of some tens of millions of degrees, the most effective energy E_0 is usually 10 to 30 kev. This energy is greater than $kT = 0.086T_6$ kev, reflecting the fact that the barrier-penetration factor has favored the selection of particles on the high-energy tail of the Maxwell-Boltzmann energy distribution.

The method of steepest descent is equivalent to the replacement of a sharply peaked exponential function by a gaussian function having a maximum of the same size and the same curvature at the maximum, in this case at $E = E_0$. That is, the integral will be evaluated by the replacement

$$\exp\left(-\frac{E}{kT} - bE^{-\frac{1}{2}}\right) \approx C \exp -\left(\frac{E - E_0}{\Delta/2}\right)^2 \tag{4-48}$$

where clearly

$$C = \exp\left(-\frac{E_0}{kT} - bE_0{}^{-\frac{1}{2}}\right) \tag{4-49}$$

and where the $1/e$ width, $\Delta/2$, is estimated by the requirement that the second derivatives match at E_0.

Problem 4-10: Show that the constant C is also equal to

$$C = \exp -\frac{3E_0}{kT} \tag{4-50}$$

and the full width at $1/e$ is

$$\Delta = \frac{4}{\sqrt{3}}(E_0kT)^{\frac{1}{2}} \tag{4-51}$$

$$\Delta = 0.75(Z_1{}^2Z_2{}^2AT_6{}^5)^{\frac{1}{6}} \qquad \text{kev} \tag{4-52}$$

It is apparent from Eq. (4-51) that the full width is approximately twice the geometric mean of kT and the peak energy E_0 but is still smaller than E_0 itself.

Problem 4-11: Show that, for the reaction $C^{12}(p,\gamma)N^{13}$,

$$E_0 = 3.93T_6{}^{\frac{2}{3}} \qquad \text{kev}$$

$$\Delta = 1.35T_6{}^{\frac{5}{6}} \qquad \text{kev}$$

and evaluate at the center of the sun, where $T = 15 \times 10^6$ °K. The numerical value of the exponential factor for $T_6 = 30$ is plotted in Fig. 4-7. The energies at $1/e$ of maximum are shown.

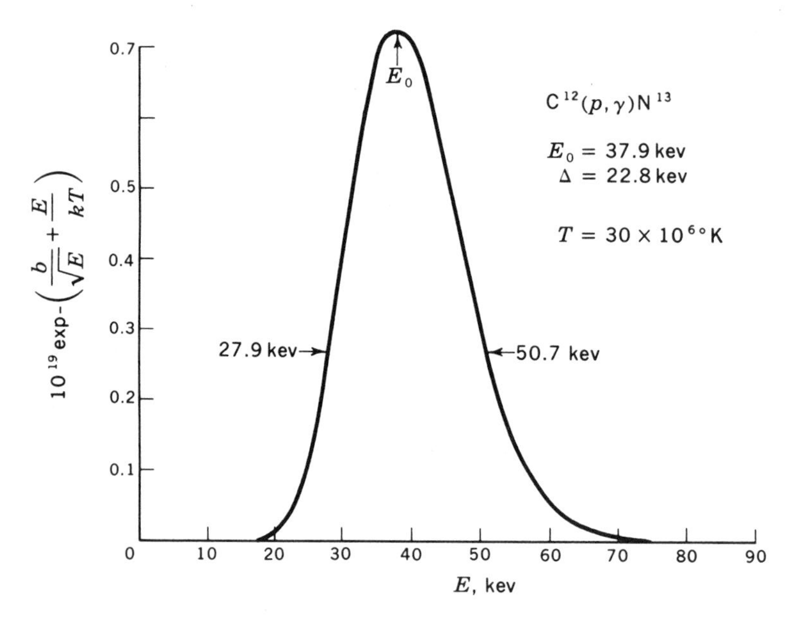

Fig. 4-7 The Gamow peak for the reaction $C^{12}(p,\gamma)N^{13}$ at $T = 30 \times 10^6$ °K. The curve is actually somewhat asymmetric about E_0, but it is nonetheless adequately approximated by a gaussian.

The most effective particles have energies ranging only about 10 kev from the most effective energy E_0. This range of energies is quite small compared to the average energy separation of quasistationary nuclear states in the light nuclei and accounts for the fact that the effects of nuclear forces, which are lumped into $S(E)$, may often be considered to be constant. The factor $S(E)$ will generally change by a large percentage of its value over the range Δ only if there is a nuclear resonance near the range of effective stellar energies, but in that case the resonant reaction rates must be employed. These will be discussed later.

Suffice it to say, then, that the experimentally measured cross-section factor can be plotted, as in Fig. 4-5 for the $C^{12}(p,\gamma)N^{13}$ reaction, and extrapolated to the range of stellar energies. This extrapolation can be made with considerably greater accuracy than could the extrapolation of the cross-section itself. In fact the solid line of Fig. 4-5 is a semitheoretical fit to the points, made in a manner to be explained later. From this analysis one can describe $S(E)$ at low energies by its intercept and slope. In the particular case of the $C^{12}(p,\gamma)N^{13}$ reaction, for instance, one finds that $S(E = 0) = 1.20$ kev barns and $dS/dE = 5.81 \times 10^{-3}$ barn. In like manner, the cross-section factor has been determined with varying degrees of accuracy for most of the important energy-generating reactions in stellar interiors.

Making the approximate substitution

$$\exp\left(-\frac{E}{kT} - bE^{-\frac{1}{2}}\right) \approx e^{-\tau} \exp - \left(\frac{E - E_0}{\Delta/2}\right)^2 \tag{4-53}$$

where the quantity $3E_0/kT$ has been designated by τ, the reaction rate per pair of particles may be written from Eq. (4-44) as

$$\lambda = \left(\frac{8}{\mu\pi}\right)^{\frac{1}{2}} \left(\frac{1}{kT}\right)^{\frac{3}{2}} e^{-\tau} \int_0^\infty S(E) \exp\left[-\left(\frac{E - E_0}{\Delta/2}\right)^2\right] dE \tag{4-54}$$

Once again it is obvious that if $S(E)$ is nearly constant, most of the value of the integral comes from values of E near E_0. The first approximation for well-behaved cross-section factors is to treat $S(E)$ as a constant S_0 defined as the value of $S(E_0)$. It is also evident that negligible error will be committed by extending the lower limit of the integral to minus infinity.[1]

Problem 4-12: Show that when S_0 is expressed in cgs units of erg cm² and when the approximations indicated above are performed, one obtains for the reaction rate per pair

$$\lambda = \frac{4.50 \times 10^{14}}{AZ_1Z_2} S_0\tau^2 e^{-\tau} \qquad \text{cm}^3/\text{sec} \tag{4-55}$$

Since kev is a more appropriate energy unit than ergs, and barns a more appropriate cross section than cm², it is more common to express S_0 in units of kev barns, e.g., Fig. 4-5. We shall follow that practice throughout this book. The reaction rate per pair then becomes numerically

$$\lambda = \frac{7.20 \times 10^{-19}}{AZ_1Z_2} S_0 \text{ (kev barns)} \tau^2 e^{-\tau} \qquad \text{cm}^3/\text{sec} \tag{4-56}$$

whereas the reaction rate is obtained by multiplying by the number of pairs per unit volume:

$$r_{12} = (1 + \delta_{12})^{-1} N_1 N_2 \lambda_{12} \tag{4-57}$$

The convenience of writing the reaction rate in this form is that the all-important temperature dependence of the rate is entirely contained in the parameter τ.

Problem 4-13: Show that

$$\tau = 42.48 \left(\frac{Z_1^2 Z_2^2 A}{T_6}\right)^{\frac{1}{3}} \tag{4-58}$$

For any given reaction, τ is proportional to $T^{-\frac{1}{3}}$; thus one can write

$$\tau = BT_6^{-\frac{1}{3}} \tag{4-59}$$

[1] See, for instance, a table of the normal probability integral for a characteristic value of $2E_0/\Delta$.

where $B = 42.48(Z_1{}^2Z_2{}^2A)^{\frac{1}{3}}$. Then the temperature dependence is explicitly displayed in a simple analytic form. It is this feature of the approximate integration which makes it so useful: one can quickly and simply determine the manner in which a chosen reaction rate depends upon the temperature.

Problem 4-14: Show that for temperature T near some value T_1 the reaction rate varies with temperature as

$$r_{12}(T) \approx r_{12}(T_1)\left(\frac{T}{T_1}\right)^n \tag{4-60}$$

where

$$n = \frac{\tau - 2}{3} \tag{4-61}$$

It is also convenient for purposes of stellar-structure computations to explicitly introduce the local density into the reaction rate. Since

$$N_i = \rho N_0 \frac{X_i}{A_i}$$

where X_i is the fraction by mass of species i, we have

$$r_{12} = \frac{2.62 \times 10^{29}}{(1 + \delta_{12})AZ_1Z_2}\,\rho^2\,\frac{X_1X_2}{A_1A_2}\,S_0(\text{kev barns})\tau^2 e^{-\tau} \qquad \text{cm}^{-3}\ \text{sec}^{-1} \tag{4-62}$$

These equations, along with the definitions of the associated quantities, comprise the basic nonresonant stellar reaction-rate formulas in first approximation.

Three corrections to the above expressions for nonresonant thermonuclear reaction rates are important enough to be mentioned in this book. Having used approximations in the integration, we must concern ourselves with the extent of the error introduced. For one thing, a better approximation to $S(E)$ than a simple constant would seem to be

$$S(E) = S(E_0) + \left(\frac{dS}{dE}\right)_{E_0}(E - E_0) \tag{4-63}$$

Furthermore, even if a constant is used for $S(E)$, some error is introduced into the reaction-rate integral by substituting a gaussian for the sharply peaked exponential. And finally it will be noted that the high density of free electrons near the nuclei in stellar interiors increases the reaction rate somewhat by reducing the coulomb repulsion. The first two corrections can be considered at this time, but discussion of the last effect will be postponed until the physics of barrier penetration has been introduced.

Perhaps the most interesting question is that of the extent of the error introduced by replacing the factor $\exp(-E/kT - bE^{-\frac{1}{2}})$, the area under which is proportional to the rate, by the gaussian $e^{-\tau}\exp\{-[(E - E_0)/(\Delta/2)]^2\}$, the area under which is $e^{-\tau}\sqrt{\pi}\,\Delta/2$. It is clear that the reaction rate, as derived, should be multiplied by a correction factor which is simply the ratio of those two areas.

Let that correction factor be denoted by $F(\tau)$:

$$F(\tau) = \frac{\int_0^\infty \exp\left(-\frac{E}{kT} - bE^{-\frac{1}{2}}\right) dE}{e^{-\tau}\sqrt{\pi}\,\Delta/2} \tag{4-64}$$

Problem 4-15: It is not obvious that F is a function only of τ. Show that the correction factor may be written

$$F(\tau) = \left(\frac{\tau}{\pi}\right)^{\frac{1}{2}} \frac{e^\tau}{2} \int_0^\infty \exp\left[-\frac{\tau}{3}(\epsilon + 2\epsilon^{-\frac{1}{2}})\right] d\epsilon \tag{4-65}$$

where the newly defined dimensionless energy is $\epsilon = E/E_0$. Hence $F = F(\tau)$.

Perhaps the simplest procedure would be to numerically integrate Eq. (4-65) on a computer for a large number of values of τ. The value for any specific value of τ could then be interpolated. For pedagogical and historical reasons we outline instead an approximate technique for evaluating Eq. (4-65).

From Eq. (4-58) it is evident that the parameter τ is usually a large number. Thus another simple procedure will be to define some new parameter that varies inversely with τ, and is hence small, and then expand F in powers of that small parameter.

Problem 4-16: Show that in terms of a new variable $y = \sqrt{\epsilon} - 1$,

$$F(\tau) = \left(\frac{\tau}{\pi}\right)^{\frac{1}{2}} \int_{-1}^\infty (1 + y)e^{-\tau y^2} \exp\left[\frac{2\tau y^3}{3(1 + y)}\right] dy \tag{4-66}$$

Then show that

$$F(\beta) = \left(\frac{3}{\pi}\right)^{\frac{1}{2}} \int_{-\beta^{-1}}^\infty \exp\left(-3\zeta^2 + \frac{2\zeta^3\beta}{1 + \beta\zeta}\right)(1 + \beta\zeta)\, d\zeta \tag{4-67}$$

where $\beta = (3/\tau)^{\frac{1}{2}}$ is the new small parameter and $\zeta = y/\beta$.

The point of the previous manipulative problem is that the expression for $F(\beta)$ may be expanded by a Taylor series:

$$F(\beta) = F(0) + \left(\frac{\partial F}{\partial \beta}\right)_0 \beta + \frac{1}{2}\left(\frac{\partial^2 F}{\partial \beta^2}\right)_0 \beta^2 + \cdots \tag{4-68}$$

which gives the correction factor to the integral when $\beta = (3/\tau)^{\frac{1}{2}}$ is reintroduced. The first term of that expansion is

$$F(0) = \left(\frac{3}{\pi}\right)^{\frac{1}{2}} \int_{-\infty}^\infty \exp(-3\zeta^2)\, d\zeta = 1 \tag{4-69}$$

which simply expresses the fact that the reaction rate is correct in zeroth order:

$$F(\beta) = 1 + \left(\frac{\partial F}{\partial \beta}\right)_0 \beta + \frac{1}{2}\left(\frac{\partial^2 F}{\partial \beta^2}\right)_0 \beta^2 + \cdots \tag{4-70}$$

Problem 4-17: Show that

$$\left(\frac{\partial F}{\partial \beta}\right)_0 = 0 \qquad \left(\frac{\partial^2 F}{\partial \beta^2}\right)_0 = \frac{5}{18} \tag{4-71}$$

It follows immediately from Eq. (4-71) and the definition of β that

$$F(\tau) = 1 + \frac{5}{12\tau} + \text{order}\,\frac{1}{\tau^2} \tag{4-72}$$

is the correction factor to be applied to the reaction-rate equations to account for the error due to the gaussian approximation. Because τ is a large number, we see that the correction is generally small. Of course, one must also demonstrate rapid convergence of the series.

The second correction is attendant to the case of a cross-section factor that changes linearly with energy, as in Eq. (4-63). In almost all cases where accurate experimental data exist or where an accurate theoretical calculation is possible, $S(E)$ is found not to be precisely constant over the 25 kev or so of energy that are most important in stellar interiors. It is again possible to write a correction factor that can be expanded in powers of $1/\tau$. In doing so, one cannot use the gaussian approximation because it is symmetric about E_0.

Problem 4-18: (This is a lengthy problem.) Show that the explicit introduction of

$$S(E) = S(E_0) + \left(\frac{\partial S}{\partial E}\right)_{E_0} (E - E_0)$$

into the reaction-rate integral causes the basic reaction-rate formula to be multiplied by the correction factor

$$G(\tau) = 1 + \frac{5}{2}\frac{E_0}{S(E_0)}\left(\frac{\partial S}{\partial E}\right)_{E_0}\frac{1}{\tau} + \text{order}\,\frac{1}{\tau^2} \tag{4-73}$$

Show also that the original asymmetric (about E_0) integrand must be retained.

This correction factor is subject to a very simple interpretation. When the definition $\tau = 3E_0/kT$ is inserted, there remains

$$G(T) = 1 + \frac{5}{6}\frac{kT}{S(E_0)}\left(\frac{\partial S}{\partial E}\right)_{E_0} \tag{4-74}$$

This correction factor indicates how $S(E)$ is to be averaged to obtain the appropriate constant S_0. That is, one should define S_0 such that

$$S_0 \equiv S(E_0)G(T) = S(E_0) + \frac{5}{6}\left(\frac{\partial S}{\partial E}\right)_{E_0} kT \tag{4-75}$$

Equation (4-75) may be immediately interpreted as prescribing that the constant cross-section factor S_0 be the value of $S(E)$, not at the energy E_0, but rather at the energy $E_0 + \frac{5}{6}kT$. In the remainder of the book we shall adopt the convention for nonresonant reaction rates that

$$S_0 \equiv S(E_0 + \tfrac{5}{6}kT) \tag{4-76}$$

This convention will allow us to ignore explicit use of the correction factor $G(\tau)$.

The third refinement is that the penetration of the particles through their mutual coulomb barrier is aided by the dense electron gas surrounding the nuclei. Laboratory experiments measure $S(E)$ for penetration through a pure coulomb barrier (except at very large distances, where atomic electrons provide shielding, but with a negligible enhancement of the penetration factors). For this reason the reaction rates must be multiplied by an electron-shielding enhancement factor, which is traditionally designated by the letter f. Thus, for example, Eq. (4-62) becomes

$$r_{12} = \frac{2.62 \times 10^{29}}{1 + \delta_{12}} \rho^2 \frac{X_1 X_2}{A_1 A_2 A Z_1 Z_2} f S_0 \left(1 + \frac{5}{12\tau}\right) \tau^2 e^{-\tau} \tag{4-77}$$

where $S_0 = S(E_0 + \frac{5}{6}kT)$ in units of kev barns and f is the screening factor. The evaluation of f must wait until the penetration factors have been calculated.

Problem 4-19: The cross-section factor for $C^{12}(p,\gamma)N^{13}$ shown in Fig. 4-5 can be characterized by the value of its zero-energy intercept $S(0) = 1.20$ kev barns and its slope $dS/dE = 5.81 \times 10^{-3}$ barn. Compute the lifetime of a C^{12} nucleus against proton capture in a stellar interior containing 80 percent hydrogen by weight and having a density of 15 gm/cm³ and a temperature of 30×10^6 °K. (Ignore electron screening.)

Ans: About 160 years.

This discussion more or less sums up the calculation of nonresonant reaction rates for which experimental data are available, an empirical approach to nonresonant reaction rates. If all nuclear reactions occurring in stellar interiors were similar to the $C^{12}(p,\gamma)N^{13}$ reaction, no more discussion would be necessary. Two important practical considerations render the foregoing discussion inadequate for some nuclear reactions, however: (1) for many nuclear reactions the nonresonant cross section at low energies is too small to be measured with sufficient accuracy to allow a small extrapolation of $S(E)$ to stellar energies; and (2) for many nuclear reactions there exist resonances in the range of effective stellar energies that invalidate the assumption of a slowly varying $S(E)$. In fact, a different approach must be developed to calculate the reaction rate when resonances occur in the range of effective stellar energies. For both of these reasons it will be necessary to take a small detour into elements of nuclear physics before proceeding onward.

4-4 NUCLEAR STATES

The nucleus bears many resemblances to the atom. Both are many-body problems that are interpreted in terms of interactions between constituent particles. Both show sets of bound states characterized by discrete binding energies and auxiliary quantum numbers characteristic of the symmetries of the collective states. More success has been encountered in the interpretation of the atom because the hamiltonian is almost entirely due to well-known terms:

$$H_{\text{atom}} \approx \sum_{\text{electrons}} \left(\frac{p^2}{2m} + V \right)$$

where V is dominated by the well-understood coulomb potential between charges, although a relatively small $\mathbf{L} \cdot \mathbf{S}$ interaction must be added to account for the fine-structure splitting between nearly degenerate states. The largest coulomb interaction, furthermore, exists between the electrons and the massive nucleus, which provides an essentially stationary center of mass for the total structure. Because the problem is well defined, the properties of atomic states can be well calculated with the necessary computational labor.

The interpretation of the nucleus is comparatively obscured by several features. The potential representing the dominant force between nucleons is not well known in detail, other than that it is a strong force which acts over relatively small distances, characteristically represented by a deep potential well of 25 Mev or so for a distance of a few centimeters $\times 10^{-13}$. The analogous $\mathbf{l} \cdot \mathbf{s}$ spin-orbit interaction for nucleons is relatively much stronger than in atoms, furthermore. Even more complicated phenomenological potentials appear to be necessary to interpret the observations. Then, too, the nucleons are of equal mass, so that no single particle can represent a massive point to which the structure can be attached. To a certain extent the nucleus is somewhat like a very imperfect gas held together by the strong short-range attraction; and although the description of nuclear structure is contained in the complete wave function of the nucleus, formidable theoretical difficulties are faced in an attempt to calculate the wave functions. Basically the hamiltonian is still written as the sum of the kinetic energies and the potential energies between all constituent particles, but the uncertainty in the form of the potential and its dependence upon the state of all the nucleons renders the problem extremely intractable.

Nonetheless, the nucleons are found to cluster into bound states characterized by a discrete energy and a quantized angular momentum that may be thought of as the sum of the orbital and spin angular momenta of the constituent nucleons. These bound states are analogous to the bound energy eigenstates of the atom. Just as in the atom, however, states are not quite stationary, because interaction terms exist in the complete hamiltonian, leading to transitions between states. These transitions cause each state to have a width in energy

$$\Gamma = \frac{\hbar}{\tau} \tag{4-78}$$

where τ is the mean lifetime of the decaying state. The states therefore have an energy profile

$$P(E) = \frac{\hbar/2\pi\tau}{(E - E_i)^2 + (\hbar/2\tau)^2} \tag{4-79}$$

as argued in Eq. (3-103). But if the state lifetime τ is sufficiently long, the width Γ of the state is sufficiently small compared to the energy difference between states for one to see good approximations of stationary energy eigenstates. The discovery of schemes for interpreting these quasistationary nuclear states is one of the major objectives of the science of nuclear physics.

THE SHELL MODEL

One of the most successful models for interpreting the properties of nuclear states is the nuclear shell model, a model for which Mayer[1] and Jensen shared the 1963 Nobel Prize in physics. Anyone interested in discussing nuclear reactions in stars will be aided by a passing familiarity with this model. It is patterned after the shell model of the atom, in which electrons fill consecutively the lowest lying available bound states. It is assumed in this model that the interaction of each nucleon *with the remainder of the nucleus* is well approximated by a spherical potential $V(r)$. The motion of each nucleon is then an independent single-particle orbit with fixed angular momentum and energy. The energy of the nucleus is then given by the sum of the single-particle energies, and the total angular momentum of the nucleus is obtained by coupling the angular momenta of the independent orbits together by the usual rules of addition of angular momentum. It is no simple matter to see why this model should give a good description, although advanced theoretical arguments can provide partial justification. But the model has been highly successful, and because it is conceptually simple, it has been a very popular framework within which many of the gross properties of nuclei fall naturally into place.

The bound states of a particle in a spherically symmetric potential are described primarily in terms of two quantum numbers. In a central potential the orbital angular momentum of each nucleon is a constant of the motion. For each orbital quantum number l there is a series of energy levels, which are distinguished by a quantum number n related to the number of nodes in the radial wave function. For example, the state of lowest energy, having $l = 1$, is called the $1p$ state; the fifth lowest state, with $l = 3$, is called the $5f$ state, etc. For any spherically symmetric potential the energy of each orbital is determined by some function of n and l which depends upon the shape of the potential. For instance, for the $1/r$ potential of the hydrogen atom the energy of the states is proportional to $1/(n + l)^2$, and in that case one defines a *principal* quantum number $N = n + l$ such that $E_N = -R/N^2$ and $N > l$. For potentials other than $1/r$, however, the energy depends upon both n and l. For this reason the concept of a principal quantum number is discarded for nuclear potentials, and one deals instead with n and l themselves. The way in which $E(n,l)$ depends upon the quantum numbers depends upon the shape of the potential.

For preliminary understanding of the ordering of nuclear energy levels, the nuclear average potential is often approximated by two simple potentials for which

[1] M. Mayer's Nobel lecture, The Shell Model, reprinted in *Science*, **145**:999 (1964), makes excellent reading.

the energy levels are easily soluble. These are the *harmonic-oscillator potential*

$$V = -V_0 + \tfrac{1}{2}M\omega^2 r^2 \tag{4-80}$$

where M is the nucleon mass and ω is a number which depends upon the range of the oscillator potential, and the *infinite square well*

$$V = \begin{cases} -V_0 & r < R \\ \infty & r > R \end{cases} \tag{4-81}$$

These potentials represent opposite extremes in the sense that the oscillator potential rises smoothly from its value $V(0) = -V_0$ whereas the square well arises abruptly at the radius R. For the oscillator potential the energy levels are

$$E(n,l) = -V_0 + \hbar\omega(2n + l - \tfrac{1}{2}) \tag{4-82}$$

which depend only upon the combination $2n + l$ and differ by multiples of $\hbar\omega$. Empirically it is found that $\hbar\omega \approx 30 \text{ Mev}/A^{\frac{1}{3}}$. Because the energy depends only upon $2n + l$, these energy levels are highly degenerate, each energy consisting of several different states. For the square well, on the other hand, the energy is determined from the nodes of the spherical Bessel functions:

$$J_{l+\frac{1}{2}}\left[\sqrt{\frac{2M(E + V_0)}{\hbar^2}}\, R\right] = 0$$

The corresponding energy levels are approximately expressible as

$$E(n,l) \approx -V_0 + \frac{\hbar^2}{2MR^2}\left[\pi^2\left(n + \frac{l}{2}\right)^2 - l(l + 1)\right] \tag{4-83}$$

This potential is seen to split the degeneracy of the harmonic-oscillator potential. These energy levels are shown in Fig. 4-8. Shown adjacent to the quantum numbers of each level are the number of nucleons that may occupy that orbital without violation of the exclusion principle, in analogy to the atom. Also shown in square brackets above each level is the total number of protons (or neutrons) that may occupy that plus all lower levels. These numbers characterize the "shells" for the respective potentials.

Problem 4-20: Taking into account the spin $\frac{1}{2}$ of the nucleon, confirm the number of independent states shown in parentheses beside each level.

Problem 4-21: Confirm that 40 neutrons may occupy oscillator energy levels with $E \leq 3\hbar\omega$.

A more realistic nuclear potential is believed to lie intermediate in shape between the oscillator well and the square well. A corresponding level spacing has been arbitrarily interpolated, as demonstrated in the middle column of Fig. 4-8. This sequence of levels, then, represents the approximate ordering of bound nucleon states to be anticipated. A model for the ground states of nuclei is then constructed by placing Z protons and N neutrons in the lowest-lying levels available.

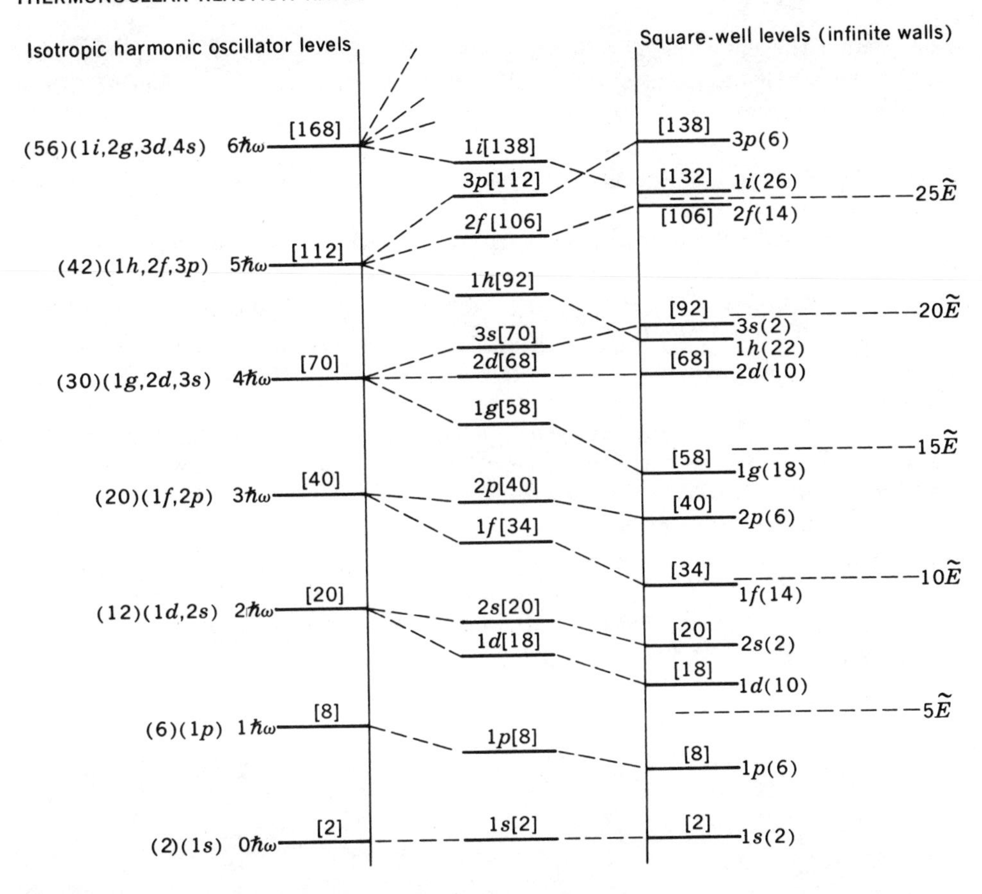

Fig. 4-8 The energy levels of a harmonic oscillator are shown on the left, and those of a square well are shown on the right. The characteristic energies $\hbar\omega$ and $\tilde{E}$ are determined by the range of the potentials. The quantum configuration of each level and the number of degenerate states within it are shown in parentheses. The nuclear potential has a shape intermediate to these two, and the corresponding sets of energy levels are indicated schematically in the middle column. The energies of bound nucleons within the nucleus would be expected to resemble this middle column were it not for the spin-orbit interaction, which has not been taken into account. (*M. A. Preston, "Physics of the Nucleus," Addison-Wesley Publishing Company, Inc., Reading, Mass.*, 1962.)

It is useful in the coupling of the angular momenta to use a representation based on the total angular momentum of each nucleon orbital (the so-called *j-j coupling*). Since the nucleon spin is one-half the total angular momentum of each orbital,[1]

$$\mathbf{j} = \mathbf{l} + \mathbf{s}$$

[1] The letters j, l, and s will be used as vectors to designate the total, orbital, and spin angular momenta, respectively, and as scalars to designate the eigenvalues of those angular momenta.

can assume the eigenvalues $j = l \pm \frac{1}{2}$ for each $l > 0$. The j of each nucleon must have a half-integral value, which is usually attached as a subscript to the orbital quantum numbers in giving the spectroscopic designation of the nucleon, that is, $(nl)_j$. For instance, an $n = 1$ nucleon with $l = 1$ can be in either state $(1p)_{\frac{3}{2}}$ or state $(1p)_{\frac{1}{2}}$, the first of which has four independent substates, the second only two.

The j representation of orbitals is useful in that there exists an important non-central force in nuclear physics which splits the degeneracy between the two j states. This interaction, based largely on phenomenological evidence, is written as a spin-orbit interaction between the $\mathbf{l}$ of the orbit and the $\mathbf{s}$ of the nucleon,

$$V = -a\mathbf{l} \cdot \mathbf{s} \tag{4-84}$$

where $a \approx 13A^{-\frac{2}{3}}$ Mev from study of nuclear-level splittings. This potential is not to be confused with the electromagnetic interaction of the same form, which leads to fine-structure splitting in atoms. This interaction is much stronger, is specifically nuclear in origin, and lowers the state with larger j ("parallel" $\mathbf{l}$ and $\mathbf{s}$) relative to the state of smaller j.

Problem 4-22: Using the fact that $\mathbf{j} \cdot \mathbf{j} = (\mathbf{l} + \mathbf{s}) \cdot (\mathbf{l} + \mathbf{s})$, show that

$$\mathbf{l} \cdot \mathbf{s} = \tfrac{1}{2}[j(j+1) - l(l+1) - s(s+1)]$$

Show also that the zeroth-order energy of the states is shifted by

$$\Delta E = \begin{cases} -\dfrac{a}{2}l & j = l + \frac{1}{2} \\[2ex] +\dfrac{a}{2}(l+1) & j = l - \frac{1}{2} \end{cases} \tag{4-85}$$

These energy splittings are very large and separate considerably the two j states; in fact, in some cases they are even greater than the zeroth-order level spacings shown for the spherically symmetric potential in Fig. 4-8 (middle column). The shell structure of the nucleus must then be recalculated by superimposing the spin-orbit energies on the energies of Fig. 4-8. When this is done, one gets an ordering of states like that shown in Fig. 4-9. Comparison of this figure with the preceding one shows the reordering introduced by the spin-orbit interaction. An s state ($l = 0$) is of course not split, and the splitting of the p states is not very large. The order of the first eight states is unchanged. The d-state splitting is sufficiently large so that the $1d_{\frac{3}{2}}$ is elevated above the $2s_{\frac{1}{2}}$. The $1d_{\frac{3}{2}}$-$2s_{\frac{1}{2}}$ reversal is the first of several such that are caused by the spin-orbit interaction. At a somewhat higher energy, one sees that the $1g_{\frac{9}{2}}$ orbital is pulled down almost to the energy of the $2p_{\frac{1}{2}}$ orbital. Similarly the $1h_{\frac{11}{2}}$ is now near the $2d_{\frac{3}{2}}$.

This ordering of states and the positions of large energy gaps between states will have very significant repercussions upon nucleosynthesis. Each of these levels represents a minor shell which may be filled with nucleons in the con-

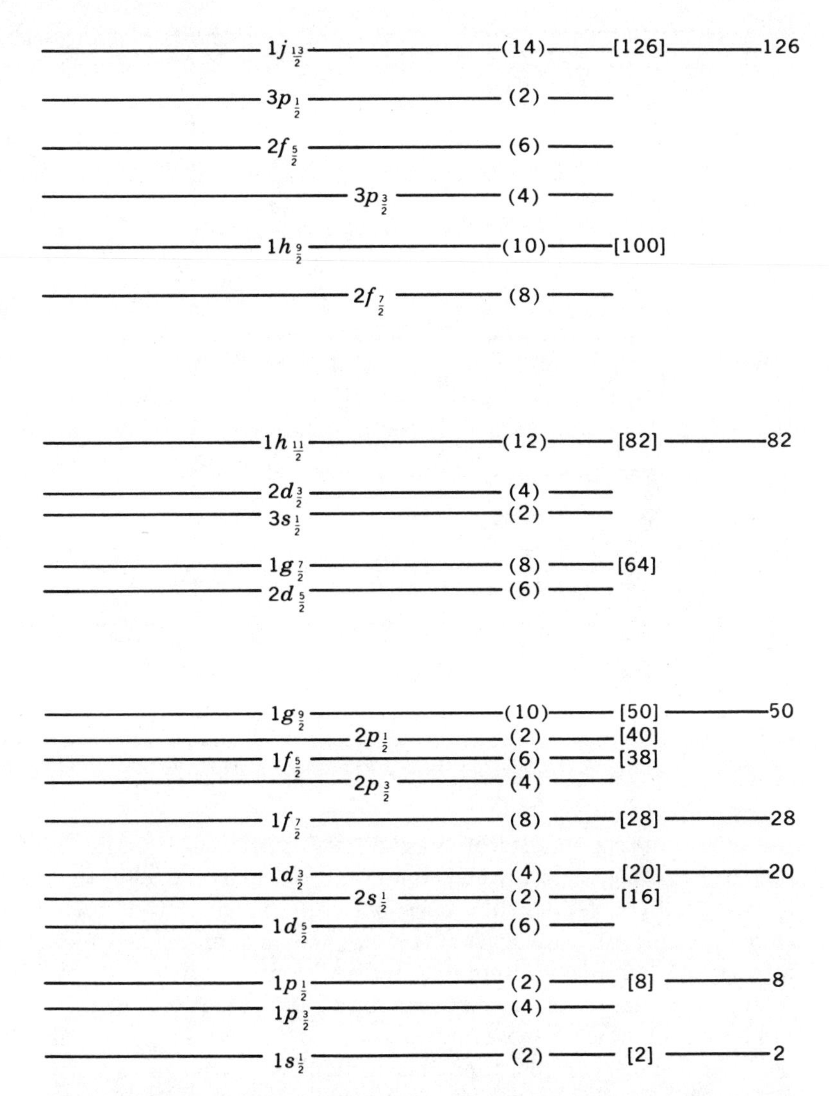

Fig. 4-9 The energy levels of the first 126 neutrons in the simple shell model. The energy shifts caused by the spin-orbit interaction have been added to the central column of Fig. 4-8. (*M. A. Preston, "Physics of the Nucleus," Addison-Wesley Publishing Company, Inc., Reading, Mass.*, 1962.)

struction of a nucleus. For instance, two neutrons and two protons may be assigned to the $1p_{\frac{1}{2}}$ subshell, etc. Of particular importance for nucleosynthesis are those shells at which the energy gap to the next shell is much larger than the average. These gaps occur at the nucleon numbers 2, 8, 20, 28, 50, 82, and 126, the so-called *magic numbers* of nuclear structure. The major effect is the high

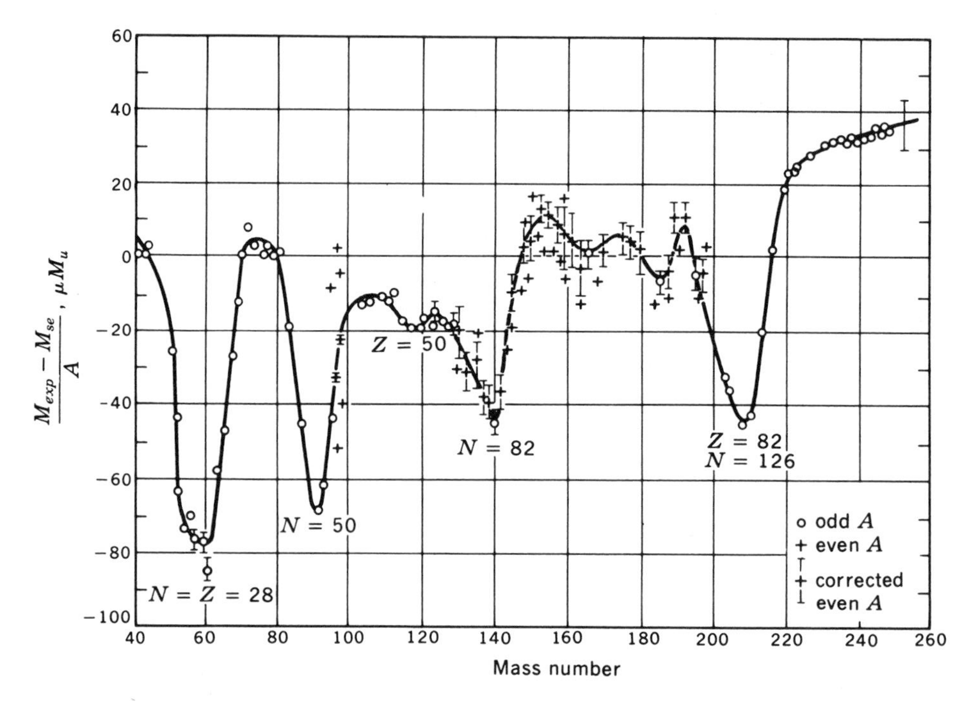

Fig. 4-10 The difference between the measured mass of the most stable nuclear isobar at each value of the atomic weight and a smooth semiempirical mass law. (*M. A. Preston, "Physics of the Nucleus," Addison-Wesley Publishing Company, Inc., Reading, Mass.*, 1962.)

stability (binding energy) of those nuclei consisting of major filled shells of protons or neutrons or both when compared to nearby nuclei in the chart of nuclides. Figure 4-10 shows the difference between the observed nuclear masses and a smooth semiempirical curve fitted to the masses. The smaller than average masses at the magic numbers reflect their relatively large binding energy. (Remember that a negative energy is the equivalent of a negative mass.) Many properties of nuclear systematics show dramatic changes at magic numbers of particles. For example, the cross section for capturing a neutron is quite small (relatively) for those nuclei having magic numbers of neutrons, a fact that will be quite important for heavy-element nucleosynthesis by neutron capture. In fact, the whole structure of the theory of heavy-element nucleosynthesis (to be discussed in Chap. 7) is dominated by the magic numbers $N = 50$, 82, and 126. Smaller effects of the same type happen near other subshells.

As examples, the following nuclei are particularly stable as a result of their closed-shell structure:

$${}_2\mathrm{He}_2^4 = (1s_{\frac{1}{2}})^2$$

$${}_8\mathrm{O}_8^{16} = (1s_{\frac{1}{2}})^2(1p_{\frac{3}{2}})^4(1p_{\frac{1}{2}})^2$$

$${}_{20}\mathrm{Ca}_{20}^{40} = (1s_{\frac{1}{2}})^2(1p_{\frac{3}{2}})^4(1p_{\frac{1}{2}})^2(1d_{\frac{5}{2}})^6(2s_{\frac{1}{2}})^2(1d_{\frac{3}{2}})^4$$

etc., where the superscript on each nucleonic state is equal to $2j + 1$, which is the number of neutrons and protons assignable to that filled shell. A massive doubly magic nucleus is Pb^{208} with 82 protons and 126 neutrons. Many properties of nuclear ground states are clarified by noting the shell-model configuration of the nucleus. As one other example, the ground-state configuration of C^{13} according to the shell model would be

$$_6C_7^{13} = \begin{cases} \text{protons: } (1s_{\frac{1}{2}})^2(1p_{\frac{3}{2}})^4 \\ \text{neutrons: } (1s_{\frac{1}{2}})^2(1p_{\frac{3}{2}})^4(1p_{\frac{1}{2}}) \end{cases}$$

The first benefit of the shell model is that it enables one to understand nuclear spins. First one observes that every closed shell of nucleons has a total angular momentum of zero. That this must be so results from the fact that the sum of j_z, taken along any z axis, for all the particles in a subshell must be zero if the shell is full. Thus the sum of all the individual j vectors in a shell sum to a total vector whose component along any axis is zero. Such a vector must itself be zero. Thus all the doubly magic nuclei in the previous listing have $J = 0$. (It also happens, because of the angular-momentum coupling of *lowest energy*, that all nuclei having even numbers of neutrons and even numbers of protons have $J = 0$, but this fact is not obvious a priori.) It follows that the angular momentum of nuclei having some numbers of nucleons in excess of closed shells is determined by the angular momentum of the extra nuclei. For example, the N^{13} nucleus has one $1p_{\frac{1}{2}}$ proton outside the C^{12} core, and hence has $J = \frac{1}{2}$ in the ground state. Similarly, O^{17} has one $1d_{\frac{5}{2}}$ neutron outside the O^{16} core and hence has $J = \frac{5}{2}$ in the ground state. An example of two nucleons outside a closed core is provided by $N^{14} = C^{12} + (1p_{\frac{1}{2}})^2$. The total angular momentum of N^{14} must be given by

$$\mathbf{J}(N^{14}) = \frac{\mathbf{1}}{\mathbf{2}} + \frac{\mathbf{1}}{\mathbf{2}} = \mathbf{1} \text{ or } \mathbf{0}$$

Because symmetric states have lower energy, the ground state of N^{14} has $J = 1$, whereas the first excited state is the $J = 0$ member of the ground-state configuration. Further examples of this system, which we only illustrate here, can be found in any textbook on nuclear physics.

PARITY

The *parity* of a nuclear state describes the behavior of the wave function of the nucleus if the space coordinates are reversed ($x, y, z \rightarrow -x, -y, -z$), the spins of particles remaining fixed. Under such a reflection, the wave function must either remain the same or change to the negative of itself. States having wave functions of the first type are called *even-parity* states, whereas states of the second type are called *odd-parity* states. A quick outline of the reasoning is as follows. The basic hamiltonian

$$H = \sum_i \frac{p_i^2}{2m} + V_i(r) = \sum_i \frac{\hbar^2}{2m} \nabla_i^2 + V_i(r) \tag{4-86}$$

is invariant to reflection of coordinate system, so that the eigenfunctions can simultaneously be eigenfunctions of the hamiltonian operator and of the parity operator P, which changes $x \rightarrow -x$, etc. That this is so comes from the fact that the time rate of change of any operator that does not depend explicitly on time is proportional to the commutator of the operator with the hamiltonian. If H is invariant under space reflection, then $HP - PH = 0$, and P is a constant. Thus for such a hamiltonian the space-inverted wave function

$$P\psi(x,y,z) \equiv \psi(-x,-y,-z) \tag{4-87}$$

is also an energy eigenfunction of the problem. Two applications of P must give back the same function:

$$P^2\psi(x,y,z) = P\psi(-x,-y,-z) = \psi(x,y,z) \tag{4-88}$$

Thus the eigenvalues of P are given by $P^2 = 1$, and $P = \pm 1$. That is, inverting the space coordinates must change ψ to $\pm\psi$. Thus every nuclear state has either even parity or odd parity.[1]

It is not difficult to see what the parity of a nuclear state is in terms of the shell model. Each nucleon is described by a wave function

$$\psi = f(r) Y_l^m(\theta,\phi) \tag{4-89}$$

The parity operator changes $\theta \rightarrow \pi - \theta$ and $\phi \rightarrow \pi + \phi$. Since

$$P_l^m(\pi - \theta) = (-1)^{l+m} P_l^m(\theta)$$

and

$$e^{im(\pi-\phi)} = (-1)^m e^{im\phi} \tag{4-90}$$

it follows that

$$Y_l^m(\pi - \theta, \pi + \phi) = (-1)^l Y_l^m(\theta,\phi) \tag{4-91}$$

Thus the parity of each nucleon orbital is $(-1)^l$. Since the wave function for the nucleus is, in this model, just the product of the wave functions for the individual nucleons in the nucleus, the parity of a nuclear state containing n nucleons is

$$P = (-1)^{l_1+l_2+l_3+\cdots+l_n} \tag{4-92}$$

The first and most obvious conclusion from Eq. (4-92) is that a nucleus consisting of closed shells must have even parity, since there is an even number of nucleons characterized by each value of l. Thus the parity of a complex nucleus is determined by the product of the parities for the extra nucleons outside closed-

[1] If terms are included in the hamiltonian which are not invariant to space inversion, the result does not follow. Such a force comes from the $(\bar{n}p)(\bar{p}n)$ term in the weak-interaction current discussed in Chap. 3. Thus the weak interaction may introduce a very small parity impurity into nuclear states. F. C. Michel, *Phys. Rev.*, **133**:B329 (1964).

shell configurations. Reconsider the same examples used for the angular momentum of ground states. Using the standard notation in which the parity (even/odd) is designated by placing a superscript (+/−, respectively) on the angular momentum, the ground state of N^{13} must be

$$J^\pi(N^{13}) = \tfrac{1}{2}^-$$

since there is one extra $1p_{\frac{1}{2}}$ proton outside the C^{12} core. Similarly for O^{17}

$$J^\pi(O^{17}) = \tfrac{5}{2}^+$$

since O^{17} has one extra $l = 2$ neutron.

Problem 4-23: Show that for the N^{14} ground state

$$J^\pi(N^{14}) = 1^+$$

Problem 4-24: The first two negative-parity states of N^{14} have $J^\pi = 0^-$ and 2^-. To what configuration do these states likely belong?
Ans: $(1p_{\frac{1}{2}})(2s_{\frac{1}{2}})$ for 0^-, $(1p_{\frac{1}{2}})(1d_{\frac{5}{2}})$ for 2^-.

Further discussion of the role of angular momentum and parity in nuclear states will be found when we consider the problem of resonances in nuclear reaction rates.

COMPOUND NUCLEAR STATES

If the reaction involving particles a and X proceeds through a resonant state of the compound nucleus W, it will be convenient to think of the process

$$a + X \to W \to Y + b$$

as a two-body interaction between particles a and X that is described by a potential $V(r)$ which characterizes the interaction of these two particles. Indeed the chance that the interaction can happen is directly proportional to the probability that the state of W can be so described. Then the motion of the pair of particles in a stationary state will be given by the Schrödinger equation

$$-\frac{\hbar^2}{2\mu}\nabla^2\psi + V(r)\psi = E\psi \tag{4-93}$$

where μ is the reduced mass of a and X and the spatial coordinates are relative ones. When $V(r)$ is, as assumed, only a function of separation, one usually writes the laplacian in spherical coordinates:

$$\nabla^2 = \frac{\partial^2}{\partial r^2} + \frac{2}{r}\frac{\partial}{\partial r} + \frac{1}{r^2}\left(\frac{1}{\sin^2\theta}\frac{\partial^2}{\partial\phi^2} + \frac{1}{\sin\theta}\frac{\partial}{\partial\theta}\sin\theta\frac{\partial}{\partial\theta}\right) \tag{4-94}$$

The angular-momentum operator is

$$\mathbf{L} = \mathbf{r} \times \mathbf{p} = \frac{\hbar}{i}\mathbf{r} \times \nabla \tag{4-95}$$

When **L** is expressed in spherical coordinates, it is

$$L_z = \frac{\hbar}{i}\frac{\partial}{\partial\phi} \qquad L_x = -\frac{\hbar}{i}\left(\sin\phi\frac{\partial}{\partial\theta} + \cot\theta\cos\phi\frac{\partial}{\partial\phi}\right)$$
$$L_y = \frac{\hbar}{i}\left(\cos\phi\frac{\partial}{\partial\theta} - \cot\theta\sin\phi\frac{\partial}{\partial\phi}\right) \tag{4-96}$$

The sum of the squares of these components reveals

$$\begin{aligned} L^2 &= L_x^2 + L_y^2 + L_z^2 \\ &= -\hbar^2\left(\frac{1}{\sin^2\theta}\frac{\partial}{\partial\phi^2} + \frac{1}{\sin\theta}\frac{\partial}{\partial\theta}\sin\theta\frac{\partial}{\partial\theta}\right) \end{aligned} \tag{4-97}$$

which is seen to be proportional to the angular part of the laplacian. Thus the kinetic-energy term of the Schrödinger equation is

$$\begin{aligned} -\frac{\hbar^2}{2\mu}\nabla^2 &= -\frac{\hbar^2}{2\mu}\left(\frac{\partial^2}{\partial r^2} + \frac{2}{r}\frac{\partial}{\partial r}\right) + \frac{L^2}{2\mu r^2} \\ &= \frac{p_r^2}{2\mu} + \frac{L^2}{2\mu r^2} \end{aligned} \tag{4-98}$$

This result parallels the classical one that the kinetic energy of a central-field problem can be written as the sum of the energy of radial motion plus an energy associated with an angular momentum.

Problem 4-25: Confirm that the kinetic energy of a classical particle moving in a circular orbit is

$$\mathrm{KE} = \frac{L^2}{2\mu r^2}$$

Problem 4-26: Show by taking the root of p_r^2 that

$$p_r = \frac{\hbar}{i}\left(\frac{\partial}{\partial r} + \frac{1}{r}\right)$$

The conventional procedure for central-field problems is to separate the radial and angular parts of the wave functions by

$$\psi(r,\theta,\phi) = f(r)Y_l^m(\theta,\phi) \tag{4-99}$$

where functions $Y_l^m(\theta,\phi)$, the so-called *spherical harmonics*, are eigenfunctions of L^2 and L_z, viz.,

$$Y_l^m(\theta,\phi) = A_l^m e^{im\phi}P_l^m(\cos\theta) \tag{4-100}$$

and A_l^m is a constant such that $\int|Y_l^m|\,d\Omega = 1$. Because they are eigenfunctions,

$$\begin{aligned} L^2Y_l^m(\theta,\phi) &= l(l+1)\hbar^2Y_l^m(\theta,\phi) \\ L_zY_l^m(\theta,\phi) &= m\hbar Y_l^m(\theta,\phi) \end{aligned} \tag{4-101}$$

where l and m are restricted to integers such that $m \leq l$. The angular momentum is a constant of the motion in central-field problems. Thus the radial part of the wave function describing an encounter must bear a subscript designating it to be the solution for a particular value of l. Insertion of the separated wave function into the Schrödinger equation gives the radial equation

$$-\frac{\hbar^2}{2\mu}\left(\frac{\partial^2}{\partial r^2} + \frac{2}{r}\frac{\partial}{\partial r}\right) f_l(r) + \left[\frac{l(l+1)\hbar^2}{2\mu r^2} + V(r)\right] f_l(r) = E f_l(r) \tag{4-102}$$

Problem 4-27: Show that the substitution $f_l(r) = \chi_l(r)/r$ leads to

$$-\frac{\hbar^2}{2\mu}\frac{d^2\chi_l}{dr^2} + \left[\frac{l(l+1)\hbar^2}{2\mu r^2} + V(r) - E\right]\chi_l(r) = 0 \tag{4-103}$$

Although it is not possible to describe in detail the complete potential representing the forces between the two particles, certain general factors can be stated with confidence. The nuclear force is a very strong attractive one that operates only at short distances (on the order of the pi-meson Compton wavelength, $\hbar/m_\pi c = 1.4 \times 10^{-13}$ cm). Until the surfaces of the nuclear particles come that close to each other, there is no nuclear force. For greater separations the potential is just that of the coulomb repulsion, $V(r) = Z_1Z_2e^2/r$.

One other feature is plainly evident from Eq. (4-102). As far as the radial motion is concerned, the kinetic energy associated with angular motion appears only as an additive term to $V(r)$. Indeed the quantity $l(l+1)\hbar^2/2\mu r^2$ is often interpreted as a *centrifugal potential*, in direct analogy to its classical counterpart $L^2/2\mu r^2$. That is, since L^2 is a constant of the motion, the kinetic energy $L^2/2\mu r^2$ must be supplied by the initial energy E to reach a separation r. Viewed that way, $L^2/2\mu r^2$ can be interpreted as a repulsive potential as far as the radial motion is concerned. Thus we can define a radial potential

$$\begin{aligned} V_l(r) &= \frac{l(l+1)\hbar^2}{2\mu r^2} + V(r) \\ &= \frac{l(l+1)\hbar^2}{2\mu r^2} + \frac{Z_1Z_2e^2}{r} \qquad \text{for } r > R \end{aligned} \tag{4-104}$$

where R is the radial separation outside of which the nuclear forces cannot be felt.

It is apparent that the potential $V_l(r)$ can be schematically represented as in Fig. 4-11. The potential is the classical repulsive one for $r > R$. R is the interaction radius, sometimes called the nuclear radius, and must be approximately the sum of the radii of the two nuclear particles plus the range of nuclear forces. In practice it has been found that

$$R = 1.4(A_1^{1/3} + A_2^{1/3}) \times 10^{-13} \text{ cm} \tag{4-105}$$

is a very good approximation to the interaction radius, where A_1 and A_2 are the atomic weights of the interacting particles. At $r = R$ the force becomes strongly attractive as a result of the nuclear pi-meson field, and the potential drops precip-

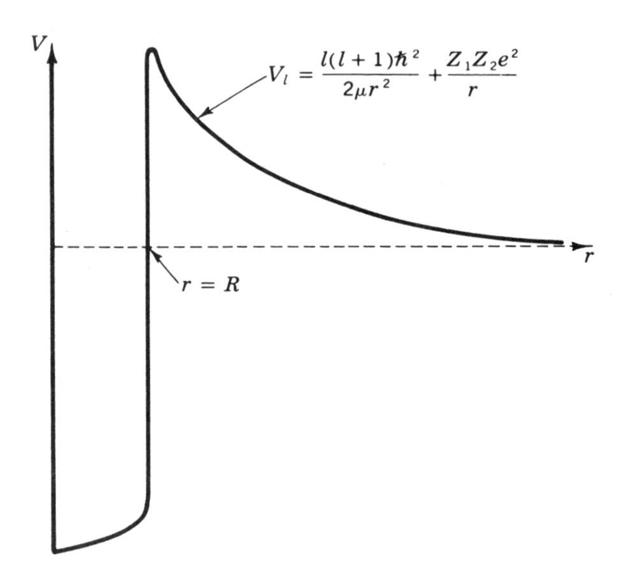

Fig. 4-11 The potential governing the motion of one nucleus relative to another. For $r < R$ the nuclei are essentially in contact, and the strongly attractive short-range nuclear force results in a deep negative potential. For $r > R$ the nuclear force can no longer be felt, and the coulomb potential dominates. When one considers the radial motion of the two nuclei, the angular momentum adds an effective centrifugal potential. The total extranuclear radial potential is designated by V_l.

itously. Near $r = 0$ the potential becomes relatively flatter (in some unknown manner) since the nucleon-nucleon forces tend to produce little net radial force near the center of the nuclear matter. In practice one often approximates $V_l(r)$ by a square well for $r < R$. For considerations relevant to understanding nuclear reactions in stellar interiors, it will not be necessary to have a good understanding of the form of $V_l(r)$ for $r < R$.

Resonances can occur in nuclear reactions if the kinetic energy E of the particles at infinity is just such that the total energy coincides with one of the quasistationary states of the compound nucleus. The situation is illustrated on a suggestive energy diagram in Fig. 4-12. Quasistationary states having positive energy like the one labeled E_n can exist because the high potential walls give the state a long lifetime against breakup. Classically the energy E at infinity would be expended against the radial energy $V_l(r)$ by the time the particle has reached a distance R_0 such that

$$V_l(R_0) = E \tag{4-106}$$

Classically the particle would rebound at R_0, which is called the *classical turning point,* and move back outward to a kinetic energy of separation at infinity equal to E. In quantum mechanics, however, there is a probability that the particle can penetrate the potential wall and reach the nuclear force at $r = R$. In Fig. 4-12 this energy E *does not* coincide with the energy of a quasistationary state, and the reaction is *nonresonant,* but in practice one of the states E_n will often lie close enough to the energy E so that a resonant reaction rate will have to be employed.

If a state of the compound nucleus is to mediate a reaction, the formation of the state must satisfy the conservation laws. That the center-of-mass kinetic energy of a and X should be such as to coincide with the energy E_n of an excited state of the compound system ensures only that energy and momentum can be conserved in the formation of that excited state. Additional requirements for the formation or breakup of the compound system are that angular momentum and nuclear symmetries (parity and isotopic spin) must also be conserved. The angular-momentum requirement is the most elementary. If $\mathbf{J}_n$ is the angular momentum of the excited state of energy E_n in the compound nucleus W, $\mathbf{J}_a$ and $\mathbf{J}_X$ the spins of particles a and X, and $\mathbf{L}$ the orbital angular momentum of a rela-

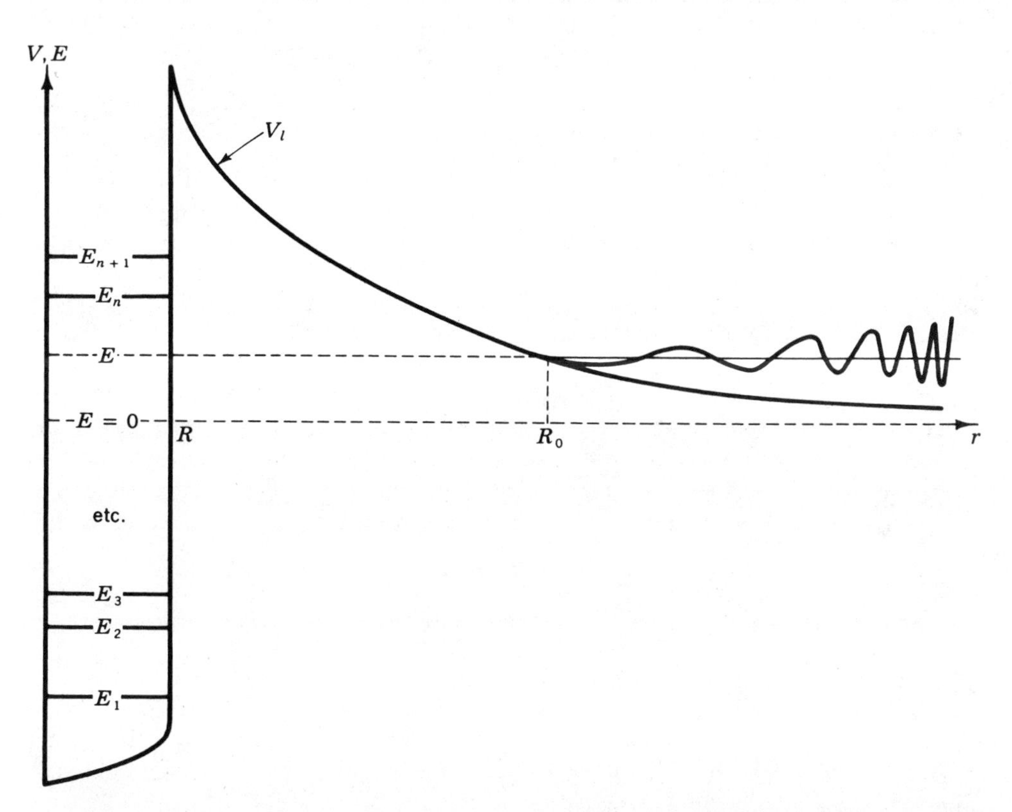

Fig. 4-12 The stationary nuclear states in the compound nucleus formed by the coalescence of the two colliding particles are designated by E_1, E_2, ... The increasing wavelength of the incoming wave reflects the loss of momentum as the kinetic energy E is expended against the repulsive extranuclear potential V_l. Within the context of classical mechanics the incoming particle would be expected to rebound from the potential at the classical turning point R_0, but in the quantum treatment the wave has a nonzero probability of tunneling through the potential barrier to the interaction radius R. The compound nucleus formed has an energy E that, in this case, falls between the natural resonances of the compound nucleus, so that the cross section will have a slowly varying dependence on the energy. The zero of energy is determined relative to the ground state E_1 of the compound nucleus by the extra mass of the colliding particles.

tive to X, then one demands that

$$\mathbf{J}_n = \mathbf{J}_a + \mathbf{J}_X + \mathbf{L} \tag{4-107}$$

where the vector addition must be made by the standard quantum rules for the addition of angular momenta.

The major importance of the concept of parity on nuclear reactions in stellar interiors is that the character of the possible resonances is limited by the demand that parity be conserved. The nuclear force is of a type that does not change the parity of a wave function by an interaction. This means that if there is to be a resonance in the interaction of $a + X$, the parity of the resonant state of the compound nucleus W must be the product of the parity of a, the parity of X, and the parity of the relative motion of a and X. Symbolically,

$$\pi(W) = \pi(a)\pi(X)(-1)^l \tag{4-108}$$

where $\pi(a)$ is the parity of a, etc., and l is the relative angular momentum of a and X when they interact. It should be added that the parities of both proton and neutron are positive by definition.

Problem 4-28: In the reaction $C^{12} + p$, what are the values of the relative angular momenta that can form the following resonant states: (*a*) $\frac{1}{2}^-$ (*b*) $\frac{1}{2}^+$ (*c*) $\frac{3}{2}^+$ (*d*) $\frac{3}{2}^-$ (*e*) $\frac{5}{2}^-$?
Ans: *p*, *s*, *d*, *p*, *f*.

The foregoing problem demonstrates that the knowledge of spin and parity of a resonant state will often determine the l wave of a reaction. Knowledge of l is necessary when the penetration factor must be calculated. This need arises often for resonances so low in energy that the cross section cannot be measured in the laboratory.

Problem 4-29: What are the possible spins and parities of states in Ne^{20} that can serve as resonances for the reaction $O^{16}(\alpha,\gamma)Ne^{20}$?
Ans: $0^+, 1^-, 2^+, 3^-, 4^+$, etc.

In the previous two problems at least one of the particles had zero spin. For reactions in which both of the particles have nonzero spin, the J^π of the resonance does not uniquely determine the l wave of the reaction.

Problem 4-30: What l waves can form a 2^+ resonance in N^{14} in the reaction $C^{13}(p,\gamma)N^{14}$?
Ans: $l = 1$, $l = 3$.

Before leaving the subject, two other observations on nuclear parity seem in order. The first is that the strongest nuclear electromagnetic transitions (gamma rays) are between states of opposite parity that differ in angular momentum by one (vector) unit. This situation is analogous to the atomic case, where electric-dipole transitions are almost all that are seen. A favorite laboratory technique of nuclear physicists for measuring the spin and parity of some state is to measure the angular distributions (relative to some other particles) of gamma rays from that state. Finally, it should be added that Yang and Lee shared a Nobel prize for their prediction that parity of wave functions is not an invariant in beta decay, a prediction that was dramatically confirmed by Wu and coworkers. Although parity is not an inviolable quantum number, it is conserved to very high accuracy in interactions of nuclei by the nuclear force.

An even more fundamental nuclear symmetry has important applications in nuclear astrophysics. The foregoing discussion has been applicable to all reactions except those between identical particles; but pairs of identical particles obey a more complete and fundamental symmetry that depends upon their spin. The total wave function (space $\times$ spin) of a pair of identical fermions (half-integral spin) must have the property that it changes sign when the particles are interchanged. This property is called *total antisymmetry*. The total wave function of a pair of identical bosons, on the other hand, must be unchanged by switching the particles. This is a fundamental rule of quantum mechanics, total antisymmetry for a pair of identical fermions and total symmetry for a pair of identical bosons. It is this fundamental rule which lies at the heart of quantum statistics. Although we shall not prove it here, *if the total wave function describing the spin and space coordinates of two identical particles is symmetric to exchange of the particles, then collections of the particles obey Bose-Einstein statistics, whereas if the wave function is antisymmetric to exchange, collections of the particles obey Fermi-Dirac statistics.*

Rather than dwelling on this principle, we shall illustrate its application to interactions of importance to nuclear astrophysics. For a pair of particles, the total wave function can be factored into the product of a space function describing the relative motion of the particles and a spin function describing the vector coupling of the two spins into a total spin. The symmetry of the total wave function is equal to the product of the symmetry of the space function times the symmetry of the spin function. But the space symmetry is just the parity of the wave function, whereas the spin symmetry is given by the angular-momentum coupling rules.

Consider, for example, the proton-proton interaction. This interaction results in the first nuclear reaction in a star of pure hydrogen. Because of the centrifugal barrier, the easiest interaction at low energy (temperature) is between s-wave protons. That wave function has even parity, so that the spin function must be antisymmetric for the whole wave function to be antisymmetric. The spin function of the two protons is symmetric if the spins are parallel, giving $S = 1$, and antisymmetric if the spins are antiparallel, giving $S = 0$. *Thus the s-wave interaction of two protons must occur with antiparallel spins,* the so-called *singlet interaction.*

Problem 4-31: What is the spin wave function for the p-wave interaction of two protons that has no angular-momentum component in the direction of relative motion (taken to define the z axis)?

Ans: $\frac{1}{\sqrt{2}}[\uparrow(1)\downarrow(2) + \downarrow(1)\uparrow(2)]$.

This peculiar feature manifests itself in the laboratory cross section for proton-proton scattering.

An alternative example of this principle, featuring bosons instead of fermions, is given by the interaction of two alpha particles. Since they have zero spin, the total spin must also be zero, a necessarily symmetric situation. Since the total symmetry of two bosons must be symmetric, it follows that the space function of the two alpha particles must have even parity. Thus the interaction of two alpha particles is limited to the states $J^\pi = 0^+, 2^+, 4^+, \ldots$.

The effects of nuclear states on nuclear reactions in stars are intimately related to problems of barrier penetration. In Fig. 4-12 it was apparent that all resonant states by which the incoming particles can interact are *unbound* nuclear states; since they can be formed by a and X with positive kinetic energy at infinity, they can decay into a and X with positive kinetic energy. The resonant states for a and X have positive energy compared to separated a and X particles at infinity. These quasistationary resonant states can live for relatively long times, however, because the potential barrier $V_l(r)$ inhibits the breakup into a and X in just the same way that it inhibits the formations of the states by those particles. Nuclear states of increasingly higher excitation energy are increasingly uninhibited by $V_l(r)$, and hence they break apart increasingly rapidly, and into more and more possible combinations of final particles, i.e., other than a and X. The larger the rate of breakup, the larger the energy width of the state according to the relationship $\Gamma = \hbar/\tau$, where τ is the mean lifetime of the state. For sufficiently high excitation energies, the lifetimes become so short (hence the width Γ so great) that neighboring quasistationary states overlap each other to large degree. At this point, the notion of quasistationary states must be discarded in favor of a continuum theory; but for the low kinetic energies in stellar interiors, the lifetimes for breakup into charged particles are long enough for the widths to be much less than the energy separation of nuclear states in the light nuclei. (Excited states that can decay into neutrons will tend to be broad compared to those which cannot, since there exists no coulomb barrier for neutrons.)

Several of these features of the decay of nuclear states are well illustrated by a conventional energy-level diagram of the nucleus B^{11}, illustrated in Fig. 4-13.

(*1*) The ground state of B^{11} is stable: $\tau = \infty$, $\Gamma(0) = 0$.

(*2*) All excited states of B^{11} can decay to a lower-lying state with the emission of a gamma ray. They are also energetically capable of beta decay to the ground state of C^{11}, but the beta decays of excited nuclear states occur so slowly that they are never seen in practice. Therefore $\Gamma_\gamma = \hbar/\tau_\gamma \gg \Gamma_{\beta-} = \hbar/\tau_\beta$, and the energy width of the states is determined by their gamma lifetime. The widths

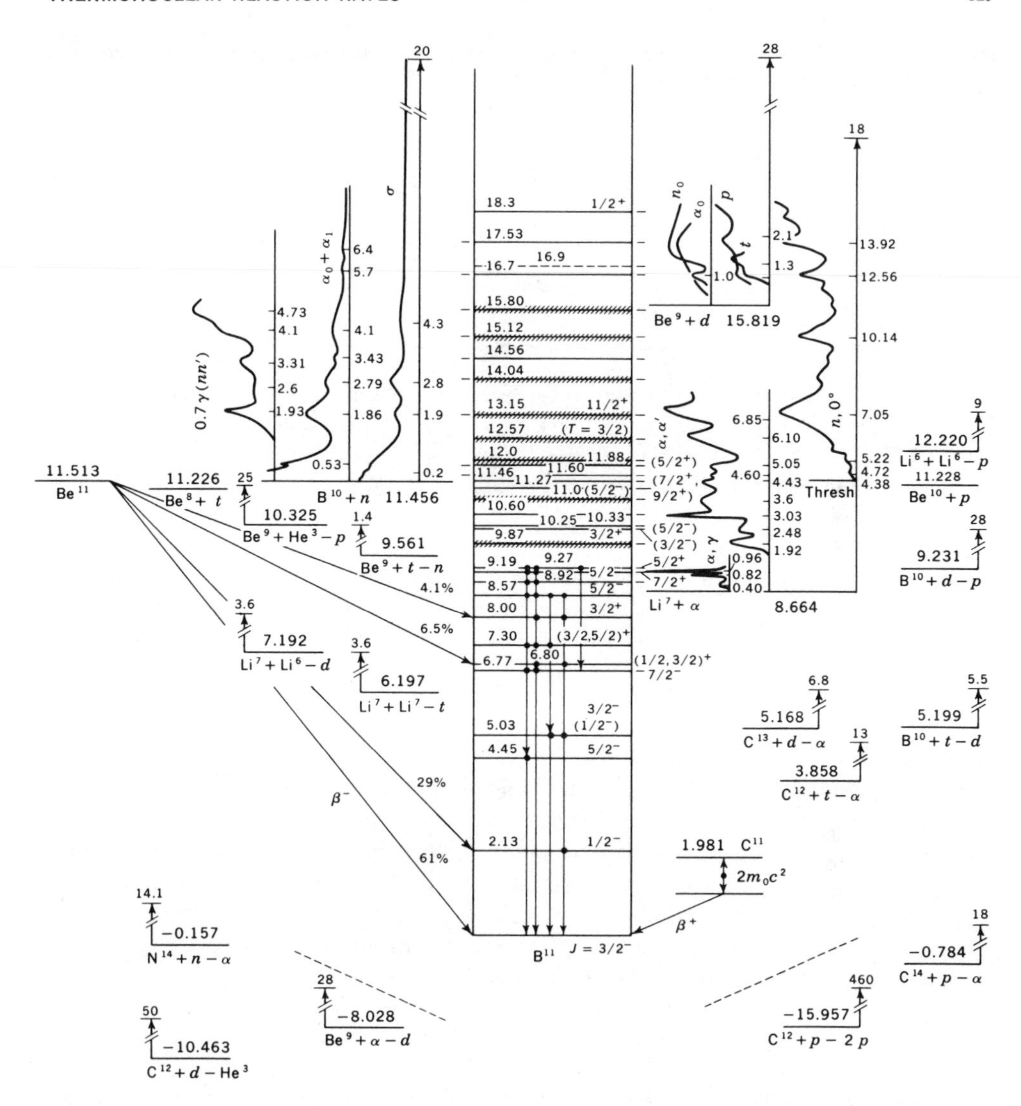

Fig. 4-13 Energy-level diagram of the B^{11} nucleus. Energy is plotted vertically on this diagram in units of Mev, with the ground state of B^{11} taken as the zero of the energy scale. Each known bound state of B^{11} is labeled by its excitation energy and by the spin and parity of the state. Also shown is the energy required to separate B^{11} into two particles, for example, $Li^7 + \alpha$, and the mass-energy released in the formation of B^{11} by specific reactions, for example, $Be^9 + He^3 - p$. This energy-level diagram is typical of those encountered in the nuclear literature, e.g., T. Lauritsen and F. Ajzenberg-Selove, Energy Levels of Light Nuclei VII, $A = 5$–10, *Nucl. Phys.*, **78**:1 (1966).

of states which may only decay by gamma emission are characteristically on the order of 1 ev.

Problem 4-32: The second excited state of B^{11} has an excitation energy of 4.45 Mev and decays by gamma emission with a mean lifetime of 1.2×10^{-15} sec. What is the natural width of that state? ($\hbar = 6.582 \times 10^{-16}$ ev sec.)
Ans: $\Gamma_\gamma = 0.55$ ev.

Problem 4-33: Compute from the table of atomic mass excesses the excitation energy of B^{11} such that all higher-lying states can decay in the channel $B^{11} \to Li^7 + \alpha$.
Ans: 8.664 Mev.

(*3*) States with excitation energy greater than 8.664 Mev can also decay by breaking up into Li^7 and an alpha particle. For instance, the width of the state at 9.27 Mev, which is observed to be $\Gamma(9.27) = 5$ kev, must be the sum of the partial widths; $\Gamma(9.27) = \Gamma_\gamma(9.27) + \Gamma_\beta(9.27) + \Gamma_\alpha(9.27) \approx \Gamma_\gamma(9.27) + \Gamma_\alpha(9.27)$ since the beta-decay width is negligible.

Problem 4-34: What is the mean lifetime $\tau(9.27)$ of B^{11}? One can compute theoretically that the width contributed to this state by gamma decays is about $\Gamma_\gamma(9.27) = 5$ ev. What is the probability that this excited state will emit a gamma ray in preference to alpha-particle breakup?

This last problem has indicated that $\Gamma_\alpha(9.28) \approx 5$ kev. Lower-lying states will have a smaller alpha width, however, since the coulomb barrier will then offer greater inhibition to the alpha decay. In fact, there exists an energy in the $Li^7 + \alpha$ system below which Γ_α would be smaller than Γ_γ, and most of the decays would be by gamma emission. (There need not actually *be* such a state in B^{11}, of course.) The immediate physical point is that the partial width is to be proportional to the rate at which particles can escape through the barrier to infinity.

Many of the principles of nuclear reactions are well illustrated by the following problem, which consists of a schematic summary of laboratory data on the interaction $Li^7 + p$, plus questions regarding the interpretation of those data.

Problem 4-35: This is a special problem on $Li^7 + p$ data. Questions follow the numbered statements.

(*1*) A schematic graph of the cross section for the reaction

$Li^7 + p \to He^4 + He^4$

is shown in Fig. 4-14 in terms of the *laboratory proton energy.*

(*2*) The increase of the cross section with increasing energy is due to the greater ease of penetration of the coulomb barrier at higher energies. By dividing the observed cross section by the penetration factor, exp $(-bE^{-\frac{1}{2}})$, a much smoother curve is obtained, shown in Fig. 4-15.

(*3*) There is another channel through which a reaction can proceed, viz., $Li^7 + p \to Be^{8*} \to Be^8 + \gamma$. The corresponding cross section, σ_γ, is shown in Fig. 4-16. At $E_p = 441$ kev the scattering cross section is $500\sigma_\gamma$.

(*4*) The resonance shown in σ_γ is distorted by the ease of penetrating the coulomb barrier at the higher energies. The product exp $(bE^{-\frac{1}{2}})\sigma_\gamma$ yields the symmetric dispersion curve of Fig. 4-17. The width of the Be^{8*} compound nuclear state is measured by the width of this

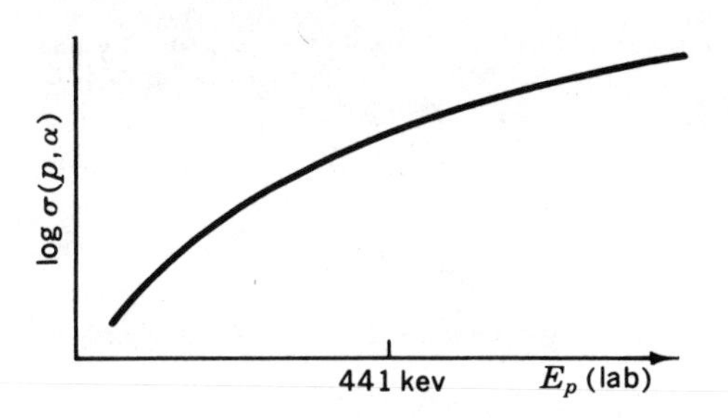

Fig. 4-14 A schematic representation of the measured cross section for the reaction $Li^7(p,\alpha)He^4$. The scale is logarithmic and encompasses an increase of several orders of magnitude between 100 and 1,000 kev.

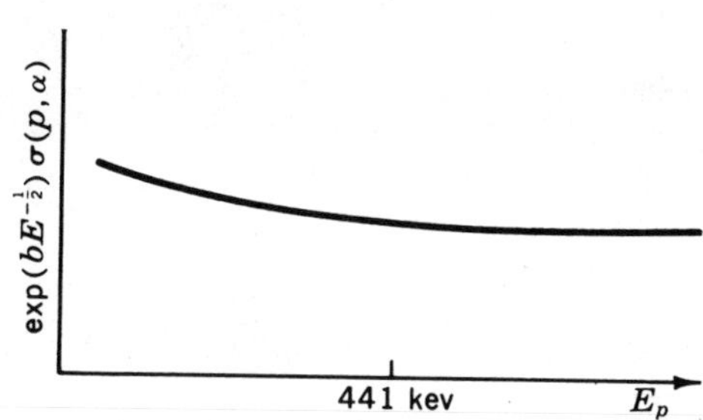

Fig. 4-15 The product of the cross section for the reaction $Li^7(p,\alpha)He^4$ and the factor exp $(bE^{-\frac{1}{2}})$ is approximately constant, showing that the nuclear mechanism changes very slowly with energy.

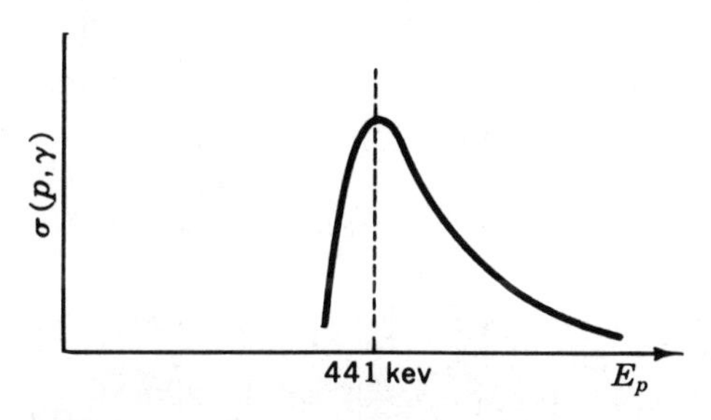

Fig. 4-16 A schematic representation of the measured cross section for the reaction $Li^7(p,\gamma)Be^8$ shows a large resonance at $E_p = 441$ kev. The cross section for scattering the protons shows the same resonance.

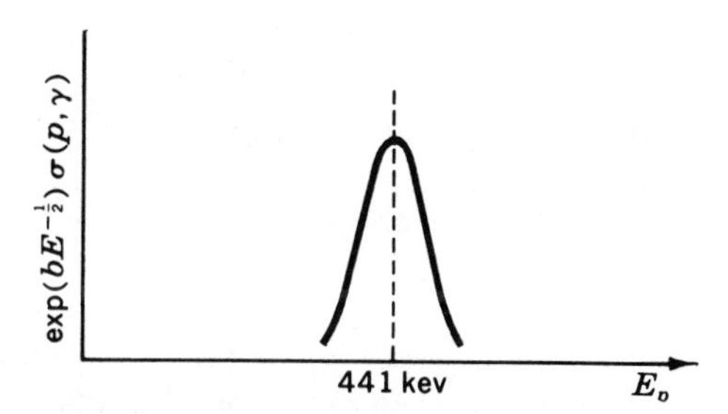

Fig. 4-17 The product of the cross section for the reaction $Li^7(p,\gamma)Be^8$ and the factor exp $(bE^{-\frac{1}{2}})$ is symmetric about the resonance energy. The full width at half maximum is $\Gamma = 12$ kev.

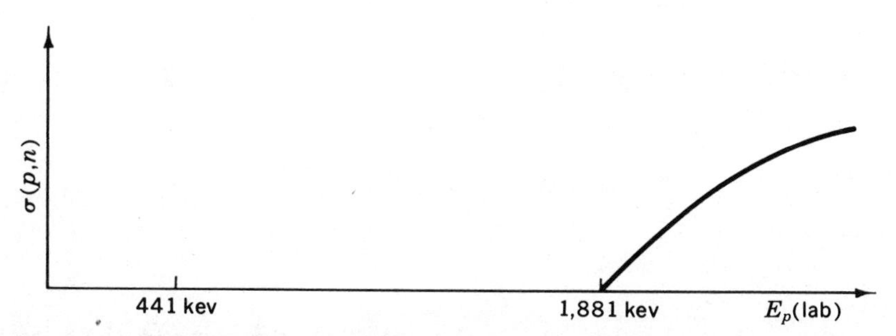

Fig. 4-18 A schematic representation of the measured cross section for the reaction $Li^7(p,n)Be^7$. The yield of neutrons appears to be zero for laboratory proton energy less than 1,881 kev.

curve to be $\Gamma = 12$ kev. The angular distribution of scattered protons shows that the angular momentum J of Be^{8*} is less than 2.

(5) Another channel of the possible reactions is $Li^7 + p \rightarrow Be^7 + n$. The neutron cross section σ_n is observed to be zero below 1,881 kev and rises abruptly with energy above that value, as shown in Fig. 4-18.

Questions

(1) What is the excitation energy in Be^8 of the Be^{8*} resonance at 441 kev proton energy?
Ans: 17.64 Mev.

(2) Why are no neutrons made for $E_p < 1{,}881$ kev?

(3) What are likely causes for the absence of a resonance in σ_α although one does appear in σ_γ?

(4) What is the gamma-ray lifetime of the 441-kev resonance?
Ans: 3×10^{-17} sec.

(5) Assume that the 441-kev resonance occurs by the simple addition of an $l = 1$ proton to the $J^\pi = \frac{3}{2}^-$ ground state of Li^7. What are the spin and parity J^π of Be^{8*}?

The discussion of particle widths is usually carried on with the aid of two concepts, the *penetration factor* and the *dimensionless reduced width.* The definitions of these quantities are motivated by the following considerations.

The Schrödinger equation

$$i\hbar \frac{\partial \psi}{\partial t} = -\frac{\hbar^2}{2\mu} \nabla^2 \psi + V\psi \tag{4-109}$$

is physically interpreted to give the probability density for the particle according to $\rho = \psi^*\psi$. If the Schrödinger equation is multiplied on the left by ψ^*, if the complex conjugate of the Schrödinger equation is multiplied on the left by ψ, and if the difference is taken between those two equations, the result is

$$\begin{aligned} i\hbar \frac{\partial}{\partial t} (\psi^*\psi) &= -\frac{\hbar^2}{2\mu} (\psi^* \nabla^2 \psi - \psi \nabla^2 \psi^*) \\ &= -\frac{\hbar^2}{2\mu} \nabla \cdot (\psi^* \nabla \psi - \psi \nabla \psi^*) \end{aligned} \tag{4-110}$$

which has the form of the equation of continuity

$$\frac{\partial \rho}{\partial t} + \nabla \cdot \mathbf{J} = 0 \tag{4-111}$$

if

$$\begin{aligned} \rho &= \psi^*\psi \\ \mathbf{J} &= \frac{\hbar}{2i\mu} (\psi^* \nabla \psi - \psi \nabla \psi^*) \end{aligned} \tag{4-112}$$

In the same sense that $\psi^*\psi$ is the probability density of the particle, $\mathbf{J}$ is the probability current of particles. To obtain a simple physical feeling of these ideas, consider a flux of particles with momentum $\mathbf{p}$ moving in the x direction.

Then since $\mathbf{p} = -i\hbar\nabla$, the wave function is a plane wave of the form

$$\psi(x) \propto \exp\left(i\frac{px}{\hbar}\right) \tag{4-113}$$

In fact if there are n particles per unit volume, the normalized plane wave will be

$$\psi(x) = \sqrt{n}\exp\left(i\frac{px}{\hbar}\right)$$

for then $\psi^*\psi = n$, the particle density.

Problem 4-36: Show that the probability current for the same plane wave is just

$$J_x = n\frac{p}{m} = nv$$

which is the classical particle flux. For such a simple case the flux is just $\mathbf{J} = \mathbf{v}\psi^*\psi$.

Suppose that the wave function for the relative motion of particles a and X has been obtained. To first approximation the particles are initially thought of as bound by the nuclear potential in the quasistationary excited state of the compound nucleus. There is an exponential decrease of the wave function through the potential barrier which turns into an outgoing wave of particles at infinity. The decay rate of that excited state (for particle emission) is

$$\begin{aligned}\lambda = \frac{1}{\tau} &= \text{probability/sec for particle from decaying system to cross large spherical shell}\\ &= \lim_{r\to\infty} v \iint_{\theta,\phi} |\psi(r\theta\phi)|^2 r^2 \sin\theta\, d\theta\, d\phi\\ &= \lim_{r\to\infty} v \iint_{\theta,\phi} \left|\frac{\chi_l}{r}\right|^2 |Y_l^m(\theta,\phi)|^2 r^2 \sin\theta\, d\theta\, d\phi\\ &= v|\chi_l(\infty)|^2\end{aligned} \tag{4-114}$$

The penetration factor for particles of relative angular momentum l is defined in this case by

$$P_l = \frac{\chi_l^*(\infty)\chi_l(\infty)}{\chi_l^*(R)\chi_l(R)} \tag{4-115}$$

whereupon the decay rate may be written

$$\lambda = vP_l|\chi_l(R)|^2 \tag{4-116}$$

It is obvious by the same manipulations that

$$|\chi_l(R)|^2 = \iint_{\theta,\phi} |\psi_l(R,\theta,\phi)|^2 R^2 \sin\theta\, d\theta\, d\phi \tag{4-117}$$

gives the probability per unit radial distance that the particle a is to be found at the interaction radius R, where the potential switches from the nuclear potential to the external potential $V_l(r)$. Thus the decay rate is thought of as a product of three factors:

$$\lambda = (\text{velocity at infinity}) \times (\text{penetration factor}) \times (\text{probability/unit } dr \text{ that particle is at nuclear radius})$$

The velocity at infinity is determined simply by the energy of the excited state relative to the combined masses of a and X. The penetration factors can be calculated with sufficient accuracy by a separate wave-mechanical calculation that is independent of the uncertainty in the nuclear forces provided one knows the nuclear radius R. The major uncertainty in the particle widths arises from the last factor, the probability of finding the particle at the nuclear surface. That probability depends very much on the detailed nature of the nuclear state involved and hence on the potential inside the nuclear radius. It can only be calculated for certain specific models of a particle in a potential well. This uncertainty is usually concentrated into a dimensionless number, called the *dimensionless reduced width*, which turns out to vary in a rather predictable manner for nuclear states. The simplified *raison d'être* of the usual definition of the dimensionless reduced width is as follows.

The attractive nuclear force tends to produce the largest probability density near the center of the nucleus, although the combination of the exclusion principle plus orbital angular momentum makes that conclusion not strictly true. But for the most part one expects that $|\chi_l(R)|^2$ should not exceed its value for a uniform probability density. For uniform probability density,

$$\chi_l^*(R)\chi_l(R)\,dr = \frac{4\pi R^2\,dr}{\frac{4}{3}\pi R^3} = \frac{3}{R}\,dr \tag{4-118}$$

With this result as a guide, the dimensionless reduced width θ_l^2 is defined by

$$\chi_l^*(R)\chi_l(R) = \theta_l^2\,\frac{3}{R} \tag{4-119}$$

For various simple potentials that can be calculated it is found that θ_l^2 is something somewhat less than unity, and empirically it is found for nuclear states that usually

$$0.01 < \theta_l^2 < 1 \tag{4-120}$$

In a sense θ_l^2 turns out to be a measure of the degree to which the actual quasi-stationary nuclear state can be described by the relative motion of a and X in a potential. Even though the quantum numbers of the nuclear state may be correct, its actual wave function may bear little similarity to a bound state of particles a and X, in which case the reduced width is small.

Problem 4-37: Show that with the above definitions, the partial width of a state to particle decay can be written

$$\Gamma_l = \frac{3\hbar v}{R} P_l \theta_l^2 \tag{4-121}$$

The advantage of such an expression for Γ_l is that once the interaction radius R has been estimated, everything in Γ_l with the exception of θ_l^2 can be evaluated. Empirical rules from nuclear systematics and carefully chosen experiments can independently provide good estimates of θ_l^2 in many cases. The really useful feature is that the dependence of Γ_l on the energy, that is, v, can be made explicit, since the explicit dependence of P_l on v can be calculated. Before considering the effects of resonances due to nuclear states on the thermonuclear reaction rates, we shall want to calculate expressions for the penetration factors.

4-5 PENETRATION FACTORS

The wave function describing the relative motion of particles a and X with relative angular momentum l has been found to be expressible as

$$\psi_l(r\theta\phi) = \frac{\chi_l(r)}{r} Y_l^m(\theta,\phi) \tag{4-122}$$

where $\chi_l(r)$ satisfies the differential equation

$$\left[-\frac{\hbar^2}{2\mu}\frac{d^2}{dr^2} + \frac{l(l+1)\hbar^2}{2\mu r^2} + V(r) - E\right]\chi_l(r) = 0 \tag{4-123}$$

This second-order differential equation has the form of a wave equation

$$\frac{d^2\chi_l}{dr^2} + \frac{2\mu}{\hbar^2}[E - V_l(r)]\chi_l(r) = 0$$

where $V_l(r)$ is the effective radial potential for the lth partial wave:

$$V_l(r) = \begin{cases} \dfrac{l(l+1)\hbar^2}{2\mu r^2} + \dfrac{Z_1 Z_2 e^2}{r} & r > R \\[2ex] \dfrac{l(l+1)\hbar^2}{2\mu r^2} + V_c + V_{\text{nuc}} & r < R \end{cases}$$

This potential was shown schematically in Fig. 4-11. The calculation of the penetration factors demands solution of the radial wave equation to give the ratio

$$P_l = \frac{\chi_l^*(\infty)\chi_l(\infty)}{\chi_l^*(R)\chi_l(R)} \tag{4-124}$$

The exact solution of the radial wave equation is impossible because the nuclear potential is not well known. Fortunately, the penetration factors can be calculated without detailed knowledge of the nuclear force, because the equation is well defined for $r > R$. Therefore the equation defines precisely the ratio of

$|\chi_l(\infty)|^2$ to $|\chi_l(R)|^2$ although the absolute value of the wave function cannot be calculated without knowledge of the nuclear potential. The wave function for $r < R$ must be matched (value and derivative at $r = R$) to the wave function for $r > R$, so that the absolute value of $\chi_l(r)$ depends upon knowing the potential everywhere. As far as the penetration factor is concerned, one could pick any value of $\chi_l(R)$ and integrate Eq. (4-123) numerically from $r = R$ to $r = \infty$ to obtain the penetration factor. Such a program has been carried out by many investigators.

Accurate calculations of penetration factors are also derivable from the analytic solutions of the radial Schrödinger equation in a coulomb field. This method depends upon the fact that the two independent solutions to Eq. (4-123) are well known.[1] Those functions are called *coulomb wave functions* for angular momentum l or *coulomb partial waves*. The *regular* coulomb wave function is called $F_l(r)$, and it has the asymptotic values

$$\begin{aligned} &F_l(0) \to 0 \\ &F_l(r) \underset{r\to\infty}{\longrightarrow} \sin\left(kr - \frac{l\pi}{2} - \eta \ln 2kr + \sigma_l\right) \end{aligned} \tag{4-125}$$

It is the only allowed solution for problems in which the origin is not excluded since $\chi_l(0)$ must vanish. The other independent solution is called the *irregular* coulomb wave $G_l(r)$, and it has the asymptotic values

$$\begin{aligned} &G_l(0) \to \infty \\ &G_l(r) \underset{r\to\infty}{\longrightarrow} \cos\left(kr - l\frac{\pi}{2} - \eta \ln 2kr + \sigma_l\right) \end{aligned} \tag{4-126}$$

In these equations $k = p/\hbar$ is the wave number ($k = 2\pi/\lambda$), $\eta = Z_1Z_2e^2/\hbar v$, and σ_l is the argument of a complex gamma function:

$$e^{i\sigma_l}|\Gamma(l + 1 + i\eta)| = \Gamma(l + 1 + i\eta) \tag{4-127}$$

Since the origin is excluded in the present problem (the coulomb waves are solutions only for $r > R$), the solution for $r > R$ is a linear combination of regular and irregular solutions:

$$\chi_l(r) = AF_l(r) + BG_l(r) \tag{4-128}$$

Since the time dependence of the wave is $\exp[-i(E/\hbar)t]$, the demand that the linear combination represent an outgoing wave at infinity is equivalent to demanding $A = iB$. Only then is the asymptotic form of $\chi_l(r) = \exp(+ikr)$. It follows immediately that

$$P_l = \frac{|\chi_l(\infty)|^2}{|\chi_l(R)|^2} = \frac{1}{F_l^2(R) + G_l^2(R)} \tag{4-129}$$

[1] An exhaustive discussion of coulomb wave functions may be found in M. H. Hull and G. Breit, "*Handbuch der Physik*," S. Flügge (ed), vol. 41, p. 408, Springer-Verlag OHG, Berlin, 1959; also C. E. Fröberg, *Rev. Mod. Phys.*, **27**:399 (1955).

An exact solution of the problem is then to have recourse to tabulated values of F_l and G_l. This approach, though exact, would require a very extensive set of tables, since P_l is not only a continuous function of E but is parametrized in l, Z_1Z_2, μ, and R, the last of which cannot easily be prescribed exactly. Not only is such an approach tedious, but for discussions of this type, it is also not very illuminating. For bombarding energies much smaller than $V_l(R)$ it is more instructive to have approximate expressions for the penetration factors that show in functional form the explicit dependence upon all the variables and parameters of the problem. Fortunately, accurate approximations exist for $E \ll V_l(R)$. One approach is based upon an expansion of G_l^2 (which is much greater than F_l^2 for $E \ll V$) in terms of modified Bessel functions. These formulas are the ones used by Burbidge et al. in their historic paper.[1] A more illuminating approach pedagogically is based on the WKB technique (named for the inventors, Wentzel, Kramers, and Brillouin) for obtaining approximate solutions.

The WKB method is an approximate treatment of equations of the type

$$\frac{d^2y}{dx^2} \equiv y'' = -f(x)y \tag{4-130}$$

If $f(x)$ is a constant, the solution is obvious: it is a sinusoidal oscillation if $f(x)$ is positive and an exponential if $f(x)$ is negative. Even if $f(x)$ is not a constant, we expect solutions similar to sin $(\sqrt{f}\,x)$ or exp $(\sqrt{f}\,x)$ if $f(x)$ is a slowly varying function. Since the characteristic distance over which the approximate solution undergoes large changes is of the order $\Delta x \approx 1/\sqrt{f}$, the possibility of writing an approximate solution should be good if the change $\Delta f(x)$ over the distance Δx is much less than $f(x)$ itself. That is, we expect an approximate solution if

$$\Delta f(x) \approx f'(x)\,\Delta x \approx \frac{f'(x)}{\sqrt{f(x)}} \ll f(x) \tag{4-131}$$

With this thought in mind, let us seek a solution to Eq. (4-130) of the form $y(x) = e^{i\phi(x)}$. The sign of $f(x)$ has not been specified, so that $\phi(x)$ may be real, imaginary, or complex. Substitution of the trial solution into Eq. (4-130) yields

$$i\phi'' - (\phi')^2 + f(x) = 0 \tag{4-132}$$

If we are to obtain a solution, $\phi(x)$ must be slowly varying, so that we shall temporarily discard ϕ'' as being negligibly small, with the reservation that the validity of this assumption must eventually be checked when a solution is obtained. When this step is made, $\phi' \approx \pm\sqrt{f(x)}$ and $\phi'' = \frac{1}{2}f'/\sqrt{f}$. From Eq. (4-132) we see that the neglect of ϕ'' is valid only if $\phi'' \ll f(x)$, which is equivalent to the condition anticipated by Eq. (4-131).

The solution for $\phi(x)$ may be exactly specified by letting

$$\phi'(x) \equiv \pm\sqrt{f(x)} + \eta(x) \tag{4-133}$$

[1] E. M. Burbidge, G. R. Burbidge, W. A. Fowler, and F. Hoyle, *Rev. Mod. Phys.*, **29**:547 (1957).

where the additional function $\eta(x)$ represents whatever must be added to $\pm\sqrt{f(x)}$ in order to reproduce ϕ' exactly. The advantage of this representation is that $\eta(x)$ is very small if $f(x)$ is slowly varying, going to zero in the limit of constant $f(x)$. Of course, nothing has been gained in the search for an approximate solution if $\eta(x)$ is not small compared to $\sqrt{f}$, an assumption that must be checked after the solution is obtained. Differentiating a second time and substituting into Eq. (4-132) gives

$$\pm\frac{i}{2}\frac{f'}{\sqrt{f}} + i\eta' \mp 2\eta\sqrt{f} - \eta^2 = 0 \qquad (4\text{-}134)$$

If η is to be small compared to $\sqrt{f}$ and slowly varying as well, then $i\eta' - \eta^2$ may tentatively be neglected in Eq. (4-134). This neglect will be justified if

$$|\eta^2 - i\eta'| \ll \left|\frac{f'}{2\sqrt{f}}\right| \qquad (4\text{-}135)$$

Assuming momentarily that this condition is met, we see from Eq. (4-134) that

$$\eta \approx \frac{i}{4}\frac{f'}{f} \qquad (4\text{-}136)$$

Problem 4-38: Show from this last result that Eq. (4-135) is satisfied if

$$\left|-\frac{1}{4}\frac{f''}{f} + \frac{5}{16}\left(\frac{f'}{f}\right)^2\right| \ll \left|\frac{f'}{2\sqrt{f}}\right| \qquad (4\text{-}137)$$

The results of the previous problem provide the condition that must be satisfied if the approximate solution is to be valid. It is apparent that this condition will be met by well-behaved functions for which $f \gg f'$. However, Eq. (4-137) indicates clearly that the approximate solution will be of no value where $f(x) \to 0$. Without any of these difficulties, however, Eq. (4-133) may be integrated to give

$$\phi(x) = \pm\int^x \sqrt{f(z)}\, dz + i\ln|f(x)|^{\frac{1}{4}} \qquad (4\text{-}138)$$

The absolute-magnitude sign is used in the logarithm to render the integral of f'/f independent of the sign of f. Therefore by the original assumption

$$y = e^{i\phi} \approx A|f(x)|^{-\frac{1}{4}} \exp\left[\pm i\int^x \sqrt{f(z)}\, dz\right] \qquad (4\text{-}139)$$

A general solution consists of an arbitrary linear combination of terms containing positive and negative exponentials. The two arbitrary constants are to be established by boundary conditions.

Problem 4-39: What is the solution to the equation

$$y'' + e^{x/100}y = 0$$

with boundary conditions $y(0) = 1$ and $y'(0) = 0$, and over what range of x is the solution valid?

As was mentioned previously, the approximate solution becomes incorrect as $f(x)$ approaches zero. Equation (4-139) blows up, but this effect is spurious, since the result is invalidated by the violation of the requirement of Eq. (4-137) in the same region. It is often the case in problems of physical interest, however, that the function $f(x)$ passes through zero at some value $x = x_0$ but is sufficiently well behaved on both sides of x_0 so that the approximate solution can be validly written down in both regions except in the vicinity of x_0. On one side of x_0 the solution is approximately exponential, and on the other side of x_0 the solution is approximately sinusoidal. The situation is illustrated in Fig. 4-19.

The function $f(x)$ is slowly varying and is shown as changing sign at x_0. The curve labeled $y(x)$ is intended to represent the exact solution to the differential equation, whereas the dashed curves represent WKB approximations to the solutions in the two asymptotic regions. Both of these approximate solutions, $y_-(x)$ for $x \ll x_0$ and $y_+(x)$ for $x \gg x_0$, are adequate solutions far from x_0, but since both solutions are invalid near x_0, special considerations are required to determine which solution $y_-(x)$ is to be joined onto a given solution $y_+(x)$. The prescription for identifying the corresponding solutions in the two asymptotic regions was first discovered by Kramers. These are commonly called the *WKB con-*

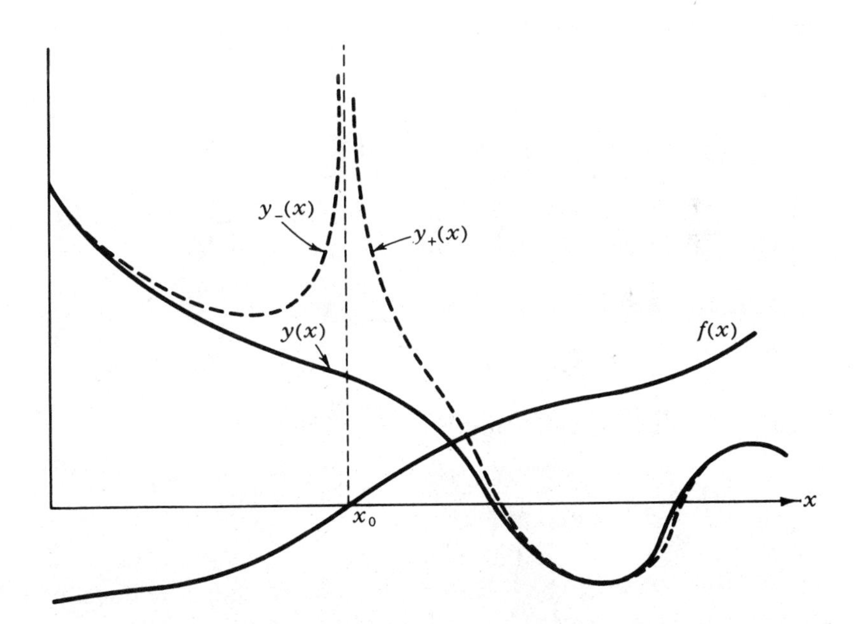

Fig. 4-19 The correct solution to the differential equation (4-130) is $y(x)$. The approximate WKB solutions are $y_-(x)$ and $y_+(x)$ for $f(x) < 0$ and $f(x) > 0$, respectively. The approximate solutions are quite good except near x_0. With knowledge of only two boundary conditions the difficult analytical problem is deciding which approximate solution y_- corresponds to the approximate solution y_+.

nection formulas. The way of establishing them and the types of solutions for which connections may validly be made are somewhat complicated and need not be discussed here.[1] Instead we shall quote the connection for the case of an outgoing wave at infinity. The solution

$$(i) \quad y_+(x) = Af(x)^{-\frac{1}{4}} \exp\left(i \int_{x_0}^{x} \sqrt{f}\, dx\right) \qquad \begin{matrix} x \gg x_0 \\ f(x) > 0 \end{matrix} \tag{4-140}$$

transforms to

$$(ii) \quad y_-(x) \approx Ae^{i\pi/4}|f(x)|^{-\frac{1}{4}} \exp\left[\int_{x}^{x_0} \sqrt{|f(x)|}\, dx\right. \qquad \begin{matrix} x \ll x_0 \\ f(x) < 0 \end{matrix} \tag{4-141}$$

Actually, if the first solution is a pure outgoing wave, the second solution is a linear combination of increasing and decreasing exponentials. But regardless of their relative admixture near the turning point x_0, the decreasing exponential is all that remains for $x \ll x_0$, whereupon the connection indicated by Eq. (4-141) becomes valid. This special result quickly gives a good approximation to the penetration factors, because the Schrödinger equation is of such a form as to be an appropriate candidate for the WKB method in the region $r > R$, where we have the equation

$$\chi_l''(r) + f(r)\chi_l(r) = 0 \tag{4-142}$$

with the function $f(r)$ identified as

$$f(r) = \frac{2\mu}{\hbar^2}\left[E - \frac{Z_1 Z_2 e^2}{r} - \frac{l(l+1)\hbar^2}{2\mu r^2}\right] \tag{4-143}$$

The function $f(r)$ changes sign at the classical turning point R_0, shown in Fig. 4-12. By the connection formulas just quoted we have

$$\chi_l(r \gg R_0) = A[E - V_l(r)]^{-\frac{1}{4}} \exp\left[i \int_{R_0}^{r} \sqrt{f(r)}\, dr\right] \tag{4-144}$$

$$\chi_l(R < r \ll R_0) = Ae^{i\pi/4}[V_l(r) - E]^{-\frac{1}{4}} \exp\left[\int_{r}^{R_0} \sqrt{-f(r)}\, dr\right] \tag{4-145}$$

Then the WKB value of the penetration factor is

$$P = \frac{\chi_l^*(\infty)\chi_l(\infty)}{\chi_l^*(R)\chi_l(R)} = \left[\frac{V_l(R) - E}{E}\right]^{\frac{1}{2}} \exp\left\{-\frac{2\sqrt{2\mu}}{\hbar}\int_{R}^{R_0} [V_l(r) - E]^{\frac{1}{2}}\, dr\right\} \tag{4-146}$$

A first step in evaluation of Eq. (4-146) is to introduce some notation and numerical values for the heights of the coulomb and centrifugal potentials at the nuclear surface. Let E_c be the height of the coulomb barrier:

$$E_c = \frac{Z_1 Z_2 e^2}{R} = 1.44(\text{Mev fm})\,\frac{Z_1 Z_2}{R} \tag{4-147}$$

[1] See, for instance, J. Mathews and R. Walker "Mathematical Methods of Physics," W. A. Benjamin, Inc., New York, 1964.

The quantity R, which we have called the nuclear radius, really represents the separation of the respective centers of mass of particles 1 and 2 at that point where the attractive nuclear force overcomes the repulsive barriers. This separation R is of necessity somewhat "fuzzy," but it is intuitively appealing and supported by experimental results to adopt R as being about equal to the sum of the radii of the two interacting nuclei. Data for the radii of nuclei come both from experimental nuclear reactions and from the scattering of relativistic electrons from nuclei. The effects in nuclear reactions indicate that the radius of any one nucleus is approximately given by

$$R_A = 1.4A^{\frac{1}{3}} \qquad \text{fm} \tag{4-148}$$

A good rule of thumb for the height of the coulomb barrier may be obtained by using this result to evaluate the coulomb energy of two touching spheres:

$$E_c \approx \frac{Z_1 Z_2}{A_1^{\frac{1}{3}} + A_2^{\frac{1}{3}}} \qquad \text{Mev} \tag{4-149}$$

Let E_l be the height of the centrifugal barrier at the nuclear surface. Then

$$E_l = \frac{l(l+1)\hbar^2}{2\mu r^2} = 20.9(\text{Mev fm}^2)\,\frac{l(l+1)}{AR^2} \tag{4-150}$$

Finally, the barrier energy E_B is defined to be the sum of these two potential barriers, $E_B \equiv E_c + E_l$. Using these definitions, the WKB solution for the penetration factors becomes

$$P_l = \left(\frac{E_B - E}{E}\right)^{\frac{1}{2}} \exp\left[-\frac{2\sqrt{2\mu}}{\hbar}\int_R^{R_0}\left(\frac{E_c R}{r} + \frac{E_l R^2}{r^2} - E\right)^{\frac{1}{2}} dr\right] \tag{4-151}$$

Clearly both E_c and E_l are of the order Mev. The energies of interest in stars, however, are of the order $E \approx E_0$, viz., Eq. (4-47), which is characteristically some tens of kilovolts in magnitude. Thus it is common in most astrophysical problems to neglect E in comparison to E_B in Eq. (4-151). In fact, it has been found that better agreement with exact computation is obtained by replacing $(E_B - E)^{\frac{1}{2}}$ with $E_c^{\frac{1}{2}}$. We make that replacement from here on. It will also be convenient to denote the exponent in Eq. (4-151) by $-W_l$, to which we now turn our attention.

(1) $l = 0$: The case of zero angular momentum is of interest because it is the simplest and because most reactions occur through $l = 0$ (s-wave) interactions when there is no resonance in the region of effective stellar energies. The dominance of s-wave interactions comes about because the penetration factor is smallest for $l = 0$.

Problem 4-40: Show by integrating with $l = 0$ that

$$W_0 = \frac{4Z_1Z_2e^2}{\hbar v}\left[\frac{\pi}{2} - \sin^{-1}\left(\frac{E}{E_c}\right)^{\frac{1}{2}} - \left(\frac{E}{E_c}\right)^{\frac{1}{2}}\left(1 - \frac{E}{E_c}\right)^{\frac{1}{2}}\right] \tag{4-152}$$

In cases of stellar interest, E/E_c is a small number. Equation (4-152), when expanded in terms of this ratio, becomes

$$W_0 = \frac{2\pi Z_1 Z_2 e^2}{\hbar v} \left[1 - \frac{4}{\pi} \left(\frac{E}{E_c} \right)^{\frac{1}{2}} + \frac{2}{3\pi} \left(\frac{E}{E_c} \right)^{\frac{3}{2}} \cdots \right] \tag{4-153}$$

Problem 4-41: The second term in Eq. (4-153) is independent of energy. Show that it is

$$-\frac{2\pi Z_1 Z_2 e^2}{\hbar v} \frac{4}{\pi} \left(\frac{E}{E_c} \right)^{\frac{1}{2}} = -4 \left(\frac{Z_1 Z_2 e^2}{R} \frac{2\mu R^2}{\hbar^2} \right)^{\frac{1}{2}}$$

$$= -0.88(AR^2 E_c)^{\frac{1}{2}} = -1.05(ARZ_1 Z_2)^{\frac{1}{2}} \tag{4-154}$$

where, as before, E_c is in Mev and R in fermis.

The first term of W_0 is just the exponent $bE^{-\frac{1}{2}}$ that was adopted without justification in the discussion of the nonresonant cross-section factors in Sec. 4-3. The second term, considered in the previous problem, vanishes in the limit as $R \to 0$ and may be thought of as a correction term due to nonzero radius in the penetration factors. It is physically clear that the size of the penetration factor must depend strongly upon the separation R to which the particles must penetrate. In the limit of vanishing R the particles would have to penetrate the complete potential, whereas only a small portion of the barrier must be penetrated for large values of R.

Problem 4-42: The correction term for nonzero radius gave the first good measurement of the radius of radioactive alpha emitters. It has been found that the dimensionless reduced width for heavy alpha emitters is near unity. The decay

$${}_{88}\mathrm{Ra}^{226} \to {}_{86}\mathrm{Rn}^{222} + \alpha$$

with $\tau_{\frac{1}{2}} = 1{,}620$ years and $E_\alpha = 4.78$ Mev, is such a decay. Show that the interaction radius of Rn^{222} with an alpha particle is $R = 10$ to 11 fm. (Because ground states of even-even nuclei have $J^\pi = 0^+$, we have $l = 0$.)

The third term in Eq. (4-153) is again energy-dependent. It represents a significant correction factor to the first term when the energy becomes a significant fraction of the coulomb barrier. Therefore

$$W_0 = -4 \left(\frac{Z_1 Z_2 e^2}{R} \frac{2\mu R^2}{\hbar^2} \right)^{\frac{1}{2}} + bE^{-\frac{1}{2}} \left[1 + \frac{2}{3\pi} \left(\frac{E}{E_c} \right)^{\frac{3}{2}} \right] \tag{4-155}$$

where we have once again used $b = 31.28 Z_1 Z_2 A^{\frac{1}{2}} (\mathrm{kev})^{\frac{1}{2}}$.

The s-wave particle width, on the other hand, is given by Eq. (4-121) as

$$\Gamma_0 = 6\theta_0{}^2 \left(\frac{\hbar^2}{2\mu R^2} E_c \right)^{\frac{1}{2}} \exp(-W_0) \tag{4-156}$$

If the correction in E/E_c is ignored, it is clear that the energy dependence is approximately

$$\Gamma_0 \propto \exp(-bE^{-\frac{1}{2}})$$

Nonresonant reactions discussed in Sec. 4-3 occur via the wings of broad s-wave resonances whose central energy is far from the bombarding energy. Thus nonresonant reaction rates are proportional to the s-wave particle width. It is now clear why the cross-section factor $S(E)$ was defined as in Eq. (4-37), i.e., in order to be a slowly varying function of energy. We note from Eq. (4-155) that this definition of $S(E)$ factors out almost the entire energy dependence as long as

$$\frac{2}{3\pi}\left(\frac{E}{E_c}\right)^{\frac{3}{2}} \ll 1$$

As the bombarding energy becomes a significant fraction of the coulomb barrier (a situation encountered in the late high-temperature stages of evolution of the stellar interior), the penetration factors take on a more complicated energy dependence:

$$\Gamma_0 \propto \exp\left[-bE^{-\frac{1}{2}}\left(1 + \frac{2}{3\pi}\frac{E}{E_c}\right)^{\frac{3}{2}}\right] \tag{4-157}$$

Some authors maintain that this more complete energy dependence should be employed for the definition of $S(E)$, so that it becomes advisable to make a mental note regarding the correction term.

(2) $l \neq 0$: The expression for W_l was found to be

$$W_l = \frac{2\sqrt{2\mu}}{\hbar}\int_R^{R_0}\left[E_c\frac{R}{r} + E_l\left(\frac{R}{r}\right)^2 - E\right]^{\frac{1}{2}} dr \tag{4-158}$$

The integration can be performed with the aid of some lengthy algebra.[1] The result is cumbersome, however, and can be easily interpreted only by an expansion. A much simpler approach which yields the leading terms is to expand the integrand before integration. Suppose that E_c is greater than E_l; both, of course, are much greater than E. Since the range of integration is $R < r < R_0$, we see that the ratio R/r is everywhere less than unity. In fact, since $R/R_0 \approx 10^{-3}$, the ratio R/r is quite small over most of the range of integration. As a consequence the second term in the square-root bracket never dominates, and the integrand may be expanded, whereupon the two leading terms become

$$W_l \approx \frac{2\sqrt{2\mu}}{\hbar}\int_R^{R_0}\left(E_c\frac{R}{r} - E\right)^{\frac{1}{2}} dr + \frac{\sqrt{2\mu}}{\hbar}\int_R^{R_0}\frac{E_l R^{\frac{3}{2}}}{E_c^{\frac{1}{2}}}\frac{dr}{r^{\frac{3}{2}}} \tag{4-159}$$

The first term is just equal to W_0, the exponent for $l = 0$, whereas the second term reflects the additional effects of the centrifugal barrier.

Problem 4-43: Show that Eq. (4-159) becomes

$$W_l = W_0 + 2\left[\frac{l(l+1)E_l}{E_c}\right]^{\frac{1}{2}}\left[1 - \left(\frac{E}{E_c}\right)^{\frac{1}{2}}\right]$$

[1] See, for instance, H. B. Dwight, "Tables of Integrals and Other Mathematical Data," formula 380.311, The Macmillan Company, New York, 1957.

The correction in $(E/E_c)^{\frac{1}{2}}$ may be neglected in this approximation, giving the following approximate expression for the penetration factor:

$$P_l \approx \left(\frac{E_c}{E}\right)^{\frac{1}{4}} \exp\left[-\frac{2\pi Z_1 Z_2 e^2}{\hbar v} + 4\left(\frac{E_c}{\hbar^2/2\mu R^2}\right)^{\frac{1}{2}} - 2l(l+1)\left(\frac{\hbar^2/2\mu R^2}{E_c}\right)^{\frac{1}{2}}\right] \qquad (4\text{-}160)$$

The correction of order $(E/E_c)^{\frac{3}{2}}$ to the first term in the exponent of W_0 has also been dropped. For numerical work one finds better agreement with more accurate calculations by replacing the product $l(l+1)$ with $(l+\frac{1}{2})^2$. Making this substitution and using nuclear numerical units, we have

$$P_l \approx \left(\frac{E_c}{E}\right)^{\frac{1}{4}} \exp\left[-bE^{-\frac{1}{2}} + 1.05(ARZ_1Z_2)^{\frac{1}{2}} - 7.62(l+\tfrac{1}{2})^2(ARZ_1Z_2)^{-\frac{1}{2}}\right] \qquad (4\text{-}161)$$

This approximation is adequate for most applications in nuclear astrophysics, although more accurate expressions can be obtained if the nuclear knowledge warrants the improvement. It can be seen that the penetration factor is a product of three exponential factors: (Gamow velocity factor) × (finite-radius factor) × (angular-momentum factor). The angular-momentum factor shows that for a given reaction

$$\frac{P_l}{P_0} \approx \exp\left[-2l(l+1)\left(\frac{\hbar^2/2\mu R^2}{E_c}\right)^{\frac{1}{2}}\right] = \exp\left[-7.62l(l+1)(ARZ_1Z_2)^{\frac{1}{2}}\right] \qquad (4\text{-}162)$$

The corresponding particle width from Eq. (4-121) is

$$\Gamma_l \approx 6\theta_l^2\left(\frac{\hbar^2}{2\mu R^2}E_c\right)^{\frac{1}{2}} \exp -W_l$$

$$= 3.33 \times 10^4\theta_l^2\left(\frac{Z_1Z_2}{AR^3}\right)^{\frac{1}{2}} \exp(-W_l) \qquad \text{kev} \qquad (4\text{-}163)$$

where R is again expressed in fermis. We have replaced the penetration factor with its approximate value based on the WKB method. The student should be aware, however, of the fact that other methods of computing P_l are in common use, in which case they may be used in Eq. (4-121).

Problem 4-44: The peak shown in Fig. 4-4 is due to a $J^\pi = \frac{1}{2}^+$ resonance in $C^{12} + p$ at 424 kev center-of-mass energy. The resonance has a full width at half maximum $\Gamma = 40$ kev. This width is essentially the proton width, since the only other channel is Γ_γ, which is much less than Γ_p. What is the value of the dimensionless reduced width θ_p^2 for that state?

4-6 MAXIMUM CROSS SECTION AND RESONANT REACTIONS

When resonances (quasistationary states in the compound nucleus) occur in the range of effective stellar energies, the stellar reaction rate is usually dominated by them. Resonant cross sections are many orders of magnitude greater than nonresonant cross sections at energies near the resonance, so that a resonance can dominate the reaction rate in spite of the required integration over the particle energy spectrum. At resonances the reaction cross section may approach the geometrical limit imposed by quantum mechanics. Since this limit increases linearly with the relative orbital angular momentum of the resonance, the resonant reaction rate may dominate even for large values of l required to form the resonant state. The geometrical limit comes about as follows.

The radial Schrödinger equation can be written

$$\left[\frac{d^2}{dr^2} + \frac{2}{r}\frac{d}{dr} - \frac{l(l+1)}{r^2} - \frac{2\mu}{\hbar^2}V(r) + k^2\right] f_l(r) = 0 \tag{4-164}$$

where $k^2 = 2\mu E/\hbar^2$. The solutions to this equation for a *free particle* ($V = 0$) are well known. They are

$$f_l(r) = j_l(kr) = \left(\frac{\pi}{2kr}\right)^{\frac{1}{2}} J_{l+\frac{1}{2}}(kr) \tag{4-165}$$

where the functions j_l are called *spherical Bessel functions*. They are related in the indicated manner to ordinary Bessel functions of half-integral order. Explicit forms for the first three solutions are

$$\begin{aligned} j_0(kr) &= \frac{\sin kr}{kr} \\ j_1(kr) &= \frac{\sin kr}{(kr)^2} - \frac{\cos kr}{kr} \\ j_2(kr) &= -\left[\frac{1}{kr} - \frac{3}{(kr)^3}\right]\sin kr - \frac{3}{(kr)^2}\cos kr \end{aligned} \tag{4-166}$$

Furthermore, all spherical Bessel functions have the asymptotic behavior

$$j_l(kr) \underset{r\to\infty}{\longrightarrow} \frac{\sin(kr - l\pi/2)}{kr} \tag{4-167}$$

Thus for a free particle moving with angular momentum $L^2 = l(l+1)\hbar^2$, $L_z = m\hbar$, and energy $E = (\hbar k)^2/2m$, the wave function is proportional to

$$\psi(r,\theta,\phi) \propto j_l(kr) Y_l^m(\theta,\phi) \tag{4-168}$$

If the direction of particle motion is defined to be the z direction, then $L_z = 0$, and

$$\psi(r,\theta) \propto j_l(kr) P_l(\cos\theta)$$

On the other hand, the free-particle wave equation could have been written in rectangular, rather than spherical, coordinates: $(\nabla^2 + k^2)\psi = 0$, with solution $\psi \propto e^{ikz}$. This solution is just the plane wave and contains all values of orbital angular momentum. Since both solutions are for free particles, and since the value of the wave function does not depend on the type of coordinate system used for description of the motion, it follows that a plane wave must be a well-defined sum of motions characterized by a definite l value. That is,

$$e^{ikz} = \sum_{l=0}^{\infty} A_l j_l(kr) P_l(\cos\theta) \tag{4-169}$$

where A_l are the coefficients that can be shown to be $A_l = (2l + 1)i^l$. Suppose further that the plane wave is examined at large values of r, that is, in the asymptotic region:

$$e^{ikz} \underset{r\to\infty}{\longrightarrow} \sum_{l=0}^{\infty} (2l + 1)i^l P_l(\cos\theta) \frac{\sin(kr - l\pi/2)}{kr}$$

$$= \frac{1}{2kr} \sum_{l=0}^{\infty} (2l + 1)i^{l+1} P_l(e^{-i(kr-l\pi/2)} - e^{i(kr-l\pi/2)}) \tag{4-170}$$

Accordingly, one sees that when viewed from the origin, a plane wave traveling in the z direction looks like a superposition of incoming and outgoing spherical waves for each value of l.

Now suppose a short-range potential exists about the origin. How can such a potential change the asymptotic form of Eq. (4-170)? Because the potential is assumed short range, it will not alter the incoming wave (the negative exponential) but can only affect the wave leaving the origin (the positive exponential). Thus the outgoing wave may be multiplied by a complex number $\tilde{\eta}_l$, where $|\eta_l|^2 \le 1$ is required in order that the outgoing flux not exceed the ingoing flux. Then,

$$\psi(r,\theta) = \frac{1}{2kr} \sum_{l=0}^{\infty} (2l + 1)i^{l+1} P_l(e^{-i(kr-l\pi/2)} - \tilde{\eta}_l e^{i(kr-l\pi/2)}) \tag{4-171}$$

The scattered wave is the difference between this wave function and the wave function in the absence of a potential:

$$\psi_{sc} = \psi - e^{ikz}$$

$$= \frac{1}{2kr} \sum_{l=0}^{\infty} (2l + 1)i^{l+1} P_l(1 - \tilde{\eta}_l) e^{i(kr-l\pi/2)} \tag{4-172}$$

Now from Eq. (4-112) the scattered flux is

$$J_{sc}(\theta) = \frac{\hbar}{2i\mu} (\psi_{sc}^* \nabla\psi_{sc} - \psi_{sc} \nabla\psi_{sc}^*) \tag{4-173}$$

The total rate of particle scattering, furthermore, is

$$N_{sc} = \int_0^{\pi} J_{sc}(\theta) r^2 2\pi \sin\theta \, d\theta \tag{4-174}$$

Since the only term in J_{sc} of order $1/r^2$ comes from the radial derivative acting on the exponential, the calculation of the rate of particle scattering is elementary.

Problem 4-45: Show that

$$N_{sc} = \frac{\hbar\pi}{\mu k} \sum_{l=0} (2l + 1)|1 - \tilde{\eta}_l|^2 \tag{4-175}$$

Recall that $\int_{-1}^{1} [P_l(u)]^2 \, du = 2/(2l + 1)$.

The scattering cross section is defined as the ratio of the number of particles scattered per unit time to the incident flux. Since the incident flux is $\hbar k/\mu$, the cross section is

$$\sigma_{sc} = \frac{\pi}{k^2}\sum_{l=0}^{\infty}(2l+1)|1-\tilde{\eta}_l|^2 = \pi\lambda\!\!\!^-{}^2\sum_{l=0}^{\infty}(2l+1)|1-\tilde{\eta}_l|^2 \tag{4-176}$$

The cross section contains a sum over all values of l. Each term is usually interpreted as a partial-wave cross section $\sigma_{sc,l}$, where

$$\sigma_{sc,l} = \pi\lambda\!\!\!^-{}^2(2l+1)|1-\tilde{\eta}_l|^2 \tag{4-177}$$

In like manner, the reaction cross section can be calculated. If a sphere of large radius is drawn about the origin, the number of reactions per unit time must equal the difference between the rate at which the incident particles enter the sphere and the rate at which they come back out; but this difference is just the total flux integrated over the sphere:

$$N_{in} - N_{out} = -\frac{\hbar}{2i\mu}\int_0^{\pi}\left(\psi^*\frac{\partial\psi}{\partial r} - \psi\frac{\partial\psi^*}{\partial r}\right)2\pi r^2 \sin\theta\, d\theta \tag{4-178}$$

where ψ is the complete wave function.

Problem 4-46: Show that the reaction cross section for the lth partial wave is

$$\sigma_{r,l} = \pi\lambda\!\!\!^-{}^2(2l+1)(1-|\tilde{\eta}_l|^2) \tag{4-179}$$

It is this last equation which contains the most important information. It is clear from its derivation that the restriction $|\tilde{\eta}_l|^2 \leq 1$ corresponds to the demand that the flux out of the bombarded sphere not exceed the flux into the sphere. With this restriction it is also evident that the maximum value of the l-wave reaction cross section is

$$\sigma_{r,l}(\max) = (2l+1)\pi\lambda\!\!\!^-{}^2 \tag{4-180}$$

This maximum occurs for $\tilde{\eta}_l = 0$, which clearly corresponds to a situation in which all the particles are absorbed and there is no outgoing wave.

The fact that the partial wave cross section increases with l does not mean that the physics of reactions favors large l but rather that there are more particles of large l in a plane wave. In fact the result can be understood in the following semiclassical way. Since the momentum of the incoming particles is $p = \hbar/\lambda\!\!\!^- = \hbar k$, it follows that the angular momentum is $L = bp = b\hbar/\lambda\!\!\!^-$, where b is the classical impact parameter. But since the angular momentum is restricted to the quantized values $L = l\hbar$, it follows that in a semiclassical picture the impact parameters are quantized; that is,

$$L = \frac{b\hbar}{\lambda\!\!\!^-} = l\hbar$$

$$b = l\lambda\!\!\!^-$$

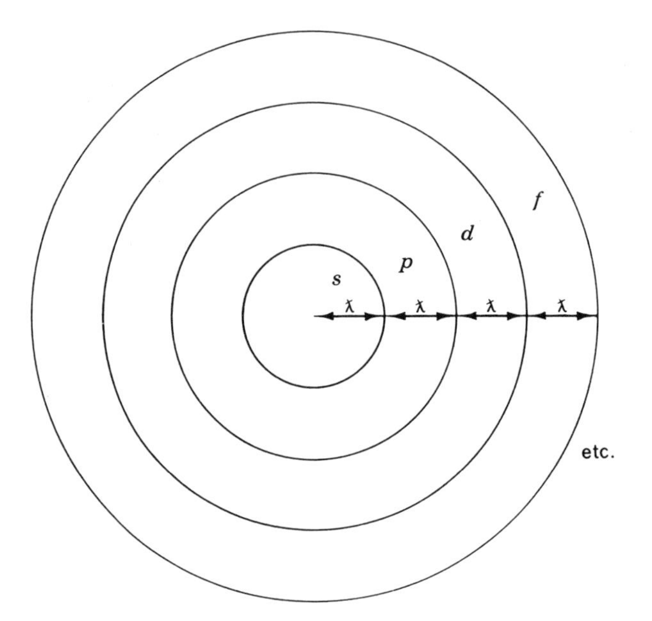

Fig. 4-20 A classical particle with an angular momentum $l\hbar$ has an impact parameter equal to $l\lambda\!\!\!^-$. In the quantum calculation the maximum cross section for the partial wave having angular momentum $l\hbar$ is equal to the area of the lth zone of this diagram. This correspondence is no more than a mnemonic device, however, because interactions having a given angular momentum are not in reality restricted to one specific zone.

The semiclassical interpretation is that the plane of impact parameters is divided into zones of differing values of l, as indicated in Fig. 4-20. The area of the lth zone, which is the maximum possible cross section if all particles react, is just equal to that of Eq. (4-180):

$$\sigma_{r,l}(\max) = \pi[(l+1)\lambda\!\!\!^-]^2 - \pi(l\lambda\!\!\!^-)^2 = (2l+1)\pi\lambda\!\!\!^-{}^2 \tag{4-181}$$

This maximum can be attained in the reaction

$$\sigma(a + X \rightarrow W \rightarrow b + Y)$$

only if the compound state is physically similar, both $a + X$ moving in a common potential and $b + Y$ moving in a common potential. For instance, the maximum cross section could be attained in the reaction

$$C^{12} + p \rightarrow N^{13} + \gamma$$

for an energy coinciding with a quasistationary state of N^{13} that is physically like a proton moving in the nuclear potential of a C^{12} nucleus.

However, every state has a natural width

$$\Gamma = \Gamma_p + \Gamma_n + \Gamma_\alpha + \Gamma_\gamma + \Gamma_\beta + \cdots$$

and the rate at which the state decays (once formed) to each mode is proportional to the partial width of that mode. The total width is associated with the fact that the energy of the state is indeterminate, the probability of the state's having energy E being given by Eq. (4-79). The rate of forming the state with particles of energy E must be proportional to the probability that the state has energy E, that is, proportional to $P(E)$. If the reaction is to be initiated by an incoming particle of type a, furthermore, the cross section must be proportional to Γ_a. A reaction occurs if the excited state breaks up *in any other channel* the rate of which is proportional to $\Gamma - \Gamma_a$. Thus the reaction cross section at resonance ($E = E_r$) must be proportional to

$$\sigma_r \propto \Gamma_a(\Gamma - \Gamma_a)P(E_r)$$

$$\propto \frac{\Gamma_a(\Gamma - \Gamma_a)}{(\Gamma/2)^2} \tag{4-182}$$

Problem 4-47: Show that the maximum occurs in Eq. (4-182) if $\Gamma_a = \Gamma/2$.

Since the maximum value of the factors in Eq. (4-182) is unity, they should be multiplied by the maximum value of the reaction cross section. Thus the reaction cross section for the lth partial wave is

$$\sigma_{r,l} = (2l + 1)\pi\lambdabar^2 \frac{\Gamma_a(\Gamma - \Gamma_a)}{(E - E_r)^2 + (\Gamma/2)^2} \tag{4-183}$$

The factor $\Gamma - \Gamma_a$ represents the sum of all of the partial widths for breakup of the compound state into a channel different from the one which formed this state. The reaction cross section for the $b + Y$ products is obtained by using only the partial width for that channel:

$$\sigma_{r,l}(a,b) = (2l + 1)\pi\lambdabar^2 \frac{\Gamma_a\Gamma_b}{(E - E_r)^2 + (\Gamma/2)^2} \tag{4-184}$$

This result is called the *Breit-Wigner single-level formula.* In the example of the resonance described at $E_p = 441$ kev in the special problem on $Li^7 + p$ (Prob. 4-35), we have $\Gamma_a = \Gamma_p$ and $\Gamma_b = \Gamma_\gamma$ for the $Li^7(p,\gamma)$ reaction and $\Gamma_b = \Gamma_\alpha = 0$ for the $Li^7(p,\alpha)$ reaction. The case in which the outgoing particle equals the incoming particle corresponds to *resonant scattering* rather than to a reaction. Although it might at first seem that the maximum reaction cross section would be unaccompanied by scattering, such is not the case. It is impossible in quantum mechanics to have a reaction without also having scattering. In fact, a comparison of Eq. (4-177) with Eq. (4-179) shows that the maximum reaction cross section, which occurs whenever $\tilde{\eta}_l = 0$, is accompanied by an equal scattering cross section.

Problem 4-48: Show that the maximum possible resonant-reaction cross section obtainable from the Breit-Wigner formula is accompanied by an equal resonant scattering cross section. In the case of the $E_p = 441$ kev resonance in $Li^7 + p$, the scattering cross section is 500 times the reaction cross section. One concludes from that fact that $\Gamma_p = 500\Gamma_\gamma$.

One simplification in Eq. (4-184) is that it has ignored the spins of the particles. Consider the example of a $J = \frac{3}{2}$ resonance formed by $l = 1$ protons interacting with C^{12}. The possible values of the total angular momentum that can be formed by $l = 1$ protons are

$$\mathbf{J} = \mathbf{J}(\mathrm{C}^{12}) + \mathbf{J}(p) + \mathbf{L}$$
$$= \mathbf{0} + \frac{\mathbf{1}}{\mathbf{2}} + \mathbf{1} = \frac{\mathbf{3}}{\mathbf{2}} \text{ or } \frac{\mathbf{1}}{\mathbf{2}}$$

Since the statistical weight of each J is $2J + 1$, it is evident that two-thirds of the $l = 1$ protons form $J = \frac{3}{2}$ states whereas one-third form $J = \frac{1}{2}$ states. If the resonance in question has $J = \frac{3}{2}$, the maximum possible cross section is two-thirds of Eq. (4-184), since only two-thirds of the $l = 1$ protons can be used. Thus, in general one writes

$$\sigma_l(1,2) = \pi\lambda\!\!\!^{-2}\omega \frac{\Gamma_1\Gamma_2}{(E - E_r)^2 + (\Gamma/2)^2} \tag{4-185}$$

$$\omega = \frac{2J + 1}{(2J_1 + 1)(2J_2 + 1)} \tag{4-186}$$

where J is the angular momentum of the resonance and J_1 and J_2 are the spins of particles 1 and 2.

Problem 4-49: Show that the numerical value of the geometrical cross-section factor is

$$\pi\lambda\!\!\!^{-2} = \frac{656.6}{AE} \quad \text{barns} \tag{4-187}$$

where E is the energy in kev. The fact that the omnipresent factor $\pi\lambda\!\!\!^{-2}$ is proportional to $1/E$ explains why the factor $1/E$ was removed from the cross section for the definition of $S(E)$ in Sec. 4-3.

4-7 RESONANT REACTION RATES IN STARS

The principles of resonances in nuclear reactions and of the particle widths have widespread application to the calculation of reaction rates in stellar interiors. These applications are of considerable importance to both stellar evolution and nucleosynthesis. In this section we demonstrate how knowledge of the properties of resonances greatly enlarges our capacity to calculate thermonuclear reaction rates.

REACTIONS IN THE WINGS OF RESONANCES

It often happens that it is impossible to measure a nonresonant cross section over a sufficiently wide range of energies to allow extrapolation of $S(E)$ experimentally to stellar energies. Such an unfortunate experimental situation could occur, for instance, if the reaction in the laboratory were accompanied by a large background of counts from undesired reactions that hopelessly contaminate the reaction of interest. It still is often possible to calculate S_0 from the knowledge that the reaction at stellar energies is proceeding by compound nuclear capture into the wing of a known resonance at higher energies. For instance, it is an easy matter for a nuclear physicist to argue that the $C^{12}(p,\gamma)N^{13}$ reaction in stars proceeds through the low-energy tail of the *s*-wave resonance at $E = 424$ kev (see Fig. 4-5) even though the effective stellar energy E_0 is only about 25 kev. A look at the N^{13} energy-level diagram shown in Fig. 4-21 reveals no other resonance in N^{13} anywhere near the energy of $C^{12} + p$, *so that the most likely nuclear structure for the proton to be captured into* is the structure of the state at 424 kev. The likelihood of this surmise is strengthened by the added knowledge that $J^\pi = \frac{1}{2}^+$ for the resonance, assuring that it is the $(2S_{\frac{1}{2}})$ state which will have a reasonably large $\theta_p{}^2$ for *s*-wave protons. Since the properties of the resonance can be measured at the energy of the resonance itself, much better than at low energy, these properties could be used to calculate a value of S_0 even if it happened that it was experimentally impossible to measure $S(E)$ at low energies.

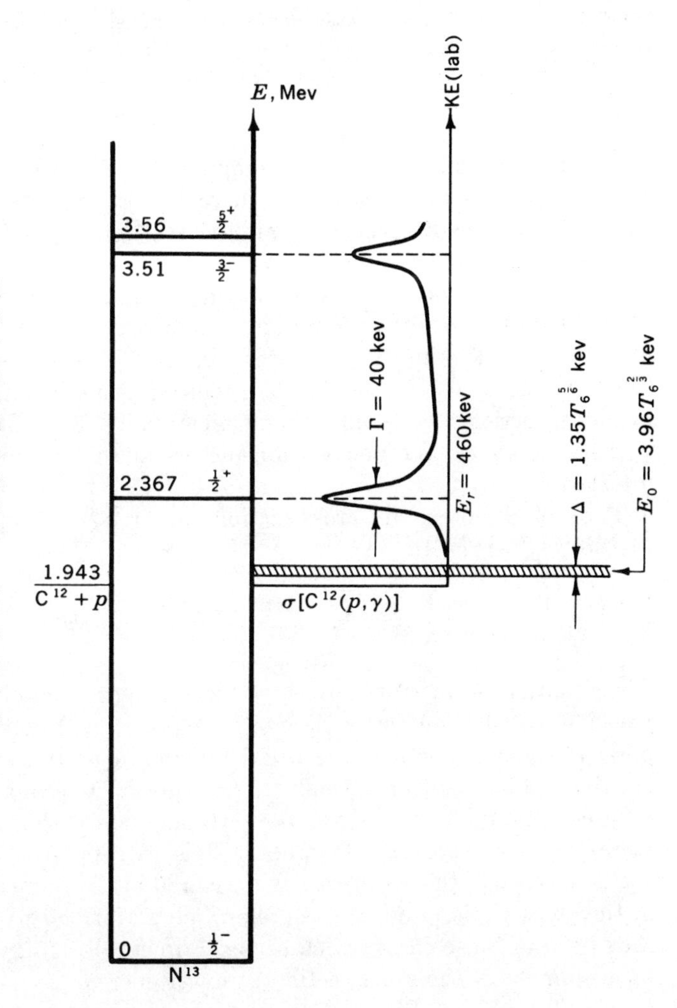

Fig. 4-21 An energy-level diagram of N^{13}, which serves as the compound nucleus for the reaction $C^{12}(p,\gamma)N^{13}$. The measured cross section for that reaction is displayed vertically just to the right of the discrete-level structure of N^{13}. The cross section is dominated at low energy by an *s*-wave resonance at $E_p = 460$ kev. Also shown as a hatched band is the range of effective stellar energies near E_0. At that energy the bombarding protons are captured into the wing of the resonance at 460 kev.

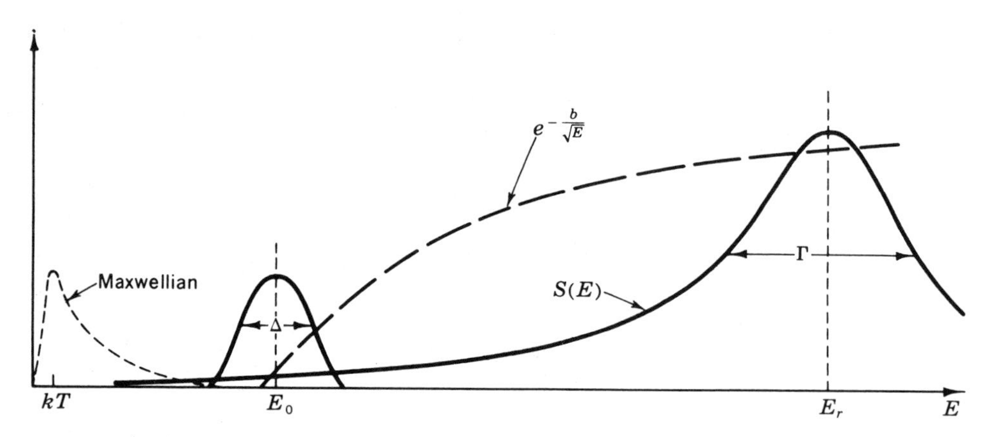

Fig. 4-22 A schematic representation of the major energy-dependent factors for a reaction, like that in Fig. 4-21, which proceeds through the wing of a broad distant resonance. In such a case the nonresonant-reaction formulation is used, and $S(E)$ is calculable by Eq. (4-188).

The extrapolation would be accomplished with the use of the Breit-Wigner single-level resonance formula, where for $C^{12}(p,\gamma)N^{13}$ we have $\Gamma_1 = \Gamma_p$ and $\Gamma_2 = \Gamma_\gamma$. In this reaction, a characteristic one, $\Gamma_2 = \Gamma_\gamma$ is a number independent of bombarding energy, whereas $\Gamma_1 = \Gamma_p$ is strongly dependent upon E through the penetration factor. That is, Γ_p varies markedly over the resonance, whereas Γ_γ is constant. Another common type of reaction is a (p,α) reaction, for example, $N^{15}(p,\alpha)C^{12}$. In these cases the reaction is usually strongly exothermic, such that $\Gamma_2 = \Gamma_\alpha$ is also independent of relatively small changes of the energy of the incident particle. The incident-particle width Γ_1 is once again a strong function of energy via the penetration factors at the low incident energies of stellar interiors.

The definition of the cross-section factor $S(E)$ leads, with the aid of Eqs. (4-185) to (4-187), to

$$S(E) = \frac{657}{A} \frac{\omega\Gamma_1(E)\Gamma_2}{(E - E_r)^2 + (\Gamma/2)^2} \exp\,(31.28Z_1Z_2A^{\frac{1}{2}}E^{-\frac{1}{2}}) \qquad \text{kev barns} \qquad (4\text{-}188)$$

The situation is illustrated in Fig. 4-22. The product of the Maxwell-Boltzmann velocity distribution times the factor $\exp\,(-bE^{-\frac{1}{2}})$ common to all the incident-particle widths produces the usual maximum at the energy E_0 in the a priori capability of causing reactions. The wing of a resonance occurring at quite a different energy than E_0 can be extrapolated to the energy E_0 by use of the Breit-Wigner single-level formula. This extrapolation may be made using only the properties of the resonance as measured in the immediate vicinity of the peak in the cross section and can, therefore, be performed with reasonably good accuracy even in those cases which are experimentally unfavorable to the laboratory measurement of the cross section at lower energy.

Problem 4-50: Confirm Eq. (4-188) and show that when the expression for Γ_1 is used, it reduces to

$$S_0 = 3.95 \times 10^3 \frac{\theta_1{}^2}{A} \left(\frac{\hbar^2}{2\mu R^2} E_c \right)^{\frac{1}{2}} \frac{\omega \Gamma_2}{(E_0 - E_r)^2 + (\Gamma/2)^2} \times \exp \left[4 \left(\frac{E_c}{\hbar^2/2\mu R^2} \right)^{\frac{1}{2}} - (l + \tfrac{1}{2})^2 \left(\frac{\hbar^2/2\mu R^2}{E_c} \right)^{\frac{1}{2}} \right] \text{ kev barns} \quad (4\text{-}189)$$

Problem 4-51: Calculate S_0 for the $C^{12}(p,\gamma)N^{13}$ reaction from Eq. (4-189) using the value of $\theta_1{}^2 = \theta_p{}^2$ computed in Prob. 4-44 and $\Gamma_\gamma = 0.77$ ev. Assume $E_0 = 24$ kev for this calculation. Compare this value of S_0 with the measured one shown in Fig. 4-5. Although the percentage difference between the two values of S_0 is appreciable, only a small adjustment of the temperature would be needed to produce equal reaction rates from these two values of S_0.

The calculation of the previous problem has indicated the manner in which S_0 can be calculated from the parameters of a resonance when there is confidence that the reaction is occurring in the wing of the resonance. The curve through the points of Fig. 4-5 is a theoretical one computed from an equation like (4-189) with, however, one additional theoretical parameter that allows the points to be fitted by a resonance curve.[1]

RESONANCES IN THE RANGE OF STELLAR ENERGIES

In some cases the compound nucleus provides a resonance in the range of effective stellar energies. Just what the range of stellar energies is, is somewhat hard to define. It is clear that if the resonance is very far from E_0, the reaction must proceed through the wings of the resonance, and the nonresonant-reaction-rate formalism must be used. For resonances in the vicinity of E_0, on the other hand, the full height of the resonant cross section must be employed.

Performing the following thought experiment will illustrate that a different type of calculation is required for resonances in the vicinity of E_0. Imagine that the resonance energy E_r in Fig. 4-22 can be continuously lowered toward E_0. What changes come about? The first thing that comes to mind is that the value of S_0 is increased as the two peaks come nearer. Although that conclusion is correct, it is hardly the whole story. Equally important effects come about as a result of the shrinkage of the incident-particle width Γ_1 at the resonance energy and the translation of the main peak of the resonance up the high-energy tail of the Maxwell-Boltzmann energy distribution. A point is eventually reached, thanks to the great size of resonant cross sections at their peak, where the product of the area under the resonant cross section times the Maxwell-Boltzmann probability at E_r exceeds the product of the cross section in the wing of the resonances times the Maxwell-Boltzmann probability integrated over the peak at E_0. At energies lower than this, more reactions occur into the main peak of the resonance than occur near E_0. The total width of the resonance is usually very small for resonances near E_0. Suppose, for example (as is common), that

[1] See D. F. Hebbard and J. L. Vogl, *Nucl. Phys.*, **21**:652 (1960).

the reaction is a (p,γ) reaction and that the low-energy resonance can only break up into two channels $\Gamma_1 = \Gamma_p$ and $\Gamma_2 = \Gamma_\gamma$. Then the total width of the resonance would be $\Gamma = \Gamma_p + \Gamma_\gamma$. Gamma widths are usually on the order of electron volts, which are, therefore, usually small compared to particle widths *if the breakup energy of the state is not extremely low*. At very low energies, however, the penetration factors reduce Γ_p markedly, until it becomes much smaller than the already narrow gamma width. This type of situation often occurs, and its end result is that the resonance at low energy is very narrow. [This statement is invalid if other exothermic particle channels are open, such as (p,α) reactions, for instance.] Thus the resonance becomes a narrow spike which interacts strongly with particles only in a narrow window of the Maxwell-Boltzmann distribution.

Problem 4-52: Suppose there had been a resonance with $J^\pi = \frac{3}{2}^-$ in $C^{12} + p$ at $E_r = 50$ kev with a dimensionless reduced width $\theta_p^2 = 0.2$. What would be the value of the proton width Γ_p at the resonance? It is clear that the total width of such a state would be given by Γ_γ.

To see how the correct calculation should be made, it is helpful to return to the initial expression for the reaction rate,

$$r_{10} = N_1 N_0 \int_0^\infty \psi(E) v(E) \sigma(E)\, dE \tag{4-190}$$

where the maxwellian energy distribution is

$$\psi(E)\, dE = \frac{2}{\sqrt{\pi}} \frac{E}{kT} \exp\left(-\frac{E}{kT}\right) \frac{dE}{(kTE)^{\frac{1}{2}}} \tag{4-191}$$

The circumstance that makes the integral easy to perform is the narrow-width Γ of the low-energy resonances. Since Γ will, in most cases, be about 1 ev, and since $\psi(E)$ and $v(E)$ change only minutely over an energy range corresponding to Γ, the values of $\psi(E)$ and $v(E)$ in the integral may be replaced by their values at resonance:

$$r_{10} \approx N_1 N_0 \psi(E_r) v(E_r) \int_0^\infty \sigma(E)\, dE \tag{4-192}$$

The value of $\int \sigma(E)\, dE$ is easy to compute if we once more assume that because of the narrowness of the resonance, λ^2 and Γ_1 may be replaced by their values at E_r.

Problem 4-53: Show that with the assumptions of the previous paragraph

$$\int \sigma(E)\, dE = 2\pi^2 \lambda_r^2 \omega f \frac{\Gamma_1 \Gamma_2}{\Gamma} \tag{4-193}$$

Notice that the area under the cross-section curve is $(\pi/2)\Gamma$ times the peak value of the cross section. The factor f again represents an enhancement due to the collective effects of the dense electron gas in stars and will be discussed shortly.

Pulling the factors together gives

$$r_{10} = N_1 N_0 \frac{2}{\sqrt{\pi}} E_r^{\frac{1}{2}} (kT)^{-\frac{3}{2}} \exp\left(-\frac{E_r}{kT}\right) \sqrt{\frac{2E_r}{\mu}}\, 2\pi^2 \lambda\!\!\!^{-}{}_r^2 \omega f \frac{\Gamma_1 \Gamma_2}{\Gamma}$$

$$= N_1 N_0 \hbar^2 \left(\frac{2\pi}{\mu kT}\right)^{\frac{3}{2}} \omega f \frac{\Gamma_1\Gamma_2}{\Gamma} \exp -\frac{E_r}{kT} \qquad (4\text{-}194)$$

$$r_{10} = 8.10 \times 10^{-12} N_1 N_0 \frac{\omega f}{(AT_6)^{\frac{3}{2}}} \frac{\Gamma_1\Gamma_2}{\Gamma} \exp\left(-11.61 \frac{E_r}{T_6}\right) \qquad \text{cm}^{-3}\ \text{sec}^{-1} \qquad (4\text{-}195)$$

$$r_{10} = 2.94 \times 10^{36} \rho^2 \frac{X_1 X_0}{A_1 A_0} \frac{\omega f}{(AT_6)^{\frac{3}{2}}} \frac{\Gamma_1\Gamma_2}{\Gamma} \exp\left(-11.61 \frac{E_r}{T_6}\right) \qquad \text{cm}^{-3}\ \text{sec}^{-1} \qquad (4\text{-}196)$$

where in the last two of these equations, the widths Γ are in units of kev, as is E_r, and the temperature T_6 is measured in millions of degrees Kelvin.

Problem 4-54: Show that the resonant reaction rate varies with temperature as T^n, where

$$n = \frac{11.61 E_r}{T_6} - \frac{3}{2}$$

Problem 4-55: Return to Prob. 4-52, where it was imagined that there is a $\frac{3}{2}^-$ resonance in the $C^{12} + p$ system at $E_r = 50$ kev having a dimensionless reduced width $\theta_p^2 = 0.2$. Assume further that at that resonance, $\Gamma_\gamma = 1$ ev. Under these circumstances, what would be the lifetime of a C^{12} nucleus against proton capture in a stellar interior containing 80 percent H by weight, having a density of 15 g/cm³, and with a temperature of 30×10^6 °K? It was found in an earlier problem in this chapter that under the same conditions the lifetime of C^{12} is about 160 years for the nonresonant conditions that actually correspond to the experimental facts for $C^{12} + p$. What this present problem demonstrates is how the effect of resonances can be handled if θ_1^2 can be determined and that lifetimes may be vastly shortened by resonances occurring near E_0. (For this problem again set $f = 1$.)

From the previous problem it will be clear that in those light-element reactions of importance to stellar evolution one must locate all resonances near the range of effective stellar energies. This investigation must often be performed with the aid of a reaction different from the one of interest in stars. An example can be seen in the energy-level diagram of N^{14} shown in Fig. 4-23, which can conveniently be discussed within the historical context. The nucleus N^{14} serves as the compound nucleus for the reaction $C^{13}(p,\gamma)N^{14}$, which participates in the CN cycle of hydrogen burning. The cross section for the reaction could be measured in the laboratory to energies slightly below 100 kev, yielding nonresonant behavior quite similar to that shown in Fig. 4-5. The corresponding $S(E)$ could be extrapolated downward to E_0 to obtain the nonresonant reaction rate, but that rate would be quite inappropriate if there existed a resonance in N^{14} between the mass of $C^{13} + p$ and the first 90 kev of laboratory proton energy. That region of N^{14} was not explorable by $C^{13} + p$ directly because of the small cross section at such low bombarding energy, and so it had to be explored by other

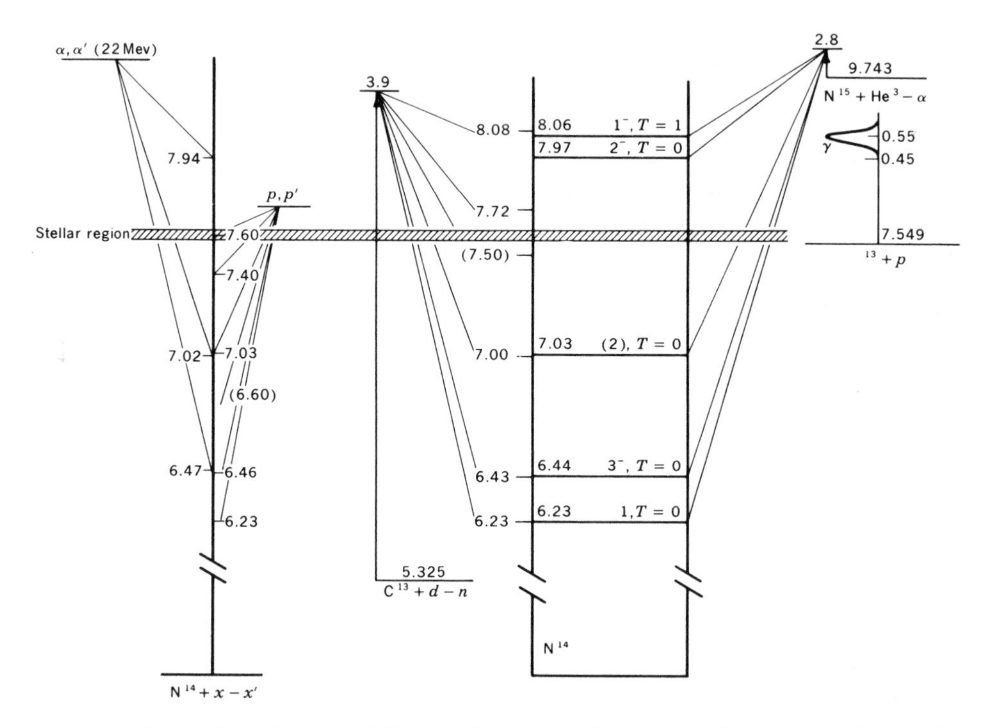

Fig. 4-23 Energy-level diagram of N^{14}, which serves as the compound nucleus for the reaction $C^{13}(p,\gamma)N^{14}$. It was necessary to perform the reaction $N^{15}(He^3,\alpha)N^{14}$ to see whether the N^{14} nucleus possessed undetected states near the mass of $C^{13} + p$ which would provide resonances near the effective stellar energy for the reaction $C^{13}(p,\gamma)N^{14}$. The existence of such states had erroneously been suggested by inelastic-scattering experiments and by the reaction $C^{13}(d,n)N^{14}$, as shown on the left. Such laboratory experiments are an essential part of the calculation of thermonuclear reaction rates. [*D. D. Clayton, Phys. Rev.*, **128**:2254 (1962).]

reactions. The inelastic scattering of protons from C^{13} had indicated the existence of N^{14} states at 7.40 and 7.60 Mev of excitation, the latter falling directly in the region of interest. To check for the existence of those states with higher accuracy, the experiment $N^{15}(He^3,He^4)N^{14}$ was performed. It found no evidence of those states, whereupon the inelastic proton scattering was repeated with greater precision and was found also not to have truly indicated such states. With this assurance, the nonresonant treatment of Sec. 4-3 could be safely applied. This parable of nuclear astrophysics illustrates the close tie of the nuclear laboratory to the astrophysical theory.

One can easily see that Eqs. (4-194) to (4-196) have two limiting cases of considerable interest. The different cases correspond to the possibilities for the ratio $\Gamma_1\Gamma_2/\Gamma$, which is common to the three expressions. The most frequent situation is that of having only two possible channels of decay for the resonance, that is, $\Gamma = \Gamma_1 + \Gamma_2$. Then the two limiting cases are:

$$(i) \quad \Gamma_2 \gg \Gamma_1\colon \quad \frac{\Gamma_1 \Gamma_2}{\Gamma} \to \Gamma_1$$

$$(ii) \quad \Gamma_1 \gg \Gamma_2\colon \quad \frac{\Gamma_1 \Gamma_2}{\Gamma} \to \Gamma_2$$

In case (*i*) the reaction rate is independent of Γ_2. The physical reason for this fact is clear. Each time the state is formed, a reaction occurs, since Γ_2 (the reaction channel) is much greater than Γ_1 (the scattering channel). Thus the reaction rate depends only on the rate for forming the resonant state. That rate is proportional to Γ_1. Inserting the value of Γ_1 from (4-163) into Eq. (4-195) shows that for the case (*i*) limit:

(*i*) $\Gamma_2 \gg \Gamma_1$:

$$r_{10} = 2.70 \times 10^{-7} N_1 N_0 \left(\frac{Z_1 Z_0}{R^3}\right)^{\frac{1}{2}} \frac{\omega}{A^2} \theta_1^2 f T_6^{-\frac{2}{3}} \exp\left(-W_l - 11.61 \frac{E_r}{T_6}\right) \tag{4-197}$$

$$r_{10} = 9.80 \times 10^{40} \rho^2 \frac{X_1 X_0}{A_1 A_0} \left(\frac{Z_1 Z_0}{R^3}\right)^{\frac{1}{2}} \frac{\omega}{A^2} \theta_1^2 f T_6^{-\frac{2}{3}} \exp\left(-W_l - 11.61 \frac{E_r}{T_6}\right) \tag{4-198}$$

The applications of these formulas for the reaction rate are usually found with resonances in the lowest energy range where penetration factors reduce the incident-particle width to a value much below the natural width of the state. This situation is the usual one.

Case (*ii*), on the other hand, is valid only for resonances of energy sufficiently high for the incident-particle width still to dominate the natural width of the state; i.e., the state predominantly decays by the reemission of the incident particle. For this limit, we have:

(*ii*) $\Gamma_1 \gg \Gamma_2$:

$$r_{10} = 8.10 \times 10^{-12} N_1 N_0 \frac{f \omega \Gamma_2}{(A T_6)^{\frac{3}{2}}} \exp\left(-11.61 \frac{E_r}{T_6}\right) \tag{4-199}$$

$$r_{10} = 2.94 \times 10^{36} \rho^2 \frac{X_1 X_0}{A_1 A_0} \frac{f \omega \Gamma_2}{(A T_6)^{\frac{3}{2}}} \exp\left(-11.61 \frac{E_r}{T_6}\right) \tag{4-200}$$

An observation of significance for the limit of case (*ii*) can be established by forgoing the numerical expressions and returning to Eq. (4-194), simultaneously introducing $\Gamma_2 = \hbar/\tau_2$. Then

$$(ii) \quad r_{10} = \frac{1}{\tau_2} N_1 N_0 \frac{h^3}{(2\pi\mu kT)^{\frac{3}{2}}} \omega f \exp - \frac{E_r}{kT} \tag{4-201}$$

Now let N_{10}^* be the number density *of excited nuclei existing in the resonant state of particles* 1 *and* 0. It then follows that the number of reactions per cubic centimeter per second must also be given by the formula

$$r_{10} = \frac{N_{10}^*}{\tau_2} \tag{4-202}$$

Equations (4-201) and (4-202) can both be correct only if

$$N_{10}^* = N_1 N_0\, \omega f \frac{h^3}{(2\pi\mu kT)^{3/2}} \exp - \frac{E_r}{kT} \tag{4-203}$$

Equation (4-203) is reminiscent of the Saha ionization equation. If we view N_{10}^* as a bound state of 1 and 0, then the binding energy is negative: $\chi_r = -E_r$. The laboratory value of E_r, valid at low density, needs modification for electron screening. The coulomb interactions in the gas change the effective energy of the excited state to $E_r + U_0$, where U_0 is the interaction energy (negative) of the two shielding charge clouds, and so the binding energy (negative) at high density is $\chi_r = -(E_r + U_0)$. Since, as we will show in the next section, the electron shielding factor is $f = \exp(-U_0/kT)$, the two exponentials can be combined. By further introducing the explicit expression for the statistical factor ω, Eq. (4-203) can be reshuffled to read

$$\frac{N_1 N_0}{N_{10}^*} = \frac{(2S_1 + 1)(2S_0 + 1)}{2S_{10} + 1} \frac{(2\pi\mu kT)^{3/2}}{h^3} \exp - \frac{\chi_r}{kT} \tag{4-204}$$

which is exactly Saha's equation, written in this case for a completely different process than the ionization of an atom. Naturally, the physical nature of the binding force nowhere appeared in the considerations leading to Saha's equation, which is statistical in its content. It is really not surprising, then, that in this case (*ii*) an equation of the same type is obtained. The condition $\Gamma_2 \ll \Gamma_1$ ensures that the decay of the compound state into the reaction channel occurs so seldom that the process of formation and dissolution of N_{10}^* into 1 and 0 is essentially uninfluenced by the possibility of the reaction. Viewed this way, it is clear that the population of the resonant state should be determined by statistical factors alone.

Problem 4-56: The ground state of Be^8 with spin and parity $J^\pi = 0^+$ is unstable by 94 kev against breakup into two alpha particles. In a stellar interior composed entirely of helium at a density of 10^5 g/cm^3 and a temperature of 10^8 °K, what is the number of Be^8 nuclei per cubic centimeter? (Ignore the coulomb interactions of the gas.)

The previous problem is an example of the fact that some reaction rates of interest in stars can be computed from applications of statistical mechanics. One need only ascertain that the conditions are suitable for equilibrium. It should be clear that an equilibrium calculation is entirely inappropriate to case (*i*) because a reaction occurs every time the resonance is formed. Under such conditions there can clearly be no equilibrium between absorption and reemission. But whenever equilibrium does apply, it can be very useful. The most important application of this technique is to the reaction $3He^4 \rightarrow C^{12}$, where the small equilibrium concentration of Be^8 nuclei serves as a target for the capture of the third alpha particle. A generalization from two particles to many particles provides the calculational tool for nuclear statistical equilibrium, which will be discussed in Chap. 7.

Finally, it should be mentioned that in the medium and heavy nuclei the density of resonances becomes large, and all reactions are resonant. Many resonances contribute to the thermal reaction cross section. In analogy with Eq. (4-194) for a single resonance we have

$$\langle \sigma v \rangle = \left(\frac{2\pi\hbar^2}{\mu kT}\right)^{\frac{3}{2}} \sum_r \left(\frac{\omega\Gamma_1\Gamma_2}{\hbar\Gamma}\right)_r \exp - \frac{E_r}{kT} \tag{4-205}$$

where the sum over r designates a sum over the resonances in the compound nucleus. In some cases the properties of the contributing resonances can be measured in the laboratory. In other cases one can determine only the density of resonances, in which case one often expresses the average cross section in terms of the level density of compound nuclear states. The student may find such treatments in the literature.[1]

4-8 ELECTRON SHIELDING

Thermonuclear reaction rates in stars are increased over their laboratory analogs because of the presence of the dense electron gas. The net negative charge surrounding each nucleus reduces the coulomb repulsion to a value smaller than $Z_1Z_2e^2/r$. This reduction makes the penetration of the coulomb barrier easier, which in turn increases the cross section in comparison with the cross section between bare nuclei having the same relative velocity at infinity. In the previous material, the cross section was evaluated as if there were a pure coulomb law between Z_1 and Z_2. The present section discusses briefly this modification of the reaction rates.[2]

Each nucleus, even though completely ionized, attracts neighboring electrons somewhat. The subject of the ionized real gas was introduced in Sec. 2-3. There will exist, on the average, some sphere around each nucleus Z which contains enough negative charge to neutralize the cloud. This sphere should not be thought of as containing just Z free electrons, for it will usually contain other positive nuclei with sufficient electrons to neutralize them as well (the Debye-Hückel ion sphere). Only if the average coulomb energy between neighboring particles is greater than kT (usually not the case) will the cloud tend to reduce to Z free electrons. When two nuclei Z_1 and Z_2 approach each other, the shielding charge density introduces a perturbing potential on the coulomb one. We write the total coulomb interaction energy as

$$U_{\text{tot}}(r_{12}) = \frac{Z_1Z_2e^2}{r_{12}} + U(r_{12}) \tag{4-206}$$

where $U(r_{12})$ obviously represents the added interaction due to shielding.

The shielding cloud must be at least as large as the average interparticle dis-

[1] Particularly recommended to students who have mastered the principles of this chapter is W. A. Fowler and F. Hoyle, *Astrophys. J. Suppl. Ser.*, **9**:201 (1965), app. C.

[2] The discussion is based upon a paper by E. E. Salpeter, *Australian J. Phys.*, **7**:373 (1954).

tance and perhaps much larger, and $U(r_{12})$ will change by large amounts only over distances on the order of the radius of the shielding cloud. Equation (4-146), on the other hand, shows explicitly that the penetration factors depend upon the integral of $(U_{\text{tot}} - E)^{\frac{1}{2}}$ between the classical turning point R_0 and the nuclear radius R. Since reactions are most favored for the energy E_0, a characteristic classical turning radius is of the order

$$R_0 \approx \frac{Z_1 Z_2 e^2}{E_0} \tag{4-207}$$

Problem 4-57: Calculate R_0 for the $C^{12}(p,\gamma)N^{13}$ reaction at $T_6 = 30$ and compare to average interparticle distances near $\rho = 50$.

This turning radius is usually much less than the radii of the shielding clouds. Thus the shielding interaction $U(r_{12})$ must be essentially constant over the relevant range of interparticle distances for which both particles are near the center of the shielding clouds. To good approximation $U(r_{12})$ in the penetration factor can be replaced by U_0, the shielding potential at the origin. The potentials are shown schematically in Fig. 4-24.

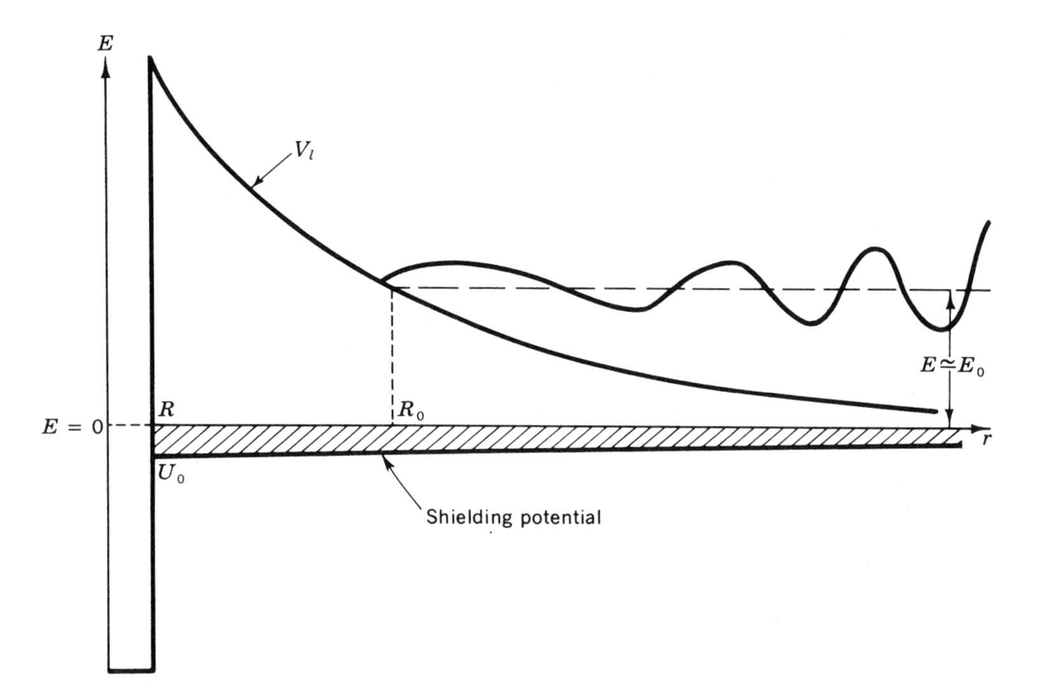

Fig. 4-24 The effective radial potential V_l modified by the screening potential. The polarization of the electron-ion plasma results in a small attractive potential, which is here drawn beneath the $E = 0$ axis. This small negative potential has the effect of reducing V_l and thereby increasing the penetrability of the barrier.

The reaction-rate integral is

$$r_{12} = N_1N_2 \int_0^\infty \psi(E)v(E)\sigma(E)\,dE \tag{4-208}$$

where $\psi(E)$ is the probability that the center-of-mass kinetic energy at large separation is E. The cross section $\sigma(E)$ can be written as the product of a penetration factor $P(E)$ times a nuclear factor $\sigma_{\text{nuc}}(E)$:

$$\sigma(E) = P(E)\sigma_{\text{nuc}}(E) \tag{4-209}$$

The s-wave penetration factor is, for instance,

$$P_0(E) = \left(\frac{E_c}{E}\right)^{\frac{1}{2}} \exp\left[-\frac{2\sqrt{2\mu}}{\hbar}\int_R^{R_0}\left(\frac{Z_1Z_2e^2}{r} + U_0 - E\right)^{\frac{1}{2}} dr\right] \tag{4-210}$$

The potential U_0 has the effect of making the penetration factor in a star for energy E equal to the laboratory penetration factor for the energy $E - U_0$. Since U_0 is negative, the appropriate penetration factor is that for a slightly increased energy. By the same token, the nuclear part of the cross section is that for the energy $E - U_0$, since the potential $U(r_{12})$ has effectively increased the kinetic energy by the magnitude of U_0. Thus the corrected form of Eq. (4-208) is

$$r_{12} = N_1N_2 \int_0^\infty \psi(E)v(E)P(E - U_0)\sigma_{\text{nuc}}(E - U_0)\,dE \tag{4-211}$$

$$r_{12} \propto \int_0^\infty E^{\frac{1}{2}}e^{-E/kT}P(E - U_0)\sigma_{\text{nuc}}(E - U_0)\,dE \tag{4-212}$$

The simplest way of relating this result to the reaction rate computed without consideration of electron screening is to define a new energy variable $E' = E - U_0$. Then

$$r_{12} \propto \int_{-U_0}^\infty (E' + U_0)^{\frac{1}{2}}e^{-(E'+U_0)/kT}P(E')\sigma_{\text{nuc}}(E')\,dE' \tag{4-213}$$

In practice, it turns out that $-U_0$ is small compared to the most effective energy E_0 (but not necessarily small compared to kT). Thus $(E' + U_0)^{\frac{1}{2}} \approx (E')^{\frac{1}{2}}$. The penetration factor, furthermore, is so small for $E' = -U_0$ that the lower limit of the integral can be extended to zero without changing the value of the integral significantly. It follows that the corrected reaction rate is approximately

$$r_{12} \approx e^{-U_0/kT}N_1N_2 \int_0^\infty \psi(E)v(E)\sigma(E)\,dE \tag{4-214}$$

so that electron shielding has increased the rate by the factor

$$f = e^{-U_0/kT} \tag{4-215}$$

Such a simple form results from the fact that the shielding potential increases the energy at small distances relative to that at large separations by the amount $-U_0$. This energy enters the rate exponentially because the particles involved are on the high-energy tail of the maxwellian distribution. Thus the calculation of the effect of electron screening on thermonuclear reaction rates is reduced to the computation of U_0, the value of the shielding interaction energy at the origin.

Most thermonuclear reactions in stars occur at sufficiently high temperature and sufficiently low density for kT greatly to exceed the average coulomb energy between adjacent ions. In that case the Debye-Hückel treatment gives a fair representation of the potential around each ion. From Eq. (2-232) the potential around the ion Z_1 is

$$V_1 \approx \frac{eZ_1}{r} \exp - \frac{r}{R_D} \tag{4-216}$$

where the Debye radius is

$$R_D = \left(\frac{kT}{4\pi e^2 \rho N_0 \zeta}\right)^{\frac{1}{2}} \tag{4-217}$$

and

$$\zeta = \sum_Z \frac{(Z^2 + Z)X_Z}{A_Z} \tag{4-218}$$

If one of the ions Z_2 of this ion sphere passes very close to Z_1, the interaction is

$$eZ_2V_1 \approx \frac{Z_1Z_2e^2}{r} - \frac{Z_1Z_2e^2}{R_D} + \cdots \tag{4-219}$$

By analogy to Eq. (4-206) we have in this limit

$$-\frac{U_0}{kT} = \frac{Z_1Z_2e^2/R_D}{kT} \tag{4-220}$$

and

$$-\frac{U_0}{kT} = 0.188 Z_1 Z_2 \rho^{\frac{1}{2}} \zeta^{\frac{1}{2}} T_6^{-\frac{3}{2}} \tag{4-221}$$

This result can be accurate only if $-U_0/kT$ is considerably smaller than unity, inasmuch as the Debye-Hückel treatment fails otherwise. As a result, the enhancement of the reaction rate in this case is also approximately expressible as

$$f \approx 1 - \frac{U_0}{kT} + \cdots$$

This limit has come to be called *weak screening*, following Salpeter, who discussed its conditions of applicability and the effect of electron degeneracy on the result. The weak-screening formula is the one applicable to the majority of the thermonuclear reactions encountered in astrophysics.

The most difficult region for screening calculations is that in which the coulomb energy of neighboring ions is roughly comparable to kT. At even higher density, where the coulomb energy dominates, the situation again simplifies. In that case, the large coulomb repulsive energy squeezes all other positive ions out of the ion sphere, leaving only a sphere of electrons of sufficient size to neutralize the charge of the ion. The energetics of such ion spheres was estimated in the discussion of the zero-temperature ionized gas in Chap. 2. The total electrostatic energy of an ion sphere is given by Eq. (2-262),

$$U = -\frac{9}{10}\frac{(Ze)^2}{R_Z} \tag{4-222}$$

where

$$R_Z = \left(\frac{3}{4\pi}\frac{Z}{n_e}\right)^{\frac{1}{3}} \tag{4-223}$$

Problem 4-58: Show that in this limit we have

$$-U = 17.6Z^{\frac{5}{3}}\left(\frac{\rho}{\mu_e}\right)^{\frac{1}{3}} \quad \text{ev} \tag{4-224}$$

Equation (4-224) gives the electrostatic energy of each ion when they are separated. Since the penetration enhancement is determined by the difference in potential energies at the origin and at large distances, we have

$$-U_0 = 17.6\left(\frac{\rho}{\mu_e}\right)^{\frac{1}{3}}[(Z_1 + Z_2)^{\frac{5}{3}} - Z_1^{\frac{5}{3}} - Z_2^{\frac{5}{3}}] \quad \text{ev} \tag{4-225}$$

or

$$-\frac{U_0}{kT} = 0.205\,[(Z_1 + Z_2)^{\frac{5}{3}} - Z_1^{\frac{5}{3}} - Z_2^{\frac{5}{3}}]\left(\frac{\rho}{\mu_e}\right)^{\frac{1}{3}} T_6^{-1} \tag{4-226}$$

It turns out that this result, commonly called *strong screening*, does not have great significance for astrophysics, because it is applicable only at very high density, where its value exceeds unity. At ultrahigh density reactions proceed even at zero temperature, and a more careful evaluation of the inter-ion potential is necessary.[1] We close the discussion by repeating that Eq. (4-221) has been the one used in most studies of nuclear reactions during stellar evolution.

Problem 4-59: By what factor does electron screening enhance the rate of the $C^{12}(p,\gamma)$ reaction in a hydrogen gas at $\rho = 15$, $T_6 = 30$?

Problem 4-60: Convince yourself that for the case (*ii*) resonant reactions that can be treated by statistical mechanics ($\Gamma_1 \gg \Gamma_2$) the enhancement due to interactions with electrons is identical with that obtained by the argument presented in this section. This may seem surprising at first, because the penetration factor in the incident channel never enters the discussion if $\Gamma_1 \gg \Gamma_2$, so that the treatment of this section is not relevant for such reactions. For this case the name electron shielding is particularly inappropriate. The reason for modification of the rates is that the polarizability of the plasma changes the energetics of the reaction. By what factor is the density of Be^8 nuclei increased in a helium gas at $\rho = 10^5$ and $T = 10^8$?

[1] R. A. Wolf, *Phys. Rev.*, **137**:B1634 (1965); another thorough analysis has been made by H. Van Horn, Ph.D. thesis, Cornell University, Ithaca, N.Y., 1965.

chapter 5

MAJOR NUCLEAR BURNING STAGES IN STELLAR EVOLUTION

Stars are formed from interstellar gas by a gravitational instability of that gas. Whenever a sufficiently large mass of gas is compressed to a small enough volume, its force of self-gravitation becomes sufficiently great to cause gravitational collapse. Because the acceleration of gravity at the edge of a given gas cloud is inversely proportional to the square of the radius of the cloud, the collapse will be accelerated by the decreasing size of the cloud. For a short time after the onset of the instability, the cloud may be nearly in a condition of free fall. Eventually pressure forces begin to restrict the collapse. The directed motion of free fall is converted into random thermal energy in the gas, and its temperature begins to rise. By the virial theorem, discussed in Chap. 2, it can be concluded that when the acceleration of the moment of inertia has become small, one-half of the gravitational energy must have been converted to internal thermal energy. The combination of increasing temperature and increasing density causes the pressure to rise so rapidly that the collapse is decelerated to a slow quasistatic one. As the collapse slows down, the virial theorem becomes quite accurate and, in fact, is the dominant principle governing the subsequent evolution of the star. Half of each increment in the magnitude of the gravitational energy is converted to kinetic energy of the particles, and half is consumed in the production of radiation. Although the radiation escapes at first, the stellar matter becomes opaque as its collapse is decelerated by the pressure buildup. The situation of thermodynamic equilibrium is then established for the first time, the internal radiation field being describable by a local temperature equal to the kinetic temperature of the gas. Much of the radiation still escapes. A great deal of it is consumed in ionizing the constituent matter, which was probably initially neutral. Because about 90 percent of the atoms are hydrogen, the interior temperature cannot rise above 10^4 °K until hydrogen has been ionized. Hydrostatic equilibrium can be established when the hydrogen and helium in the bulk of the interior have been ionized. In the case of the sun, insufficient gravitational energy had been released for this purpose until the radius had shrunk to about 60 times the present radius of the sun, or about one-fourth the distance from the sun to the orbit of the earth. Once hydrostatic equilibrium has been established, the subsequent slow contraction to the main sequence can be calculated with reasonable confidence.

It was shown by Hayashi that a star cannot achieve hydrostatic equilibrium if its surface is too cool. It follows that stars contracting toward the main sequence have very large luminosity as a consequence of the required high surface temperature and the large radius. The combination of the large luminosity and large opacity during the contraction phase demands that the star be convective to get the energy out fast enough. What Hayashi showed is that a star contracts along a nearly vertical path in the H-R diagram until reaching the vicinity of the main sequence. Figure 5-1 shows such a contraction path calculated for the sun and also the time required for the star to reach each point on the track. The early contraction is seen to be rapid and very luminous. The energy source for this luminosity is entirely due to the gravitational work of the contraction. As the star approaches its final luminosity on the main sequence, the evolution becomes slow. The star is fully convective for a few million years, after which time a central core in radiative equilibrium begins to grow. That radiative core slowly moves outward in the star until it achieves its final main-sequence size. Correspondingly, the subsurface convection zone shrinks to its final main-sequence size. During the entirety of this process the central temperature has been increasing. And there is the main point. The temperature rises until it becomes sufficient to cause thermonuclear reactions to occur at a rate adequate to supply the power radiated from the surface. When that balance is achieved, the stellar configuration becomes almost perfectly static. It would be perfectly static were it not for the fact that the thermonuclear reactions cause a very slow change in the composition of the interior gas. The task of the present chapter is to outline the major static burning phases encountered in stellar evolution.

Which nuclear reactions are of importance in any particular environment depends in an obvious way upon the composition of the gas. The results of the last chapter have indicated the ways in which reaction rates depend upon the properties of the particles and the temperature. Most energy-generating reactions involving light particles are nonresonant, in which case Eqs. (4-56) to (4-58) indicate that the reaction rate contains, among other things, the factors

$$r_{12} \propto N_1 N_2 \exp\left[-42.48\left(\frac{Z_1{}^2 Z_2{}^2 A}{T_6}\right)^{\frac{1}{3}}\right] \tag{5-1}$$

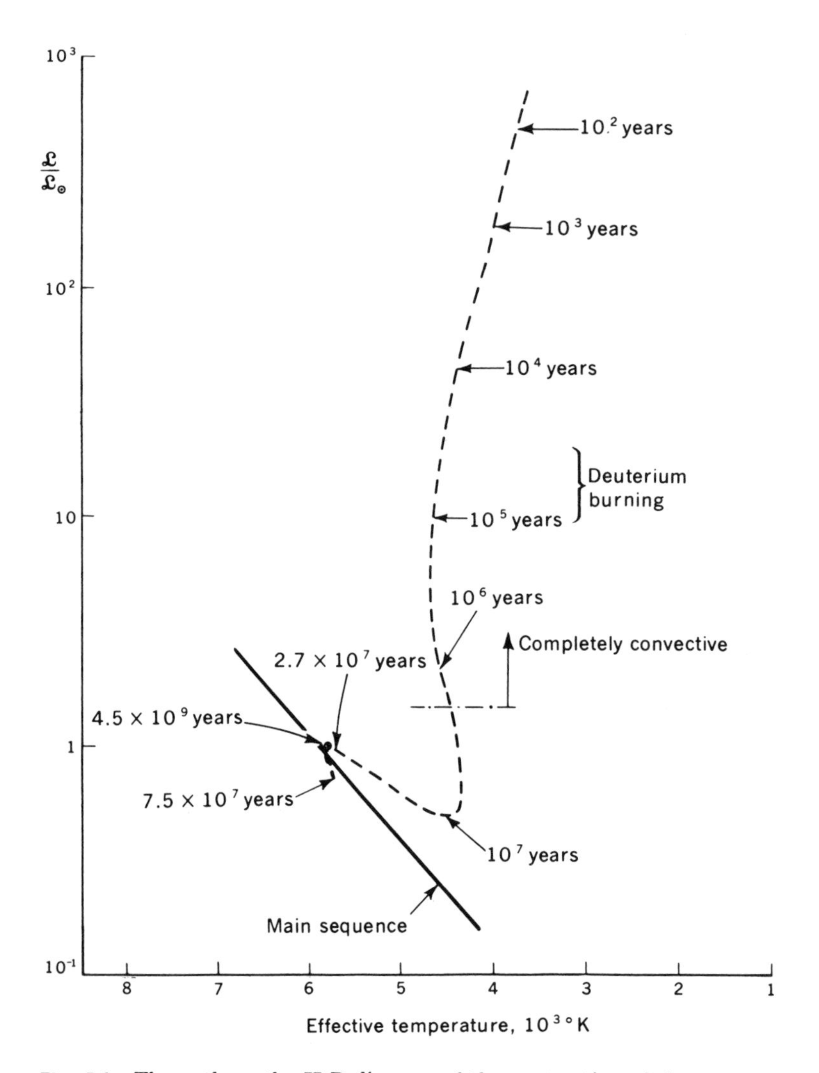

Fig. 5-1 The path on the H-R diagram of the contraction of the sun to the main sequence. The interior has become sufficiently hot to burn deuterium after about 10^5 years. The contraction ceases near the main sequence when the core has become hot enough to replenish the solar luminosity with the thermonuclear power generated by the fusion of hydrogen into helium. [*After D. Ezer and A. G. W. Cameron, The Contraction Phase of Stellar Evolution, in R. F. Stein and A. G. W. Cameron (eds.), "Stellar Evolution," Plenum Press, New York,* 1966.]

It is clear that as the temperature rises during gravitational contraction, the rates increase rapidly. Because the exponent is large, however, the major reaction tends to be that one for which the product Z_1Z_2 is as small as possible. In the vast majority of stars the initial abundances are dominated by hydrogen and helium, so that one is led to search for reactions involving these nuclei. Historically a major impasse was connected with the fact that all the major two-particle combinations have unstable ground states:

$$\begin{aligned} &p + p \rightarrow \mathrm{He}^2 \text{ (unstable)} \rightarrow p + p \\ &p + \mathrm{He}^4 \rightarrow \mathrm{Li}^5 \text{ (unstable)} \rightarrow p + \mathrm{He}^4 \\ &\mathrm{He}^4 + \mathrm{He}^4 \rightarrow \mathrm{Be}^8 \text{ (unstable)} \rightarrow \mathrm{He}^4 + \mathrm{He}^4 \end{aligned} \tag{5-2}$$

That is, the nuclear force produces no two-particle exothermic reactions in a gas of protons and alpha particles. Thus one is led to look either for more peculiar reactions between those particles or for reactions with rarer constituents of the gas.

The primary candidates would seem to be the minor isotopes of hydrogen and helium. The first thermonuclear reaction to proceed at a significantly rapid rate occurs between the two isotopes of hydrogen:

$$\mathrm{H}^1 + \mathrm{D}^2 \rightarrow \mathrm{He}^3 + \gamma \tag{5-3}$$

This reaction converts deuterium to He^3 during the pre-main-sequence contraction of the star; but there is probably so little deuterium that it is quickly exhausted, and its only effect on the star is to slow the contraction somewhat during the phase marked "deuterium burning" in Fig. 5-1. It is worth noting, however, that this phase may have provided an important source of He^3 in the early sun if D^2 was initially more abundant than He^3. The He^3 made in this way may survive to this day in the outer layers of the sun.

Similar results apply to the elements Li, Be, and B, which are destroyed effectively by reactions with protons at temperatures of a few million degrees. It is a matter of current interest to see how much of these elements can survive the deep surface convection zones of the contraction phase, but again one finds that their abundances are so small that they are quickly consumed in the interior, with only a small brief effect on the evolution of the star.

The most important reactions needed are those capable of converting hydrogen into helium. The two major ways by which this can be accomplished are the so-called *proton-proton chains* and the CNO cycle. From the table of atomic mass excess it is evident that whatever reactions result in the conversion of four hydrogen atoms into an He^4 atom liberate 26.731 Mev of energy. The sequence of reactions must also change two protons into two neutrons, however, so that each alpha particle created in the sun must be accompanied by the emission of two neutrinos. Thus the internal heat energy liberated per alpha particle fused from protons is 26.731 Mev minus the average kinetic energy of the two neutrinos. The amount of energy carried away by the two neutrinos will depend upon the detailed sequence of reactions by which the alpha particle is assembled.

Problem 5-1: Assuming that the energy for the sun's luminosity is provided by the conversion of $4H \rightarrow He^4$ and that the neutrinos carry off only about 3 percent of the energy liberated, how many neutrinos are liberated each second from the sun? What is the neutrino flux at the earth from the sun?
Ans: $9.4 \times 10^{37}\ \nu$/sec; $6.6 \times 10^{10}\ \nu\ \text{cm}^{-2}\ \text{sec}^{-1}$.

5-1 THE PROTON-PROTON REACTION

It was Hans Bethe[1] who first realized that the weak nuclear interaction was capable of converting a proton into a neutron during the brief encounter of a scattering event. Since the neutron is more massive than a hydrogen atom, such a decay would require energy (be endothermic) except for the fact that the neutron can appear in a state bound to the proton in the form of the deuterium nucleus. The binding energy of the deuteron is sufficient (2.2245 Mev) to make the reaction exothermic.

Problem 5-2: Show that the reaction

$$H^1 + H^1 \rightarrow D^2 + \beta^+ + \nu \tag{5-4}$$

liberates 0.420 Mev of kinetic energy to the positron and neutrino. The subsequent annihilation of the positron and an electron brings the total energy release to 1.442 Mev.

The beta decays that proceed at the greatest rate are the so-called *allowed decays*, in which the two leptons (e, $\bar{\nu}$ or $\bar{e}$, ν) are emitted without orbital angular momentum. The leptons may or may not carry off spin angular momentum in allowed decays, depending upon whether the electron spin ($s_e = \frac{1}{2}$) and the neutrino spin ($s_\nu = \frac{1}{2}$) add vectorially to zero or to unity. If $\mathbf{s}_e + \mathbf{s}_\nu = \mathbf{0}$, the allowed decays are called *Fermi transitions*, for which the nuclear spin cannot change. If $\mathbf{s}_e + \mathbf{s}_\nu = \mathbf{1}$, the allowed decays are called *Gamow-Teller transitions*, for which the nuclear spin must change by one *vector* unit.

The allowed decays do not change the orbital angular momenta of nucleons in nuclei, and they do not change the spatial arrangement of nucleons. One of the nucleons in the composite wave function changes from proton to neutron (or vice versa), and in the Gamow-Teller case flips its spin, but the spatial wave function remains the same, and to the extent that the initial and final wave functions are dissimilar the transition is suppressed. The quantum statement of this requirement is that the rate be proportional to the square of an *overlap integral* over all nucleon coordinates,

$$\lambda \propto \left|\int \psi_f^* \psi_i \, dV\right|^2 \tag{5-5}$$

which quantitatively measures the degree to which the final nucleon wave function ψ_f is similar to the initial wave function ψ_i. There are several examples of beta decay in nuclear physics wherein the initial and final wave functions are

[1] *Phys. Rev.*, **55**:103, 434 (1939).

believed to be nearly identical; i.e., the overlap integral is unity. The best known of these are the decays

$$He^6(J^\pi = 0^+) \rightarrow Li^6(J^\pi = 1^+) + \beta^- + \bar{\nu}$$

which is used to determine the Gamow-Teller coupling constant, and

$$O^{14}(J^\pi = 0^+) \rightarrow N^{14*}(J^\pi = 0^+) + \beta^+ + \nu$$

which is used to determine the Fermi coupling constant.[1] These decays are called *superallowed* because not only are they allowed decays, but the overlap integral is essentially unity.

The beta-decay rates are also proportional to the volume of phase space into which the leptons can be emitted. The combined momentum space into which the leptons can be emitted is given by

$$n(p_\beta,p_\nu)\, dp_\beta\, dp_\nu = \frac{1}{h^6} (4\pi p_\beta{}^2\, dp_\beta)(4\pi p_\nu{}^2\, dp_\nu) \tag{5-6}$$

subject to the restriction that the combined energy equal the energy release in the decay, viz.,

$$E_\beta + E_\nu = W \tag{5-7}$$

When this restriction is imposed on the differential momentum-space volume, and when the result is integrated over all possible ways in which the pair can share the energy, the rate becomes proportional to a function called $f(W)$, which measures the phase-space volume into which the decay products can go.[2] It turns out that the rate of the proton-proton beta decay can be factored into

$$\lambda = g|M_{\text{sp}}|^2\, |\int \psi_f^* \psi_i\, dV|^2 f(W) \tag{5-8}$$

where $g = (6.9 \pm 0.3) \times 10^{-4}$ sec^{-1} is the appropriate beta-decay coupling constant, $|M_{\text{sp}}|^2$ is a spin matrix element that includes the sum over the possible projections of the involved spin states, and the momentum-space volume $f(W)$ is

$$f(W) = (W^2 - 1)^{\frac{1}{2}} \left(\frac{W^4}{30} - \frac{3W^2}{20} - \frac{2}{15} \right) + \tfrac{1}{4} W \log [W + (W^2 - 1)^{\frac{1}{2}}] \tag{5-9}$$

where W is the total decay energy (including the electron rest mass) in units of m_ec^2, that is, the kinetic energy released is $(W - 1)m_ec^2$. This factor contains the empirical rule of thumb $\lambda \propto W^5$ observed for decays having $W \gg 1$.

The reaction $p + p \rightarrow d + \beta^+ + \nu$ is a rather special kind of positron decay because the initial wave function is that for the scattering of protons. Since the

[1] See, for instance, R. K. Bardin, C. A. Barnes, W. A. Fowler, and P. A. Seeger, *Phys. Rev.*, **127**:583 (1962).

[2] It would be somewhat out of place to develop the theory of beta decay in this book. Qualified readers will find a thorough discussion in M. A. Preston, "Physics of the Nucleus," chap. 15, Addison-Wesley Publishing Company, Inc., Reading, Mass., 1962.

final wave function is contained in the volume of the deuterium nucleus, the overlap integral $\int \psi_d^* \psi_{pp}\, dV$ will be a small number. Only those parts of the initial scattering wave at small separation can contribute to the overlap integral. As the deuteron is a surprisingly large nucleus (because it is weakly bound), very small separations in the scattering wave are not required, and the largest contribution to the overlap integral comes from the tail of the deuteron wave function that extends outside the potential well. Although there is some difficulty in defining an interaction radius for this reaction, it is nonetheless clear that the rate must contain the penetration factor for two charge-1 particles. Salpeter[1] has discussed the uncertainties in the reaction rate introduced by the uncertainties in the deuteron wave function.

The ground state of the deuteron is predominantly describable as a bound s state (zero orbital angular momentum) of the neutron and proton with the nucleon spins aligned to give a total $J = 1$ to the nucleus. Since the allowed beta decays do not alter orbital angular momentum, the protons must scatter in an s-wave ($l = 0$) state for the decay to occur. From the symmetry principles of identical fermions, the proton spins must be antiparallel ($S = 0$) before the decay. The decay must therefore be of the Gamow-Teller type, which flips the spin of one of the protons as it changes to a neutron in the deuterium ($S = 1$) ground state. These observations allow unambiguous evaluation of the spin matrix element in Eq. (5-8).

The inclusion of the factor $f(W)$ produces little complication. The kinetic energy released (0.420 Mev) is so much greater than the kinetic energies of the protons before the collision that $f(W)$ is nearly constant. The major temperature dependence is in the penetration factor. Suffice it to say that all these factors can be pulled together and evaluated with considerable confidence in the calculation of the stellar reaction rate. The resulting cross section is about 10^{-47} cm^2 at 1 Mev of laboratory energy, which is much too small for detection. A thick hydrogen target would have to be bombarded for 10 years with 1 amp of 1-Mev protons to obtain one reaction! But the theoretical understanding of this process is so good that the cross section can be computed with great confidence. It is in fact the weakness of the beta-decay interaction that allows stars as we know them to exist at all. Because of the low coulomb barrier in the p-p reaction, a star would consume its hydrogen quickly if it were not slowed down by the weakness of the beta interaction.

To preserve the similarity to the other thermonuclear reaction-rate formulas, the resulting cross section is usually expressed in terms of its cross-section factor $S(E)$. It suffices to give $S(0)$ and dS/dE, the values of which are

$$S(0) = 3.78 \pm 0.15 \times 10^{-22} \text{ kev barn}$$

$$\frac{dS}{dE} = 4.2 \times 10^{-24} \text{ barn} \qquad (5\text{-}10)$$

[1] E. E. Salpeter, *Phys. Rev.*, **88**:547 (1952). See also E. Frieman and L. Motz, *Phys. Rev.*, **83**:202 (1951), and J. N. Bahcall and R. M. May, *Astrophys. J.*, **152** (April, 1968).

Problem 5-3: Show that the thermonuclear rate of the *p-p* reaction (without the electron-screening correction) is

$$r_{pp} = 3.09 \times 10^{-37} n_p{}^2 T_6{}^{-\frac{2}{3}} \exp(-33.81 T_6{}^{-\frac{1}{3}}) \times (1 + 0.0123 T_6{}^{\frac{1}{3}} + 0.0109 T_6{}^{\frac{2}{3}} + 0.00095 T_6) \quad \text{cm}^{-3}\,\text{sec}^{-1} \qquad (5\text{-}11)$$

The factor $3.09 \times 10^{-37} n_p{}^2$ can be replaced by the factor $11.05 \times 10^{10} \rho^2 X_H{}^2$.

Problem 5-4: Calculate the lifetime of protons against the *p-p* reaction at a temperature $T = 15 \times 10^6$ °K, density $\rho = 100$ g/cm^3, and composition $X_{He} = X_H = 0.5$. This partial lifetime is defined as the quantity which satisfies

$$\frac{dn_H}{dt} = -\frac{n_H}{\tau_p(H)}$$

Ans: About 10^{10} years.

Not all the 1.442 Mev released by this reaction is converted into local heat. The neutrinos carry away some kinetic energy (nominally about one-half of the kinetic energy of the leptons). It takes a detailed calculation of the lepton spectrum to determine the average energy of the neutrinos. The result of such a calculation is that $\bar{E}_\nu = 0.262$ Mev. The average heat input from each reaction is then equal to $1.442 - 0.262 = 1.180$ Mev.

5-2 PPI CHAIN

The conversion of hydrogen into helium involves a chain of reactions which have that conversion as their net product. The determination of the actual chain requires the calculation of the rates of all possible reactions between all the pairs of particles present. Experience shows, however, that such complexity is unnecessary in practice. Many reactions proceed at a negligible rate, either because the cross section is too small or because the product of the abundances of the interacting species is too small. As an example, after the deuterium has been formed, one could imagine that He^4 might be produced by the reaction $D + D \rightarrow He^4 + \gamma$. This reaction, however, suffers from a small cross section and, more importantly, from the fact that the deuterium abundance is kept very small by its interaction with protons. By that type of analysis one can pick out the reactions responsible for the conversion. The simplest chain is the only one that occurs in a pure hydrogen gas, and is called PPI. The reactions and their respective rates are

Reaction	*Rate*
$H^1 + H^1 \rightarrow D^2 + \beta^+ + \nu$	$r_{pp} = \lambda_{pp} \frac{H^2}{2}$
$D^2 + H^1 \rightarrow He^3 + \gamma$	$r_{pd} = \lambda_{pd} HD$
$He^3 + He^3 \rightarrow He^4 + 2H^1$	$r_{33} = \lambda_{33} \frac{(He^3)^2}{2}$

(5-12)

where the number densities are designated by the chemical symbols in the rate equations. That these are the major reactions comes about because the reactions between $H + He^3$, $D + D$, and $D + He^3$ all proceed at a negligible rate, the major reasons being that Li^4 is unstable and that D can build up only to a very small abundance. A rather complete discussion of all the possibilities can be found in a paper by Parker et al.[1]

Once the relevant reactions are restricted to those of PPI, the differential equations for the abundances can be written. The deuterium abundance is given by

$$\frac{d\mathrm{D}}{dt} = \lambda_{pp} \frac{\mathrm{H}^2}{2} - \lambda_{pd}\mathrm{HD} \tag{5-13}$$

If the deuterium abundance is very small, $d\mathrm{D}/dt$ is positive, and the D abundance builds up until the right-hand side of Eq. (5-13) vanishes; alternatively, if D is initially large, $d\mathrm{D}/dt$ is initially negative, and the D abundance decays until the right-hand side vanishes. Equation (5-13) is self-regulating in the sense that the D abundance seeks an equilibrium value, $(\mathrm{D/H})_e$:

$$\left(\frac{\mathrm{D}}{\mathrm{H}}\right)_e = \frac{\lambda_{pp}}{2\lambda_{pd}} = \frac{\tau_p(\mathrm{D})}{2\tau_p(\mathrm{H})} \tag{5-14}$$

The deuterium burning reaction is so fast that the lifetime of deuterium inside a star may be on the order of seconds. The cross-section factor is characterized by

$$S(0) = 2.5 \times 10^{-4} \text{ kev barn} \qquad \text{and} \qquad \frac{dS}{dE} = 7.9 \times 10^{-6} \text{ barn}$$

The form of this reaction is similar to that of the *p-p* reaction. The much greater cross section occurs because the transition-causing interaction is electromagnetic rather than weak.

Problem 5-5: Compute the equilibrium ratio $(\mathrm{D/H})_e$ for $T = 15 \times 10^6$ °K. This ratio is actually not very dependent upon the temperature, as shown in Fig. 5-2.
Ans: $(\mathrm{D/H})_e = 2.8 \times 10^{-18}$.

Because the deuterium lifetime is so short, it takes very little time for D to reach its equilibrium value at high temperature. In fact, Eq. (5-13) can to excellent approximation be solved with the assumption that H is constant. This assumption is justifiable on the grounds that D achieves equilibrium in a time so short that the H abundance cannot change.

[1] P. D. Parker, J. N. Bahcall, and W. A. Fowler, *Astrophys. J.*, **139**:602 (1964). The nuclear data on thermonuclear reaction rates have been thoroughly reviewed by W. A. Fowler, G. R. Caughlan, and B. A. Zimmerman, *Ann. Rev. Astron. Astrophys.*, **5**:525 (1967).

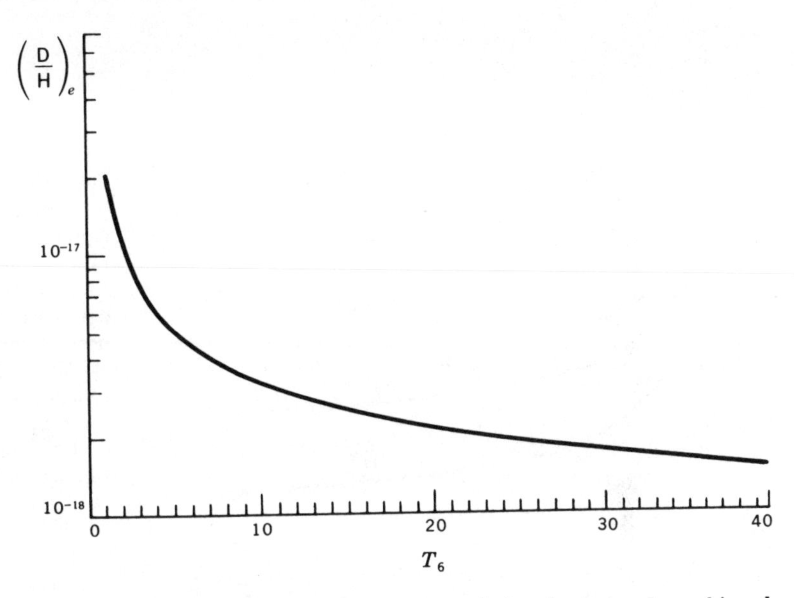

Fig. 5-2 The equilibrium ratio $(\mathrm{D/H})_e$ expected after deuterium has achieved a steady-state abundance during hydrogen burning.

Problem 5-6: Show that with the assumption of constant H the D abundance is given by

$$\left(\frac{\mathrm{D}}{\mathrm{H}}\right)_t = \left(\frac{\mathrm{D}}{\mathrm{H}}\right)_e - \left[\left(\frac{\mathrm{D}}{\mathrm{H}}\right)_e - \left(\frac{\mathrm{D}}{\mathrm{H}}\right)_0\right] \exp - \frac{t}{\tau_p(\mathrm{D})} \tag{5-15}$$

Thus the initial D decays exponentially to its equilibrium value with a $1/e$ time equal to $\tau_p(\mathrm{D})$.

The lifetimes of deuterium against the equilibrium amount of deuterium, against the equilibrium amount of He^3, and against protons are shown in Fig. 5-3. The lifetime of deuterium against protons is by far the shortest, showing that the assumption of Eq. (5-13) [that the $\mathrm{D}(p,\gamma)\mathrm{He}^3$ reaction is the only important one for deuterium destruction] is justified.

The short lifetime of deuterium has important astrophysical implications that have by no means been deciphered. The terrestrial abundance, as determined from sea water, for instance, is $\mathrm{D/H} = 1.5 \times 10^{-4}$. This ratio is very large compared to the value $(\mathrm{D/H})_e \approx 10^{-17}$ to be expected from the remnants of hydrogen burning in stars and has necessitated auxiliary investigations of the source of the deuterium abundance. Burbidge et al.[1] ascribed synthesis to spallation reactions in stellar surfaces. The high-energy particles would presumably be accelerated by flares, as in the case of solar cosmic rays. Their interaction with heavier nuclei could knock out deuterons as products, thereby contaminating the surface with deuterium. The problem is further complicated by the fact

[1] E. M. Burbidge, G. R. Burbidge, W. A. Fowler, and F. Hoyle, *Rev. Mod. Phys.*, **29**:547 (1957).

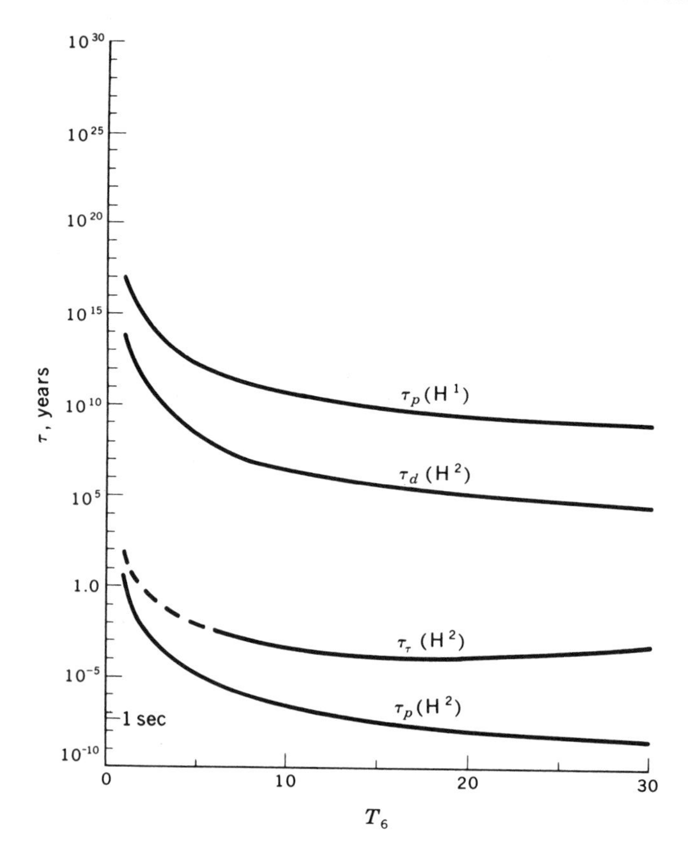

Fig. 5-3 The lifetimes of deuterium (H^2) against the equilibrium amount of deuterium, against the equilibrium amount of He^3 (here designated by τ), and against protons for a proton density of 50 g/cm^3. By far the shortest lifetime is the one against protons. Also shown is the lifetime of protons against protons. [*After P. D. Parker, J. N. Bahcall, and W. A. Fowler, Astrophys. J.,* **139**:602 (1964). *By permission of The University of Chicago Press. Copyright 1964 by The University of Chicago.*]

that the deuterium abundances in the interstellar medium and in the sun are unknown. Fowler et al.[1] subsequently proposed that terrestrial deuterium is not the product of stars at all but rather has resulted from spallation in small planetesimals bombarded with high-energy protons from the young sun. It has, therefore, become a crucial scientific question whether the interstellar D/H ratio is as large as the terrestrial ratio. The question is very much open. If the interstellar D/H ratio is as large as the terrestrial ratio, it must follow that the first thermonuclear energy source to become operative in a newly contracting star is that of deuterium burning, $D(p,\gamma)He^3$. There is a resulting uncertainty in the time scale with which a newly formed star can contract through the deuterium-burning phase shown in Fig. 5-1.

A convenient practical consequence of the short deuterium lifetime is that the deuterium abundance may always be safely assumed to be in equilibrium. The differential equation for the He^3 abundance in the PPI chain,

$$\frac{dHe^3}{dt} = \lambda_{pd}HD - 2\lambda_{33}\frac{(He^3)^2}{2} \qquad (5\text{-}16)$$

[1] W. A. Fowler, J. L. Greenstein, and F. Hoyle, *Geophys. J.*, **6**:148 (1962).

can therefore be simplified to

$$\frac{dHe^3}{dt} = \lambda_{pp}\frac{H^2}{2} - 2\lambda_{33}\frac{(He^3)^2}{2} \tag{5-17}$$

This equation is also of a self-regulating type, in that He^3 builds toward an equilibrium abundance given by

$$\left(\frac{He^3}{H}\right)_e = \left(\frac{\lambda_{pp}}{2\lambda_{33}}\right)^{\frac{1}{2}} \tag{5-18}$$

The cross-section factor for the $He^3(He^3,2p)He^4$ reaction is not well known, because of experimental difficulties with gas targets. A frequently quoted value has been $S = 1.1 \times 10^3$ kev barns at all low energies, but recent measurements[1] indicate that $S \approx 5 \times 10^3$ kev barns is more nearly correct. With the aid of Eq. (5-18), Eq. (5-17) can be written in a more suggestive form,

$$\frac{dHe^3}{dt} = \lambda_{33}H^2\left[\left(\frac{He^3}{H}\right)_e^2 - \left(\frac{He^3}{H}\right)^2\right] \tag{5-19}$$

which clearly demonstrates the tendency of He^3/H to seek its equilibrium value. Figure 5-4 shows the equilibrium He^3 concentration.

[1] At the time of writing H. C. Winkler and M. R. Dwarakanath at the California Institute of Technology are improving their measurements of this important reaction; see, for example, *Bull. Am. Phys. Soc.*, **12**:16 (1967).

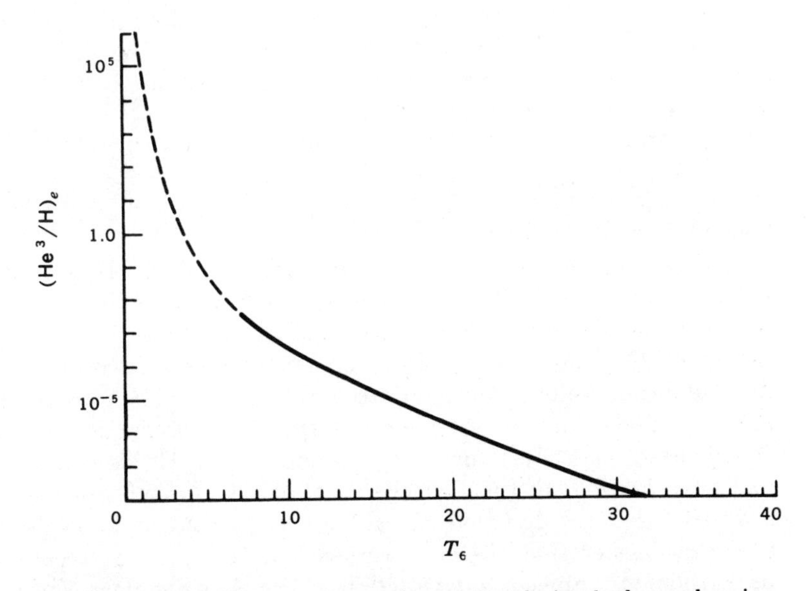

Fig. 5-4 The equilibrium concentration of He^3 during hydrogen burning. The curve is dashed for $T_6 < 8$ because the length of time required for He^3 to achieve equilibrium at such low temperatures is unreasonably long.

The question of the equilibrium of He^3 is an important one. It will shortly be demonstrated that the energy generation from PPI may be expressed very simply if He^3 has achieved equilibrium in the chain. Because of the larger coulomb barrier ($Z_1Z_2 = 4$) for this reaction, the lifetime of He^3 against itself is much longer than the lifetime of deuterium against protons. Consequently He^3 must build up to a much larger abundance than D must in order to achieve equilibrium. An important question for the PP chain is the length of time required for that buildup. The lifetime of He^3 against He^3 is

$$\tau_3(3) \equiv \tau_{He^3}(He^3) = (\lambda_{33}He^3)^{-1} \tag{5-20}$$

which obviously depends not only on the temperature (via λ_{33}) but upon the He^3 abundance itself. Early in the burning, when the He^3 abundance may be very small, the lifetime $\tau_3(3)$ is extremely long, and to the extent that He^3 is burned at all, it is burned not by interactions with itself but by the following interaction with deuterium:

$$He^3(d,p)He^4$$

$$S(0) = 6.7 \times 10^3 \text{ kev barns} \tag{5-21}$$

$$\frac{dS}{dE} = 27 \text{ barns}$$

Since the deuterium abundance may be assumed to be in equilibrium after very short times on the order of seconds, the lifetime of He^3 against deuterium may be unambiguously calculated for a given temperature and hydrogen density. That lifetime, $\tau_d(He^3)$, is shown in Fig. 5-5, along with the lifetime $[\tau_3(3)]_e$ of He^3 against He^3 *after* He^3 *has reached its equilibrium value.* The latter lifetime is given from Eqs. (5-20) and (5-18) as

$$[\tau_3(3)]_e = \left(\frac{2}{\lambda_{33}\lambda_{pp}}\right)^{\frac{1}{2}} \frac{1}{H} \tag{5-22}$$

The lifetime $\tau_3(3)$ falls from infinity (for zero initial He^3 content) to the value in Eq. (5-22) as He^3 builds up to its equilibrium value. It is clear from Fig. 5-5 that since $[\tau_3(3)]_e \ll \tau_d(He^3)$, the He^3 is predominantly destroyed by interactions with itself *after it has reached equilibrium.* In fact, the destruction of He^3 is predominantly by interactions with itself after it has built up to only 1 percent of its equilibrium value. Only for very low He^3 abundance (compared to its equilibrium value) will the destruction of He^3 occur predominantly by interactions with deuterium. But under those circumstances, He^3 is being produced by the first two reactions of the PPI chain at a rate so much greater than it is being destroyed that the destruction reactions may probably be neglected. As far as the energy-generation rate is concerned, only the primary reactions of PPI need be considered. Since the destruction of He^3 is negligible except when its abundance is high enough for it to interact with itself, the destruction-term approximation in Eq. (5-19) is justified.

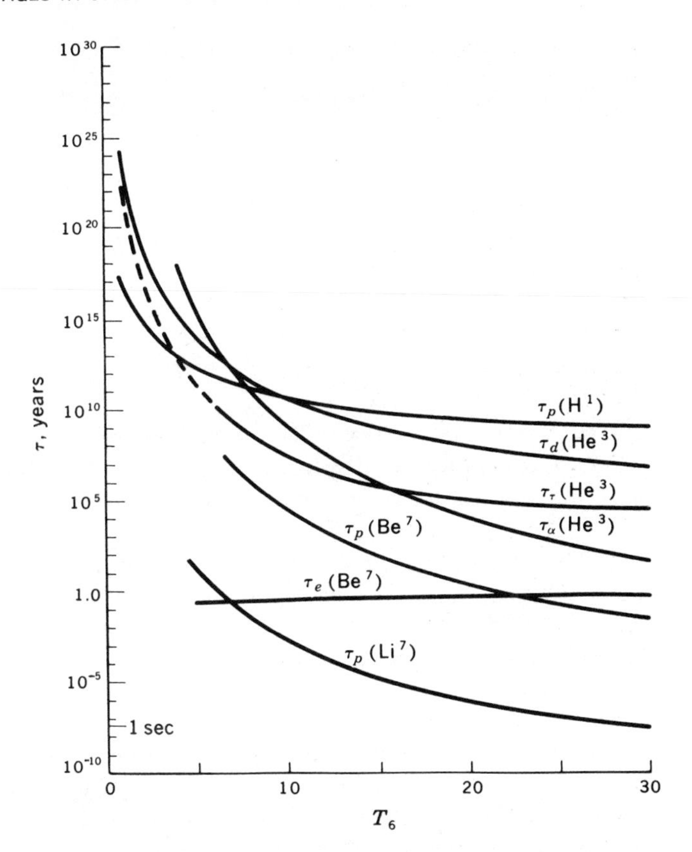

Fig. 5-5 Comparison of the lifetimes of He^3 against the equilibrium concentrations of He^3 and D shows that destruction of He^3 by reactions with deuterium is unimportant when the chain is in equilibrium. The proton density for this calculation was taken to be 50 g/cm³. If the alpha-particle density is high, however, the He^3 may be destroyed by interactions with He^4, a reaction that leads to the chains PPII and PPIII. The lifetime against He^4 shown above is computed for $Y = X$. [*After P. D. Parker, J. N. Bahcall, and W. A. Fowler, Astrophys. J.,* **139**:602 (1964). *By permission of The University of Chicago Press. Copyright* 1964 *by The University of Chicago.*]

There still exists the important question of the length of time required for He^3 to achieve its equilibrium value. This time is a very steep function of temperature. Figure 5-4 shows that the equilibrium abundance $(He^3/H)_e$ is reasonably small for temperatures greater than about 10^7 °K but rapidly rises toward unity at temperatures of a few million degrees. This last part of the curve is shown dashed because, as we shall see, the time required to actually achieve such high equilibrium abundances is excessive. This time may be estimated in the following way. If $(He^3/H)_e$ is small, say less than 0.01, very little hydrogen must be consumed in achieving equilibrium, so that Eq. (5-19) can be solved assuming H to be constant.

Problem 5-7: If the ratio He^3/H is defined to be W, and if it is assumed that H is constant, Eq. (5-19) is

$$\frac{dW}{dt} = \lambda_{33}\mathrm{H}(W_e^2 - W^2) \tag{5-23}$$

Integrate this equation to show that if the He^3 abundance is initially zero, its abundance at time t is

$$\left(\frac{\text{He}^3}{\text{H}}\right) = \left(\frac{\text{He}^3}{\text{H}}\right)_e \tanh\left[\lambda_{33}\text{H}\left(\frac{\text{He}^3}{\text{H}}\right)_e t\right]$$
$$= \left(\frac{\text{He}^3}{\text{H}}\right)_e \tanh \frac{t}{[\tau_3(3)]_e} \tag{5-24}$$

From Eq. (5-24) the time required for He^3 to build its abundance to some fraction f of its equilibrium value is seen to be

$$t_f = [\tau_3(3)]_e \tanh^{-1} f \tag{5-25}$$

The time required to achieve 99 percent of the equilibrium abundance is displayed in Fig. 5-6. Specifically it can be seen that the required time exceeds 10^9 years for temperatures less than 8×10^6 °K. For this reason it is seldom relevant to use arguments based on equilibrium amounts of He^3 at $T_6 < 8$. The energy generation from PPI must be handled in two different ways depending upon whether He^3 has achieved equilibrium or not, and it is Fig. 5-6 that divides the time-temperature plane into equilibrium and nonequilibrium regions. For stellar ages less than the curve, the equilibrium energy-generation rate cannot be used. For example, since it may take a few times 10^7 years for a newly formed star to settle onto the main sequence, only those regions at temperatures higher than 10^7 °K will have achieved He^3 equilibrium at that time. Of course, there may exist some initial amount of He^3 in the nebula from which a star formed that would modify the calculations leading to Fig. 5-6.

The rate of energy generation from PPI must in general be split into two parts. Since the proton-proton reaction is rapidly followed by the $D(p,\gamma)He^3$ reaction, the net effect is

$$3\text{H} \rightarrow \text{He}^3 + \nu$$

at the rate r_{pp}. The energy liberated is 6.936 Mev, which must be reduced by the 0.263 Mev carried off by the neutrino. Since 1 Mev = 1.602×10^{-6} erg, this part of the PPI chain liberates energy at the rate

$$\rho\epsilon(3\text{H} \rightarrow \text{He}^3) = 1.069 \times 10^{-5} r_{pp} \qquad \text{erg cm}^{-3}\ \text{sec}^{-1} \tag{5-26}$$

Likewise, the $He^3(He^3,2p)He^4$ reaction liberates 12.858 Mev of kinetic energy. The energy-generation rate is expressible as the sum of both contributions:

$$\rho\epsilon_{\text{PPI}} = 1.069 \times 10^{-5} r_{pp} + 2.060 \times 10^{-5} r_{33} \qquad \text{erg cm}^{-3}\ \text{sec}^{-1} \tag{5-27}$$

The expression for r_{pp} was given in Eq. (5-11), and r_{33} can be computed from the cross-section factor $S = 5.0 \times 10^3$ and the He^3 abundance. In light of the previous discussion, it is apparent that the He^3 abundance depends upon the length of time the star has been burning. In numerical computations of stellar structure one must follow the He^3 abundance numerically from time step to time step until it has built up to its equilibrium value (if it ever does). A stellar model is com-

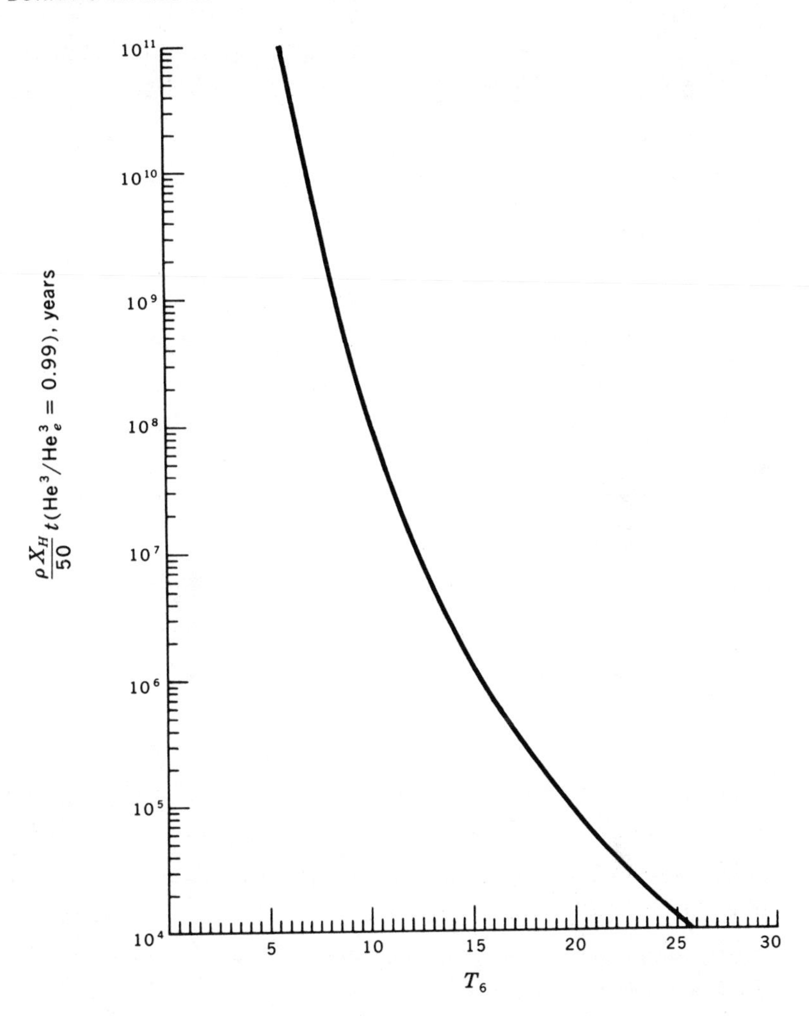

Fig. 5-6 The time required for He^3 to build up to 99 percent of its equilibrium abundance. Because this time exceeds 1 billion years for $T_6 < 8$, it is seldom reasonable to assume that He^3 has achieved equilibrium at such low temperatures.

puted at $t = 0$ from a known composition. Suppose for the sake of argument that initially $He^3(0) = 0$. That model will have a calculable rate r_{pp} for the production of He^3 at each point in the star. The next model, say at time $t = \Delta t$, will be constructed with a helium abundance $He^3(\Delta t) = r_{pp}\,\Delta t$, unless that amount of He^3 is already of the same order of magnitude as He^3_e. In the next model, say at time $t = 2\Delta t$, the He^3 will be calculated from

$$\mathrm{He}^3(2\Delta t) = \mathrm{He}^3(\Delta t) + r_{pp}\,\Delta t - 2r_{33}\,\Delta t$$

etc. And in each model the energy generation is calculated from Eq. (5-27) (provided, of course, that PPI is the dominant energy source). An alternative if the star is already static is to use Eq. (5-24) for the He^3 abundance. The assumptions of Eq. (5-24), viz., $(He^3/H)_e < 0.1$ and the temperature is constant, are often met with sufficient accuracy for it to be a useful approximation. After a period of burning indicated by Fig. 5-6 the He^3 abundance may safely be assumed to be in equilibrium; i.e., the rate of He^3 destruction has become as large as its rate of synthesis.

Once equilibrium has been attained, the energy-generation rate can be simplified. Since $d\text{He}^3/dt \approx 0$ at equilibrium, it follows from Eq. (5-17) that $2r_{33} = r_{pp}$. The rate of production of He^4 is then

$$\frac{d\text{He}^4}{dt} = r_{33} = \frac{r_{pp}}{2} \tag{5-28}$$

where the second equality is true only at He^3 equilibrium. Since two proton-proton reactions are required for the production of each alpha particle, the rate of production of alpha particles is exactly half of the rate of production of deuterons. Substituting into Eq. (5-27) results in

$$\rho\epsilon_{\text{PPI}} = 2.099 \times 10^{-5} r_{pp} \qquad \text{erg cm}^{-3}\ \text{sec}^{-1} \tag{5-29}$$

at equilibrium.

Problem 5-8: Show that at He^3 equilibrium

$$\epsilon_{\text{PPI}} = 2.32 \times 10^6 \rho X_{\text{H}}^2 T_6^{-2/3} \exp(-33.81 T_6^{-1/3})(1 + 0.0123 T_6^{1/3} + 0.0109 T_6^{2/3} + 0.00095 T_6) \qquad \text{erg g}^{-1}\ \text{sec}^{-1} \tag{5-30}$$

Problem 5-9: Show that the correction for electron screening in the weak-screening limit (appropriate to main sequence) is

$$f = \exp 0.27 \rho^{1/2} T_6^{-3/2} \tag{5-31}$$

in a predominantly hydrogen gas.

5-3 PPII AND PPIII CHAINS

The discussion of PPI has ignored the role of He^4 in the completion of the hydrogen fusion process, but the reaction $He^3(He^4,\gamma)Be^7$ has been found to have a sufficiently large cross-section factor that He^4 may, if it is abundant and if the temperature is not too low, provide the main source for consuming He^3. This possibility leads to two new cycles for converting hydrogen into helium. These cycles are commonly called PPII and PPIII, and they correspond to the two possible fates for the Be^7 nucleus. The atom Be^7 is radioactive, and in the laboratory its only energetically allowed decay involves the capture of an atomic electron:

$$\text{Be}^7 + e \rightarrow \text{Li}^7 + \nu$$

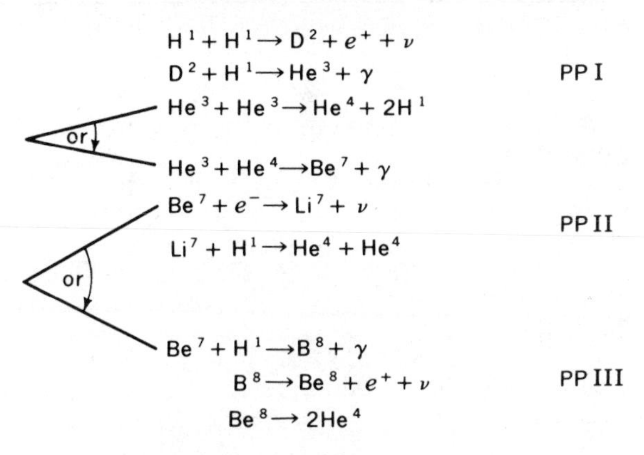

Fig. 5-7 The alternative PP chains. When He^3 is destroyed by the capture of an alpha particle, the chain is completed either through PPII or PPIII, depending upon the fate of the Be^7 nucleus.

At the center of a star the Be^7 nucleus is ionized, but the decay can still occur by the capture of free electrons. Once the decay occurs, the Li^7 is quickly destroyed by the reaction $Li^7(p,\alpha)He^4$.

The Be^7 nucleus may instead react with protons to form the nucleus B^8 by the reaction $Be^7(p,\gamma)B^8$. Which fate accrues to the Be^7 nucleus will depend upon the relative lifetimes of Be^7 against electron capture and against proton capture. Once B^8 is formed, it decays to Be^8,

$$B^8 \rightarrow Be^8 + \beta^+ + \nu$$

and the Be^8, which is unstable, breaks up into two alpha particles, $Be^8 \rightarrow He^4 + He^4$. These various possibilities are diagramed in Fig. 5-7, which reveals that the He^4 nucleus initiating these alternate cycles serves only as a catalyst which allows He^3 and a proton to be converted to He^4 and a neutrino. Thus the total energy released is the same for either cycle, but there will be a difference in the energy carried away by the neutrino in the two cases. Another look will reveal an important difference from PPI; viz., an alpha particle is produced with only one proton-proton reaction. Thus, if all of the subsequent reactions come into equilibrium, the rate of production of alpha particles can be as great as r_{pp} rather than the value $r_{pp}/2$, as in PPI. This fact will increase the energy generation by a significant amount (up to a factor of 2 minus the increased neutrino loss). Of course, PPI, PPII, and PPIII all operate *simultaneously* in a hydrogen-burning star containing significant amounts of He^4, and the details of the cycle will depend upon the density, temperature, and composition. The computation of these details requires the various reaction rates. The cross-section factors are listed in Table 5-1, along with other numerical information for each reaction. It should also be noted that other reactions than the ones indicated are possible and in fact do occur to some extent. For example, He^3 could react with Li^7, but in this case Li^7 interacts much more rapidly with protons both because of the lower coulomb barrier and because protons greatly outnumber He^3 nuclei. Similar

Table 5-1 Reactions of the PP chains

Reaction	*Q value, Mev*	*Average ν loss, Mev*	S_0, *kev barns*	$\frac{dS}{dE}$, *barns*	*B*	τ_{12}, *years*†
$H^1(p,\beta^+\nu)D^2$	1.442	0.263	3.78×10^{-22}	4.2×10^{-24}	33.81	7.9×10^9
$D^2(p,\gamma)He^3$	5.493		2.5×10^{-4}	7.9×10^{-6}	37.21	4.4×10^{-8}
$He^3(He^3,2p)He^4$	12.859		5.0×10^3		122.77	2.4×10^5
$He^3(\alpha,\gamma)Be^7$	1.586		4.7×10^{-1}	-2.8×10^{-4}	122.28	9.7×10^5
$Be^7(e^-,\nu)Li^7$	0.861	0.80				3.9×10^{-1}
$Li^7(p,\alpha)He^4$	17.347		1.2×10^2		84.73	1.8×10^{-5}
$Be^7(p,\gamma)B^8$	0.135		4.0×10^{-2}		102.65	6.6×10^1
$B^8(\beta^+\nu)Be^{8*}(\alpha)He^4$	18.074	7.2				3×10^{-8}

† Computed for $X = Y = 0.5$, $\rho = 100$, $T_6 = 15$ (sun).

analyses of all such possibilities indicate that the reactions listed in Fig. 5-7 are the significant ones.

The reactions of the PP chains are the most carefully studied of all the major sequences in nuclear astrophysics. Tombrello[1] has provided a vivid account of the interplay of theory and experiment in the study of these reactions. They are unusual in that the mechanisms of the reactions involve direct transitions from scattering states to bound final states rather than the mediation of a compound nuclear state. That feature has given these reactions extraordinary interest to nuclear physicists.

The functioning of the complete chain is described by the set of differential equations governing the abundances. If the only sources of abundance change are nuclear reactions, i.e., no expansions, mixing, etc., the time derivatives are

$$\frac{dH}{dt} = -2\lambda_{pp}\frac{H^2}{2} - \lambda_{pd}HD + 2\lambda_{33}\frac{(He^3)^2}{2} - \lambda_{17}HBe^7 - \lambda'_{17}HLi^7 \tag{5-32a}$$

$$\frac{dD}{dt} = \lambda_{pp}\frac{H^2}{2} - \lambda_{pd}HD \tag{5-32b}$$

$$\frac{dHe^3}{dt} = \lambda_{pd}HD - 2\lambda_{33}\frac{(He^3)^2}{2} - \lambda_{34}He^3He^4 \tag{5-32c}$$

$$\frac{dHe^4}{dt} = \lambda_{33}\frac{(He^3)^2}{2} - \lambda_{34}He^3He^4 + 2\lambda_{17}HBe^7 + 2\lambda'_{17}HLi^7 \tag{5-32d}$$

$$\frac{dBe^7}{dt} = \lambda_{34}He^3He^4 - \lambda_{e7}n_eBe^7 - \lambda_{17}HBe^7 \tag{5-32e}$$

$$\frac{dLi^7}{dt} = \lambda_{e7}n_eBe^7 - \lambda'_{17}HLi^7 \tag{5-32f}$$

[1] T. A. Tombrello, Astrophysical Problems, in J. B. Marion and D. M. Van Patter (eds.), "Nuclear Research with Low Energy Accelerators," Academic Press Inc., New York, 1967.

The notation of these equations is largely self-explanatory. The reaction rates per pair for $Be^7 + p$ and for $Li^7 + p$ are designated respectively by λ_{17} and λ'_{17}, and the rate of the Be^7 electron capture is proportional to the free-electron density n_e, the proportionality constant being λ_{e7}. One simplification has been made in writing this set of equations: the B^8 abundance has been eliminated. This is possible because the B^8 lifetime against positron decay is so short (0.78 sec) that the sequential reactions $Be^7(p,\gamma)\ B^8(\beta^+\nu)\ Be^8(\alpha)\ He^4$ can be considered as a simple step $Be^7 + H \rightarrow 2He + \nu$. Since Eqs. (5-32*a*) to (5-32*f*) are complicated and nonlinear, their solution can best be accomplished by using approximations valid under a prescribed set of circumstances. Consider the following sequential simplifications.

(*1*) Deuterium equilibrium occurs in times comparable to $\tau_p(D)$, which, as discussed previously, is a matter of seconds to hours. Thus dD/dt vanishes after short times, the deuterium equation (5-32*b*) can be eliminated, and the term $\lambda_{pd}HD$ in (5-32*a*) and (5-32*c*) can be replaced by $\lambda_{pp}H^2/2$.

(*2*) Li^7 and Be^7 equilibrium occurs on a relatively short time scale because at temperatures and densities of interest both nuclei have lifetimes of a year or less. This fact effectively means that those two abundances quickly come into equilibrium with He^3, because the sum of Eqs. (5-32*e*) and (5-32*f*) gives

$$\frac{d(Be^7 + Li^7)}{dt} = \lambda_{34}He^3He^4 - \lambda_{17}HBe^7 - \lambda'_{17}HLi^7 \approx 0 \tag{5-33}$$

Since this equation is true after times on the order of years, it means that Be^7 and Li^7 will thereafter follow the buildup of He^3. Equation (5-33) is satisfied long before He^3 achieves equilibrium. This result produces a greatly simplified equation for the rate of He^4 production:

$$\frac{dHe^4}{dt} = \lambda_{33}\frac{(He^3)^2}{2} + \lambda_{34}He^3He^4 \tag{5-34}$$

The fact that the last term appears with a positive sign even though the corresponding reaction actually *consumes* an alpha particle reflects the fact that *two* alpha particles are returned at the end of the cycle. The hydrogen equation becomes

$$\frac{dH}{dt} = -3\lambda_{pp}\frac{H^2}{2} + 2\lambda_{33}\frac{(He^3)^2}{2} - \lambda_{34}He^3He^4 \tag{5-35}$$

With the approximations made to this point, the equations involve only H, He^3, and He^4 and are accurate after a few years of high-temperature burning. The equations still have no simple solution. The procedure that must in general be followed is analogous to that in PPI: the three differential equations must be followed from time step to time step in the sequence of stellar models until He^3 has achieved equilibrium. An alternative if the star is static is to integrate Eq. (5-32*c*) with the aid of the assumptions that both H and He^4 are constants.

Actually, these assumptions are not bad, since the He^3 equilibrium abundance is generally so small that little H is destroyed and little He^4 synthesized in the time required for He^3 to approach equilibrium. This integration will not be presented explicitly here. The solution is much like the one obtained in PPI, and the time required for He^3 to achieve 99 percent of its equilibrium value is not greatly different from that shown in Fig. 5-6.

(*3*) The next simplification occurs when He^3 reaches equilibrium. The equilibrium abundance of He^3 will be somewhat less than in PPI because there is an additional reaction destroying He^3. The solution to the energy generation will again be simplified when He^3 achieves equilibrium. Then $dHe^3/dt \approx 0$, and from Eq. (5-32*c*)

$$\lambda_{pp}\frac{H^2}{2} - \lambda_{33}(He_e^3)^2 - \lambda_{34}He_e^3He^4 \approx 0 \tag{5-36}$$

which has the solution

$$He_e^3 = (2\lambda_{33})^{-1}\{-\lambda_{34}He^4 + [(\lambda_{34}He^4)^2 + 2\lambda_{pp}\lambda_{33}H^2]^{\frac{1}{2}}\} \tag{5-37}$$

Problem 5-10: Show that Eq. (5-37) has the proper limits as $He^4 \to 0$ and as $H \to 0$. Show also that in terms of the fractions by weight

$$\left(\frac{He^3}{H}\right)_e = (8\lambda_{33})^{-1}\left\{\frac{-\lambda_{34}Y}{X} + \left[\left(\frac{\lambda_{34}Y}{X}\right)^2 + 32\lambda_{pp}\lambda_{33}\right]^{\frac{1}{2}}\right\} \tag{5-38}$$

It is actually Eq. (5-38) with $X = Y = 0.5$ that was plotted in Fig. 5-5, rather than the PPI approximation. The two expressions differ under those circumstances only for $T_6 > 14$, however, because for $T_6 < 14$ and $X = Y$, the He^3 is primarily destroyed by He^3. Only for $T_6 > 14$ (if $X = Y$) does the lifetime of He^3 against He^4 become shorter than the lifetime against itself.

It is important to have a good physical appreciation of the competition for the He^3 nuclei, because it is that competition which determines whether it is PPI or PPII and PPIII that dominate. The ratio of the rate of alpha-particle production by PPI to the rate of alpha-particle production by PPII and PPIII is

$$\frac{\text{PPI}}{\text{PPII} + \text{PPIII}} = \frac{r_{33}}{r_{34}} = \frac{\lambda_{33}(He^3)^2/2}{\lambda_{34}He^3He^4} = \frac{1}{2}\frac{\lambda_{33}}{\lambda_{34}}\frac{He^3}{He^4} \tag{5-39}$$

One can see right away that this branching ratio (1) varies inversely with He^4 (or Y), and thus will change with age even if ρ and T remain constant; (2) is zero initially ($He^3 = 0$) and rises to a maximum as He^3 builds to its equilibrium concentration; and (3) decreases with temperature since $(He^3/H)_e$ decreases rapidly with increasing temperature. Once He^3 equilibrium has been achieved, the ratio is

$$\left(\frac{\text{PPI}}{\text{PPII} + \text{PPIII}}\right)_e = \frac{1}{2}\frac{\lambda_{33}}{\lambda_{34}}\frac{He_e^3}{He^4} \tag{5-40}$$

which by Eq. (5-37) becomes

$$\left(\frac{\text{PPI}}{\text{PPII} + \text{PPIII}}\right)_e = \frac{-\lambda_{34} + [\lambda_{34}^2 + 2\lambda_{pp}\lambda_{33}(\text{H}/\text{He}^4)^2]^{\frac{1}{2}}}{4\lambda_{34}} \tag{5-41}$$

For analysis of this and other related problems, it is convenient to define a function of temperature and composition:

$$\alpha\left(T, \frac{\text{He}^4}{\text{H}}\right) \equiv \frac{\lambda_{34}^2}{\lambda_{33}\lambda_{pp}}\left(\frac{\text{He}^4}{\text{H}}\right)^2 \tag{5-42}$$

Problem 5-11: Show that with this definition of α

$$\frac{\text{PPI}}{\text{PPII} + \text{PPIII}} = \frac{(1 + 2/\alpha)^{\frac{1}{2}} - 1}{4} \tag{5-43}$$

Show also that

$$\alpha\left(T, \frac{\text{He}^4}{\text{H}}\right) \approx \frac{S_{34}^2}{S_{11}S_{33}}\left(\frac{\text{He}^4}{\text{H}}\right)^2 \exp\,(-100T_6^{-\frac{1}{3}})$$

$$= 1.2 \times 10^{17}\left(\frac{\text{He}^4}{\text{H}}\right)^2 \exp\,(-100T_6^{-\frac{1}{3}}) \tag{5-44}$$

From the value of α it is a simple matter to see the relative numbers of alpha particles made by the $\text{He}^3 + \text{He}^3$ interaction and the $\text{He}^3 + \text{He}^4$ interaction. Figure 5-8 shows the fraction of He^3 nuclei destroyed by He^4 and by He_e^3 in the special case $X = Y$, that is, $\text{He}/\text{H} = \frac{1}{4}$. It can be seen in this special case that the crossover occurs near $T_6 = 14$.

Problem 5-12: Show that the rate of production of alpha particles in PPI equals that in PPII and PPIII for a value $\alpha = \frac{1}{12}$. Show that if $X = Y$, this value of α occurs near $T_6 = 14$.

The system of differential equations is further simplified when He^3 achieves equilibrium:

$$\frac{d\text{He}^4}{dt} = \tfrac{1}{4}\lambda_{pp}\text{H}^2 + \tfrac{1}{2}\lambda_{34}\text{He}_e^3\text{He}^4 \tag{5-45}$$

$$\frac{d\text{H}}{dt} = -\lambda_{pp}\text{H}^2 - 2\lambda_{34}\text{He}_e^3\text{He}^4 \tag{5-46}$$

Thus when the entire PP chains operate in equilibrium,

$$\frac{d\text{He}^4}{dt} = -\frac{1}{4}\frac{d\text{H}}{dt} \tag{5-47}$$

as it must for $4\text{H} \rightarrow \text{He}^4$. It is conventional to write the He^4 production rate as its value for PPI times a correction factor brought about by the operation of PPII and PPIII:

$$\frac{d\text{He}^4}{dt} = \tfrac{1}{2}\lambda_{pp}\frac{\text{H}^2}{2}\left(1 + \frac{2\lambda_{34}\text{He}_e^3\text{He}^4}{\lambda_{pp}\text{H}^2}\right) \tag{5-48}$$

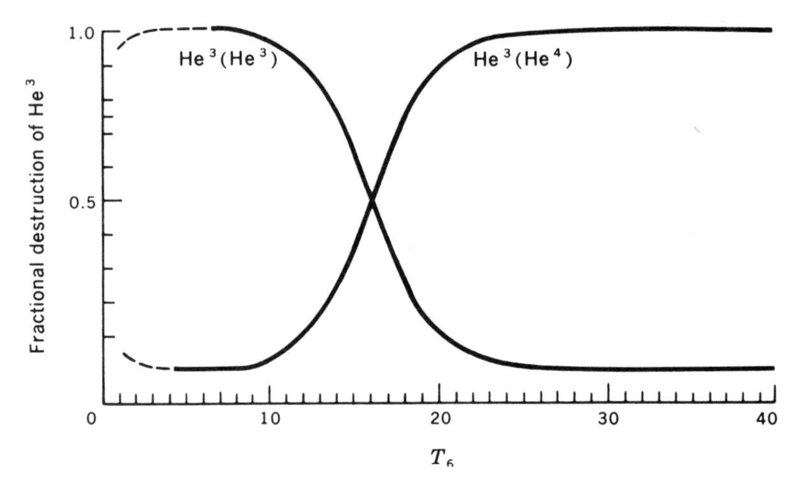

Fig. 5-8 The fraction of He^3 nuclei destroyed by reactions with an equilibrium concentration of He^3 and by reactions with He^4, respectively. For purposes of this calculation the mass fractions of hydrogen and helium are taken to be equal, so that the relative rate of destruction due to He^4 must be moved either up or down by the factor Y/X. For any common mixture, however, destruction of He^3 by reactions with He^4 dominates for $T_6 > 14$. [*After P. D. Parker, J. N. Bahcall, and W. A. Fowler, Astrophys. J.*, **139**:602 (1964). *By permission of The University of Chicago Press. Copyright* 1964 *by The University of Chicago.*]

Problem 5-13: Show by inserting the value of He^3_e into Eq. (5-48) that the He^4 production rate can be written

$$\frac{dHe^4}{dt} = \tfrac{1}{2}\lambda_{pp}\frac{H^2}{2}\Phi(\alpha) \tag{5-49}$$

where

$$\Phi(\alpha) = 1 - \alpha + \alpha\left(1 + \frac{2}{\alpha}\right)^{\frac{1}{2}} \tag{5-50}$$

Notice that $\Phi(\alpha) \to 1$ as $\alpha \to 0$, that is, $He^4 \to 0$, which corresponds to the operation of PPI by itself. As $\alpha \to \infty$, which implies $Y/X \to \infty$ and high temperature, $\Phi(\alpha) \to 2$. This last feature reflects the fact that only one *p-p* reaction is required for each alpha particle when the chain is completed exclusively by PPII and PPIII. Figure 5-9 displays the function $\Phi(\alpha)$ for the specific composition $X = Y$, a restriction which allows Φ to be displayed as a function only of temperature. It is clear that for $X = Y$ the transition from PPI to PPII and PPIII occurs smoothly between $T_6 = 13$ and $T_6 = 20$.

The energy-generation rate would be solved by the foregoing were it not for the fact that the neutrino losses are different in each PP chain. If there were no neutrino losses, the energy-generation rate would be the product of Eq. (5-49) and the energy released in the fusion of each He^4 atom from four H atoms. From the third column of Table 5-1, however, it is apparent that the neutrino losses differ markedly:

(*i*) PPI: $\frac{2 \times 0.263}{26.73} = 2.0\%$

(*ii*) PPII: $\frac{0.263 + 0.80}{26.73} = 4.0\%$

(*iii*) PPIII: $\frac{0.263 + 7.2}{26.73} = 27.9\%$

Thus the *total rate* of energy liberation, i.e., including neutrino energy, must be multiplied by $0.980F_{\text{PPI}} + 0.960F_{\text{PPII}} + 0.721F_{\text{PPIII}}$, where F_{PPI} is the fraction of the alpha particles produced by the PPI set of reactions, F_{PPII} is the fraction of the alpha particles produced by the PPII set of reactions, and F_{PPIII} is the fraction via PPIII. The rate of energy generation is

$$\rho\epsilon = \frac{d\text{He}^4}{dt}(4M_{\text{H}} - M_{\text{He}^4})c^2(0.980F_{\text{PPI}} + 0.960F_{\text{PPII}} + 0.721F_{\text{PPIII}}) \tag{5-51}$$

Since the case $F_{\text{PPI}} = 1$ reduces to PPI, it follows that

$$\epsilon = \frac{\epsilon_{\text{PPI}}}{0.980}\Phi(\alpha)(0.980F_{\text{PPI}} + 0.960F_{\text{PPII}} + 0.721F_{\text{PPIII}}) \tag{5-52}$$

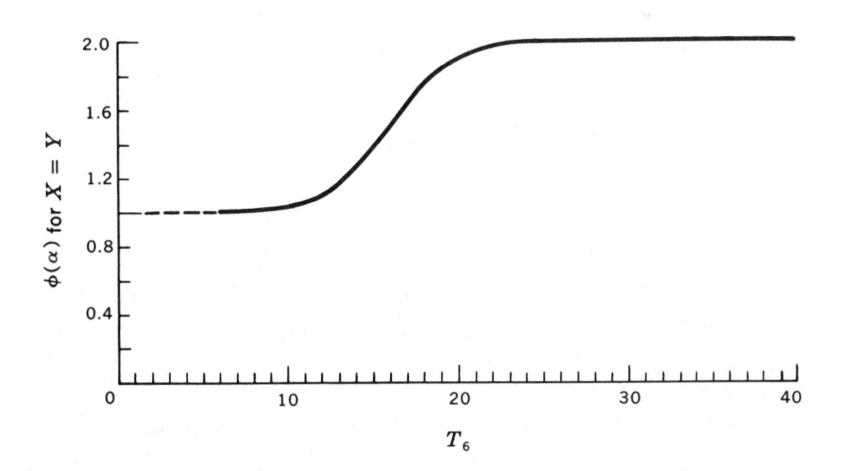

Fig. 5-9 The rate of production of He^4 is increased over its rate in PPI by a factor $\Phi(\alpha)$, which is shown here as a function of temperature for the particular composition $X = Y$. [*After P. D. Parker, J. N. Bahcall, and W. A. Fowler, Astrophys. J.,* **139**:602 (1964). *By permission of The University of Chicago Press. Copyright* 1964 *by The University of Chicago.*]

Thus the energy generation for the complete PP chains operating in equilibrium can be written with the aid of Eq. (5-30) as

$$\epsilon = 2.36 \times 10^6 \rho X_H{}^2 T_6{}^{-2/3} \exp(-33.81 T_6{}^{-1/3}) \psi_\epsilon(\alpha)(1 + 0.0123 T_6{}^{1/3} + 0.0109 T_6{}^{2/3} + 0.00095 T_6) \qquad \text{erg g}^{-1}\text{ sec}^{-1} \tag{5-53}$$

where

$$\psi_\epsilon(\alpha) = \Phi(\alpha)(0.980 F_{\text{PPI}} + 0.960 F_{\text{PPII}} + 0.721 F_{\text{PPIII}}) \tag{5-54}$$

It is apparent that the function $\psi_\epsilon(\alpha)$ represents the correction of the PPI energy-generation rate necessitated by the simultaneous occurrence of PPII and PPIII.

Problem 5-14: Show that Eq. (5-53) reduces to Eq. (5-30) for $Y = 0$.

The F factors must be calculated from the branching in the nuclear reactions. From the discussion of the competition between He^3_e and He^4 for the He^3 nuclei leading up to Eq. (5-43) it follows that

$$\frac{F_{\text{PPI}}}{1 - F_{\text{PPI}}} = \frac{(1 + 2/\alpha)^{1/2} - 1}{4}$$

Problem 5-15: Show that

$$F_{\text{PPI}} = \left[\left(1 + \frac{2}{\alpha}\right)^{1/2} - 1\right]\left[\left(1 + \frac{2}{\alpha}\right)^{1/2} + 3\right]^{-1} \tag{5-55}$$

The competition between PPII and PPIII rests upon the fate of the Be^7 nucleus created by $He^3(He^4,\gamma)Be^7$ reaction. In the laboratory Be^7 decays by capturing a $1S$ electron from its innermost atomic shell:

$$Be^7 + e^- \rightarrow Li^7 + \nu \qquad \tau_{1/2} = 53 \text{ days}$$

The rate of this reaction is, among other things, proportional to the density of $1S$ electrons at the nucleus, viz., $|\psi_{1S}(r = 0)|^2$. In the stellar interior the Be^7 nucleus is ionized and would be stable were it not for its encounters with the free electrons in the gas. Since the other factors in the reaction are approximately the same, the decay rate is multiplied by the ratio of the free-electron density at the nucleus in a star to the bound-electron density at the nucleus in the atom. The calculation is relatively straightforward. Rather than assume a uniform density of free electrons, however, one must use the coulomb wave functions appropriate to an electron scattering from a $4e$ charge. These coulomb waves introduce a $1/v$ factor in the rate. When this $1/v$ is averaged over the maxwellian electron distribution (assuming nondegeneracy), the $1/v$ becomes a factor $T^{-1/2}$.[1] The result of the calculation is

$$\frac{1}{\tau_e(Be^7)} = \lambda_{e7} n_e = 7.05 \times 10^{-33} n_e T_6{}^{-1/2} \qquad \text{sec}^{-1} \tag{5-56}$$

[1] This type of average occurred also in the free-free opacity in Chap. 3. The calculation was made by J. N. Bahcall, *Phys. Rev.*, **128**:1297 (1962).

Near the centers of characteristic main-sequence stars the Be^7 lifetime against electron capture is about 1 year. It is also evident that this lifetime is, in contrast with nuclear lifetimes against positive particles, only very weakly dependent upon the temperature. Specifically, the lifetime of Be^7 against protons, which is the competing reaction leading to the PPIII branch, is strongly temperature-dependent.

Problem 5-16: Show that

$$\frac{1}{\tau_p(\mathrm{Be}^7)} = \lambda_{17} n_p = 6.3 \times 10^{-17} n_p T_6^{-\frac{2}{3}} \exp\left(-102.65 T_6^{-\frac{1}{3}}\right) \tag{5-57}$$

This lifetime is greater than τ_e for $T_6 < 23$ but becomes considerably shorter at higher temperature.

Problem 5-17: Show that

$$F_{\mathrm{PPII}} = (1 - F_{\mathrm{PPI}}) \frac{\tau_p(\mathrm{Be}^7)}{\tau_p(\mathrm{Be}^7) + \tau_e(\mathrm{Be}^7)} \tag{5-58}$$

and $F_{\mathrm{PPIII}} = 1 - F_{\mathrm{PPI}} - F_{\mathrm{PPII}}$.

An example of F_{PPI}, F_{PPII}, and F_{PPIII} calculated for the composition $X = Y$ is shown in Fig. 5-10. This example shows the characteristic feature of the PP

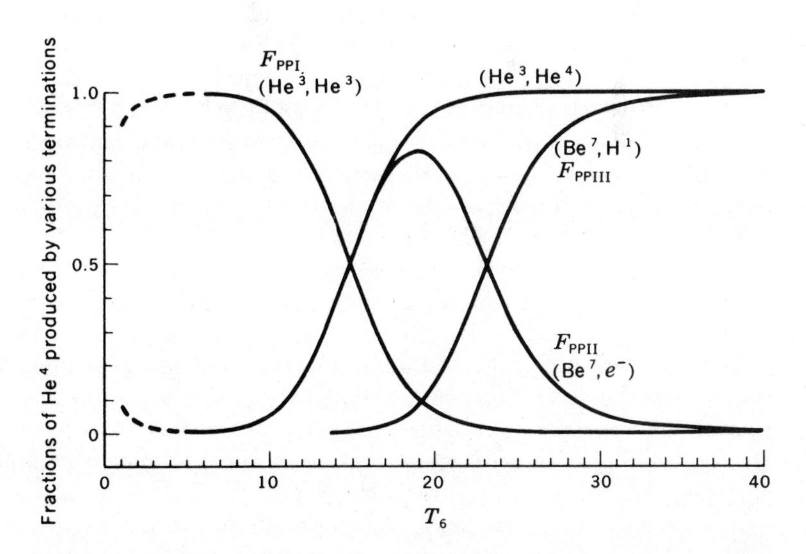

Fig. 5-10 The fraction of the He^4 production due to PPI, PPII, and PPIII, respectively. The chains are assumed to be in equilibrium, and for the purpose of this figure it was assumed that $Y = X$. [*After P. D. Parker, J. N. Bahcall, and W. A. Fowler, Astrophys. J.*, **139**:602 (1964). *By permission of The University of Chicago Press. Copyright* 1964 *by The University of Chicago.*]

chains. The alphas are produced predominantly by PPI at low temperatures. Near $T_6 = 14$, depending moderately upon the value Y/X, PPII takes over from PPI. Near $T_6 = 23$, depending moderately on the hydrogen mass fraction X, PPIII takes over from PPII. Of course all three modes operate simultaneously. Bahcall and Wolf[1] have considered other modes of completion that involve electron captures and have shown them to be important only for $\rho > 10^4$ g/cm³.

Considerable interest currently exists in the operation of PPIII in the sun, because the decay $B^8(\beta^+\nu)Be^8$, which occurs only in PPIII, is the major solar source of neutrinos of sufficiently high energy to be absorbed efficiently by a Cl^{37} target. This reaction, $Cl^{37} + \nu \rightarrow Ar^{37} + e^-$, is endothermic ($Q = -0.81$ Mev) and thus can detect only the $Be^7 + e^-$ neutrinos and the B^8 neutrinos. Since the central temperature of the sun is about 16 million °K, the energy generation is predominantly via PPII. For this reason, the neutrino flux from Be^7 is much greater than that from B^8. These neutrino fluxes have been calculated from solar models. The results of Sears[2] are

$$\Phi_\nu(Be^7) = (1.2 \pm 0.5) \times 10^{10} \; \nu \; cm^{-2} \; sec^{-1}$$

$$\Phi_\nu(B^8) = (2.25 \pm 1) \times 10^{7} \; \nu \; cm^{-2} \; sec^{-1}$$

Although the Be^7 neutrinos are more abundant by a factor of about 500, the B^8 neutrinos are actually more capable of producing the Cl^{37} absorption reaction. This curious fact comes about because only the B^8 neutrinos have sufficient energy to make the more favorable transitions to excited states of the Ar^{37} nucleus. Bahcall[3] has computed the cross sections and shown that about 90 percent of the absorptions are due to the B^8 neutrinos and about 10 percent to the Be^7 neutrinos. But even so, the cross sections are very small, and the rate of absorption of neutrinos from all transitions is quoted as $(4 \pm 2) \times 10^{-35}$ sec^{-1} per Cl^{37} atom. Thus it is that any detector of these neutrinos must contain a lot of chlorine. Nonetheless Davis[4] has undertaken the sizable task of building a detector with 100,000 gal of C_2Cl_4 at the bottom of a mine. From his experience with a similar 1,000-gal tank, Davis believes the solar neutrino flux can be measured. This measurement not only would be an experimental tour de force but would produce an astrophysical datum of great importance. There is no way known other than by neutrinos to see into a stellar interior. There is no other known direct experimental observation of nuclear reactions occurring at high temperatures in the center of a star. The rather elaborate theoretical structure of stars is built upon inference from known physical principles rather than from measured facts. Of course, the inferences are cogent ones and are generally accepted as being correct. There can be no doubting, however, the philosophical importance of direct measurement. It flies in the face of history to say there will be no surprises.

[1] J. N. Bahcall and R. A. Wolf, *Astrophys. J.*, **139**:622 (1964).

[2] R. L. Sears, *Astrophys. J.*, **140**:477 (1964). This paper also discusses uncertainties in the neutrino fluxes.

[3] J. N. Bahcall, *Phys. Rev. Letters*, **12**:300 (1964).

[4] R. Davis, Jr., *Phys. Rev. Letters*, **12**:302 (1964); R. Davis, Jr., D. S. Harmer and K. C. Hoffman, *Phys. Rev. Letters*, **20**:1205 (1968).

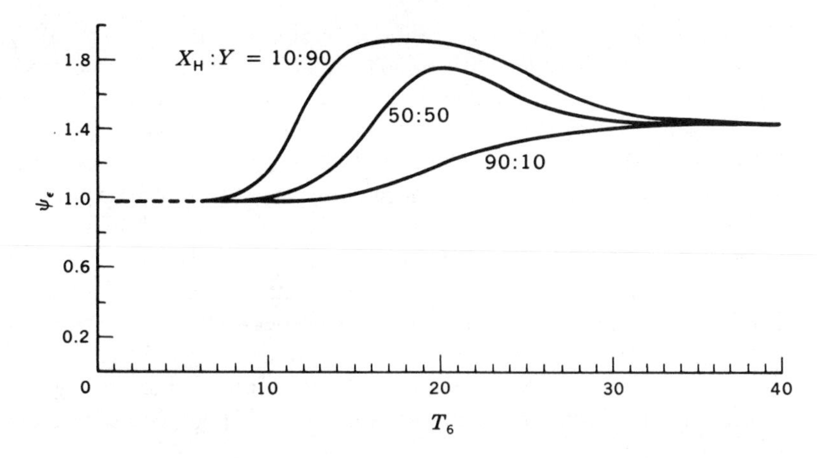

Fig. 5-11 The function ψ_ϵ, which measures the rate of thermal-energy release relative to the rate of the proton-proton reaction, is plotted for three different choices of composition. [*After P. D. Parker, J. N. Bahcall, and W. A. Fowler, Astrophys. J.*, **139**:602 (1964). *By permission of The University of Chicago Press. Copyright* 1964 *by The University of Chicago.*]

Even if our general understanding is correct, a good measurement can yield the central temperature of the sun. The branching between PPII and PPIII is strongly temperature-dependent. Bahcall has stated that a measurement of the B^8 neutrino flux accurate to ± 50 percent would fix the central solar temperature to better than ± 10 percent.

Returning to the energy generation for a moment, it is apparent that $\psi_\epsilon(\alpha)/0.981$ is the factor that multiplies the rate ϵ_{PPI} to convert it to the energy-generation rate for the complete chain. This function $\psi_\epsilon(\alpha)$ is a function of the temperature and the composition. Figure 5-11 displays it as a function of temperature for three different ratios of He^4/H. The main features of the function ψ_ϵ are easily understandable:

(*1*) $\psi \rightarrow 0.98$ at $T_6 = 8$ to 10, where only PPI is effective.
(*2*) $\psi \rightarrow 1.44$ at high T, since the alpha-particle production rate doubles $[\Phi(\alpha) \rightarrow 2]$ but only 72 percent of the energy release is converted to heat in PPIII, which dominates.
(*3*) The intermediate maximum in ψ_ϵ corresponds to domination by PPII, for which the relative rate of alpha-particle production $\Phi(\alpha)$ is approaching the value 2, but the neutrino losses are not large.

It is apparent that the computation of the energy-generation rate even when He^3 is in equilibrium is somewhat involved. In the construction of stellar models on electronic computers a fairly elaborate subroutine would be required to calculate ϵ as accurately as possible. Such a subroutine may or may not be desirable in all cases. One may prefer to approximate the energy generation by simpler expressions that are accurate over a limited range of temperature and composition. When a small error in ϵ is introduced into a structure computation, the end result is generally a star having a much smaller percentage error in the temperature at each point. Any such approximations can easily be constructed for specific problems when the more complete solution has been appreciated.

A final comment on the PP chains concerns the abundance of Li^7. Although it is produced in PPII, the Li^7 is destroyed so rapidly that its abundance at any point in hydrogen burning is almost vanishingly small. It seems certain that the Li^7 in the universe is not the result of hydrogen burning. The Li^7 abundance is rather linked to the whole problem of the origin of the isotopes of Li, Be, and B. The nuclei are generally believed to be the results of high-energy spallation reactions in locations other than stellar interiors or of poorly understood circumstances in supernova explosions.

5-4 THE CNO BI-CYCLE

The PP chains must be invoked to synthesize He^4 from hydrogen in a gas consisting only of helium and hydrogen. If any stellar systems are formed of essentially pure hydrogen, their main-sequence stars can obtain energy only from those PP chains. With the exception of the extreme population II objects (globular clusters, high-velocity subdwarfs, etc.), however, most stars have apparently formed from gas having a healthy admixture of the heavier elements. It then becomes necessary to consider other reactions as possible sources of energy. Because lifetimes rise rapidly with increasing coulomb barrier, the reactants must have nuclear charges such that the product Z_1Z_2 is as small as possible. To provide significant energy generation, moreover, the reactants must also be abundant. It was independently suggested by Bethe and by von Weizsäcker in 1938 that reactions of protons with carbon and nitrogen nuclei would provide competition with PP chains. They showed that a series of reactions, called the CN cycle, had the property that the CN nuclei served only as catalysts for the conversion of hydrogen to helium but were not themselves destroyed. The basic CN cycle is as follows:

$$\begin{aligned} &C^{12}(p,\gamma)N^{13}(\beta^+\nu)C^{13} \\ &C^{13}(p,\gamma)N^{14} \\ &N^{14}(p,\gamma)O^{15}(\beta^+\nu)N^{15} \\ &N^{15}(p,\alpha)C^{12} \end{aligned} \tag{5-59}$$

By summing the particles before and after the cycle one obtains

$$C^{12} + 4H \rightarrow C^{12} + He^4 + 2\beta^+ + 2\nu$$

The C^{12} nucleus only plays the role of catalyst. It is also evident that the same cycle occurs with any of the four nuclei C^{12}, C^{13}, N^{14}, or N^{15} as catalysts, or any

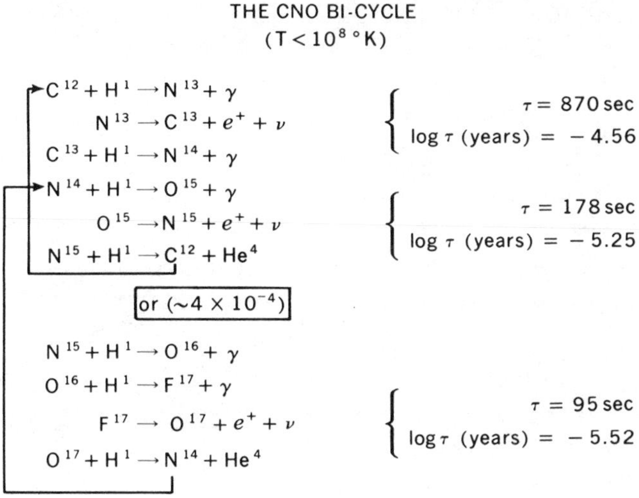

THE CNO BI-CYCLE
($T < 10^8\,°K$)

$C^{12} + H^1 \rightarrow N^{13} + \gamma$
$N^{13} \rightarrow C^{13} + e^+ + \nu$ — $\tau = 870$ sec, $\log \tau$ (years) $= -4.56$
$C^{13} + H^1 \rightarrow N^{14} + \gamma$
$N^{14} + H^1 \rightarrow O^{15} + \gamma$
$O^{15} \rightarrow N^{15} + e^+ + \nu$ — $\tau = 178$ sec, $\log \tau$ (years) $= -5.25$
$N^{15} + H^1 \rightarrow C^{12} + He^4$

or ($\sim 4 \times 10^{-4}$)

$N^{15} + H^1 \rightarrow O^{16} + \gamma$
$O^{16} + H^1 \rightarrow F^{17} + \gamma$
$F^{17} \rightarrow O^{17} + e^+ + \nu$ — $\tau = 95$ sec, $\log \tau$ (years) $= -5.52$
$O^{17} + H^1 \rightarrow N^{14} + He^4$

Fig. 5-12 The reactions of the CNO bi-cycle.

mixture of them. In fact a mixture of those nuclei must soon result regardless of the initial composition. It is also clear that the energy generation per catalyst will be related to the reciprocal of the time it takes to go around the cycle multiplied by $(4M_H - M_{He^4})c^2$ minus the energy loss in the two neutrinos.

It was later realized that oxygen plays a role in the CN cycle. If oxygen is initially present, the reactions

$$O^{16}(p,\gamma)F^{17}(\beta^+\nu)O^{17}$$
$$O^{17}(p,\alpha)N^{14} \tag{5-60}$$

feed nuclei into the CN subcycle at N^{14}. It was also discovered that the bombardment of N^{15} with protons does not always result in $C^{12} + \alpha$: about 4 times in 10^4 the result is $O^{16} + \gamma$. Thus CN nuclei are slowly drained into O^{16}, and O^{16} nuclei are reinjected into the CN cycle at N^{14}. There are really two cycles, and the combination has been christened the CNO bi-cycle. Figure 5-12 shows the full set of reactions and the positron decay lifetimes. The other lifetimes depend, of course, on the proton density and the temperature. The cross-section factors and other relevant data for each reaction are listed in Table 5-2. The quantity B in the last column is the same one as defined in Eq. (4-59).

The key to understanding the CNO-cycle reactions lies in appreciating the lifetimes of the nuclei against protons. From Eq. (4-22) the lifetime of species 2 against protons is $\tau_p(2) = (\lambda_{p2}N_p)^{-1}$.

Problem 5-18: Show that the lifetime in years for nonresonant reactions is given by

$$\frac{1}{\tau_p(2)} = 2.45 \times 10^{16} \rho X_H f S_0 \left[\frac{(A_2+1)Z_2}{A_2}\right]^{\frac{1}{3}} T_6^{-\frac{2}{3}} \left(1 + \frac{5}{12BT_6^{\frac{1}{3}}}\right) \exp\left(-BT_6^{-\frac{1}{3}}\right) \quad \text{yr}^{-1} \tag{5-61}$$

Table 5-2 The CNO reactions

Reaction	*Q value, Mev*	*Average ν loss, Mev*	*S(E = 0), kev barns*	$\frac{dS}{dE}$, *barns*	*B*
$C^{12}(p,\gamma)N^{13}$	1.944		1.40	4.26×10^{-3}	136.93
$N^{13}(\beta^+\nu)C^{13}$	2.221	0.710			
$C^{13}(p,\gamma)N^{14}$	7.550		5.50	1.34×10^{-2}	137.20
$N^{14}(p,\gamma)O^{15}$	7.293		2.75		152.31
$O^{15}(\beta^+,\nu)N^{15}$	2.761	1.00			
$N^{15}(p,\alpha)C^{12}$	4.965		5.34×10^4	8.22×10^2	152.54
$N^{15}(p,\gamma)O^{16}$	12.126		2.74×10^1	1.86×10^{-1}	152.54
$O^{16}(p,\gamma)F^{17}$	0.601		1.03×10^1	-2.81×10^{-2}	166.96
$F^{17}(\beta^+\nu)O^{17}$	2.762	0.94			
$O^{17}(p,\alpha)N^{14}$	1.193		Resonant reaction		167.15

where

$$S_0 = S(E=0) + \frac{dS}{dE}(E_0 + \tfrac{5}{6}kT) \qquad \text{kev barns}$$

$$= S(E=0) + \frac{dS}{dE}\left[1.220\left(\frac{Z_2{}^2A_2T_6{}^2}{A_2+1}\right)^{\frac{1}{3}} + 0.072T_6\right] \tag{5-62}$$

and f is the electron-screening factor.

From Table 5-2 it can be seen that all the CNO reactions, with the exception of $O^{17}(p,\alpha)N^{14}$, are presently believed to be nonresonant. This belief is based upon research into the structure of the compound nuclei of the several reactions.[1] Only the $O^{17} + p$ reaction has been found to have resonant states of the compound nucleus (F^{18} in that case) so close in energy to the range of stellar energies, i.e., near E_0, that the resonant-reaction-rate formula must be used. For others, the data are much like those for the $C^{12}(p,\gamma)N^{13}$ reaction, which was used as a nonresonant example throughout Chap. 4. Equation (5-61) can therefore be used with the nuclear information of Table 5-2 to calculate the lifetimes of all the species except O^{17}. Since the lifetimes are inversely proportional to the hydrogen abundance, the product $\tau\rho X_H$ is a function only of the temperature (except for a weak density and composition dependence in the electron-screening factor f, which is, however, not very important in hydrogen burning). Since ρX_H may be of order of magnitude 100 g/cm³ at the centers of main-sequence stars, the logarithm of the product

$$\log \frac{\rho X_H}{100} \tau_p$$

has been tabulated for each reaction in Table 5-3, with τ_p expressed in years.

[1] Much of this research has been conducted at the Kellogg Radiation Laboratory of the California Institute of Technology, which, under the guidance of W. A. Fowler, has made a specialty of the CNO reactions. Most of the reactions proceed via a compound nucleus, exceptions being found in the important captures by N^{14} and O^{16}.

Table 5-3 Dependence of $\log\ (\tau\rho X_{\mathrm{H}}/100)$ on temperature†

Temperature, T_6	*Reaction‡* $C^{12}(p,\gamma)N^{13}$	$C^{13}(p,\gamma)N^{14}$	$N^{14}(p,\gamma)O^{15}$	$N^{15}(p,\alpha)C^{12}$	$10^4\gamma$	$O^{16}(p,\gamma)F^{17}$	$O^{17}(p,\alpha)N^{14}$
5	16.32	15.73	19.79	15.53	4.649	22.95	21.92
6	14.32	13.73	17.57	13.29	4.598	20.51	20.02
7	12.72	12.13	15.79	11.50	4.551	18.56	18.26
8	11.41	10.81	14.32	10.03	4.508	16.95	16.50
9	10.29	9.69	13.08	8.78	4.468	15.59	15.10
10	9.33	8.73	12.02	7.70	4.431	14.42	14.05
11	8.50	7.90	11.09	6.76	4.396	13.39	13.15
12	7.75	7.15	10.26	5.93	4.363	12.49	12.38
13	7.09	6.49	9.52	5.18	4.332	11.68	11.68
14	6.49	5.89	8.86	4.51	4.303	10.95	11.02
15	5.95	5.35	8.26	3.90	4.275	10.29	10.32
16	5.45	4.85	7.71	3.34	4.248	9.68	9.55
17	5.00	4.39	7.20	2.83	4.223	9.13	8.70
18	4.58	3.97	6.73	2.35	4.198	8.61	7.86
19	4.18	3.58	6.30	1.91	4.175	8.14	7.01
20	3.82	3.21	5.89	1.50	4.152	7.69	6.18
22	3.16	2.55	5.16	0.75	4.110	6.89	4.78
24	2.57	1.97	4.51	0.09	4.071	6.18	3.63
25	2.30	1.70	4.21	−0.21	4.052	5.85	3.10
26	2.05	1.44	3.93	−0.50	4.034	5.54	2.62
28	1.58	0.97	3.41	−1.03	4.000	4.97	1.75
30	1.15	0.54	2.93	−1.51	3.967	4.45	1.05
35	0.23	−0.38	1.91	−2.55	3.893	3.33	−0.42
40	−0.53	−1.14	1.07	−3.42	3.829	2.41	−1.50
45	−1.18	−1.78	0.36	−4.14	3.771	1.64	−2.33
50	−1.73	−2.33	−0.25	−4.77	3.719	0.97	−2.99
55	−2.21	−2.82	−0.78	−5.32	3.673	0.39	−3.53
60	−2.64	−3.24	−1.25	−5.81	3.630	−0.12	−3.97
65	−3.02	−3.63	−1.67	−6.24	3.590	−0.58	−4.33
70	−3.37	−3.97	−2.05	−6.63	3.554	−0.99	−4.65
75	−3.68	−4.28	−2.39	−6.99	3.521	−1.37	−4.91
80	−3.97	−4.57	−2.71	−7.32	3.489	−1.71	−5.14
85	−4.23	−4.83	−2.99	−7.62	3.460	−2.02	−5.35
90	−4.48	−5.08	−3.26	−7.90	3.433	−2.31	−5.52
95	−4.70	−5.30	−3.51	−8.15	3.407	−2.58	−5.68
100	−4.91	−5.51	−3.74	−8.39	3.383	−2.83	−5.82

† Adapted from G. R. Caughlan and W. A. Fowler, *Astrophys. J.*, **136**:453 (1962). By permission of The University of Chicago Press. Copyright 1962 by The University of Chicago.

‡ The lifetimes against protons are expressed in years, and the density ρ is in grams per cubic centimeter.

Also listed in Table 5-3 is the parameter $\gamma = 1 - \alpha$, which is the fraction of the $N^{15} + p$ reactions which occurs in the $O^{16} + \gamma$ channel. This quantity is approximately $\gamma = 4 \times 10^{-4}$ but has a weak temperature dependence due to slight changes with temperature of the excitation energy of the compound nucleus. From the point of view of the compound nucleus, there is a probability proportional to Γ_α that $N^{15} + p \rightarrow O^{16*}$ will decay into $O^{16*} \rightarrow C^{12} + \alpha$ and a probability proportional to Γ_γ that it will decay into $O^{16*} \rightarrow O^{16} + \gamma$. The first probability is far greater in this particular case, but both quantities depend somewhat upon the energy of O^{16*}. Then $\gamma/\alpha = \langle \Gamma_\gamma/\Gamma_\alpha \rangle$, where the brackets indicate an average of the ratio over the energies corresponding to the maxwellian distribution of proton energies responsible for the state. The role of the quantity γ will be discussed later. From Table 5-3 it can be seen that near $T_6 = 25$, which is characteristic of most CNO burning temperatures, the sequence of lifetimes is, in increasing order,[1] τ_{15}, τ_{13}, τ_{12}, τ_{17}, τ_{14}, and τ_{16}. That is, the fastest-burning species is N^{15}, and the slowest is O^{16}. In performing the calculations resulting in Table 5-3, Caughlan and Fowler treated the electron screening in an approximate way by arguing that for main-sequence stars one has $\log f \approx 0.025 Z_1 Z_2$. If more exact expressions for f are used, the entries in Table 5-3 should therefore be decreased by the amount $\log f - 0.025 Z_1 Z_2$. The lifetimes serve as a guide to physical approximations necessary for subsequent analysis of the CNO bi-cycle.

The differential equations for the CNO nuclei are

$$\frac{dC^{12}}{dt} = -\lambda_{p12} HC^{12} + \alpha \lambda_{p15} HN^{15} = -\frac{C^{12}}{\tau_{12}} + \alpha \frac{N^{15}}{\tau_{15}} \tag{5-63a}$$

$$\frac{dN^{13}}{dt} = \frac{C^{12}}{\tau_{12}} - \frac{N^{13}}{\tau_\beta(13)} - \frac{N^{13}}{\tau_p(N^{13})} \qquad \tau_\beta(N^{13}) = 870 \text{ sec} \tag{5-63b}$$

$$\frac{dC^{13}}{dt} = \frac{N^{13}}{\tau_\beta(13)} - \frac{C^{13}}{\tau_{13}} \tag{5-63c}$$

$$\frac{dN^{14}}{dt} = \frac{C^{13}}{\tau_{13}} - \frac{N^{14}}{\tau_{14}} + \frac{O^{17}}{\tau_{17}} \tag{5-63d}$$

$$\frac{dO^{15}}{dt} = \frac{N^{14}}{\tau^{14}} - \frac{O^{15}}{\tau_\beta(15)} - \frac{O^{15}}{\tau_p(O^{15})} \qquad \tau_\beta(O^{15}) = 178 \text{ sec} \tag{5-63e}$$

$$\frac{dN^{15}}{dt} = \frac{O^{15}}{\tau_\beta(15)} - \frac{N^{15}}{\tau_{15}} \tag{5-63f}$$

$$\frac{dO^{16}}{dt} = \gamma \frac{N^{15}}{\tau_{15}} - \frac{O^{16}}{\tau_{16}} \tag{5-63g}$$

$$\frac{dF^{17}}{dt} = \frac{O^{16}}{\tau_{16}} - \frac{F^{17}}{\tau_\beta(17)} - \frac{F^{17}}{\tau_p(F^{17})} \qquad \tau_\beta(F^{17}) = 95 \text{ sec} \tag{5-63h}$$

$$\frac{dO^{17}}{dt} = \frac{F^{17}}{\tau_\beta(17)} - \frac{O^{17}}{\tau_{17}} \tag{5-63i}$$

[1] Here, and in what follows, we shall use the less cumbersome notation

$$\tau_p(C^{12}) \equiv \tau_{12} \qquad \tau_p(C^{13}) \equiv \tau_{13} \qquad \text{etc.}$$

These equations include explicitly the possibility of the reactions $N^{13}(p,\gamma)O^{14}$, $O^{15}(p,\gamma)F^{16}$, and $F^{17}(p,\gamma)Ne^{18}$, although the equations for O^{14}, F^{16}, and Ne^{18} are not displayed. If those nuclei are formed, they quickly decay to N^{14}, O^{16}, and O^{18}, respectively. But below 10^8 °K these complications are unwarranted and allow the first simplification of the set of equations to be made. For $T < 10^8$, the lifetimes of N^{13}, O^{15}, F^{17} against protons are much longer than their beta-decay lifetimes. Thus, for instance, Eq. (5-63*b*) becomes

$$\frac{dN^{13}}{dt} \approx \frac{C^{12}}{\tau_{12}} - \frac{N^{13}}{\tau_\beta(13)} \qquad \text{for } \tau_p(N^{13}) \gg 870 \text{ sec} \tag{5-64}$$

The branch to O^{14} and the equation for O^{14} are then unnecessary. A similar situation occurs (even more strongly) for the other two short-lived positron emitters. Since in hydrogen-burning main-sequence stars the hydrogen temperature is never so high as 10^8 °K, for most applications the approximation of Eq. (5-64) is entirely justifiable. One should keep in mind, however, that there may be special astrophysical circumstances in which protons may interact with carbon at temperatures higher than 10^8 °K and that in those circumstances the interaction of N^{13} with protons may be important. Such a situation will occur, for instance, when protons are liberated during carbon burning. The remainder of the discussion will be limited to the approximation $T < 10^8$ (the range of Table 5-3).

Problem 5-19: Show that the abundance of N^{13} is

$$N^{13}(t) = \frac{\tau_\beta(13)}{\tau_{12}} C^{12}(1 - e^{-t/\tau_\beta(13)}) \tag{5-65}$$

for times short enough for C^{12} and τ_{12} to be essentially constant.

Equation (5-65) indicates the next approximation to the CNO bi-cycle equations, for it shows that N^{13} approaches an equilibrium value $(N^{13}/C^{12})_e = \tau_\beta(13)/\tau_{12}$ in times on the order of τ_β. Thus after times on the order of minutes Eqs. (5-63*b*), (5-63*e*), and (5-63*h*) may be set equal to zero. Then N^{13}, O^{15}, and F^{17} can be eliminated from the system of equations. There results

$$\frac{dC^{12}}{dt} = -\frac{C^{12}}{\tau_{12}} + \alpha\frac{N^{15}}{\tau_{15}} \tag{5-66a}$$

$$\frac{dC^{13}}{dt} = \frac{C^{12}}{\tau_{12}} - \frac{C^{13}}{\tau_{13}} \tag{5-66b}$$

$$\frac{dN^{14}}{dt} = \frac{C^{13}}{\tau_{13}} - \frac{N^{14}}{\tau_{14}} + \frac{O^{17}}{\tau_{17}} \tag{5-66c}$$

$$\frac{dN^{15}}{dt} = \frac{N^{14}}{\tau_{14}} - \frac{N^{15}}{\tau_{15}} \tag{5-66d}$$

$$\frac{dO^{16}}{dt} = \gamma\frac{N^{15}}{\tau_{15}} - \frac{O^{16}}{\tau_{16}} \tag{5-66e}$$

$$\frac{dO^{17}}{dt} = \frac{O^{16}}{\tau_{16}} - \frac{O^{17}}{\tau_{17}} \tag{5-66f}$$

This much simplified set of equations, then, is adequate to describe the CNO abundances after a few minutes of burning at a temperature of less than 10^8 °K.

Problem 5-20: Show that the sum of the CNO abundances is a constant.

The shortest proton lifetime among the CNO nuclei belongs to N^{15}. At temperatures near 25 million °K, which is roughly characteristic of CNO burning on the upper main sequence, τ^{15} is on the order of years. Equation (5-66*d*) shows that the N^{15} abundance seeks the equilibrium value

$$\left(\frac{N^{15}}{N^{14}}\right)_e = \frac{\tau_{15}}{\tau_{14}} \tag{5-67}$$

with $1/e$ time equal to τ_{15}. After a very short time in hydrogen burning, therefore, the N^{15} equation may also be eliminated from the set of equations. This ratio leaves the observed N^{15} abundance as inexplicable in terms of hydrogen burning, however, for $(N^{15}/N^{14})_e \approx 4 \times 10^{-5}$ at all temperatures, whereas the observed isotopic ratio on the earth is $N^{15}/N^{14} = 3.7 \times 10^{-3}$. That is, the N^{15} abundance is two orders of magnitude greater than the residue expected from the CNO cycle. It has been suggested many times that there may be an undetected resonance in the $N^{14} + p$ reaction, in which case $(N^{15}/N^{14})_e$ would be a larger number. The present nuclear evidence seems to point firmly to the conclusion used here, however, that the $N^{14} + p$ reaction is nonresonant, with the cross-section factor listed in Table 5-2. The observed N^{15} abundance is so small, however, that it may easily be the result of the relatively rare nonthermal reactions (such as spallation) that are invoked in the solution of the Li-Be-B problem.

In spite of the linearity of the set of equations, they are somewhat difficult to solve even with the aid of the simplifications made to this point. To reduce the complexity to a manageable level it is helpful to notice that the complete bi-cycle can be separated with high precision into two cycles, the CN cycle and the ON cycle. The CN cycle is just the set of reactions in Eq. (5-59), the ones originally proposed by Bethe and von Weizsäcker. This would be the complete set of reactions if the branching ratios were $\gamma = 0$, $\alpha = 1$ (instead of $\gamma \approx 4 \times 10^{-4}$, $\alpha \approx 0.9996$) and if there were no oxygen. Of course these two *ifs* are untrue, but a little thought about the comparative lifetimes shows that the CN cycle is independent anyway. The motivation is as follows. Since $\gamma \approx 4 \times 10^{-4}$, it would take on the order of 10^3 complete CN cycles before a significant fraction of the CN nuclei could be shunted off into the ON cycle. The time required would be of order $10^3\tau_{14}$. On the other hand, the time required for a significant amount of O^{16} to be shunted into the CN cycle is of the order τ_{16}. Both of these times, $10^3\tau_{14}$ and τ_{16}, are several thousand times greater than the time required for the CN cycle to come to equilibrium. To understand this requires an analysis of the CN cycle.

APPROACH TO EQUILIBRIUM OF THE CN CYCLE

The basic equations of the CN cycle with N^{15} in equilibrium with N^{14} and with $\alpha \approx 1$ are

$$
\begin{aligned}
\frac{dC^{12}}{dt} &= -\frac{C^{12}}{\tau_{12}} + \frac{N^{14}}{\tau_{14}} \\
\frac{dC^{13}}{dt} &= \frac{C^{12}}{\tau_{12}} - \frac{C^{13}}{\tau_{13}} \\
\frac{dN^{14}}{dt} &= \frac{C^{13}}{\tau_{13}} - \frac{N^{14}}{\tau_{14}}
\end{aligned}
\tag{5-68}
$$

which may conveniently be written as a matrix equation

$$
\frac{d}{dt}\begin{bmatrix} C^{12} \\ C^{13} \\ N^{14} \end{bmatrix} = \begin{bmatrix} -\frac{1}{\tau_{12}} & 0 & \frac{1}{\tau_{14}} \\ \frac{1}{\tau_{12}} & -\frac{1}{\tau_{13}} & 0 \\ 0 & \frac{1}{\tau_{13}} & -\frac{1}{\tau_{14}} \end{bmatrix} \begin{bmatrix} C^{12} \\ C^{13} \\ N^{14} \end{bmatrix} \tag{5-69}
$$

which is of the form

$$
\frac{d}{dt}\mathbf{U} = [\Lambda]\mathbf{U} \tag{5-70}
$$

where the three components of the vector $\mathbf{U}$ are C^{12}, C^{13}, and N^{14}, and $[\Lambda]$ is the 3×3 matrix in Eq. (5-69). The solution consists in finding the three eigenvectors of $[\Lambda]$, defined as those vectors satisfying the equations

$$
\begin{aligned}
[\Lambda]\mathbf{U}_1 &= \lambda_1\mathbf{U}_1 \\
[\Lambda]\mathbf{U}_2 &= \lambda_2\mathbf{U}_2 \\
[\Lambda]\mathbf{U}_3 &= \lambda_3\mathbf{U}_3
\end{aligned}
\tag{5-71}
$$

where the quantities λ_1, λ_2, and λ_3 are the three eigenvalues of $[\Lambda]$. From Eq. (5-70) it follows that if $\mathbf{U}(t)$ is expressed as a linear combination of the eigenvectors with exponential time dependence

$$
\mathbf{U}(t) = Ae^{\lambda_1 t}\mathbf{U}_1 + Be^{\lambda_2 t}\mathbf{U}_2 + Ce^{\lambda_3 t}\mathbf{U}_3 \tag{5-72}
$$

then Eq. (5-70) is exactly satisfied.

Problem 5-21: Confirm that Eq. (5-72) is a solution of Eq. (5-70). The constants A, B, and C are evidently determined by the initial abundances: $\mathbf{U}(0) = A\mathbf{U}_1 + B\mathbf{U}_2 + C\mathbf{U}_3$.

Before proceeding with this solution, we note that it is correct only if the individual nuclear lifetimes are constant. This condition is not strictly met because of a gradual depletion in hydrogen and the possibility of changes in temperature. In most cases when a star has settled into a static configuration and is obtaining energy from the CNO cycle, however, the lifetimes do not change very much over the length of time required to achieve equilibrium in the CN portion of the cycle, in which case this solution is physically meaningful. What must always be done with analytic solutions of coupled equations in nuclear astrophysics is to make an ex post facto self-consistency check, i.e., to see whether the assumptions leading to the analytic solution are consistent with the conditions of the environment in which the solutions are presumed to hold. The present case is no exception to this rule.

From elementary algebraic theory it is known that solutions for eigenvalues as in Eq. (5-71) can be obtained only if the eigenvalues themselves are such that the determinant of the matrix $[[\Lambda] - \lambda[1]]$ vanishes. That so-called *secular equation* is, in this case, the cubic equation

$$\begin{vmatrix} -\left(\frac{1}{\tau_{12}} + \lambda\right) & 0 & \frac{1}{\tau_{14}} \\ \frac{1}{\tau_{12}} & -\left(\frac{1}{\tau_{13}} + \lambda\right) & 0 \\ 0 & \frac{1}{\tau_{13}} & -\left(\frac{1}{\tau_{14}} + \lambda\right) \end{vmatrix} = 0 \tag{5-73}$$

Problem 5-22: Show that the eigenvalues are

$$\lambda_1 = 0 \qquad \lambda_2 = \frac{-\Sigma + \Delta}{2} \qquad \lambda_3 = \frac{-\Sigma - \Delta}{2} \tag{5-74}$$

where

$$\Sigma \equiv \frac{1}{\tau_{12}} + \frac{1}{\tau_{13}} + \frac{1}{\tau_{14}}$$

and

$$\Delta \equiv \left[\Sigma^2 - 4\left(\frac{1}{\tau_{12}\tau_{13}} + \frac{1}{\tau_{12}\tau_{14}} + \frac{1}{\tau_{13}\tau_{14}}\right)\right]^{\frac{1}{2}}$$

Problem 5-23: The eigenvector belonging to the first eigenvalue ($\lambda_1 = 0$) is determined from the equation

$$[\Lambda]\mathbf{U}_1 = 0 \tag{5-75}$$

Show that the first eigenvector, normalized such that the sum of the components is unity, is

$$\mathbf{U}_1 = \frac{1}{\tau_{12} + \tau_{13} + \tau_{14}} \begin{bmatrix} \tau_{12} \\ \tau_{13} \\ \tau_{14} \end{bmatrix} \tag{5-76}$$

The first eigenvector has the property that if C^{12}, C^{13}, and N^{14} exist in this ratio, each of those abundances is constant in time. These are the equilibrium abundance ratios in the CN cycle, and the fact that $\lambda_1 = 0$ corresponds to the fact that abundances in this ratio do not change in time. Since the sum of the abundances is constant, $\mathbf{U}_1$ may also be written

$$\mathbf{U}_1 = \frac{1}{C^{12} + C^{13} + N^{14}} \begin{bmatrix} C_e^{12} \\ C_e^{13} \\ N_e^{14} \end{bmatrix} \tag{5-77}$$

where C_e^{12} designates the equilibrium abundance of C^{12}. It is the other two eigenvectors that show how the equilibrium abundances are approached.

Problem 5-24: Show that the other two eigenvectors are

$$\mathbf{U}_2 = \begin{bmatrix} 1 \\ \dfrac{1/\tau_{12}}{1/\tau_{13} - (\Sigma - \Delta)/2} \\ -1 - \dfrac{1/\tau_{12}}{1/\tau_{13} - (\Sigma - \Delta)/2} \end{bmatrix} \qquad \mathbf{U}_3 = \begin{bmatrix} 1 \\ -1 - \dfrac{1/\tau_{12} - (\Sigma + \Delta)/2}{1/\tau_{14}} \\ \dfrac{1/\tau_{12} - (\Sigma + \Delta)/2}{1/\tau_{14}} \end{bmatrix} \tag{5-78}$$

No particular normalization is needed, because the sum of the three components is identically zero. Why should that be so?

By grouping these eigenvectors together into a solution of the form of Eq. (5-72) we obtain

$$\begin{bmatrix} C^{12}(t) \\ C^{13}(t) \\ N^{14}(t) \end{bmatrix} = \begin{bmatrix} C_e^{12} \\ C_e^{13} \\ N_e^{14} \end{bmatrix} + Be^{\lambda_2 t}\mathbf{U}_2 + Ce^{\lambda_3 t}\mathbf{U}_3 \tag{5-79}$$

The eigenvalues λ_2 and λ_3 are negative, corresponding to the fact that the abundances decay exponentially to the equilibrium values. It is also evident that the sum of the components of $\mathbf{U}_2$ and of $\mathbf{U}_3$ must vanish, or the number of CN nuclei would not be constant with time. The constants B and C are determinable from the initial abundances. It follows almost by inspection that

$$B + C = C^{12}(0) - C_e^{12}$$

and

$$\frac{1/\tau_{12}}{1/\tau_{13} - (\Sigma - \Delta)/2} B + \frac{1/\tau_{12} - (\Sigma + \Delta)/2}{1/\tau_{14}} C = C^{13}(0) - C_e^{13} \tag{5-80}$$

These formulas represent the formal solution to the CN abundances when the nuclear lifetimes are constant. To appreciate the time scale for the approach to equilibrium it is best to consider a specific numerical example.

Problem 5-25: Evaluate the eigenvalues and eigenvectors of the CN cycle for $T_6 = 25$ and $\rho X_H = 25$ and write the solution to the CN abundances if $C^{12}(0) = C^{13}(0) = N^{14}(0) = N/3$.
Ans:

$$\begin{bmatrix} C^{12} \\ C^{13} \\ N^{14} \end{bmatrix} = N \begin{bmatrix} 0.0122 \\ 0.00305 \\ 0.985 \end{bmatrix} + 0.320N \begin{bmatrix} 1 \\ 0.336 \\ -1.336 \end{bmatrix} \exp(-1.274 \times 10^{-3}t)$$

$$+ \frac{0.222}{240.2} N \begin{bmatrix} 1 \\ 240.5 \\ -241.5 \end{bmatrix} \exp(-4.983 \times 10^{-3}t)$$

where t is expressed in years.

Several interesting observations can be made from the numerical results of the previous problem. First note that for the environment of that problem, $1/\tau_{12} = 1.25 \times 10^{-3}$ and $1/\tau_{13} = 4.99 \times 10^{-3}$. Thus λ_2 and λ_3 are very nearly equal to the reaction rates of C^{12} and of C^{13} in the proton bath. The second decay mode corresponds primarily to the distribution of the initial C^{13}. It is apparent from the equilibrium vector that the end product of the CN cycle is primarily N^{14}, and so the only things that must be accomplished to reach equilibrium involve the conversion of C^{12} and C^{13} to N^{14}. Obviously that conversion must proceed at rates determined by the C^{12} and C^{13} lifetimes. The slowest mode decays away with a $1/e$ time that is approximately equal to τ_{12}. From the table of lifetimes it is apparent that the crucial point has been demonstrated; viz., the CN cycle reaches equilibrium in a time of order τ_{12}, which is much faster than any significant interchange of nuclei between the CN cycle and the ON cycle. Those two interchange times are characteristically $10^3\tau_{14}$ and τ_{16}.

APPROACH TO EQUILIBRIUM OF THE ON CYCLE

Because of its long lifetime, O^{16} is the slowest species to come into equilibrium in the CNO bi-cycle. The establishment of its equilibrium involves significant interchange of nuclei between the ON and the CN portions of the cycle. Once equilibrium has been achieved, of course, all the abundances remain constant thereafter.

It might be mentioned at this time that the seed nuclei for the CNO cycle are believed to be predominantly C^{12} and O^{16}, since these are the nuclei produced in helium burning. Only after C^{12} and O^{16} have been synthesized in a previous generation of stars can the CNO cycle operate. Of course, a young star that has formed recently from the interstellar medium will also contain N^{14}, N^{15}, C^{13}, and O^{17}. But the solar abundance ratio seems to be

$$C^{12}:N^{14}:O^{16} = 5.5:1:9.6$$

and much smaller amounts of the other nuclei. These ratios are probably fairly representative of the entire population I. The values of C^{13}, N^{15}, and O^{17} are quite small (though uncertain) in interstellar gas, and so they clearly are not significant seed nuclei for the cycle. The major seed nuclei are C^{12} and O^{16}, with

N^{14} a distant third. We have just seen that in the CN portion of the cycle, the nuclei are converted predominantly ($\sim$98 percent) to N^{14} at equilibrium, which is achieved in several C^{12} lifetimes. The O^{16} will also be converted predominantly to N^{14} at equilibrium but on a much slower time scale. In fact O^{16} equilibrium is often not achieved at all in the CNO cycle.

The essential points required for achieving the simplest understanding of the CNO bi-cycle are the following: (1) *for short times when the* CN *cycle is approaching equilibrium the oxygen can be ignored, because only vanishingly small amounts of oxygen can burn in the time required for* CN *equilibrium (except for* O^{17} *at high temperature, but* O^{17} *is not a significant seed); and* (2) *for times long enough to consider oxygen burning, as well as the transfer of* CN *nuclei to the* ON *cycle via the* γ *branch, it may safely be assumed that the* CN *nuclei have already achieved their equilibrium distribution.* With these thoughts in mind, the relevant differential equations can be written from a new point of view. Picture in your mind's eye the CN cycle racing around and around its track, while every now and then ($\gamma = 4 \times 10^{-4}$) the $N^{15} + p$ reaction, which in equilibrium is proceeding at the same rate as the $N^{14} + p$ reaction, produces an O^{16} nucleus. The same reaction removes one nucleus from the CN cycle, which means it removes $\tau_{14}/(\tau_{14} + \tau_{12} + \tau_{13})$ nuclei from N^{14}. At the same time, there is a slow addition of nuclei into the CN cycle by $O^{17}(p,\alpha)N^{14}$. Each such reaction adds $\tau_{14}/(\tau_{14} + \tau_{12} + \tau_{13})$ nuclei to N^{14}. Thus

$$\frac{dN^{14}}{dt} = -\frac{\tau_{14}}{\tau_{14} + \tau_{12} + \tau_{13}} \gamma \frac{N^{14}}{\tau_{14}} + \frac{\tau_{14}}{\tau_{14} + \tau_{12} + \tau_{13}} \frac{O^{17}}{\tau_{17}} \tag{5-81}$$

Although this equation has been carefully formulated to allow for the redistribution among the CN nuclei, the care is warranted only in principle. The factor $\tau_{14}/(\tau_{14} + \tau_{12} + \tau_{13})$, which is approximately equal to 0.985, might as well be set equal to unity, considering that τ_{14} and τ_{17} are not known well enough (1.5 percent) to make retention of this factor meaningful. Thus to a high degree of accuracy, the equations for N^{14}, O^{16}, and O^{17} are

$$\frac{d}{dt}\begin{bmatrix} N^{14} \\ O^{16} \\ O^{17} \end{bmatrix} \approx \begin{bmatrix} -\dfrac{\gamma}{\tau_{14}} & 0 & \dfrac{1}{\tau_{17}} \\ \dfrac{\gamma}{\tau_{14}} & -\dfrac{1}{\tau_{16}} & 0 \\ 0 & \dfrac{1}{\tau_{16}} & -\dfrac{1}{\tau_{17}} \end{bmatrix} \begin{bmatrix} N^{14} \\ O^{16} \\ O^{17} \end{bmatrix} \tag{5-82}$$

Since Eq. (5-82) is exactly the same as Eq. (5-69), it has exactly the same solution. All the eigenvalues and eigenvectors are calculable by the substitution $1/\tau_{12} \rightarrow \gamma/\tau_{14}$, $1/\tau_{13} \rightarrow 1/\tau_{16}$, and $1/\tau_{14} \rightarrow 1/\tau_{17}$ in the equations for the solution of the CN cycle. The solution therefore takes the form

$$\begin{bmatrix} N^{14}(t) \\ O^{16}(t) \\ O^{17}(t) \end{bmatrix} \approx \begin{bmatrix} N_e^{14} \\ O_e^{16} \\ O_e^{17} \end{bmatrix} + Be^{\lambda_2 t}\mathbf{U}_2 + Ce^{\lambda_3 t}\mathbf{U}_3 \tag{5-83}$$

Since this solution applies after CN equilibrium, in the evaluation of B and C the value of $N^{14}(0)$ must be taken to represent the initial abundance of all the CN nuclei. The equilibrium distribution (the eigenvector belonging to $\lambda = 0$) has components in the ratios

$$\left(\frac{O^{17}}{O^{16}}\right)_e = \frac{\tau^{17}}{\tau^{16}} \qquad \left(\frac{O^{16}}{N^{14}}\right)_e = \frac{\gamma\tau_{16}}{\tau_{14}} \tag{5-84}$$

The second ratio is particularly interesting since its value shows that only 1 percent or so of the O^{16} remains as O^{16}; the bulk of it is converted to N^{14}. This result emphasizes the fact that *the* CNO *bi-cycle, if it has time to achieve equilibrium, essentially converts all the* CNO *nuclei to* N^{14}.

Examination of the lifetimes in Table 5-3 shows that in the higher-temperature regions another approximation exists that further reduces the complexity of Eq. (5-83). For $T_6 > 25$ the O^{17} lifetime becomes as short as the C^{12} lifetime, although for $T_6 < 25$ the τ_{17} is much greater than τ_{12}. This fact means the O^{17} will come into equilibrium with O^{16} as fast as the CN cycle comes to equilibrium for $T_6 > 25$. Moreover, for $T_6 > 22$ the ratio $(O^{17}/O^{16})_e$ is less than 10^{-2}, which means that to excellent approximation, all the oxygen is O^{16} and all the CN nuclei are N^{14}. Thus the interchange between the two cycles reduces to a single equation:

$$\frac{dN^{14}}{dt} \approx -\frac{\gamma}{\tau_{14}} N^{14} + \frac{O^{16}}{\tau_{16}} \approx -\frac{dO^{16}}{dt} \tag{5-85}$$

Problem 5-26: Derive Eq. (5-85).

To 1 percent accuracy, moreover,

$$N^{14}(t) + O^{16}(t) \approx \text{const} = N_{CN}(0) + N_O(0) \tag{5-86}$$

where $N_{CN}(0)$ and $N_O(0)$ represent the initial CN abundances and oxygen abundances, respectively.

Problem 5-27: Show that the solution of Eq. (5-85) when τ_{14} and τ_{16} are assumed constant is

$$N^{14}(t) = \frac{N_{CN}(0) + N_O(0)}{(\gamma\tau_{16}/\tau_{14}) + 1}\left[1 - \exp\left(-\frac{\gamma}{\tau_{14}} - \frac{1}{\tau_{16}}\right)t\right] + N_{CN}(0)\exp\left(-\frac{\gamma}{\tau_{14}} - \frac{1}{\tau_{16}}\right)t \tag{5-87}$$

Derive the formal solution if it is assumed instead that the temperature remains constant but that the hydrogen density decreases exponentially as $\exp(-t/\tau_H)$. Estimate the value of τ_H.

Equation (5-87) shows in a simple way that $N^{14}(t)$, which represents the number of nuclei participating in the CN cycle, rises from its initial value $N_{CN}(0)$ to its equilibrium value N_e^{14}, because

$$N^{14} \underset{t\to\infty}{\longrightarrow} \frac{N_{CN}(0) + N_O(0)}{(\gamma\tau_{16}/\tau_{14}) + 1} = \frac{N_{CN}(0) + N_O(0)}{(O^{16}/N^{14})_e + 1} = N_e^{14}$$

An alternative form for Eq. (5-87) is

$$N^{14}(t) = N_e^{14} + (N_{CN}(0) - N_e^{14}) \exp\left(-\frac{\gamma}{\tau_{14}} - \frac{1}{\tau_{16}}\right) t \tag{5-88}$$

This equation presents a much simpler solution than Eq. (5-83). Of course, it is not exact, nor is Eq. (5-83), but they are both accurate in the case of constant τ to about 1 percent for times greater than several τ_{12}. It seems that Eq. (5-88) is to be preferred for those circumstances for which it is valid; viz., for $T_6 > 22$ and $t > \tau_{17}$. With some thought about these solutions, the student should acquire a good physical feeling for the operation of the CNO bi-cycle.

The equilibrium abundances themselves may, of course, be obtained from the initial set of differential equations (5-66a) to (5-66f) by setting all time derivatives

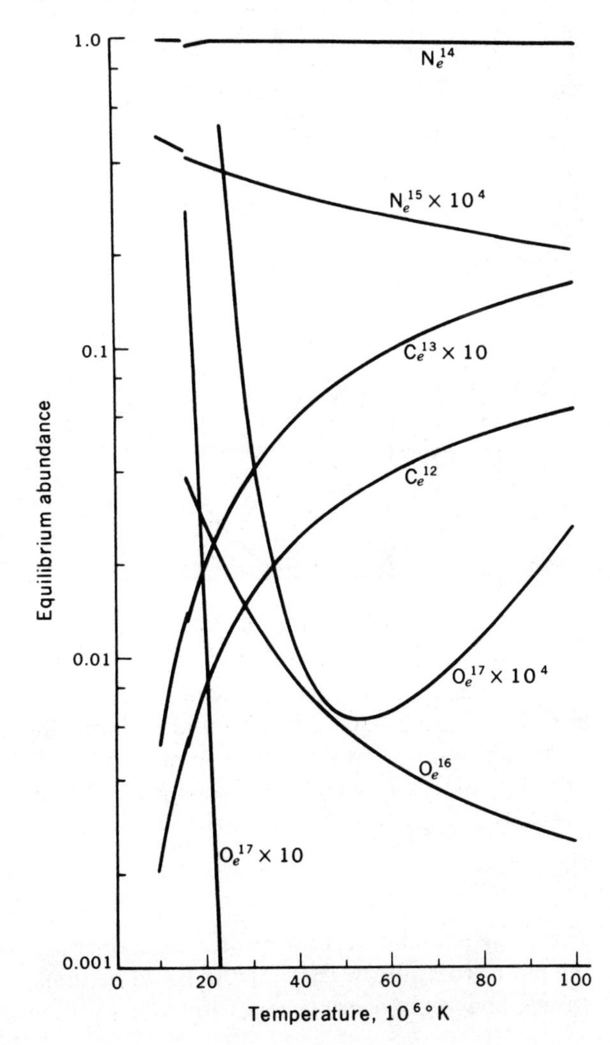

Fig. 5-13 The fractional abundance of each CNO nucleus when the cycle is operating in equilibrium. [*After G. R. Caughlan and W. A. Fowler, Astrophys. J.*, **136**:453 (1962). *By permission of The University of Chicago Press. Copyright* 1962 *by The University of Chicago.*]

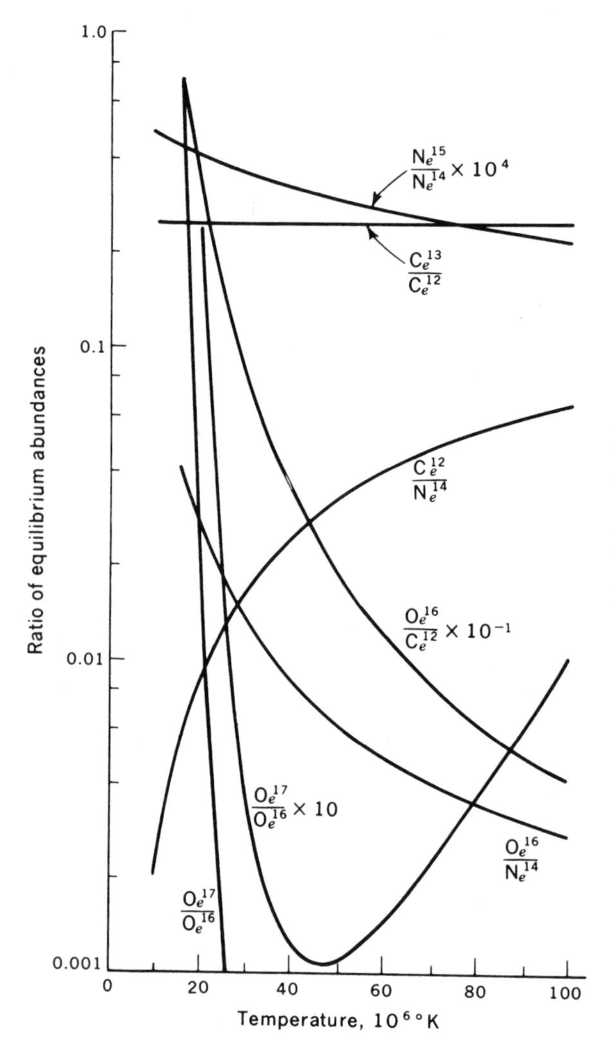

Fig. 5-14 Ratios of abundances of CNO nuclei when the cycle is operating in equilibrium. [*After G. R. Caughlan and W. A. Fowler, Astrophys. J.*, **136**:453 (1962). *By permission of The University of Chicago Press. Copyright* 1962 *by The University of Chicago.*]

equal to zero. The fractional equilibrium abundances are displayed in Fig. 5-13 and the ratios of equilibrium abundances in Fig. 5-14.

At this point a few summary comments regarding the abundances in the CNO bi-cycle seem in order. Since the basic CNO equations are linear in the abundances, the entire set could be written as one vector equation

$$\frac{d\mathbf{U}}{dt} = [\Lambda]\mathbf{U}$$

where the complete vector has six components and $[\Lambda]$ is a 6×6 matrix. With the aid of computers, that matrix could be diagonalized by solving for the eigenvalues and eigenvectors, and complete solution could be expressed as a linear combination of the equilibrium eigenvector plus five exponentially decaying eigenvectors.[1] In this chapter we have sacrificed 1 percent or so in accuracy to make a sequential treatment that is more physically instructive and adequate for any application. For times of order τ_{12} one need only consider the CN portion of the bi-cycle, which approaches equilibrium like Eq. (5-79). For longer times, in which some interchange between the CN and ON cycles can occur, it is adequate to regard the CN portion as having already achieved equilibrium. Then the solution is accurately represented by Eq. (5-83) for $T_6 < 22$ and Eq. (5-88) for $T_6 > 22$. These are rules of thumb. The bi-cycle is sufficiently complicated so that the user should make his own approximations consistent with the use to which the cycle is to be put. This critical judgment can be developed from the material presented here. As a graphical example to sum up the way the abundances develop, Fig. 5-15 indicates the abundances as a function of time and of protons consumed by the CNO nuclei at $T_6 = 20$ for an initial abundance that is

[1] This solution has been performed by G. R. Caughlan, *Astrophys. J.*, **141**:688 (1965).

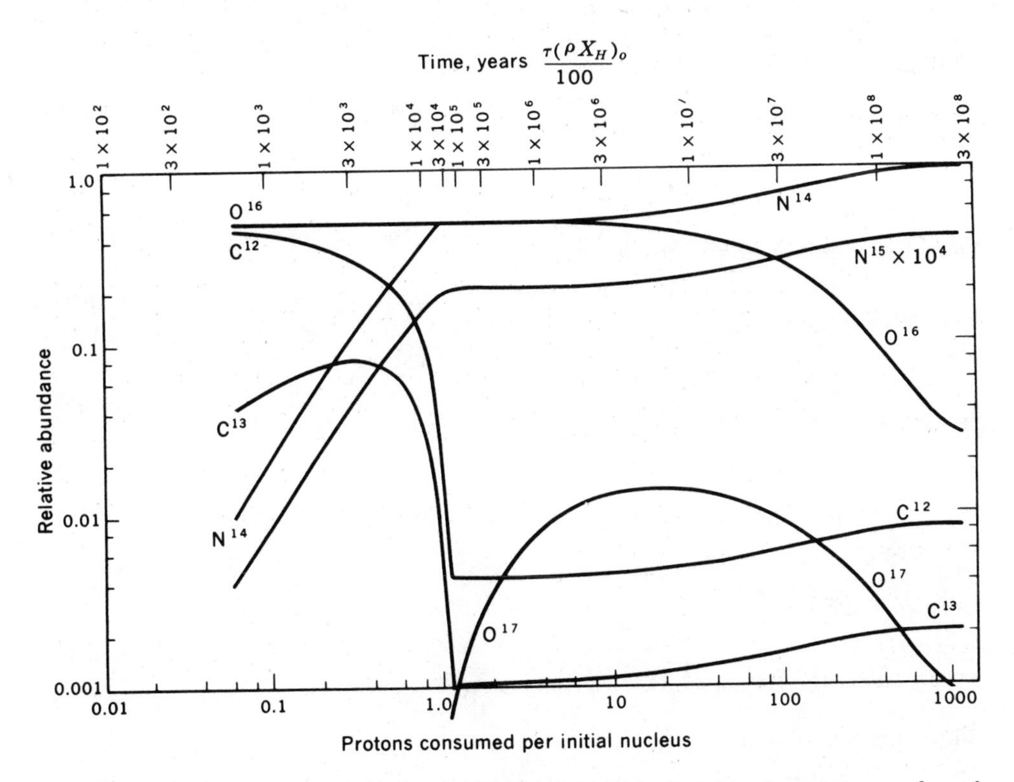

Fig. 5-15 The approach to equilibrium in the CNO bi-cycle as a function of the number of protons captured per initial CNO nucleus. This particular calculation started with equal concentrations of C^{12} and O^{16}. [*After G. R. Caughlan, Astrophys. J.*, **141**:688 (1965). *By permission of The University of Chicago Press. Copyright* 1964 *by The University of Chicago.*]

50 percent O^{16} and 50 percent C^{12}. Particularly note that O^{16} has not yet achieved equilibrium after even 1,000 proton captures. For population I stars, therefore, there may well be insufficient hydrogen ever to drive oxygen to equilibrium. The hydrogen is consumed so rapidly by the CN cycle that it vanishes before the oxygen can be depleted. The formal solutions valid for constant τ have been presented only to develop physical insight into the operation of the cycles, and they are not correct for a real star in which X_H and T both change with time. The student will be well advised to review the second part of Prob. 5-27.

There is considerable observational evidence in stars confirming the rates of the CNO cycle. As a result of mixing or mass loss or both many stars have apparently exposed interior matter that was burning on the CNO cycle. Both McKellar and Climenhaga have studied the molecular bands of the C_2 molecule in carbon stars, and from the relative intensity of the lines of the C^{12}-C^{13} molecule they have reported C^{13}/C^{12} ratios as great as $\frac{1}{4}$, approximately the value expected in the CNO cycle. Note that on the surfaces of stars in general this ratio will be expected to be much smaller than $\frac{1}{4}$, because the star will have formed from matter much enriched in C^{12} by the remnants of helium-burning nucleosynthesis.

Wallerstein and coworkers[1] have studied the hydrogen-poor star HD 30353, whose surface ratio $H/He \approx 10^{-4}$ shows that the hydrogen has been almost completely consumed in the layer now exposed. At the same time they find $N/C \approx 10^3$ and $N/O \approx 50$, a composition so nitrogen-rich that it can easily be understood only as the remnants of the CNO cycle.

ENERGY GENERATION BY THE CNO BI-CYCLE

The rate of energy generation is the sum over all reactions of the product of the reaction rate and the difference of the energy release (Q value) and the neutrino loss. The relevant quantities are all listed in Table 5-2.

For the CN portion of the cycle it follows from Eq. (5-79) that at constant temperature and constant hydrogen density

$$\epsilon_{CN} = \epsilon_{CNe} + \epsilon_{CN2}e^{\lambda_2 t} + \epsilon_{CN3}e^{\lambda_3 t} \tag{5-89}$$

In this computation, the energy release is considered as occurring in three separate pieces:

$$\begin{array}{ll} C^{12}(p,\gamma)N^{13}(\beta^+\nu)C^{13} & 5.534 \times 10^{-6} \text{ erg/reaction} \\ C^{13}(p,\gamma)N^{14} & 12.093 \times 10^{-6} \text{ erg/reaction} \\ N^{14}(p,\gamma)O^{15}(\beta^+\nu)N^{15}(p,\alpha)C^{12} & 22.453 \times 10^{-6} \text{ erg/reaction} \end{array} \tag{5-90}$$

The numerical value of ϵ_{CNe} is

$$\rho\epsilon_{CNe} = \left(5.534\,\frac{C_e^{12}}{\tau_{12}} + 12.093\,\frac{C_e^{13}}{\tau_{13}} + 22.453\,\frac{N_e^{14}}{\tau_{14}}\right) \times 10^{-6} \qquad \text{erg cm}^{-3}\,\text{sec}^{-1}$$

[1] G. Wallerstein, T. F. Greene, and L. J. Tomley, *Astrophys. J.*, **150**:245 (1967).

But from Eqs. (5-76) and (5-77)

$$\begin{bmatrix} C^{12}_e \\ C^{13}_e \\ N^{14}_e \end{bmatrix} = \frac{N_{CN}}{\tau_{12} + \tau_{13} + \tau_{14}} \begin{bmatrix} \tau_{12} \\ \tau_{13} \\ \tau_{14} \end{bmatrix}$$

and so the energy generation by the equilibrated cycle reduces to

$$\rho\epsilon_{CNe} = 4.008 \frac{N_{CN}}{\tau_{12} + \tau_{13} + \tau_{14}} \times 10^{-5} \qquad \text{erg cm}^{-3}\text{ sec}^{-1} \tag{5-91}$$

The sum $\tau_{12} + \tau_{13} + \tau_{14}$ is [neglecting τ_{15}, $\tau_\beta(N^{13})$, and $\tau_\beta(O^{15})$] called the *cycle time* and is approximately equal to τ_{14}, since that lifetime is so much greater than the other two.

Problem 5-28: Show that

$$\epsilon_{CNe} \approx 8 \times 10^{27} \rho X_H X_{CN} f_N T_6^{-\frac{2}{3}} \exp(-152.31 T_6^{-\frac{1}{3}}) \qquad \text{erg g}^{-1}\text{ sec}^{-1} \tag{5-92}$$

The uncertainty in this expression is about ± 20 percent, which is a reasonable measure of the uncertainty in S_0 for $N^{14}(p,\gamma)O^{15}$. The value of the temperature required to produce a given value of ϵ_{CNe}, however, is certain to much better than 20 percent accuracy.

Equation (5-91) displays a point that has been made many times in the literature, viz., in a specific environment the energy-generation rate would be greatly increased if there should exist an undetected resonance in the $N^{14} + p$ reaction. If that were the case, it would probably follow that $\tau_{14} < \tau_{12}$, and the energy-generation rate would be determined by the C^{12} lifetime. All the available nuclear evidence indicates that the reaction is nonresonant, however.

The second term in Eq. (5-89) is given by

$$\rho\epsilon_{CN2} = B\left(5.534 \frac{U_{2,1}}{\tau_{12}} + 12.093 \frac{U_{2,2}}{\tau_{13}} + 22.453 \frac{U_{2,3}}{\tau_{14}}\right) \times 10^{-6} \qquad \text{erg cm}^{-3}\text{ sec}^{-1} \tag{5-93}$$

where $U_{2,1}$, $U_{2,2}$, and $U_{2,3}$ are the three components of the second CN eigenvector given in Eq. (5-78) and B is one of the two initial-composition constants to be obtained from solution of Eq. (5-80). A similar expression exists for $\rho\epsilon_{CN3}$ with B replaced by C and $\mathbf{U}_2$ replaced by $\mathbf{U}_3$.

Problem 5-29: Return to the problem of the CN cycle at $T_6 = 25$, $\rho X_H = 25$, and $C^{12}(0) = C^{13}(0) = N^{14}(0) = N_{CN}/3$. Write the numerical expression for the energy generation if $N_{CN}/H = 5 \times 10^{-4}$ and $\rho = 50$.
Ans: $\epsilon_{CN} = 3.0 \times 10^3 + 4.1 \times 10^4 e^{-1.27 \times 10^{-3} t(\text{yr})} + 6.3 \times 10^4 e^{-4.98 \times 10^{-3} t(\text{yr})}$.

Although the preceeding problem is of little practical value, it does reveal an interesting feature qualitatively. The initial rate of energy generation is some 30 times as great as ϵ_{CNe}, and after 4×10^3 years ϵ is within 10 percent of ϵ_{CNe}.

These times are to be compared to $\tau_{12} \approx 200$ years. Thus $\epsilon_{\rm CN}$ is more than twice as great as $\epsilon_{{\rm CN}e}$ for burning times of order $10\tau_{12}$. Although this time seems fairly short at $T_6 = 25$, it becomes significantly long for $T_6 < 15$ to be comparable to contraction times onto the main sequence. Iben[1] in particular has considered in detail the effects associated with the approach to CN equilibrium that occur as a star settles onto the main sequence. In such a calculation it is usually necessary to follow the changes of the abundances in the computer program, however. The integral solution for the CN abundances is not correct unless the various lifetimes are constant. As stars settle into their static main-sequence configuration, the internal temperatures (and hence the lifetimes) change as the CN cycle moves toward equilibrium. Thus the lifetimes have to be recomputed at each time step of the evolutionary sequence.

For the longer static-burning periods it is sufficient, so far as energy generation is concerned, to regard the CN cycle as being in equilibrium. In the lower range of temperatures, where a long time is required for CN equilibrium, the major energy generation comes from the PP chains, whereas for those temperatures high enough for $\epsilon_{\rm CN}$ to dominate the energy generation the CN cycle reaches equilibrium by the time the star is fully settled on the main sequence. This is not to say that there is no interest in the effects of nonequilibrium CN abundances persisting over long periods of time at relatively low temperature; rather, those abundances are not significant sources of energy generation. On the longer time scales, the only feature that need be watched relative to energy generation is the distribution of CNO nuclei between the CN cycle (operating in equilibrium) and the oxygen isotopes.

The solution to the energy-generation problem is derived in this case from Eq. (5-83). In this equation $\rm N^{14}$ stands for $N_{\rm CN}$, the number of nuclei circulating in the CN portion of the bi-cycle. The energy generation due to that abundance is obtained by inserting $\mathrm{N}^{14}(t)$ for $N_{\rm CN}$ in Eq. (5-91). The remainder of the energy generation is contained in the following two steps:

$$\begin{array}{ll} \mathrm{O}^{16}(p,\gamma)\mathrm{F}^{17}(\beta^+\nu)\mathrm{O}^{17} & 3.87 \times 10^{-6} \text{ erg/reaction} \\ \mathrm{O}^{17}(p,\alpha)\mathrm{N}^{14} & 1.91 \times 10^{-6} \text{ erg/reaction} \end{array} \tag{5-94}$$

It is quite evident by comparison with Eq. (5-90) that these two reactions are relatively low with respect to Q values. This small energy release from the oxygen isotopes means that the energy-generation rate can be approximated by

$$\rho\epsilon_{\rm CNO} = \left(40.08 \frac{N_{\rm CN}}{\tau_{14} + \tau_{13} + \tau_{12}} + 5.78 \frac{N_{\rm CNO} - N_{\rm CN}}{\tau_{16}}\right) \times 10^{-6} \qquad \text{erg cm}^{-3}\text{ sec}^{-1} \tag{5-95}$$

which is employable in conjunction with Eq. (5-85). If we make the rather poor assumption in this case that the lifetimes are constant during the burning, the

[1] I. Iben, Jr., *Astrophys. J.*, **141**:993 (1965).

auxiliary condition is simply

$$N_{CN}(t) \approx N_{CNO} - [N_{CNO} - N_{CN}(0)] \exp\left(-\frac{\gamma}{\tau_{14}} - \frac{1}{\tau_{16}}\right) t \tag{5-96}$$

It will be found that in practical applications all the energy generation effectively comes from the first term in Eq. (5-95). Only if the initial oxygen abundance is overwhelmingly greater than carbon and nitrogen will this conclusion be incorrect.

Problem 5-30: A newly formed star with $C^{12}:N^{14}:O^{16} = 4:1:10$ begins burning at the center with $T_6 = 25$ and $\rho X_H = 25$. Assuming constant lifetimes, write an expression for the energy generation valid for $t > 10^4$ years. Assume $N_{CNO}/H = 1.5 \times 10^{-3}$. It will be seen that in this example the energy generation triples over a period of time of about 10^7 years.
Ans: $\epsilon_{CNO} \approx 9.0 \times 10^3 - 6.0 \times 10^3 \exp[-3.6 \times 10^{-7}t(\text{yr})]$ erg g^{-1} sec^{-1}.

Whether the complications of the time changes of energy-generation rates as the various cycles approach equilibrium should be incorporated into models of stars is a question requiring some judgment. As has been emphasized many times, an error in the energy-generation rate may be transferred to a temperature error in the construction of a stellar model. It is quite incorrect to think that doubling the rates of all nuclear reactions will double the luminosity of a star, because a star is a self-regulating machine that selects temperatures such that the energy generation is balanced by the energy flow down the temperature gradient.

Problem 5-31: Show that the temperature dependence of the CN energy-generation rate near $T_6 = 25$ is

$$\epsilon_{CN}(T_6) \approx \epsilon_{CN}(25) \left(\frac{T_6}{25}\right)^{16.7} \tag{5-97}$$

From Eq. (5-97) it can be calculated that a factor of 3 in ϵ can be compensated by a 7 percent change in the temperature. When faced with an effect like that in the previous problem, therefore, anyone constructing stellar models must ascertain whether an error of a few percent in the temperature is important to his calculation. It is more important to include all the major energy-generating effects in the PP chains than it is in the CNO bi-cycle, because the temperature dependence of the proton-proton reaction is much weaker than the temperature dependence of the $N^{14}(p,\gamma)$ reaction.

How far in error a naïve instinct about the effect of nuclear reactions of stars may be is illustrated by a demonstration of Donald Morton's.[1] He showed that if the $N^{14}(p,\gamma)$ reaction is resonant, which would cause ϵ_{CN} to be greater by a factor of 10^2 since τ_{12} would then dominate the cycle, the luminosity of upper-main-sequence stars would actually be *smaller* by about 30 percent. The "reverse" effect occurs because the star must have a major readjustment in structure. The great increase in ϵ would generate more energy than could be carried out along the existing temperature gradient. To reduce ϵ to compatible proportions, the central temperature must decrease, which causes the star to expand to remain in

[1] *Astrophys. J.*, **129**:20(1959).

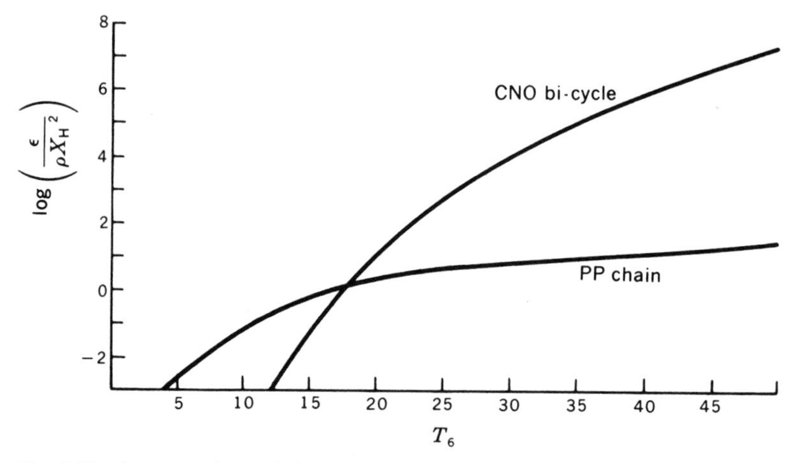

Fig. 5-16 A comparison of thermonuclear power from the PP chains and the CNO cycle. Both chains are assumed to be operating in equilibrium. The calculation was made for the choice $X_{CN}/X_H = 0.02$, which is representative of population I composition.

hydrostatic equilibrium. The attendant decrease in the temperature gradient in radiative zones causes the luminosity to decrease. The point to be emphasized is that considerable experience and physical insight must be brought to bear on the question of how accurately ϵ must be represented in stellar-structure calculations to achieve a good stellar model.

In most hydrogen-burning stars the PP chain and the CNO bi-cycle operate simultaneously. The question of which source dominates the energy generation depends on the relative abundances of hydrogen and the CN nuclei and on the temperature. Since

$$\epsilon_{pp} \propto \rho X_H{}^2 \qquad \epsilon_{CN} \propto \rho X_H X_{CN} \tag{5-98}$$

the quantity $\epsilon/\rho X_H{}^2$ is a function only of temperature for PPI and is X_{CN}/X_H times a function of temperature for the CN cycle. To present a rough idea of the relative importance of these sources, Fig. 5-16 displays $\epsilon_{PPI}/\rho X_H{}^2$ when the cycle is in equilibrium and $\epsilon_{CN}/\rho X_H{}^2$ for the specific value $X_{CN}/X_H = 0.02$. For other values of X_{CN}/X_H the ϵ_{CN} curve can be moved up or down by the logarithm of the ratio. The value $X_{CN}/X_H = 0.02$ was chosen because that value is fairly characteristic of population I composition. We note here that for $X_{CN}/X_H = 0.02$ the CN cycle takes over from the PP chains near $T_6 = 18$.

This concludes the discussion of hydrogen burning, which has been considered in some detail, not only because of its prominence in studies of stellar structure, but also because it illustrates many principles encountered in any nuclear burning stage. Hydrogen burning in stars is found to occur as the central energy source for main-sequence stars and as a shell source in later stages of stellar evolution.

5-5 HELIUM BURNING

Some of the historical puzzles in nuclear astrophysics are very interesting. One was discussed in the last section, the problem of initiating nuclear reactions in stars in the face of the fact that there exist no stable nuclei composed only of protons. With the solution of that problem it became possible to synthesize nuclei up to $A = 4$ (the small number of $A = 7$ nuclei synthesized in PPII and PPIII are very rare and do not survive the exhaustion of hydrogen). The clarification of this nuclear physics suggested a solution to the problem of why He^4 should be the second most abundant nucleus in the universe. At least, there exists the possibility that the present ratio $He^4/H \sim 0.1$ reflects the results of hydrogen burning in early cosmological stages followed by about 13 billion years of star formation, death, and remixing. When scientific attention turned to the next two most abundant nuclei, C^{12} and O^{16}, another temporary stumbling block was encountered; viz., there are no stable nuclei at $A = 5$ and $A = 8$. In particular, the latter fact seemed to forbid the fusion of two alpha particles into an $A = 8$ nucleus. Careful analysis showed that there exist no chains of light-particle reactions that efficiently hurdle these gaps. The fact that C^{12} and O^{16} are composed of numbers of protons and neutrons numerically equal to three and four He^4 nuclei, respectively, led to the idea that these nuclei may be the results of more-than-two-body alpha-particle collisions. As appealing as this idea was, it encountered numerical difficulty from the very low probability of many-body collisions. The quantum treatment of collisions showed that some special resonant interactions between alpha particles would be required to achieve a sufficient stellar rate. The resolution of this problem took shape from an interesting interplay between experiment, inference, and theory.[1]

THE 3α REACTION

The heart of the reaction by which $3He^4 \rightarrow C^{12} + \gamma$ is the temporary formation of Be^8 from two alpha particles. Although it was shown in the late 1940's that Be^8 is unstable against breakup into two alpha particles, it is unstable by only 92 kev. The ground state has a width of 2.5 ev, which corresponds to a natural Be^8 lifetime of 2.6×10^{-16} sec. Although short, this lifetime is much longer than the time required for two alpha particles to scatter past each other in some nonresonant way. That is, each time a Be^8 nucleus is formed from two alphas, it sticks together much longer than it would if the two alphas simply scattered. As a result, a small concentration of Be^8 nuclei builds up in the helium gas until the rate of breakup of Be^8 is equal to its rate of formation; that is, Be^8 comes into equilibrium (Fig 5-17):

$$He^4 + He^4 \rightleftarrows Be^8$$

[1] E. E. Salpeter, *Astrophys. J.*, **115**:326 (1952), and *Ann. Rev. Nucl. Sci.*, **2**:41 (1953); E. J. Opik, *Proc. Roy. Irish Acad.*, **A54**:49 (1951), and *Mem. Soc. Roy. Sci. Liege*, **14**:131 (1954); F. Hoyle, *Astrophys. J. Suppl.*, **1**:121 (1954); W. A. Fowler and J. L. Greenstein, *Proc. Natl. Acad. Sci. U.S.*, **42**:173 (1956); C. Cook, W. A. Fowler, C. C. Lauritsen, and T. Lauritsen, *Phys. Rev.*, **107**:508 (1957); and E. E. Salpeter, *Phys. Rev.*, **107**:516 (1957).

This equilibrium can be calculated from the resonant cross-section rate and the breakup time of Be^8. The prototype calculation was described in Sec. 4-7. This is a case (*ii*) resonance inasmuch as $\Gamma_2 \ll \Gamma_1$. In the laboratory $\Gamma_2 = 0$, since there is no other mode of decay for Be^8 than into the initial channel of two alphas, but in a dense helium gas the interaction of Be^8 with a third alpha particle constitutes the counterpart of Γ_2. Therefore from Eq. (4-203) we have

$$N(Be^8) = N_\alpha{}^2 \omega f \frac{h^3}{(2\pi\mu kT)^{\frac{3}{2}}} \exp - \frac{E_r}{kT}$$

$$\approx 1.87 \times 10^{-33} N_\alpha{}^2 f T_8{}^{-\frac{3}{2}} \times 10^{-4.64/T_8} \qquad (5\text{-}99)$$

where T_8 is the temperature in units of 10^8 °K. At $T_8 = 1$ and $\rho = 10^5$ there exists about 1 Be^8 nucleus for 10^9 He^4 nuclei. Although 1 part in 10^9 may seem small, it is quite sufficient to allow a third alpha particle to interact with the Be^8 nuclei:

$$Be^8 + He^4 \rightarrow C^{12} + \gamma$$

Even after the recognition by Salpeter of the two-stage nature of the 3α reaction, Hoyle pointed out that the overall reaction would still not be sufficiently fast unless the $Be^8 + He^4$ reaction were also resonant at stellar energies. Since Be^8 and He^4 are both $J^\pi = 0^+$ nuclei, an *s*-wave resonance in stars demands that C^{12} have a 0^+ state with energy near $E_0 \pm 2\Delta E_0$ above the mass of $Be^8 + He^4$. The work of Cook, Fowler, Lauritsen, and Lauritsen in the Kellogg Radiation Laboratory at the California Institute of Technology demonstrated the existence of such a state.

Problem 5-32: Calculate E_0 and ΔE_0 for $Be^8 + He^4$.
Ans: $E_0 = 146 T_8{}^{\frac{2}{3}}$ kev, $\Delta E_0 = 82 T_8{}^{\frac{5}{6}}$ kev.

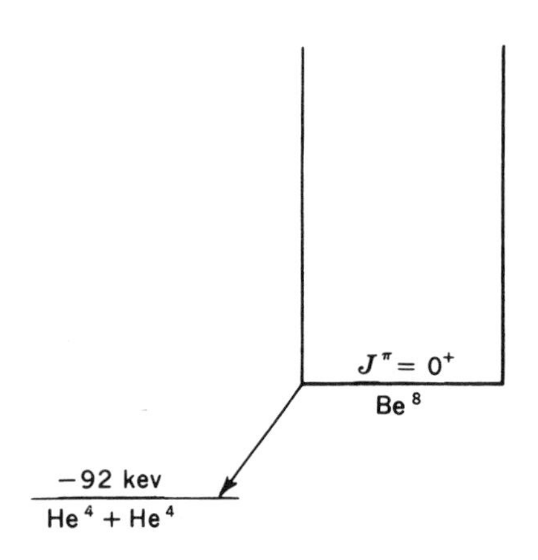

Fig. 5-17 The ground state of Be^8 is unstable against breakup into two alpha particles.

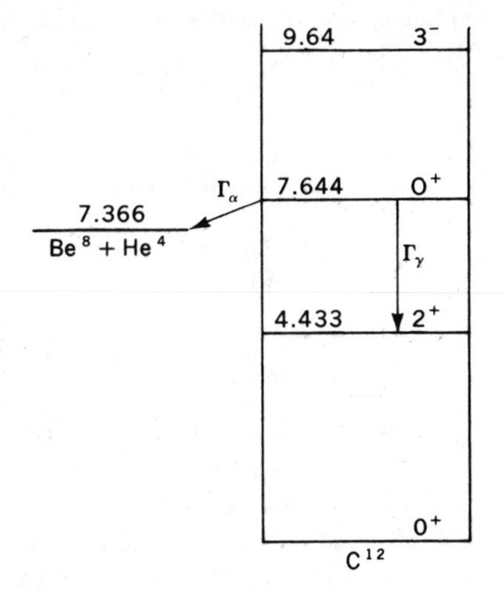

Fig. 5-18 The energy-level diagram of C^{12}. Alpha particles may fuse with the transient Be^8 nuclei to form the 7.644-Mev state of C^{12}. This state usually breaks up by rejecting the alpha particle, but with a smaller probability it also decays electromagnetically to the 4.433-Mev state.

The $J^\pi = 0^+$ excited state of C^{12} is found to lie at a resonance energy $E_r = 278$ kev above the combined mass of $Be^8 + He^4$. Thus the reaction $Be^8 + He^4 \rightarrow C^{12}$ is also a resonant reaction, and its rate in stellar interiors is also calculable from the resonant-reaction-rate formulas.

The energy-level diagram of C^{12} is shown in Fig. 5-18. Before applying the reaction-rate formulas it is necessary to examine the partial widths of the C^{12} excited state. It so happens that this state of C^{12} breaks up almost every time it is formed into $Be^8 + He^4$ rather than by decay to the C^{12} ground state via two successive gamma rays. (The gamma decay cannot go directly to the ground state because $0^+ \rightarrow 0^+$ gamma transitions are forbidden.) Thus in a star the state usually decays back into $He^4 + Be^8$. The best present estimates of the widths are $\Gamma_\alpha \approx \Gamma = 8.3$ ev and $\Gamma_\gamma = (2.8 \pm 0.5) \times 10^{-3}$ ev.[1] From the point of view of resonant reaction rates, therefore, this resonant reaction also falls into the case (*ii*) limit of Sec. 4-7 ($\Gamma_2 \ll \Gamma_1$). In the spirit of that discussion, therefore, it is possible to calculate the equilibrium concentration of C^{12*} and obtain the reaction rate by multiplying $N(C^{12*})$ by the gamma-decay rate. It follows that

$$N(C^{12*}) = N(Be^8)N_\alpha f \frac{h^3}{(2\pi\mu kT)^{\frac{3}{2}}} \exp - \frac{E_r}{kT} \tag{5-100}$$

where $\omega = 1$ since the particles are all spin zero and the f, μ, and E_r refer to the $Be^8 + \alpha$ reaction.

[1] Fowler, Caughlan, and Zimmerman, *loc. cit.* These widths are measured relative to the width for decay of the 7.64-Mev state to the ground state by the emission of a positron-electron pair. That rate is calculated in turn by measuring the monopole matrix element for inelastic electron scattering from C^{12}. Thus even an electron linear accelerator enters nuclear astrophysics in an important way.

Problem 5-33: Show that

$$N(\mathrm{C}^{12*}) = N_\alpha{}^3 f_{\alpha\alpha} f_{\alpha\mathrm{Be}} \frac{h^6(3)^{\frac{3}{2}}}{(2\pi M_\alpha kT)^3} \exp - \frac{\chi}{kT} \tag{5-101}$$

where $f_{\alpha\alpha}$ and $f_{\alpha\mathrm{Be}}$ are the electron-screening factors for the two successive reactions and χ is the energy difference between C^{12*} and the alpha particles: $\chi = 92 + 278 = 370$ kev.

Since the decay rate for $\mathrm{C}^{12*} \to \mathrm{C}^{12} + \gamma$ is $\Gamma_\gamma/\hbar$, the number of reactions per cubic centimeter per second is (using $\Gamma_\gamma = 2.8 \times 10^{-3}$ ev)

$$r_{3\alpha\to\mathrm{C}^{12}} = \frac{N(\mathrm{C}^{12*})\Gamma_\gamma}{\hbar} = 9.8 \times 10^{-54} \frac{N_\alpha{}^3}{T_8{}^3} f \exp - \frac{42.94}{T_8} \qquad \mathrm{cm}^{-3}\ \mathrm{sec}^{-1} \tag{5-102}$$

The numerical uncertainty in Eq. (5-102) is about 60 percent, most of which is due to the uncertainty of the total width of the C^{12} excited state. [The total width Γ did not enter into the calculation leading to Eq. (5-102), but it does enter linearly into the experimental determination of Γ_γ.] Since each reaction consumes three alpha particles, the lifetime of alpha particles against the 3α reaction is given by

$$\frac{N_\alpha}{\tau_{3\alpha}} \equiv 3r_{3\alpha}$$

which reduces numerically to

$$\frac{1}{\tau_{3\alpha}} = 6.7 \times 10^{-7} \frac{(\rho X_\alpha)^2}{T_8{}^3} f \exp - \frac{42.94}{T_8} \qquad \mathrm{sec}^{-1} \tag{5-103}$$

The energy generation from the 3α reaction is given by the product of the rate and the energy liberated per reaction: $Q_{3\alpha} = [3M_\alpha - M(\mathrm{C}^{12})]c^2 = 7.274$ Mev. Therefore

$$\epsilon_{3\alpha} = \frac{r_{3\alpha}Q_{3\alpha}}{\rho} = 3.9 \times 10^{11} \frac{\rho^2 X_\alpha{}^3}{T_8{}^3} f \exp - \frac{42.94}{T_8} \qquad \mathrm{erg\ g^{-1}\ sec^{-1}} \tag{5-104}$$

Problem 5-34: Show that the weak-electron-screening formula for the 3α reaction is approximately $f_{3\alpha} \approx \exp\,(2.76 \times 10^{-3}\rho^{\frac{1}{2}}T_8{}^{-\frac{3}{2}})$.

One of the most dramatic features of the 3α reaction is the very strong temperature dependence. Near some value of the temperature T_0 the energy-generation rate is

$$\epsilon(T) = \epsilon(T_0)\left(\frac{T}{T_0}\right)^n$$

Problem 5-35: Show that $n = 42.9/T_8 - 3$.

Thus near $T_8 = 1$, for instance,

$$\epsilon_{3\alpha} \approx 4.4 \times 10^{-8}\rho^2 X_\alpha{}^3 f T_8{}^{40} \qquad \mathrm{erg\ g^{-1}\ sec^{-1}} \tag{5-105}$$

This very strong energy dependence means that the energy generation in a star will be very strongly peaked toward the regions of highest temperature. It also means that an error of a factor of 2 in ϵ corresponds to a very small error in the temperature. At sufficiently high temperature and density a helium gas is very explosive in the sense that a small temperature rise greatly accelerates the rate of energy liberation. In a stellar center supported by electron degeneracy, the onset of helium burning is believed to be accompanied by just such an explosive reaction, the so-called *helium flash*.

NUCLEOSYNTHESIS DURING HELIUM BURNING

The three-alpha-particle reaction suggests (from the point of view that the elements were synthesized from hydrogen in stars) why it is that C^{12} is the fourth most abundant nuclear species. The third most abundant, O^{16}, may logically be formed by the capture of yet another alpha particle. The fifth most abundant is apparently Ne^{20} (its abundance is hard to determine), which presumably can be synthesized by the capture of another alpha particle by O^{16}. Continued successive alpha-particle captures can occur in principle, but calculation shows that the increasing coulomb barrier severely limits the number of alpha-particle captures at temperatures low enough for some helium still to remain. The main line of nucleosynthesis during helium burning is found to consist of the reactions

$$3He^4 \rightarrow C^{12}$$

$$C^{12}(\alpha,\gamma)O^{16}$$

and perhaps

$$O^{16}(\alpha,\gamma)Ne^{20}$$

The $C^{12}(\alpha,\gamma)O^{16}$ is a nonresonant reaction; i.e., there are no O^{16} states near E_0. The reaction is presumed to occur in the tail of an O^{16} state lying just below the mass of $C^{12} + He^4$ which spreads into the positive-energy region by virtue of its natural width. The 7.115-Mev state of O^{16} has a width of only $\Gamma_\gamma = 0.066$ ev and has $J^\pi = 1^-$, so that the capture of an alpha particle into its tail must occur by an $l = 1$ alpha wave. An additional complication is that the rate is propor-

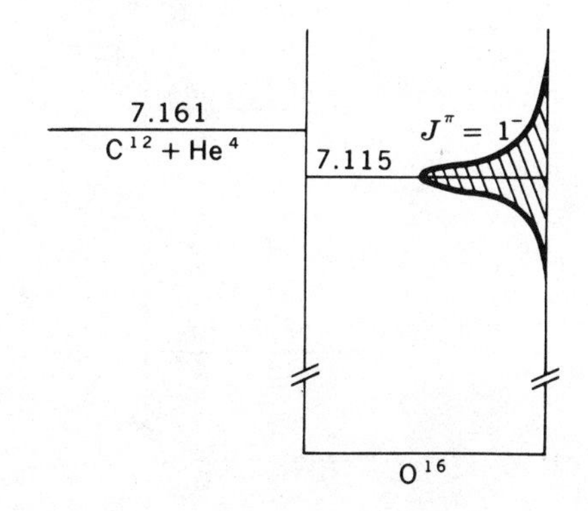

Fig. 5-19 A portion of the energy-level diagram of O^{16}. The 7.115-Mev state is stable by 46 kev against breakup into $C^{12} + He^4$. The C^{12} nucleus may nonetheless capture an alpha particle to form the high-energy wing of this state.

tional to θ_α^2 for the 7.115-Mev state, and θ_α^2 is unknown for that state (it is not actually observed to break up into $C^{12} + He^4$ because it has insufficient energy except on the very high-energy tail of the state). Most workers agree that it should be safe to assume $\theta_\alpha^2 = 0.1$ within a factor of 2. Then the Breit-Wigner resonance shape can be extrapolated into the positive-energy region for $C^{12} + He^4$ and an S_0 calculated. The resulting rate

$$\lambda_{\alpha 12} \approx \frac{3.6 \times 10^{-14}}{T_8{}^2} \exp - \frac{69.18}{T_8{}^{\frac{1}{3}}} \qquad \text{cm}^3\,\text{sec}^{-1} \tag{5-106}$$

should be within a factor of 2 of the correct value. The corresponding lifetime in years is

$$\frac{1}{\tau_\alpha(C^{12})} \approx 1.7 \times 10^{17} \frac{\rho X_\alpha}{T_8{}^2} \exp - \frac{69.18}{T_8{}^{\frac{1}{3}}} \qquad \text{year}^{-1} \tag{5-107}$$

An interesting situation occurs in the subsequent $O^{16}(\alpha,\gamma)Ne^{20}$ reaction that illustrates several physical features of the physics of thermonuclear reactions.

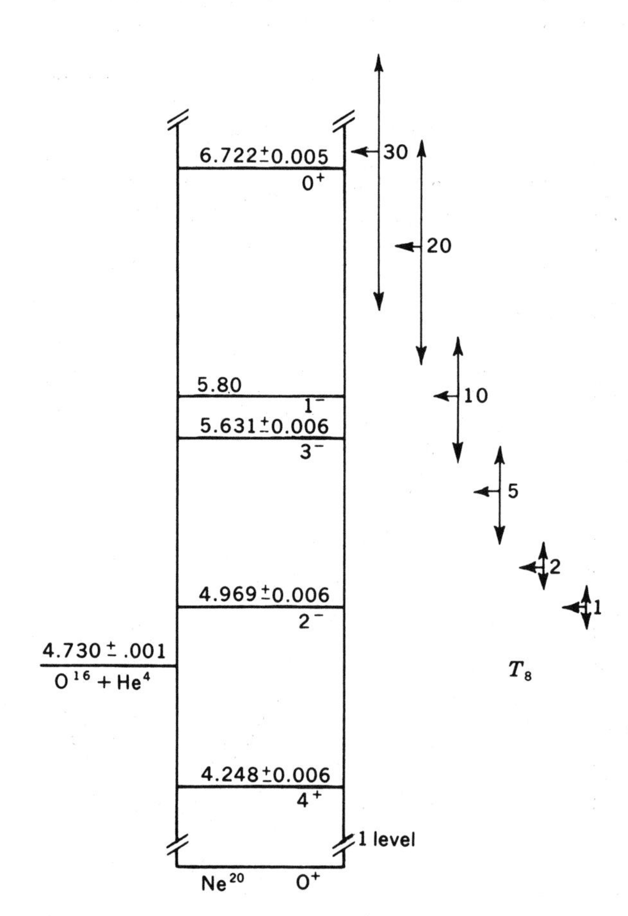

Fig. 5-20 The energy-level diagram of Ne^{20}. The arrows to the right indicate the most effective stellar energy E_0 for the $C^{12}(\alpha,\gamma)O^{16}$ reaction for several different values of T_8. It can be seen that the question of which states of Ne^{20} are important for this reaction depends upon the temperature.

The relevant energy-level diagram for Ne^{20} is shown in Fig. 5-20. Also designated by arrows to the right of the diagram is the energy range $E_0 \pm \Delta E_0$ that is the most favored energy range for various values of the temperature of helium burning.

Since helium burning starts at about $T_8 = 1$, for which $E_0 = 246$ kev, it is evident that the 4.969-Mev state falls almost at the center of the Gamow peak. Thus that state would serve as the dominant resonance in the $O^{16} + He^4$ reaction if it were allowed to do so by the nuclear quantum numbers. But the spin and parity of that state ($J^\pi = 2^-$) cannot be formed from O^{16} ($J^\pi = 0^+$) and He^4 ($J^\pi = 0^+$) with any l-wave capture contributing a factor of $(-1)^l$ to the parity. Thus that state is "invisible" to the $O^{16} + He^4$ system. By an analogous argument, the states at 5.63, 5.80, and 6.72 Mev are allowed in the $O^{16} + He^4$ system.

Problem 5-36: Confirm the availability of the 5.63-, 5.80-, and 6.72-Mev states to the $O^{16} + He^4$ system and determine the required l wave of the alpha-particle capture.
Ans: $l = 3$, $l = 1$, and $l = 0$, respectively.

Thus near $T_8 = 1$ the $O^{16}(\alpha,\gamma)Ne^{20}$ reaction will be nonresonant; the S_0 for the reaction will be provided by the tails of the higher-lying resonances extending downward to the energy E_0. From the form of the Breit-Wigner cross section it is apparent that far from the peak of the resonance, where $(E - E_r)^2 \gg (\Gamma/2)^2$, the cross section (hence S_0) will be proportional to

$$S_0 \propto \frac{\Gamma_\alpha \Gamma_\gamma}{(E_0 - E_r)^2} \tag{5-108}$$

Because the widths of the 5.63-Mev state are so much smaller than the widths of the other two states, it turns out that the nonresonant value of S_0 is largely due to the 5.80- and 6.72-Mev states. According to Reeves,[1] the nonresonant lifetime is valid for $1 < T_8 < 2.1$ and is given by (within a factor of 10)

$$\frac{1}{\tau_\alpha(O^{16})} = 1.4 \times 10^{17} \frac{\rho X_\alpha f}{T_8^{2/3}} \exp - \frac{85.66}{T_8^{1/3}} \qquad \text{year}^{-1} \tag{5-109}$$

From Fig. 5-20, however, it is evident that as T_8 increases, a point will be reached where the resonant reaction rate will come into play. The higher-lying states will be important when this reaction occurs in some later stage of nuclear burning.

Problem 5-37: Assuming that for the 5.63-Mev state $\Gamma_\alpha = 6 \times 10^{-3}$ ev and $\Gamma_\gamma = 4 \times 10^{-4}$ ev, show that the lifetime in years due to alpha-particle capture into this state is given by

$$\frac{1}{\tau_\alpha(O)^{16}} = 1.9 \times 10^{10} \frac{\rho X_\alpha f}{T_8^{3/2}} \exp - \frac{104.6}{T_8} \qquad \text{yr}^{-1} \tag{5-110}$$

[1] Good discussions of the nuclear physics problems can be found in H. Reeves, Stellar Energy Sources, in L. H. Aller and D. B. McLaughlin (eds.), "Stellar Structure," The University of Chicago Press, Chicago, 1965, and in A. G. W. Cameron, Yale Lecture Notes in Nuclear Astrophysics (unpublished).

Comparison shows that Eqs. (5-110) and (5-109) become equal near $T_8 = 2.1$. Actually the 5.80-Mev level begins to dominate when $T_8 > 8$, but such high temperatures are apparently of little importance in helium burning. This reaction is a good example of the way a nonresonant reaction rate changes into a resonant rate as temperature increases bring new compound states into play.

The lifetimes of the three nuclear species He^4, C^{12}, and O^{16} against alpha particles are listed in Table 5-4 as a function of temperature. The density dependence is noted explicitly, and the entries in the table are calculated without electron screening. The numbers can be corrected for electron screening by dividing each entry by the appropriate value of f. It can be seen that for the rather dense case $\rho X_\alpha = 10^5$ g/cm^3, the lifetime of He^4 is the shortest of the three over the entire temperature range listed. As the helium is depleted at $\rho = 10^5$, that is, $X_\alpha \to \ll 1$, or for helium burning at lower density, however, the C^{12} lifetime may be shorter than the He^4 lifetime. The effect of the resonance in $O^{16} + He^4$ can also be plainly seen, because $\tau_\alpha(O^{16})$ becomes shorter than $\tau_\alpha(C^{12})$ at the higher temperatures in spite of the larger coulomb barrier. This table of lifetimes will be of assistance in understanding the properties of helium burning.

The lifetime of Ne^{20} against the next reaction in the chain, $Ne^{20}(\alpha,\gamma)Mg^{24}$, has not been included in Table 5-4. That reaction is not of much importance in helium burning in low-mass stars, because almost no Ne^{20} can be synthesized there as a result of the long O^{16} lifetime. For those cases when considerable Ne^{20} is produced in helium burning, however, one should also add on the next reaction. The lifetime of Ne^{20} against alpha particles is (from Reeves)

$$\log\left[\tau_\alpha(Ne^{20})\left(\frac{\rho X_\alpha}{10^5}\right)\text{yr}\right] = -32.3 + \tfrac{2}{3}\log T_8 + 43.73 T_8^{-\frac{1}{3}} - 0.09 T_8^{\frac{2}{3}} \qquad (5\text{-}111)$$

Since this lifetime is actually shorter than $\tau_\alpha(O^{16})$, the reaction should be included if much Ne^{20} is produced. It prevents Ne^{20} from ever being the major product of helium burning.

The interesting question for nucleosynthesis is that of the abundances at the end of helium burning, i.e., when the helium is exhausted. Early in the burning the main synthesis must be $3He \to C^{12}$, but since the rate of that reaction is proportional to the cube of the helium density, the helium nuclei will mainly be captured by C^{12} and O^{16} as the helium abundance becomes small. To explore this problem it is simplest (and reasonably realistic because very little Ne^{20} will be synthesized) to temporarily ignore all captures past Ne^{20} by assuming that the capture chain ends there. Then the three reactions and their rates are

$$r(3He^4 \to C^{12}) = \lambda_{3\alpha}(He^4)^3 \qquad (5\text{-}112a)$$

$$r[C^{12}(\alpha,\gamma)O^{16}] = \lambda_{\alpha 12} He^4 C^{12} \qquad (5\text{-}112b)$$

$$r[O^{16}(\alpha,\gamma)Ne^{20}] = \lambda_{\alpha 16} He^4 O^{16} \qquad (5\text{-}112c)$$

and the differential equations for the abundances are (again using chemical

Table 5-4 Helium-burning lifetimes

T_8	$\tau_{3\alpha}(\mathrm{He}^4)\left(\frac{\rho X_\alpha}{10^5}\right)^2$, years	$\tau_\alpha(\mathrm{C}^{12})\frac{\rho X_\alpha}{10^5}$, years	$\tau_\alpha(\mathrm{O}^{16})\frac{\rho X_\alpha}{10^5}$, years
0.8	1.0×10^{12}		
0.9	3.9×10^{9}	9.3×10^{8}	
1.0	4.2×10^{7}	9.6×10^{7}	1.1×10^{15}
1.1	1.1×10^{6}	1.3×10^{7}	7.7×10^{13}
1.2	5.2×10^{4}	2.3×10^{6}	7.7×10^{12}
1.3	4.1×10^{3}	4.9×10^{6}	9.6×10^{11}
1.4	4.6×10^{2}	1.2×10^{5}	1.5×10^{11}
1.5	7.2×10	3.3×10^{4}	2.6×10^{10}
1.6	1.4×10	1.0×10^{4}	5.8×10^{9}
1.7	3.4	3.6×10^{3}	1.4×10^{9}
1.8	1.0	1.4×10^{3}	3.8×10^{8}
1.9	3.3×10^{-1}	5.4×10^{2}	1.1×10^{8}
2.0	1.2×10^{-1}	2.4×10^{2}	3.6×10^{7}
2.1	4.9×10^{-2}	1.0×10^{2}	1.0×10^{7}
2.2	2.3×10^{-2}	5.1×10	1.2×10^{6}
2.3	1.1×10^{-2}	2.6×10	1.5×10^{5}
2.4	5.5×10^{-3}	1.3×10	2.4×10^{4}
2.5	3.2×10^{-3}	7.2	4.4×10^{3}
2.6	1.8×10^{-3}	4.0	9.3×10^{2}
2.8	6.6×10^{-4}	1.4	5.6×10
3.0	2.9×10^{-4}	5.2×10^{-1}	5.0
3.2	1.4×10^{-4}	2.1×10^{-1}	6.2×10^{-1}
3.4	7.5×10^{-5}	9.6×10^{-2}	9.6×10^{-2}
3.6	4.5×10^{-5}	4.5×10^{-2}	1.9×10^{-2}
3.8	2.9×10^{-5}	2.3×10^{-2}	4.3×10^{-3}
4.0	1.9×10^{-5}	1.2×10^{-2}	1.2×10^{-3}

symbols for number density)

$$\frac{d\mathrm{He}^4}{dt} = -3\lambda_{3\alpha}(\mathrm{He}^4)^3 - \lambda_{\alpha 12}\mathrm{He}^4\mathrm{C}^{12} - \lambda_{\alpha 16}\mathrm{He}^4\mathrm{O}^{16} \tag{5-113a}$$

$$\frac{d\mathrm{C}^{12}}{dt} = \lambda_{3\alpha}(\mathrm{He}^4)^3 - \lambda_{\alpha 12}\mathrm{He}^4\mathrm{C}^{12} \tag{5-113b}$$

$$\frac{d\mathrm{O}^{16}}{dt} = \lambda_{\alpha 12}\mathrm{He}^4\mathrm{C}^{12} - \lambda_{\alpha 16}\mathrm{He}^4\mathrm{O}^{16} \tag{5-113c}$$

$$\frac{d\mathrm{Ne}^{20}}{dt} = \lambda_{\alpha 16}\mathrm{He}^4\mathrm{O}^{16} \tag{5-113d}$$

Actually the symbol "Ne^{20}" represents here the sum of the number densities of Ne^{20} and of any nuclei formed by subsequent alpha capture (Mg^{24}, Si^{28}, etc.). This set of equations may be followed numerically in the time steps of an evolving stellar model.

To develop an appreciation of the likely products of helium burning and the dependence of those results on the rates it is more instructive to examine them in a convenient representation devised by Cameron.[1] His particularly illuminating discussion runs as follows. To obtain the abundances as a function of the He^4 consumed it is appropriate to change the independent variable from time to the He^4 abundance. For instance,

$$\frac{dC^{12}}{dHe^4} = \frac{dC^{12}/dt}{dHe^4/dt} = \frac{\lambda_{3\alpha}(He^4)^3 - \lambda_{\alpha 12}He^4C^{12}}{-3\lambda_{3\alpha}(He^4)^3 - \lambda_{\alpha 12}He^4C^{12} - \lambda_{\alpha 16}He^4O^{16}}$$

$$= \frac{1 - \dfrac{\lambda_{\alpha 12}}{\lambda_{3\alpha}}\dfrac{C^{12}}{(He^4)^2}}{-3 - \dfrac{\lambda_{\alpha 12}}{\lambda_{3\alpha}}\dfrac{C^{12}}{(He^4)^2} - \dfrac{\lambda_{\alpha 16}}{\lambda_{3\alpha}}\dfrac{O^{16}}{(He^4)^2}} \tag{5-114}$$

and likewise for the other two equations. These equations are simplified by the definition of two dimensionless temperature-dependent parameters

$$R_{12} \equiv \frac{\lambda_{\alpha 12}}{\lambda_{3\alpha}He^4(0)} \qquad R_{16} \equiv \frac{\lambda_{\alpha 16}}{\lambda_{3\alpha}He^4(0)} \tag{5-115}$$

where $He^4(0)$ is the *initial* He^4 number density, and by the definition of four dimensionless variables

$$x \equiv \frac{He^4}{He^4(0)} \qquad u \equiv \frac{C^{12}}{He^4(0)} \qquad v = \frac{O^{16}}{He^4(0)} \qquad w = \frac{Ne^{20}}{He^4(0)}$$

Problem 5-38: Show that with these dimensionless quantities the differential equations are

$$\begin{aligned}\frac{du}{dx} &= \frac{1 - R_{12}u/x^2}{-3 - R_{12}u/x^2 - R_{16}v/x^2}\\ \frac{dv}{dx} &= \frac{R_{12}u/x^2 - R_{16}v/x^2}{-3 - R_{12}u/x^2 - R_{16}v/x^2}\\ \frac{dw}{dx} &= \frac{R_{16}v/x^2}{-3 - R_{12}u/x^2 - R_{16}v/x^2}\end{aligned} \tag{5-116}$$

Then the abundances may be obtained by numerical integration of the coupled set of equations. When the helium is exhausted, for instance, the carbon abundance is given by

$$u(0) = \int_1^0 \frac{du}{dx}\,dx$$

In a completely realistic calculation one more equation would be needed, one relating the helium abundance x to the temperature. Since the quantities R_{12} and R_{16} are functions only of the temperature [although inversely proportional to $He^4(0)$ also], they will change in general as the He^4 is exhausted. The relationship of x to T can be computed only by a complete set of evolving stellar models, but a good estimate of the results should be obtainable by considering helium

[1] A. G. W. Cameron, *op. cit.* and *Atomic Energy of Canada Limited Rept. AECL* 454, 1957.

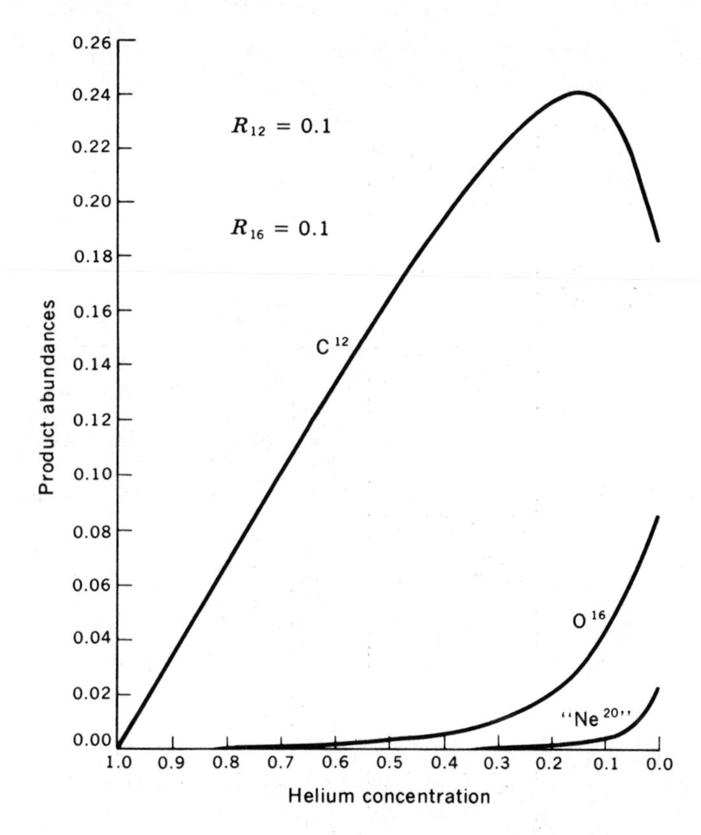

Fig. 5-21

Figs. 5-21 to 5-28 Abundances of C^{12}, O^{16}, and "Ne^{20}" produced by helium burning as a function of the fraction of He^4 remaining. These eight figures represent calculations with values of $R_{16} = 0.1$ and 1.0 for each of the values $R_{12} = 0.1$, 1.0, 10, and 100. As the value of R_{12} is increased, the amount of C^{12} that survives is decreased. For $R_{16} = 0.1$ or less, the O^{16} produced exceeds that of "Ne^{20}," whereas "Ne^{20}" is the greater for $R_{16} = 1.0$ or greater. For these calculations the alpha-particle captures were artificially terminated after capture by O^{16}, so that the "Ne^{20}" symbolizes the abundance of Ne^{20} plus heavier alpha nuclei. It does not appear to be possible that Ne^{20} can actually be a major product of helium burning, because when "Ne^{20}" is abundant it is largely Mg^{24} rather than Ne^{20}. [*After A. G. W. Cameron, Atomic Energy of Canada Limited Rept. AECL* 454, 1957 *and Yale Lecture Notes in Nuclear Astrophysics* (*unpublished*).]

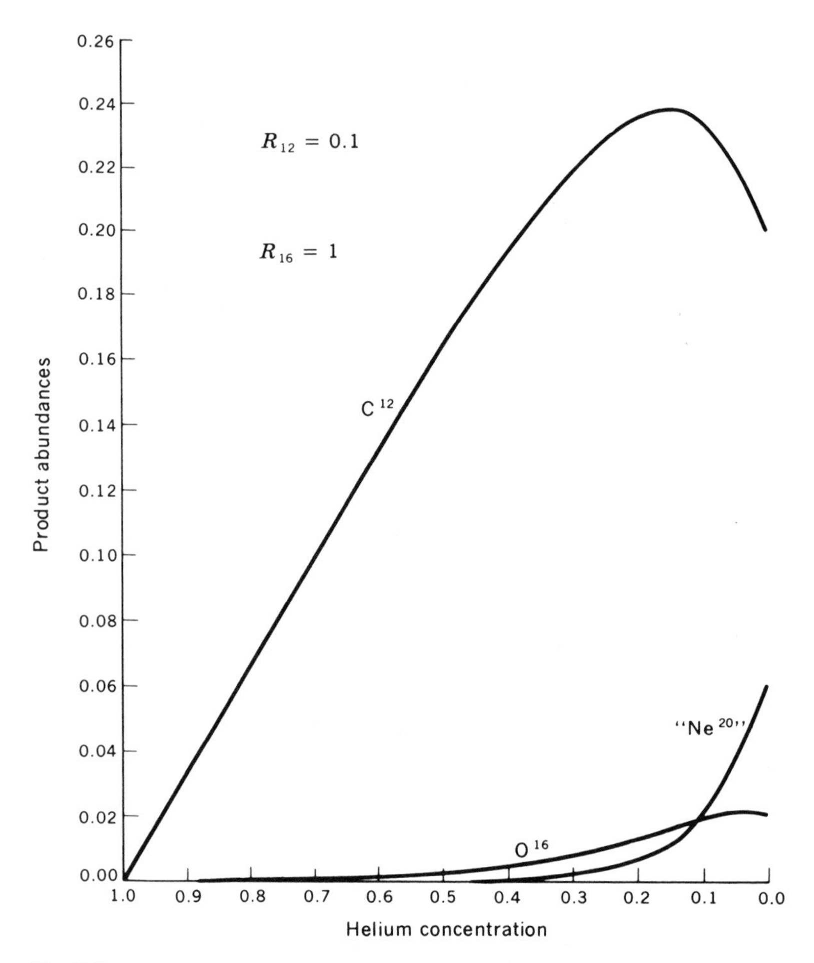

Fig. 5-22

burning at constant temperature, in which case R_{12} and R_{16} are constants and the equations can be numerically integrated.

Cameron has performed such integrations. The results of constant-temperature helium burning are computed for $R_{12} = 0.1$, 1.0, 10, and 100, coupled with two values of $R_{16} = 0.1$ and 1.0, and are plotted (per initial He^4) in Figs. 5-21 to 5-28. In Fig. 5-21, with the smallest values of the parameters, the C^{12} abundance is initially linear but passes through a maximum as the He^4 is depleted, and the final He^4 nuclei are scoured out by production of small amounts of O^{16} and Ne^{20}. When R_{16} is increased to unity, the C^{12} production is hardly changed, as illustrated in Fig. 5-22, but the final concentration of "Ne^{20}" exceeds that of O^{16}. In both cases the end product of helium burning is primarily C^{12}.

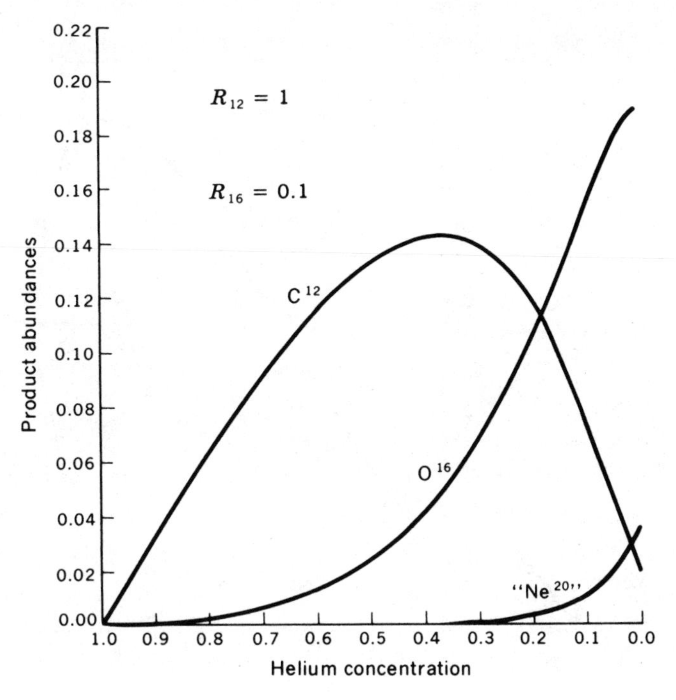

Fig. 5-23

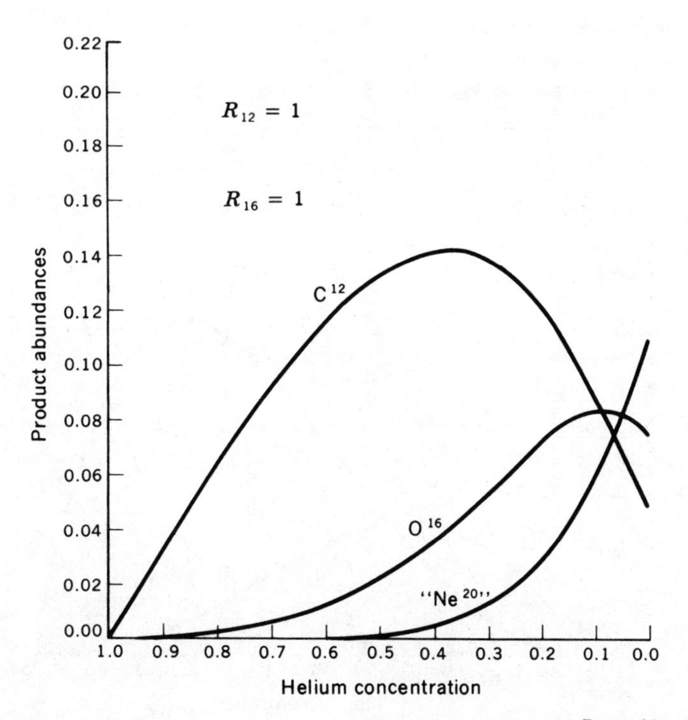

Fig. 5-24

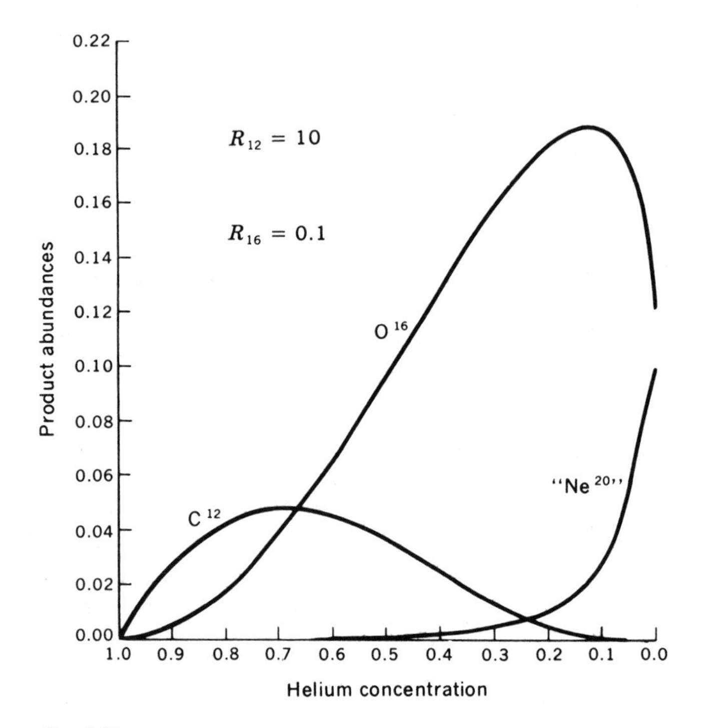

Fig. 5-25

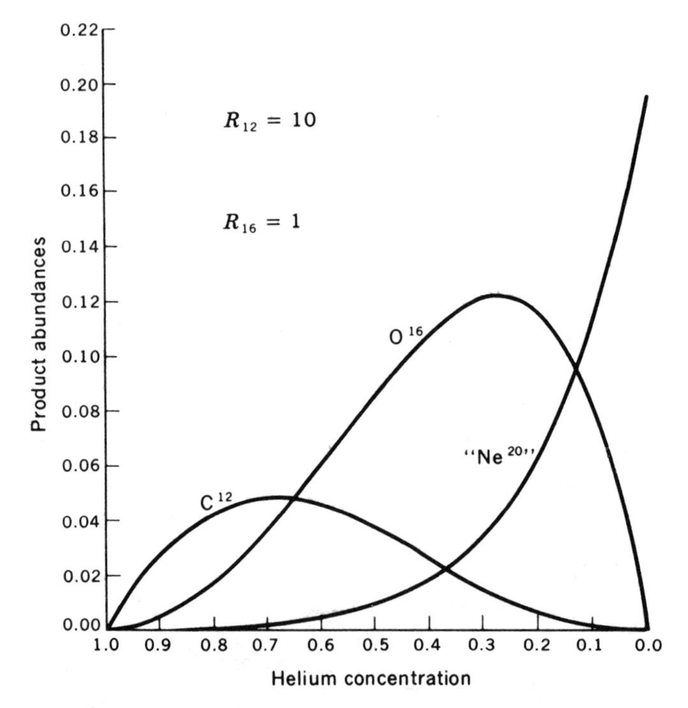

Fig. 5-26

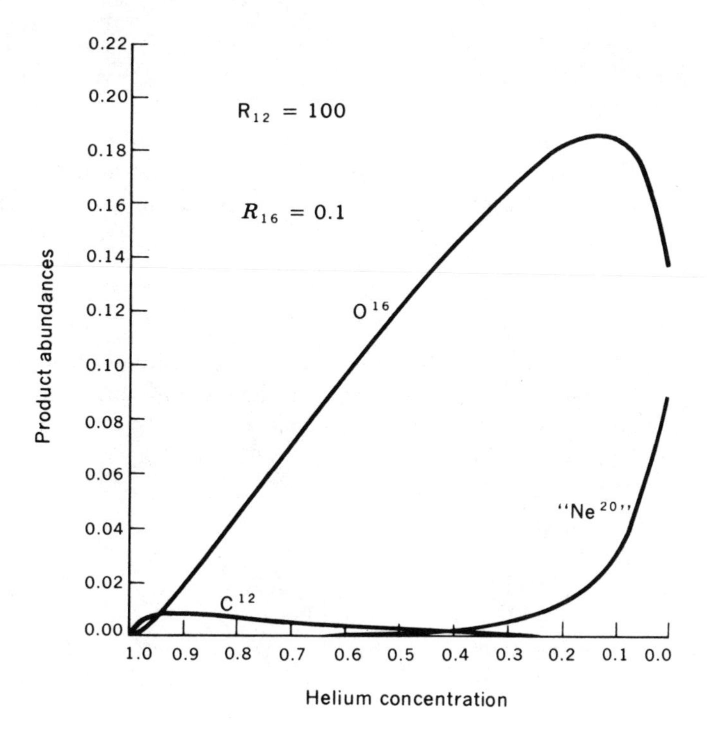

Fig. 5-27

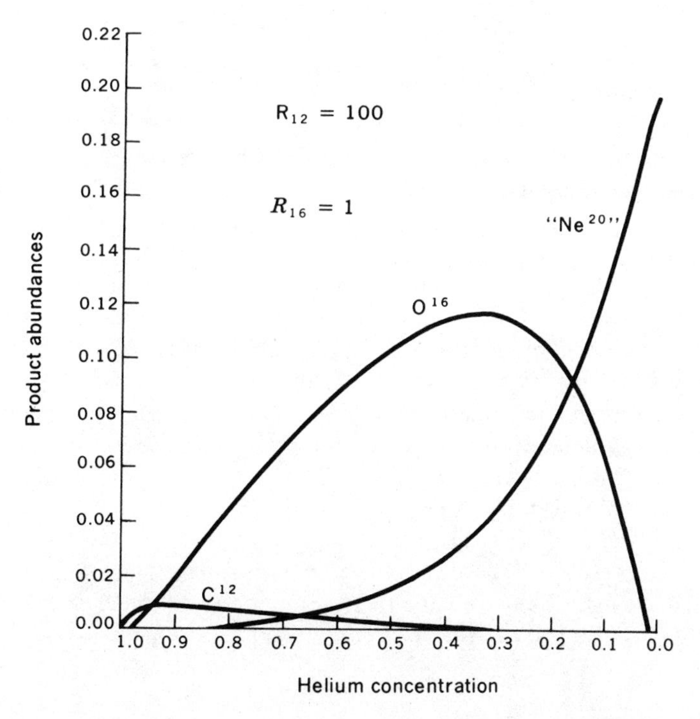

Fig. 5-28

In Figs. 5-23 and 5-24 the parameter R_{12} is increased to unity. This change increases the rate of $C^{12}(\alpha,\gamma)O^{16}$ relative to $r_{3\alpha}$ such that the final product is not predominantly C^{12}: with $R_{16} = 0.1$ it is predominantly O^{16}, and with $R_{16} = 1.0$ is mostly "Ne^{20}," with O^{16} and C^{12} not far behind.

When R_{12} is increased again to 10, the C^{12} is only the dominant product early in the burning, and none of it survives to the end. With $R_{16} = 0.1$ the results are nearly equally O^{16} and "Ne^{20}," but with $R_{16} = 1.0$ the results are entirely "Ne^{20}." In these cases, of course, the $Ne^{20}(\alpha,\gamma)Mg^{24}$ reaction should be included. *Whenever the final* "Ne^{20}" *exceeds* O^{16}, *the* "Ne^{20}" (*which really stands for the sum* $Ne^{20} + Mg^{24} + Si^{28} + \cdots$) *is predominantly* Mg^{24}, since the Ne^{20} lifetime is shorter than the O^{16} lifetime. It is doubtful whether Ne^{20} itself can ever be the major product of helium burning. The major end products are C^{12}, O^{16}, or Mg^{24}, depending upon the values of the R's. The final Ne^{20} abundance probably cannot exceed 15 percent by mass. This is a very interesting result, because the Ne^{20} abundance is very near the C^{12} abundance in most astronomical objects.

The results of increasing R_{12} to 100 are self-evident in Figs. 5-27 and 5-28. By interpolating among these eight figures a good appreciation of the dependence of the end products on the parameters R_{12} and R_{16} can be obtained. When temperature changes are included in the calculation, one finds that the appropriate values of R_{12} and R_{16} are those nearer the end of helium burning than the beginning—but not too near the end. (When the alpha particles are almost gone, the temperature does not matter, because there cannot be enough additional captures to alter the composition further.) In summary, C^{12} will be the major final abundance if R_{12} is much less than unity, and the major uncertainty in R_{12} lies in the uncertainty of the dimensionless reduced width θ_α^2 for the 7.12-Mev state of O^{16}. If R_{12} is of order unity or greater, the final products are predominantly O^{16}, Ne^{20}, and Mg^{24}. If R_{16} is about unity or greater, the final products are mostly Mg^{24} with some Ne^{20}. A particularly good discussion of this problem from a slightly different approach can be found in an account of the burning of pure helium stars by Deinzer and Salpeter.[1]

The values of R_{12} and R_{16} can be computed for any temperature and density from their definitions. Of course, almost any values may be obtained for arbitrary choices of temperature and density. To understand what the likely results of helium burning in stars are requires some restriction to the range of temperatures and densities relevant to helium burning. Cameron has found an interesting way to display the relevant range. From the luminosities of stars believed to be burning helium (giants) and from models of helium-burning stars it is known that energy-generation rates on the order of 10^2 to 10^8 ergs g^{-1} sec^{-1} are required in the various burning phases. The exact value of ϵ depends, of course, upon the mass of the star and the location of the interior point. Equation (5-104) has been plotted in Fig. 5-29 in such a way as to show the locus of temperature versus density with $X_\alpha = 1$, resulting in a given energy generation. Stars near one solar mass must have $\epsilon \sim 10^2$ to 10^4 ergs g^{-1} sec^{-1} and burn at relatively low tempera-

[1] W. Deinzer and E. E. Salpeter, *Astrophys. J.*, **140**:499 (1964).

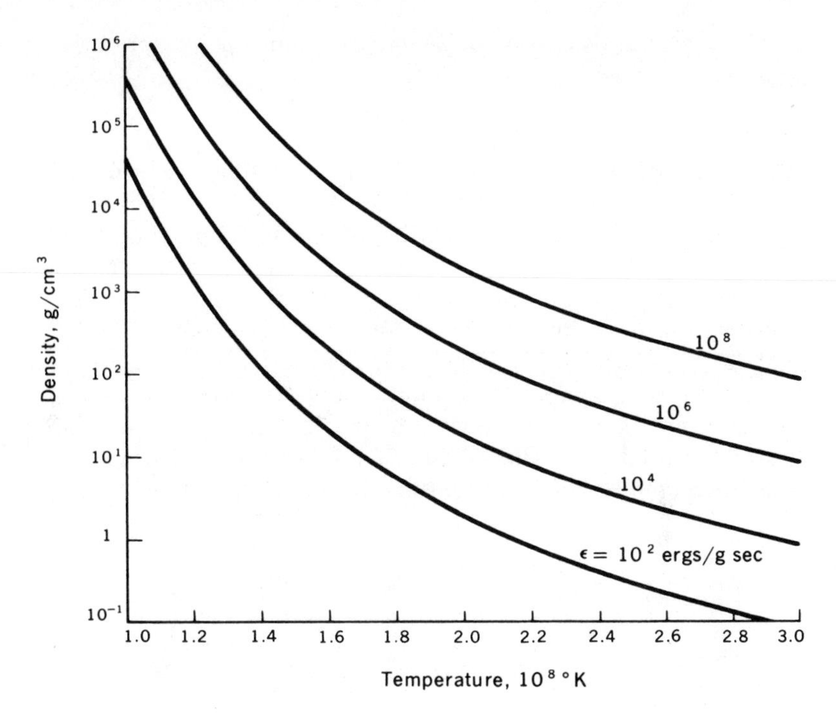

Fig. 5-29 The locus of points in the temperature-density plane resulting in a given value of the energy generation. The values of ϵ labeling the curves span those of primary significance for helium-burning stars.

ture $T_8 \sim 1.0$ to 1.2 and relatively high density $\rho \sim 10^4$ to 10^5. On the other hand, stars of 10 solar masses and more require $\epsilon \sim 10^6$ to 10^8 and burn at higher temperatures $T_8 = 2$ to 3 and lower densities $\rho \sim 10^2$ to 10^3. *For a given rate of energy generation*, the parameters R_{12} and R_{16} are functions only of the temperature, and they are so displayed in Figs. 5-30 and 5-31. For a one-solar-mass star with $\epsilon \approx 10^4$ and $T_8 \approx 1$ the value of R_{12} is near unity. The earlier curves show that this value of R_{12} is near enough to the dividing line between C^{12} and O^{16} production to make it clear that the final product cannot be emphatically stated. It appears likely that both C^{12} and O^{16} are produced in significant amounts by stars of moderate mass but that the final product is almost entirely O^{16} for $\mathfrak{M} > 10\mathfrak{M}_\odot$. Figure 5-31 shows that for the same star the value of R_{16} is almost vanishingly small, which means that no Ne^{20} can be produced. From the same figure it can be seen that the large value $R_{16} > 0.1$ required for appreciable production of Ne^{20} and heavier nuclei occurs only at high temperatures, which automatically restricts their synthesis in helium burning to massive stars (say $\mathfrak{M} \gg 20\mathfrak{M}_\odot$).[1]

[1] In detailed calculations of the evolution of a $15\mathfrak{M}_\odot$ star through He^4 burning, Iben finds that the final weight fraction of Ne^{20} at the center is only 1 percent; *Astrophys. J.*, **143**:516 (1966).

These general results are quite important for the interpretation of the following abundance ratios:

$$C^{12}:N^{14}:O^{16}:Ne^{20}:Mg^{24} = 5.5:1.0:9.6:5:0.3$$

The N^{14} is included because its synthesis is presumed to be due to the operation of the CNO bi-cycle upon the C^{12} and O^{16} which were synthesized in helium burning in another star. That interpretation of the N^{14} abundance can be correct if, from the above ratios, about 6 percent of the C^{12} and O^{16} has been reincorporated into another star burning on the CNO cycle and ultimately rejected (without destruction) back into the interstellar medium. If 6 percent of the $C^{12} + O^{16}$ is to have that fate, a considerably larger amount must actually have been remixed into later-generation stars, for the bulk of the N^{14} produced in the CNO cycle will probably not be ejected intact into the interstellar medium; most of it will be destroyed in later burning phases of the same star in which it was synthesized. The abundance of N^{14} is regarded (from the point of view of stellar nucleo-

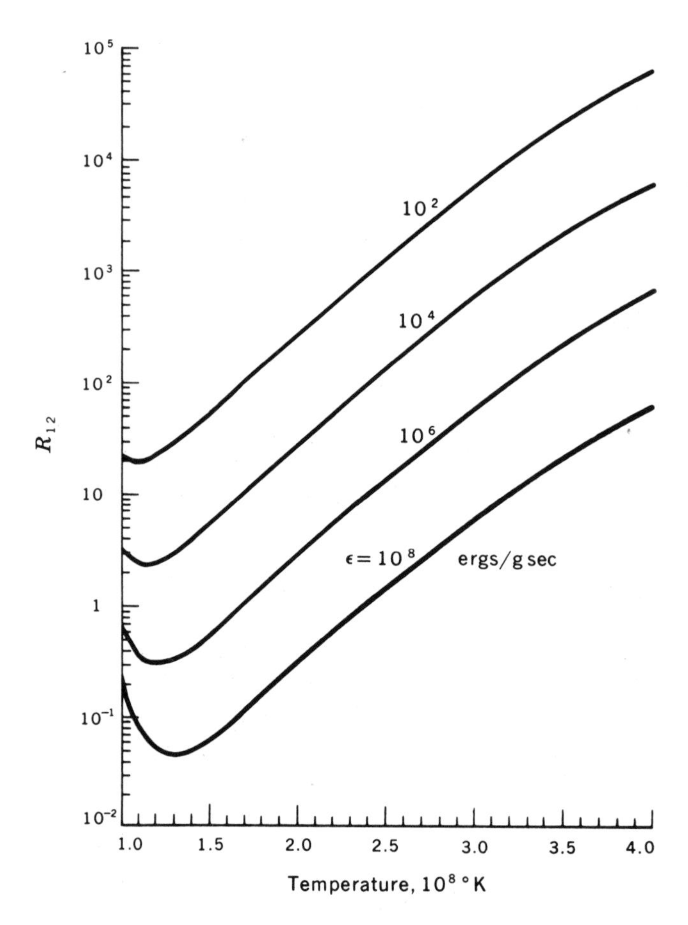

Fig. 5-30 The value of the parameter R_{12} as a function of temperature for several values of the energy generation. The important value $R_{12} = 1$ lies near the center of the expected requirements for helium-burning stars. [*After A. G. W. Cameron, Yale Lecture Notes in Nuclear Astrophysics (unpublished).*]

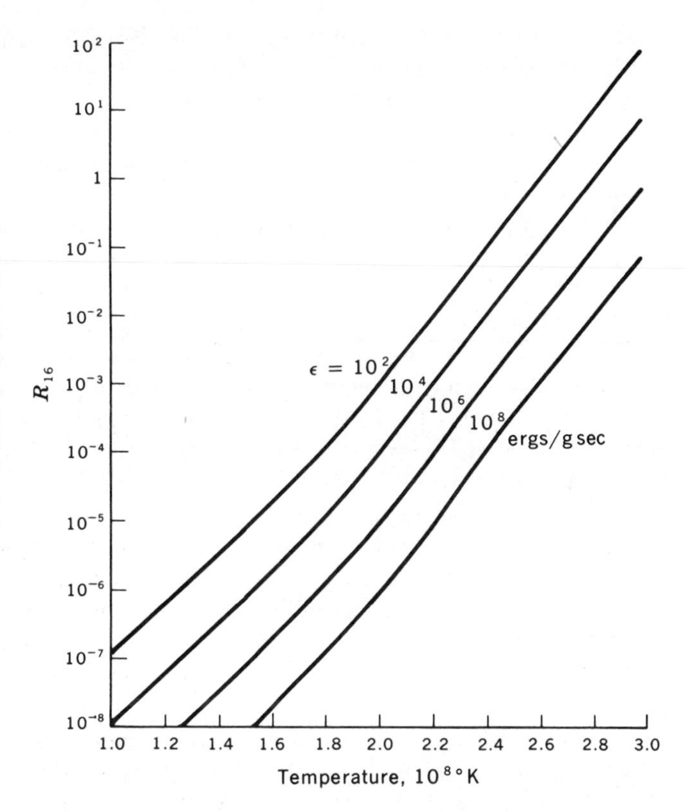

Fig. 5-31 The value of the parameter R_{16} as a function of temperature for several values of the energy generation. The expected values are almost entirely in the domain $R_{16} \ll 1$, indicating that O^{16} will not generally be a significant capturer of alpha particles during helium burning. [*After A. G. W. Cameron, Yale Lecture Notes in Nuclear Astrophysics (unpublished).*]

synthesis) as a remixing problem in the chemical evolution of the galaxy. It is worth adding at this point that the same remixing ratio is not adequate to account for the abundances of C^{13} and N^{15}, which are *terrestrially* found to be $C^{13}/C^{12} = \frac{1}{90}$ and $N^{15}/N^{14} = \frac{1}{275}$.

Problem 5-39: Show that the remixing ratio necessary to account for the N^{14} abundance results in a ratio $C^{13}/C^{12} = 10^{-3}$, a full factor of 10 less than the terrestrially observed ratio. (The solar ratio appears to be less but is not well known, so that the solar C^{13}/C^{12} ratio may be consistent with the simple remixing idea, in which case a special source for enhancing the terrestrial C^{13} by a factor of 10 must exist.[1]

One may try to interpret the remainder of the abundances as remnants of helium burning. No particular difficulty results from the ratio $O^{16}/C^{12} = 1.7$, especially if θ_α^2 (which enters linearly in R_{12}) for the 7.12-Mev state of O^{16} is smaller than 0.1. Such an O^{16}/C^{12} ratio may well be the logical consequence of helium burning in the spectrum of stellar masses. The abundances of Ne^{20} and Mg^{24} present severe problems, however. There seems to be no way to produce Ne^{20} with roughly half of the O^{16} abundance without simultaneously creating more Mg^{24} than Ne^{20}. It would seem to require that some reaction rates be in

[1] See, for example, Fowler, Greenstein, and Hoyle, *loc. cit.*, for a detailed suggestion.

error if the abundances are to be the result of He burning. The difficulty lies with the high abundance of Ne^{20}. Either the Ne^{20} abundance is an overestimate (by a factor of 10 or so), or one must conclude that it is probably not the result of helium burning. The situation could be eased by a large increase in the $O^{16}(\alpha,\gamma)Ne^{20}$ reaction rate, but there is no indication in the nuclear data on Ne^{20} suggesting such a possibility. In summary one can only say that these abundances are not easily explainable in terms of helium burning, but the knowledge of nuclear physics and of stellar evolution is not secure enough to claim a definite conflict.

The uncertainty in the products of nucleosynthesis during helium burning causes a corresponding uncertainty in the subsequent evolution of the star. If C^{12} is a substantial remnant, the next nuclear burning phase will be from interactions of carbon with itself. If little C^{12} is produced, that burning phase will be omitted, and the star will progress directly from helium burning to oxygen burning. If substantial Ne^{20} can be produced in helium burning, moreover, there will be a special burning phase involving the photodisintegration of Ne^{20}. At the time of writing the indications are that C^{12} and O^{16} are both produced but that Ne^{20} is not.

The question of energy generation in helium burning is a complicated one. In principle it is easy enough, for all one must do, as usual, is multiply the rate of each reaction by the energy release and then sum. By far the best technique is to actually do this in a computer subroutine in the construction of stellar models. Since all the rates have already been given, they need only be multiplied by the Q values:

$$\begin{aligned} Q(3\alpha \rightarrow C^{12}) &= 7.274 \text{ Mev} \\ Q(C^{12} \rightarrow O^{16}) &= 7.161 \text{ Mev} \\ Q(O^{16} \rightarrow Ne^{20}) &= 4.73 \text{ Mev} \\ Q(Ne^{20} \rightarrow Mg^{24}) &= 9.31 \text{ Mev} \end{aligned} \tag{5-117}$$

At the beginning of burning the energy generation is due entirely to the 3α reaction, and ϵ is as given in Eq. (5-104). After sufficient He^4 depletion and C^{12} buildup the rate of the $C^{12}(\alpha,\gamma)O^{16}$ reaction becomes as great as the rate of formation of C^{12} with the effect that $\epsilon \approx 2\epsilon_{3\alpha}$. As the helium is further exhausted, the 3α reaction becomes an insignificant portion of the energy generation. There seems to be no general way of expressing this change easily, and it seems best to compute ϵ in the computer program in terms of the composition.[1]

5-6 ADVANCED BURNING STAGES

When the helium burning ceases to provide sufficient power to the star, gravitational contraction begins again. By the virial theorem the temperature of the

[1] Deinzer and Salpeter, *op. cit.*, present a graph of the correction factor to $\epsilon_{3\alpha}$ as a function of X_α for pure helium stars.

helium-exhausted region rises during the contraction, which continues until the next nuclear fuel begins burning at an adequate rate or until electron degeneracy provides enough pressure to halt the contraction. Stars more massive than $0.7M_\odot$ must contract until the temperature is large enough for carbon to interact with itself, although less massive stars may settle into degenerate white-dwarf configurations.

Following helium burning the most abundant nuclei in the gas are expected to be C^{12} and O^{16}, although in relative amounts that are somewhat difficult to predict with certainty; but both will be abundant in most stars. Essentially all the initial hydrogen and helium has been converted into these two nuclei, which are therefore far more abundant than the traces of the original heavy elements. Because the product Z_1Z_2 is smallest for the case of carbon interacting with itself, one expects that reaction to be the next to burn at a significant rate.

The combined mass of $C^{12} + C^{12}$ falls at an excitation of about 14 Mev in the compound Mg^{24} nucleus. At this high excitation energy in Mg^{24} there exist many compound nuclear states. Furthermore, the most effective range of stellar energies, Δ, is nearly 1 Mev broad, so that a large number of resonances may contribute significantly to the reaction rate.

Problem 5-40: What are the spins and parities of Mg^{24} states that can resonate in the $C^{12} + C^{12}$ system?
Ans: $J^\pi = 0^+, 2^+, 4^+, \ldots$

Because the centrifugal barrier has relatively small effect on penetration factors for massive nuclei, even the large-angular-momentum resonances may be important. The properties of all the relevant resonances are not easily determinable, furthermore, so that some statistical analysis of the compound nucleus must be employed. Basically one assumes that the cross-section factor observed at the higher energies observable in the laboratory is not greatly different from the average over resonances in the astrophysical region of energies. Unfortunately an intensive study of this reaction at the lowest energies achievable in the laboratory has not yet been made.

One complication for this reaction that has not been encountered in the earlier burning stages is a multiplicity of energetically allowable reaction channels:

Reaction Channel	*Q, Mev*
$C^{12} + C^{12} \rightarrow Mg^{24} + \gamma$	13.930
$\rightarrow Na^{23} + p$	2.238
$\rightarrow Ne^{20} + \alpha$	4.616
$\rightarrow Mg^{23} + n$	−2.605
$\rightarrow O^{16} + 2\alpha$	−0.114

(5-118)

At the temperature of 6 to 7×10^8 °K, where the carbon begins to react with itself, the neutron-liberating reaction requires too much particle energy to be very important. Laboratory experiments at higher energy show that the electromagnetic channel, $Mg^{24} + \gamma$, and the three-particle channel, $O^{16} + 2\alpha$, have low probability in comparison with the two-particle channels, $Na^{23} + p$ and $Ne^{20} + \alpha$, which occur with nearly equal probability. Thus the direct products of the carbon reaction are believed to be Na^{23}, Ne^{20}, protons, and alpha particles. According to Reeves, the rate of C^{12} reactions per pair is

$$\log \lambda_{12,12} = \log f_{12,12} + 4.3 - \frac{36.55(1 + 0.1T_9)^{\frac{1}{3}}}{T_9^{\frac{1}{3}}} - \frac{2}{3}\log T_9 \tag{5-119}$$

where T_9 is the temperature in billions of degrees Kelvin. This value of $\lambda_{12,12}$ is judged to be within a factor of 10 of the actual value.

Arguments of nucleosynthesis and, to some extent, energy generation during carbon burning depend upon the fate of the liberated protons and alpha particles, which will certainly be consumed quickly at the temperatures of carbon burning. There are many possible alternatives,[1] solution of which requires numerical analysis of the rates in a reaction network. Most of the protons seem to be captured by C^{12}, thereby forming N^{13}, which usually beta-decays to C^{13}. The C^{13} reacts with liberated alpha particles by the reaction $C^{13}(\alpha,n)O^{16}$. The net effect of this important sequence is the conversion of the free proton into a free neutron while converting C^{12} and He^4 into O^{16}. The C^{13} builds up a rather high abundance, roughly 1 percent of the initial C^{12} concentration, shortly after the carbon reactions begin. The free neutrons are available for capture by all the species in the gas, a very important circumstance for the synthesis of heavy elements by neutron-capture chains (see Chap. 7). The alpha particles are also captured by C^{12}, O^{16}, Ne^{20}, and Mg^{24}. By the end of carbon burning the initial C^{12} nuclei have been converted primarily to O^{16}, Ne^{20}, Na^{23}, Mg^{24}, and Si^{28}. It seems likely that this burning phase is a major source of Na^{23} and perhaps also of Mg^{24}. Reeves and Salpeter found a synthesis ratio $Na^{23}:Mg^{24}$ in the range 1:2 to 1:5, rather larger than the ratio of about 1:20 observed in nature. It is a very difficult problem, however, to estimate the products, especially since only those returned to the interstellar medium are actually observed.

These secondary reactions greatly augment the energy released by the initial carbon reaction. Reeves has estimated that each $C^{12} + C^{12}$ reaction liberates an average of about 13 Mev. The corresponding energy generation is

$$\begin{aligned}\rho\epsilon_c &\approx (20.8 \times 10^{-6}\ \text{erg/reaction})r_{12,12} \\ &= 20.8 \times 10^{-6}\lambda_{12,12}\frac{(C^{12})^2}{2}\end{aligned} \tag{5-120}$$

In terms of the C^{12} weight fraction one obtains

$$\epsilon_c \approx 2.6 \times 10^{40}\rho X_{12}{}^2\lambda_{12,12} \qquad \text{ergs g}^{-1}\text{ sec}^{-1} \tag{5-121}$$

We note that the temperature is sufficiently high so that in this and all subsequent nuclear burning stages the role of neutrino emission must be examined with care.

[1] A good published discussion is H. Reeves and E. Salpeter, *Phys. Rev.*, **116**:1505 (1959); also H. Reeves, Ph.D. thesis, Cornell University, Ithaca, N.Y., 1960.

Reactions between C^{12} and O^{16} are not important. At the carbon-burning temperatures the larger coulomb barrier ($Z_1Z_2 = 48$ instead of 36) renders the rate too slow to be of importance. The carbon is nearly completely exhausted by the time the temperature has risen to a sufficient value for $C^{12} + O^{16}$ reactions. The next reactions are those between O^{16} and O^{16} (although photodisintegration rearrangement of Ne^{20} happens at about the same temperature). For reactions between two O^{16} nuclei the energetically accessible channels are

Reaction Channel	*Q, Mev*
$O^{16} + O^{16} \rightarrow S^{32} + \gamma$	16.539
$\rightarrow P^{31} + p$	7.676
$\rightarrow S^{31} + n$	1.459
$\rightarrow Si^{28} + \alpha$	9.593
$\rightarrow Mg^{24} + 2\alpha$	−0.393

(5-122)

This reaction network is similar to that of carbon burning. A slightly larger network of secondary reactions is involved, and the major nuclei are more massive. Because of the larger coulomb barrier, the reaction occurs near $T_9 = 1$, and the major final nucleus synthesized appears to be Si^{28}, although a wide range of reaction products survives. The rate is given to an order of magnitude by

$$\log \lambda_{16,16} = \log f_{16,16} + 17.9 - \frac{59.02(1 + 0.14T_9)^{\frac{1}{3}}}{T_9^{\frac{1}{3}}} - \frac{2}{3} \log T_9 \tag{5-123}$$

The nuclear energy generation is

$$\epsilon_0 \approx 2 \times 10^{10} \rho X_{16}{}^2 \lambda_{16,16} \qquad \text{ergs g}^{-1}\text{ sec}^{-1} \tag{5-124}$$

Neutrino losses will be high during oxygen-burning phases, so much so that most of the energy generated is radiated in the form of neutrinos. In fact the oxygen must burn at temperatures in excess of $T_9 = 1$ to replace the heavy neutrino losses. An extensive study of the competition between oxygen burning and neutrino losses in massive stars has been made by Fowler and Hoyle.[1] Both the time scale and the internal temperature of the oxygen-burning stage depend upon whether the $(\bar{e}\nu)(\bar{\nu}e)$ term in the weak-interaction hamiltonian exists with the same strength as is shared by the known weak interactions.

5-7 PHOTODISINTEGRATION

At the temperatures encountered in carbon and oxygen burning a new type of nuclear reaction becomes important, the disintegration of nuclei by the thermal photon bath. The process is an exact analog of the ionization of atoms with increasing temperature near 10^4 °K, the difference being that the particles are bound together by the nuclear force and with such large binding energies relative

[1] *Astrophys. J. Suppl.*, **9**:201 (1964).

to atomic binding energies that nuclear disintegration can occur only at very high temperatures.

The photodisintegration rates are related in a very simple way to the inverse radiative-capture reactions. We have written the radiative-combination rate of nuclei 1 and 2 as $r_{12} = n_1 n_2 \lambda_{12}$. Let us now designate the rate at which the composite nucleus 12 is photodisintegrated into 1 and 2 by $r_\gamma = n_{12}\lambda_\gamma$. We seek here the value of λ_γ. It could be calculated without much difficulty by a calculation similar to the one that yielded the thermonuclear reaction rates, but is is somewhat simpler to appeal to the conditions that would prevail in thermodynamic equilibrium. If the gas were allowed to sit for a sufficient length of time to establish thermodynamic equilibrium, the concentrations n_1, n_2, and n_{12} would achieve equilibrium values such that the combination and disintegration rates would be equal,

$$\lambda_\gamma (n_{12})_e = (n_1)_e (n_2)_e \lambda_{12} \tag{5-125}$$

and such that the Saha statistical relation would apply,

$$\left(\frac{n_1 n_2}{n_{12}}\right)_e = \frac{(2\pi\mu kT)^{\frac{3}{2}}}{h^3} \frac{G_1 G_2}{G_{12}} \exp - \frac{Q}{kT} \tag{5-126}$$

Here μ is the reduced mass of 1 and 2, G is the nuclear partition function, and Q is the binding energy of 1 to 2. The thermodynamic equilibrium may never be achieved in a given astrophysical environment, but the photodisintegration rate, which depends only upon the photon density, must nonetheless be given by

$$\lambda_\gamma = \frac{(2\pi\mu kT)^{\frac{3}{2}}}{h^3} \frac{G_1 G_2}{G_{12}} \exp\left(-\frac{Q}{kT}\right) \lambda_{12} \tag{5-127}$$

The advantage of writing the rate in this way is that all the nuclear properties (except the statistical weights and Q values) are contained in the calculation of λ_{12}. The previous results on radiative-capture rates may thus be inverted into photodisintegration rates. The rate per pair, λ_{12}, must of course be evaluated at the temperature for which λ_γ is desired. Numerically we have

$$\lambda_\gamma = 5.943 \times 10^{33} \frac{G_1 G_2}{G_{12}} (AT_9)^{\frac{3}{2}} \exp\left(-\frac{11.605Q}{T_9}\right) \lambda_{12} \tag{5-128}$$

where A is the reduced atomic weight and Q is in Mev.

Probably the first photodisintegration of importance is that of N^{13}, formed when C^{12} captures a proton liberated in carbon burning. If $T_9 > 0.75$, it turns out that the photodisintegration of N^{13} is faster than its beta decay, with the result that the C^{13} concentration is sharply reduced, as is its effectiveness as a source of free neutrons. It turns out that most carbon-burning stages are at $T_9 > 0.75$, moreover.

The role of photodisintegration becomes very important in oxygen burning and later phases, however, which probably occur at $T_9 > 1$ if the $(\bar{e}\nu)(\bar{\nu}e)$ neutrino terms exist. At such high temperatures many nuclei become subject to photodisintegration. The first major disintegration phase is the Ne^{20}:

$$\gamma + Ne^{20} \rightarrow O^{16} + \alpha \tag{5-129}$$

At $T_9 > 1$ the capture, and hence the photodisintegration, is dominated by the 5.63-Mev level in Ne^{20}, shown in Fig. 5-20.

Problem 5-41: Using the results of Prob. 5-37, show that the photodisintegration rate of Ne^{20} is

$$\log \lambda_\gamma(Ne^{20}) = 12.7 - \frac{28.4}{T_9} \quad \text{sec}^{-1} \tag{5-130}$$

Around $T_9 \approx 1.3$ this rate becomes greater than the rate for $O^{16}(\alpha,\gamma)Ne^{20}$, whereupon the Ne^{20} is effectively disintegrated. The liberated alpha particles are probably captured primarily by the remaining Ne^{20}, so that the net effect is

$$2Ne^{20} \rightarrow O^{16} + Mg^{24} + 4.583 \text{ Mev} \tag{5-131}$$

The corresponding energy generation is

$$\epsilon_{Ne} = 10^{17.34} X_{20}\lambda_\gamma(Ne^{20}) \qquad \text{ergs g}^{-1}\text{ sec}^{-1} \tag{5-132}$$

There may be a quite significant Ne^{20} abundance as a by-product of the previous carbon-burning state, so that a significant amount of energy may accompany this particular photodisintegration and rearrangement. The reaction also has some interesting implications for nucleosynthesis. Since the Ne^{20} is destroyed around $T_9 = 1.3$, it must have been ejected from the star before that temperature was achieved. If the Ne^{20} is in fact synthesized primarily during carbon burning, those carbon-burning remnants must be rather quickly ejected, an awkward requirement for carbon burning at the center of a star. The relevant carbon-burning stage for nucleosynthesis is probably one in a shell surrounding an even more advanced core of a star in the terminal phases of its lifetime.

At the conclusion of oxygen burning the gas continues to heat up. The subsequent nuclear reactions are primarily of a rearrangement type, in which a particle is photoejected from one nucleus and captured by another. The effect of such a rearrangement network is that of converting nuclear particles to their most stable forms. There exists a maximum in the binding energy per nucleon at the nucleus Fe^{56}; thus the rearrangement attempts to convert oxygen-burning remnants into nuclei in the vicinity of Fe^{56}. The time scale for this conversion is controlled primarily by the photodisintegration of Si^{28}. Many exciting problems are associated with this terminal stage of nuclear burning. The aftermath of that holocaust will be discussed briefly in Chap. 7.

chapter 6

CALCULATION OF STELLAR STRUCTURE

The preceding chapters have introduced the basic physical principles from which one can attempt to compute numerical models of stars. The complexity of the physical principles used in such calculations has generally been restricted to a minimum. Whereas real stars frequently occur in binary pairs, have observable magnetic fields, and rotate, almost all model studies of stellar evolution have ignored the special physics of these perturbations. This seems a reasonable procedure, especially since it can be argued that the corresponding physical effects on most stages of stellar evolution are not terribly great, and, moreover, there seems to be as yet no general way of including these effects, which are really not well understood. The clearest approach parallels the historical one, and with the exception of some comments to be found in later sections of this chapter, we shall consider the problem of an isolated nonrotating nonmagnetic spherical mass of gas held together by gravity. Such gas spheres have been shown by many investigators to have properties nearly identical to those of stars. Subsequent inclusion of additional effects can be thought of as a perturbation to a realistic model of a star.

The relationships to be satisfied by the physical variables of the stellar model can be assembled from the preceding chapters. The internal pressure is related to the local weight and acceleration by

$$\frac{dP}{dr} = -\rho\,\frac{GM(r)}{r^2} - \rho\,\frac{d^2r_\rho}{dt^2} \tag{6-1}$$

where r_ρ is the position of the mass element ρ and is thought of as moving with it.

The last term vanishes for a star in hydrostatic equilibrium. The function $M(r)$ represents the mass enclosed within a sphere of radius r:

$$\frac{dM}{dr} = 4\pi r^2 \rho \tag{6-2}$$

The mass and density are here regarded as rest masses, but relativistic effects[1] do alter the equations and the meaning of the variables in extreme astrophysical conditions.

Once the star achieves hydrostatic equilibrium and there is negligible kinetic energy in collective mass motions, the conservation of thermal energy will require

$$\frac{dL(r)}{dr} = 4\pi r^2 \rho \left[\epsilon(r) - T \frac{dS}{dt} \right] \tag{6-3a}$$

where $L(r)$ is the energy emerging each second from the sphere of radius r, $\epsilon(r)$ is the power generated per gram of matter by nuclear reactions, and S is the entropy per unit mass. The energy source ϵ may be negative if the dominant reactions are endothermic. Before applying Eq. (6-3*a*) to stellar structure one must decide how one will do the bookkeeping on energy losses in the form of neutrinos. As it stands, the power $L(r)$ in Eq. (6-3*a*) contains the neutrino power as well as the thermal transport. It is more convenient and common, however, to transfer the neutrino component of L to the right-hand side of the equation, where it plays the role of a negative ϵ, viz.,

$$\epsilon_\nu = -\frac{1}{4\pi r^2 \rho} \frac{dL_\nu(r)}{dr} \tag{6-3b}$$

This meaning will be followed from here on, viz., that *the neutrino power produced per gram will be regarded as a negative contribution to* ϵ, *whereas the symbol L will represent only the thermal luminosity.* This system is the one that has been used in most research papers in the field.

The thermal luminosity $L(r)$, being the by-product of the mechanisms of thermal-energy transport, is related to the temperature gradient. If the material is in radiative equilibrium, the temperature gradient is related to the luminosity by the diffusion approximation to radiative transfer:

$$\frac{dT}{dr} = -\frac{3}{4ac} \frac{\kappa\rho}{T^3} \frac{L(r)}{4\pi r^2} \tag{6-4a}$$

If the temperature gradient required is excessive, the matter becomes convectively unstable. The material mixes nearly adiabatically, i.e., such that $dS(r)/dr \approx 0$. The corresponding temperature gradient is

$$\frac{dT}{dr} = \frac{\Gamma_2 - 1}{\Gamma_2} \frac{T}{P} \frac{dP}{dr} \tag{6-4b}$$

Equation (6-4*b*) is used in place of (6-4*a*) whenever the magnitude of dT/dr calculated from (6-4*a*) exceeds that calculated from (6-4*b*). Stated another way,

[1] See K. S. Thorne, The General-relativistic Theory of Stellar Structure and Dynamics, in L. Gratton (ed.), "High-energy Astrophysics," Academic Press Inc., New York, 1966.

one uses whichever temperature gradient is smaller in absolute magnitude. Equation (6-4b) is often inadequate near the stellar surface, however.

These four equations govern the stellar structure, which is considered as solved when the functions $P(r)$, $\rho(r)$, $T(r)$, $M(r)$, and $L(r)$ are specified. The solution for these five structural variables involves also five specific functions of the local thermodynamic state:

$$P = P(\rho,T,\{X_z\})$$

$$\kappa = \kappa(\rho,T,\{X_z\})$$

$$\epsilon = \epsilon(\rho,T,\{X_z\})$$

$$S = S(\rho,T,\{X_z\})$$

$$\Gamma_2 = \Gamma_2(\rho,T,\{X_z\})$$

where $\{X_z\}$ represents the set of composition parameters. The specific forms of these functions depend upon the physical circumstances. Although they may be complicated, these functions are explicitly and uniquely defined in the state of local thermodynamic equilibrium. It should be noted at once that the existence of the first of these functions, the equation of state, means that three of the structural variables, P, ρ, and T, are not independent of each other. From the point of view of the differential equations, the five functions to be specified are actually only four in number.

The differential equations involve both space (structure) and time (evolution). The time enters in Eqs. (6-1) and (6-3). Fortunately the second-order time derivative in Eq. (6-1) is too small to be of significance except in stars in a state of free fall or explosion. Inasmuch as GM/r^2 is the local value of the gravitational acceleration, the mass accelerations must be comparable to g to be of importance. This term will accordingly be set equal to zero for the following discussion, although it must be reintroduced in certain dynamical situations. The first-order time derivative of Eq. (6-3) is, however, essential to calculations of stellar evolution. It can be ignored only in the long-lived static phases of stellar evolution. But when a star contracts or expands, even at a modest rate, that time derivative becomes important. It is called a *gravitational energy source* in some research papers, although that terminology is not entirely appropriate. *The importance of this time derivative is an essential complication because it means that the structure of a star cannot be computed without specific knowledge of the star's previous history.* The structure of a static star (none really exist) can be computed by setting the time derivative equal to zero, but for a star that is changing, the structure depends not only on the values of all the physical variables but also upon how fast they are changing.

6-1 BOUNDARY CONDITIONS

What is it, we might ask, that one tries to calculate in an attempt to interpret a star? First of all it should be clear that one does not try to compute the structure of a given star—Deneb, say, or Sirius A, or Capella. For one thing, the relevant physical quantities of very few stars are completely known. Even if one knew the mass, radius, luminosity, and surface composition of a star, one would not know its age. Without knowing the extent of a star's evolution, one could not hope to know the interior composition. The accepted operational program is the following. For a given stellar mass and composition, a sequence of models describing the star as it ages is computed. This procedure is repeated for a spectrum of masses and compositions, whereupon an attempt can be made to locate each observed star in the resulting mass-age grid. Many subtle features come into play in this semiempirical correlation, but the upshot is, for the present discussion, that the calculations are begun with the selection of a mass for the model. This choice immediately gives one trivial, and worthless, boundary condition, viz., $M(\infty) = \mathfrak{M}$, where $\mathfrak{M}$ is the mass of the star. (In this chapter we shall use $\mathfrak{M}$ and $\mathfrak{L}$ for the surface values of M and L, respectively, although this convention was not adhered to in Chap. 1.) This condition is not at all what is desired, for presumably the stellar atmosphere terminates rather abruptly at some value $r = R$ for which $M(R) = \mathfrak{M}$. But the value of this radius is not known in advance; in fact, the value of R is one of the desired results of the calculation.

Consider the stellar center, on the other hand. For a small sphere of radius δr located there we have

$$\begin{aligned} M(\delta r) &= \frac{4\pi}{3}(\delta r)^3\rho_c \\ L(\delta r) &= \frac{4\pi}{3}(\delta r)^3\rho_c\left(\epsilon_c - T_c\frac{dS_c}{dt}\right) \end{aligned} \tag{6-5}$$

both of which must vanish as δr approaches zero. So two boundary conditions at the center must be

$$r = 0: \quad M(0) = 0 \qquad L(0) = 0 \tag{6-6}$$

There can be no more central boundary conditions because, clearly, the values of the central pressure P_c and of the central temperature T_c are to be results of the computation. If the calculation were in fact started by specifying P_c and T_c, one would obtain the mass of the star only as one of the results of the computation, a rather hit-and-miss arrangement if one does in fact want to evolve a star of a given mass.

To achieve simultaneous solution of the four first-order differential equations for a given mass, it is necessary to prescribe the two additional boundary conditions at the surface. This is somewhat troublesome in that one does not know in advance the radius R of the photosphere or the photospheric temperature and pressure, which together constitute the real boundary conditions. Fortunately this problem is not so serious as it might seem, for it turns out that the *internal* structure of the star is not overly sensitive to errors in the photospheric values, at least for surfaces in radiative equilibrium. For such stars the internal variables converge rapidly to the correct values for any reasonable choice of photospheric temperature and pressure. This rapid convergence occurs in radiative envelopes because the opacity behaves appropriately, as can be seen by the following argument. If Eq. (6-1) is divided by Eq. (6-4a), one obtains

$$\frac{dP}{dT} = \frac{16\pi acGM(r)}{3L(r)} \frac{T^3}{\kappa} \qquad \text{radiative} \tag{6-7}$$

Near the surface M and L are constant and equal to their surface values, so that

$$\frac{dP}{dT} = A \frac{T^3}{\kappa} \qquad A = \frac{16\pi acG\mathfrak{M}}{3\mathfrak{L}} = 1.33 \times 10^{-10} \frac{\mathfrak{M}/\mathfrak{M}_\odot}{\mathfrak{L}/\mathfrak{L}_\odot} \tag{6-8}$$

Problem 6-1: Show that if the opacity can be written in the form

$$\kappa = \kappa_0 P^\alpha T^\gamma \tag{6-9a}$$

valid near the surface, then the pressure and temperature differ from their photospheric values P_e and T_e according to

$$P^{1+\alpha} - P_e^{1+\alpha} = \frac{A}{\kappa_0} \frac{1+\alpha}{4-\gamma} (T^{4-\gamma} - T_e^{4-\gamma}) \tag{6-9b}$$

From Eq. (6-9b) it will be clear that if $1 + \alpha$ and $4 - \gamma$ are both positive, the first terms on each side of the equation will dominate when P and T exceed P_e and T_e by only modest amounts. Thus the relationship between P and T rapidly becomes independent of P_e and T_e as we proceed inward from the photosphere of a star in radiative equilibrium. Because convective instability occurs in the hydrogen ionization zones because of their very great opacity, the stars with radiative surfaces are those for which hydrogen is predominately ionized ($T_e >$ 8000°K). Under such conditions, the quantities $1 + \alpha$ and $4 - \gamma$ are in fact both positive. Because the interior structure is then insensitive to the surface conditions, it is common to choose the simplest possible choices for the surface boundary condition:

$$r = R: \quad P = 0 \qquad T = 0 \qquad \text{radiative surface} \tag{6-10}$$

The choice of Eq. (6-10) is commonly referred to as the *zero boundary conditions*. The justification for its use is that it correctly describes the interior, although it is clearly no description of the observed surface. The computed model may be placed at the appropriate point in the H-R diagram by the following reasoning. Consider again the radiative temperature gradients of Eq. (6-4a). This equation requires that the temperature drop rapidly near the surface, because the combination $\kappa\rho/T^3$ becomes very large as T approaches zero in radiative surfaces. For that reason the artificial boundary condition $T = 0$ introduces no significant error in the computed radius R. Because the internal structure has been accurately computed, moreover, the luminosity $\mathfrak{L}$ is well known and is equal to the total power emerging from a sphere well below the photosphere. Since $\mathfrak{L}$ and R are both known, the effective surface temperature computed from the Planck equation is

$$\sigma T_e^4 = \frac{\mathcal{L}}{4\pi R^2} \tag{6-11}$$

which is accurately given by the zero boundary condition. Inasmuch as the calculated model yields the observable properties $\mathcal{L}$, R, $\mathfrak{M}$, and T_e, the zero boundary condition is adequate for the stars on the upper main sequence.[1]

For stars in the lower range of surface temperatures the situation is somewhat more complicated. If the hydrogen is largely neutral at the surface, the opacity is quite large beneath the surface, with the result that convection will generally be required to transport the luminosity to the surface. The temperature gradient must accordingly be determined at the surface in accordance with a physical picture of convection. For the low-density surfaces of cool stars, however, the adiabatic approximation to the convective temperature gradient, which leads to Eq. (6-4*b*), is not really adequate. The temperature gradient must often be considerably *superadiabatic* to carry a sufficient energy flux. The necessary excess in the temperature gradient can only be computed with a theory of convection, such as the mixing-length theory, for example. Once the hydrogen ionization zone has been penetrated, Eq. (6-4*b*) is probably adequate, but for the present discussion the question is how that zone is penetrated in a realistic way in setting up the equations.

The boundary condition is implicit, rather than explicit as in the radiative case.[2] A set of trial photospheric boundary conditions must be estimated from the theory of stellar atmospheres. Because of its extreme transparency, the outermost region (down to the photosphere, by definition) is radiative. The photosphere itself is approximately at an optical depth equal to $\frac{2}{3}$, that is, such that

$$\int_R^\infty \kappa\rho \, dr \approx \tfrac{2}{3} \tag{6-12}$$

For a model of mass $\mathfrak{M}$ and given composition, trial values of $\mathcal{L}$ and R can be assumed, at least provisionally, in advance. These assumed boundary conditions yield a trial value of the surface temperature T_e (from the Planck law) and a trial value of the surface gravity g (from the law of gravity). These quantities are sufficient to allow a computation of the photospheric pressure from the theory of radiative outer atmospheres. These values of T_e and P_e can then be regarded as *tentative boundary conditions*, and a detailed theory of convection can be used to integrate inward through the hydrogen ionization zone, where the basic equations of interior stellar structure become valid. In general it will happen that the

[1] The theory of model stellar atmospheres must be used to obtain the relationship of T_e to the other observed surface temperatures.

[2] The true boundary condition is implicit even for the radiative surface, but because the interior solution is independent of the boundary condition, it was possible to make the explicit choice $T_e = P_e = 0$.

assumed surface will not successfully join onto a solution in the interior, because either $\mathcal{L}$ or R or both will have been guessed incorrectly, but by repeating the procedure with alternative guesses of $\mathcal{L}$ and R, a grid of convective envelopes can be generated, and the proper one can be interpolated from the grid. This general technique will be discussed later in the description of the fitting method associated with the integration of the equations of stellar structure, but for the moment we return to the discussion of the logical structure of the overall problem.

When the four differential equations are solved in conjunction with the four boundary conditions, they yield the values of M, L, ρ, and T as functions of r from the center of the star to its surface. The corresponding tables of numbers constitute the solution to the problem. Note particularly, however, that this solution is only specified in terms of the structure of the star as it was a time Δt earlier. The time derivative dS/dt in a stellar model is generally estimated by the ratio of the difference in S between the present model and the previous model to the time difference Δt between the two models. It is for this reason that the state of a star can be calculated only by a sequence of stellar models which advance the time from the starting model to the one desired. This necessity greatly increases the time required to compute a model of an evolved star. It also means that if one chooses a poor initial model, the mathematically correct solution a short time later need not bear any resemblance to actual stars.[1]

6-2 M AS THE INDEPENDENT VARIABLE

As they have been written, the equations of stellar structure contain one awkward feature, viz., the range of the independent variable r is unspecified. This situation reflects the fact that one usually sets out to construct a model of a star of mass $\mathfrak{M}$ rather than a star of radius R. The value of R is one of the results desired from the calculation. This situation is awkward in that it is conceptually simpler in performing numerical solutions to simultaneous differential equations to have the boundary values specified at fixed end points of the range of the independent variable.

Matters are simplified by the choice of $M(r)$ as the independent variable rather than r. Then each point in the spherical structure is labeled by the value M of the mass interior to that point, and the distance scale of the structure is obtained by solving for the function $r(M)$. The solution of the differential equations then consists of specifying the values of the functions $r(M)$, $P(M)$, $T(M)$, and $L(M)$. Another advantage of this change of variable occurs in the specification of chemical composition. If a star does nothing but expand or contract, the composition parameter $X_Z(M)$ remains constant, because the coordinate M moves with the gas, whereas the function $X_Z(r)$ changes simply by virtue of describing a different mass element. A coordinate that moves with the bulk motion of a mass element is called a *lagrangian coordinate*.

[1] P. Bodenheimer has examined this point for the important case of pre-main-sequence contraction; *Astrophys. J.*, **144**:709 (1966).

This change in the differential equations results in the inversion of Eq. (6-2),

$$\frac{dr}{dM} = \frac{1}{4\pi r^2(M)\rho(M)} \tag{6-13}$$

whereas the others become

$$\frac{dP(M)}{dM} = -\frac{GM}{4\pi r^4(M)} \tag{6-14}$$

$$\frac{dL(M)}{dM} = \epsilon - T\frac{dS}{dt} \tag{6-15}$$

and

$$\frac{dT}{dM} = -\frac{3}{4ac}\frac{\kappa}{T^3}\frac{L}{16\pi^2 r^4} \qquad \text{radiative} \tag{6-16a}$$

or

$$\frac{dT}{dM} = \frac{\Gamma_2 - 1}{\Gamma_2}\frac{T}{P}\frac{dP}{dM} \qquad \text{convective} \tag{6-16b}$$

One extra advantage of this change is that the density appears only in Eq. (6-13), which defines $r(M)$.

Problem 6-2: Confirm Eqs. (6-14) to (6-16).

The independent variable now ranges from $M = 0$ to $M = \mathfrak{M}$, which is a fixed range (except for the circumstances discussed in the next paragraph), and the boundary conditions are

$$\begin{aligned} M &= 0: \quad r = 0 \qquad L = 0 \\ M &= \mathfrak{M}: \quad T = T_e \qquad P = P_e \end{aligned} \tag{6-17}$$

Although the range of the independent variable at any epoch of time is fixed, it may change as time progresses if the star loses or gains mass. It seems virtually certain that all stars lose mass at some rate, but it is not yet known which ones lose mass fast enough to affect evolutionary calculations. Observed rates of mass loss for those very few stars for which the loss is observable range from about 10^{-7} to $10^{-13} M_\odot$ per year, the first number corresponding to certain blue giants, red giants, and T Tauri stars and the second number corresponding to the stellar wind from the sun.[1] The problem here is that mass loss itself may be unobservable and yet have profound observable consequences. Several authors have suggested that important classes of observations may be explained by mass loss. The real role of mass loss is unresolved, however, largely due to the lack of observational data. If the rate of mass loss is known, it can be included in the

[1] A recent review of mass loss is R. Weymann, *Ann. Rev. Astron. Astrophys.*, **1**:97 (1963).

sequence of evolutionary models by altering the total mass at each time step:

$$\mathfrak{M}(t + \Delta t) = \mathfrak{M}(t) + \frac{d\mathfrak{M}}{dt}\,\Delta t \tag{6-18}$$

In performing such calculations, account must be made of the power appearing as mass-loss luminosity (see Chap. 1). Because of the observational uncertainties, almost all calculations of evolutionary sequences have been carried out at constant mass. Many of these calculations will no doubt require revision because of the effects of mass loss. Most of the subsequent discussion, however, will be of evolution at constant mass.

6-3 COMPOSITION CHANGES

Any evolutionary calculation is begun with an initial composition which is usually taken to be homogeneous throughout the star. In the course of time that composition changes for a variety of reasons. The corresponding development of chemical inhomogeneities is intimately tied up with the evolution of the star. A successful scheme of computation, therefore, must do a careful job of bookkeeping on the composition. The two most important mechanisms for altering composition appear to be nuclear reactions and mixing by turbulent convection.

The time derivative in Eq. (6-3) is not the only way in which time explicitly enters the calculation. The thermonuclear generation of power is proportional to the rates of key reactions among the constituent nuclei, and a local change in chemical composition necessarily occurs with the same rates. Within any isolated mass element, say a unit volume, the abundance of each nuclear species changes at constant mass density, according to

$$\left[\frac{\partial N(A,Z)}{\partial t}\right]_{\text{nuc reac}} = r_+(A,Z) - r_-(A,Z) \tag{6-19}$$

where r_+ is the rate at which nuclei (A,Z) are synthesized per unit volume by nuclear reactions and r_- is the rate at which they are consumed. If no material of differing composition is being mixed in by convection,[1] Eq. (6-19) will give the total rate of composition change. More generally the composition parameter at time $t + \Delta t$ in a radiative region can be related to those at time t,

$$X_Z(t + \Delta t,\, M) = X_Z(t,M) + \left(\frac{\partial X_Z}{\partial t}\right)_{\text{nuc reac}} \Delta t \tag{6-20}$$

where M is the lagrangian mass coordinate of the element. Notice that this equation would not be correct during an expansion if the radius r were used as the independent variable instead of the mass M. Nor would it be correct if number

[1] Convection is, of course, not the only type of mixing. Diffusion also sorts the elements. Eddington concluded that diffusive separation due to gravity is too slow to cause the star to evolve, but the phenomenon should probably be reinvestigated with particular attention to diffusion near the boundaries of abrupt changes in chemical composition. Circulation currents along meridian planes are also induced by rotation. The basic idea of this effect will be discussed later in this chapter.

density were used instead of mass fraction. The partial derivatives, i.e., the rates, are also evaluated at the point M and the time t. Equation (6-20) is clearly only the first term of an expansion and will not be correct if one takes too large a time step Δt.

The way in which convection alters this simple prescription reflects the physical nature of convection in the stellar interior. For a turbulent mixing length of order 10^9 to 10^{10} cm, it is an easy matter to estimate that turbulent velocities of order 10^3 cm/sec are required to transport most anticipated stellar heat fluxes along the slightly superadiabatic temperature gradient. It follows that mixing throughout stellar convection zones is expected to occur on time scales measured in months. This time is so short compared to the time scales of most phases of stellar evolution that it is commonly assumed that *the chemical composition in stellar convective zones is homogeneous.* This is not to say that nuclear reactions do not proceed preferentially in the hottest regions of the convective zone, but rather that the new nuclei are redistributed throughout the convective zone faster than a significant inhomogeneity can be built up.

Suppose then that a star has a convective zone whose boundaries are at $M(r) = M_1$ and $M(r) = M_2$, as shown in Fig. 6-1. Because of evolutionary changes in the structure, both zone boundaries may be changing at the rates dM_1/dt and dM_2/dt. Let X represent the mass fraction of any nuclear species, whose value at each point is being altered at the rate $[\partial X(M)/\partial t]_{\text{nuc}}$ *due to nuclear reactions alone;* let X_1 and X_2 respectively represent the values of X *on the radiative*

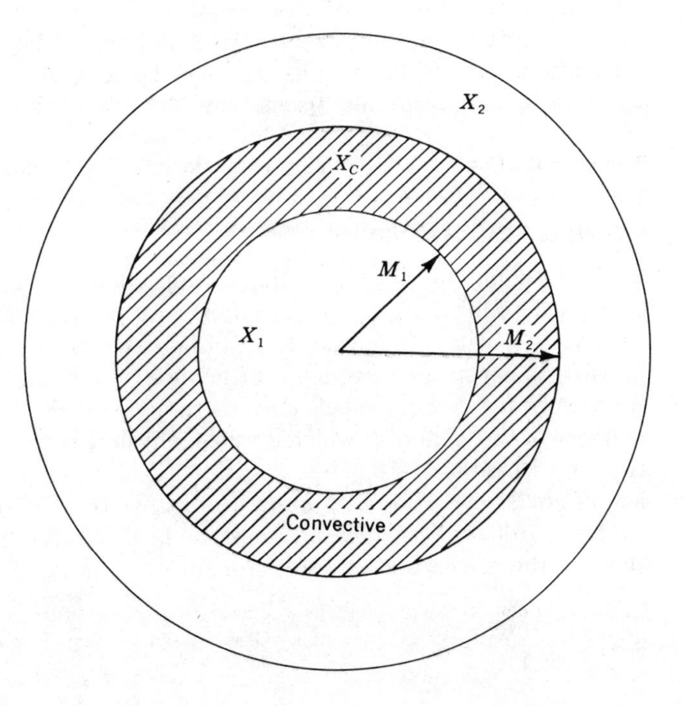

Fig. 6-1 Cross section of a star with an intermediate convection zone. The central radiative zone extends to mass coordinate M_1. The mass fraction of any nuclear species is equal to X_1 on the radiative side of that boundary and is equal to X_c throughout the convection zone, which extends from M_1 to M_2. At M_2 the star switches to radiative equilibrium, where the mass fraction in question is equal to X_2.

side of the boundaries M_1 and M_2; and let X_c represent the homogeneous value of X throughout the convective region. The value of X_c changes for two reasons: (1) nuclear reactions within the convective zone change X_c at a rate equal to the mass average of $[\partial X(M)/\partial t]_{\text{nuc}}$ over the zone; and (2) enlargement of the convective zone mixes material of different composition into the convective zone. The situation is somewhat tricky, because material of composition X_2 is admixed only if M_2 is *increasing* with time, whereas material of composition X_1 is admixed only if M_1 is *decreasing* with time. These conditions can be incorporated simply, but somewhat clumsily, into an equation expressing the rate of change of the composition in the convective zone:

$$\frac{dX_c}{dt} = \frac{1}{M_2 - M_1} \int_{M_1}^{M_2} \left[\frac{\partial X(M)}{\partial t}\right]_{\text{nuc}} dM + \frac{X_2 - X_c}{M_2 - M_1} \frac{1}{2} \left(\frac{dM_2}{dt} + \left|\frac{dM_2}{dt}\right|\right) + \frac{X_1 - X_c}{M_2 - M_1} \frac{1}{2} \left(\left|\frac{dM_1}{dt}\right| - \frac{dM_1}{dt}\right) \tag{6-21}$$

The first term gives the rate of change for a zone with static boundaries, whereas the second and third terms give the rates of change associated with motion of the boundaries. The last two terms have been constructed in such a way that they vanish if the convection zone is shrinking. If the zone does recede at either interface, however, it will leave behind a discontinuity in the composition in the radiative zone. Since any of these composition changes may prove crucial to stellar evolution and nucleosynthesis, they must be followed with care during the calculation. It will be obvious that the choice of M as the independent variable has greatly simplified this discussion.

Problem 6-3: Confirm Eq. (6-21) by simple hypothetical examples.

6-4 NUMERICAL TECHNIQUES

From the equations one would like to construct a numerical model of a star. As numerical problems go, this is a fairly extensive one. The science advanced considerably with the development of large electronic computers, which made it possible to construct a sequence of models rapidly and accurately. Within this framework a "stellar model" may be thought of as a set of tables stored in the memory of the computer which give the physical variables and chemical composition at selected discrete points and times in the stellar interior. The basic problem of stellar evolution as a numerical rather than an observational science is to compute from the set of tables describing the initial configuration assumed for the star the corresponding tables for future times.[1]

[1] This statement of the problem, as well as the logical structure of this section, has been adapted from M. H. Wrubel's chapter in L. Gratton (ed.), "Star Evolution," Academic Press Inc., New York, 1963.

The procedure for solving a problem on a computer is called the *program*. The program expresses the logical sequence of numerical manipulations that the computer must perform to arrive at the answer. Large programs are frequently constructed from small programs, called *subroutines*, each of which has a limited and specific function in the main program. Examples of subroutines in the stellar-structure program are the calculations of the equation of state, the opacity, the nuclear energy generation, the entropy, and the adiabatic exponents from the local thermodynamic conditions.

The logical recipes for solving specific problems are called *algorithms*. There are often many suitable algorithms for the solution of a specific problem, and it is important to choose a good one. A poor choice of algorithm may inordinately increase the time required to reach the answer. One of the most generally successful types of algorithm is a rapidly converging iterative procedure, in which an approximate solution is repeatedly improved until two successive approximations differ by a negligible amount.

In this section we shall discuss two commonly used algorithms for solving the stellar-evolution problem. The first of these consists of a stepwise procedure for integrating the differential equations. The second technique forces the structure, which is described by a series of concentric shells, to relax to a stable configuration satisfying the differential equations. These two techniques are commonly called by a variety of names, the first most commonly the *integration method*, the *stepwise method*, or the *fitting method* and the second the *relaxation method*, the *difference method*, the *shell method*, or the *Henyey method*.[1] We choose the first names given, simply because they seem to convey the basic idea most clearly.

THE INTEGRATION METHOD

The basic algorithm is most easily described in terms of a much simpler example. Suppose one wants to estimate the solution to the equation

$$\frac{dy}{dx} = f(x,y) \tag{6-22}$$

subject to some boundary condition, say, $y(0) = y_0$. The slope at the origin is then known to be

$$\left(\frac{dy}{dx}\right)_{x=0} = f(0,y_0) \tag{6-23}$$

so that the value of y a short distance $x = h$ from the origin is given approximately by

$$y(h) = y_0 + f(0,y_0)h \tag{6-24}$$

[1] After L. R. Henyey, one of the most active developers of this method. The numerical techniques are analyzed by R. D. Richtmeyer, "Difference Methods for Initial Value Problems," Interscience Publishers, Inc., New York, 1957.

whereupon the slope at $x = h$ can be computed and the solution continued,

$$y(2h) = y(h) + f(h,y(h))h \tag{6-25}$$

etc. This procedure, known as *Euler's method*, is quite easy to program if a subroutine is available for computing $f(x,y)$.

Although this procedure generates the solution $y(x)$ in the form of numerical values at selected discrete points, the value at each point is subject to numerical error. The first source of error is that the differential equation itself is not solved, but rather an approximation to it. In this case the approximation is a difference equation generated by the first term in a Taylor series. An inaccuracy caused by the fact that the differential equation has been approximated to some order is called *truncation error*. It is proportional to some power of the step size h and is therefore decreased by making the step size smaller and using more steps.

A second type of error inherent in numerical solutions of differential equations is due to the fact that the calculations are carried out with a finite number of significant figures. The last digit in each number is rounded off, depending upon the size of the remainder, before the calculation can be continued. The corresponding error is called *rounding error*, and it accumulates at each step, gradually decreasing the true number of significant digits. To minimize this error it is necessary to begin the calculation with the largest convenient number of significant figures and to make as few steps as possible in achieving the answer. Rounding error prohibits, for instance, allowing the step size h of the previous example to be made arbitrarily small, for eventually the rounding error exceeds the truncation error (to say nothing of the increased time involved).

The truncation error can be reduced by using more elaborate procedures in making each step h. To do so one must make the approximation to the differential equation correct to higher order. Each step forward in Euler's method requires that the subroutine for $f(x,y)$ be evaluated once, and the corresponding truncation error is proportional to the first power of h. But the approximation is easily improved. It is easy to verify, for instance, that the error is reduced to second order in the foregoing example merely by evaluating $f(x,y)$ in the middle of each step rather than at the beginning, as illustrated by the following problem.

Problem 6-4: Show that steps are correct to order h^2 (and hence that the truncation error is proportional to h^2) if one uses

$$y(x + h) = y(h) + f\left(x + \frac{h}{2}, y(h) + f(x,y)\frac{h}{2}\right)h$$

The most widely used of these improved methods is the Runge-Kutta method. To cross the interval h requires four evaluations of the subroutine, and the corresponding truncation error is proportional to h^4. The procedure is that the increment Δy is given by

$$\Delta y = \tfrac{1}{6}(k_1 + 2k_2 + 2k_3 + k_4) \tag{6-26}$$

where

$$k_1 = f(x,y)h$$

$$k_2 = f\left(x + \frac{h}{2}, y + \frac{k_1}{2}\right) h$$

$$k_3 = f\left(x + \frac{h}{2}, y + \frac{k_2}{2}\right) h$$

$$k_4 = f(x + h,\ y + k_3)h$$

The stellar problem, as it has been defined, consists of four first-order differential equations and four boundary conditions. These boundary conditions are at two different points, however, so that it is not possible to generate the complete solutions by progressing away from any single point at which the solution is known. The central thermodynamic state, embodied in the values of P_c and T_c, is initially unknown, whereas at the surface the radius and luminosity are initially unknown. (We continue to assume that the given problem is to compute a model of a star of mass $\mathfrak{M}$ rather than one of radius R, or one of luminosity $\mathfrak{L}$, etc. Such a change in the logical object of the calculation would necessarily alter the logistical basis of the following discussion.) The existence of two-point boundary conditions rather than single-point boundary conditions requires that one assume values of P_c and T_c to begin integrations outward from the center or alternatively that one assume values of R and $\mathfrak{L}$ to begin integrations inward from the surface. In most instances, integrations begun at one boundary will lead to values of variables at the other boundary which do not satisfy the correct boundary conditions there.

Problem 6-5: Using Eqs. (6-1) to (6-4), show that the variables near the center have the values

$$
\begin{aligned}
M &= \frac{4\pi}{3}\rho_c r^3 \\
P &= P_c - \frac{2\pi}{3} G\rho_c^2 r^2 \\
L &= \frac{4\pi}{3}\rho_c r^3 \left(\epsilon_c - T_c \frac{dS_c}{dt}\right) \\
T &= \begin{cases} T_c - \dfrac{\kappa_c \epsilon_c \rho_c^2}{8acT_c^3} r^2 & \text{radiative} \\[2ex] T_c - \dfrac{2\pi}{3} G \dfrac{\Gamma_2 - 1}{\Gamma_2} \dfrac{T_c}{P_c} \rho_c^2 r^2 & \text{convective} \end{cases}
\end{aligned}
\tag{6-27}
$$

The boundary-value problem has generally been handled by integrating away from both boundaries to a predetermined intermediate point where the two solutions are joined. In general, some type of fitting procedure is required at the intermediate point, which is hereafter designated by M_f. The outward integrations are begun by assuming values for T_c and P_c, whereupon the values of the dependent variables are specified out to M_f. The inward integrations are begun by assuming values for R and $\mathfrak{L}$, whereupon the value of the dependent variables are specified inward to M_f. If the starting conditions were correctly guessed, the dependent variables will match at the fitting point, and a correct model will have been achieved. In general, of course, the solutions will not match. The usual procedure in that case is to vary one of the parameters, R or $\mathfrak{L}$, characterizing the inward integration and one of the parameters, T_c or P_c, characterizing the outward integration and to repeat the integration. By observing the extent to which the discontinuities at the fitting point have been reduced, it is possible to interpolate new starting conditions which give a satisfactory model. Usually six trial integrations, three inward and three outward, are run with the trial values:

Outbound Integrations	*Inbound Integrations*
P_c, T_c	$\mathfrak{L}$, R
$P_c + \Delta P_c$, T_c	$\mathfrak{L} + \Delta\mathfrak{L}$, R
P_c, $T_c + \Delta T_c$	$\mathfrak{L}$, $R + \Delta R$

From these trial integrations one computes the derivatives of the discontinuities at the fitting point with respect to the values of the trial parameters. Then the new trial values can be interpolated from those derivatives. The procedure may be visualized as a search in a four-dimensional space for that point, with coordinates P_c, T_c, R, and $\mathfrak{L}$, which characterizes that pair of integrations having continuity in the dependent variables at the fitting point. The entire procedure can be executed automatically on a computer.

The discussion to this point has ignored the time derivative in Eq. (6-15), and so we shall now briefly consider how the calculation of the model can be made. It bears repeating at this point that a model of a star can be computed only in terms of the model a short time earlier. Therefore the first problem is that of how the calculation is begun. The initial model is obtained by assuming a value of $T(dS/dt)$ at each point M. The resulting model, calculated by the algorithm just described, will clearly depend upon that initial assumption, so that the first model can be no better than one's guess regarding the value of the time derivatives. Because of this dependence, one must be sure that the desired goal of the calculation is independent of the starting assumptions. Fortunately this problem is not serious, because it can be shown in most cases of interest that the third or fourth model is nearly independent of the first model; i.e., the model has relaxed to a solution nearly independent of the starting model. If this relaxation has occurred in a time short compared to the lifetime of the evolutionary stage under investigation, the calculation will have been successful. Because the most frequent calculations have been of main-sequence stars, which are nearly static structures for a very long time, the most common assumption for the starting model has been that the time derivative is equal to zero. Although this assumption yields very good models of all but the most massive main-sequence stars, it clearly will not give a satisfactory initial model of an evolving star, one contracting to the main sequence, say. For the particular investigation of contraction to the main sequence, one must choose an initial model that has contracted so little that the subsequent contraction quickly becomes independent of the initial model.

Once an initial model has been constructed by assumption, the subsequent models can be calculated explicitly. The values of the temperature and the entropy per unit mass for the initial model are stored in the memory of the computer. One then chooses a time step Δt that is to be the age of the first evolved model relative to the initial model. From the value of Δt and the structure of the initial model the change in composition due to nuclear reactions and due to mixing by convection is computed. This calculation gives the run of the composition for the first evolved model. One then calculates the new model in the same manner as the previous one, with one exception: at each point M in the integration the new value of the entropy must be computed and compared with its value in the initial model as an estimate of the time derivative:

$$T \frac{dS}{dt} \approx \overline{T(M)} \, \frac{S(M,\Delta t) - S(M,0)}{\Delta t} \tag{6-28}$$

This estimate must be inserted in Eq. (6-15) during the construction of the model. Once the proper set of values $(P_c,T_c,R,\mathfrak{L})_1$ has been found for the first evolved model, the procedure is repeated with a new time step. A simplified flow diagram is shown in Fig. 6-2. It is clear that calculation of an evolving star requires a more complicated algorithm than calculation of a static star, because not only do the values of the variables enter into the coefficients of the differential equations, but their differences from those of the previous model enter as well. A new dimension of self-consistency is imposed on the equations. For rapidly evolving stars the algorithm has been found somewhat awkward by many workers. The relaxation method has proved more popular for such stars, but the integration technique is rapid and accurate for those stages of stellar evolution for which $\epsilon \gg |T(dS/dt)|$.

THE RELAXATION METHOD

The idea of this method is as follows: one first assumes a structure for the star and notices that the differential equations are not satisfied; corrections to the physical variables are calculated by a specific prescription at each point, thereby rendering the differential equation more nearly satisfied; the corrective process is repeated until the magnitude of the corrections becomes insignificant. The algorithm is best illustrated by the example

$$\frac{dy}{dx} = f(x,y) \qquad y(0) = y_0$$

Suppose that y is to be computed for values of x between $x = 0$ and $x = x_N$.

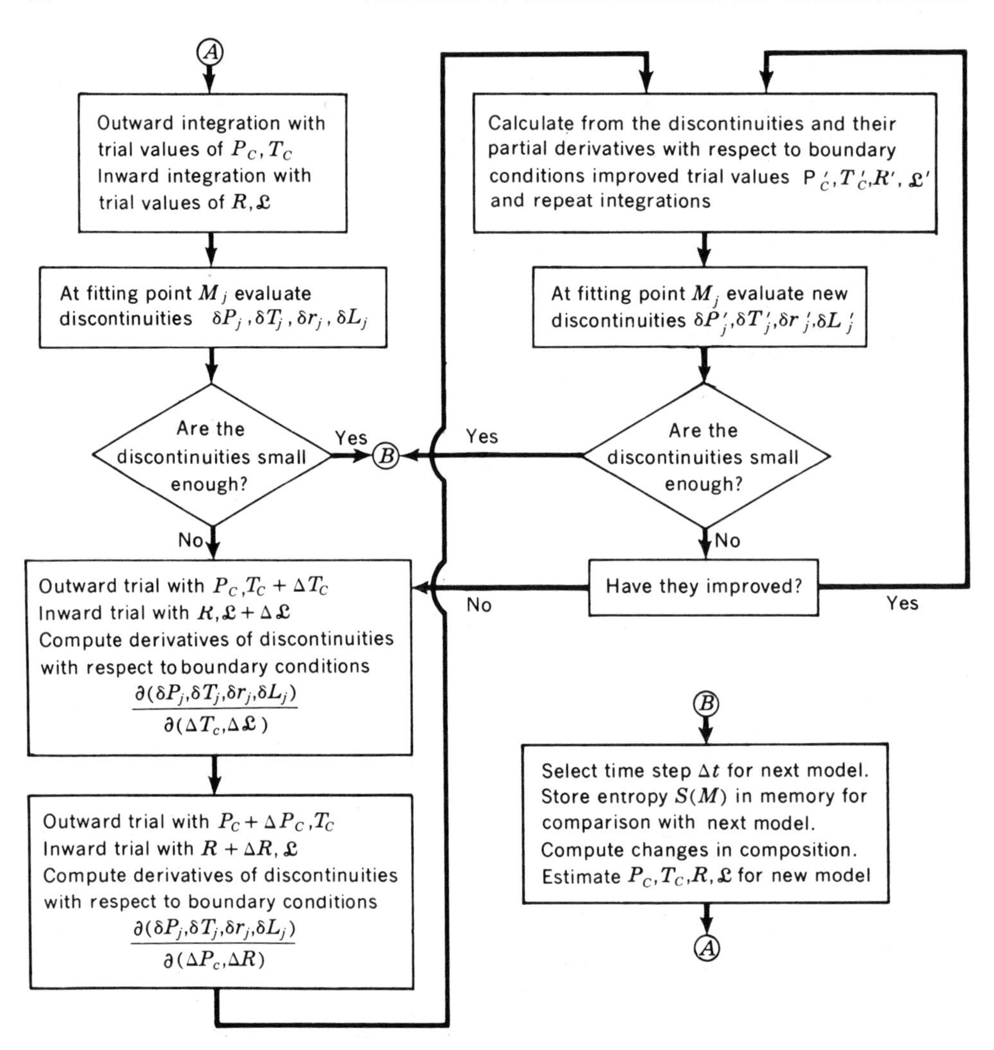

Fig. 6-2 Flow diagram for the integration method. The sequence of logical decisions and steps employed in the compution of the stellar structure is indicated by the arrows. [*Adapted from R. L. Sears and R. R. Brownlee, Stellar Evolution and Age Determinations, in L. H. Aller and D. B. McLaughlin (eds.), "Stellar Structure," The University of Chicago Press, Chicago,* 1965, *by permission of The University of Chicago Press. Copyright* 1965 *by The University of Chicago.*]

Divide that range of x into zones by selecting specific points $x_1, x_2, \ldots, x_N$, as illustrated in Fig. 6-3. These points will remain fixed throughout the calculation. Suppose one now makes a guess of the form of $y(x)$ and labels it $y^0(x)$. The object will be to calculate corrections to $y^0(x)$ making it more nearly equal to $y(x)$, the function being sought. The functions are represented in the computer memory by a table giving their numerical value at each of the selected discrete points $x_1, x_2, \ldots, x_N$.

The differential equation is first approximated by a series of difference equations, which in this case can be quite simple:

$$\left(\frac{dy}{dx}\right)_{i+\frac{1}{2}} \approx \frac{y_{i+1} - y_i}{x_{i+1} - x_i} \approx f\left(\frac{x_{i+1} + x_i}{2}, \frac{y_{i+1} + y_i}{2}\right) \equiv f_{i+\frac{1}{2}} \tag{6-29}$$

i.e., the derivatives are replaced by the ratio of differences across each interval and the functions occurring in the differential equation are evaluated at the middle of the interval. The known function could be evaluated at x_i instead, but evaluation in the middle of the interval makes the truncation error second order rather than first order in the interval size. Thus, except for the truncation error, which can be made small by using enough points, the single differential equation is replaceable by N difference equations of the form

$$(y_{i+1} - y_i) - f_{i+\frac{1}{2}}(x_{i+1} - x_i) = 0 \qquad 0 \leq i \leq N - 1 \tag{6-30}$$

For the trial function these equations will not in general be satisfied:

$$(y^0_{i+1} - y_i{}^0) - f^0_{i+\frac{1}{2}}(x_{i+1} - x_i) \neq 0 \tag{6-31}$$

Inasmuch as the points x_i are predetermined and fixed, the problem reduces to finding those corrections to each $y_i{}^0$ which bring Eq. (6-31) closer to the desired null value.

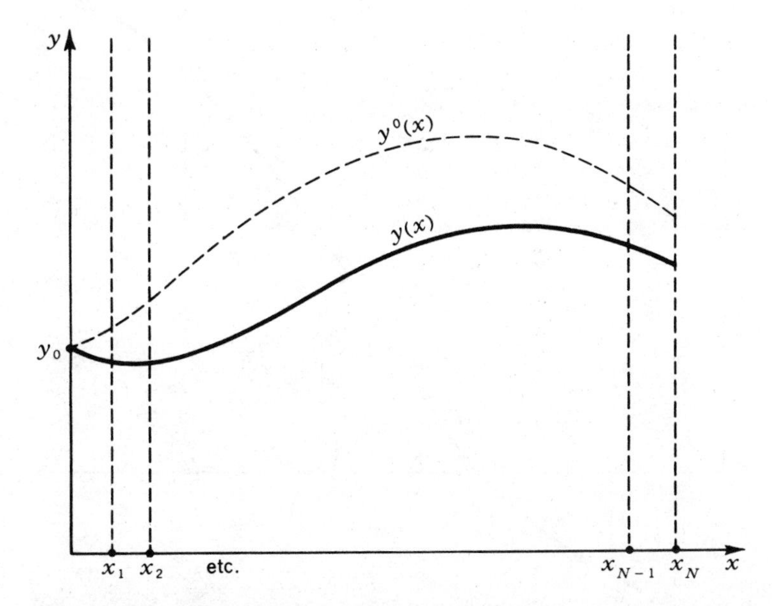

Fig. 6-3 A coordinate grid for application of the relaxation method to the simple differential equation $dy/dx = f(x,y)$. The true solution is designated by $y(x)$, and the initial trial solution is designated by $y^0(x)$.

The problem can be solved by an iterative procedure for converging to the root of a function if the function can be differentiated. Newton first derived the following geometrical conception of the location of that value of x for which a known function $g(x)$ vanishes. Select a test value $x = x^0$ and evaluate $g(x^0) = g^0$, which will in general not vanish. The tangent to $g(x)$ through the point (x^0,g^0) intercepts the x axis at point x^1, which is taken as the new trial value of x:

$$x^1 = x^0 - \frac{g(x^0)}{(dg/dx)_{x^0}}$$

The process is continually repeated,

$$x^{i+1} = x^i - \frac{g(x^i)}{(dg/dx)_{x^i}}$$

and the root is approached along a saw-toothed path, illustrated in Fig. 6-4.

Problem 6-6: Use Newton's method on the function $g(x) = x^2 - c$ to find an algorithm for computing $\sqrt{c}$.

Ans: $x^{i+1} = \frac{1}{2}\left(x^i + \frac{c}{x^i}\right)$.

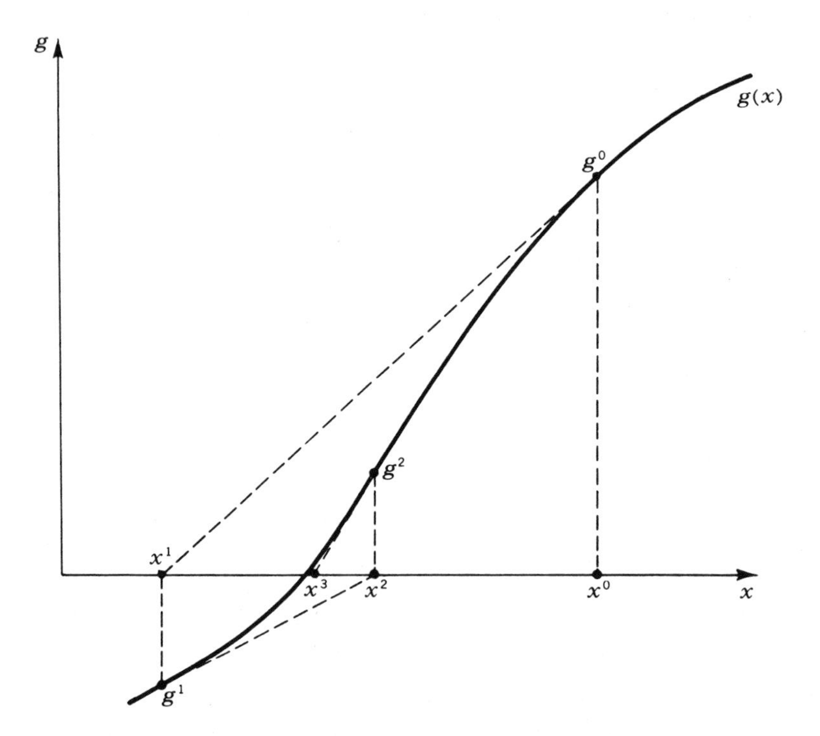

Fig. 6-4 Newton's method for finding the root of a function $g(x)$. By successive iterations the root is approached along a saw-toothed path.

For the solution of the differential equation, we have simultaneously to find the roots of the N functions of the form of Eq. (6-30):

$$g_i(y_1,y_2, \ldots, y_N) \equiv (y_{i+1} - y_i) - f_{i+\frac{1}{2}}(x_{i+1} - x_i) \tag{6-32}$$

This simultaneous root is approached by an analogous prescription. The functions g_i^0 evaluated at the trial values of the variables $y_1^0, y_2^0, \ldots, y_N^0$ are not generally zero, so that increments are computed to make them more nearly zero:

$$g_i^0 + \sum_{j=1}^{N} \left(\frac{\partial g_i}{\partial y_j}\right)^0 \delta y_j^0 = 0 \tag{6-33}$$

This linear system is solved for the values of the δy_j^0, which are added to the trial values to give the first approximation to the solution:

$$\begin{aligned} y_0^1 &= y_0^0 = y_0 \qquad \text{boundary condition} \\ y_1^1 &= y_1^0 + \delta y_1^0 \\ y_2^1 &= y_2^0 + \delta y_2^0 \\ &\ldots\ldots\ldots \end{aligned} \tag{6-34}$$

The entire process is then repeated until the corrections δy_j become as small as desired or until truncation error limits the convergence.

Problem 6-7: Show that for this special example one obtains for the kth iteration

$$y_{i+1}^k - y_i^k - f_{i+\frac{1}{2}}^k h_i + \left(1 - \frac{\partial f_{i+\frac{1}{2}}}{\partial y_{i+1}} h_i\right)^k \delta y_{i+1}^k - \left(1 + \frac{\partial f_{i+\frac{1}{2}}}{\partial y_i}\right)^k \delta y_i^k = 0$$

where $h_i = x_{i+1} - x_i$ is the size of the ith interval. This set of equations is very simple in that only two increments appear in each equation. Thus for any stage k of iteration, the increment δy_{i+1}^k to y_{i+1}^k can be found in terms of the increment δy_i^k to y_i^k. Inasmuch as $\delta y_0 = 0$ because of the boundary condition, each δy_i can be solved for in turn. Then the new set of variables can be constructed and the procedure repeated.

Problem 6-8: Set up the numerical algorithm for solution of the differential equation

$$\frac{dy}{dx} = x^{\frac{1}{2}} y$$

and show that it can be written in the form

$$(1 - Z_i)y_{i+1}^k - (1 + Z_i)y_i^k + (1 - Z_i)\,\delta y_{i+1}^k - (1 + Z_i)\,\delta y_i^k = 0$$

where Z_i is defined to be

$$Z_i = \frac{h_i}{2}\left(x_i + \frac{h_i}{2}\right)^{\frac{1}{2}}$$

Establish a five-point grid between $x = 0$ and $x = 1$ and find the numerical approximation to the differential equation subject to the boundary condition $y(0) = 1$. That is, choose a trial solution and iterate the equations until the solution ceases to improve. Compare the results to the exact integral of the differential equation. You will note that after rapid early convergence toward the solution it ceases to improve. This is a manifestation of truncation error and could be improved by increasing the number of zones in the grid. Students with access to digital electronic computers may wish to program this or other equations for automatic solution.

This example has shown that a trial solution to a differential equation can be made increasingly like the exact solution by successive correction. In this self-correction the trial solution "relaxes" to its correct value. This aspect of the technique gives it its name. Consider now its application to the stellar problem.

The natural selection for the independent variable is the lagrangian coordinate M. It has a prescribed range from $M = 0$ at the center to $M = \mathfrak{M}$ at the surface. Thus the star will be subdivided into a number (somewhere between 10 and 1,000, generally, depending upon the problem) of concentric shells whose boundaries (in the M coordinate) are preselected and remain fixed during the calculation. For this calculation the lagrangian coordinate has the obvious advantage of labeling each shell with a coordinate which remains fixed during expansions and contractions. The total mass is by no means divided equally among the zones. The amount of mass in each zone is something that can be selected only on the basis of experience in such a way that the computing time and accuracy of the calculation are optimized. The following general rules apply. In the energy-generating regions, where the temperature and density must be known with relative accuracy, the mass fraction must be small, say about 10^{-4} of the total mass. For main-sequence stars this demand requires a small central zone, but for stars with inert (and therefore nearly isothermal) cores the central zone may be much larger, whereas an energy-generating shell around the core must be small. Because of the steep gradients and low density near the surface, the outermost zone should probably contain no more than 10^{-4} of the total mass. These small zones are compensated by large zones (as much as $0.1\mathfrak{M}$) in interior regions without energy generation. Larger zones may generally be used in static stars than in rapidly evolving ones. These general rules are best amplified by actual experience in the construction of stellar models.

There is no unique lagrangian coordinate, and because the mass is unevenly distributed among the zones, it may be preferable to employ a dummy variable ξ chosen as a specific explicit function of the mass M such that[1]

$$M = M(\xi) \qquad 0 \leq \xi \leq 1 \tag{6-35}$$

Such a function must be monotonic and have as end points

$$M(0) = 0 \qquad \text{and} \qquad M(1) = \mathfrak{M} \tag{6-36}$$

It must be clearly understood that such a change is merely a change in represen-

[1] The subsequent discussion draws heavily on the papers of L. G. Henyey and his collaborators. See especially L. G. Henyey, J. E. Forbes, and N. L. Gould, *Astrophys. J.*, **139**:306 (1964), and other papers cited there.

tation and enumeration and does not introduce any new unknowns into the problem, because $M(\xi)$ is a function selected before the problem is begun. The advantage of the change can be a more nearly linear dependence of the variables upon coordinate. The method is unchanged, but the equations are more generally flexible if allowance for this transformation is included. The problem then is to compute the dependent variables P, T, r, and L as a function of ξ.

The basic equations (6-13) to (6-16) of stellar structure then are

$$\frac{dP}{d\xi} + \frac{GM\rho}{r^2}\frac{dr}{d\xi} = 0 \tag{6-37}$$

$$M' - 4\pi r^2 \rho \frac{dr}{d\xi} = 0 \tag{6-38}$$

$$\frac{dL}{d\xi} - M'\left(\epsilon - T\frac{dS}{dt}\right) = 0 \tag{6-39}$$

and either

$$\frac{dT}{d\xi} + \frac{3\kappa\rho L}{16\pi acT^3r^2}\frac{dr}{d\xi} = 0 \qquad \text{radiative} \tag{6-40a}$$

or

$$\frac{dS}{d\xi} = 0 \qquad \text{convective} \tag{6-40b}$$

The symbol M' represents the derivative of $M(\xi)$ and is a known function of ξ. The derivatives all have been written as ordinary derivatives although each physical variable also depends upon time.

When approximating the differential equations by difference equations it is necessary to use both differences and averages of the variables at neighboring points in the grid. To minimize truncation error it is desirable to work with variables that are as nearly linear as possible. The pressure and density vary by so many orders of magnitude through the star that they may not be the best variables for performing differences and averages. To minimize that difficulty, Henyey and his coworkers replace them with the artificial variables

$$p = P^{\frac{1}{4}} \tag{6-41}$$

and

$$q = \rho^{\frac{1}{3}} \tag{6-42}$$

These particular exponents are chosen by analogy to the standard-model polytrope of index 3, for which p is proportional to q. And because the luminosity L usually rises abruptly near the stellar center, where the energy is generated, it is also convenient to moderate its variation by the use of pseudoflux F defined by

$$L = \xi^2 F \tag{6-43}$$

Problem 6-9: Show that in terms of these variables the structural equations become

$$\frac{dp}{d\xi} + \frac{GMq^3}{4p^3r^2}\frac{dr}{d\xi} = 0 \tag{6-44}$$

$$M' - 4\pi r^2 q^3 \frac{dr}{d\xi} = 0 \tag{6-45}$$

$$\xi^2 \frac{dF}{d\xi} + 2\xi F - M'\left(\epsilon - T\frac{dS}{dt}\right) = 0 \tag{6-46}$$

and either

$$\frac{dT}{d\xi} - \frac{16K\xi^2 F}{M}\frac{dp}{d\xi} = 0 \qquad \text{radiative} \tag{6-47a}$$

or

$$\frac{dS}{d\xi} = 0 \qquad \text{convective} \tag{6-47b}$$

where a new function of state playing the role of a pseudopacity is defined as

$$K = \frac{3\kappa p^3}{64\pi acGT^3} \tag{6-48}$$

Show also that an increment in the entropy per unit mass may be written in terms of these variables as

$$dS = dU - \frac{3p^4}{q^4}\,dq \tag{6-49}$$

where U is the internal energy per unit mass. (Many papers use the symbol E for internal energy.)

Imagine the star to be divided into J shells, whose boundaries are at the points ξ_j ($j = 0, 1, \ldots, J$) arranged in order of position proceeding from the center ($\xi_0 = 0$) to the surface ($\xi_J = 1$). For simplicity the values of the physical variable at those points will be designated by the corresponding subscripts; for example, $T(\xi_j) = T_j$, $q(\xi_j) = q_j$, etc. The mass and its derivative will often be evaluated in the center of a zone, in which case a half-integral subscript is attached:

$$m_{j+\frac{1}{2}} \equiv m(\xi_{j+\frac{1}{2}}) \equiv m\left(\frac{\xi_{j+1} + \xi_j}{2}\right) \tag{6-50}$$

There is no unique prescription for preparing a difference equation from a differential equation. What is usually done, however, is to use variables as nearly linear as possible so that their value in the center of a zone can be approximated by the average of their values at the two boundaries. Consider, for example, Eq. (6-44) evaluated at the center of the $j + 1$ zone as a difference equation. There results

$$p_{j+1} - p_j + \frac{GM_{j+\frac{1}{2}}(q_{j+1} + q_j)^3}{(p_{j+1} + p_j)^3(r_{j+1} + r_j)^2}(r_{j+1} - r_j) \tag{6-51}$$

Note that five factors of $\frac{1}{2}$ times the 4 in the denominator have canceled the three factors of $\frac{1}{2}$ occurring in the numerator. The remainder of the differential equations are similarly cast as the following difference equations:

For Eq. (6-45):

$$\frac{8}{\pi} M'_{j+\frac{1}{2}}(\xi_{j+1} - \xi_j) - (q_{j+1} + q_j)^3(r_{j+1} + r_j)^2(r_{j+1} - r_j) = 0 \tag{6-52}$$

For Eq. (6-46):

$$\begin{aligned} F_{j+1}(\xi_{j+1} + \xi_j)(3\xi_{j+1} - \xi_j) + F_j(\xi_{j+1} + \xi_j)(\xi_{j+1} - 3\xi_j) \\ - 2M'_{j+\frac{1}{2}}(\xi_{j+1} - \xi_j)\left[\epsilon_{j+1} + \epsilon_j - \frac{U_{j+1} + U_j - U_{j+1}^{(-\Delta t)} - U_j^{(-\Delta t)}}{\Delta t}\right. \\ \left. + 3\left(\frac{p_{j+1} + p_j}{q_{j+1} + q_j}\right)^4 \frac{q_{j+1} + q_j - q_{j+1}^{(-\Delta t)} - q_i^{(-\Delta t)}}{\Delta t}\right] = 0 \end{aligned} \tag{6-53}$$

For Eq. (6-47*a*):

$$T_{j+1} - T_j - \frac{(K_{j+1} + K_j)(\xi_{j+1} + \xi_j)^2(F_{j+1} + F_j)}{M_{j+\frac{1}{2}}}(p_{j+1} - p_j) = 0 \tag{6-54a}$$

For Eq. (6-47*b*):

$$U_{j+1} - U_j - 3\left[\frac{p_{j+1} + p_j}{q_{j+1} + q_j}\right]^4 (q_{j+1} - q_j) = 0 \tag{6-54b}$$

Equation (6-53) requires special comment because it contains the nuclear energy generation and because it contains the time derivative linking one model in the evolutionary sequence to the previous one.

The energy generation is very temperature-dependent, varying as $\epsilon \propto T^n$, where $4 < n < 40$ in most stars. For such steep temperature dependence, the average of ϵ at the two boundaries of the zone need not be a good estimate of its average throughout the zone unless that zone is sufficiently small. This fact emphasizes the general comment made earlier that the energy-generating zones must have small mass. It has been suggested that a geometric mean, $\epsilon_{j+\frac{1}{2}} = (\epsilon_{j+1}\epsilon_j)^{\frac{1}{2}}$, might better represent the average energy generation within a shell than $\epsilon_{j+\frac{1}{2}} = (\epsilon_{j+1} + \epsilon_j)/2$ does, but no prescription is in fact safe unless ϵ is not greatly different at opposite ends of the zone.

The time derivative is estimated as the ratio of the increment in the quantity from its value at the same coordinate in a model a time Δt earlier to the time interval Δt. Thus the construction of a model requires knowledge of the values of the entropy at the zone boundaries in the previous model. Quantities associated with the previous model are designated by the superscript $(-\Delta t)$. It will also be noticed that although the space derivatives are evaluated symmetrically, i.e., in terms of variables on both sides of the point where the derivative is evaluated, the time derivative is not. It involves only a "backward difference" in time. This scheme is actually preferred over a time-centered one for most stellar problems because it quickly damps out short-period oscillations in the structure.

Whether Eq. (6-54*a*) or (6-54*b*) determines the temperature structure depends upon the usual considerations of stability against convection. The structure is computed with the provisional assumption that Eq. (6-54*a*) applies, but at each point the temperature gradient must be compared against that implicit in Eq. (6-54*b*), and if the former is the greater of the two in absolute magnitude, Eq. (6-54*b*) must replace Eq. (6-54*a*). Of course, the transition from the radiative to the convective gradient will generally lie somewhere between two of the preselected shell boundaries, ξ_k and ξ_{k+1}, say. The location of the boundary may be made by interpolating between two models, one for which the radiative equation is terminated at ξ_k and one for which it is terminated at ξ_{k+1}. The excess of the provisional radiative temperature gradient over the adiabatic one is of opposite algebraic sign at the provisional boundaries of the radiative zones of the two models, and the magnitude of the change allows the proper boundary to be interpolated. When there is either an interior convection zone or both a central and a surface convection zone, the interpolation procedure is more complicated. In those cases there are two boundaries between radiative and convective zones, and a minimum of three models is required to simultaneously interpolate the positions of both boundaries. It may be preferable to use such fine zoning near the interface that interpolation becomes unnecessary.

As mentioned in the previous section, the surface introduces an added complication when it is convective. In that case the adiabatic approximation leading to Eq. (6-54*b*) is invalid. As far as the logical structure of this discussion is concerned, it is convenient to relegate these matters to a separate program which treats the entire stellar atmosphere down through the helium ionization zone as the outermost zone of the star from $j = J - 1$ to $j = J$. Because the mass and composition of the star are preselected, the assumption of values of R and $\mathfrak{L}$ is sufficient to calculate the variables at the base of the atmospheric zone:

$$
\begin{aligned}
r_{J-1} &= f_1(R,\mathfrak{L}) \\
F_{J-1} &= f_2(R,\mathfrak{L}) \qquad (6\text{-}55)\\
T_{J-1} &= f_3(R,\mathfrak{L}) \\
q_{J-1} &= f_4(R,\mathfrak{L})
\end{aligned}
$$

Another scheme of interpolation must be established for obtaining these boundary values from a small but representative collection of model-atmosphere calculations. Once this scheme has been established, the values of the variables at ξ_{J-1} are set by the self-consistency requirement that the atmosphere match the rest of the stellar model.[1]

Equations (6-51) to (6-54) are then solved by the same technique illustrated in the introductory example. The variables whose values are to be found are r_j,

[1] A full description of the interpolation procedure may be found in Henyey, Forbes, and Gould, *loc. cit.*

F_j, T_j, and q_j at each point ξ_j. All other functions in the equations, including the pressure p_j, are functions of state calculable in terms of T, q, and the chemical composition. The total number of variables is given by the product of the four variables times the number of grid points and is equal to $4(J + 1)$. The four difference equations are evaluated in the center of each shell, so that the number of equations is equal to $4J$. The existence of four boundary conditions, two explicit ones at the center and two implicit ones from the theory of stellar atmospheres at the surface, allows the problem to be solved. The four equations of each shell can be thought of as four functions of the variables, each of which equals zero for the proper values of the variables. Therefore the problem is to find the simultaneous roots of $4J$ equations in $4J$ unrestricted variables, and the method is the Newton-Raphson scheme. The $4J$ equations

$$g_k(x_1,x_2, \ldots ,x_{4J}) = 0 \qquad k = 1, 2, \ldots , 4J \tag{6-56}$$

are solved by assuming trial values $x_1{}^0$, $x_2{}^0$, . . . , $x_{4J}{}^0$ of the set of variables, which are then corrected by amounts $\delta x_1{}^0$, $\delta x_2{}^0$, . . . determined by

$$g_k{}^0 + \sum_{i=1}^{4J} \left(\frac{\partial g_k}{\partial x_i}\right)^0 \delta x_i{}^0 = 0 \tag{6-57}$$

where the $g_k{}^0$'s will not generally vanish because the initial assumption of the values of the variables was not correct. The increments are added to the initial values, and the process is repeated. For a reasonable initial guess the method converges rapidly to the solution.

Problem 6-10: Set up the incremental equation (6-57) for the second of the structural equations (6-52).
Ans:

$$\frac{8}{\pi} M'_{j+\frac{1}{2}}(\xi_{j+1} - \xi_j) - (q^0_{j+1} + q_j{}^0)^3(r^0_{j+1} + r_j{}^0)^2(r^0_{j+1} - r_j{}^0)$$
$$- 3(q^0_{j+1} + q_j{}^0)^2(r^0_{j+1} + r_j{}^0)^2(r^0_{j+1} - r_j{}^0)(\delta q^0_{j+1} + \delta q_j{}^0)$$
$$- (q^0_{j+1} + q_j{}^0)^3(r^0_{j+1} + r_j{}^0)[(3r^0_{j+1} - r_j{}^0)\,\delta r^0_{j+1} + (r^0_{j+1} - 3r_j{}^0)\,\delta r_j{}^0] = 0 \tag{6-58}$$

This equation has a simple geometrical interpretation if $\delta r^0_{j+1} = \delta r_j{}^0$. Can you derive this meaning and state it verbally?

The other structural equations, although somewhat more complicated than this one, are incremented in much the same way. Notice that when the iteration algorithm is employed to find the values of the variables, the identity of the unknowns shifts from the variables themselves to the increments to be applied to the assumed values of the variables, a subtle but important difference. Equation (6-58) illustrates another important feature of the problem; viz., although there are $4J$ unrestricted variables, only a few at adjacent coordinates occur in any single equation. In this particular case the only unknowns appearing are δr and δq at the adjacent points ξ_j and ξ_{j+1}. It is this general feature that allows the construction of a logic ladder in which the values of the increments at one point are expressed in terms of the values of the increments at the previous point. Because four of the increments at the boundaries are tied down by the boundary conditions, the increments at successive points are evaluated in turn. Henyey, Forbes, and Gould[1] give a proof by mathematical induction of the existence of such a solution. The principle is identical to that illustrated by the example of the single differential equation $y' = f(x,y)$.

When the values of all the increments are smaller than some predetermined value, the model is regarded as satisfactory. A time step Δt is chosen for the relative age of the next model. The composition changes due to nuclear reactions and mixing are calculated and added to the existing composition to obtain the composition of the next model. For evolving stars, a good trial value for the variables in the next model can be obtained by a linear extrapolation of the rates of change of the variables from the last two models. It is for just those stages of rapid evolution that the relaxation method handles the problem most elegantly.

Lest confusion occur, it should perhaps be emphasized that both of these numerical methods treat only stars in hydrostatic equilibrium. Many *dynamic* stars exist for which a hydrodynamic treatment including both the momentum and acceleration of mass elements must be used. The most notable examples are to be found in pulsating stars and in supernovas. In such hydrodynamic problems the lagrangian shell coordinates are used but in such a way that the structure relaxes *simultaneously* in space and in time, rather than in space at a given time as outlined in the previous material.

6-5 CONTRACTION TO THE MAIN SEQUENCE

A brief discussion of the contraction of a star to the main sequence served as an introduction to Chap. 5, and will not be repeated here. The dominant feature of contraction to the main sequence is that it occurs on a nearly vertical path in the H-R diagram. At any given luminosity hydrostatic equilibrium is unachievable for stars of a given mass whose outer layers are too cool. For a given mass, therefore, there exists a forbidden region to the right of some critical curve, as illustrated in Fig. 6-5. A star to the right of this boundary collapses rapidly until the critical surface temperature for that mass is reached, whereupon it follows the quasistatic evolutionary sequence of a star in hydrostatic equilibrium. The boundary of the forbidden region is somewhat difficult to locate accurately because it depends upon the opacity and the detailed theory of convective energy transport.

The static structure of a star of mass $\mathfrak{M}$ is the main-sequence structure. At the time when the star first achieves hydrostatic equilibrium at the boundary of the forbidden region, its luminosity is much greater than the main-sequence

[1] *Ibid.*

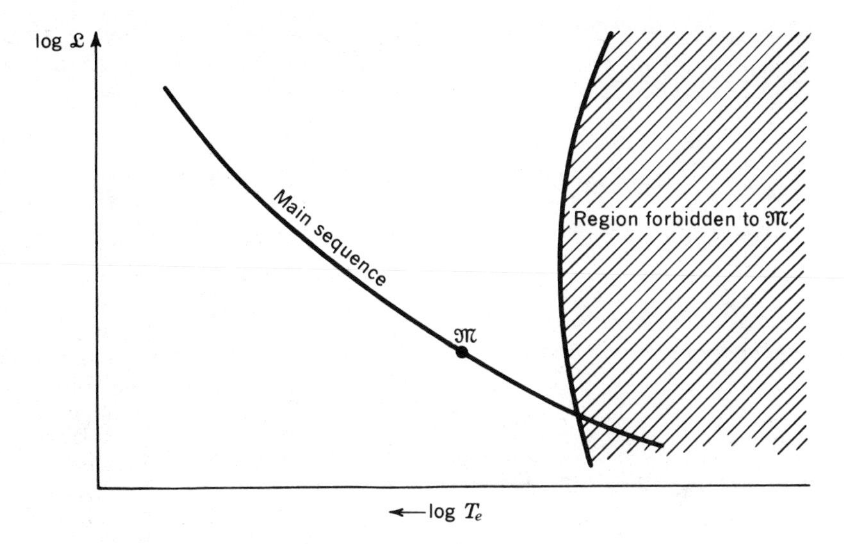

Fig. 6-5 The forbidden region in the H-R diagram. For a star of given mass and composition, hydrostatic equilibrium cannot be established for a surface temperature that is, for each $\mathcal{L}$, less than a critical value. The boundary of the forbidden zone is commonly called the *Hayashi track* because Hayashi first argued that a star contracting toward the main sequence should descend that boundary.

luminosity for the same mass. At that time the subphotospheric layers have such high opacity and the total luminosity is so great that the star is convective throughout. The star then contracts, maintaining the largest radius consistent with hydrostatic equilibrium. This means that it approaches the main sequence by descending the boundary of the forbidden zone.[1] Such a track computed for the sun was shown in Fig. 5-1.

As the star descends this track, the central opacity drops rapidly by virtue of the increasing central temperature. Eventually the radiative temperature gradient becomes smaller than the adiabatic one, and a central core in radiative equilibrium develops. As the contraction slows its pace, the luminosity reaches a minimum, but it then begins to rise again for the following reason. As the hot radiative core, which has low opacity, encompasses more and more of the star, the star becomes less opaque. Thus more energy can flow out radiatively as the opacity continues to fall (recall that the fairly representative Kramers opacity varies as $\rho T^{-3.5}$). Because the star is still shrinking as its luminosity rises, its surface is becoming hotter, and the star moves up and to the left in the H-R diagram but at a considerably slower rate of evolution than in the descent.

A feature peculiar to the pre-main-sequence evolution is characteristic of deep

[1] The existence of the forbidden zone was first demonstrated by Hayashi. See C. Hayashi, *Ann. Rev. Astron. Astrophys.*, **4**:(1966).

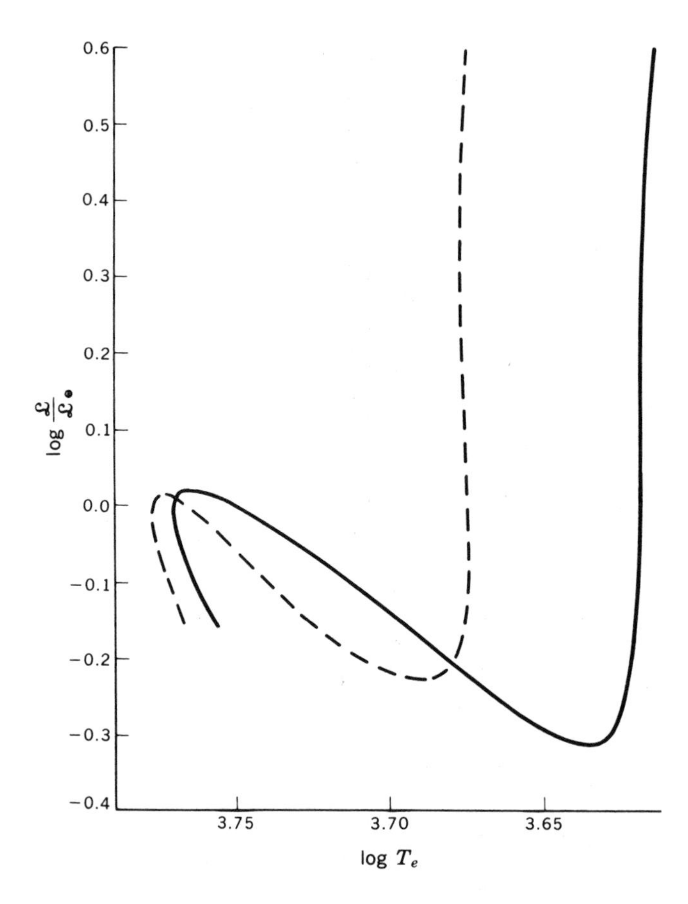

Fig. 6-6 Computed paths in the H-R diagram for the contraction of a one-solar-mass star to the main sequence. The two tracks shown differ only in the metal abundance, which affects the structure of such stars by virtue of its effect on the surface opacity. The solid curve corresponds to a metallic mass fraction $X_M = 5.4 \times 10^{-5}$, and the dashed curve corresponds to $X_M = 5.4 \times 10^{-6}$. [*After I. Iben, Jr., Astrophys. J.*, **141**: 993 (1965). *By permission of The University of Chicago Press. Copyright* 1965 *by The University of Chicago.*]

convection zones; viz., the structure of the whole star is sensitive to the photospheric opacity. This situation reflects the fact that in any model atmosphere the relationship between photospheric pressure and temperature shows strong dependence upon photospheric opacity. Because the photospheric values serve as boundary conditions for an inward development based upon a theory of convection (roughly $P = KT^{\frac{5}{2}}$), the internal values in the convective zone depend strongly on the photospheric values. If the convection zone is deep, the whole star is affected. Figure 6-6 shows a contraction path of a normal $1\mathfrak{M}_{\odot}$ star and, for comparison, the corresponding track when the abundances of the metals of low ionization potential are reduced by a factor of 10. The tracks are similar, but the more transparent photosphere results in a track of higher surface temperature. Low-temperature opacity remains an important problem in stellar structure.

It can be seen from Fig. 6-6 that the luminosity and surface temperature pass through a maximum before they both decrease to the main-sequence position. This decrease occurs, paradoxically enough, when the central temperature becomes great enough to burn nuclear fuel. The nuclear fuels of importance at that stage are the onset of the PP chain and the conversion of the initial C^{12} into N^{14} by two successive proton captures. These processes happen simultaneously, and the second deserves a brief explanatory comment. As discussed in Chap. 5, the equilibrium abundance of the nuclei in the CN cycle is almost entirely N^{14}, because the reaction $N^{14}(p,\gamma)O^{15}$ is about two orders of magnitude slower than the other reactions of the CN cycle. But the initial abundance of C^{12} may be large, and its conversion to N^{14} proceeds at temperatures that are competitive with those of the PP chain. When the C^{12} has been scoured out by conversion to N^{14}, however, that energy source disappears and is replaced by the slower equilibrium CN cycle. Now the significance of the temporary C^{12} energy source is that the corresponding value of ϵ has strong temperature dependence.

Problem 6-11: Show that near $T_6 = 12$, where the $C^{12}(p,\gamma)$ reaction begins at a significant rate, the corresponding energy generation varies as $\epsilon \propto T^{19}$.

It is almost a general rule that *when a large fraction of a stellar luminosity is provided by nuclear reactions having a strong temperature dependence, the energy-generating zone is convective.* The reason for this can be made physically clear. Because energy is flowing out of the zone, the zone itself must possess a temperature gradient. (An inert central zone on the other hand tends to be isothermal.) Because of the strong temperature dependence of the nuclear rates, the nuclear power is strongly concentrated into the hottest portion of the zone, in this case the center of the star. Thus a very temperature-dependent central source is somewhat like a point source at the center.

Problem 6-12: Show that if a highly temperature-dependent energy source exists at the center of a star, there exists some minimum distance from the center for which the star can be in radiative equilibrium.

When the C^{12} burns, the change of the core from a radiative state to a convective state, which is much less centrally condensed, halts the general contraction and causes the central structure to readjust in a way that requires work against gravity. Power converted into gravitation work is unavailable for radiation, of course, so that the luminosity falls, and hence the surface temperature as well. Only about 80 percent of the energy liberated by nuclear reactions during this phase is available to be radiated. As the C^{12} is exhausted, the radiative core reappears in a low-mass star because the PP chain has a sufficiently weak temperature dependence to be distributed smoothly over the stellar center. The luminosity drops during this process to its main-sequence value. In stars of higher mass, however, the core remains convective because it becomes hot enough for the equilibrium CN cycle to dominate the PP chains in importance.

The developments in time of several of the characteristic quantities of the collapse of a $1\mathfrak{M}_\odot$ model are shown in Fig. 6-7. Shown there are the quantities $\log \mathfrak{L}$, $\log R$, $\log T_e$, $\log \rho_c/\bar{\rho}$, and the mass fraction Q_{RC} of the stellar core in radia-

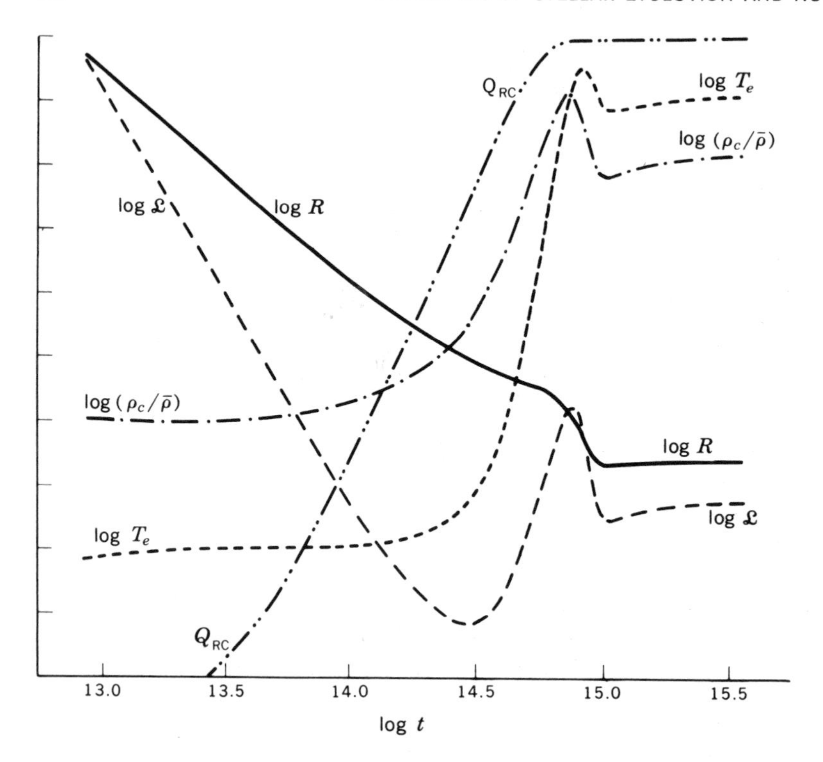

Fig. 6-7 The variation with time t (in seconds) of the surface temperature, the luminosity, the radius, the ratio of central density to mean density, and the mass fraction Q_{RC} within the radiative core during the contraction of a one-solar-mass star to the main sequence. The full-scale limits correspond to $3.78 > \log T_e > 3.58$, $0.6 > \log (\mathcal{L}/\mathcal{L}_\odot) > -0.4$, $0.6 > \log (R/R_\odot) > -0.4$, $2.0 > \log (\rho_c/\bar{\rho}) > 0$, and $1 > Q_{RC} > 0$. [*After I. Iben, Jr., Astrophys. J.*, **141**:993 (1965). *By permission of The University of Chicago Press. Copyright* 1965 *by The University of Chicago*.]

tive equilibrium. Each of these quantities has its own ordinate scale, which is stated in the legend to the figure. The abscissa is log t, with t expressed in seconds, and ranges from epochs of about 3×10^5 to 10^8 years after the initial model. The interpretation of this age at early epochs is somewhat ambiguous because it is not known at what point the dynamically collapsing protostar first joins the Hayashi track. The early evolution is so fast, however, that this initial time uncertainty represents a negligible correction to the age during the final approach to the main sequence. There is no unique specification of the time when the star reaches the main sequence, but gravitational contraction is contributing less than 1 percent of the luminosity after log t = 15.2, or about 5×10^7 years.

Notice from Fig. 6-7 that as the size of the radiative core Q_{RC} grows, the star becomes more centrally condensed. In keeping with the discussion of polytropes in Chap. 2, the gradual change in $\rho_c/\bar{\rho}$ may be thought of as a change in the effective polytropic index. The early value $\log \rho_c/\bar{\rho} = 0.78$ corresponds to a polytrope of index 1.5, appropriate to a star in convective equilibrium, whereas the late values are near $\log \rho_c/\bar{\rho} = 1.74$, which corresponds to a polytrope of index 3.0.

The final drop toward the main sequence begins near $\log t = 14.9$. At this time the central nuclear reactions begin burning rapidly, and a small convective core develops at the center of the larger radiative core (although not reflected in Q_{RC}, which indicates the interface between the convective envelope and the radiative interior). At this time the star becomes less centrally condensed, and the gravitational energy release drops markedly. Finally the central convective core disappears again near $\log t = 15.2$, when the star has effectively reached the main sequence.

Contraction tracks in the H-R diagram calculated by Iben are shown in Fig. 6-8. Each track there is labeled by the mass of the star. Each track has several points enumerated on it, and the times required for that model to approach each point on the track are listed in Table 6-1. It will be noted that the more massive stars do not come so far down the Hayashi track before achieving radiative equilibrium in their interiors. They arrive at the main sequence much faster than low-mass stars:

$$t_{\text{contraction}} \approx 8 \times 10^7 \frac{\mathfrak{M}}{\mathfrak{M}_\odot} \frac{\mathcal{L}_\odot}{\mathcal{L}} \qquad \text{years} \tag{6-59}$$

where $\mathcal{L}$ is the final main-sequence luminosity. It must be borne in mind, however, that these tracks and ages were computed with the assumption that the star maintains constant mass during the contraction. When the details of pre-main-sequence mass loss are known, it may well turn out that both the tracks and the time required are considerably in error. It is quite clear that if the mass loss is extensive, the associated power requirements (stellar-wind luminosity) will make the total contraction time much less than that of a star of equal final mass which has contracted without mass loss. And it is now known that the T Tauri stars, which are believed to be in the final stages of contraction to the main sequence, are losing mass at a very significant rate.[1]

There is one final point of great significance to be gathered from Fig. 6-8; viz., *the final states of the stars at the end of their contraction tracks reproduce the low-T_e envelope of the observational main sequence quite accurately.* Thus the zero-age main sequence is identified as those stars of initially uniform composition which have ceased gravitational contraction and which obtain their luminous power from the thermonuclear conversion of hydrogen into helium.

[1] L. V. Kuhi, *Astrophys. J.*, **140**:1409 (1964). Parenthetically it may be added that the effects of angular momentum due to rotation on the contraction may also be appreciable, and certainly warrant investigation.

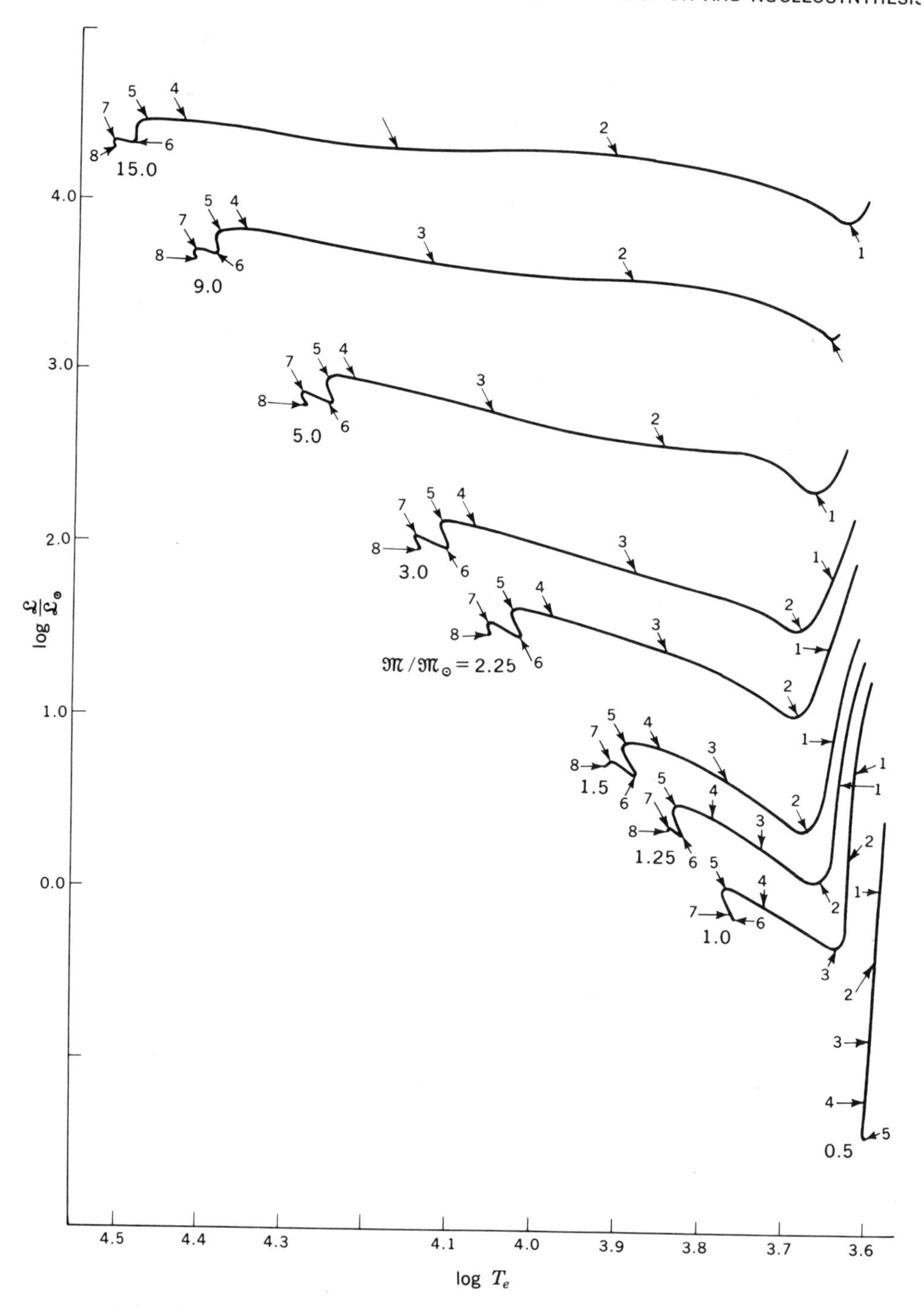

Fig. 6-8 Pre-main-sequence contraction paths for models of mass $\mathfrak{M}/\mathfrak{M}_\odot = 0.5, 1.0, 1.25, 1.5, 2.25, 3.0, 5.0, 9.0$, and 15.0. The time intervals required for the stars to reach enumerated points along the tracks are listed in Table 6-1. [*After I. Iben, Jr., Astrophys. J.*, **141**:993 (1965) *By permission of The University of Chicago Press. Copyright* 1965 *by The University of Chicago.*

Table 6-1 Evolutionary lifetimes, years†

	$\mathfrak{M}/\mathfrak{M}_\odot$								
Point	15.0	9.0	5.0	3.0	2.25	1.5	1.25	1.0	0.5
1	6.740×10^{2}	1.443×10^{3}	2.936×10^{4}	3.420×10^{4}	7.862×10^{4}	2.347×10^{5}	4.508×10^{5}	1.189×10^{5}	3.195×10^{5}
2	3.766×10^{3}	1.473×10^{4}	1.069×10^{5}	2.078×10^{5}	5.940×10^{5}	2.363×10^{6}	3.957×10^{6}	1.058×10^{6}	1.786×10^{6}
3	9.350×10^{3}	3.645×10^{4}	2.001×10^{5}	7.633×10^{5}	1.883×10^{6}	5.801×10^{6}	8.800×10^{6}	8.910×10^{6}	8.711×10^{6}
4	2.203×10^{4}	6.987×10^{4}	2.860×10^{5}	1.135×10^{6}	2.505×10^{6}	7.584×10^{6}	1.155×10^{7}	1.821×10^{7}	3.092×10^{7}
5	2.657×10^{4}	7.922×10^{4}	3.137×10^{5}	1.250×10^{6}	2.818×10^{6}	8.620×10^{6}	1.404×10^{7}	2.529×10^{7}	1.550×10^{8}
6	3.984×10^{4}	1.019×10^{5}	3.880×10^{5}	1.465×10^{6}	3.319×10^{6}	1.043×10^{7}	1.755×10^{7}	3.418×10^{7}	
7	4.585×10^{4}	1.195×10^{5}	4.559×10^{5}	1.741×10^{6}	3.993×10^{6}	1.339×10^{7}	2.796×10^{7}	5.016×10^{7}	
8	6.170×10^{4}	1.505×10^{5}	5.759×10^{5}	2.514×10^{6}	5.855×10^{6}	1.821×10^{7}	2.954×10^{7}		

† I. Iben, Jr., *Astrophys. J.*, **141**:993 (1965). By permission of The University of Chicago Press.

6-6 THE MAIN SEQUENCE

For a given chemical composition, the zero-age main sequence can be defined as the locus of points in the H-R diagram characterizing static stars of homogeneous composition which are burning hydrogen at their centers. Although the position of a given star in the diagram depends rather strongly on the composition, the main sequence itself does not. To understand why this is so, it is helpful to examine the dependence of the luminosity of the standard model upon mass and composition. From Eq. (3-193) we have

$$\mathcal{L} \propto \kappa_0^{-1}\mu^{7.5}\mathfrak{M}^{5.5} \tag{6-60}$$

Although this formula is correct only for the standard model with Kramers opacity $\kappa = \kappa_0\rho T^{-3.5}$, its qualitative features are correct. The mass dependence is exaggerated for the massive stars by the fact that the Kramers opacity is invalid at nearly complete ionization. If the opacity is given a weaker temperature dependence, the corresponding luminosity has a weaker mass dependence; for instance, $\mathcal{L} \propto \mathfrak{M}^{4.5}$ if $\kappa \propto T^{-2.5}$. Thus the observed power-law dependence upon the mass is not difficult to understand and is, in fact, accurately reproduced by the models constructed with electronic computers. Above $7\mathfrak{M}_\odot$ the dependence weakens to $\mathcal{L} \propto \mathfrak{M}^3$.

The luminosity varies inversely with the mean opacity, a fact already obvious from the luminosity formula for the standard model. It is clear simply from the equation of radiative transfer that if the opacity is changed while the temperature structure is held fixed, the rate of energy flow changes inversely with the opacity. It was pointed out in Chap. 3 that the major contribution to the Rosseland mean opacity between 10^6 and 10^7 °K is due, roughly speaking, to the elements of high atomic number (if they are present). Thus to first approximation for population I stars, κ_0 is proportional to Z, the mass fraction of the elements heavier than helium. The luminosity is more strongly dependent upon the mean molecular weight. If the mean molecular weight μ is changed, the central temperature increases by roughly the same factor. The associated reduction in interior opacity greatly increases the radiative flux. The radius R, on the other hand, varies only little with these quantities.

These general features are sufficient to understand why it is that the position of the zero-age main sequence is not very sensitive to the makeup of the composite stars. Because of the relative insensitivity of the radius of a model to composition change, the surface temperature must increase as $T_e^4 \propto \mathcal{L} \propto \mu^{7.5}$ if the radius is to be nearly constant. Thus changes in composition of main-sequence stars tend to slide the star along a line that is approximately parallel to the main sequence itself. The zero-age main sequence, then, shows only a weak dependence upon composition, although that difference may be of importance.[1]

The problems are more acute, however, if one wants to calculate the age of an

[1] For a discussion of the differences between the Hyades and Pleiades in terms of numerical models, see I. Iben, Jr., *Astrophys. J.*, **138**:452 (1963).

old cluster by calculating the time at which the most luminous member of the main sequence would begin to move up the giant branch. From the discussion of Sec. 1-7 we see that the main-sequence lifetime is roughly proportional to

$$t \propto \frac{\mathfrak{M} X_{\mathrm{H}}}{\mathfrak{L}} \tag{6-61}$$

The quantity generally observed is the luminosity, which is a function of both mass and composition. From (6-60) the mass of a star of given luminosity varies with composition according to

$$\mathfrak{M}\ (\text{given } \mathfrak{L}) \propto \frac{\kappa_0^{\frac{1}{5}}}{\mu^{1.4}} \tag{6-62}$$

so that the formula

$$t \propto \frac{\kappa_0^{\frac{1}{5}} X_{\mathrm{H}}}{\mu^{1.4}} \tag{6-63}$$

displays how the main-sequence lifetime of a star of luminosity $\mathfrak{L}$ varies with the chemical composition. It is clear that the errors in the calculation of opacity do not greatly disturb the age calculation, whereas errors in the mean molecular weight are serious. This problem has been reflected as a major uncertainty in the calculation of the ages of the globular clusters, because the helium content of the extreme population II stars has not been well fixed. But it is clear from Eq. (6-63) that increasing the initial helium content reduces X_{H} and increases μ, and both changes reduce the main-sequence lifetime.

Problem 6-13: Consider extreme population II stars composed only of hydrogen and helium ($Z = 0$). The interior opacity will then be dominated by free-free transitions in the field of protons or alpha particles. Show that the compositional dependence of the opacity is then simply

$$\kappa_0 \propto 1 + X$$

Show then that the main-sequence lifetime of a star of given $\mathfrak{L}$ varies with hydrogen composition according to

$$\frac{\partial}{\partial X} \log t = X + \frac{1}{5(X+1)} + \frac{7.0}{5X+3} \approx 2$$

for $0.5 < X < 1.0$. Thus the age t is proportional to 10^{2X}.

Probably the most characteristic features of the structure of main-sequence stars are the location and extent of convection zones. The main sequence is commonly divided into an upper main sequence, with a radiative surface and a convective core, and a lower main sequence, with a convective surface and a radiative core. To a certain extent, these properties depend upon the composition of the stars. The surface convection zone is due to the very high opacity in the subsurface layers, which are cool enough for hydrogen to be partly neutral. The depth of the convection zone, however, depends upon the opacity beneath the hydrogen ionization zone, which in turn depends upon the composition. The central convection zone, on the other hand, occurs for central temperatures high enough for energy generation from the CN cycle to dominate that from the PP chain. Its extent depends in part on the abundances of the CNO nuclei.

There are no upper-main-sequence stars remaining in population II clusters because the cluster ages exceed the stellar burning time on the main sequence. All upper-main-sequence stars in our galaxy, therefore, are of population I composition. It is not surprising then that almost all calculations of their structure have been based on the population I composition, although massive stars of pure hydrogen are of interest because of their possible role in the initial nucleosynthesis in the galaxy. The lower main sequence, on the other hand, is represented in both populations I and II, so that both types of composition have been used in the calculation of their structure.

DEPTH OF THE OUTER CONVECTION ZONE IN MAIN-SEQUENCE STARS

Important questions hinge on the depth of the outer convection zone in the stars of the lower main sequence. That depth is computed in the following way. Given the mass and composition of the star, representative values of R and $\mathcal{L}$ are specified. The equation of hydrostatic equilibrium is then applied to a model atmosphere (a separate calculation) of known surface gravity and effective temperature. The values of P and T at an optical depth $\tau = \frac{2}{3}$ then serve as boundary conditions for inward integrations. In the low-density outer layers, the adiabatic approximation to convection is inadequate, so that what is usually done is to relate the temperature gradient to the energy flux by the simple mixing-length model of convection (which may also be inadequate). Because the mixing-length model is not a fundamental theory, the effective mixing length is not given. It is usually assumed to be some multiple α of the pressure scale height H_P. That is, one takes

$$l = \alpha H_P \tag{6-64}$$

where

$$H_p^{-1} = \frac{1}{P}\left|\frac{dP}{dr}\right| = \frac{GM\rho}{Pr^2} \tag{6-65}$$

The parameter α is then regarded as a parameter of the model. Its value has been argued to be of order unity on the basis of hydrodynamic principles. With each assumed value of α the structural equations can be integrated to the base of the convection zone. Then, of course, the family of models must be fitted there to an internal structure to uniquely specify those values of R and $\mathcal{L}$ appropriate to the given mass, composition, and mixing length. The results of such calculations have important implications for astronomy.

Figure 6-9 displays the depth of the convection zone Δr in terms of the total radius R as a function of the surface temperature of the star. The four different

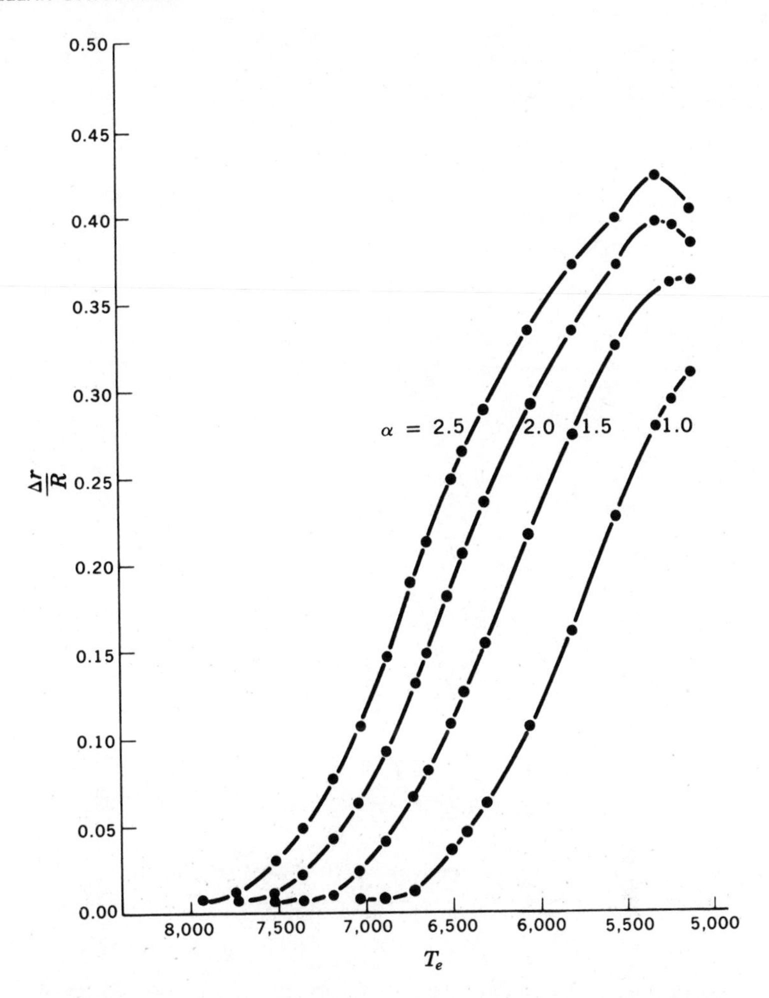

Fig. 6-9 Depth of the outer convection zone of main-sequence stars as a function of T_e. The four separate curves were computed for four different choices of the mixing-length parameter α. [*After N. Baker, The Depth of the Outer Convection Zone in Main-sequence Stars, Inst. Space Studies Rept., New York* (*undated*).]

curves shown correspond to four different choices of the mixing-length parameter α. For each value of α the convection zone is quite thin at the highest temperatures. As one considers stars of lower and lower surface temperature, a fairly well-defined *transition temperature* is reached, where $\Delta r/R$ changes from a small value to an almost linear increase with decreasing T_e. Figure 6-10 is a quite similar display of the temperature T_i at the lower boundary of the convection zone. It will be apparent from these figures that a theory of convection is needed.

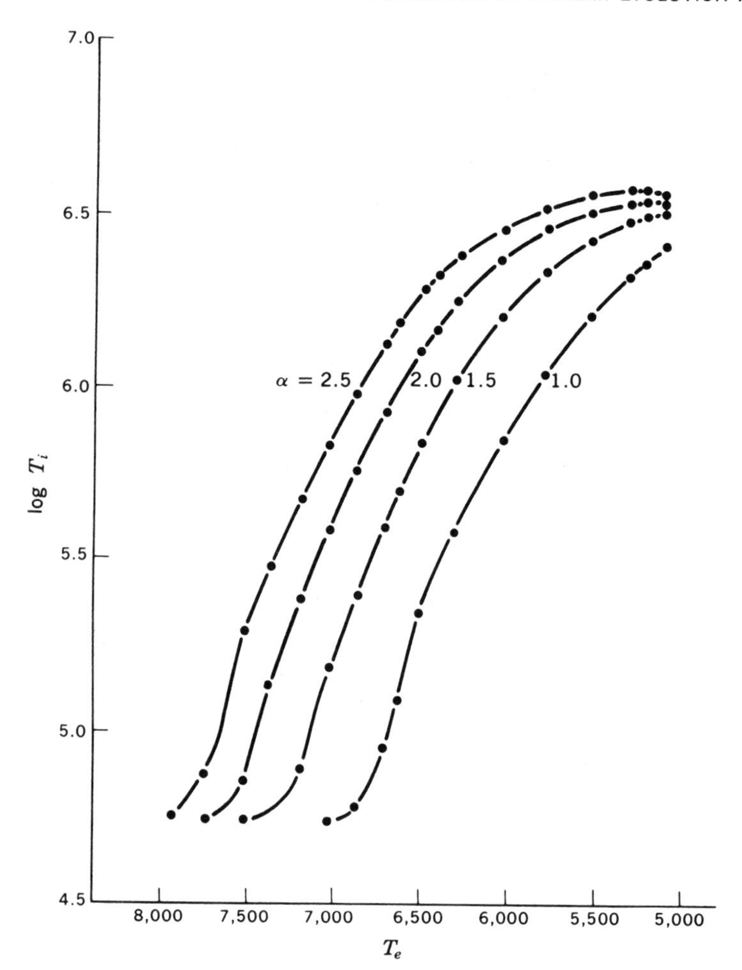

Fig. 6-10 The temperature T_i at the base of the outer convection zone as a function of T_e. [*After N. Baker, The Depth of the Outer Convection Zone in Main-sequence Stars, Inst. Space Studies Rept., New York (undated).*]

Problem 6-14: Try to understand why changes of the mixing length affect these two figures in the way that they do. Why, for instance, does the convection zone become deeper as α is increased? *Hint:* Does increasing α cause the absolute magnitude of the temperature gradient to increase or decrease?

Several interesting composition changes on the surface are related to the depth of the convective zone. It will suffice to illustrate three of them at this point. Many stars are observed to have lithium in their photosphere.[1] Because lithium

[1] G. H. Herbig, *Astrophys. J.*, **141**:588 (1965); G. Wallerstein, G. H. Herbig, and P. S. Conti, *Astrophys. J.*, **141**:610 (1965).

is destroyed by interactions with protons at sufficiently high temperature, the surface lithium will slowly be depleted in those stars having convection zones extending downward to temperatures in excess of 10^6 °K. The lithium is, of course, destroyed only in the lower portions of the convection zone, but the complete mixing throughout the zone has the effect of decreasing the surface lithium as well.

It has been estimated[1] that the surface abundances of many of the rarer nuclear species may be the results of spallation reactions in the stellar surface induced by high-energy particles accelerated in flares. If surface convection zones are present, the nuclei produced over periods long compared to the convective mixing time must be redistributed evenly throughout the convection zone. In some cases—that of lithium or deuterium, perhaps—the nuclei could conceivably be produced by spallation at the surface and destroyed by thermonuclear reactions deep in the convection zone. In that case the corresponding abundance will tend to build up to an equilibrium concentration such that the rate of production at the surface is balanced by the rate of destruction in the convective zone.

A third interesting possibility involves surface convection zones that grow deeper with time. There exists the possibility in that case of bringing nuclear products synthesized in the interior to the surface. This type of phenomenon is probably related in some way to Merrill's observation of the radioactive element technetium in red giants, which also have deep surface convection zones.

Whenever a nuclear species is to be distributed throughout the convection zone, the relevant measure of the zone is its total mass rather than its depth in radius. Because of the low density near the surface, the convection zone extends over a much smaller fraction of the stellar mass than the corresponding fraction of the stellar radius. The fraction of the stellar mass contained in the surface convection zone of main-sequence stars is illustrated in Fig. 6-11.

For mixing-length parameters in the range $1.0 < \alpha < 2.5$, the transition temperature below which deep convection zones occur is $T_e = 7300 \pm 500$°K. The surface temperature $T_e = 7300$°K corresponds to a main-sequence star of mass $\mathfrak{M} = 1.5\mathfrak{M}_\odot$ and of spectral type FO. Thus one may roughly designate spectral types of class F and cooler as lower-main-sequence stars and spectral types of class A and hotter as upper-main-sequence stars. Finally, it should be mentioned that rotation may be strongly coupled to the convection-zone problem. We shall not consider this complicated possibility except to point out the coincidence (?) that, in general, upper-main-sequence stars are fast rotators whereas lower-main-sequence stars are slow rotators.

CENTRAL STRUCTURE OF MAIN-SEQUENCE STARS

The most characteristic feature of the internal structure of main-sequence stars is the extent of the central convection zone. In stars with masses smaller than about $1.2\mathfrak{M}_\odot$, the central temperature is low enough for only the PP chains to

[1] W. A. Fowler, E. M. Burbidge, and G. R. Burbidge, *Astrophys. J. Suppl.*, **2**:167 (1955).

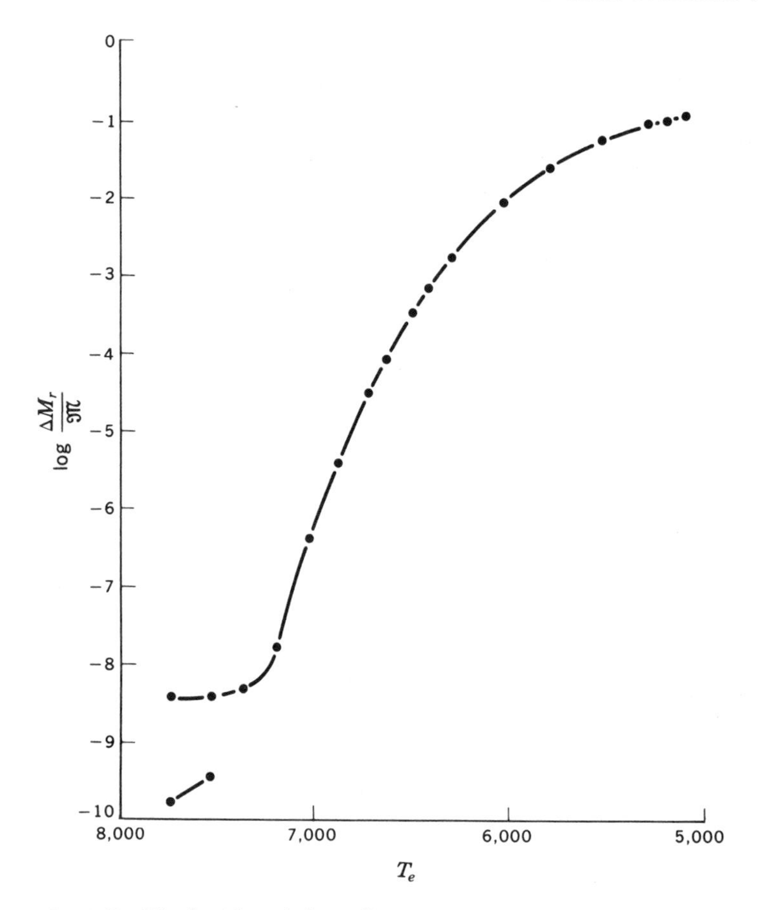

Fig. 6-11 The fraction of the stellar mass contained in the outer convection zone of a main-sequence star as a function of T_e. [*After N. Baker, The Depth of the Outer Convection Zone in Main-sequence Stars, Inst. Space Studies Rept., New York* (*undated*).]

contribute significantly to the thermonuclear power. Because of the relatively weak temperature dependence of that energy source, it is distributed over a relatively large region of the stellar center. The luminosity $L(r)$ builds up slowly at increasing distances from the center, so that the heat flux is never very large inasmuch as $L(r)$ is moderated by the inverse square of the distance from the center. It turns out that the flux can be carried adequately by radiative transfer along a temperature gradient smaller than the adiabatic one.

In stars of larger mass the central temperature is greater, with the result that the coulomb barrier for the CN cycle is more easily overcome. Provided that C, N, and O occur in representative amounts in the interior, the energy liberated by their radiative capture of protons becomes more important with increasing mass. But because of the stronger temperature dependence of these reactions, the energy from them is primarily liberated very near the center of the star. Such a pointlike source of energy produces very large fluxes near the center, fluxes that cannot be carried by underadiabatic gradients. The size of the convective core increases with mass, representing about half of the total mass at $\mathfrak{M} = 15\mathfrak{M}_\odot$.

Some of the fine details of the transition from the radiative to the convective core are displayed in Fig. 6-12 as a function of the stellar mass. The particular composition used in this case was $X_H = 0.7$, $Y = 0.28$, $Z = 0.02$ with carbon and nitrogen present in the amount $X_{CN} = 0.18Z = 0.0036$. The parameter R_c represents the ratio at the stellar center of the temperature gradient that would be required to carry the heat flux by radiative transfer to the adiabatic temperature gradient. It follows that the center is convectively unstable when $R_c > 1$. Note that R_c increases markedly with mass, being greater than 2 for $\mathfrak{M} > 1.7\mathfrak{M}_\odot$. The value of R_c also depends upon the calculation of the central opacity; i.e., the

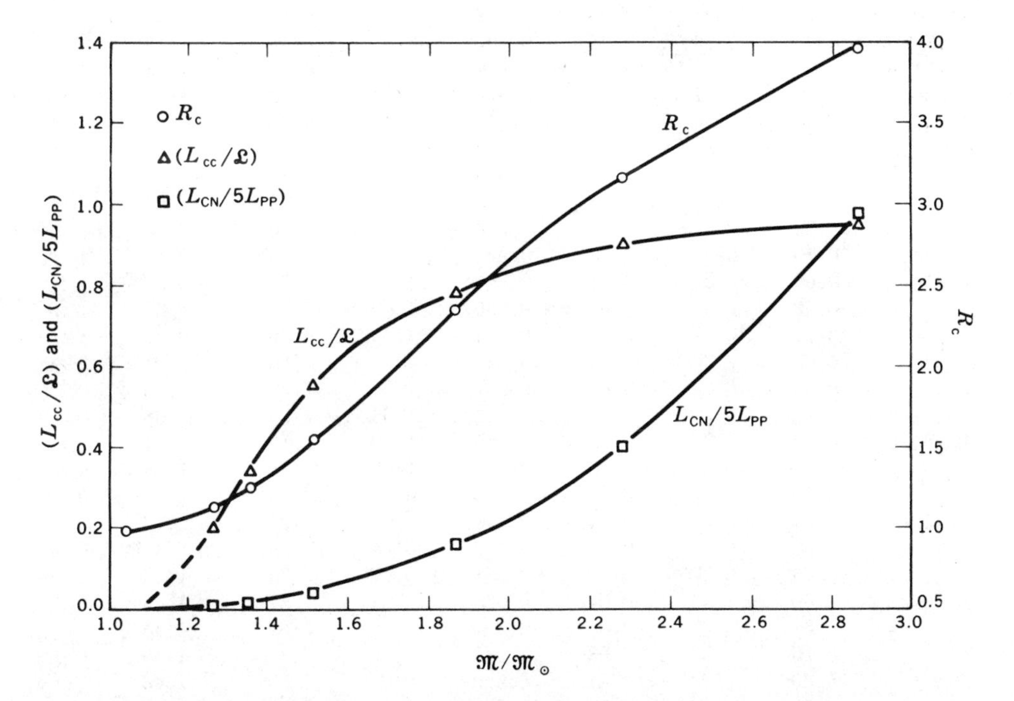

Fig. 6-12 Variation with stellar mass of the central structure of a main-sequence star. R_c is the ratio of the temperature gradient that would be required to carry the central flux radiatively to the adiabatic temperature gradient. $L_{cc}/\mathfrak{L}$ is the fraction of the total luminosity generated within the convective core. $L_{CN}/5L_{PP}$ is a measure of the nuclear power generated respectively by the CN cycle and by the PP chains, the power being summed over the entire star. [*After I. Iben, Jr., and J. R. Ehrmann, Astrophys. J.*, **135**:770 (1962). *By permission of The University of Chicago Press. Copyright* 1962 *by The University of Chicago.*]

Table 6-2 Zero-age model for three compositions with $\mathfrak{M} = 2.82\mathfrak{M}_\odot$†

	X	Y	Z	$R/R_\odot$	$\mathfrak{L}/\mathfrak{L}_\odot$	$\log T_e$
Surface	0.60	0.36	0.04	2.1	93	4.10
	0.60	0.37	0.03	2.0	110	4.12
	0.70	0.27	0.03	2.0	63	4.07

r/R	X	Y	Z	$M(r)/\mathfrak{M}$	$L(r)/\mathfrak{L}$	$\log T$	$\log \rho$	κ
0.95	0.60	0.36	0.04	1.000	1.00	5.42	−4.81	7.6
	0.60	0.37	0.03	1.000	1.00	5.44	−4.74	7.0
	0.70	0.27	0.03	1.000	1.00	5.41	−4.73	8.9
0.85	0.60	0.36	0.04	1.000	1.00	5.92	−3.04	4.1
	0.60	0.37	0.03	1.000	1.00	5.94	−2.97	3.5
	0.70	0.27	0.03	1.000	1.00	5.91	−2.97	4.4
0.75	0.60	0.36	0.04	0.998	1.00	6.20	−2.09	3.8
	0.60	0.37	0.03	0.998	1.00	6.22	−2.02	3.3
	0.70	0.27	0.03	0.998	1.00	6.19	−2.02	4.0
0.65	0.60	0.36	0.04	0.992	1.00	6.43	−1.43	3.8
	0.60	0.37	0.03	0.992	1.00	6.44	−1.36	3.2
	0.70	0.27	0.03	0.992	1.00	6.42	−1.36	4.0
0.55	0.60	0.36	0.04	0.975	1.00	6.60	−0.85	2.78
	0.60	0.37	0.03	0.973	1.00	6.61	−0.77	2.31
	0.70	0.27	0.03	0.974	1.00	6.59	−0.77	2.90
0.45	0.60	0.36	0.04	0.926	1.00	6.75	−0.26	1.78
	0.60	0.37	0.03	0.922	1.00	6.76	−0.18	1.45
	0.70	0.27	0.03	0.925	1.00	6.74	−0.19	1.86
0.35	0.60	0.36	0.04	0.806	1.00	6.90	+0.34	1.20
	0.60	0.37	0.03	0.795	1.00	6.92	0.41	1.01
	0.70	0.27	0.03	0.804	1.00	6.89	0.41	1.25
0.25	0.60	0.36	0.04	0.546	1.00	7.06	0.91	0.80
	0.60	0.37	0.03	0.527	1.00	7.08	0.96	0.72
	0.70	0.27	0.03	0.544	1.00	7.05	0.98	0.84
0.20	0.60	0.36	0.04	0.365	1.00	7.15	1.14	0.66
	0.60	0.37	0.03	0.347	1.00	7.17	1.18	0.60
	0.70	0.27	0.03	0.362	0.99	7.14	1.21	0.71
0.15	0.60	0.36	0.04	0.190	0.98	7.24	1.32	0.56
	0.60	0.37	0.03	0.179	0.98	7.25	1.35	Conv.
	0.70	0.27	0.03	0.188	0.97	7.22	1.38	0.59
0.10	0.60	0.36	0.04	0.065	0.82	7.31	1.43	Conv.
	0.60	0.37	0.03	0.062	0.80	7.32	1.45	Conv.
	0.70	0.27	0.03	0.065	0.79	7.30	1.49	Conv.
0.05	0.60	0.36	0.04	0.010	0.26	7.35	1.49	Conv.
	0.60	0.37	0.03	0.008	0.24	7.36	1.51	Conv.
	0.70	0.27	0.03	0.009	0.25	7.34	1.56	Conv.

Table 6-2 Zero-age model for three compositions with $\mathfrak{M} = 2.82\mathfrak{M}_\odot$† (Continued)

	X	Y	Z	$\frac{R}{R_\odot}$	$\frac{\mathfrak{L}}{\mathfrak{L}_\odot}$	$\log T_e$
Surface	0.60	0.36	0.04	2.1	93	4.10
	0.60	0.37	0.03	2.0	110	4.12
	0.70	0.27	0.03	2.0	63	4.07

$\frac{r}{R}$	X	Y	Z	$\frac{M(r)}{\mathfrak{M}}$	$\frac{L(r)}{\mathfrak{L}}$	$\log T$	$\log \rho$	κ
0.00	0.60	0.36	0.04	0.000	0.00	7.36	1.51	Conv.
	0.60	0.37	0.03	0.000	0.00	7.37	1.53	Conv.
	0.70	0.27	0.03	0.000	0.00	7.35	1.58	Conv.
0.148‡	0.60	0.36	0.04	0.183	0.98	7.24	1.32	0.55
0.155‡	0.60	0.37	0.03	0.194	0.98	7.24	1.33	0.53
0.147‡	0.70	0.27	0.03	0.179	0.97	7.23	1.39	0.59

† Adapted from B. Strömgren, Stellar Models for Main-sequence Stars and Subdwarfs, in L. H. Aller and D. B. McLaughlin (eds.), "Stellar Structure." By permission of The University of Chicago Press. Copyright 1965 by The University of Chicago.

‡ Boundary of the convective core.

value $R_c = 2$, for instance, means that either the core is convective or the calculated value of the central opacity is too great by a factor of 2. Thus one must examine the opacity calculation carefully when predicting the existence of marginal convection zones. In the example just cited, the central opacity near $R_c = 2$ was computed to be about three times as great as the opacity due to the scattering from free electrons, and the electron-scattering opacity clearly is a lower bound to the central opacity of such a star.

The ratio $L_{cc}/\mathfrak{L}$ in Fig. 6-12 represents the ratio of the power generated within the convective core to the total luminosity of the star. About half of the power is generated within the convective core at $\mathfrak{M} = 1.5\mathfrak{M}_\odot$, but the ratio is nearly unity at $\mathfrak{M} = 3\mathfrak{M}_\odot$. The ratio $L_{CN}/5L_{PP}$ measures the ratio for the entire star of the rate at which energy is provided by the CN cycle to the rate provided by the PP chains. The respective contributions are equal near $\mathfrak{M} = 2\mathfrak{M}_\odot$, but the CN cycle becomes much more dominant at larger masses. From this ratio it can be seen that the convective core develops *before* the CN cycle takes over as the major source of power. It must be clear that these ratios depend upon the number of CN nuclei in the initial composition. In extreme population II stars (which no longer survive in this mass range) with X_{CN} smaller by a factor of 100 or more, core convection may not appear at all.

Tables 6-2 and 6-3, respectively, show the characteristics of zero-age models of upper-main-sequence stars of $2.8\mathfrak{M}_\odot$ and $7.1\mathfrak{M}_\odot$. Each model is displayed for three sets of composition parameters: $\{X,Y,Z\} = \{0.60,0.36,0.04\}$, $\{0.60,0.37,0.03\}$, and $\{0.70,0.27,0.03\}$. The generally small effects of composition difference can be interpolated over a small range of composition changes.

Table 6-3 Zero-age model for three compositions with $\mathfrak{M} = 7.08\mathfrak{M}_{\odot}$†

	X	Y	Z	$\frac{R}{R_{\odot}}$	$\frac{\mathcal{L}}{\mathcal{L}_{\odot}}$	$\log T_e$
Surface	0.60	0.36	0.04	3.5	2,800	4.35
	0.60	0.37	0.03	3.4	2,800	4.36
	0.70	0.27	0.03	3.3	2,000	4.35

$\frac{r}{R}$	X	Y	Z	$\frac{M(r)}{\mathfrak{M}}$	$\frac{L(r)}{\mathcal{L}}$	$\log T$	$\log \rho$	κ
0.95	0.60	0.36	0.04	1.000	1.00	5.59	−4.77	1.88
	0.60	0.37	0.03	1.000	1.00	5.61	−4.70	1.72
	0.70	0.27	0.03	1.000	1.00	5.58	−4.72	2.09
0.85	0.60	0.36	0.04	1.000	1.00	6.09	−3.06	1.34
	0.60	0.37	0.03	1.000	1.00	6.10	−2.99	1.22
	0.70	0.27	0.03	1.000	1.00	6.08	−2.99	1.42
0.75	0.60	0.36	0.04	0.997	1.00	6.38	−2.18	1.33
	0.60	0.37	0.03	0.997	1.00	6.39	−2.10	1.19
	0.70	0.27	0.03	0.997	1.00	6.37	−2.10	1.43
0.65	0.60	0.36	0.04	0.987	1.00	6.59	−1.52	1.10
	0.60	0.37	0.03	0.987	1.00	6.60	−1.44	0.95
	0.70	0.27	0.03	0.987	1.00	6.58	−1.44	1.16
0.55	0.60	0.36	0.04	0.958	1.00	6.75	−0.92	0.81
	0.60	0.37	0.03	0.955	1.00	6.77	−0.85	0.70
	0.70	0.27	0.03	0.957	1.00	6.74	−0.85	0.87
0.45	0.60	0.36	0.04	0.880	1.00	6.91	−0.35	0.64
	0.60	0.37	0.03	0.872	1.00	6.93	−0.28	0.58
	0.70	0.27	0.03	0.878	1.00	6.90	−0.28	0.68
0.35	0.60	0.36	0.04	0.705	1.00	7.07	+0.18	0.50
	0.60	0.37	0.03	0.689	1.00	7.08	0.24	0.47
	0.70	0.27	0.03	0.702	1.00	7.06	0.25	0.53
0.25	0.60	0.36	0.04	0.410	1.00	7.23	0.61	0.41
	0.60	0.37	0.03	0.393	1.00	7.24	0.65	0.39
	0.70	0.27	0.03	0.407	1.00	7.22	0.67	0.44
0.20	0.60	0.36	0.04	0.252	1.00	7.31	0.75	Conv.
	0.60	0.37	0.03	0.239	0.99	7.32	0.78	Conv.
	0.70	0.27	0.03	0.249	1.00	7.30	0.82	Conv.
0.15	0.60	0.36	0.04	0.123	0.94	7.37	0.86	Conv.
	0.60	0.37	0.03	0.116	0.93	7.38	0.89	Conv.
	0.70	0.27	0.03	0.125	0.94	7.36	0.93	Conv.
0.10	0.60	0.36	0.04	0.041	0.66	7.42	0.94	Conv.
	0.60	0.37	0.03	0.038	0.64	7.43	0.96	Conv.
	0.70	0.27	0.03	0.040	0.67	7.41	1.00	Conv.
0.05	0.60	0.36	0.04	0.006	0.16	7.45	0.99	Conv.
	0.60	0.37	0.03	0.005	0.11	7.45	1.01	Conv.
	0.70	0.27	0.03	0.005	0.16	7.44	1.05	Conv.

Table 6-3 Zero-age model for three compositions with $\mathfrak{M} = 7.08\mathfrak{M}_\odot$† (Continued)

	X	Y	Z	$\frac{R}{R_\odot}$	$\frac{\mathfrak{L}}{\mathfrak{L}_\odot}$	$\log T_e$		
Surface	0.60	0.36	0.04	3.5	2,800	4.35		
	0.60	0.37	0.03	3.4	2,800	4.36		
	0.70	0.27	0.03	3.3	2,000	4.35		
$\frac{r}{R}$	X	Y	Z	$\frac{M(r)}{\mathfrak{M}}$	$\frac{L(r)}{\mathfrak{L}}$	$\log T$	$\log \rho$	κ
0.00	0.60	0.36	0.04	0.000	0.00	7.45	1.00	Conv.
	0.60	0.37	0.03	0.000	0.00	7.46	1.02	Conv.
	0.70	0.27	0.03	0.000	0.00	7.45	1.06	Conv.
0.211‡	0.60	0.36	0.04	0.282	1.00	7.29	0.73	0.39
0.218‡	0.60	0.37	0.03	0.290	1.00	7.30	0.74	0.38
0.207‡	0.70	0.27	0.03	0.270	1.00	7.29	0.80	0.42

† Adapted from B. Strömgren, Stellar Models for Main-sequence Stars and Subdwarfs, in L. H. Aller and D. B. McLaughlin (eds.), "Stellar Structure." By permission of The University of Chicago Press. Copyright 1965 by The University of Chicago.

‡ Boundary of the convective core.

It is generally believed that the hydrogen content of the population I stars, which are the only ones left at masses this great, falls somewhere in the range $0.60 < X < 0.70$. It remains a major theoretical problem to understand how it is that the whole range of population I stars share this narrow band of hydrogen concentration if the galaxy began as essentially pure hydrogen. It may well be that a major stage of galactic evolution separates population II and population I.

As the upper-main-sequence stars consume their central hydrogen, several effects occur simultaneously: (1) the convective core shrinks, thereby leaving behind a continual gradation of hydrogen concentration, (2) the radius expands, (3) the core contracts gravitationally to larger central density and temperature, and (4) the luminosity increases but in such a way that the surface temperature drops relatively little because of the expanding radius. This last effect means that the *evolved main sequence* is shifted upward to higher luminosity than the zero-age star of the same surface temperature. The principal characteristics of evolutionary sequences for models of four different masses of initial composition $X = 0.70$, $Y = 0.27$, and $Z = 0.03$ are listed in Table 6-4. Most of the entries are self-explanatory. The entry q(core) represents the fraction of the total mass in the convective core, whereas $|\Delta M_b|$ represents the increase of the luminosity *over that of the zero-age star of the same surface temperature.* The bolometric magnitudes M_b decrease as the luminosity increases, so that $|M_b|$ represents the magnitude of the decrease.

The change of the lower main sequence can be illustrated with two models of the sun, the first at zero age and the second after 4.5×10^9 years, i.e., today.

Table 6-4 Evolved main sequence for four masses†

$\mathfrak{M}/\mathfrak{M}_\odot$	X_c	Age, 10^6 years	$R/R_\odot$	M_b	log T_e	$\lvert\Delta M_b\rvert$	q(core)	T_c, 10^6 °K	ρ_c, g/cm³
1.78	0.70	0	1.54	2.1	3.93	0.0	0.12	20	68
	0.60	210	1.64	2.0	3.92	0.1	0.11	20	72
	0.50	390	1.74	2.0	3.92	0.3	0.10	21	76
	0.40	540	1.86	2.0	3.90	0.5	0.09	21	82
	0.30	670	1.99	1.9	3.89	0.7	0.08	22	89
	0.20	770	2.14	1.9	3.88	0.9	0.07	22	99
	0.10	860	2.28	1.9	3.86	1.0	0.06	24	117
2.82	0.70	0	1.96	0.2	4.07	0.0	0.18	23	38
	0.60	70	2.11	0.1	4.07	0.2	0.16	23	39
	0.50	120	2.28	0.0	4.06	0.4	0.14	24	41
	0.40	170	2.46	−0.1	4.05	0.6	0.12	24	43
	0.30	210	2.67	−0.1	4.03	0.8	0.10	25	46
	0.20	240	2.91	−0.1	4.02	1.1	0.08	25	50
	0.10	260	3.15	−0.1	4.00	1.3	0.07	27	59
4.47	0.70	0	2.54	−1.7	4.20	0.0	0.22	25	21
	0.60	23	2.75	−1.8	4.20	0.2	0.20	26	21
	0.50	42	2.99	−1.9	4.19	0.4	0.17	26	22
	0.40	56	3.26	−2.0	4.18	0.7	0.15	27	22
	0.30	68	3.57	−2.0	4.16	0.9	0.12	28	24
	0.20	78	3.91	−2.1	4.15	1.2	0.10	28	26
	0.10	86	4.27	−2.1	4.13	1.5	0.08	30	30
7.08	0.70	0	3.3	−3.5	4.32	0.0	0.27	28	12
	0.60	9	3.6	−3.6	4.32	0.2	0.24	29	12
	0.50	16	3.9	−3.7	4.31	0.4	0.21	29	12
	0.40	21	4.3	−3.8	4.30	0.7	0.18	30	12
	0.30	26	4.7	−3.9	4.29	1.0	0.16	31	13
	0.20	29	5.2	−3.9	4.27	1.3	0.13	32	14
	0.10	32	5.8	−4.0	4.26	1.5	0.11	33	16

† Adapted from B. Strömgren, Stellar Models for Main-sequence Stars and Subdwarfs, in L. H. Aller and D. B. McLaughlin (eds.), "Stellar Structure." By permission of The University of Chicago Press. Copyright 1965 by The University of Chicago. The initial composition is $X = 0.70$, $Y = 0.27$, $Z = 0.03$.

The physical properties are listed in Tables 6-5 and 6-6 as a function of the mass coordinate.

Problem 6-15: Calculate the change of the effective surface temperature T_e of the sun over the period 4.5×10^9 years.

For lower-main-sequence stars there is interest in a considerable range of compositions. The heavy-element concentration ranges between $Z = 0.04$ in young population I to $Z = 0.001$, or perhaps less, in extreme population II. The mass range $0.6 < \mathfrak{M}/\mathfrak{M}_\odot < 1.3$ is of particular interest to population II composition, because this is the mass range of the globular-cluster main sequences.

Table 6-5 Zero-age model of the sun†

$M(r)$, solar masses	r, 10^{11} cm	T, 10^6 °K	ρ, g/cm^3	$L(r)$, 10^{33} ergs/sec	ϵ, ergs g^{-1} sec^{-1}	κ, cm^2/g
0.0	0.00	13.7	90	0.00	13.9	1.38
0.05	0.07	12.3	74	0.95	7.2	1.64
0.1	0.09	11.6	65	1.54	4.8	1.82
0.2	0.11	10.4	51	2.20	2.3	2.16
0.3	0.14	9.4	40	2.53	1.1	2.50
0.4	0.16	8.5	30.5	2.68	0.5	2.87
0.5	0.18	7.6	22.4	2.75	0.2	3.3
0.6	0.20	6.8	15.7	2.77	0.04	3.8
0.7	0.23	5.9	10.0	2.78	0.01	4.4
0.8	0.26	5.0	5.5	2.78	0.00	5.2
0.9	0.32	3.8	2.09	2.78	0.00	7.0
0.95	0.37	3.0	0.87	2.78	0.00	8.6
0.99	0.46	1.73	0.142	2.78	0.00	11.1
0.99968	0.60	0.62	0.0057	2.78	0.00	Conv.
1.0	0.659			2.78		

† B. Strömgren, Stellar Models for Main-sequence Stars and Subdwarfs, in L. H. Aller and D. B. McLaughlin (eds.), "Stellar Structure." By permission of The University of Chicago Press. Copyright 1965 by The University of Chicago.

Table 6-6 Model of the sun at 4.5×10^9 years†

$M(r)$, solar masses	r, 10^{11} cm	T, 10^6 °K	ρ, g/cm^3	$L(r)$, 10^{33} ergs/sec	ϵ, ergs g^{-1} sec^{-1}	κ, cm^2/g	X_H
0.0	0.00	15.7	158	0.00	17.5	1.09	0.36
0.05	0.06	13.8	103	1.30	10.0	1.32	0.52
0.1	0.08	12.8	83	2.13	6.8	1.48	0.58
0.2	0.10	11.3	59	3.09	3.3	1.78	0.65
0.3	0.13	10.1	43	3.55	1.6	2.09	0.68
0.4	0.15	9.0	31.5	3.77	0.7	2.42	0.69
0.5	0.17	8.1	22.4	3.86	0.3	2.79	0.70
0.6	0.20	7.1	15.2	3.90	0.06	3.2	0.70
0.7	0.23	6.2	9.4	3.90	0.02	3.8	0.71
0.8	0.26	5.1	5.0	3.90	0.00	4.5	0.71
0.9	0.32	3.9	1.84	3.90	0.00	6.0	0.71
0.95	0.38	3.0	0.74	3.90	0.00	7.4	0.71
0.99	0.48	1.73	0.117	3.90	0.00	9.6	0.71
0.99955	0.62	0.66	0.0063	3.90	0.00	Conv.	0.71
1.0	0.694			3.90			0.71

† B. Strömgren, Stellar Models for Main-sequence Stars and Subdwarfs, in L. H. Aller and D. B. McLaughlin (eds.), "Stellar Structure." By permission of The University of Chicago Press. Copyright 1965 by The University of Chicago. Initial composition $X = 0.71$, $Y = 0.27$.

Since the globular clusters are believed to be the oldest objects in the galaxy, considerable effort has been expended in an attempt to compute their age by matching their color-magnitude diagrams to those calculated for stars in this mass range. A star of low heavy-element content has a greater luminosity than its population I counterpart of equal mass because its interior opacity is reduced. It is also bluer, and the zero-age main sequence for extreme population II lies about half a magnitude below that of population I. But because the population II stars have a greater fuel supply per unit mass, their main-sequence lifetimes are longer than their population I counterparts of equal luminosity.

For sufficiently low mass, the central temperature of lower-main-sequence stars will not become great enough to cause the hydrogen to burn at a rapid enough rate to stop the contraction. Kumar[1] has calculated that for $\mathfrak{M} < 0.07\mathfrak{M}_\odot$ the contraction proceeds until it is stopped by electron degeneracy, and the star cools to invisibility.

6-7 ADVANCED STELLAR EVOLUTION

The preceding sections illustrate the fact that stellar evolution is continuous. Nuclear reactions begin in the interior before the star has completed the shrinkage of its radius to its minimum value on the zero-age main sequence. This process merges continuously into evolution within the main-sequence band. This evolution is characterized by the depletion of central hydrogen accompanied by slow core contraction and heating while the radius begins to slowly expand from the zero-age value. During the depletion of the central hydrogen the star's position in the color-magnitude diagram moves somewhat upward. The overriding features of this evolution, as well as those of subsequent stages, are embodied in the principles of hydrostatic equilibrium and the virial theorem. When light nuclei are fused into heavier ones, the reduction of the mean molecular weight leads to a pressure deficiency. This deficiency in turn allows (or should we say *demands?*) the contraction of the core under the excessive weight of the overlying layers. The physics of the virial theorem implies that the gravitational contraction will be accompanied by a rise in central temperature, and in this way hydrostatic equilibrium can be reestablished. The rising central temperature tends to steepen the temperature gradient, which usually implies that energy will flow out at a faster rate than it is being provided. To maintain the energy balance, the outer layers expand to reduce the temperature gradient once again to a level consistent with the rate of energy production. All of stellar evolution is dominated by similar physical principles; only the details of the energy generation, the opacity, the equation of state, etc., change. Gravity continually sends its incessant order to contract, an order that can be delayed and compromised for varying lengths of time but not ignored. So far as is known, the contraction can come to an absolute halt only if the star can be stabilized as a ball of degenerate matter that has ceased to radiate (a black white-dwarf star or neutron star).

[1] S. S. Kumar, *Astrophys. J.*, **137**:1121 (1963).

With the application of these simple principles (and a large amount of hindsight!), it is possible to give a rough verbal description of the sequence of events subsequent to the exhaustion of hydrogen in the core of a star. The core continues to contract as the hydrogen is exhausted, leaving a central region of helium plus heavier trace elements. This helium core will tend to be isothermal because nuclear energy generation has ceased to send out energy, although a relatively small temperature gradient will be required to transport the energy released by continued contraction of the core. Because of the continuity of temperature and the increasing value of the core temperature, the temperature of a shell of hydrogen surrounding the core will be elevated to temperatures sufficient to liberate significant amounts of energy. The increased internal temperatures require the expansion of the stellar radius to keep the temperature gradient at a consistently low level. The star therefore reddens at a relatively rapid rate while the hydrogen-burning shell slowly increases the mass of the helium core. As the outer layers cool as a result of the great increase in radius, the luminosity may decrease as a result of both a decreasing rate of energy production in the hydrogen shell source and an increased radiative opacity in the portions of the star outside the shell source.

This trend is terminated by the following development. When the surface becomes cool enough for the ionized metals to recapture electrons and become neutral, the surface opacity drops. It does so because the main source of opacity is the H^- ion, which cannot be easily formed without free electrons. The smaller surface opacity allows energy to be radiated at a faster rate. To supply this energy the high-opacity subphotospheric layers become convective, thereby drawing energy directly from the low-opacity interior. The radius continues to increase with increasing luminosity, so that the outer convective zone becomes deeper and deeper as the core continues its contraction. During this phase the star moves upward on a nearly vertical path in the H-R diagram that bears many similarities to the Hayashi track that was descended during the original contraction phase.

Eventually the center becomes hot enough for the helium gas to begin fusing to C^{12}, whereupon the core becomes convective once more. For structural balance the surface contracts again as the star enters a phase characterized by a helium central energy source and a hydrogen shell source. In relatively low-mass stars the history is somewhat modified. Because of a larger central density, the helium core is supported by a degenerate electron gas at the time the 3α reaction begins at a significant rate. Because of the weak temperature dependence of degeneracy pressure (see Chap. 2), a thermal runaway occurs, the so-called *helium flash*. The manner in which the star reorganizes itself following this flash is a major unsolved problem in stellar evolution, particularly so because of its importance to the understanding of the globular-cluster H-R diagrams.

In subsequent stages of evolution, the same types of events occur. When the helium is exhausted, the core contracts until the temperature has risen to the value required for the $C^{12} + C^{12}$ reactions to begin. Such a star may then have

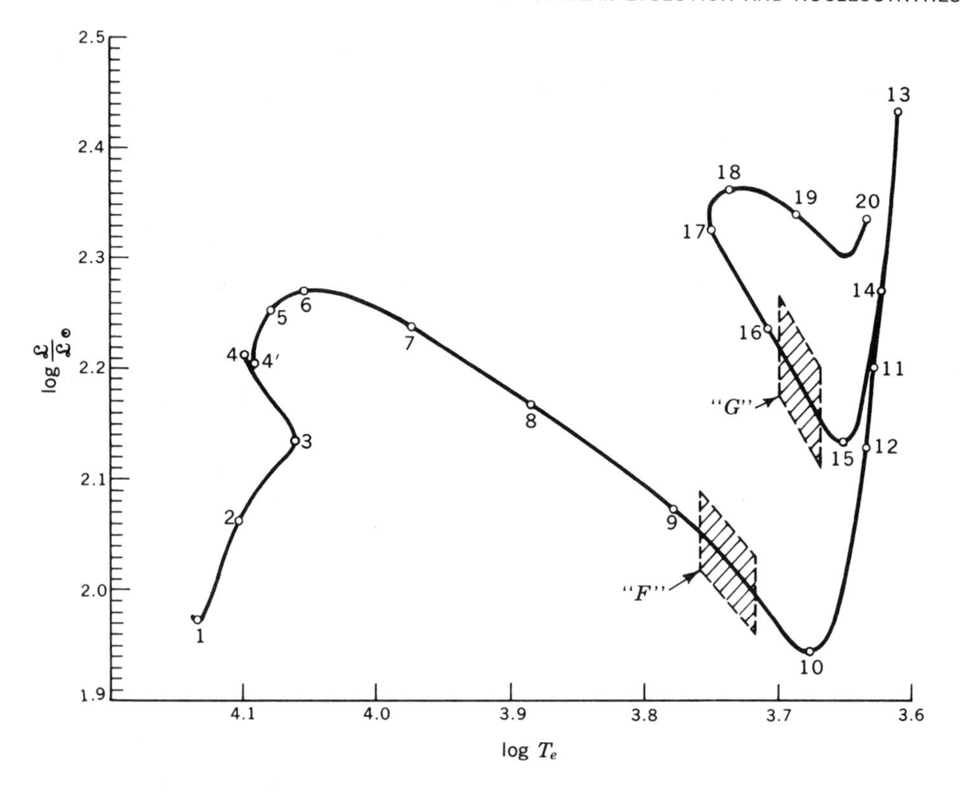

Fig. 6-13 The evolutionary track of a star of three solar masses in the H-R diagram. The time required to reach the enumerated points is given in Table 6-7. [*After I. Iben, Jr., Astrophys. J.*, **142**:1447 (1965). *By permission of The University of Chicago Press. Copyright* 1965 *by The University of Chicago.*]

a helium shell source and perhaps even a third energy source in a hydrogen shell. The initial burning of C^{12} also often occurs in a degenerate electron gas, so that the carbon flashes too. The next stage, oxygen burning, is the last one capable of providing adequate power to delay further collapse for any significant length of time. By this time the neutrino losses have become important.

Let us now illustrate some of these developments in a more quantitative way.[1] Figure 6-13 shows the evolution of a population I star of three solar masses in the

[1] The following material was taken primarily from the published papers of I. Iben, Jr., who has conducted very detailed examinations of stellar evolution. The student should also read the papers of Hayashi, Henyey, Schwarzschild, Kippenhahn, and their associates, each of whom has made major contributions to the modern science of stellar evolution. In this literature the student will find references to the early exposition of the concepts of stellar evolution. Many of the older papers are very illuminating insofar as the conceptual framework of stellar evolution is concerned.

H-R diagram. Several key points along the evolutionary track are labeled by numbers. The time required for the model to evolve to each point is listed in Table 6-7. The model points listed represent only a small fraction of the 560 models constructed in this sequence. These evolution times are not in themselves accurate to six significant digits for a real star of three solar masses, because the real star may do things that have not been included in the model, e.g., rotate, mix by circulation, or lose mass. The *difference in time* between two models is probably accurate to the degree shown, however, unless the star's structure is for some unknown reason different from that of the model at the corresponding point. The way in which many quantities vary in time is illustrated in Figs. 6-14 and 6-15. The time abscissa on these figures may be correlated with the evolutionary track

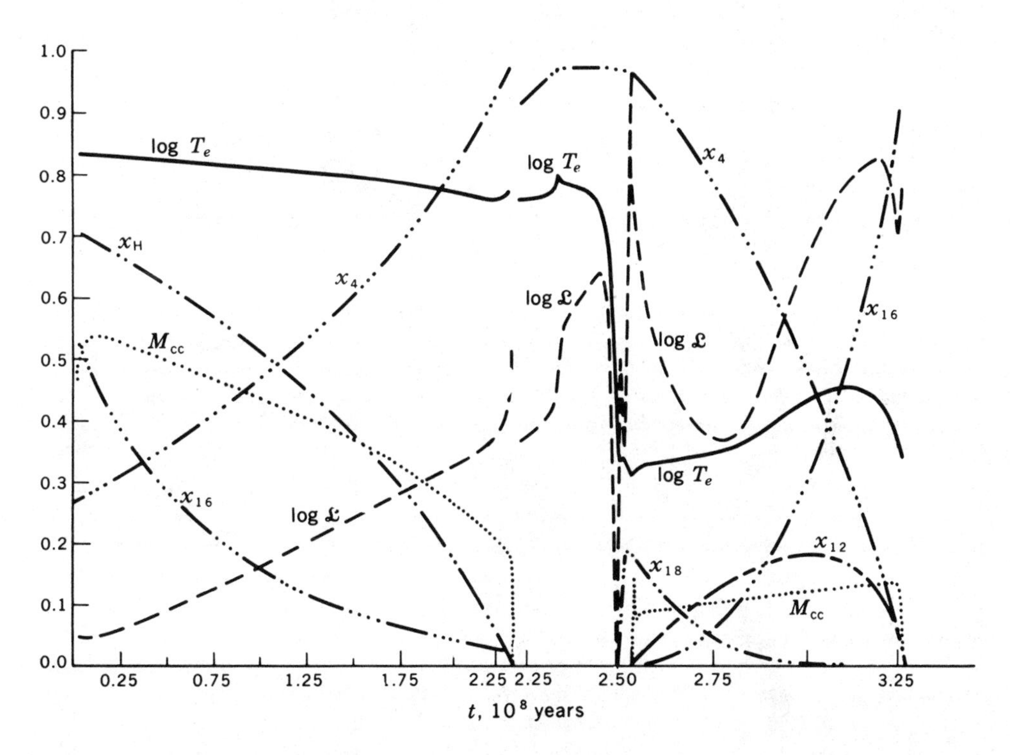

Fig. 6-14 The variation with time of the luminosity, the surface temperature, the mass fraction M_{cc} within the convective core, and the central mass fractions of H, He^4, C^{12}, O^{16}, and O^{18} during the evolution of a three-solar-mass star. The full-scale limits correspond to $2.45 > \log \mathfrak{L}/\mathfrak{L}_\odot > 1.95$, $4.3 > \log T_e > 3.3$, and $\frac{1}{3} > M_{cc} > 0$. The scale for the composition parameters changes at $t = 2.25 \times 10^8$ years. To the left of the break $0.02 > X_{16} > 0$ and $1.0 > x_H, X_4 > 0$, and to the right of the break $0.1 > X_{18} > 0$ and $1.0 > X_4, X_{12}, X_{16} > 0$. [*After I. Iben, Jr., Astrophys. J.*, **142**:1447 (1965). *By permission of The University of Chicago Press. Copyright* 1965 *by The University of Chicago.*]

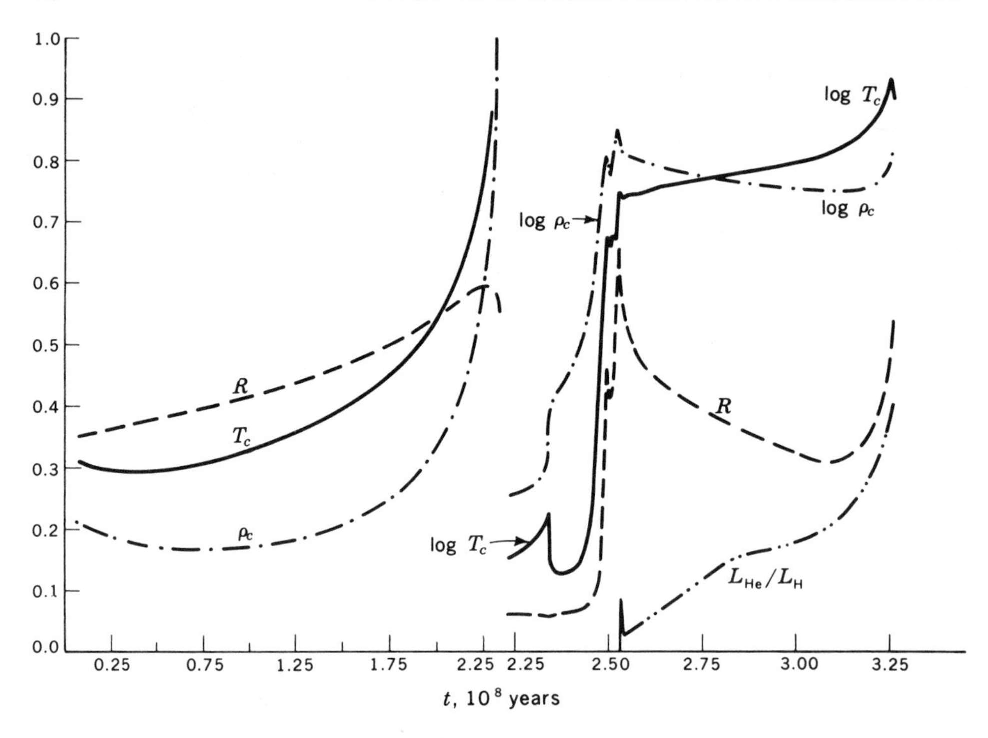

Fig. 6-15 The variation with time of the radius, the central density, the central temperature, and the ratio of the helium-burning power to the hydrogen-burning power during the evolution of a three-solar-mass star. To the left of the break at $t = 2.25 \times 10^8$ years the full-scale limits correspond to $5 > R/R_\odot > 0$, $31 > T_c/10^6 > 21$, and $80 > \rho_c > 30$. To the right of the break the full-scale limits correspond to $50 > R/R_\odot > 0$, $2.3 > \log T_c/10^6 > 1.3$, $5.5 > \log \rho_c > 0.5$, and $1.0 > L_{He}/L_H > 0$. [*After I. Iben, Jr., Astrophys. J.,* **142**:1447 (1965). *By permission of The University of Chicago Press. Copyright* 1965 *by The University of Chicago.*]

Table 6-7 Evolutionary lifetime for $3\mathfrak{M}_\odot$†

Point	*t*, 10^8 *years*	*Point*	*t*, 10^8 *years*	*Point*	*t*, 10^8 *years*
1	0.024586	7	2.47004	14	2.55850
2	1.38921	8	2.47865	15	2.78295
3	2.23669	9	2.48429	16	2.94233
4	2.34089	10	2.48925	17	3.06968
4′	2.34222	11	2.49817	18	3.19043
5	2.40119	12	2.50728	19	3.23566
6	2.44420	13	2.53163	20	3.26323

† I. Iben, Jr., *Astrophys. J.*, **142**:1447 (1965). By permission of The University of Chicago Press. Copyright 1965 by The University of Chicago.

by the use of Table 6-7. A thumbnail sketch describing the evolution between characteristic points follows:

(1–3) The track between points 1 and 3 corresponds to the main-sequence depletion of hydrogen and the reduction in size of the convective core, as illustrated in Fig. 6-14. The time required to evolve through this portion is about two-thirds of the total time represented on the entire evolutionary track. The long time spent by stars near the main sequence accounts for the fact that most observed stars are in their main-sequence phases.

(3–4) The star contracts for about 10^7 years, during which time the mass fraction in the convective core decreases rapidly.

(4′–6) A thick hydrogen-burning shell is formed near 4′. At the same time the convective core disappears, and the central hydrogen vanishes. The core contracts very rapidly and becomes nearly isothermal. The main phase of hydrogen burning in the shell lasts about 10^7 years and ends at point 6.

(6–10) Between 6 and 10 the core contracts, and the central temperature rises rapidly. The envelope expands rapidly while hydrogen burning continues in an intermediate shell. This phase requires about 4.5×10^6 years. The luminosity falls during this phase because the hydrogen shell source becomes less and less able to provide energy as a result of overall structural changes in the star. The expanding and cooling envelope is also absorbing a great deal of energy. As point 10 is approached by the $3\mathfrak{M}_\odot$ star, the contraction of the core liberates energy at the rate $L_{\text{core}} \approx 7\mathfrak{L}_\odot$, the hydrogen-burning shell contributes $L_{\text{shell}} \approx 122\mathfrak{L}_\odot$, and the expanding envelope absorbs $L_{\text{env}} \approx 42\mathfrak{L}_\odot$.

(10–13) Near point 10 the surface has become so cool that a deep outer convection zone appears. The star ascends a nearly vertical track as the core contracts, the radius expands, and the mass contained in the outer convective core increases roughly in proportion to log $\mathfrak{L}$. This process is halted temporarily at point 11, where the N^{14} at the center begins to burn rapidly by radiative capture of an alpha particle; $N^{14}(\alpha,\gamma)F^{18}$, followed by $F^{18}(\beta^+\nu)O^{18}$. The N^{14} is reasonably abundant at this stage, because almost all the original CNO nuclei were converted to N^{14} during the main-sequence phase. This burning produces a short core expansion, during which time the luminosity regresses to point 12. When the N^{14} has been converted to O^{18}, the central energy source disappears, and the core contracts again. The luminosity and radius then continue their increase to point 13, where the 3α reaction halts the contraction of the core.

(14–18) This portion of the track is characterized by helium burning in the core and hydrogen burning in a shell. This phase lasts about 7.3×10^7 years for the $3\mathfrak{M}_\odot$ star, which is about one-quarter of the total lifetime of the star. The onset of energy production in the core as the temperature there reaches 10^8 °K leads to expansion of the core, which becomes convective. Simultaneous with the core expansion, the luminosity falls because the energy production in the hydrogen-burning shell is decreased and the surface contracts, becoming bluer as it does so. During helium burning, the hydrogen shell remains the major source of power, producing as it does about 6 to 8 times as much energy as the core.

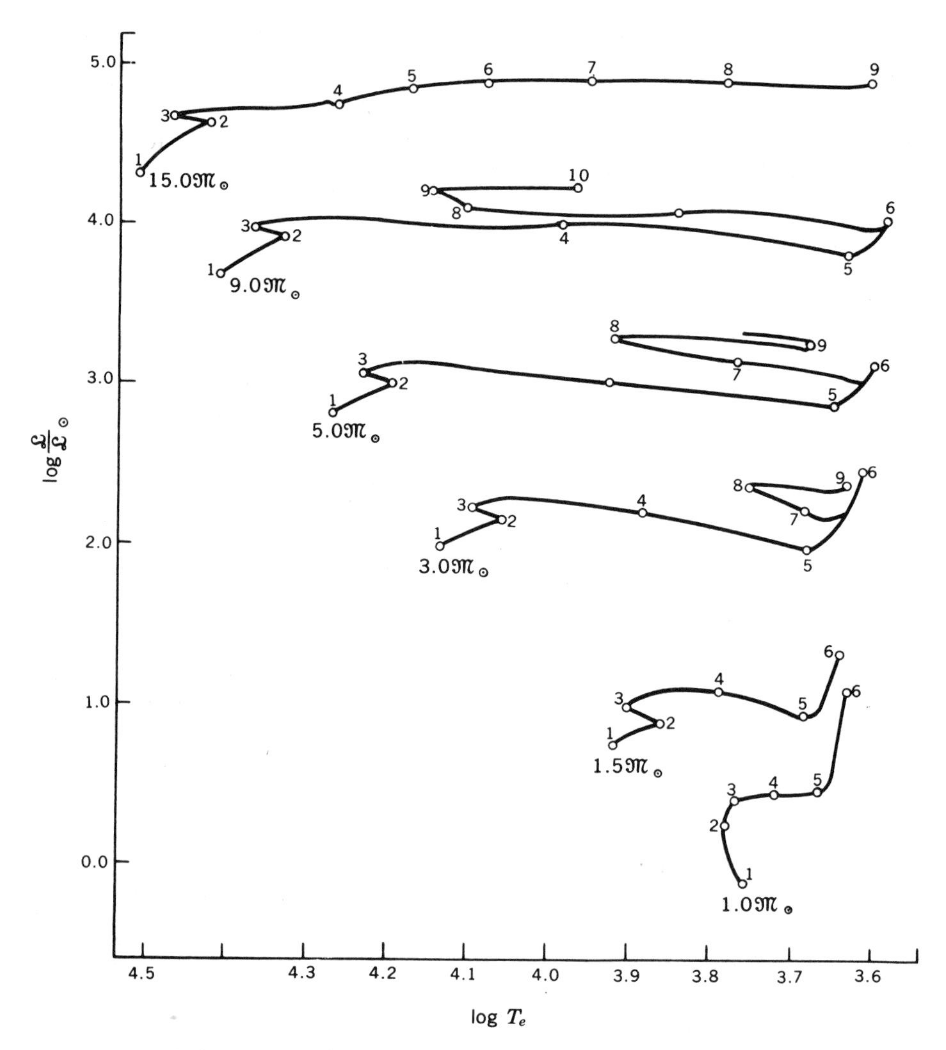

Fig. 6-16 Evolutionary paths in the H-R diagram for population I stars of mass $\mathfrak{M}/\mathfrak{M}_\odot$ = 1.0, 1.5, 3, 5, 9, and 15. The initial point is on the zero-age main sequence. The ages of the stars at the enumerated points are listed in Table 6-8. [*After I. Iben, Jr., Astrophys. J.*, **140**:1631 (1964). *By permission of The University of Chicago Press. Copyright* 1964 *by The University of Chicago.*]

(18–20) This portion of the track represents the exhaustion of helium in the core. When the helium abundance has been reduced to small values, the reaction $C^{12}(\alpha,\gamma)O^{16}$ becomes more important than the 3α reaction. The physical principles governing the exhaustion of helium are quite similar to those of the earlier exhaustion of hydrogen. The core contracts and heats up the surrounding layers until helium begins to burn in a shell around the helium-exhausted core. At point 20 the star has two major shell sources, an inner one of helium and an outer one of hydrogen, burning around a contracting core of C^{12}, O^{16}, and heavier trace elements. In the future life of this star, the C^{12} core will eventually contract enough as it grows in mass for carbon burning to begin.

The evolution of this $3\mathfrak{M}_\odot$ star affords an illustration of the interplay between photospheric abundances and nucleosynthesis within the interior. Prior to the growth of the outer convection zone, the surface abundances of the elements Li, C, and N retain the values they had when the star was on the main sequence. When the star begins its vertical climb in the H-R diagram at point 10, the surface convection zone begins reaching further into the star. As the lower boundary of the convection zone moves inward past the point where lithium has been destroyed by interactions with protons ($T \sim 2 \times 10^6$ °K), the unburnt lithium is mixed convectively with matter in which the lithium has already been destroyed. The surface abundance of Li therefore drops. By the time the star has ascended to point 11, the convective zone has reached inward to the point where C^{12} has been converted to N^{14} by the CN cycle. Thereafter C^{12} is convected inward, and N^{14} is convected outward, with the result that the surface ratio of N^{14} to C^{12} begins to increase. When the star reaches the tip of the red-giant branch (point 13), the convective envelope contains the outer 82 percent of the star's mass. The abundance of lithium at the surface has been reduced by a factor of about 60, and the ratio N^{14}/C^{12} has been increased by somewhat more than a factor of 3. These effects may be observable. The situation is somewhat clouded by the likelihood of mass loss on the subgiant branch, however. Even so, the analogous effects can be computed if the rate of mass loss is known.

Paths of evolution in the H-R diagram for a spectrum of stellar masses of population I composition ($X = 0.708$, $Z = 0.02$) are shown in Fig. 6-16. The times required for the initial pre-main-sequence model to evolve to the enumerated points are listed in Table 6-8. On each path the main phase of core hydro-

Table 6-8 Evolutionary lifetimes, years†

	$\mathfrak{M}/\mathfrak{M}_\odot$					
Point	15.0	9.0	5.0	3.0	1.5	1.0
1	6.160×10^4	1.511×10^5	5.760×10^5	2.510×10^6	1.821×10^7	5.016×10^7
2	1.023×10^7	2.129×10^7	6.549×10^7	2.273×10^8	1.567×10^9	8.060×10^9
3	1.048×10^7	2.190×10^7	6.823×10^7	2.394×10^8	1.652×10^9	9.705×10^9
4	1.050×10^7	2.208×10^7	7.019×10^7	2.478×10^8	2.036×10^9	1.0236×10^{10}
5	1.149×10^7	2.213×10^7	7.035×10^7	2.488×10^8	2.105×10^9	1.0446×10^{10}
6	1.196×10^7	2.214×10^7	7.084×10^7	2.531×10^8	2.263×10^9	1.0875×10^{10}
7	1.210×10^7	2.273×10^7	7.844×10^7	2.887×10^8		
8	1.213×10^7	2.315×10^7	8.524×10^7	3.095×10^8		
9	1.214×10^7	2.574×10^7	8.782×10^7	3.262×10^8		
10		2.623×10^7				

† I. Iben, Jr., *Astrophys. J.*, **140**:1631 (1964). By permission of The University of Chicago Press. Copyright 1964 by The University of Chicago.

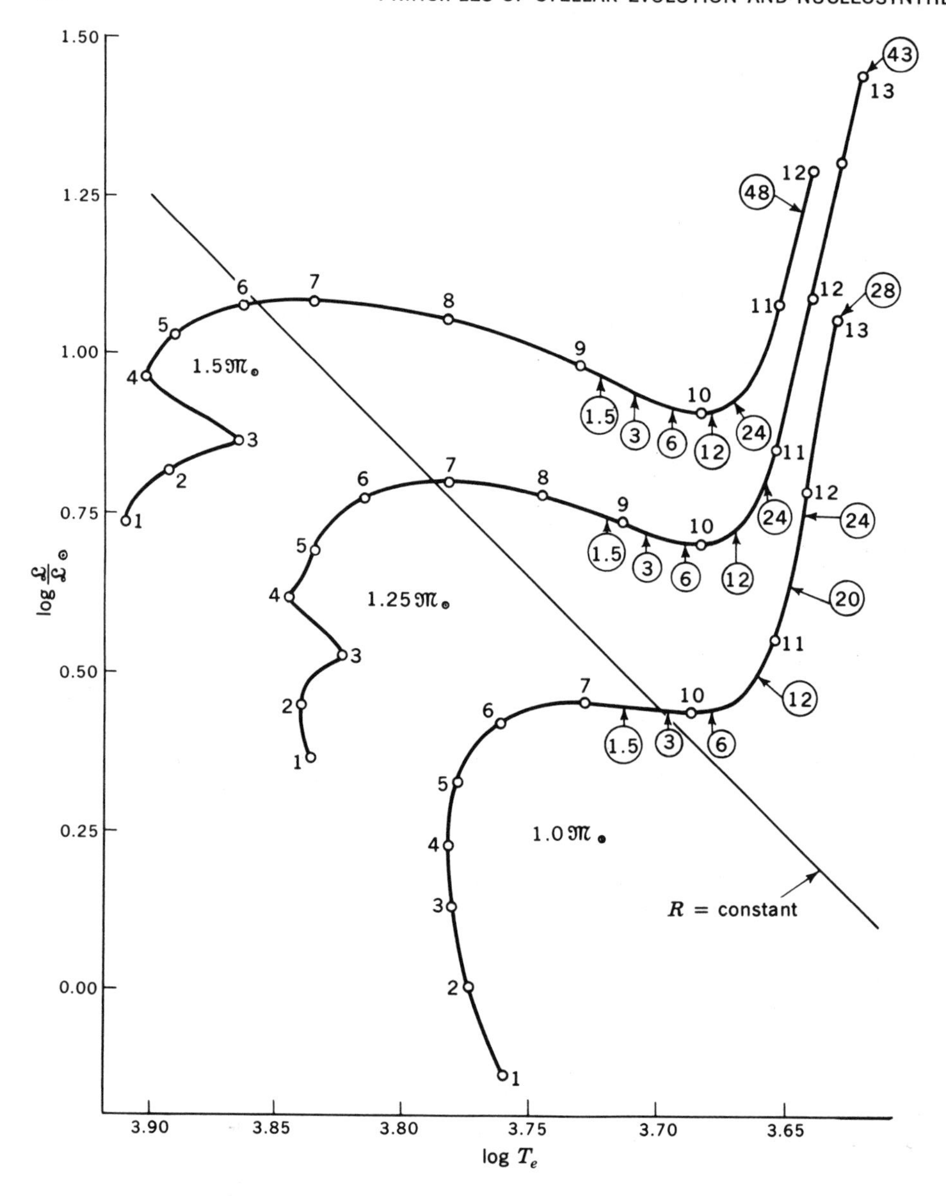

Fig. 6-17 Evolutionary track of lower-main-sequence population I stars of mass $\mathfrak{M}/\mathfrak{M}_\odot$ = 1.0, 1.25, and 1.5. The ages of the stars at the enumerated points along each track are listed in Table 6-9. The circled numbers along the tracks represent the factors by which the surface Li^7 abundance has been depleted by the deepening of the outer convection zone. A diagonal line of constant radius has been included for added physical insight. [*After I. Iben, Jr., Astrophys. J.*, **147**:624 (1967). *By permission of The University of Chicago Press. Copyright* 1967 *by The University of Chicago.*]

gen burning is represented by the segment joining points 1 and 2. The previously mentioned fact that *the main-sequence lifetime is a steeply decreasing function of stellar mass* is without doubt the single most important conclusion of the quantitative science of stellar evolution. This fact is the basic stepping stone into most discussions of galactic chronology. A clear understanding of these tracks and the time scales involved will enable the student to follow most of the semiempirical interpretations of the diagrams of clusters of young stars. It would be profitable at this point to return to the composite color-magnitude diagram for galactic clusters shown in Fig. 1-16. In the relatively young clusters, stars remain on the main sequence above zero magnitude. The red giants in such clusters have luminosities not greatly different from the most luminous main-sequence member, a fact that is explained naturally by the evolution of upper-main-sequence stars to the right as they age. In old clusters, however, the stars at the tip of the red-giant branch are considerably more luminous than those at the tip of the main sequence, a fact that can be interpreted naturally in terms of the shape of the $1\mathfrak{M}_\odot$ track.

For the stars $\mathfrak{M}/\mathfrak{M}_\odot = 3$, 5, and 9, helium burning in the core starts at the tip of the red-giant branch, point 6. In the $15\mathfrak{M}_\odot$ model the central temperatures are so great that helium burning occurs on the way to the red-giant tip. The models for $1\mathfrak{M}_\odot$ and $1.5\mathfrak{M}_\odot$ are terminated before helium burning begins, however, because of difficulties in computing the subsequent evolution.

More details of the tracks for low-mass stars of population I are shown in Fig. 6-17, and the corresponding evolution times are listed in Table 6-9. These tracks are identical to those in Fig. 6-16, but the scale has been expanded, and more ages along the track are explicitly noted. The phase of central hydrogen

Table 6.9 Evolutionary lifetimes (10^9 years)†

Point	$1.0\mathfrak{M}_\odot$	$1.25\mathfrak{M}_\odot$	$1.50\mathfrak{M}_\odot$
1	0.05060	0.02954	0.01821
2	3.8209	1.4220	1.0277
3	6.7100	2.8320	1.5710
4	8.1719	3.0144	1.6520
5	9.2012	3.5524	1.8261
6	9.9030	3.9213	1.9666
7	10.195	4.0597	2.0010
8		4.1204	2.0397
9		4.1593	2.0676
10	10.352	4.2060	2.1059
11	10.565	4.3427	2.1991
12	10.750	4.4505	2.2628
13	10.875	4.5349	

† I. Iben, Jr., *Astrophys. J.*, **147**:624 (1967). By permission of The University of Chicago Press. Copyright 1967 by The University of Chicago.

burning (points 1 to 3) shows interesting differences that reflect differences in the central structure. In all three stars energy generation from the PP chains dominates that from the CN cycle, but the CN cycle contributes sufficient central power for the cores of the $1.5\mathfrak{M}_\odot$ and $1.25\mathfrak{M}_\odot$ models to be convective, as discussed in Sec. 6-6, whereas the core of the $1\mathfrak{M}_\odot$ model is in radiative equilibrium. This difference accounts for the difference in the early evolution of these models. The $1\mathfrak{M}_\odot$ model initially evolves toward hotter surface temperatures, in spite of the fact that the radius is increasing, whereas the $1.5\mathfrak{M}_\odot$ and $1.25\mathfrak{M}_\odot$ models initially evolve toward the red as the central hydrogen is consumed. When the convective core disappears in the two more massive stars at point 3, the restructuring of the star causes the short kink toward bluer surfaces. Because the core is always radiative in the $1\mathfrak{M}_\odot$ model, however, this transition does not occur there, and the evolution proceeds smoothly. During the remainder of the tracks, the CN cycle contributes a progressively larger fraction of the energy generation in the hydrogen-burning shell source until, along the giant branches, the CN cycle dominates in all three models.

Another principle of importance to stellar evolution can be derived from the results embodied in Fig. 6-17; viz., *the ratio of the time spent by a star burning hydrogen in a shell around a hydrogen-exhausted core to the time spent in the core hydrogen-burning phase is small.* This fact is especially important near $1\mathfrak{M}_\odot$. It can be demonstrated from Table 6-9, which shows the ratio of times to be

$$\frac{t(4 \rightarrow 13)}{t(1 \rightarrow 4)} = 0.33 \qquad 1.00\mathfrak{M}_\odot$$

That is, for stars near $1\mathfrak{M}_\odot$, the time required to evolve through the subgiant branch is small compared to the time spent near the main sequence. Since the main-sequence lifetime is a strong function of mass, it follows that in a highly evolved cluster, the stars along the entire subgiant region are only slightly more massive than the most luminous members near the main sequence. In first approximation, then, the H-R diagram of the subgiant branch in an old cluster is very nearly equal to the evolutionary track of the most luminous member of the main sequence. This feature allows one to estimate the mass of the star involved by comparing the subgiant branch to the evolutionary tracks of single stars. Another important argument can be made if the number of giant stars in the cluster is large. Over the small range of masses represented by the giant branch the number of stars per unit mass interval should be nearly constant. As a result one expects the number of stars between two points on the track to be proportional to the time required for the star to evolve between the two points.

Problem 6-16: Consider a cluster whose subgiant stars are $1.25\mathfrak{M}_\odot$. What number ratios should one expect statistically for stars on the segments $2 \rightarrow 3$, $3 \rightarrow 4$, and $4 \rightarrow 5$?
Ans: $N(2 \rightarrow 3):N(3 \rightarrow 4):N(4 \rightarrow 5) = 7.8:1.0:2.9$.

These features are illustrated in Fig. 6-18, which represents a fit to the observed H-R diagrams of the old galactic clusters M 67 and NGC 188. These subgiant

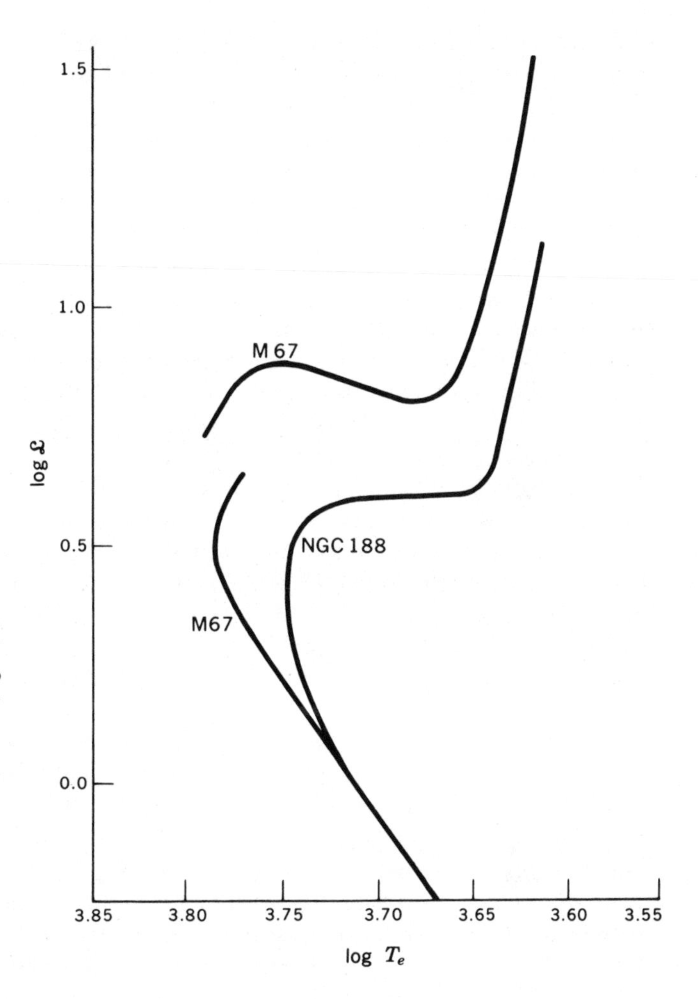

Fig. 6-18 A characterization of the observed H-R diagrams of two old galactic clusters. The ages of these clusters are estimated by the age of an ensemble of stellar models having the property that the locus of H-R positions of the individual stars within the ensemble, which differ with respect to mass only, best reproduces the observed diagram of the cluster. [*After I. Iben, Jr., Astrophys. J.*, **147**:624 (1967). *By permission of The University of Chicago Press. Copyright* 1967 *by The University of Chicago.*]

branches bear strong resemblance to the evolutionary tracks of $1.25\mathfrak{M}_\odot$ and $1.00\mathfrak{M}_\odot$ stars, respectively. The gap in the M 67 diagram corresponds to a paucity of observed stars, which may reflect the anticipated deficiency of the segment $3 \rightarrow 4$ as illustrated in Prob. 6-16. By attempting more exact matching, Iben has estimated the ages of these clusters to be:

M 67: $\tau = (5.5 \pm 1) \times 10^9$ years

NGC 188: $\tau = (11 \pm 2) \times 10^9$ years

Many details render analyses of this type uncertain, but the reader will be aware that estimating the ages of clusters is a considerably more sophisticated process than was indicated in the first chapter.

In contrast to the case of more massive stars, electron degeneracy is responsible for a major fraction of the pressure, and electron conduction is the dominant means of energy transport in the hydrogen-exhausted cores of all three stars of Fig. 6-17. The good conductivity results in an isothermal core which grows slowly over an extended period of hydrogen burning in an extremely thin shell around the core. The narrowness of this shell requires the use of small time steps in the calculation and makes the evolution hard to follow to the ignition of core helium. When the helium gets hot enough to burn, furthermore, it does so explosively. The high degree of degeneracy of the gas means that a rise in central temperature is at first not accompanied by a comparably rapid rise in central pressure. The temperature shoots up rapidly, causing the 3α reaction to proceed at ever greater rates. This runaway process is halted when the temperature has risen to the point where the electrons are no longer strongly degenerate. In this helium flash, as the event is called, the central temperature doubles, and the *instantaneous rate* of nuclear energy production reaches a very short-lived peak of about $10^{11}\mathcal{L}_\odot$. The time steps between successive models may have to be as short as seconds[1] to follow the flash! The large central power lasts only a short time and liberates only enough energy to lift the degeneracy of the electrons and to expand the core, whereupon the temperature falls again. The details of this crucial phase of low-mass evolution have been exceedingly difficult to follow, but it appears likely that the stellar structure simply readjusts to one appropriate to the burning of helium in a nondegenerate core plus the burning of hydrogen in a shell. The subsequent evolution is probably similar to that of more massive stars, for which the flash problem does not cloud the issue. The flash is a very interesting and important phenomenon and probably warrants further physical analysis. The tracks in Fig. 6-17 were terminated quite short of the flash which will occur when $\log (\mathcal{L}/\mathcal{L}_\odot) \approx 3.0$.

As the stars evolve into the giant region, the surface convection zone dips below the outer regions, where the primordial lithium has remained intact. The mixing to the surface of lithium-depleted material reduces the abundance of lithium on the surface. The circled numbers on Fig. 6-17 represent the factor by which the surface lithium has been reduced. They also represent, therefore, the growth in mass of the convective envelope.

This brief survey of the features of stellar evolution must be terminated at this point. The reader should be aware that only the general ideas of stellar evolution have been discussed. The particulars could easily fill an entire book. Some of the more important subjects that have not been discussed are (1) the evolution of extreme population II stars and the uncertainty in the helium abundance, (2) semiempirical studies of stellar evolution, (3) carbon- and oxygen-burning stars and the effect of neutrino emission on their time scales, (4) the presupernova star and supernova explosions, (5) white-dwarf structure and evolution, and (6) the problem of star formation. It is hoped that the principles outlined in this book will make these more detailed problems more accessible to the uninitiated.

[1] See R. Harm and M. Schwarzschild, *Astrophys. J.*, **145**:496 (1966), as well as their earlier papers on the subject cited therein.

The discussion has centered on nonrotating stars of constant mass. It seems quite clear, however, that the problems of mass loss and rotation, which may often be coupled to each other, must soon be injected in a natural way into studies of stellar evolution. This chapter will continue with a brief discussion of the major physical effects involved, and in the final section we consider the principles of pulsational stability.

6-8 ROTATION

Rotation can affect the evolution of a star in at least two very important ways. The most obvious consideration is that a new principle, the conservation of angular momentum, becomes relevant to any structural change. Second, we shall find that fluid circulation may be necessary to maintain the energy balance in the nonspherical star. In this section we shall introduce these ideas in the simplest possible way, not making any attempt to discuss the full complexity of this important stellar problem.

If a star is envisioned as rotating, it seems sensible that the centrifugal forces will render it nonspherical. Although we may still choose to think of an isotropic scalar pressure at each point in the interior, it no longer seems likely that the pressure gradient will be perfectly radial. The equation of hydrostatic equilibrium must be written so as to contain any additional forces:

$$\nabla P = -\rho \nabla \phi_G + \rho \mathbf{F} \tag{6-66}$$

where ϕ_G is the gravitational potential, which may be nonspherical, $\mathbf{F}$ is any additional force per unit mass acting on each mass element, and ∇P is the gradient of the pressure and is also nonradial. For the case in point, the additional force $\mathbf{F}$ will be regarded as a centrifugal force, although it could also represent forces due to magnetic fields or due to the tidal distortion by a companion star in a binary.

Although it is certainly not necessary to do so, the effect of rotation is most easily seen by imagining all parts of the star to be rotating at a uniform angular velocity Ω. Then the components of the centrifugal force on each mass element,

$$\begin{aligned} F_r &= \Omega^2 r \sin^2 \theta \\ F_\theta &= \Omega^2 r \sin \theta \cos \theta \\ F_\phi &= 0 \end{aligned} \tag{6-67}$$

can themselves be written as the gradient of the *centrifugal potential*

$$\phi_\Omega = -\tfrac{1}{2}\Omega^2 r^2 \sin^2 \theta \tag{6-68}$$

Then the hydrostatic equilibrium can be expressed as

$$\nabla P = -\rho \nabla \phi \tag{6-69}$$

where

$$\phi = \phi_G + \phi_\Omega \tag{6-70}$$

Concentrate for a moment on Eq. (6-69). Although the equipotentials are no longer spherical, the pressure gradient remains perpendicular to the surfaces of constant potential. The pressure is therefore constant on a given equipotential surface, so that the pressure at each point can be regarded as a function of the potential labeling the equipotential surface passing through the point: $P = P(\phi)$. Equation (6-69),

$$\frac{dP}{d\phi} = -\rho \tag{6-71}$$

implies that the density is also a function only of ϕ. It follows from the equation of state, then, that the equipotential surfaces are also surfaces of constant temperature.

Consider a region of a rotating star in radiative equilibrium, where the heat flux is

$$\begin{aligned}\mathbf{H} &= -\frac{4ac}{3}\frac{T^3}{\kappa\rho}\nabla T \\ &= -\frac{4ac}{3}\frac{T^3}{\kappa\rho}\frac{dT}{d\phi}\nabla\phi\end{aligned} \tag{6-72}$$

Inasmuch as all quantities are a function only of ϕ, the radiation flux must be proportional to the potential gradient, and, furthermore, *the absolute value of the heat flux along a potential surface is proportional to the magnitude of the potential gradient.* This last point is important, because in the nonspherical star, the potential gradient cannot have constant magnitude at all points on the equipotential surface. For a rotating star, the equipotential surfaces will be more widely spaced in the equatorial plane than at the corresponding values along the axis of rotation. Therefore the heat flux will be greater in the polar direction than along the equatorial planes. This asymmetry raises an important question, for although the star may adjust its temperature structure in such a way that the net heat flow out of an equipotential surface is equal to the integrated power generated within, it is not clear whether energy balance can be satisfied locally. For the sake of simplicity consider a region outside the central region of energy generation. If the thermodynamic state of that mass element is to be truly static, it must be true that

$$\nabla \cdot \mathbf{H} = 0 \tag{6-73}$$

From Eq. (6-72) and the knowledge that all functions of state are functions only of ϕ, we have, on the other hand,

$$\nabla \cdot \mathbf{H} = -\frac{4ac}{3}\frac{T^3}{\kappa\rho}\frac{dT}{d\phi}\nabla^2\phi - |\nabla\phi|^2 \frac{d}{d\phi}\left(\frac{4ac}{3}\frac{T^3}{\kappa\rho}\frac{dT}{d\phi}\right) \tag{6-74}$$

Problem 6-17: Confirm Eq. (6-74). Then evaluate the laplacian to show that its magnitude

$$\nabla^2\phi = 4\pi G\rho - 2\Omega^2 \tag{6-75}$$

is a function only of ϕ.

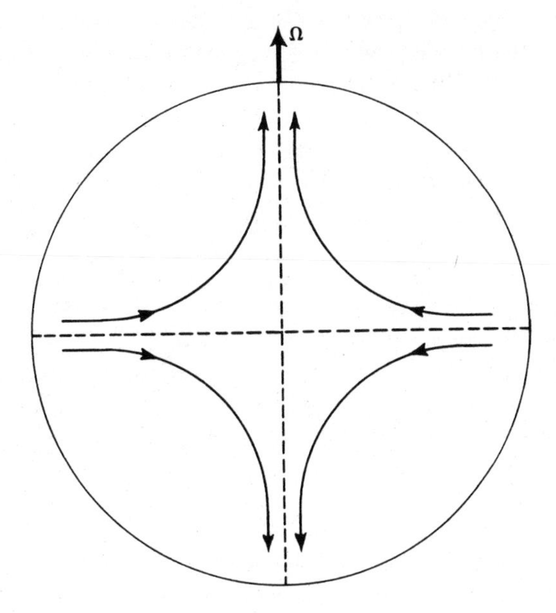

Fig. 6-19 A meridian plane cut through a rotating star. Circulation along the directions indicated is established to maintain the thermal balance of the star.

When Eq. (6-75) is reincorporated in (6-74), we see that all terms in $\nabla \cdot \mathbf{H}$ are functions only of ϕ with the exception of the factor $|\nabla\phi|$. But the value of $\nabla\phi$ depends also on θ, being greatest for a given value of ϕ on the axis of rotation and having a minimum in the equatorial plane. It follows that $\nabla \cdot \mathbf{H}$ cannot vanish everywhere on the equipotential.

Under such circumstances it may be that the best a star can do is to adjust its temperature structure so that the integrated heat flow through an equipotential surface is balanced by the energy generation from within. But in a given local region $\nabla \cdot \mathbf{H}$ will not vanish. As a result, the gas along the axis of rotation heats up and begins to rise, whereas gas elements in the equatorial plane lose heat energy and begin to fall. This pattern amounts to the establishment of fluid currents, as illustrated in Fig. 6-19. This circulation along meridian planes has the potentially important effect of mixing the composition of the gas. The speed of this circulation adjusts itself so that the heating by compression of a falling element is balancing at each point the cooling due to the divergence of the radiation flux. When Eddington first calculated the mixing speeds, he concluded that they were great enough to prevent the development of chemical inhomogeneity. Subsequent corrections by Sweet and by Mestel showed this conclusion to be a great overestimate, and the velocity of circulation in the sun was found to be of the order 10^{-9} cm/sec. Such a velocity requires about 10^{12} years to mix center and surface in the sun and therefore appears negligible. But it is by no means certain that special effects cannot occur in important thin layers of more rapidly rotating stars. (The sun is a slow rotator, compared to upper-main-sequence stars.) In particular one may wonder about the status of thin shells of nuclear burning and whether such circulations could bring lithium-depleted material to the lower bound of a surface convective zone. Although it seems likely that the chemical inhomogeneities upon which stellar evolution is based are secure, very few workers in the field believe that important applications of meridian currents do not exist anywhere.[1] Much modern research is directed toward finding a nonuniform $\Omega(\mathbf{r})$ which allows $\nabla \cdot \mathbf{H} = 0$, in which case circulation may not be required.

Because of the contribution of the centrifugal force, the internal pressure in a rotating star will be somewhat less than in an otherwise identical nonrotating star. The reduced interior temperature will liberate less thermonuclear power, so we may expect rotating stars to be slightly subluminous. For main-sequence stars constrained to rotate rigidly, the luminosity is reduced by about 10 percent.[2]

Another very important principle to be satisfied by isolated rotating bodies is the conservation of angular momentum. This principle must be considered during expansions and contractions of a star. There will be a tendency for a star to contract more easily along the axis of rotation than along the equatorial plane because of the centrifugal barrier for the latter case. The expansions may be very large. When a typical main-sequence star evolves to a giant, for example, the core may contract in radius by an order of magnitude, whereas the outer radius may expand by about the same factor. A relatively small amount of effort has been expended on the evolution of numerical models of rotating stars. The theoretical advances that have been made are only slowly digested by the field as a whole, but it may safely be said that stellar rotation is emerging as a problem of prime importance in stellar evolution.

Rotation also interacts with convection in such a way as to cause the inner portion of a star to rotate more rapidly than the outer portion. Because of the conservation of angular momentum, a falling convection cell will attempt to increase its angular velocity, whereas a rising cell will tend to decrease its angular velocity. The tendency of convection is to make the angular momentum per unit mass constant, which in turn demands greater rotational velocity in the central portion.

One of the most fascinating of the rotation problems is the apparent abrupt decrease in rotational velocity observed for main-sequence stars cooler than about type F4. Upper-main-sequence stars have long been observed to be fast rotators, whereas the dwarfs are not. One of the active areas in contemporary research is the attempt to relate these observations to the fact that the surface convection zone also begins near type F4 and deepens for cooler surfaces. The idea is that stellar winds may provide a sufficient drag on stellar surfaces to slow

[1] For a modern discussion of this problem the reader is referred to L. Mestel, Meridian Circulation in Stars, in L. H. Aller and D. B. McLaughlin (eds.), "Stellar Structure," The University of Chicago Press, Chicago, 1965. Very informative also are papers of I. Roxburgh, *Monthly Notices Roy. Astron. Soc.*, **126**:67; **128**:157, 237 (1964).

[2] J. Faulkner, I. W. Roxburgh, and P. A. Strittmatter, *Astrophys. J.*, **151**:203 (1968).

down the rotation of the surface convection zone while the radiative core continues to rotate rapidly underneath. In this case the angular momentum per unit mass of main-sequence stars would be a smoothly changing function of stellar mass rather than one having two apparently distinct branches. The question is one of cosmological importance, moreover, because Dicke's observations[1] of the solar oblateness may be consistent with a rapidly rotating solar core, in which case the argument that solar oblateness augments the precession of the perihelion of Mercury becomes more compelling.

6-9 MASS LOSS

Mass loss is a self-descriptive term that is used to describe any process by which the main body of the star, defined as the gravitationally bound mass, reduces its mass by ejecting surface layers. It is obvious that mass loss in sufficient amounts can have a large effect on evolutionary calculations. Mass loss can occur in a variety of forms and can be initiated by a variety of physical mechanisms. Any catastrophic event in which a massive outer layer is lifted off into space by some internal instability must result in a drastically new structure for the remaining core. So special are these circumstances that they will not be discussed here. The frequency of such events and whether they occur at all has not been well established.

It is not unlikely, however, that all luminous stars lose mass at some *small continuous rate* as a result of the mechanical heating of the most tenuous outer layers by the dissipation of sonic, hydromagnetic, and gravity waves.[2] For steady mass loss of this type the residual stellar structure changes quasistatically, and the modification of its evolution can be calculated if the details of the mass loss are known. Such slow rates of mass loss, which can sensibly be called *stellar winds* to contrast them with discontinuous and disruptive mass loss, can conceivably change the evolution in at least three major respects.

(*1*) If the mass of surface layer lost over the total evolutionary lifetime is a nonnegligible fraction of the total mass, the chemical composition of the evolved surface may reveal nuclear products that would not be expected on the surface of a star of constant mass.

(*2*) The rotational velocity of the surface may be reduced if the mass lost is able to carry away more angular momentum per unit mass than it had on the photosphere. A mechanism for accomplishing this involves the magnetic field. If the field is strong enough near the photosphere, the tenuous plasma corona will *corotate* out to some distance from which it will drift away freely. The requirement of corotation out to some distance from the surface provides a drag on the

[1] R. H. Dicke and H. M. Goldenberg, *Phys. Rev. Letters*, **18**:313 (1967); for a readable account of the cosmological relevance read R. H. Dicke, *Phys. Today*, **20**(1):55 (1967).

[2] For an illuminating discussion of the best-known case, the sun, the reader is referred to a review by A. J. Dessler, *Rev. Geophys.*, **5**:1 (1967), and to references listed therein.

stellar angular momentum. It is in principle possible that the angular momentum per unit mass of an evolving star may decrease by this mechanism.

Problem 6-18: Suppose that a spherical rotating star of radius R loses mass in a spherically symmetric manner in such a way that the escaping gas corotates, i.e., maintains the same angular velocity as the surface, out to some distance R', whereafter it escapes to infinity as a free body. Calculate the ratio of the average angular momentum per unit mass lost to infinity to the angular momentum per unit mass on the surface.

(*3*) If a star radiates mass quasistatically, the age of the star will be less than its age computed with the assumption of constant mass. This result follows from the fact that the core of the star must necessarily have provided power at a greater rate in the past than in the corresponding star of constant mass. The extra power demands result from the work required to deposit the ejected matter at infinity and from the greater rate of nuclear burning that will have existed in the earlier more massive progenitor of the final star.

The power required to radiate mass was described in Chap. 1 by the term *mass-loss luminosity*. It is given by

$$\mathcal{L}_{\dot{\mathfrak{M}}} = \left[G \frac{\mathfrak{M}}{R} + (U_\infty - U_e) + \tfrac{1}{2}v_\infty{}^2 \right] \dot{\mathfrak{M}} \tag{6-76}$$

where $U_\infty - U_e$ represents the amount by which the asymptotic internal energy per unit mass of the ejected gas exceeds its corresponding value at the photosphere, v_∞ is the translational velocity of the ejected matter at large distances from the surface, and $\dot{\mathfrak{M}}$ is the rate of mass loss. In the present sun this power is only equal to $2 \times 10^{-6}\mathcal{L}_\odot$ and can be neglected. Observed rates of mass loss in some other stars exceed that of the sun by as much as the factor 10^7, however, so that the possibility of significant power requirements should not be overlooked. Unfortunately the observation of mass loss is difficult because of the high degree of ionization relative to the photosphere, and so the present information is relatively meager. It is to be hoped that ultraviolet observations from space will clarify the extent of mass loss in stars of differing types.

Problem 6-19: Consider a white dwarf of one solar mass with optical luminosity $\mathcal{L} = \mathcal{L}_\odot$ and radius $R = 10^{-2}R_\odot$ that is losing mass with an asymptotic velocity no greater than the solar case. At what rate must the star be losing mass in order for there to be an error of 1 magnitude in the cooling rate?

Because the luminosity of a main-sequence star varies roughly as the fourth power of the mass, the past power expenditures of a star that has radiated a nonnegligible mass will have been greater than that of the corresponding star of constant mass. In such a case the main-sequence lifetime of the observed star will be less than would have been computed with the assumption of constant mass. Consider, for example, the evolution of a lower-main-sequence star with and

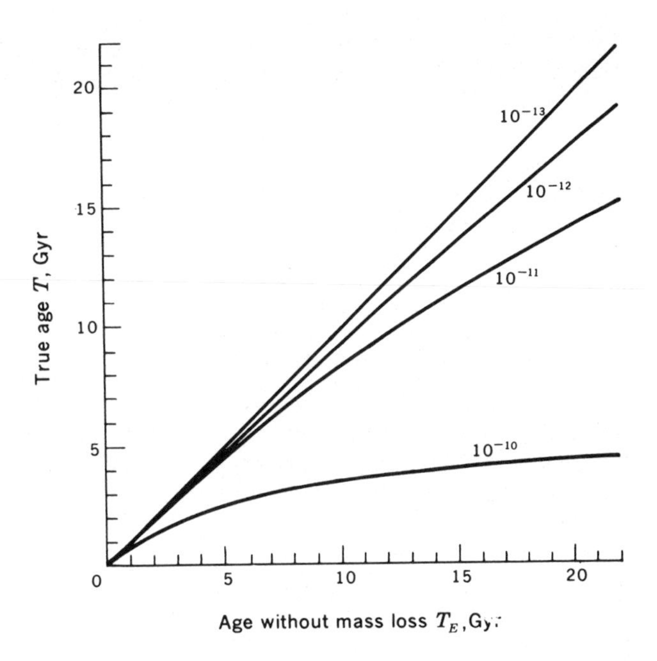

Fig. 6-20 The comparison of the true age T of an old star cluster with the evolutionary time T_E computed without mass loss. The curves are labeled with the rate of mass loss, which was taken to be constant, in units of $\mathfrak{M}_\odot$ per year. [*After D. D. Clayton, Astrophys. J.*, **140**:1604 (1964). *By permission of The University of Chicago Press. Copyright* 1964 *by The University of Chicago.*]

without mass loss. Assume that the internal structure of the two stars is essentially identical and compare the stars of *equal final mass* at a time in their evolution where they have converted equal amounts of hydrogen to helium. That is, consider identical evolved main-sequence stars but assume that one of them has evolved for a time T at constant mass whereas the other has evolved for a time T' by losing mass from an initial model of greater mass. If the evolved structures are identical, they will have transformed equal amounts of hydrogen into helium in their respective cores. Assuming for the sake of example that $\mathcal{L} = \mathfrak{M}^4$ exactly, the respective evolution times will be given by equating the time-integrated luminosities:

$$\int_0^{T'} [\mathfrak{M}(t)]^4 \, dt = \mathfrak{M}_f{}^4 T \tag{6-77}$$

where $\mathfrak{M}(T') = \mathfrak{M}_f$, the final mass of the stars compared. Because $\mathfrak{M}(t)$ has decreased from its initial value to the final value $\mathfrak{M}_f$, the true age T' will be less than the age T computed by assuming a constant mass. The rates of mass loss required to affect the computed age of globular clusters in this way are much less than the rates required to achieve a significant mass-loss luminosity. Figure 6-20 shows the comparative ages of globular clusters if it is assumed that the rate of mass loss was constant. Since the oldest globular clusters may be as old as 20×10^9 years, mass-loss rates of about $10^{-11}\mathfrak{M}_\odot$/year, or about 100 times as great as the solar rate, would be required to alter the computed ages. Mass-loss rates of this size would be very difficult to detect in globular clusters.

Problem 6-20: Consider two lower-main-sequence stars that evolved to *identical final states by losing equal amounts of mass*. Suppose, though, that the stars did not both lose mass at a *constant* rate during their lifetime. For definiteness assume that the first star lost mass faster in its early years than near the evolved final state, whereas the second star lost mass slowly at first and faster as it neared to time of comparison. If the stars are identical at the time of comparison, which star is older?
Ans: $T_1 > T_2$.

In summary it may be said that mass loss can seriously affect the course of stellar evolution in a variety of ways, but the observational evidence of its importance has not been obtained.

6-10 PULSATION

The phenomenon of pulsating, or variable, stars is one of the most intriguing in the heavens. It is one of the few cases in which man can watch a type of stellar evolution in real time, *evolution* in the sense that the observable properties change. Practically every observable feature of the regular variables, e.g., the Cepheids and RR Lyrae variables, undergoes a periodic (but not sinusoidal) variation.[1] Such variations do not represent nuclear evolution, and in fact, the time average of the observable is generally constant. These average quantities are a function of mass, composition, and state of nuclear evolution, and so they may change on time scales appropriate to the nuclear evolution of the core of the pulsating star. But the attempt to understand the pulsation of stars provides a fundamental test of the theories of stellar structure and evolution. For the purposes of this book, the physical ideas behind pulsation will illustrate significant *dynamic* principles that have been ignored in the previous discussions.

The first serious explanation of pulsation was the hypothesis that the phenomenon represented the free radial pulsations, much like a harmonic oscillator, in which the spring constant corresponds to the adiabatic compressibility of the gas. Eddington examined the theory of adiabatic radial oscillations and succeeded in showing that they must quickly die out as a result of dissipation in the gas. An early suggestion that variables represented partially eclipsing binaries was easily discounted by mounting observational evidence. By 1930 it was clear, thanks largely to the work of Eddington, that a pulsating star must in fact be some type of heat engine, in which some continuously operating mechanism transforms thermal energy into the mechanical energy of the oscillation.

Imagine that an element of gas within the star can be isolated and studied. Its thermodynamic state will be altered in the pulsation, and the first law of thermodynamics requires the heat absorbed in a small change to be equal to the sum of the rise in internal energy and the work done *by the gas* element on the

[1] For a review of the properties of variable stars, see P. Ledoux and T. Walraven, in S. Flugge (ed.), "Handbuch der Physik," vol. 51, pp. 353–604, Springer-Verlag OHG, Berlin, 1958; also L. Plaut, in A. Blaauw and M. Schmidt (eds.), "Galactic Structure," The University of Chicago Press, Chicago, 1965.

surroundings:[1]

$$dQ = dU + dW \tag{6-78}$$

If the mass element is followed through a complete cycle, the value of the internal energy U returns to its initial value, so that the work done during the cycle is

$$W = + \oint dQ \tag{6-79}$$

The element does positive work on the surroundings, i.e., drives oscillations, only if it absorbs a net amount of heat.

The way in which the absorption must be accomplished to be effective is dictated by the second law of thermodynamics. Because the gas returns to its initial state at the end of the cycle, the entropy must also have returned to its initial value:

$$\oint dS = 0 = \oint \frac{dQ}{T} \tag{6-80}$$

This well-known theorem demands that in a cyclic process absorbed heat must be partially given back up, the heat exchange being moderated by $1/T$. If, for example, the pulsation were isothermal, we would have immediately that $\oint dQ = 0$, and no work could be done. Suppose then that we imagine the cycle progressing in time and represent the heat gain by $dQ(t)$ and the temperature variation of the cycle by

$$T = T_0 + \Delta T(t) \tag{6-81}$$

where $\Delta T(t)$ is a small cyclic modulation of the temperature. Then from Eq. (6-80) we have

$$0 = \oint \frac{dQ(t)}{T_0 + \Delta T(t)} \approx \oint \frac{dQ}{T_0}\left(1 - \frac{\Delta T}{T_0}\right) \tag{6-82}$$

It follows that

$$W = \oint dQ \approx \oint \frac{\Delta T}{T_0}\, dQ \tag{6-83}$$

From this result follows the important physical principle that *if positive work on the surroundings is to be done, heat should be absorbed while the temperature is high and reemitted when the temperature falls.* Because the entire star oscillates to some extent, and because some portions of the star may absorb work, the entire star can drive the oscillation if

$$W = \int_M \oint \frac{\Delta T}{T_0(M)}\, dQ(M)\, dM > 0 \tag{6-84}$$

[1] Note the sign of W used in this discussion.

where the cyclic integral is over each mass element and the mass integral is over the entire star.

This principle is used in the simplest gasoline engine, where the heat is added by the burning of fuel while the gas is heated by compression, and the heat is partially reejected after the gas is cooled by expansion. Eddington suggested two similar principles that could work in the stellar case, the *nuclear-energy mechanism* and the *valve mechanism*.

Nuclear reaction rates are proportional to positive powers of density and temperature. In a central pulsation, therefore, one will expect heat to be liberated during the compression. Although the timing is proper to drive oscillations, the magnitude of the effect is much too small. The central portions, where the energy is liberated, oscillate with very small amplitude, and the work derivable from the central heat engine is quite inadequate to overcome the dissipation in layers farther out from the center. (We shall see later that the regions of Kramers opacity and $\Gamma = \frac{5}{3}$ are dissipative.)

The valve mechanism calls for the modulation of the radiant energy flux by the stellar material. Eddington described this mechanism in an analogy to the combustion engine:

> *Suppose that the cylinder of the engine leaks heat and that the leakage is made good by a steady supply of heat. The ordinary method of setting the engine going is to vary the* supply *of heat, increasing it during compression and diminishing it during expansion. That is the first alternative we considered. But it would come to the same thing if we varied the* leak, *stopping the leak during compression and increasing it during expansion. To apply this method we must make the star more heat-tight when compressed than when expanded; in other words, the opacity must increase with compression.*[1]

Many of the modern discussions of stellar pulsation have concentrated only on the attempt to determine the conditions under which pulsation will occur rather than on the attempt to calculate the full-blown details of the pulsation. The former problem can be analyzed by a simpler technique, the stability analysis of linearized equations of motion. An equilibrium configuration is assumed to exist, and the oscillations of infinitesimal amplitude about that configuration are studied. For small amplitudes, all the equations are linear in the perturbations of the equilibrium values of the variables. These infinitesimal amplitudes will either grow in time, in which case they are assumed to exist, or die out in time, in which case the equilibrium is stable. The technique is a very general one used in stability analysis. It can identify the modes of instability, but the linearized solution becomes invalid when the amplitudes become too large. Although the full solution of the pulsation problem requires the analysis of the

[1] A. S. Eddington, "The Internal Constitution of the Stars," p. 202, Dover Publications, Inc., New York, 1959.

nonlinear equations, we shall introduce the linear analysis because it clarifies many important features of stability.[1]

Consider r, P, T, and L to be the dependent functions of M, and let r_0, P_0, T_0, and L_0 represent the values of a satisfactory stellar model in equilibrium. The procedure is to consider small perturbations of the variables to see whether they grow or decay. To do so, define *infinitesimal fractional changes* of these variables as r', p', t', and l', and regard these as functions of time. Specifically

$$\begin{aligned} r(M,t) &= r_0(M)[1 + r'(M,t)] \\ P(M,t) &= P_0(M)[1 + p'(M,t)] \\ T(M,t) &= T_0(M)[1 + t'(M,t)] \\ L(M,t) &= L_0(M)[1 + l'(M,t)] \end{aligned} \tag{6-85}$$

The equations of stellar structure are then expanded to first order in these increments. The radial equation of motion, for example,

$$\frac{\partial P}{\partial M} = -\frac{1}{4\pi r^2}\left(\frac{GM}{r^2} + \frac{\partial^2 r}{\partial t^2}\right) \tag{6-86}$$

becomes

$$\frac{\partial P_0(1 + p')}{\partial M} = -\frac{1}{4\pi r_0^2(1 + r')^2}\left[\frac{GM}{r_0^2(1 + r')^2} + \frac{\partial^2 r_0(1 + r')}{\partial t^2}\right] \tag{6-87}$$

Expansion to first order in the increments gives

$$\frac{\partial P_0}{\partial M} + \frac{\partial P_0 p'}{\partial M} = -\frac{(1 - 2r')}{4\pi r_0^2}\left[\frac{GM(1 - 2r')}{r_0^2} + r_0\frac{\partial^2 r'}{\partial t^2}\right] \tag{6-88}$$

Because the initial model is in equilibrium,

$$\frac{\partial P_0}{\partial M} = -\frac{1}{4\pi r_0^2}\frac{GM}{r_0^2} \tag{6-89}$$

With the aid of Eq. (6-89), Eq. (6-88) reduces to

$$-\frac{GM}{4\pi r_0^4}p' + P_0\frac{\partial p'}{\partial M} = \frac{4r'GM}{4\pi r_0^4} - \frac{1}{4\pi r_0}\frac{\partial^2 r'}{\partial t^2} \tag{6-90}$$

or alternatively

$$\frac{\partial p'}{\partial M} = \frac{1}{4\pi r_0 P_0}\left[\sigma_0^2(4r' + p') - \frac{\partial^2 r'}{\partial t^2}\right] \tag{6-91}$$

[1] The subsequent discussion follows that of N. Baker in A. G. W. Cameron and R. F. Stein (eds.), "Stellar Evolution," Plenum Press, New York, 1966. See also articles by J. Cox and R. Christy in the same volume and the bibliographies of those papers. Other important reviews are R. Christy, *Ann. Rev. Astron. Astrophys.*, **4**:353 (1966) and S. Zhevakin, *ibid.*, **1**:367 (1963). For a numerical scheme for the nonlinear problem, see R. Christy, *Rev. Mod. Phys.*, **36**:555 (1964).

where we define

$$\sigma_0^2 = \frac{GM}{r_0^3} \tag{6-92}$$

Problem 6-21: Show in a similar manner that the continuity equation

$$\frac{\partial r}{\partial M} = \frac{1}{4\pi r^2 \rho}$$

becomes when linearized

$$\frac{\partial r'}{\partial M} = -\frac{1}{4\pi r_0^3 \rho_0}(3r' + \alpha p' - \delta t') \tag{6-93}$$

where the bulk moduli are defined as

$$\alpha = \frac{P}{\rho}\left(\frac{\partial \rho}{\partial P}\right)_T \qquad \delta = -\frac{T}{\rho}\left(\frac{\partial \rho}{\partial T}\right)_P \tag{6-94}$$

Show that the radiative-diffusion equation

$$L = -\frac{64\pi^2 acr^4T^3}{3\kappa}\frac{\partial T}{\partial M}$$

becomes when linearized

$$\frac{\partial t'}{\partial M} = \frac{1}{T_0}\frac{\partial T_0}{\partial M}[l' - 4r' + \kappa_p p' + (\kappa_T - 4)t'] \tag{6-95}$$

where the logarithmic derivatives of the opacity are defined as

$$\kappa_p = \frac{P}{\kappa}\left(\frac{\partial \kappa}{\partial P}\right)_T \qquad \kappa_T = \frac{T}{\kappa}\left(\frac{\partial \kappa}{\partial T}\right)_P \tag{6-96}$$

For the linearization of the energy equation

$$\frac{\partial L}{\partial M} = \epsilon - T\frac{dS}{dt} = \epsilon - \frac{dQ}{dt} \tag{6-97}$$

we must proceed in such a way that the heat gained is expressed in terms of the differentials of the variables T and P with which we are working. Because the pulsation is largely due to the outer layers, let us further simplify the problem by ignoring nuclear energy generation and by assuming that the unperturbed model is truly static. Then

$$\frac{\partial L_0}{\partial M} = 0 \tag{6-98}$$

as is appropriate to the outer layers of a static star.

Problem 6-22: Show that in terms of dT and dP

$$dQ = C_p\,dT + \left[\left(\frac{\partial U}{\partial V}\right)_T + P\right]\left(\frac{\partial V}{\partial P}\right)_T dP \tag{6-99}$$

Then show that the linearized energy equation is

$$L_0 \frac{\partial l'}{\partial M} = -C_p T_0 \frac{dt'}{dt} + \left[\left(\frac{\partial U}{\partial V}\right)_T + P_0\right] \frac{\alpha}{\rho_0} \frac{dp'}{dt} \tag{6-100}$$

It will be convenient to write Eq. (6-100) in the form

$$\frac{\partial l'}{\partial M} = -\left[\left(\frac{\partial U}{\partial V}\right)_T + P_0\right] \frac{\alpha}{\rho_0 L_0} \left(C \frac{dt'}{dt} - \frac{dp'}{dt}\right) \tag{6-101}$$

where the new parameter of state is

$$C = C_p \frac{\rho_0 T_0}{\alpha\,[(\partial U/\partial V)_T + P_0]} \tag{6-102}$$

We shall see later that C is related simply to the second adiabatic exponent. This completes the set of linearized equations. From Eqs. (6-91), (6-93), (6-95), and (6-101), along with the definitions of the auxiliary parameters of state, we must ascertain the stability of the equilibrium model.

The test of stability of these four coupled equations for an entire star is still difficult, because all the coefficients of the perturbations are functions of the coordinate M. The stability of the star depends upon the stability analysis of each mass element and how they couple together. But the rudiments of the physics can be seen by taking a simplified model consisting of a single thin zone. If the zone is thin enough, the coefficients in the linearized equations may be thought of as constants throughout the zone. That is, the question of the stability of the star is replaced with the question of the stability of a thin zone. Once the zone is understood, the star can be thought of as a succession of thin zones of differing stability.

If the fluctuations are assumed constant throughout the thin layer, we have

$$\frac{\partial r'}{\partial M} = \frac{\partial p'}{\partial M} = \frac{\partial t'}{\partial M} = 0 \tag{6-103}$$

If the same assumption were made for l', however, we would lose the essential physics of the heat engine, because the interaction to be studied is the modulation of the photon flux by its interaction with a mechanical system. Let the flux variation of the lower boundary of the shell be l'_L, and that of the upper boundary l'_U, as illustrated in Fig. 6-21. Then for the thin shell we may sensibly take

$$l' = \frac{l'_U + l'_L}{2} \tag{6-104}$$

and

$$\frac{\partial l'}{\partial M} = \frac{l'_U - l'_L}{m} \tag{6-105}$$

In the spirit of analyzing the simplest of shells with an eye toward the physical effects, assume that the flux at the lower boundary is constant; that is, $l'_L = 0$.

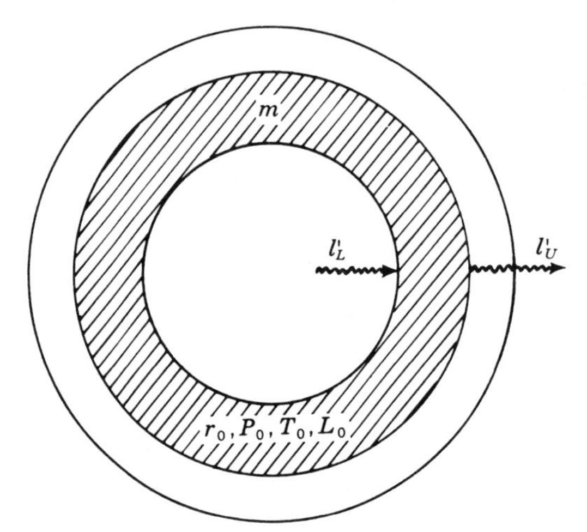

Fig. 6-21 A thin shell of mass m within the interior isolated for stability against pulsation. The zone is chosen to be small enough to permit taking the unperturbed quantities r_0, P_0, T_0, and L_0 as constant throughout the shell. The flux variations at the lower and upper boundaries of the shell are designated by l'_L and l'_U, respectively.

Then from the previous two equations we have

$$\frac{\partial l'}{\partial M} \approx \frac{2l'}{m} \tag{6-106}$$

When these simplifications are applied to the linearized equations, there results this set of four equations:

$$\begin{aligned}
&\frac{\partial^2 r'}{\partial t^2} = \sigma_0^2(4r' + p') \\
&3r' + \alpha p' - \delta t' = 0 \\
&l' - 4r' + \kappa_P p' + (\kappa_T - 4)t' = 0 \\
&C\frac{\partial t'}{\partial t} - \frac{\partial p'}{\partial t} = -K\sigma_0 l'
\end{aligned} \tag{6-107}$$

where the parameter K is defined as

$$K = \frac{2L_0\rho_0}{m\sigma_0\alpha[(\partial U/\partial V)_T + P]} \tag{6-108}$$

Problem 6-23: Show that by elimination and differentiation the four equations (6-107) can be combined into a single third-order equation

$$\frac{\partial^3 r'}{\partial t^3} + K\sigma_0 A\frac{\partial^2 r'}{\partial t^2} + \sigma_0^2 B\frac{\partial r'}{\partial t} + K\sigma_0^3 D r' = 0 \tag{6-109}$$

where the coefficients are

$$A = -\frac{\alpha(\kappa_T - 4) + \delta\kappa_P}{\alpha C - \delta}$$

$$B = \frac{3C - 4(\alpha C - \delta)}{\alpha C - \delta}$$

$$D = \frac{(4\alpha - 3)(\kappa_T - 4) + 4\delta(\kappa_P + 1)}{\alpha C - \delta}$$

Problem 6-24: Using the definition of Chap. 2 for the adiabatic exponents in a partially ionized gas, show that

$$\Gamma_1 = \frac{C}{\alpha C - \delta} \qquad \Gamma_2 = \frac{C}{C - 1} \qquad \Gamma_3 - 1 = \frac{1}{\alpha C - \delta} \tag{6-110}$$

The stability test can be performed on Eq. (6-109) by assuming that the fluctuation has exponential time dependence

$$r'(t) = \xi e^{st} \tag{6-111}$$

whereupon we obtain a cubic equation for s:

$$s^3 + K\sigma_0 A s^2 + \sigma_0{}^2 B s + K\sigma_0{}^3 D = 0 \tag{6-112}$$

Consider first the adiabatic oscillations. It is evident from the manner in which the equations were established that the demand $dQ = 0$ is equivalent to

$$C\frac{dt'}{dt} - \frac{dp'}{dt} = 0 \qquad \text{adiabatic} \tag{6-113}$$

From Eqs. (6-107) this may be formally accomplished by setting $K = 0$, whereupon Eq. (6-112) becomes quadratic, with the solutions

$$s_{\text{ad}} = \pm i\sqrt{B}\,\sigma_0 = \pm i\sqrt{3\Gamma_1 - 4}\,\sigma_0 \tag{6-114}$$

The case $\Gamma_1 > \frac{4}{3}$ allows for sinusoidal adiabatic oscillations. These will damp out relatively quickly, however, when the total solution is considered, and they do not account for the variable phenomenon. It is also apparent that the case $\Gamma_1 < \frac{4}{3}$ allows for an exponentially growing perturbation, i.e., instability. This type of dynamic instability phenomenon was mentioned in Chap. 2 in connection with the adiabatic exponents.

In order that the nonadiabatic case ($K \neq 0$) be stable, however, all three roots of the cubic equation must have negative real parts. This can happen only if

$$\begin{aligned} &\tau_0{}^2 B > 0 \\ &K\sigma_0{}^3 D > 0 \\ &K\sigma_0{}^3(AB - D) > 0 \end{aligned} \tag{6-115}$$

The first condition simply ensures dynamic stability, as outlined in the previous paragraph. For our present purposes it will suffice to assume that the first two conditions are met and to examine the third.

Problem 6-25: By multiplying out the terms in the quantity $AB - D$, as given after Eq. (6-109), show that the zone is *stable if*

$$\frac{4}{C} - \left(\frac{\kappa_T}{C} + \kappa_P\right) - \frac{4}{3\Gamma_1} > 0 \tag{6-116}$$

The examination of Eq. (6-116) shows clearly the physical effects contributing to the status of the stability of the zone. The first term always contributes to stability because it is positive. On the other hand, the quantity

$$C = \frac{\Gamma_2}{\Gamma_2 - 1} \tag{6-117}$$

becomes very large as Γ_2 approaches unity. It was illustrated in Chap. 2 that Γ_2 is only slightly greater than unity in ionization zones. Thus we see that the stabilizing term is diminished in ionization zones. This effect has been called the Γ *mechanism.* The physical reason behind the presence of this stabilizing term is the tendency of gas to lose heat by radiation when its temperature rises and to gain heat when it falls, just the opposite of a pulsation driving mechanism.

The second term reflects the way in which the opacity varies during the pulsation. Positive values of κ_T and κ_P would imply that the opacity increases upon contraction, which would remove more energy from the radiation flux (reduce Eddington's leak) at the proper time to drive mechanical work. The destabilizing effect of positive κ_T and κ_P has been called the κ *mechanism* by Baker and Kippenhahn.

Problem 6-26: Show that in the inner regions where Kramers opacity applies and where $\Gamma_2 = \frac{5}{3}$, the κ mechanism is stabilizing because

$$-\left(\frac{\kappa_T}{C} + \kappa_P\right) = +\tfrac{4}{5}$$

This fact contributes to the pulsational stability of the stellar interior.

In the ionization zones, the larger value of C reduces the importance of κ_T, whereas κ_P is always positive. There are also regions where κ_T is positive, but all these destabilizing features are related to the major ionization zones.

The third term is always destabilizing and reflects the spherical geometry of the star. It is not present in the analysis of a plane atmosphere. The total stability of the zone is, of course, determined by the sum of the three terms. The total stability of the star depends upon the cumulative effect of all the zones.

The most elaborate calculations to date of stellar pulsation have been performed by Christy. In order to solve the entire nonlinear problem, he has divided the outer portion of the star into a large number (about 40) of concentric mass shells. The differential equations were replaced by difference equations and followed in real time by a technique similar to the relaxation techniques of following stellar evolution. The relaxation technique described earlier, however, relaxes only in space at a given moment of time and is followed by a time step to the next model. For the study of pulsation, which is a very fast phenomenon in terms of stellar time scales, the relaxation in space and time must be followed together. There are many numerical pitfalls lurking in such a complicated task, and so continued improvements in the computational scheme are to be expected.

The results of Christy's models of RR Lyrae variables are very illuminating. Figure 6-22 shows the temperature variation in specific mass shells. Mass shells 32 and 33 are, in the static model, in the region of hydrogen ionization and, less importantly, of the first ionization of helium. Zones 26 and 27 are correspond-

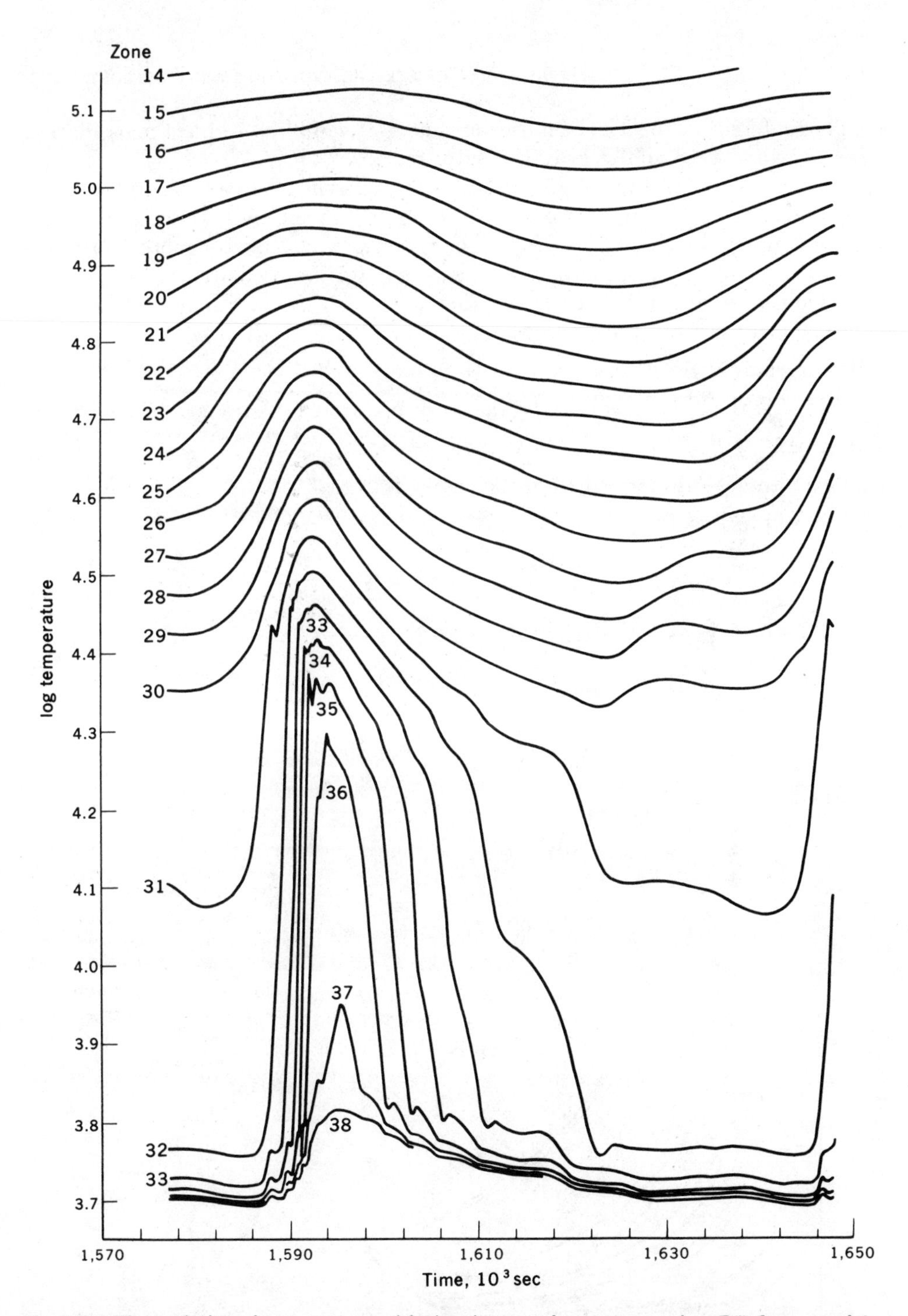

Fig. 6-22 The variation of temperature with time in several mass zones of an RR Lyrae model. The hydrogen ionization zone centers at shell 33, and the HeII ionization zone centers at shell 27. The amplitude of the temperature oscillations in these zones is even larger than the quiescent temperature in the static model, and the overall problem is nonlinear. [*After R. F. Christy, Astrophys. J.,* **144**:108 (1966). *By permission of The University of Chicago Press. Copyright* 1966 *by The University of Chicago.*]

ingly centered on the HeII ionization zone. It can be seen that the oscillations are large and very nonlinear in the outer regions.

It is not clear from Fig. 6-22 what the major driving source of the pulsation is. To determine this it is necessary to compute the mechanical work per cycle generated by each mass element. Some zones will generate more work than others, and some will absorb work, i.e., dissipate the oscillations. If the total mechanical work generated exceeds that absorbed, the star may pulsate. The mechanical work per cycle per gram is just

$$\text{Work/gram} = \oint P\, dV \tag{6-118}$$

where V is the specific volume. The work performed by any given shell is then

$$W_j = \Delta M_j \oint P_j\, dV_j \tag{6-119}$$

and the total work derived from the engine is

$$W = \sum_i W_i \tag{6-120}$$

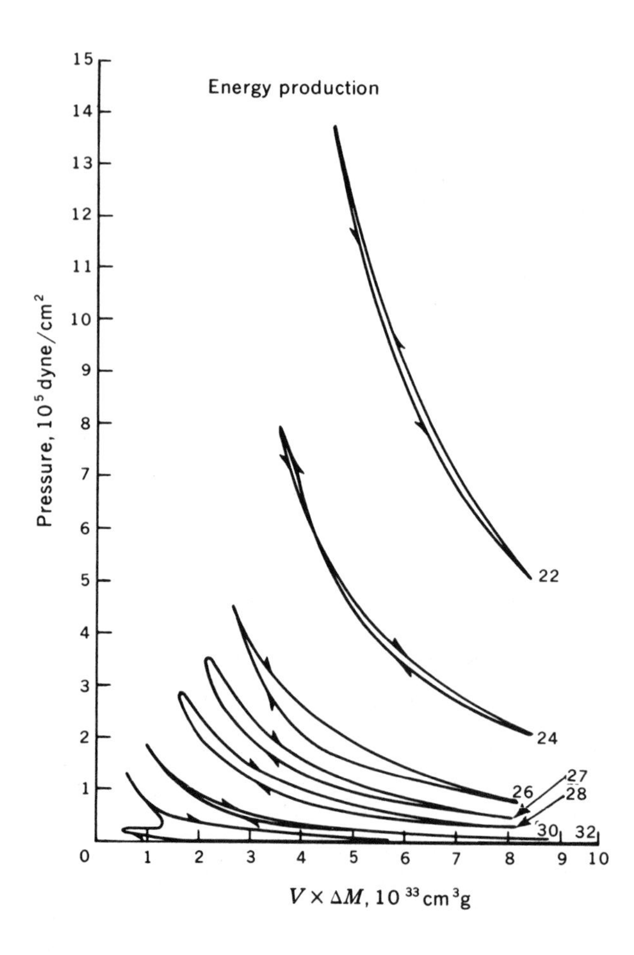

Fig. 6-23 The PV cycles for selected mass shells within a model of an RR Lyrae variable. A clockwise cycle performs positive mechanical work equal to the enclosed area. [*After R. F. Christy, Rev. Mod. Phys.*, **36**:555 (1964).]

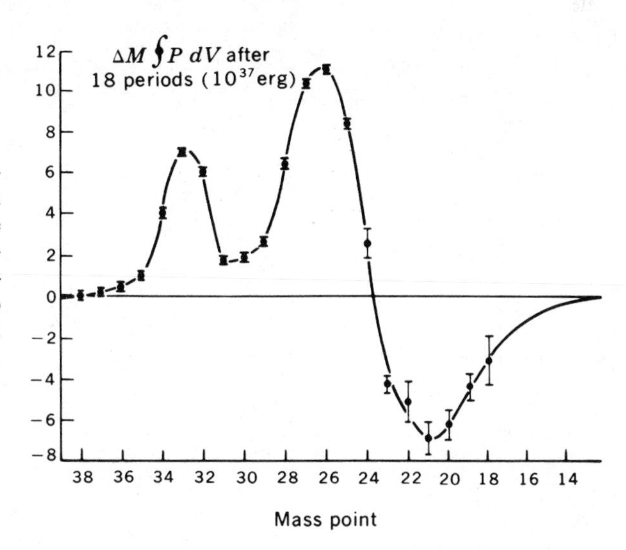

Fig. 6-24 The work done per period by the outer mass shells of an RR Lyrae model. Most of the positive work comes from the ionization zones of hydrogen and helium, whereas the regions interior to those zones are largely dissipative. [*After R. F. Christy, Rev. Mod. Phys.*, **36**:555 (1964).]

The PV cycles for several key zones are shown in Fig. 6-23. Note that a clockwise cycle does positive work equal to the enclosed area, whereas a counterclockwise cycle is dissipative by the same amount. The work per period done by the several zones is shown in Fig. 6-24. The first peak is due to hydrogen ionization, the major peak is due to HeII ionization, and the internal region is dissipative, although the amplitude quickly dies out. Hydrogen contributes one-third, and HeII contributes two-thirds of the positive work, which is 6.3×10^{38} ergs/cycle, or about 7 percent of the luminosity. At this point in the calculation, the dissipation was only 4.9×10^{38} ergs/cycle, so that the kinetic energy of the pulsation is still growing. The motion eventually saturates when the amplitude is large enough for the work done and the dissipation to cancel. It should also be remarked that Christy's calculations show many details apparent in the real phenomenon.

Our introduction to calculations of stellar structure must end here, although there are countless interesting applications that have not been discussed at all. It should perhaps be restated in conclusion that the object is to learn something of stars by comparing them to calculated models. The interplay between observation and calculation is the heart of the science and is the hardest of the tasks. By careful study of the peculiar properties displayed by stars we may hope to slowly refine our theoretical conception of their structure. For the science of nucleosynthesis we must learn how stars expel matter and be able to follow the entire thermonuclear history of that matter, a challenging task indeed. The key stellar event in this regard is probably the supernova, theoretically a more difficult dynamical problem than the variable stars. The supernovas will not be discussed here, but the nuclear events leading to the presupernova state will be outlined in the next chapter.

chapter 7

SYNTHESIS OF THE HEAVY ELEMENTS

The elementary scheme of nucleosynthesis of the light elements assigns the sources of the light elements to those energy-generating stages of thermonuclear fusion in stars which terminate with significantly increased abundances of the species in question. An abbreviated summary of the main line is as follows:

(*1*) He^4 from hydrogen burning; He^3 from incomplete PP chain.
(*2*) D, Li, Be, B are bypassed, and their small abundances relative to the main line are interpreted as being due to nonthermal processes.
(*3*) C^{12} and O^{16} from helium burning; O^{18} and Ne^{22} partially due to alpha captures by N^{14} present during helium-burning processes.
(*4*) N^{14} from conversion of CNO to N^{14} in hydrogen burning; some C^{13}, N^{15}, and O^{17} also results from conversion of CNO catalysts.
(*5*) Ne^{20}, Na, Mg, Al, Si^{28} partly due to carbon burning.
(*6*) Mg, Al, Si, P, S partly due to oxygen burning.

One can, with a fair share of success, understand the relative abundances of the light elements in terms of the ashes of these burning stages. The success is hardly an unqualified one, however. Many abundance ratios make little sense in terms of a simple picture. A quantitative theory of the evolution of the galactic abundances of the light elements is badly clouded by fundamental uncertainties in the structural evolution of galaxies and in the details of mass loss in highly evolved stars. Years ago, for example, it was somewhat naïvely assumed that the relative abundances of He^4 and C^{12} reflected in a natural way the fact that $4H \rightarrow He^4$ represents the primary stage of nuclear fusion, whereas $3He^4 \rightarrow C^{12}$ represents a secondary stage that can only follow the synthesis of He^4. But there remains the fundamental problem of why it should be so, that the hydrogen-burning mass expelled from stars so greatly exceeds the helium-burning mass that is expelled. The fusion into helium at a stellar center can hardly be retained as a source of interstellar helium unless one can find a way of removing the helium from the star without allowing it to progress through the helium-burning stage. A similar problem exists in the interpretation of each of the light-element abundances. It seems that most of the nucleosynthesis is more easily imagined as happening in shells surrounding the cores of highly evolved stars, but it is not at all clear how much may ever be expelled from the star in its death throes.

It is likely that the conventional concept of stellar nucleosynthesis is not complete. The large abundance of He^4 may reflect instead the residue of a primordial fireball in the early condensed universe. Or perhaps one or many supermassive stars (or quasistars?) have exploded at our galactic center.[1] May they not even have contributed the lion's share of the population I elements? Such questions have not been resolved. But it seems certain that the thermonuclear burning stages are still relevant even if conventional stars should turn out to be only a secondary source of nucleosynthesis. The principles of nuclear astrophysics may be regarded as a physical probe of astrophysical circumstances.

The nuclear principles involved in the synthesis of heavier elements differ in many respects from those discussed previously. In this chapter we shall outline the major conceptual features surrounding the nucleosynthesis of heavy elements, and we shall emphasize the astrophysical clues hidden in the heavy-element abundances.

7-1 PHOTODISINTEGRATION REARRANGEMENT AND SILICON BURNING

The stages of thermonuclear energy generation share the feature that light nuclei are fused into heavier ones. Each epoch of nuclear burning comes to a close when the light fuels have been scoured out, leaving only the heavier ashes. Each stage discussed so far,

$$4H \rightarrow He^4 \qquad 3He^4 \rightarrow C^{12} \qquad 2C^{12} \rightarrow Mg^{24} \qquad 2O^{16} \rightarrow S^{32}$$

where the reactions are schematic rather than literal, has been of this type. These fusion reactions liberate energy as long as the *binding energy per nucleon* of the compound system exceeds the binding energy per nucleon of the constitu-

[1] For a delightful and opinionated discourse on the relevance of cosmological astrophysics to nucleosynthesis, the reader may profitably turn to Fred Hoyle's short book, "Galaxies, Nuclei, and Quasars," Harper and Row, Publishers, Incorporated, New York, 1965. The technical basis for nucleosynthesis in fireballs was elaborated by P. J. E. Peebles, *Astrophys. J.*, **146**:542 (1966), and by R. V. Wagoner, W. A. Fowler, and F. Hoyle, *Astrophys. J.*, **148**:3 (1967).

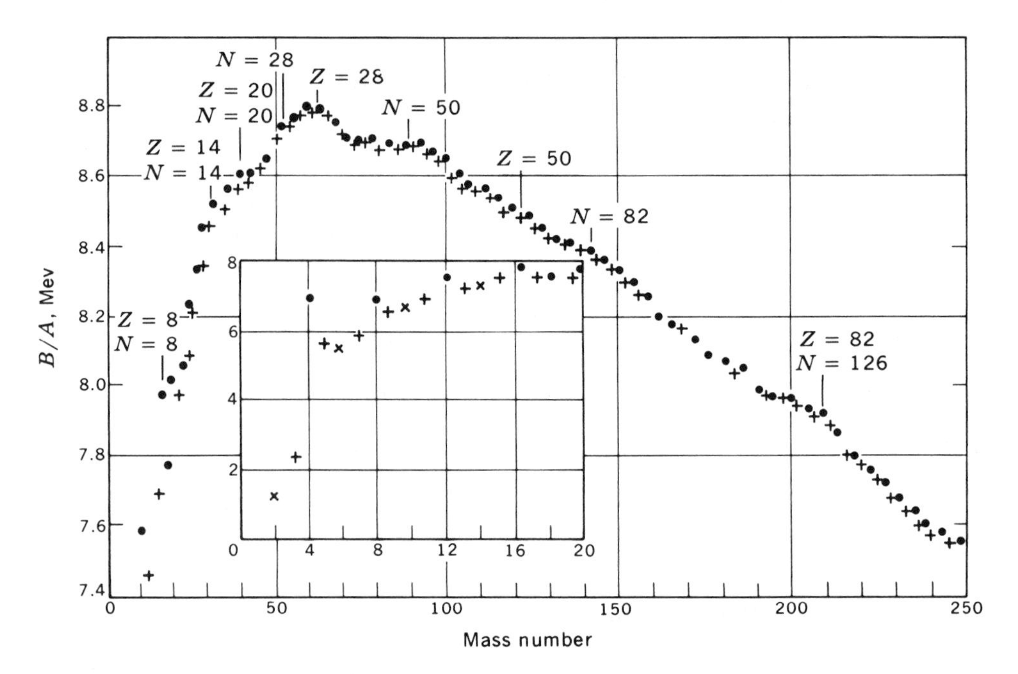

Fig. 7-1 The binding energy per nucleon of the most stable isobar of atomic weight A. The solid circles represent nuclei having an even number of protons and an even number of neutrons, whereas the crosses represent odd-A nuclei. *(M. A. Preston, "Physics of the Nucleus," Addison-Wesley Publishing Company, Inc., Reading, Mass.,* 1962.)

ents. To illustrate the peculiarities of nuclear binding, a graph of the nuclear binding energy per nucleon is displayed in Fig. 7-1 as a function of the atomic weight (the binding energy being that of the most stable isobar at each value of A). These binding energies may be related by a brief calculation to the atomic mass excesses that were listed in Table 4-1, and the student may profitably attempt such a calculation. One finds in the light nuclei that the binding is markedly increased by a fusion of nuclei. The most tightly bound nuclei per nucleon are those with $50 < A < 60$, the maximum occurring at the Fe^{56} nucleus. It might well be noted here, however, that the binding per nucleon decreases again as one progresses to yet heavier nuclei. It would appear that thermal fusion is energetically favorable until the mean atomic weight reaches the iron group.[1] The reader may already surmise that this feature of nuclear structure is closely related to the large natural-abundance peak near iron. For the time being we discuss the way in which the products of carbon and oxygen burning are transmuted to the iron group.

[1] The term *iron group* is commonly used to designate the nuclei in the abundance peak centered on Fe^{56}, primarily Cr, Mn, Fe, Co, and Ni. The utility of the term comes from the apparently common origin of these nuclei in nucleosynthesis.

At the conclusion of carbon and oxygen burning, the dominant nuclei are Si^{28} and S^{32}, with significant amounts of Mg^{24} and, to a lesser degree, the non-alpha-particle nuclei. One might imagine, by analogy with earlier burning phases, that the material again gravitationally contracts until these nuclei begin to fuse with each other. The first important reaction might be sensibly assumed to be $Mg^{24} + Mg^{24}$. Quantitative examination shows that the true sequence of events follows a quite different course, however. The coulomb barrier between two magnesium nuclei is so great that very high thermal temperatures would be required to effect their fusion. Before such high temperatures can be achieved in the contraction, the intense thermal gamma-ray flux begins the photodisintegration of key nuclei. The important reactions are (γ,p), (γ,n), and (γ,α), and the first nuclei to be stripped down are those with the smallest binding energies for protons, neutrons, and alpha particles.[1] The photodisintegration rate is dominated by the factor

$$\lambda_\gamma \propto \exp - \frac{Q}{kT}$$

where Q is the binding energy of the particle (p, n, or α). Consider what happens to the *least tightly bound particles* after their photoejection. They are quickly recaptured (because of the small coulomb barrier), perhaps to be photoejected once more. The inverse reactions attempt to establish an equilibrium such that the number of photoejections per second is balanced by the number of recaptures per second. But many of the photoejected particles will be captured by nuclei in which they are more tightly bound than in the nuclei from which they were liberated. In fact, that is to be expected, since the photoejections of the least tightly bound nucleons are the first to occur. What happens then, as the temperature rises, may be described as a redistribution of loosely bound nucleons into more tightly bound states. We choose to call this process by the descriptive term *photodisintegration rearrangement*.

Problem 7-1: Charged-particle reactions in this intermediate mass range proceed through one or several important resonances in the compound nucleus, and so the value of $\langle\sigma v\rangle$ for the radiative reaction is a sum over resonances of the resonant reaction rate. Show that in this case the rate of the inverse photodisintegration reaction is given by

$$\lambda_\gamma = \frac{\exp(-Q/kT)}{G_{12}} \sum_r \frac{(2J_r + 1)\Gamma_{1r}\Gamma_{\gamma r}}{\hbar\Gamma_r} \exp - \frac{E_r}{kT} \tag{7-1}$$

where the sum is over the resonances of the compound system of the two particles. Each resonance r is characterized by a spin J_r, by the center-of-mass energy E_r, and by Γ_{1r}, $\Gamma_{\gamma r}$, and Γ_r, which are respectively the particle width, the radiative width, and the total width of the resonant state. G_{12} is the partition function for the compound nucleus, and Q is the binding energy of the two particles in the ground state of the compound nucleus. Draw a figure showing

[1] The binding energy for a proton in the nucleus (Z,A) is given by

$$Q_p(Z,A) = [M(Z-1, A-1) + M_p - M(Z,A)]c^2$$

and similarly for the binding of other particles.

the various energies in an energy-level diagram. Note that the resonance energies E_r are generally small compared to the nucleon binding energies, which are of order $Q \approx 8$ Mev.

Nuclear energy may or may not be liberated as a result of the photodisintegration rearrangement. The relocation of nucleons into more tightly bound nuclei certainly liberates energy, but the high photon flux also maintains an equilibrium density of free nucleons at great cost to the total binding energy. The rearrangement of Si^{28} and S^{32} into the iron group liberates energy if it occurs at sufficiently low temperature for the vast majority of the nucleons to be bound to some iron-group nucleus, but if it occurs at high temperature, $T_9 > 5$, say, the large free-nucleon density will render the energy generation small. The energy balance during this stage of stellar evolution is thus somewhat uncertain. The neutrino losses above 3×10^9 °K are quite large, moreover, and they must be reimbursed by the nuclear energy generation and the gravitational work.

Many beta-radioactive nuclei are produced during the rearrangement process, and their decays tend to slowly reduce the total proton-to-neutron ratio $\bar{Z}/\bar{N}$ of the nuclear gas. Later we shall see that a reduction of only a few percent from the initial ratio of unity has a major influence on the composition of the iron-group products. The number of beta decays is determined in first order by the length of time of the rearrangement, which will in turn depend upon the temperature of the rearrangement. As a rule of thumb it will be found that durations in excess of 10^4 sec for the rearrangement will force one to keep track of the beta decays, whereas faster rearrangements, i.e., higher-temperature rearrangements, will occur quickly enough for $\bar{Z}/\bar{N}$ to retain its initial value of unity. It is difficult to predict the time available for the rearrangement because it is very likely imposed by the rapid evolution of the stellar structure in the face of severe neutrino losses. In their 1964 investigation,[1] Fowler and Hoyle concluded that the nucleons rearranged at $T_9 \approx 3.5$ to ${}_{28}Ni^{56}_{28}$ as the dominant constituent, with the release of about 10 Mev of thermal energy per Ni^{56} nucleus formed. Truran, Cameron, and Gilbert,[1] on the other hand, found at $T_9 = 5$ that the final composition is

[1] W. A. Fowler and F. Hoyle, *Astrophys. J. Suppl.*, **9**:201 (1964). These authors followed their original *Rev. Mod. Phys.*, **29**:547 (1957) article in designating the photodisintegration rearrangement by the name *α process*. The name originally arose because the first photodisintegrations of importance were shown by them to be (γ,α) reactions, and it was assumed that recapture of the alpha particles synthesized the heavier $Z = N$ nuclei. In their 1964 paper, Fowler and Hoyle showed this idea to be an oversimplification for rearrangement between Si^{28} and the iron group, but they nonetheless chose to label the rearrangement process, which does in fact synthesize mainly alpha-particle nuclei, by the name α process (see pp. 240–241 of their paper). J. W. Truran, A. G. W. Cameron, and A. A. Gilbert, *Can. J. Phys.*, **44**:576 (1966), performed detailed numerical integrations of the differential equations of the reaction network and showed that the non-alpha-particle reactions were important. D. Bodansky, D. D. Clayton, and W. A. Fowler [to be published: a preliminary report of their results may be found in *Phys. Rev. Letters*, **20**:161 (1968)] greatly clarified the nature of the process and presented many numerical results by analyzing the burning in terms of a partial-equilibrium model. All three papers are important for the serious student, and this entire section is a digest of their results.

dominated by ${}_{26}Fe^{54}_{28} + 2p$, in which case the conversion is endothermic by 1.3 Mev per Fe^{54} nucleus produced. Until one has selected the specific circumstance of an astrophysical event, it cannot be said whether the phase of photodisintegration rearrangement liberates nuclear energy or not. This question has a secondary effect on the final nuclear composition, moreover, by virtue of its effect on the time scale of gravitational contraction. That is, does the star "hold" for hours, say, while the rearrangement occurs at nearly constant temperature and nuclear energy makes good the neutrino loss, or does the star contract and heat up to make good the neutrino loss? In the latter case, the iron group is reached at higher temperature, and the final abundance distribution in the iron peak is somewhat broader than at a lower temperature. We shall return to this question of the final iron-group composition in Sec. 7-2. For the time being it will suffice to assume simply that the nuclei near $A = 28$ are rearranged into nuclei near $A = 56$.

With the foregoing overview of the process we may profitably turn to some of the details of how it occurs. Near the conclusion of oxygen burning, the dominant nuclear constituents are Si^{28} and S^{32}. Without doing violence to the physical idea we may consider those to be the only two nuclei. The binding energies of protons, alpha particles, and neutrons are smaller in S^{32} than in Si^{28}, so that as the temperature rises, the S^{32} will be the first to be photodisintegrated. But before calculating the different photodisintegration rates, we must consider a simplification of Eq. (7-1).

For the sake of an explicit illustration, concentrate for a moment on the example reaction $S^{32}(\gamma,p)P^{31}$. In this case $Q_p = 8.86$ Mev is the binding energy of the proton in the S^{32} nucleus and is the value of Q to be used in Eq. (7-1). The system $P^{31} + p$ has many resonances E_r, however, and we may sensibly wonder which ones will dominate in the sum appearing in Eq. (7-1). Recall that this formula was derived by balancing the photodisintegration rate against the recombination rate $\langle\sigma v\rangle$ for the $P^{31} + p$ system. It was shown in Chap. 4 that the most effective bombarding energy for stellar reactions is

$$E_0 = 0.122(Z_1^2Z_2^2AT_9^2)^{\frac{1}{3}} \qquad \text{Mev}$$

If there are a large number of resonances, one may be sure that the ones dominating the average $\langle\sigma v\rangle$ are those with E_r near E_0. The same thing must be true for the photodisintegration reaction. To good approximation, then, Eq. (7-1) may be rewritten as

$$\lambda_\gamma = \frac{\exp\left[-(Q + E_0)/kT\right]}{G_{12}} \sum_r \left[\frac{(2J + 1)\Gamma_1\Gamma_\gamma}{\hbar\Gamma}\right]_r \exp - \frac{E_r - E_0}{kT} \tag{7-2}$$

The advantage in writing the equation this way, rather than in the simpler form of Eq. (7-1), is that the largest temperature-dependent term, which must include E_0, is displayed in front of the summation. The major contributions to the sum occur for resonances with E_r near E_0, so that the temperature dependence of the sum is relatively small. Without this conceptual aid, one cannot easily understand the relative magnitudes for the photoejection rates of differing light-nuclear constituents, because the value of E_0 depends upon the type of particle ejected. In the case of S^{32}, for example, the binding energies of alpha particles, protons, and neutrons are $Q_\alpha = 6.95$, $Q_p = 8.86$, and $Q_n = 15.09$ Mev, so that one might expect, on the basis of the appearance of Eq. (7-1), that the alpha photodisintegration rate will much exceed the others. It turns out, on the other hand, that near $T_9 = 2.5$, for example, the effective energies E_0 for the reactions (γ,α), (γ,p), and (γ,n) are respectively 3.14, 1.34, and 0 Mev, with the result that the proton photodisintegration rate is the greatest. What is reflected by the value of E_0 is the increasing coulomb barrier as the charge of the photoejected particle is increased.

Let us examine this point with somewhat more precision by understanding once more why it is that the sum in Eq. (7-2) is dominated by terms with E_r near E_0. For $E_r \gg E_0$ the charged particle width Γ_1 becomes much greater than Γ_γ, and we have

$$\frac{\Gamma_1\Gamma_\gamma}{\Gamma} \to \Gamma_\gamma \approx \text{const} \qquad E_r \gg E_0$$

At high resonance energy the exponential therefore truncates the sum. At low energy, on the other hand, the penetration factor renders Γ_1 much smaller than Γ_γ,

$$\frac{\Gamma_1\Gamma_\gamma}{\Gamma} \to \Gamma_1 \propto \exp - \frac{b}{E_r^{\frac{1}{2}}} \qquad E_r \ll E_0$$

so that Γ_1 truncates the sum at low energy. The situation is exactly analogous to that of charged-particle reactions as described in Chap. 4. The energy that most effectively compromises these two effects is $E_r \approx E_0$. From the point of view of experimental nuclear astrophysics a nice feature is that the yield of a radiative-capture reaction measures the factor $(2J + 1)\Gamma_1\Gamma_\gamma/\Gamma$ directly. The excited states of the compound nucleus are usually far enough removed from the ground state, moreover, to permit approximation of G_{12} by $2J_{12} + 1$, where J_{12} is the spin of the ground state of the compound nucleus. Many of these features are quantitatively illustrated by the following problem.

Problem 7-2: Laboratory measurements show that the sums

$$\langle\Gamma_{\text{eff}}\rangle \equiv \sum_r \left[(2J + 1)\frac{\Gamma_1\Gamma_\gamma}{\Gamma}\right]_r \exp - \frac{E_r - E_0}{kT}$$

over the resonances in S^{32} appropriate for $T_9 = 2.5$ and for the three modes of photodisintegration of S^{32} are given by

Reaction	*Q, Mev*	$\langle\Gamma_{\text{eff}}\rangle$, *Mev*
$S^{32}(\gamma,\alpha)Si^{28}$	6.95	10^{-5}
$S^{32}(\gamma,p)P^{31}$	8.86	2×10^{-4}
$S^{32}(\gamma,n)S^{31}$	15.09	10^{-3}

Calculate the photodisintegration lifetimes at $T = 2.5$ billion °K. You may also wish to check these Q values from Table 4.1.
Ans: $\tau_{\gamma\alpha} \approx 10^4$ sec, $\tau_{\gamma p} \approx 10^3$ sec, $\tau_{\gamma n} \approx 10^{12}$ sec.

Near the end of oxygen burning the temperature reaches the neighborhood of $T_9 = 2.5$. Thus the S^{32} will begin to be disintegrated at that time, primarily by the photoejection of protons. The binding energy of a proton in P^{31} is only 7.29 Mev, and so the reaction $P^{31}(\gamma,p)Si^{30}$ rapidly follows. This in turn is rapidly followed by two photoneutron reactions $Si^{30}(\gamma,n)Si^{29}$ and $Si^{29}(\gamma,n)Si^{28}$. The liberated protons and neutrons effectively regroup into alpha particles, leaving Si^{28} as the overwhelmingly dominant constituent. Little more of importance happens until the temperature rises to the point where photoejection from Si^{28} becomes possible. For this reason the process is often called *silicon burning*, and to illustrate the main features most clearly we shall assume the gas to be pure Si^{28} at this point.

The Si^{28} nucleus is very tightly bound and is the last one in this intermediate mass range to be photodisintegrated. The binding energies of α, p, and n are 9.99, 11.58, and 17.18 Mev, respectively. The coulomb barrier renders the (γ,α) reaction slower than the (γ,p) reaction, just as in Prob. 7-2, so that the major initial reaction of importance is $Si^{28}(\gamma,p)Al^{27}$. The lifetime in seconds of the Si^{28} nucleus is easily found to be approximately

$$\log \tau_\gamma(Si^{28}) \approx -17.5 + \frac{3.36}{T_9^{1/3}} + \frac{58.3}{T_9}$$

This lifetime is shown in Fig. 7-2 for a relevant range of temperatures for the rearrangement process.

The flurry of nuclear activity following the photodisintegration of Si^{28} is difficult to describe. Each disintegration is rapidly followed by the photoejection of less tightly bound nucleons from the isotopes of Al, Mg, Ne, and O, and so at first glance it appears that the Si gas is decomposed into an alpha-particle gas. But the Si^{28} nuclei do not all break apart at the same time. Therefore the protons, neutrons, and alpha particles liberated by the disintegration of one Si^{28} nucleus may, in part, be captured by another Si^{28} nucleus; but those nuclei too are subject to reasonably rapid photodisintegration. Thus a great profusion of (α,γ), (p,γ), (n,γ) reactions and their inverses occur simultaneously. Each pair of reactions attempts to strike an equilibrium. In some cases, those corresponding to the fastest reactions, the equilibrium is achieved, but in the slower reactions it is not. It is Fig. 7-1 that contains the clue to what happens. Notice again that matter is more tightly bound in the form of an iron-group nucleus than it is in the form of two Si^{28} nuclei. It follows that the buildup of heavier nuclei in the reaction network is accompanied by an increase in nuclear binding. Superimposed on the relatively rapid pairs of inverse reactions, therefore, will be a slow leakage of nuclei from the intermediate-mass region toward the iron group. The nucleons liberated from the Si^{28} nuclei are thereby fused with other intermediate-mass

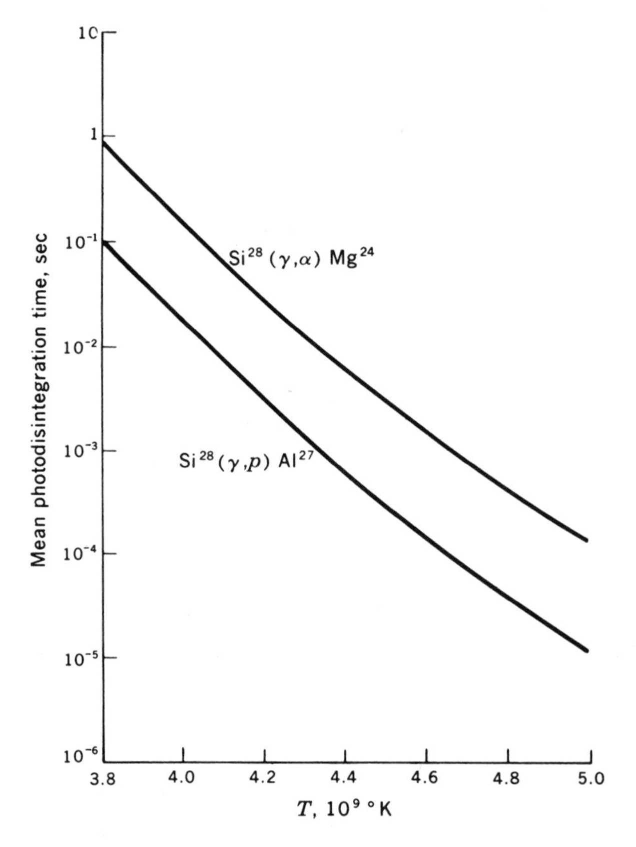

Fig. 7-2 The lifetime of the Si^{28} nucleus against photodisintegration. The lifetime has a strong temperature dependence, and the photoejection of protons is found to be somewhat faster than the photoejection of alpha particles. [*After J. W. Truran, A. G. W. Cameron, and A. A. Gilbert, Can. J. Phys.*, **44**:576 (1966).]

nuclei in a slow progression toward greater atomic weight. It should not be assumed that all the reactions are radiative, however. One must also include the strictly nuclear reactions (p,n), (α,n), (α,p) and their inverses, many of which are of importance in the total reaction network.

To systematically investigate what happens in the reaction network one must calculate the rates of all the reactions indicated, at least until it is determined which ones are of practical importance. The nuclear information is quite incomplete in this mass range, and it has been necessary in many cases to calculate the rates from the binding energies, which are mostly well known, plus estimates of nuclear level densities and level widths derived from semiempirical nuclear systematics. Fortunately, the well-known Q values are the nuclear quantities of greatest importance for this problem, and so the trend of events is not badly obscured by the relatively meager nuclear measurements.[1]

[1] The appendix to the paper by Truran, Cameron, and Gilbert, *op. cit.*, tabulates $\langle \sigma v \rangle$ for each reaction of importance in the network. This listing is particularly useful, but each entry must be multiplied by the factor 10^{-24} to convert to cgs units.

The number density $N(A,Z,t)$ of each nucleus is given by the differential equation linking all reactions that create or destroy that nucleus

$$\begin{aligned}
\frac{dN(A,Z)}{dt} = &-\lambda_\gamma(A,Z)N(A,Z) + \lambda_{\gamma,n}(A+1,Z)N(A+1,Z) \\
&+ \lambda_{\gamma,p}(A+1,Z+1)N(A+1,Z+1) \\
&\qquad + \lambda_{\gamma,\alpha}(A+4,Z+2)N(A+4,Z+2) \\
&+ n_n(t)[-\langle\sigma(A,Z)v\rangle_{n,\gamma}N(A,Z) + \langle\sigma(A-1,Z)v\rangle_{n,\gamma}N(A-1,Z)] \\
&+ n_p(t)[-\langle\sigma(A,Z)v\rangle_{p,\gamma}N(A,Z) \\
&\qquad + \langle\sigma(A-1,Z-1)v\rangle_{p,\gamma}N(A-1,Z-1)] \\
&+ n_\alpha(t)[-\langle\sigma(A,Z)v\rangle_{\alpha,\gamma}N(A,Z) \\
&\qquad + \langle\sigma(A-4,Z-2)v\rangle_{\alpha,\gamma}N(A-4,Z-2)] \\
&+ n_n(t)[-\langle\sigma(A,Z)v\rangle_{n,p}N(A,Z) + \langle\sigma(A,Z+1)v\rangle_{n,p}N(A,Z+1)] \\
&+ (p,n) + (n,\alpha) + (p,\alpha) + (\alpha,n) + (\alpha,p) \cdots \\
&- \lambda_\beta(A,Z)N(A,Z) + \lambda_\beta(A,Z-1)N(A,Z-1) \\
&- (\text{electron capture}) - (\text{positron decay}) \cdots
\end{aligned}$$

The student may profitably write out the other reaction-rate terms that have been suggested by the parenthetical notations. Note that each quantity λ_γ and $\langle\sigma v\rangle$ is a function only of the temperature.

Each differential equation appears rather formidable, even when the reactions are restricted to those involving α, n, and p. Further progress involves deciding which rates are big enough to warrant keeping track of. Considerable experience and thought are needed to see what approximations may safely be made, but, for example, one finds that for the light alpha-particle nuclei only the reactions (α,γ) and (γ,α) need be followed explicitly. It also happens that almost all beta-decay rates are insignificant, at least until most of the nuclei effectively reach the iron group. This last conclusion is extremely important, because if it is correct, it means that the total number of neutrons and protons remains equal. Of course, that cannot be strictly true, because some radioactive nuclei are produced, but for $T_9 > 3.5$ negligibly few beta decays occur during the time required for Si^{28} to be processed to the iron group by photodisintegration rearrangement. The abundances of radioactive nuclei are just too small. It should be added, however, that this conclusion rests on the existence of the neutrino emission processes implied by the universal Fermi interaction, as discussed in Chap. 3. If there should be no $(\bar{e}\nu)(\bar{\nu}e)$ interaction term, the photodisintegration rearrangement will take longer to occur because the temperature rise associated with gravitational contraction will occur more slowly. Or if the conversion occurs at $T_9 \leq 3$, it may occur so slowly that beta decays are significant. The alternatives are quite important for nucleosynthesis, as we shall soon see, because the final nuclear composition of the iron group depends markedly on whether beta decays occur during

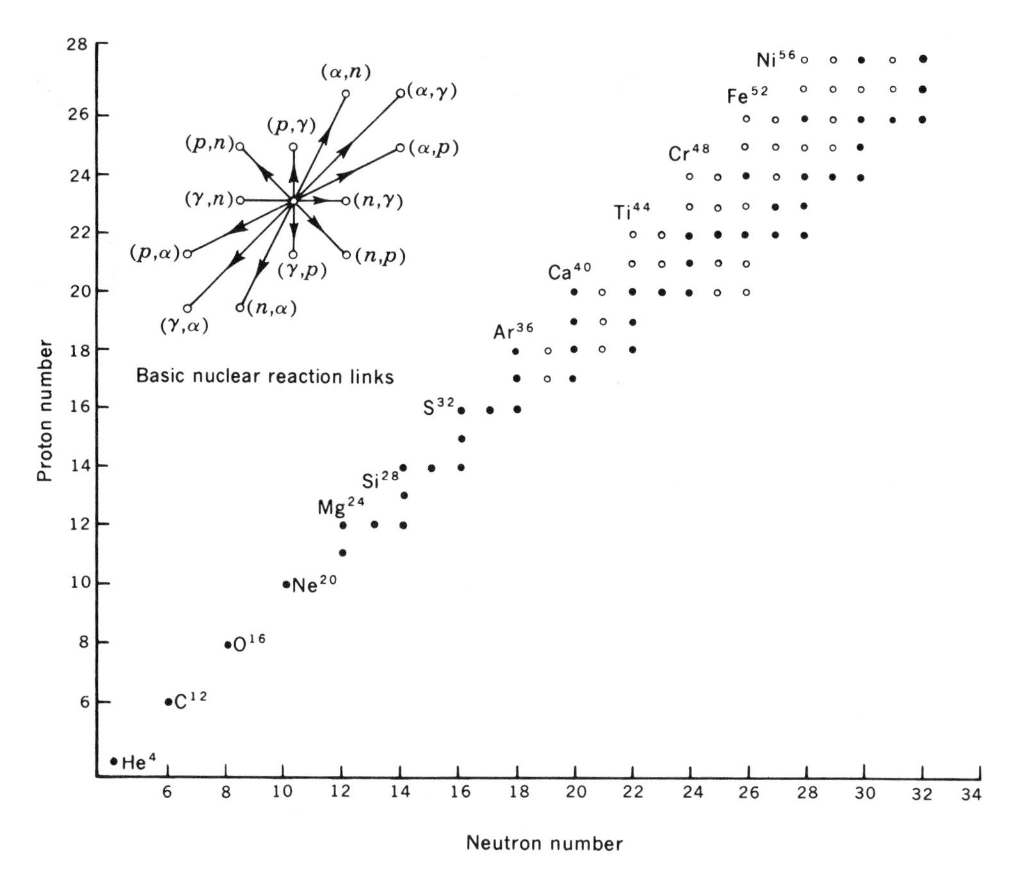

Fig. 7-3 The nuclides which participate in the reaction network established during silicon burning. The solid dots designate stable nuclei, whose relative abundances during silicon burning must be scrutinized for correlations with natural abundances. The open dots designate unstable nuclei. Their relatively low abundances may be very important for determination of the overall rate of beta decays. [*After J. W. Truran, A. G. W. Cameron, and A. A. Gilbert, Can. J. Phys.*, **44**:576 (1966).]

the time interval of the rearrangement. The nuclear-rearrangement time is governed by the rate of photodisintegration of Si^{28}, a rate that is strongly temperature-dependent. In the hottest environments, the time interval is much too short for beta decays to play any role, and the gas is constrained to maintain $\bar{Z} = \bar{N}$.[1] If the rearrangement occurs with $T_9 < 3$, however, it will take at least 10^6 sec for

[1] The statement $\bar{Z} = \bar{N}$ does not mean that only nuclei having $Z = N$ exist but rather that the density of protons (free plus bound) is equal to the density of neutrons (free plus bound). Should beta decay occur during the rearrangement, the iron group will be reached by a gas having $\bar{Z}/\bar{N} < 1$. The nuclear equilibrium of the iron group is strongly dependent on the value of $\bar{Z}/\bar{N}$. F. E. Clifford and R. J. Tayler, *Mem. Roy. Astron. Soc.*, **69**:21 (1965), have thoroughly discussed the way in which the equilibrium composition of nuclear matter depends on $\bar{Z}/\bar{N}$. We shall refer to this work again in the next section.

most of the Si^{28} to be converted to the iron group. During time intervals as long as this, the beta decays lowering the value of $\bar{Z}/\bar{N}$ will be quite important.

Analysis of the possible reactions between the nuclei shown in Fig. 7-3 and neutrons, protons, and alpha particles shows that many of the reactions can be neglected, some for energetic reasons and others because the nuclei involved are hopelessly rare. Each dot in Fig. 7-3 corresponds to a nucleus included in the analysis of the reaction network by Truran et al. The solid dots are stable, whereas the open ones correspond to the potentially important beta emitters. These unstable nuclei must be included in the reaction network even if the rearrangement is so rapid that beta decays are unimportant, because many of the important reaction links involve these nuclei; and, of course, the abundances of the unstable nuclei are needed to ascertain the extent of the beta decays. The upper left-hand corner of Fig. 7-3 shows the translation in charge and mass caused by each of the nuclear reactions involving protons, neutrons, and alpha particles.

The early phase of the nuclear rearrangement at $T_9 = 5$ is shown in Fig. 7-4. The composition at $t = 0$ was taken to be pure Si^{28}. The abundances of Mg^{24},

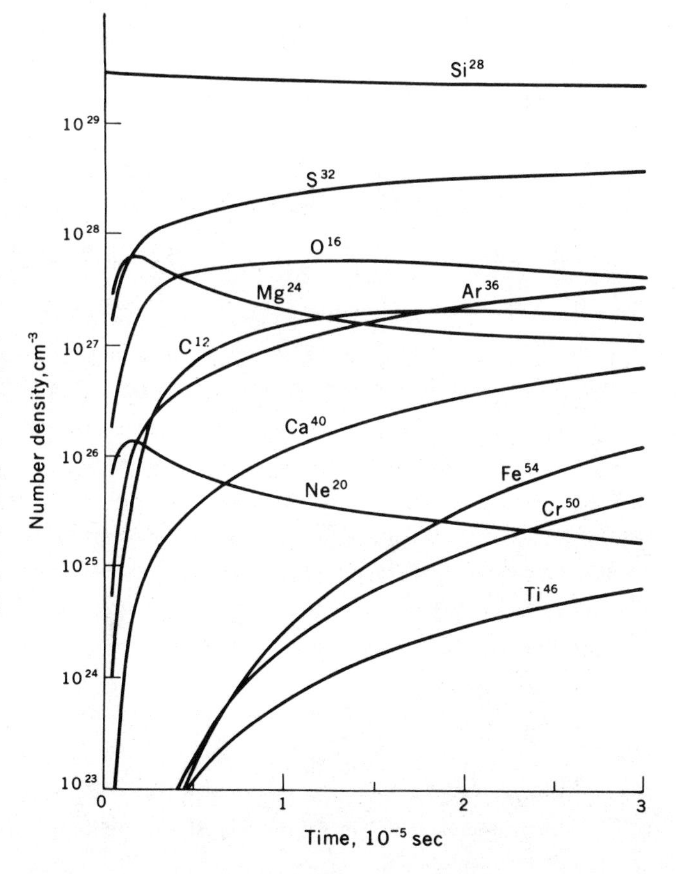

Fig. 7-4 The early phase of the nuclear rearrangement of initially pure Si^{28} at the temperature 5×10^9 °K and density 1.3×10^7 g/cm^3. The abundances grow very rapidly at first, when the liberated alpha particles are being consumed in a rapid flow toward the iron group. [*After J. W. Truran, A. G. W. Cameron, and A. A. Gilbert, Can. J. Phys.,* **44**:576 (1966).]

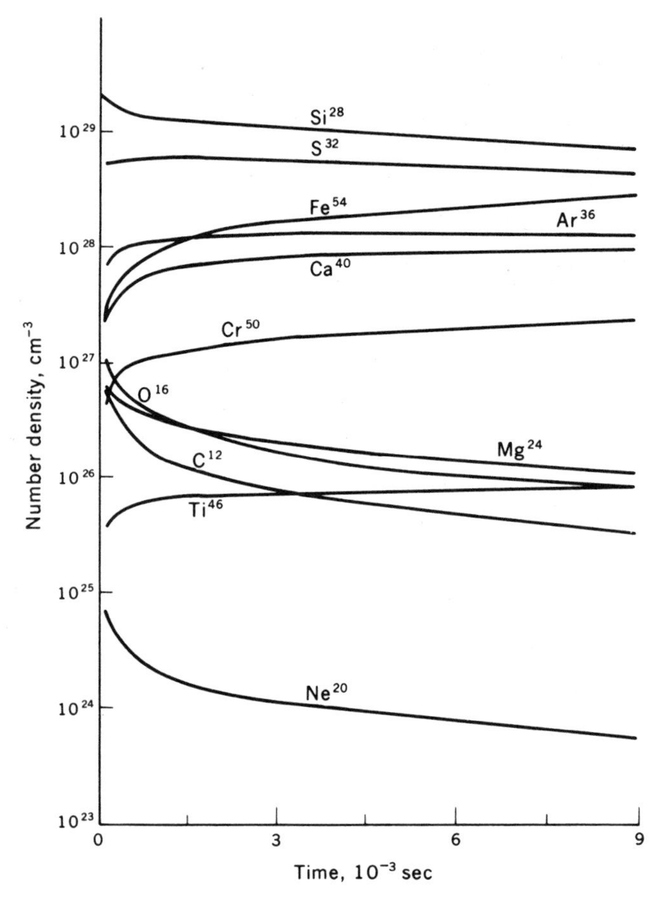

Fig. 7-5 Number densities of key nuclei during silicon burning at 5×10^9 °K at later times than those shown in Fig. 7-4. The densities of free protons and neutrons have become nearly quasistatic as a result of the near equilibration of their rates of capture and photoejection in the nuclei heavier than Si^{28}. All the heavier nuclei have approached a quasiequilibrium with Si^{28} and the pool of free nucleons, so that their abundances also change very slowly. [*After J. W. Truran, A. G. W. Cameron, and A. A. Gilbert, Can. J. Phys.*, **44**:576 (1966), *and D. Bodansky, D. D. Clayton, and W. A. Fowler (to be published).*]

Ne^{20}, O^{16}, and C^{12} build up rapidly as alpha particles are liberated. As the density of free alpha particles, protons, and neutrons is established, the rearrangement into heavier nuclei is facilitated, and S^{32}, Ar^{36}, Ca^{40}, and heavier nuclei in the iron group begin to build up. The number densities of key nuclei at later times are shown in Fig. 7-5. A very interesting thing becomes evident here. The secondary nuclei build up to an equilibrium with the residual Si^{28} and the light particles. To very good approximation, the rates of inverse reactions

$$Si^{28} + \alpha \rightleftharpoons S^{32} + \gamma$$
$$S^{32} + \alpha \rightleftharpoons Ar^{36} + \gamma$$
$$\cdots\cdots\cdots\cdots$$

become equal to each other. The abundances of the alpha-particle nuclei are then determined by the abundance of Si^{28} and the density of free alpha particles.

Problem 7-3: Show that the equality of the first pair of inverse reactions implies that the number densities satisfy the relationship

$$\frac{N(\mathrm{Si}^{28})N(\alpha)}{N(S^{32})} = \left(\frac{7}{2}\right)^{\frac{3}{2}} \left(\frac{2\pi M_u kT}{h^2}\right)^{\frac{3}{2}} \exp - \frac{Q_\alpha}{kT} \tag{7-3}$$

where M_u is the atomic mass unit and Q_α is the energy released by the reaction $\mathrm{Si}^{28}(\alpha,\gamma)\mathrm{S}^{32}$. This result has correctly assumed that the particles are spinless and have no important excited states. Similar statistical relationships can be written for the other inverse reactions that have achieved balance. A useful quantity appearing in all such equations is numerically

$$\theta \equiv \left(\frac{2\pi M_u kT}{h^2}\right)^{\frac{3}{2}} = 5.94284 \times 10^{33} T_9^{\frac{3}{2}} \qquad \mathrm{cm}^{-3} \tag{7-4}$$

Problem 7-4: Suppose that the density of alpha particles has risen to 2.7×10^{28} cm^{-3} during the time required for 75 percent of an initially pure Si^{28} gas at $\rho = 1.3 \times 10^7$ g/cm^3 to be rearranged at $T_9 = 5$ into heavier nuclei. Compute the number densities of S^{32}, Ar^{36}, and Ca^{40}, which have alpha-particle binding energies of 6.946, 6.640, and 7.044 Mev, respectively. Compare these results with Fig. 7-5 at $t = 9 \times 10^{-3}$ sec, which corresponds to the conditions assumed.

Problem 7-5: Show that if the heavy alpha nuclei are in equilibrium with the residual Si^{28} and the free-alpha-particle density, then as Si^{28} is burned, the value of N_α increases slowly as

$$N_\alpha = \left(\frac{\mathrm{Ni}^{56}}{\mathrm{Si}^{28}}\right)^{\frac{1}{7}} \frac{8\theta}{2^{\frac{3}{14}}} \exp\left(-\frac{1}{7}\sum_{A=32}^{56} \frac{Q_\alpha}{kT}\right) \tag{7-5}$$

where $\Sigma Q_\alpha = 49.382$ Mev is the sum of alpha-particle binding energies in the alpha nuclei. Show also that in the neighborhood of $T_9 = 3$ this result is numerically

$$N_\alpha \approx 10^{23.5} \left(\frac{T_9}{3}\right)^{28.8} \left(\frac{\mathrm{Ni}^{56}}{\mathrm{Si}^{28}}\right)^{\frac{1}{7}}$$

This result shows that N_α is strongly dependent on the temperature but only weakly dependent upon how much Si^{28} has been converted to Ni^{56}.

The many inverse reactions also maintain a steady-state density of the protons and free neutrons. The non-alpha-particle nuclei quickly come into equilibrium with the density of alpha-particle nuclei and this density of free nucleons, but (with the possible exception of the iron group) their abundances are much smaller than those of the alpha-particle nuclei. This situation is very reminiscent of the abundances found in nature, and it may be that silicon burning truncated before it is complete is the major source of nucleosynthesis between $A = 30$ and $A = 50$.

It may be helpful to provide a verbal description of this equilibrium that emphasizes the rates of key reactions. The intermediate nuclei are synthesized by alpha capture on Si^{28}, forming a chain of alpha-induced reactions. These reactions would exhaust the supply of free alpha particles were they not replenished by photoejection from nuclei heavier than Si^{28}. The Si^{28} itself is so tightly bound that its slow rate of disintegration allows that equilibrium to be established. Perhaps *silicon melting* is a more appropriate term than silicon burning because it emphasizes that the rate of the process is determined by the rate at which the silicon can be decomposed. Thus the reactions above Si^{28} come into equilibrium with the residual Si^{28} and the free densities of alpha particles, protons, and neutrons, whose gradual assimilation into heavier nuclei is replenished by the slow breakdown of Si^{28}.

Equilibrium is established in a few of the reactions below Si^{28}. Especially important is the fact that Mg^{24} comes into alpha-particle equilibrium with Si^{28}, an equilibrium in which $(Mg^{24})_{e\alpha} \approx 10^{-3}\ Si^{28}$. When this equilibrium is established, the rate of $Mg^{24}(\alpha,\gamma)Si^{28}$ very nearly cancels the rate of photodisintegration of Si^{28}. In this case the effective rate of photodisintegration of Si^{28} is given by the *leakage* from Mg^{24} by photodisintegration of the tightly bound Mg^{24}; that is, if alpha capture by Ne^{20} can be neglected, we have

$$\text{Photodisintegration rate of } Si^{28} \approx \lambda_\gamma(Mg^{24})(Mg^{24})_{e\alpha}$$

In the higher range of temperature, say $T_9 \approx 5$, many of the other reactions come into equilibrium also, but it is not too difficult to see how to compute the rate of flow downward from Mg^{24}. Let

$$J(A) = \lambda_\gamma(A)N_A - \langle\sigma_{\alpha,\gamma}v\rangle_{A-4}N_\alpha N_{A-4} \qquad (7\text{-}6)$$

be the effective alpha-particle flow downward from A. Because the Mg^{24} achieves equilibrium to high accuracy with Si^{28}, the effective rate of photodisintegration of Si^{28} is

$$\lambda_{\gamma,\text{eff}}Si^{28} = J(24) \qquad (7\text{-}7)$$

where we now have reverted to the convention of designating the number density by the chemical symbol. All the light alpha nuclei are much less abundant than Si^{28}, and so their abundances come to a steady state in which the downward photodisintegration current from them is replenished by the flow in from above. This demand means that the J's are equal:

$$J = J(24) = \lambda_\gamma(24)Mg^{24} - \langle\sigma v\rangle_{20}N_\alpha Ne^{20} \qquad (7\text{-}8a)$$

$$J = J(20) = \lambda_\gamma(20)Ne^{20} - \langle\sigma v\rangle_{16}N_\alpha O^{16} \qquad (7\text{-}8b)$$

$$J = J(16) = \lambda_\gamma(16)O^{16} - \langle\sigma v\rangle_{12}N_\alpha C^{12} \qquad (7\text{-}8c)$$

$$J = J(12) = \lambda_\gamma(12)C^{12} - r_{3\alpha} \qquad (7\text{-}8d)$$

where $r_{3\alpha}$ is the rate of the 3α reaction, which depends upon $N_\alpha{}^3$. This system of equations can be solved for J in terms of a ladder logic. For example, from Eq. (7-8*d*) we have

$$C^{12} = \frac{J + r_{3\alpha}}{\lambda_\gamma(12)}$$

which can be inserted into Eq. (7-8*c*) and solved for O^{16}:

$$O^{16} = \frac{1}{\lambda_\gamma(16)}\left[J + \frac{\lambda_\alpha(12)}{\lambda_\gamma(12)}(J + r_{3\alpha})\right]$$

Problem 7-6: Show by repeating this process to the top of the ladder that

$$J = \frac{\lambda_\gamma(24)\mathrm{Mg}^{24} - \dfrac{\lambda_\alpha(20)\lambda_\alpha(16)\lambda_\alpha(12)}{\lambda_\gamma(20)\lambda_\gamma(16)\lambda_\gamma(12)} r_{3\alpha}}{1 + \dfrac{\lambda_\alpha(20)}{\lambda_\gamma(20)}\left\{1 + \dfrac{\lambda_\alpha(16)}{\lambda_\gamma(16)}\left[1 + \dfrac{\lambda_\alpha(12)}{\lambda_\gamma(12)}\right]\right\}} \tag{7-9}$$

This result gives the effective photodisintegration rate of Mg^{24} in terms of the alpha-particle density and the temperature. Because Mg^{24} is always in near equilibrium with Si^{28}, it is also the rate of breakdown of Si^{28}. This result is the major one required to be able to calculate the time scale for the nuclear burning at a given set of conditions.

It turns out that $J \approx \lambda_\gamma(24)\mathrm{Mg}^{24}$ near $T_9 = 3$, but it falls to less than one-tenth of that value near $T_9 = 5$, because the radiative alpha captures become more significant corrections to the photodisintegration rates as the temperature increases.

As a result of this and similar cases of near balance, the effective rate of consumption of Si^{28} falls much below the photodisintegration rate of Si^{28}. The net flow downward, J, from Si^{28} is the excess of the photodisintegration rates over the inverse-fusion rates. Such a net leakage persists as long as Si^{28} remains a dominant member of the gas. As it is consumed, however, the ratio of the net downward flow rate to the photodisintegration rate decreases, as shown in Fig. 7-6. Inasmuch as the disintegrated nuclei are being used to build up toward the iron group, the rate of upward flow is also decreasing as the Si^{28} burns. Eventually all flow rates go to zero when the Si^{28} has been consumed, and nuclear statistical equilibrium is approached.

Figure 7-7 shows the net rates of nuclear flow in a typical case of rearrangement where Si^{28} is still the dominant nucleus. The intensity of the flow is designated by the intensity of the lines. The continued breakup of Si^{28}, Mg^{24}, Ne^{20}, O^{16}, and C^{12} into alpha particles produces a net flow downward. The recapture of the light nuclei produces a net flow upward. At the lower end the (α,γ) and (γ,α) reactions produce the major net flows, whereas the (α,p) reactions produce the major net flows above Ca^{40}. Figure 7-8 shows the abundance distribution after 10 sec at $T_9 = 4.2$. This figure, and others like it for different pairings of time and temperature, is remarkable in its similarity to the distribution of element abundances found in nature above Si^{28}. The alpha-particle nuclei dominate the abundances between Mg^{24} and Ca^{40}, there is a general abundance minimum between $45 < A < 50$, and there is an abundance peak in the iron group. This last feature occurs because the general fusion processes do not carry the matter to atomic weights much in excess of 60. This fact reemphasizes the physical content of Fig. 7-1; viz., the most tightly bound nucleons are those in nuclei in the iron group. The composition of the iron group is a sensitive function of the temperature and the extent of the beta decays. In the next section we shall see that the end product of the photodisintegration rearrangement is Ni^{56} over a wide range of temperatures and densities, in which case the rate of energy generation for the symbolic process $2\mathrm{Si}^{28} \rightarrow \mathrm{Ni}^{56}$ can be written in cgs units as[1]

$$\log \epsilon_{\mathrm{Si}} \approx 30.47 + \log X_{\mathrm{Si}} - \tfrac{1}{7} \log \frac{1 - X_{\mathrm{Si}}}{X_{\mathrm{Si}}} + 6.31 \log \frac{T_9}{3} - \frac{61.67}{T_9} \tag{7-10}$$

If, on the other hand, the temperature is too high, the major product may be $\mathrm{Fe}^{54} + 2p$, in which case

$$\epsilon_{\mathrm{Si}} \approx 0$$

To gain further insight into this unusual situation, we must consider the problem of nuclear statistical equilibrium.

[1] For a clear discussion of energy generation, the reader must consult Bodansky et al., *op. cit.*

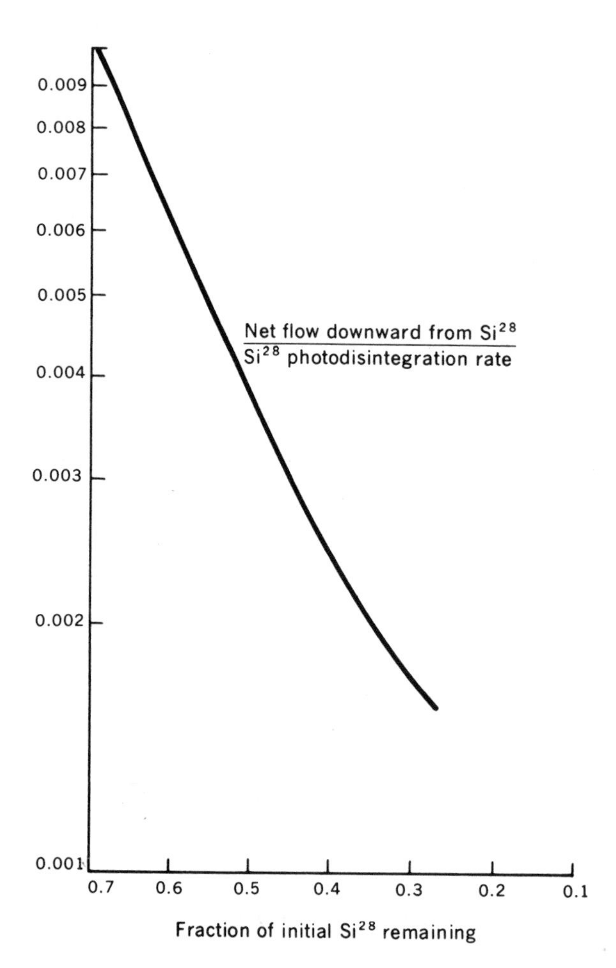

Fig. 7-6 The ratio of the disintegration current J to the rate of Si^{28} photodisintegrations. The much smaller value for J reflects the fact that the Si^{28} photodisintegration is almost exactly balanced by an upward current from Mg^{24}. As the burning progresses, the disintegration current falls markedly. [*After J. W. Truran, A. G. W. Cameron, and A. A. Gilbert, Can. J. Phys.*, **44**:576 (1966), *and D. Bodansky, D. D. Clayton, and W. A. Fowler (to be published)*.]

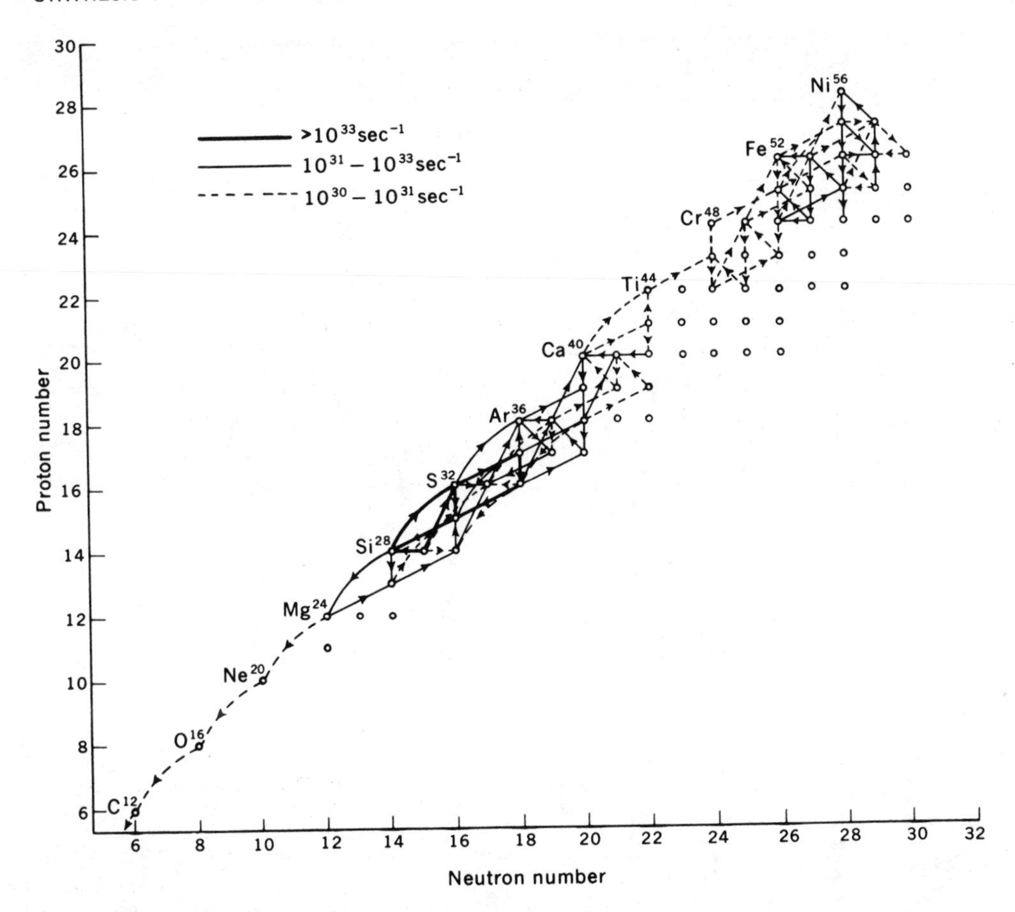

Fig. 7-7 A typical example of the net flow due to major reactions in the silicon-burning network. Each reaction is partially balanced by the inverse reaction, and the magnitude of the excess is here designated by the intensity of the lines connecting the nuclei linked by that reaction. As the burning progresses, each flow becomes a relatively small difference between two large and nearly equal opposing rates, in which case the composition of the gas may be approximated by an equilibrium. [*After J. W. Truran, A. G. W. Cameron, and A. A. Gilbert, Can. J. Phys.*, **44**:576 (1966).]

7-2 NUCLEAR STATISTICAL EQUILIBRIUM AND THE *e* PROCESS

Understanding of the composition of the iron group of elements is greatly aided by the study of nuclear statistical equilibrium. Suppose that the rates of all nuclear reactions (excepting beta decays) are exactly equal to the rates of the inverse reactions. During silicon burning these rates are only very nearly equal, because there is a small net flow of intermediate-mass nuclei toward the iron group superimposed upon the much faster pairs of inverse reactions. But even

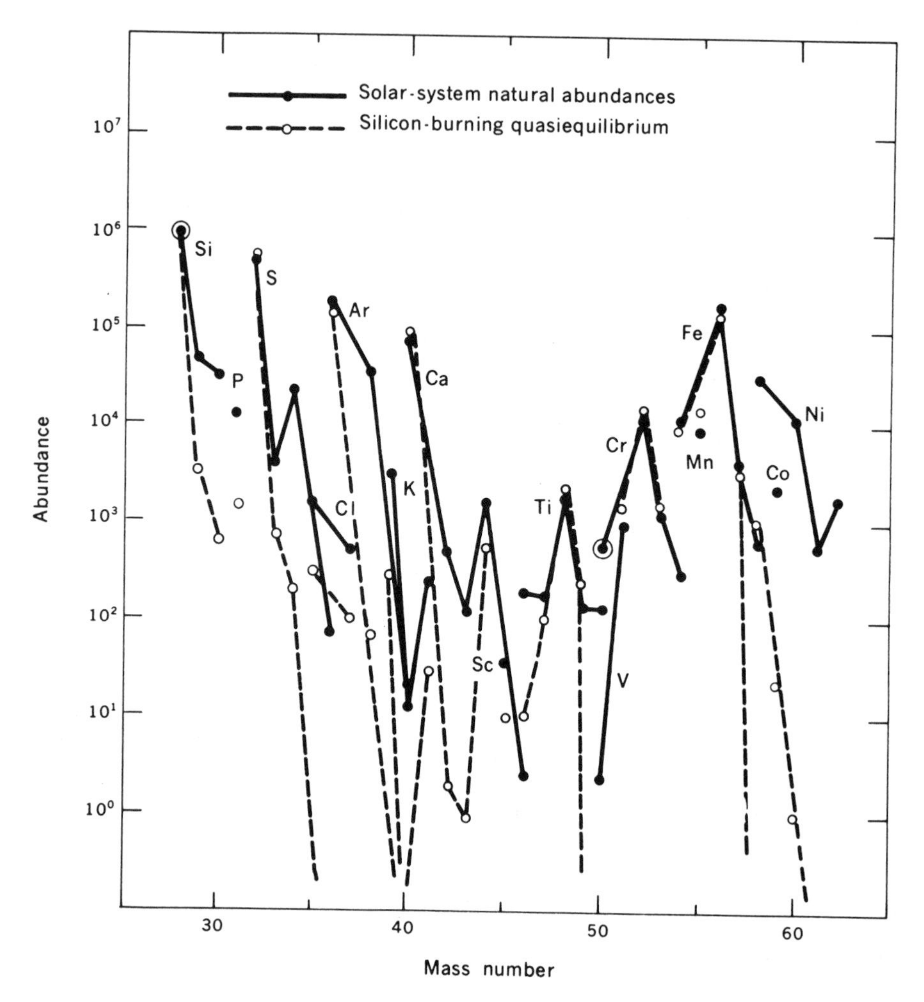

Fig. 7-8 A comparison of the abundances during silicon burning with the natural solar-system abundances. In this case, which is typical, the silicon has been burning for 10 sec at the temperature $T_9 = 4.2$, and about 35 percent of the original silicon remains. The solid dots represent the natural solar abundances, and isotopes of the same element are connected by a solid line. The open dots represent the quasiequilibrium abundances of the silicon-burning process, and isotopes of the same element are connected by dashed lines. The natural abundance of Fe^{56} is compared with the quasiequilibrium abundance of Ni^{56}, which decays after expulsion. The great similarity between the abundances of the most abundant nuclei strongly suggests that truncated silicon burning has been important in nucleosynthesis. [*After D. Bodansky, D. D. Clayton, and W. A. Fowler, Phys. Rev. Letters,* **20**:161 (1968)].

then the rates of inverse reactions within the iron peak are balanced to considerable accuracy. As the Si^{28} disappears, the gas is more and more describable as being in a steady state with respect to the nuclear reactions. In such a state, the nuclear abundances are determinable from statistical principles. The situation is exactly analogous to that of the ionization equilibrium of an atomic gas. The ratio $N(A-1, Z)/N(A,Z)$ is determined by the free-neutron density and the temperature,

$$\frac{N(A-1, Z)n_n}{N(A,Z)} = \frac{2G(A-1, Z)}{G(A,Z)} \frac{(2\pi\mu kT)^{\frac{3}{2}}}{h^3} \exp - \frac{Q_n}{kT}$$

$$= \frac{2G(A-1, Z)}{G(A,Z)} \left(\frac{A-1}{A}\right)^{\frac{3}{2}} \theta \exp - \frac{Q_n}{kT} \quad (7\text{-}11)$$

where Q_n is the binding energy of a neutron in the nucleus (A,Z) and $\theta = (2\pi M_\mu kT)^{\frac{3}{2}}/h^3$. By the same reasoning,

$$\frac{N(A-2, Z-1)n_p}{N(A-1, Z)} = \frac{2G(A-2, Z-1)}{G(A-1, Z)} \left(\frac{A-2}{A-1}\right)^{\frac{3}{2}} \theta \exp - \frac{Q_p}{kT} \quad (7\text{-}12)$$

where Q_p is the proton binding energy in $(A-1, Z)$. Note that if one takes the product of Eqs. (7-11) and (7-12), all the properties of the nucleus $(A-1, Z)$ divide out except its binding energy. In fact a whole series of such equations can be written, until the nucleus (A,Z) is related to the number densities of free neutrons and protons only.

Problem 7-7: Show by repeated action of this nuclear Saha equation that

$$N(A,Z) = G(A,Z)A^{\frac{3}{2}} \frac{n_p{}^Z n_n^{A-Z}}{2^A} \theta^{1-A} \exp \frac{Q(A,Z)}{kT} \quad (7\text{-}13)$$

where

$$Q(A,Z) = c^2[ZM_H + (A-Z)M_n - M(A,Z)] \quad (7\text{-}14)$$

is the binding energy of the nucleus (A,Z). Note that the abundance of every nuclide is given in terms of its properties and the densities of free protons and neutrons. It may also be noted that the principles of statistical mechanics yield Eq. (7-13) by a more direct route than the chain argument applied here.

Problem 7-8: Show that when Eq. (7-13) is applied to the two nuclei (A,Z) and $(A-1, Z)$, the ratio of results is Eq. (7-11).

Equation (7-13) is in itself inadequate to yield the equilibrium abundance of each nucleus, because n_p and n_n are not given; but if, as is usually the case, nuclear equilibrium is achieved faster than any significant number of beta decays can occur, we must impose an auxiliary constraint: the total number densities (free plus bound) of protons and neutrons must preserve the ratio

$$\frac{\bar{Z}}{\bar{N}} = \frac{\Sigma ZN(A,Z) + n_p}{\Sigma(A-Z)N(A,Z) + n_n} \quad (7\text{-}15)$$

Inasmuch as the number density of nucleons (free plus bound) is determined by the mass density, it follows that the values of ρ, T, and $\bar{Z}/\bar{N}$ prescribe the equilibrium abundance of each nucleus.[1]

Beta decays play a special role in nuclear equilibrium. They cannot participate in a true equilibrium because the inverse reaction to a beta decay requires the absorption of a neutrino. For example, the inverse of the neutron decay $n \rightarrow p + e^- + \bar{\nu}$ is the reaction $p + e^- + \bar{\nu} \rightarrow n$. Because neutrinos escape from the local environment (at least in the astrophysical circumstances under discussion here), their density is much too low for the inverse reaction to have a significant rate. Thus true equilibrium is impossible. The manner of including beta decays is as follows. The equilibrium with respect to the strong (nuclear) interactions is computed with the constraint $\bar{Z}/\bar{N} = (\bar{Z}/\bar{N})_t$. The abundances of the radioactive nuclei and their decay rates yield the rate at which $\bar{Z}/\bar{N}$ is changing. Then the equilibrium a short time Δt later is computed with the constraint

$$\left(\frac{\bar{Z}}{\bar{N}}\right)_{t+\Delta t} \approx \left(\frac{\bar{Z}}{\bar{N}}\right)_t + \Delta t \frac{\partial}{\partial t}\left(\frac{\bar{Z}}{\bar{N}}\right)_t \tag{7-16}$$

This approximation is justifiable only because the beta rates are sufficiently slow so that equilibrium at any specific value of $\bar{Z}/\bar{N}$ can be established in a time short compared to the time required for a significant change in the value of $\bar{Z}/\bar{N}$. Needless to say, this assumption must be checked for self-consistency, because environments may be imagined where the approximation is invalid.

The nature of the beta decays is somewhat different than in the laboratory. Because of the high temperatures involved in the approach to nuclear equilibrium, the beta decays of thermally populated excited states make a significant contribution. It is necessary to estimate the beta-decay rate of each excited state in order to assess its contribution to the total decay rate. This can usually be done on the basis of the spins and parities of the states and the systematics of nuclear beta decay. Even more important is the capture of continuum electrons from the gas. The most tightly bound nuclei in the iron group are not those with $Z = N$ but those with neutron excesses of two or four. The nuclear binding is increased, therefore, by those decays which change protons into neutrons, positron decay or electron capture. In the process under discussion, the free-electron density is about 100 times greater than it is at the nucleus of a terrestrial iron-group nucleus. As a result, the electron capture in the star in many cases significantly shortens the nuclear lifetime.[2] As we shall see, these decay rates

[1] A particularly thorough study of nuclear statistical equilibrium under conditions likely to prevail during formation of the iron group has been provided by Clifford and Tayler, *op. cit.* For results at extreme density, see S. Tsuruta and A. G. W. Cameron, *Can. J. Phys.*, **43**:2056 (1965).

[2] A thorough discussion of beta-interaction rates within this context is given in Fowler and Hoyle, *op. cit.*, app. A, and in C. J. Hansen, Ph.D. Thesis, Yale University, New Haven, Conn., 1967. Hansen's very complete work has also been issued as "Neutrino Emission from Dense Stellar Interiors," Goddard Institute for Space Studies, New York, 1967.

must be followed with care, because even a small decrease in $\bar{Z}/\bar{N}$ greatly affects the composition of the iron group.

For values of $\bar{Z}/\bar{N}$ very near unity, it turns out that the most abundant iron-group nucleus is either Ni^{56} or Fe^{54}. Important questions hinge upon the identity of the dominant nucleus. In the first place, the conversion $2Si^{28} \rightarrow Ni^{56}$ is exothermic by 10.9 Mev and may stall collapse for a short while, whereas the conversion $2Si^{28} \rightarrow Fe^{54} + 2p$ is slightly endothermic by -1.3 Mev and can provide no hindrance to the collapse. In the absence of beta decays ($\bar{Z}/\bar{N} = 1$), the dominant nucleus at moderate temperatures ($T_9 < 5$, say) is Ni^{56}, because it has the greatest binding energy per nucleon of the $Z = N$ nuclei. At very high temperatures, however, the intense photon flux keeps two protons driven off, and one obtains $Fe^{54} + 2p$ as the dominant component. For values of $\bar{Z}/\bar{N}$ near 0.97, there is a sufficient neutron excess for Fe^{54} to be the dominant nucleus even at moderate temperatures. In this case the effective reaction can be summarized as $54Si^{28} \rightarrow 28Fe^{54}$, which has been accompanied by 28 beta decays. Relatively little energy is associated with the beta decays, and this process releases about 15 Mev per Fe^{54} nucleus produced. At even smaller values of $\bar{Z}/\bar{N} \approx 0.90$, the dominant nucleus is generally Fe^{56}, which has the maximum binding energy per nucleon of all nuclei.

Problem 7-9: The binding energy of Si^{28} is 8.448 Mev per nucleon, that of Ni^{56} is 8.644 Mev per nucleon, and that of Fe^{54} is 8.736 Mev per nucleon. Calculate the increase in nuclear binding per iron-group nucleus produced associated with the reactions $2Si^{28} \rightarrow Ni^{56}$, $2Si^{28} \rightarrow Fe^{54} + 2p$, $54Si^{28} \rightarrow 28Fe^{54}$.

Another issue at stake here is the large natural abundance of Fe^{56}. Reference to Fig. 1-22 shows that iron in the solar system is much more abundant than its immediate neighbors. Observations of stellar spectra confirm this result to be a general abundance phenomenon. From the point of view of nucleosynthesis it is comfortable to incorporate this large abundance of Fe^{56} into the theory if it can be interpreted as a natural consequence of nuclear equilibrium. Burbidge et al.[1] tried, with encouraging results, to match the solar abundances to nuclear equilibrium in a gas having a sufficiently small value of $\bar{Z}/\bar{N}$ for Fe^{56} to be the dominant nucleus. They dubbed this nuclear process the *e process* as a mnemonic for *equilibrium*. In their 1964 investigation, Fowler and Hoyle found that the rearrangement of Si^{28} into iron-group nuclei occurs too quickly at $T_9 = 3.5$ for beta decays to reduce $\bar{Z}/\bar{N}$ from its initial value of unity to a value low enough for Fe^{56} to dominate. In their model Si^{28} is converted to Ni^{56}, which decays later into Fe^{56}. Not only is the explanation of the large natural Fe^{56} abundance retained thereby, but elegant auxiliary consequences emerge from the model. If the conversion results largely in Fe^{54}, however, it will be somewhat more of a problem to understand the large Fe^{56} abundance.

Because of the importance of this problem of the competition between Ni^{56}

[1] E. M. Burbidge, G. R. Burbidge, W. A. Fowler, and F. Hoyle, *Rev. Mod. Phys.*, **29**:547 (1957).

and Fe^{54}, it is advisable to examine the physical principles of the competition in terms of a slightly simplified model. The ratio of Fe^{54} to Ni^{56} in nuclear statistical equilibrium may be expressed by the balance of the reactions $Fe^{54} + p \rightleftharpoons Co^{55} + \gamma$ and $Co^{55} + p \rightleftharpoons Ni^{56} + \gamma$. By combining the two equations of the form of Eq. (7-12) we have

$$\frac{Fe^{54}}{Ni^{56}} n_p^2 = \frac{4G(Fe^{54})}{G(Ni^{56})} \left(\frac{54}{56}\right)^{\frac{3}{2}} \theta^2 \exp - \frac{Q_p(55) + Q_p(56)}{kT} \tag{7-17}$$

Problem 7-10: The proton binding energies in Ni^{56} and Co^{55} can be calculated from Table 4-1. The ratio of partition functions is near unity because both Fe^{54} and Ni^{56} are even-even nuclei with $J = 0$ whose excited states lie too high to contribute much. Show that

$$\frac{Fe^{54}}{Ni^{56}} n_p^2 = T_9^3 10^{68.13-62.09/T_9} \tag{7-18}$$

The results of the previous problem can be easily employed to investigate the competition between the Ni^{56} and Fe^{54} phases of the nuclear gas. First consider the case $\bar{Z}/\bar{N} = 1$, which will correspond to a conversion of Si^{28} to iron-group nuclei in a time short compared to an effective beta-decay lifetime of the gas. Near those thermodynamic conditions where Ni^{56} and Fe^{54} are of comparable abundance, the gas may be regarded as having to good approximation only two nuclear components, ${}_{28}Ni_{28}^{56}$ and ${}_{26}Fe_{28}^{54} + 2p$. Thus in those circumstances we have $n_p \approx 2Fe^{54}$ to preserve the charge ratio. Then Eq. (7-18) becomes

$$\frac{(Fe^{54})^3}{Ni^{56}} \approx T_9^3 10^{67.53-62.09/T_9} \tag{7-19}$$

What we wish to know is the location of the strip in the ρT plane for which $Fe^{54} \approx Ni^{56}$. Substituting that equality, expressing the number density of Fe^{54} in terms of the mass density, and setting $X_{54} = X_{56} = \frac{1}{2}$, we have

$$\rho^2 \approx T_9^3 10^{24.09-62.09/T_9} \qquad X_{54} \approx X_{56} \approx \tfrac{1}{2} \qquad \frac{\bar{Z}}{\bar{N}} = 1 \tag{7-20}$$

Equation (7-20) locates the transition strip separating regions where Ni^{56} or $Fe^{54} + 2p$ dominate the composition. It is displayed in Fig. 7-9 for values of the density between $\rho = 10^6$ and 10^8 g/cm^3. In the lower-temperature region Ni^{56} is the dominant equilibrium nucleus, whereas $Fe^{54} + 2p$ dominates at temperatures high enough to keep two protons driven off the Ni^{56} nucleus. At very high temperatures, the equilibrium shifts to 14 alpha particles, a fact that will be of importance in triggering supernovas.

Problem 7-11: Use Eq. (7-13) to obtain the equilibrium densities of He^4 and of Ni^{56}. Then show they are related to each other as

$$\frac{N_\alpha^{14}}{Ni^{56}} = \frac{4^{21}}{56^{\frac{3}{2}}} \theta^{13} \exp \frac{14Q(\alpha) - Q(Ni^{56})}{kT} \tag{7-21}$$

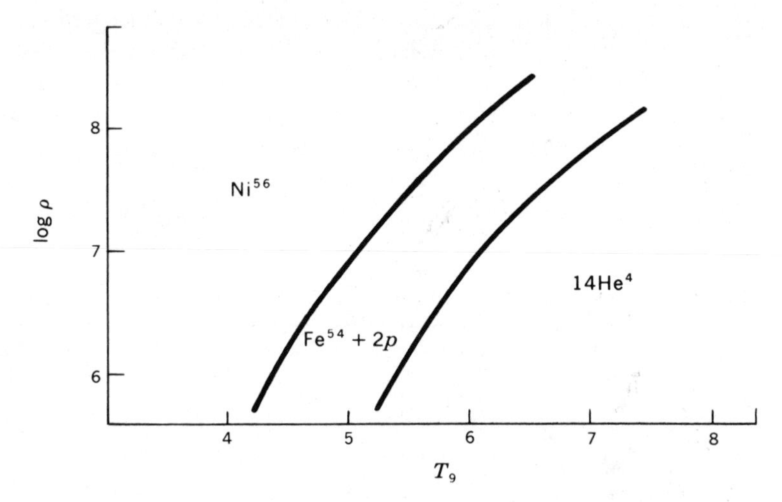

Fig. 7-9 The dominant nuclear constituent in a gas in nuclear statistical equilibrium when $\bar{Z}/\bar{N} = 1$.

where $Q(\alpha)$ and $Q(\mathrm{Ni}^{56})$ are the binding energies of He^4 and Ni^{56} in the sense of Eq. (7-14); that is, $Q(\mathrm{Ni}^{56}) - 14Q(\alpha)$ represents the energy required to dissociate Ni^{56} in 14 alphas. Using the masses of Table 4-1, show that the fractions by weight of Ni^{56} and He^4 are equal when

$$X_\alpha{}^{13}\rho^{13} = \left(\frac{T_9}{3}\right)^{\frac{39}{2}} 10^{157.47-442.9/T_9} \tag{7-22}$$

This relationship has been plotted for X_α near 0.5 in Fig. 7-9 as a rough indication of the transition to an alpha-particle gas. The calculations displayed in Fig. 7-9 are illustrative of the physical ideas involved. Neither boundary is sharp. Near the second boundary, particularly, many secondary nuclei exist.

In summarizing the physical ideas illustrated in Fig. 7-9, we may say that the photodisintegration of Si^{28} maintaining $\bar{Z} = \bar{N}$ will lead to Ni^{56} with energy release for ρT in the left-hand region, will lead to Fe^{54} without energy release in the middle region, and will decompose into alpha particles in the right-hand region. Many other nuclei exist in smaller abundance in each zone. In particular $\mathrm{Fe}^{54} + 2p$ is accompanied by $\mathrm{Co}^{55} + p$, $\mathrm{Fe}^{55} + 3p$, $\mathrm{Ni}^{58} + 2p$, and others.

Now suppose that the rearrangement occurs slowly enough for the number of beta decays not to be negligible. At densities greater than 10^6, for example, the electron-capture lifetimes of Ni^{56} and Co^{55} are on the order of 1 min or less. Since the abundance of these and other radioactive nuclei are a few percent or less during the rearrangement process, we may expect a significant number of beta decays (1 percent or so) to occur if the time of the rearrangement is 10^4 to 10^5 sec, say, or greater. Because the rearrangement time is a strongly decreasing function of the temperature, rearrangement times this long will of necessity be restricted to temperatures less than $T_9 \approx 3.5$, say. One cannot be more precise at the time of writing because the details of the problem are still unsolved, but it is easy enough to see what the results of a few beta decays will be. A small neutron excess will allow the dominant nucleus to be one with a neutron excess. In the iron region, the most stable nuclei are those having excess neutrons, the most tightly bound Fe^{56} having $N = Z + 4$. At low temperatures, neutron-rich nuclei cannot dominate unless $\bar{Z}/\bar{N}$ is reduced to a value near that of Z/N for the nucleus in question. On the basis of this simple principle alone, the equilibrium switches from Ni^{56} to Fe^{54} near

$$\frac{\bar{Z}}{\bar{N}} \approx \frac{26}{28} = 0.93 \qquad Fe^{54} \text{ maximum}$$

and it switches to Fe^{56} near

$$\frac{\bar{Z}}{\bar{N}} \approx \frac{26}{30} = 0.87 \qquad Fe^{56} \text{ maximum}$$

Because the transition to Fe^{54} occurs with many fewer decays, it is the one of overriding importance for nuclear equilibrium in the important temperature region $3 < T_9 < 4$. It is instructive to consider this situation further by returning to Eq. (7-18), which expresses the competition between Ni^{56} and Fe^{54}, which we may safely assume to be the dominant iron-group nuclei for values of $\bar{Z}/\bar{N}$ slightly less than unity. Near the transition strip, we no longer have $n_p \approx 2Fe^{54}$, however, because beta decays will have greatly reduced the free-proton density. The approximate condition of the two-component gas may be expressed by assuming that essentially all the neutrons are bound in either Ni^{56} or Fe^{54} nuclei. Then for the nucleon densities (free plus bound),

$$\begin{aligned} \bar{Z} &\approx 28Ni^{56} + 26Fe^{54} + n_p \\ \bar{N} &\approx 28Ni^{56} + 28Fe^{54} \end{aligned} \tag{7-23}$$

It should be noted that these equations are incorrect if there exists a large component of unburned silicon; therefore, the two-component model can be an appropriate approximation only for complete equilibrium.

Problem 7-12: Show that in the two-component model the proton density near the transition strip can be estimated as

$$n_p \approx \frac{\rho N_0}{2} X_{56} \left[\frac{Fe^{54}}{Ni^{56}} \left(\frac{\bar{Z}}{\bar{N}} - 0.929 \right) + \frac{\bar{Z}}{\bar{N}} - 1 \right] \tag{7-24}$$

It is now possible to make the following argument. Suppose we concentrate on a combination of ρT having the property that if $\bar{Z} = \bar{N}$, the equilibrium would have strongly favored Ni^{56}. That is, consider a ρT that would have given from Eq. (7-19) the result $X_{54}{}^3/X_{56} \ll 1$. It is then possible to show that one can have X_{54} and X_{56} of the same order of magnitude only if

$$\left[\frac{\mathrm{Fe}^{54}}{\mathrm{Ni}^{56}}\left(\frac{\bar{Z}}{\bar{N}} - 0.929\right) + \left(\frac{\bar{Z}}{\bar{N}} - 1\right)\right]^2 \approx 0 \tag{7-25}$$

Problem 7-13: Confirm the above assertion.

From Eq. (7-25) we have at once for the two-component gas that near the transition strip between Fe^{54} and Ni^{56}

$$\frac{\mathrm{Fe}^{54}}{\mathrm{Ni}^{56}} \approx \frac{1 - \bar{Z}/\bar{N}}{\bar{Z}/\bar{N} - 0.929} \tag{7-26}$$

in a ρT region where Ni^{56} would be strongly favored if $\bar{Z} = \bar{N}$. This result is plotted in Fig. 7-10, where it can be seen that the transition from Ni^{56} to Fe^{54} occurs when the beta decays have reduced $\bar{Z}/\bar{N}$ to a value near 0.97. This implies that if only 1.5 percent of the protons have made beta transitions into neutrons during the rearrangement, the equilibrium can be shifted from Ni^{56} to Fe^{54}. The switch occurs at values of $\bar{Z}/\bar{N}$ much nearer to unity, however, if there is a large component of unburned silicon. If the rearrangement occurs slowly enough to allow Fe^{54} to dominate, there is an energy release of about 15 Mev per Fe^{54} nucleus formed, but it becomes impossible to account for the large natural abundance of Fe^{56} by such rearrangements. The dominance of Fe^{56} in nature allows the following conclusion: *the e process in stars has either occurred so rapidly that the dominant nucleus produced is* Ni^{56} *(which decays later to* Fe^{56}*) or has proceeded so slowly that beta decays have reduced* $\bar{Z}/\bar{N}$ *to a value so low (near* 0.87*) that* Fe^{56} *is the dominant equilibrium nucleus.*

On the basis of the foregoing principles, one can easily understand the results of full calculations of equilibrium composition.[1] Figure 7-11 displays the frac-

[1] Extensive tables and discussion have been published by Clifford and Tayler, *op. cit.* Their results are strongly recommended for the serious student.

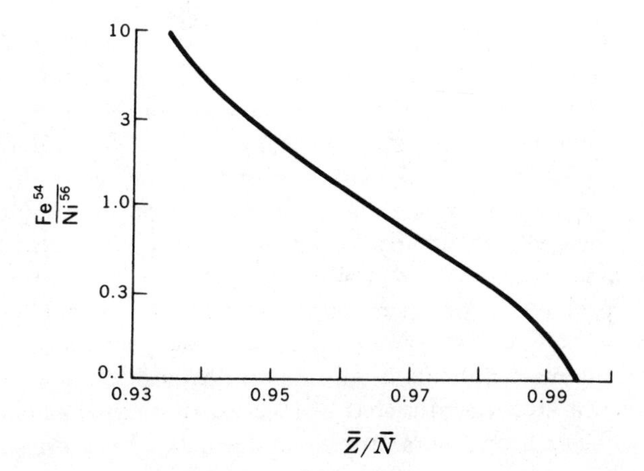

Fig. 7-10 The ratio of Fe^{54} to Ni^{56} in a gas in complete nuclear statistical equilibrium at a temperature and density such that Ni^{56} would be the dominant nucleus at $\bar{Z}/\bar{N} = 1$. This result is not valid in a silicon-burning quasiequilibrium, however, if there is a large component of unburned silicon.

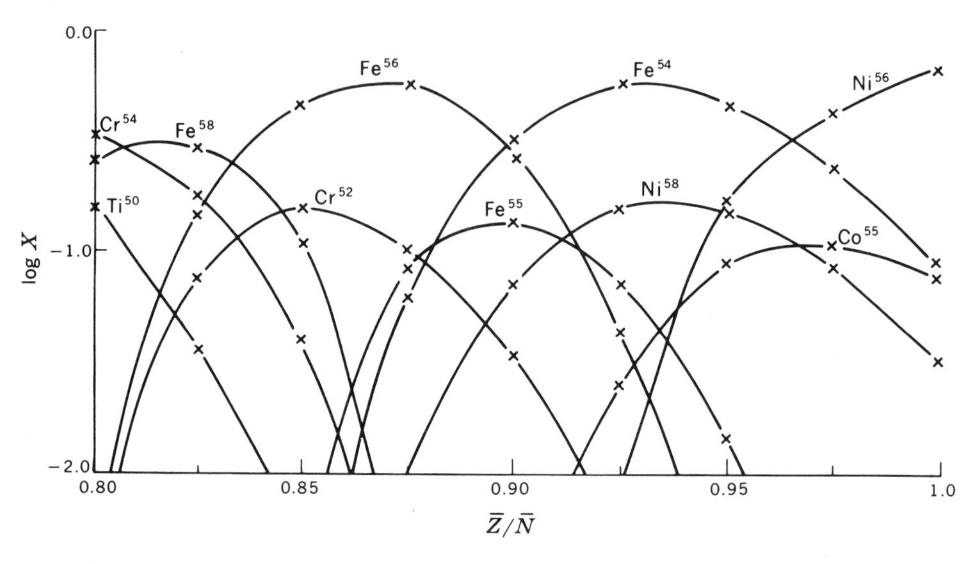

Fig. 7-11 Mass fractions of the most abundant nuclear species in nuclear statistical equilibrium at $T_9 = 4$ and $\rho = 10^6$ g/cm³. As the value of $\bar{Z}/\bar{N}$ is reduced, the equilibrium shifts to more neutron-rich nuclei. [*After F. E. Clifford and R. J. Tayler, Mem. Roy. Astron. Soc.*, **69**:21 (1965).]

tional equilibrium abundances by weight of several key nuclei as a function of $\bar{Z}/\bar{N}$. The particular thermodynamic circumstances of this example are $\rho = 10^6$ and $T_9 = 4$, but the qualitative nature of the results is much more general. The transition $Ni^{56} \rightarrow Fe^{54} \rightarrow Fe^{56}$ in the identity of the dominant nucleus is evident, along with a corresponding transition in the identity of the major secondary nucleus.

The extent of the reduction of $\bar{Z}/\bar{N}$ from its initial value of unity will clearly depend upon the length of time required for the rearrangement of a major portion of the Si^{28} into iron-group nuclei. This time depends upon the temperature of the rearrangement. In Fig. 7-12 the final mass fractions of these three dominant nuclei are schematically displayed as a function of temperature. It is clear that an important problem confronting astrophysics is that of sufficient understanding of the astrophysical environment to determine the temperature of the *e* process. In the evolution of the star, the temperature and time scale will be dominated by the energy-balance equations. Neutrino losses are extensive at the temperatures involved, and the dynamics of the late evolutionary stages is poorly understood. When it is finally understood, we shall be faced with the problem of which mass zones are reejected to the interstellar medium and with what final history of density and temperature. This last problem remains one of the most difficult in all of astrophysics.

A stellar center that has passed through the fiery furnace of the *e* process seems headed for a catastrophic holocaust. No more nuclear energy remains, but the neutrino losses are large and insistent, so that the core contracts and the temperature continues to rise. Perhaps there may exist stars at this stage with sufficiently small mass to be stabilized by electron degeneracy, but at least for those stars with $\mathfrak{M} > 1.5\mathfrak{M}_\odot$ the contraction and heating continues. In massive stars the sequence of events seems inescapable. As the temperature marches upward, the iron-group nuclei are photodisintegrated into alpha particles and neutrons. Figure 7-9 showed the approximate location of the onset of $Ni^{56} \rightarrow 14He^4$, but similar curves exist for other iron-group nuclei, $Fe^{54} \rightarrow 13He^4 + 2n$, $Fe^{56} \rightarrow 13He^4 + 4n$, etc. In any case, each photodisintegration absorbs about 100 Mev from the photon gas and constitutes a highly endoergic undoing of the whole history of fusion at the center of the star. That drain on the internal thermal energy (to say nothing of the additional neutrino losses) can only be provided by one remaining source, the compressional work accompanying gravitational collapse. As the collapse accelerates, the temperature rises further, and the alpha particles give way to photodisintegration at the high price of 28 Mev per alpha particle. The contraction must be so rapid that the core is nearly in a state of free fall! As the Fermi energy of the electron gas is squeezed past the neutron-proton mass difference, the protons capture free electrons, further robbing the gas of energy and pressure and promoting further collapse. Solid resistance is encountered only when the neutrons become degenerate near the nuclear density

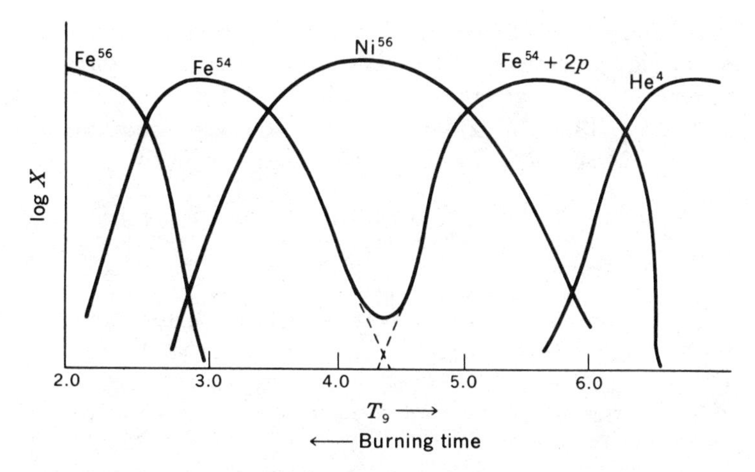

Fig. 7-12 A schematic diagram of the principal nucleus that results from the burning of silicon. For $T_9 > 4$ the burning is so fast that $\bar{Z}/\bar{N} = 1$ throughout the burning, leading to a composition dependence upon the temperature like that in Fig. 7-9. The time required for the burning increases markedly as the temperature is decreased, with the result that the relatively slow beta decays lower the value of $\bar{Z}/\bar{N}$ during the burning. Near $T_9 = 3$ the neutron-rich nucleus Fe^{54} has sufficient time to appear, and at even lower temperatures the beta decays drive the equilibrium to Fe^{56}.

of 10^{14} to 10^{15} g/cm³. The core has collapsed to a giant nucleus of neutrons.[1] Because of the extensive neutrino losses, the imploded core cannot have sufficient energy to expand back outward but seems trapped as a *neutron star*. The intense neutrinos do not flow out so easily in these extreme conditions, and a considerable amount of the energy of the collapse is believed to be deposited in the noncentral layers by neutrino interactions. The outer portions are believed to be dramatically ejected into space by the heating caused by the deposition of neutrino energy plus possible nuclear explosions in the outer layers. This sequence of events is the generally accepted model of the type II supernova explosion.[2]

Problem 7-14: Devise some reasonable way of giving an approximate numerical answer to the following question. Suppose a central core of $1\mathfrak{M}_\odot$ collapses within a $10\mathfrak{M}_\odot$ star converting 10 percent of the energy gained into neutrinos, which are subsequently absorbed with 1 percent efficiency in the outer layers. To what radius must the core collapse to provide the outer nine solar masses sufficient energy to escape from the star? To what radius must the core collapse to reach a density of 10^{15} g/cm³?

A large amount of stellar nucleosynthesis (maybe even *most of it*) surely accompanies the type II supernova. Proceeding outward from the imploded core, one passes through mass shells representing each nuclear burning state in stellar evolution: a very dense nucleon gas near the center; equilibrium nuclei somewhat farther out; then, in turn, Si burning, C and O burning, He burning, and H burning—all capped by an envelope where no nuclear burning has occurred. In the dramatic shock heating that may accompany the expulsion of this material, many significant nuclear reactions may still occur, but the dominant products should be those same nuclei which lie on the main line of the nucleosynthesis sequence. Figure 7-13 shows a schematic conception of a $30\mathfrak{M}_\odot$ presupernova model. Much more work remains to be done on the physics of this model. The

[1] Perhaps the collapse cannot even be halted by nuclear degeneracy. If the mass of the core is too large, in analogy to the Chandrasekhar limit for electron degeneracy, the nuclei are squeezed onward toward an exciting event, not well understood, in relativistic astrophysics. For a profound account with extensive bibliography, see B. K. Harrison, K. S. Thorne, M. Wakano, and J. A. Wheeler, "Gravitation Theory and Gravitational Collapse," The University of Chicago Press, Chicago, 1965.

[2] The thermal instability associated with iron-to-helium phase change was suggested by Hoyle in 1946 to be the triggering mechanism of a supernova explosion. The idea was described in some detail by Burbidge, Burbidge, Fowler, and Hoyle, *loc. cit.*, in 1957, and further elaborated by Fowler and Hoyle, *op. cit.*, sec. 8. Colgate and his collaborators made the first working numerical models of the implosion-explosion phenomenon [S. A. Colgate and H. J. Johnson, *Phys. Rev. Letters*, **5**:235 (1960); S. A. Colgate and R. H. White, *Astrophys. J.*, **143**:626 (1966)]. See also W. D. Arnett, *Can. J. Phys.*, **44**:2553 (1966), and Z. Barkat, G. Rakavy, and N. Sack, *Phys. Rev. Letters*, **18**:379 (1967). H. Y. Chiu has questioned the mechanism, believing rather that the instability is triggered by electron capture and the associated formation of very neutron-rich matter (Pre-supernova Evolution, in A. G. W. Cameron and R. F. Stein (eds.), "Stellar Evolution," Plenum Press, New York, 1966). Chiu was also largely responsible for reviving the concept of neutron stars as the final state of the supernova core; *Ann. Phys.*, **26**:364 (1964).

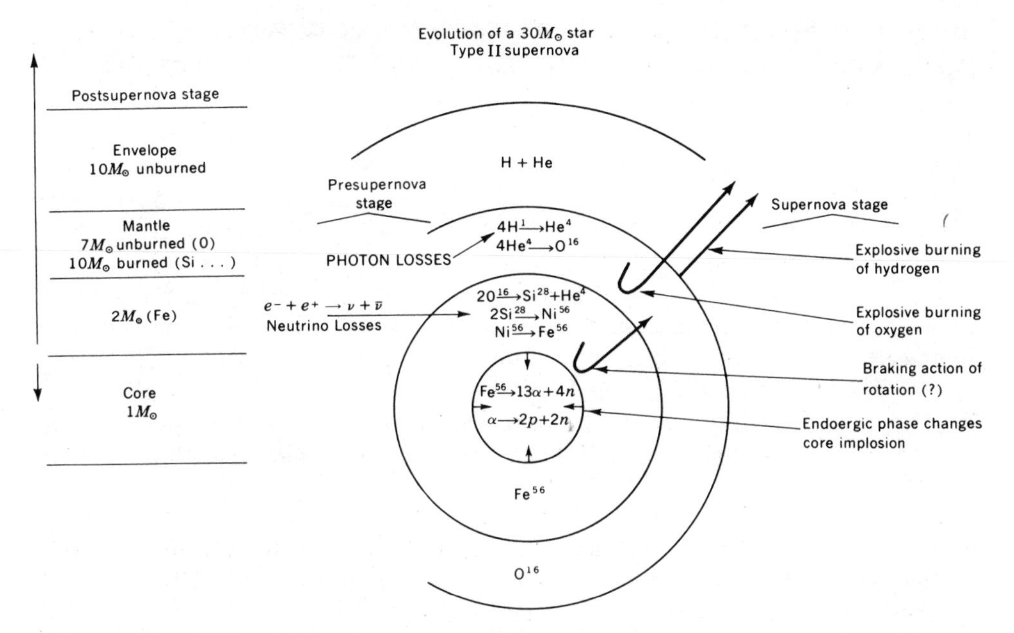

Fig. 7-13 The final evolution of a $30\mathfrak{M}_{\odot}$ star illustrating the presupernova in the center, the type II supernova events on the right, and the postsupernova results on the left. It has been assumed that a braking action due to rotation or some other mechanism ultimately leads to a mantle-envelope explosion following the core implosion caused by the endoergic nuclear phase changes. The explosive burning of previously unburned oxygen is taken to be the source of energy in the explosion as it is envisioned here. The explosion results in the ejection of unburned material plus products of hydrogen burning, helium burning, oxygen burning, silicon burning, and equilibrium material. Nuclear explosions may not be essential, however, if neutrino deposition in the middle layers is sufficient. [*After W. A. Fowler and F. Hoyle, Astrophys. J. Suppl.*, **9**:201 (1964). *By permission of The University of Chicago Press. Copyright* 1964 *by The University of Chicago.*]

effects of rotation on the core collapse must be studied, and the numerical problem of neutrino energy transport in a dynamic situation probably warrants reexamination with great care. These are hard problems, but the supernova is probably the major frontier for stellar evolution during the 1970's.

Finally, we note that the physics of the fusion reactions that proceed as portions of the dense nucleon gas near the center expand outward may differ somewhat from the circumstances along the main line of presupernova synthesis. At densities greater than 10^9 g/cm^3, and temperatures of several billion degrees, the compound nucleus may interact with a third particle in a time shorter than the natural lifetime of the compound nuclear state.[1] The corresponding problems for nucleosynthesis are both interesting and important.

[1] P. B. Shaw and D. D. Clayton, *Phys. Rev.*, **160**:1193 (1967).

7-3 NUCLEOSYNTHESIS OF HEAVY ELEMENTS BY NEUTRON CAPTURE

For the science of nucleosynthesis the term *heavy elements* is generally taken to designate those nuclei more massive than the iron-group nuclei. Their natural abundances are far greater than can be produced in nuclear statistical equilibrium, so that nuclear interpretation of their abundances seems to require a nonequilibrium mechanism. There is no apparent way to make the interpretation in terms of charged particles. The lifetime of a heavy nucleus against charged-particle reactions increases rapidly with increasing nuclear charge. For moderate temperatures, these lifetimes greatly exceed the lifetime of the corresponding epoch of stellar evolution. For high temperatures, on the other hand, the nuclear composition tends toward a nuclear equilibrium that either favors the region in the binding-energy maximum shown in Fig. 7-1 or the photodisintegration into light-nuclear particles. In short, there seems to be no conceivable way to synthesize heavy-element abundances as great as those shown in Fig. 1-22 by means of charged-particle reactions.

This limitation does not exist for the capture of neutrons. With no coulomb barrier to overcome, heavy elements capture neutrons easily even at extremely low energies. Neutron cross sections, in fact, generally *increase* with decreasing energy. Accordingly one concludes that heavy elements could be synthesized at relatively moderate temperatures by exposing lighter nuclei to a flux of neutrons. This suggestion has proved very useful because, as we shall see, the abundances of the heavy elements are dramatically correlated with the nuclear systematics of neutron capture. The most important of these systematic effects will be seen to be an enhanced stability of nuclei with the magic numbers $N = 50$, 82, and 126 neutrons, which were introduced in Sec. 4-4.

The difficulty with the idea is that free neutrons are not normally thought to be abundant in the major phases of nuclear burning. The main line of nuclear energy generation does not involve the liberation of neutrons until the carbon-burning epoch is reached. Neutrons are liberated to some extent by secondary reactions during helium burning, however, so that it will be necessary to follow the course of many of these nuclear reactions as an auxiliary calculation to the major problem of the evolution of the star. But for the time being it will be advantageous to ignore the problem of the sources of the free neutrons until the principles of synthesis by neutron capture have been surveyed.

The abundances of the nuclear species seem to suggest that two different types of neutron-capture chains have participated in the synthesis of the heavy elements. To better understand how this distinction is made operationally, it will be helpful to examine Fig. 7-14, which is a schematic section of a nuclide chart. The charge of each nucleus is plotted as ordinate, so that rows correspond to isotopes of a given element. The columns are nuclei having the same neutron number N, so that the atomic weight $A = Z + N$ is the sum of the ordinate and the abscissa. Thus, isobars, or nuclei of equal atomic weight, lie along a diagonal line from upper left to lower right. Stable nuclei are distinguished by boxes

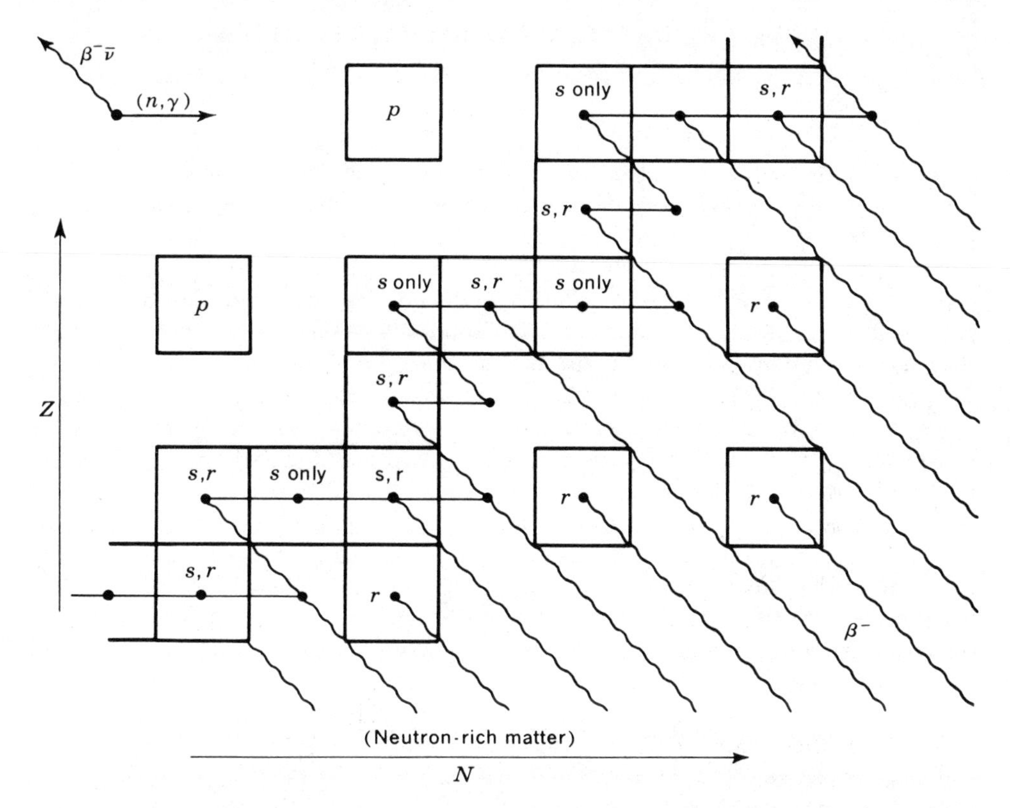

Fig. 7-14 A characterization of a portion of the chart of nuclides showing the assignment of nuclei to the classes *s*, *r*, and *p*. The *s*-process path of (n,γ) reactions followed by quick beta decays enters at the lower left and passes through each nucleus designated by the letter *s*. Neutron-rich matter undergoes a chain of beta decays terminating at the most neutron-rich of the stable isobars, which are designated by the letter *r*. Those nuclei on the *s*-process path which are shielded from *r*-process production are labeled "*s only*." The rare proton-rich nuclei which are bypassed by both neutron processes are designated by the letter *p*.

containing the letters *s*, *r*, *p* (to be explained shortly). Even values of Z are found to have many stable isotopes, whereas odd values of Z usually have only one stable isotope, and at most two. Along the diagonal lines of constant A (isobars), one finds that odd atomic weights have only one stable isobar, whereas even atomic weights may have one, two, or three stable isobars. These generalizations, obtained from the laboratory study of heavy nuclei, can be simply understood in terms of nuclear systematics, but it would take us somewhat too far afield to explore the reasons here.

Imagine the sequence of events if nuclei are placed in a neutron flux. When a nucleus captures a neutron, it becomes an isotope of the same element with one unit more of atomic weight:

$$(Z,A) + n \rightarrow (Z, A + 1) + \gamma$$

If the nucleus $(Z, A + 1)$ is stable, it waits until it captures another neutron, and so on. If, however, the nucleus $(Z, A + 1)$ is radioactive, the question whether it beta-decays to $(Z + 1, A + 1)$ or captures a second neutron depends upon the relative lifetimes of $(Z, A + 1)$ against beta decay and against capture of neutrons. It is this question of relative lifetimes that distinguishes the two major types of neutron-capture chains. The overwhelming majority of beta-decay lifetimes are of order *hours*, within perhaps a factor of 10^3 either way. Suppose then that the neutron flux is so weak that it characteristically requires long times, say 10^4 *years*, within perhaps a factor 10^3 either way, for neutron captures to occur. Then in first approximation one could say that neutron capture is *slow* compared to beta-decay rates. In that case the neutron-capture chain will march through the stable isotopes of an element until it reaches a radioactive species, at which point beta decay will occur, and the capture chain will resume in the element $(Z + 1)$. This type of neutron-capture chain was named the *s process* by Burbidge, Burbidge, Fowler, and Hoyle. The letter *s* is a mnemonic for *slow*, which in turn refers to the rate of neutron capture compared to the rate of beta decay for radioactive nuclei. In Fig. 7-14 the slow-neutron capture path is indicated by a solid line, and the stable nuclei through which the chain passes are labeled with the letter *s*. The *s*-process path is frequently said to pass along the *valley of beta stability*, which means that it sticks closely to the most tightly bound isobar at each atomic weight. It will be noticed that the chain does not pass through several of the most neutron-rich isotopes of the elements, labeled *r* for reasons soon to be explained, nor does it pass through the most proton-rich isotopes, labeled *p* as a mnemonic for *proton*. The upshot is that if the *s* process as described occurs in nature, it will synthesize only those isotopes labeled *s*. It is also clear that the abundances synthesized by this process will depend in part on the neutron-capture cross sections of the nuclei along the chain. Those nuclei with small cross sections will capture neutrons with such difficulty that the abundance may be expected to pile up at that point, whereas those with large cross sections will be so easily destroyed that they will not be expected to achieve a very large abundance.

The nuclear abundances reflect the operation of another neutron-capture sequence, however. Some astrophysical environment has produced very neutron-rich nuclei lying diagonally below and to the right of the valley of beta-decay stability displayed in Fig. 7-14. After the synthesizing event, these nuclei undergo chains of beta decay which terminate at the most neutron-rich of the stable isobars at each value of atomic weight. These beta-decay chains are schematically identified by wavy lines in Fig. 7-14, and the stable neutron-rich isobar which terminates the decay chain is labeled by the letter *r*. This mechanism seems to be responsible for the synthesis of the neutron-rich isotopes of the elements. It should also be noted, however, that this mechanism contributes to the abundances of many nuclei on the *s*-process chain as well. Those nuclei which can be synthesized by both mechanisms have been labeled with the letters *s* and *r*. Many nuclei lying on the *s*-process path, almost invariably even isotopes of even-Z elements, are *shielded* from the *r* process, as it is called, by a stable neutron-rich isobar. That is, a nucleus (Z,A) will not have received *r* contributions to its abundance if the nucleus $(Z-2, A)$ is stable. It occasionally happens for odd-A isobars that the nucleus $(Z-1, A)$, although it cannot be stable if (Z,A) is stable, nonetheless shields the nucleus (Z,A) from *r* contributions because its half-life is so long as to be virtually stable. Important examples are ${}_{37}Rb^{87}$ and ${}_{75}Re^{187}$, each with half-lives near 40 billion years, which shield the absolutely stable ${}_{38}Sr^{87}$ and ${}_{76}Os^{187}$ from *r* contributions. Shielded nuclei have played a key role in the development of the theory, because their abundances apparently reflect the yield of the slow-neutron-capture process alone.

Finally, it may be noted from the schematic Fig. 7-14 that some proton-rich nuclei cannot be synthesized by either slow-neutron-capture chains or by the decay of neutron-rich matter. These nuclei are shielded from both processes and are labeled with the letter *p* in deference to their proton-rich character. One thing comes to mind immediately regarding the abundances of the *p* nuclei. If it is to be true that the heavy-element abundances may be interpreted primarily in terms of two processes, the *s* and *r* processes, then it must also be true that the abundances of the *p* nuclei are small compared to those of the *s* and *r* nuclei. This expectation is borne out by detailed comparisons, so that one can sensibly assume that whatever mechanism has been responsible for the synthesis of the *p* nuclei has synthesized only a small fraction of the abundances of the *s* and *r* nuclei. This conclusion allows the latter nuclei to be analyzed in terms of two processes only, although it should be remembered always that this simplification has been made. In summary, one attempts to regard the heavy nuclei as being of four completions, which are respectively *p*, *s*, *sr*, and *r*. The *p* nuclei are separable as a class, which allows the possibility of interpreting the remainder of the heavy nuclei as a superposition of contributions from the *s* and *r* processes. Schematically the decomposition may be written

$$
\begin{aligned}
N(Z,A) &= N_s(Z,A) + N_r(Z,A) + N_p(Z,A) \\
&\approx N_s(Z,A) + N_r(Z,A) \qquad (7\text{-}27)
\end{aligned}
$$

For those *s*-only nuclei shielded from the *r* process we have $N_r = 0$, whereas we often have $N_s = 0$ for the heaviest isotopes of the elements.[1]

Several of these features are illustrated quantitatively in Fig. 7-15, which

[1] Each of these processes was named and quantitatively analyzed by Burbidge, Burbidge, Fowler, and Hoyle, *loc. cit.* Major discussions of the neutron-capture processes have been made by D. D. Clayton, W. A. Fowler, T. E. Hull, and B. A. Zimmerman, *Ann. Phys.*, **12**:331 (1961) and by P. A. Seeger, W. A. Fowler, and D. D. Clayton, *Astrophys. J. Suppl.*, **11**:121 (1965). All three papers are recommended to the serious student. Their point of view and results are the basis for this section.

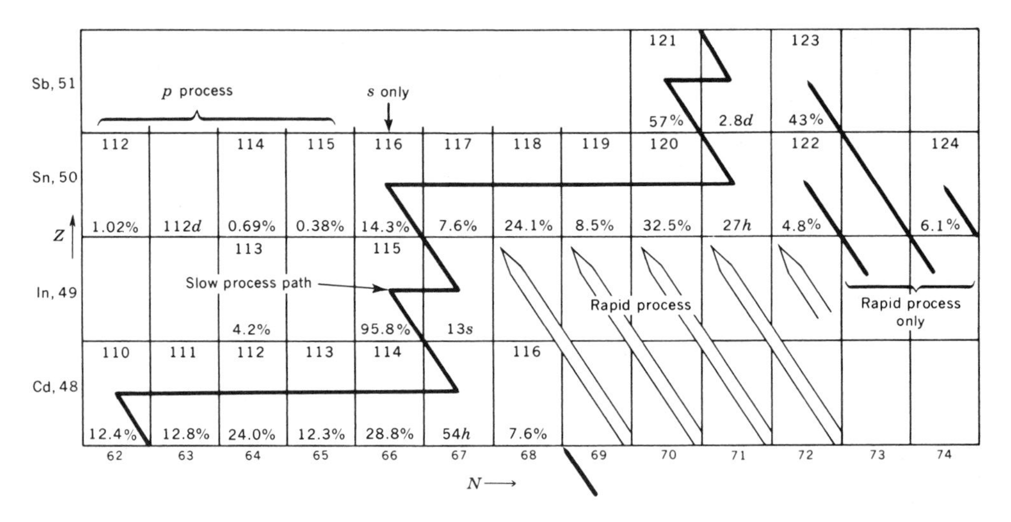

Fig. 7-15 The *s*-process path through the isotopes of cadmium, indium, tin, and antimony The element Sn is particularly interesting in that it has 10 stable isotopes, of which numbers 116 to 120 lie on the *s*-process path. The *r*-process decay chains enter from the lower right. Each stable isotope contains its solar-system abundance by percentage of the total element abundance, and each unstable nucleus on the *s*-process path contains its terrestrial lifetime against beta decay. [*After D. D. Clayton, W. A. Fowler, T. E. Hull, and B. A. Zimmerman, Ann. Phys.*, **12**:331 (1961).]

traces the *s*-process capture path through the elements Cd, In, Sn, and Sb. The isotopic composition of each element is given by percentage of the total element abundance. The element Sn is particularly interesting in that it has more stable isotopes than any other element, a fact that reflects its magic number of protons $Z = 50$. First notice the *p* nuclei. In^{113}, which is bypassed by both *s* and *r* processes, is only 4 percent as abundant as In^{115}, which lies on the main line. Going on to the next element, we find that the three light isotopes of Sn have a combined abundance of only 2 percent, less than that of any other single isotope. To good approximation, therefore, the abundances of the heavier isotopes can be analyzed in terms of *s* and *r* processes. The *s*-process path through Sn includes $A = 116$, 117, 118, 119, and 120, and the *r* process contributes to $A = 117$, 118, 119, 120, 122, and 124. Thus Sn^{116}, shielded as it is by Cd^{116}, is an *s*-only isotope, whereas Sn^{122} and Sn^{124} are due to the *r* process. The challenge is to interpret these abundances as a natural consequence of the two-neutron-capture mechanisms. The *s*-process abundance of each Sn isotope, which is the difference between its abundance and the characteristic *r*-process yield in its mass region, is expected to be anticorrelated with its neutron-capture cross section. Thus the increasing abundances of the even isotopes, Sn^{116}, Sn^{118}, and Sn^{120}, reflect decreasing neutron-capture cross sections. The smaller abundances of the odd isotopes, Sn^{117} and Sn^{119}, reflect the larger cross sections generally encountered for odd atomic weight. We shall return to the details of this correlation after discussing the s process.

It is clear that the r-process abundances cannot correlate with the neutron-capture cross sections, or any other nuclear property of the stable daughter nuclei, because the r nuclei were presumed to have been formed far on the neutron-rich side of the valley of beta stability. There is no obvious correlation between the nuclear properties of the neutron-rich parents and the nuclear properties of the first stable isobar reached in a chain of subsequent beta decays. We conclude that the abundances of the r nuclei must reflect the nuclear properties of neutron-rich matter and the process for forming it. The generally accepted mechanism for forming neutron-rich nuclei is the rapid capture of neutrons. Suppose the neutron density is great enough for the neutron-capture rate of radioactive nuclei to be greater than the beta-decay rate. Then the radioactive nuclei will quickly make one successive neutron capture after another, the capture chain terminating when no more neutrons can be captured because of insufficient binding. Very heavy isotopes of the elements would be produced by such a rapid capture chain. This is exactly the result desired, and the mechanism has been called the r process, where the mnemonic r indicates the rapid rate of neutron capture compared to the rate of beta decays.

NEUTRON-CAPTURE CROSS SECTIONS

Because the neutron-capture lifetimes depend upon the magnitude of neutron-capture cross sections, we may profitably digress into a brief discussion of the determination of neutron-capture cross sections.[1] The neutron binding energy in an average heavy nucleus is near 8 Mev, so that the excited compound nucleus $(Z, A + 1)^*$ formed by the addition of a low-energy neutron to the nucleus (Z,A) generally has an excitation of some 8 Mev in the nucleus $(Z, A + 1)$. The density of nuclear states at that excitation energy in a heavy nucleus is commonly very large, there being on the average an energy of only 1 kev or so separating adjacent resonances. The resonances are generally broad, moreover, because the neutron widths Γ_n are large. This fact reflects the absence of a coulomb barrier and is especially true for s-wave neutrons. The neutron capture in such a heavy element occurs, therefore, through many wide overlapping levels of the compound nucleus. The Maxwell-Boltzmann distribution of neutron velocities in a star yields a weighted average of σv over resonances primarily in the vicinity of kT. It is experimentally difficult to obtain a neutron energy resolution comparable to the level separation, so that the experimental measurements already yield σ averaged over an energy region sufficiently large for it to vary smoothly with energy. It is found that σ varies as v^{-1} at low energies, i.e., thermal cross sections, and

[1] An outstanding program of this type has been pursued at the Oak Ridge National Laboratory. The reader is referred to R. L. Macklin and J. H. Gibbons, Neutron Capture Data at Stellar Temperatures, *Rev. Mod. Phys.*, **37**:166 (1965). See also their account in *Science*, **156**:1039 (1967). A very interesting discussion is G. I. Bell, Cross Sections for Nucleosynthesis in Stars and Bombs, *Rev. Mod. Phys.*, **39**:59 (1967).

that this dependence changes over to v^{-2} in the kilovolt region as long as *s*-wave capture alone is effective. However, in the heavy elements *p*-wave capture, which is proportional to v^{+1}, begins to contribute significantly just in that region where the *s* wave begins to decrease more rapidly than v^{-1}. The result is that to a crude approximation $\langle \sigma v \rangle$ is very nearly constant. For the thermal problem, it is convenient to define an average cross section $\langle \sigma \rangle$ such that its product with the average thermal velocity yields $\langle \sigma v \rangle$. That is,

$$\langle \sigma \rangle \equiv \frac{\langle \sigma v \rangle}{v_T} \tag{7-28}$$

where

$$\begin{aligned} \langle \sigma v \rangle &= \int_0^\infty \sigma v \phi(v)\, dv \\ \phi(v)\, dv &= \frac{4}{\pi^{\frac{1}{2}}} \left(\frac{v}{v_T}\right)^2 \exp\left[-\left(\frac{v}{v_T}\right)^2\right] \frac{dv}{v_T} \\ v_T &\equiv \left(\frac{2kT}{\mu_n}\right)^{\frac{1}{2}} \end{aligned} \tag{7-29}$$

and $\mu_n = M_n M_A/(M_n + M_A)$ is the reduced neutron mass. It then conveniently happens for nuclei of the type under discussion that $\langle \sigma \rangle$ is numerically very nearly equal to σ_T, the cross section measured at the velocity v_T.

Problem 7-15: As an example of what is involved here, consider capture cross sections $\sigma(v)$ proportional to v^{-2}, v^{-1}, v^0. Show that the relationship between $\langle \sigma \rangle$ as defined in Eq. (7-28) and $\sigma_T = \sigma(v_T)$ is

$$\langle \sigma \rangle = \begin{cases} 1.13\sigma_T & \sigma(v) \propto v^{-2} \\ \sigma_T & \sigma(v) \propto v^{-1} \\ 1.13\sigma_T & \sigma(v) \propto v^0 \end{cases}$$

Thus for smoothly falling cross sections it appears that a measurement of σ near v_T already yields a good value for $\langle \sigma \rangle$.

An example of the application of this technique can be seen in Fig. 7-16, which shows as points measurements of the neutron-capture cross section of Ta^{181}, a nucleus with properties similar to those being discussed. A common way of characterizing the cross sections between, say, 5 and 100 kev of neutron energy, which includes the energies of astrophysical interest, is by the average radiative width Γ_γ of the compound nucleus, an *s*-wave *strength function* S_0 defined as an average ratio of *s*-wave neutron width to the average energy spacing between *s*-wave levels, a *p*-wave strength function S_1 defined analogously for *p*-wave neutrons, and an average observed level spacing $\bar{D}_{\text{obs}}$. The capture cross section can be expressed in terms of these quantities as long as *s* and *p* waves are alone important and as long as there are a large number of broad overlapping levels that moderate the capture process as compound nuclear intermediate states.

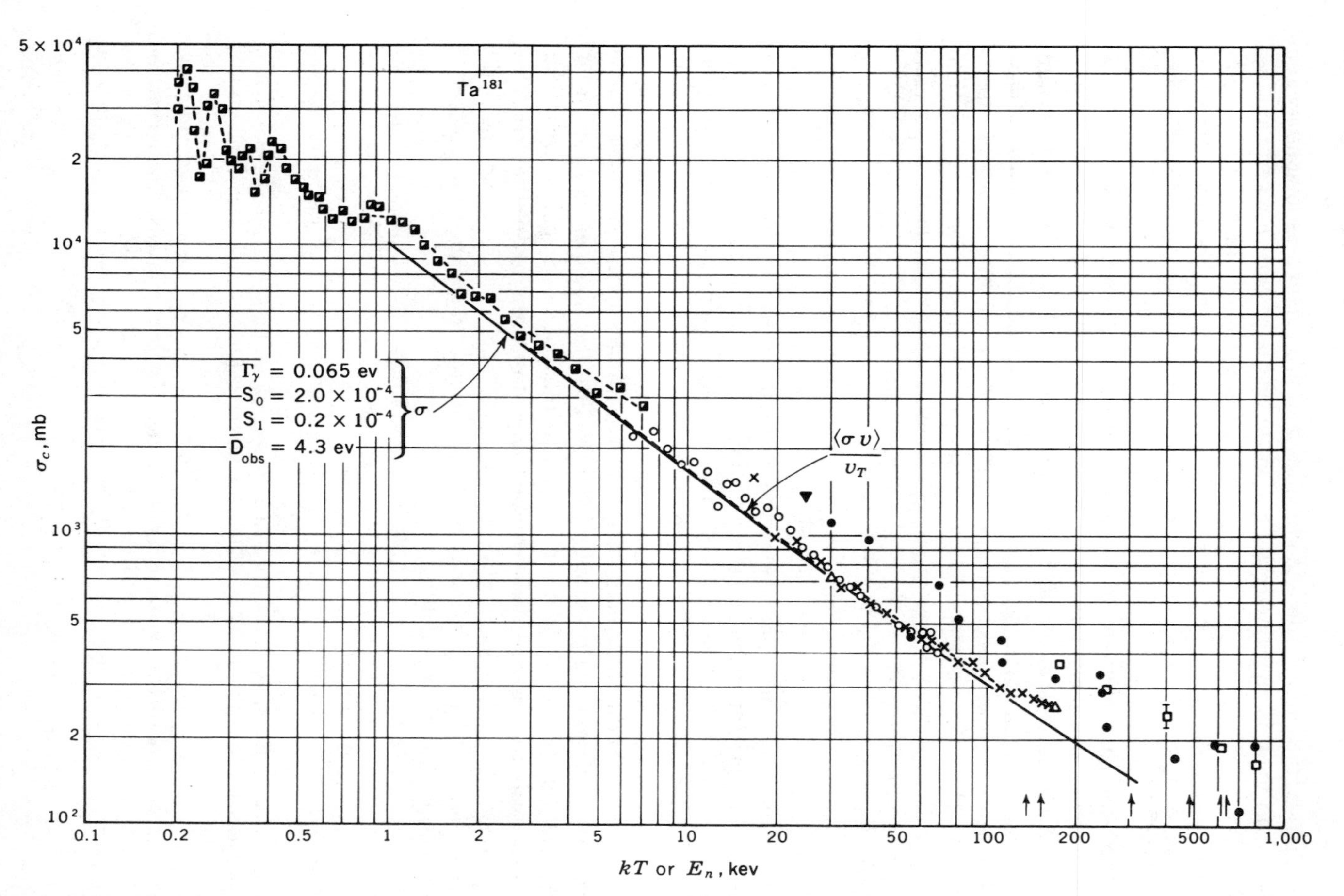

Fig. 7-16 The neutron-capture cross section of tantalum. The points represent experimental measurements at selected neutron energies E_n. The solid curve is a fit to $\sigma(E_n)$ made with the average resonance parameters and strength functions shown on the figure. The dashed curve, which is almost indistinguishable from the solid one, shows the value of $\langle\sigma\rangle$ computed for various temperatures. For nuclei of this type one sees that $\langle\sigma(kT)\rangle \approx \sigma(E_n = kT)$. [*R. L. Macklin and J. H. Gibbons, Rev. Mod. Phys.*, **37**:166 (1965).]

Such a characterization of the experimental points is shown as a solid line in Fig. 7-16. From this characterization of σ the value of $\langle\sigma\rangle$ may be computed from Eq. (7-28). The result in this case is shown as a dashed line in Fig. 7-16, but the curve is practically indistinguishable from σ, showing that to good accuracy

$$\langle\sigma\rangle = \sigma(v_T) = \sigma_T$$

Another general characteristic of the capture cross sections of nuclei of this type is that they are relatively large. The Ta^{181} cross section, for example, has a value near 1,000 millibarns, or 10^{-24} cm^2, near $E_n = 20$ kev. This cross section is some ten orders of magnitude greater than characteristic charged-particle cross sections encountered in stellar evolution, and it is two to three orders of magnitude greater than the corresponding neutron-capture cross sections in the light elements or in nuclei with magic numbers of neutrons. It is as large as the cross section for scattering a photon from a free electron!

Neutron capture by light nuclei or by magic-numbered heavy nuclei cannot be treated so simply. The distinguishing feature in either case is that the number of resonances is not so large for such nuclei. In either case the shell structure of the nucleus becomes an important feature in reducing the level density of resonances. Consider, for example, the addition of a neutron to a magic closed-neutron shell, which may be thought of in simplest approximation as an inert core with tight binding. The next neutron, like the valence electrons of the alkali metals, is not so tightly bound and is accordingly received at a smaller excitation energy in the compound nucleus. The number of resonances is smaller, because the excitation energy is lower, and because the single extra neutron has only a small number of single-particle states compared to the polyneutron configurations away from closed shells, where the host of couplings of the extra neutrons provide a multiplicity of states. In short, one finds that the thermally averaged cross section is dominated by a few discrete resonances, although the thermal s-wave tail is also significant.

Problem 7-16: Show that the contribution of a single resonance to the effective cross section is

$$\langle\Delta\sigma\rangle = \frac{2\pi^{\frac{3}{2}}\hbar^2}{(kT)^2M_n}\frac{\omega\Gamma_\gamma\Gamma_n}{\Gamma}\exp - \frac{E_r}{kT} \tag{7-30}$$

Except at extremely low energies, we find that $\Gamma_n \gg \Gamma_\gamma$, so that $\Gamma_n/\Gamma \approx 1$, and the contribution is proportional to $\omega\Gamma_\gamma \exp(-E_r/kT)$. It becomes important in this case to locate all the low-energy resonances.

Problem 7-17: The cross section for the reaction $Na^{23}(n,\gamma)Na^{24}$ is dominated by two resonances: (1) $J^\pi = 1^+$, $E_r = 2.85$ kev, $\Gamma_\gamma = 0.34$ ev and (2) $J^\pi = 2^-$, $E_r = 54$ kev, $\Gamma_\gamma = 1$ ev. For Na^{23} the spin $J = \frac{3}{2}$. Calculate the contribution of each of these two resonances to $\langle\sigma\rangle$ at $T = 3 \times 10^8$ °K.
Ans: (1) 0.91 millibarn; (2) 0.59 millibarn.

Because of the small and strongly energy-dependent cross sections of the light and magic nuclei, more experimental information is required and is harder to

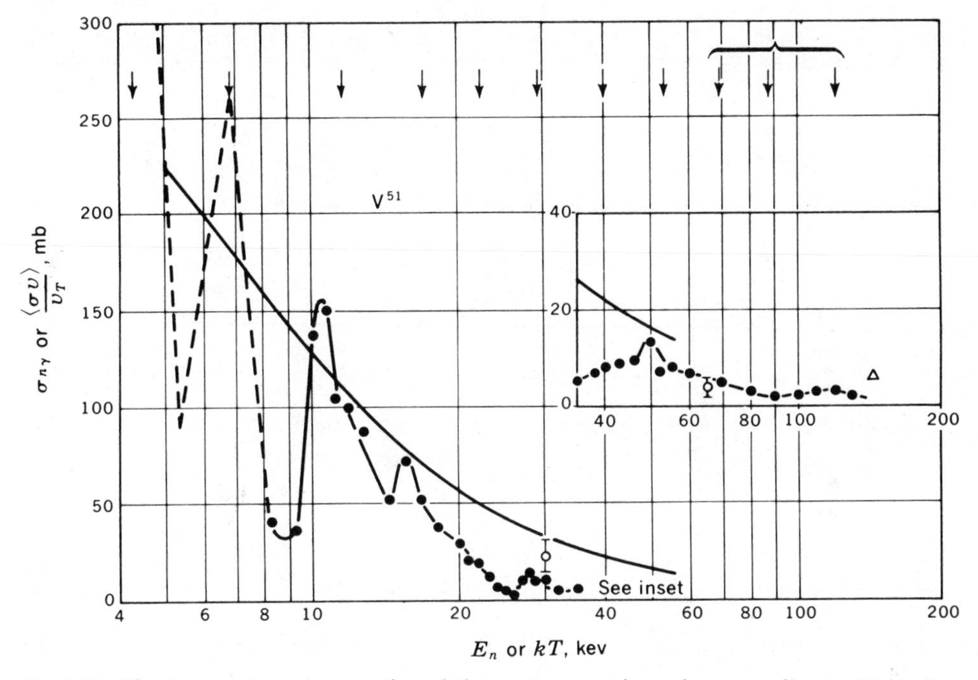

Fig. 7-17 Neutron-capture cross section of the neutron-magic nucleus vanadium. The points and the irregular curve through them represent the measured $\sigma(E_n)$. Arrows at the top indicate the positions of known resonances. The smoothly falling solid curve represents the thermally averaged cross section $\langle\sigma\rangle$. For nuclei of this type it is not true that $\langle\sigma(kT)\rangle \approx \sigma(E_n = kT)$. [*R. L. Macklin and J. H. Gibbons, Rev. Mod. Phys.*, **37**:166 (1965).]

obtain. Figure 7-17 shows the data for neutron capture in vanadium, which has the magic neutron number $N = 28$. The dots indicate measured values of the capture cross section. At least four resonances are obvious in these data, and several others are suggested. The arrows across the top of the diagram indicate the positions of resonances that are known to exist from the total cross-section data, largely $V^{51}(n,n)V^{51}$, which are much more easily detected because of the much greater size of the resonant scattering cross section. It is clear from Eq. (7-30) that the known resonances near 4 and 7 kev play an important role in the evaluation of $\langle\sigma\rangle$ even though they are undetected in the radiative-capture data themselves. The solid curve represents the calculated value of $\langle\sigma\rangle$, which is seen to be much greater than the value of the cross section itself near $E_n = kT$ between 20 and 50 kev. About 50 percent of $\langle\sigma\rangle$ at $kT = 30$ kev, for example, is due to the resonances at $E_n = 4$, 7, and 11 kev. Unlike the example of Ta^{181}, a cross section measured at one energy may be a very poor approximation to $\langle\sigma\rangle$ for a magic-neutron nucleus. Because the cross sections of the closed-shell nuclei play an important role in the s process, one would ideally like to have the ther-

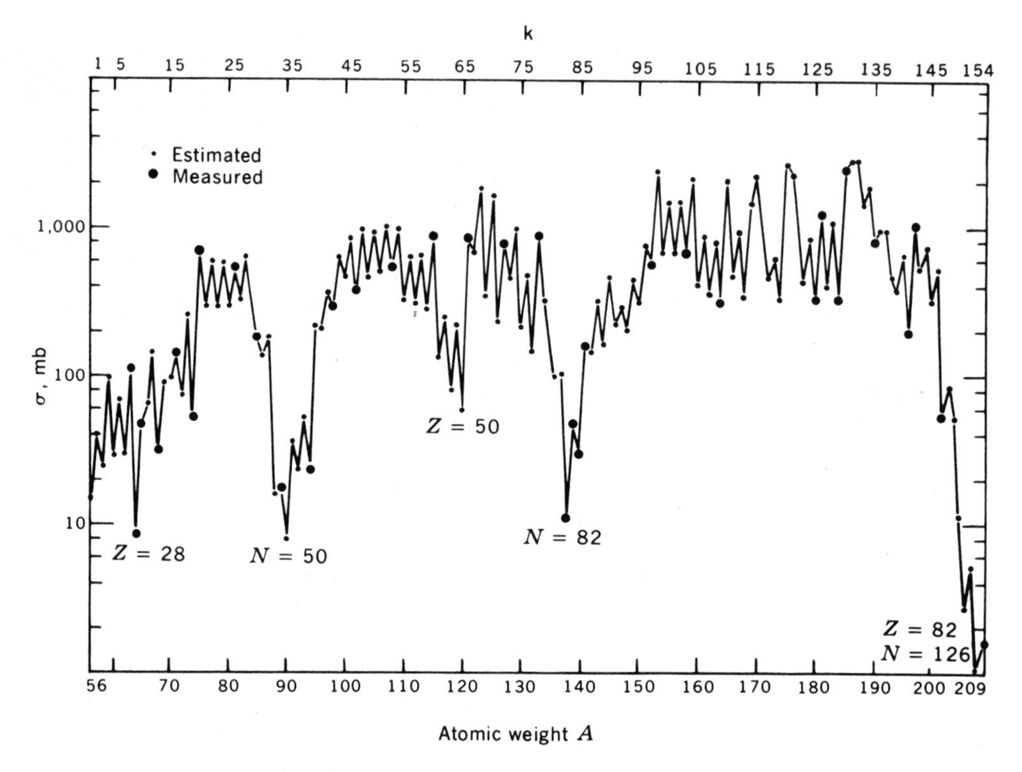

Fig. 7-18 Measured and estimated neutron-capture cross sections of nuclei on the s-process path. The neutron energy is near 25 kev. The cross sections show a strong odd-even effect reflecting average level densities in the compound nucleus. Even more obvious is the strong influence of the closed nuclear shells, or magic numbers, which are associated with precipitous drops in the cross section. Nucleosynthesis of the s-process nuclei is dominated by the small cross sections of the neutron-magic nuclei.

mally averaged values $\langle\sigma\rangle$ tabulated as a function of kT for these crucial nuclei. Macklin and Gibbons have initiated such a program at Oak Ridge, but the procedure is a painstakingly laborious one because each calculated value must depend on the measured properties of the neutron resonances—measurements not easily obtained. Their efforts have, as we shall see, provided the major nuclear experimental evidence of the correctness of neutron capture as the major mode of heavy-element nucleosynthesis.

In Fig. 7-18 the capture cross sections of the nuclei on the s-process path are displayed against the atomic weight. Several systematic features are clearly evidenced there. There is an odd-even effect of something like a factor 2 or 3 between average adjacent points. The even-A nuclei, which all have even Z and even N, have smaller cross sections because the average density of resonant states in the compound nucleus is less than that of an odd-A nucleus, where the captured neutron can couple in several ways to the unpaired nucleons. The smallest cross sections are those in nuclei with a magic number of neutrons $N = 50$, which are Sr^{88}, Y^{89}, Zr^{90}, or those having $N = 82$, which are Ba^{138}, La^{139}, Ce^{140}, Pr^{141}, Nd^{142}, or those having $N = 126$, which are Pb^{208} and Bi^{209}. These nuclei play an important role in the *s* process.

The general order of magnitude of the capture cross sections can be used to estimate the order of magnitude of the relevant neutron densities for the *s* and *r* processes. A characteristic cross section is apparently $\langle\sigma\rangle \approx 100$ millibarns, and a characteristic thermal velocity is $v_T \approx 3 \times 10^8$ cm/sec, so that we may crudely take $\langle\sigma v\rangle \approx 3 \times 10^{-17}$ cm³/sec. The lifetime of a nucleus against neutron capture is

$$\tau_n = \frac{1}{n_n\langle\sigma v\rangle} \approx \frac{3 \times 10^{16}}{n_n} \text{ sec} \approx \frac{10^9}{n_n} \text{ years} \tag{7-31}$$

If we are to have times of order $\tau_n \approx 10^4$ years for the *s* process, we should encounter free-neutron densities within of few orders of magnitude of

$$n_n \approx 10^5 \text{ cm}^{-3} \qquad s \text{ process}$$

To form sufficiently neutron-rich matter in the *r* process, on the other hand, will require that neutrons be captured in fractions of a microsecond, requiring neutron fluxes within a few orders of magnitude of

$$n_n \approx 10^{23} \text{ cm}^{-3} \qquad r \text{ process}$$

In principle one might expect to encounter astrophysical neutron fluxes in the large region between these two densities and have thereby a process intermediate to the *s* and *r* processes. Such events are apparently not common, and it is one of the fortunate simplifications in the applied theory of synthesis by neutron capture that the most common fluxes are either quite small or quite large. It may be of interest to note by way of comparison that the thermal neutron density in a reactor is of order $n_n \approx 10^7$ cm⁻³. With this background in mind we may turn to the details of these processes.

DETAILS OF THE *s* PROCESS

For preliminary analysis of the details of the *s* process it is helpful to rigorously accept the assumption of the process; viz., all beta decays of radioactive nuclei are assumed to be quite rapid compared to the rate for capturing neutrons. This assumption will be examined later, and it will be found not to hold in every case, with, however, no general damage to the theory. The assumption makes possible a great simplification in the equations because the nuclear chain generated is in that case unique, and one can let N_A represent the abundance at atomic weight A on that *s*-process path; i.e., it is not necessary to specify the charge Z of the nucleus of atomic weight A on the path because that charge is uniquely fixed by the assumption. Because the beta-decay rates are assumed fast, moreover, one can neglect the abundances of the radioactive species and assume that the nucleus $(Z, A + 1)$ produced by

$$(Z,A) + n \rightarrow (Z, A + 1) + \gamma$$

immediately decays, if it is radioactive, to the nucleus $(Z + 1, A + 1)$, which is generally stable:

$$(Z, A + 1) \rightarrow (Z + 1, A + 1) + \beta^- + \bar{\nu}$$

Presume, then, that a group of heavy elements exists in some interior region of a star. The region chosen is characterized by a temperature T and a free-neutron density $n_n(t)$ which is uniform over the region but may depend upon the time. The differential equation for the abundance of the heavy nucleus is then

$$\frac{dN_A(t)}{dt} = -\langle\sigma v\rangle_A n_n(t) N_A(t) + \langle\sigma v\rangle_{A-1} n_n(t) N_{A-1}(t) \tag{7-32}$$

where $\langle\sigma v\rangle$ is the quantity defined by Eq. (7-29) for the neutron-capture cross sections, and it is dependent upon the time only to the extent that the temperature depends upon the time. The s process no doubt occurs at several different temperatures in astrophysics corresponding to the several different sources of free neutrons that may come into play during the sequence of nuclear burning epochs, but during any single irradiation the temperature will probably not change greatly with time, as the mass zone burns at nearly constant temperature until the neutron source and/or the burning epoch is exhausted. In the last section we saw that the average $\langle\sigma v\rangle$ is very insensitive to temperature changes, especially for the nonmagic nuclei, so that we may with considerable justification make the replacement

$$\langle\sigma v\rangle = \langle\sigma\rangle v_T \tag{7-33}$$

in the sense of Eq. (7-28) and additionally assume that $\langle\sigma\rangle$ and v_T are constants corresponding to the nearly constant temperature of the neutron-liberation phase. Because the reduced mass of a neutron and heavy nucleus is so nearly equal to the neutron mass M_n, moreover, the velocity may be taken to be equal to the absolute velocity of the neutron independent of the target mass. Then Eq. (7-32) may be written

$$\frac{dN_A}{dt} = v_T n_n(t)[-\sigma_A(kT)N_A(t) + \sigma_{A-1}(kT)N_A(t)] \tag{7-34}$$

where for convenience we have discarded the brackets about the thermally averaged cross sections. In all that follows we shall assume that the unbracketed symbol σ represents the average cross section. Under these assumptions it is possible to define a new independent variable which will measure the progress of the neutron captures uniquely. We define the *neutron exposure* τ by

$$d\tau = v_T n_n(t)\, dt \qquad \tau = v_T \int n_n(t)\, dt \tag{7-35}$$

whereupon Eq. (7-34) becomes simply

$$\frac{dN_A}{d\tau} = -\sigma_A N_A + \sigma_{A-1} N_{A-1} \tag{7-36}$$

Even if Eq. (7-36) does correctly depict the generation of that component of the heavy-element abundances due to the *s* process, it cannot be solved without boundary conditions. The suitable boundary conditions are by no means obvious, because the *s*-process heavy nuclei can be synthesized by neutron-capture chains beginning on any nucleus with $A > 8$. One is forced to remember that the *s* process is not itself a major phase of nuclear burning but is a set of auxiliary reactions accompanying such a phase. Some neutrons are liberated, for example, by the reactions $C^{13}(\alpha,n)O^{16}$, $O^{17}(\alpha,n)Ne^{20}$, $Ne^{21}(\alpha,n)Mg^{24}$, and $Ne^{22}(\alpha,n)Mg^{25}$ that occur to a certain extent during helium burning, which is itself primarily characterized by He^4 fusion into C^{12} and O^{16}. These other reactions occur because C^{13}, O^{17}, Ne^{21}, and Ne^{22}, in this case, are naturally present in trace amounts in the composition of the star. In the same spirit, the liberated neutrons may be captured by any of the trace abundances present in the gas as well as by the very abundant C^{12} and O^{16}. The most efficient way of producing the heavy elements, however, is the capture of the neutrons by the iron-group nuclei, which constitute the last major abundance peak synthesized without the aid of neutron capture.

The neutron-capture cross sections for the light nuclei are on the average much smaller than the cross sections for nuclei above the iron group. Therefore a much larger integrated neutron flux would be required to convert silicon, say, into a heavy element than would be required to convert iron into the same heavy element. Yet reference to Fig. 1-22 shows that the iron group is comparable in natural abundance to the more abundant lighter elements and thus will constitute an efficient seed. Because the number of free neutrons is distinctly limited, it is much more profitable to synthesize heavy elements from the iron group as seed nuclei than to synthesize them from lighter elements. These seed nuclei must of course have been present in the original gas of the star, inasmuch as the nuclear burning site of the *s* process occurs much earlier in the scheme of nuclear evolution than the production of the iron group. The upshot is that we may realistically assume as a boundary condition that the seed nuclei for the *s* process are the iron-group nuclei.

The analysis of solar-system material shows the abundances of the iron group to be strongly peaked at Fe^{56}. One roughly finds $N_{55} \approx N_{57} \approx 0.1N_{56}$ coupled with an irregular but rapid dropoff at smaller and greater atomic weights. The abundances at large atomic weight resulting from neutron irradiation of this abundance peak will differ negligibly from that produced by the exposure of Fe^{56} alone. Thus the boundary conditions may be stated approximately as

$$N_A(0) = \begin{cases} N_{56}(0) & A = 56 \\ 0 & A > 56 \end{cases} \tag{7-37}$$

and we shall concentrate on this approximation to the actual problem.

The differential equation (7-36) becomes incorrect for large values of the atomic weight. The s-process chain terminates at $A = 209$ because Bi^{209} is the most massive stable nucleus. The material at $A = 210$ resulting from neutron capture by Bi^{209} decays by alpha-particle emission to Pb^{206}, and so a small cycle of capture and decay terminates the s process.[1] The details of this termination affect the abundance of Pb, an important element for nucleosynthesis, but it will take us too far afield to discuss them here. Suffice it to say that the system of coupled equations reduces to

$$\frac{dN_{56}}{d\tau} = -\sigma_{56}N_{56}$$

$$\frac{dN_A}{d\tau} = -\sigma_A N_A + \sigma_{A-1}N_{A-1} \qquad \begin{matrix} 57 \leq A \leq 209 \\ A \neq 206 \end{matrix} \tag{7-38}$$

$$\frac{dN_{206}}{d\tau} = -\sigma_{206}N_{206} + \sigma_{205}N_{205} + \sigma_{209}N_{209}$$

which are to be solved subject to the boundary conditions of Eq. (7-37). In the discussion to follow, however, we shall omit the equations $206 \leq A \leq 209$ and concentrate on the simpler set $56 \leq A \leq 205$. The reader is referred to the paper by Clayton and Rassbach[2] for the termination solution.

Important features of the solution can be seen directly from the differential equation

$$\frac{dN_A}{d\tau} = -\sigma_A N_A + \sigma_{A-1}N_{A-1}$$

It is obvious that

$$\frac{dN_A}{d\tau} < 0 \qquad \text{if } N_A > \frac{\sigma_{A-1}}{\sigma_A} N_{A-1}$$

$$\frac{dN_A}{d\tau} > 0 \qquad \text{if } N_A < \frac{\sigma_{A-1}}{\sigma_A} N_{A-1} \tag{7-39}$$

so that the solution is self-regulating in the sense that N_A decreases if it is too large with respect to N_{A-1} and increases if it is too small. The coupled equations have the property that they attempt to minimize the difference between the product $\sigma_A N_A$ and the product $\sigma_{A-1}N_{A-1}$. At atomic weights removed from the closed shells we shall find that the cross sections are so large that the difference between these two products is much smaller than the magnitude of either one of them, so that

$$\sigma_A N_A \approx \sigma_{A-1}N_{A-1} \qquad \text{nonmagic} \tag{7-40}$$

This result is only satisfied locally in regions between magic neutron numbers and has been called the *local approximation*. This very simple and powerful

[1] D. D. Clayton and M. E. Rassbach, *Astrophys. J.*, **148**:69 (1967).

[2] *Ibid.*

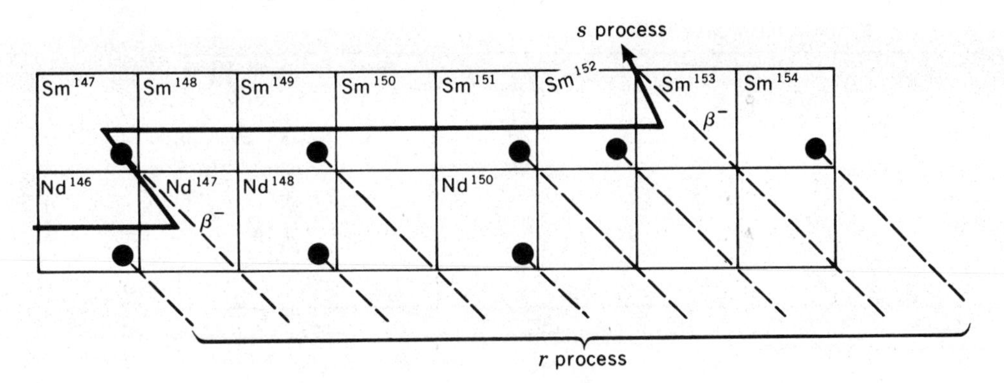

Fig. 7-19 The *s*-process path through the isotopes of samarium. The *r*-process yield contributes to the abundances of the nuclei containing the solid dots. Two of the isotopes of samarium, Sm^{148} and Sm^{150}, are *s*-only nuclei.

result has provided the most striking demonstrations of the correctness of the *s*-process idea. An outstanding example occurs in the isotopes of samarium. The *s*-process path passes through the isotopes $Sm^{147-150}$. It may also pass through Sm^{151} and Sm^{152}, because radioactive Sm^{151} has the long terrestrial half-life of 80 years, which may well be greater than the neutron-capture lifetime of Sm^{151}. But the direction of the path at $A = 151$ will not enter into the present discussion, which will concentrate primarily on those nuclei which definitely lie on the *s*-process path. Of these Sm^{148} and Sm^{150} are *s*-only isotopes because they are shielded from *r*-process production by Nd^{148} and Nd^{150}. If we assume the local approximation to be valid, we then expect

$$\sigma_{147}(N_{147})_s \approx \sigma_{148}N_{148} \approx \sigma_{149}(N_{149})_s \approx \sigma_{150}N_{150} \tag{7-41}$$

where the subscript *s* indicates that only the *s*-process part of the abundances is to be used. The subscript is not required on N_{148} and N_{150} because they are *s*-only nuclei. If Eq. (7-41) is to be satisfied, we should find in forming the σN products that those for $A = 148$ and $A = 150$ will be equal to each other and smaller than the products for $A = 147$ and $A = 149$, which are overabundant because of their *r*-process components. How well these ideas are borne out is indicated by Table 7-1, which lists the isotopic composition by percentage, the measured neutron-capture cross sections, and the product of the two. Although the expectations are impressively fulfilled, the agreement is even more convincing when the data for the *ratio* of gamma-ray yields are examined, because the ratio has a smaller error than the absolute error in the cross-section measurements. The result is

$$\frac{\sigma_{148}N_{148}}{\sigma_{150}N_{150}} = 1.02 \pm 0.06$$

Table 7-1 Samarium isotopes at 30 kev†

A	N_A, %	*Class*	σ_c, *millibarns*	$N\sigma$
144	2.87	p	119 ± 55	342
147	14.94	rs	1,173 ± 192	17,600 ± 2,900
148	11.24	s	258 ± 48	2,930 ± 540
149	13.85	rs	1,622 ± 279	22,500 ± 3,900
150	7.36	s	370 ± 72	2,770 ± 535
152	26.90	r or rs	411 ± 71	11,100 ± 1,900
154	22.84	r	325 ± 61	7,430 ± 1,400

† Data of R. L. Macklin, J. H. Gibbons, and T. Inada, *Nature*, **197**:369 (1963).

This equation and other results like it, obtained at great expense and effort, have firmly established the operation in nature of the s process.

Problem 7-18: From the data of Table 7-1 estimate the r-process contributions to the abundances of Sm^{147} and Sm^{149}.
Ans: $(N_{147})_r = 12.5 \pm 0.4$, $(N_{149})_r = 12.1 \pm 0.3$.

Problem 7-19: Estimate by the local approximation the abundance that the p nucleus Sm^{144} would have were it also to lie on the s-process path.

Over a large range of atomic weights the local approximation becomes a poor one. Equation (7-39) implies a large degree of coherence between adjacent values of atomic weight, however, such that one will expect the values of $\sigma_A(N_A)_s$ to vary smoothly with atomic weight. The values deemed most reliable in the year 1965 are shown in Fig. 7-20, along with a calculated curve that need not concern us at this juncture except, perhaps, to point out that it will result from the very simplest of theoretical assumptions, to be described later. Although a certain amount of irregularity can be seen, the points nonetheless approximate a smooth and monotonically decreasing function of atomic weight. A quick glance back to Fig. 1-22 shows that these σN points are much more smoothly behaved than are the abundances themselves. Probably the most dramatic demonstration of the reality of this correlation can be made by examining the corresponding figure for r-process abundances. Here one expects no correlation between the cross section of the first stable isotope reached in a beta-decay chain from neutron-rich matter and the abundance, and Fig. 7-21 shows that there is none. We may regard the smooth coherence of σN for the s-process nuclei as another confirmation of the correctness of the s-process idea.

The task then is to solve the differential equations to see whether the observed abundances can be simply interpreted in terms of the solution. The solutions are not hard to find, but they are somewhat tricky to use. Equations of this type are simply analyzed in terms of their Laplace transforms

$$\bar{N}_k(s) = \int_0^\infty e^{-s\tau} N_k(\tau)\, d\tau \tag{7-42}$$

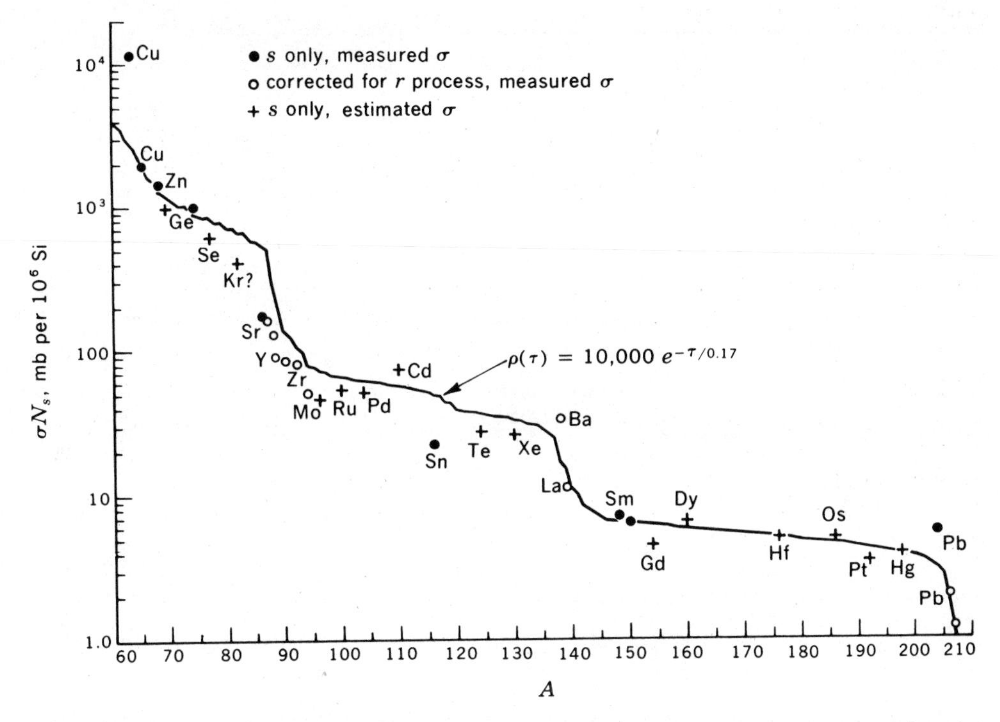

Fig. 7-20 The solar-system σN_s curve. The product of the neutron-capture cross sections for $kT = 30$ kev times the nuclide abundance per 10^6 silicon atoms is plotted versus the atomic mass number A. The solid curve is the calculated result of an exponential distribution of neutron exposures. [*P. A. Seeger, W. A. Fowler, and D. D. Clayton, Astrophys. J. Suppl.*, **11**:121 (1965). *By permission of The University of Chicago Press. Copyright* 1965 *by The University of Chicago.*]

where for convenience from here on we use the index $k = A - 55$ in order for the chain to start from $k = 1$. Using the fact that the Laplace transform of a derivative is

$$\int_0^\infty e^{-s\tau} \frac{dN_k(\tau)}{d\tau}\, d\tau = s\bar{N}_k(s) - N_k(0) \tag{7-43}$$

the transformed equations coupled with the boundary conditions chosen are

$$\begin{aligned} s\bar{N}_1(s) &= -\sigma_1\bar{N}_1(s) + N_1(0) \\ s\bar{N}_2(s) &= -\sigma_2\bar{N}_2(s) + \sigma_1\bar{N}_1(s) \\ s\bar{N}_k(s) &= -\sigma_k\bar{N}_k(s) + \sigma_{k-1}\bar{N}_{k-1}(s) \end{aligned} \tag{7-44}$$

Problem 7-20: Solve these equations algebraically to show

$$\bar{N}_k(s) = N_1(0) \frac{\sigma_{k-1}\sigma_{k-2} \cdots \sigma_2\sigma_1}{(s + \sigma_k)(s + \sigma_{k-1}) \cdots (s + \sigma_2)(s + \sigma_1)} \tag{7-45}$$

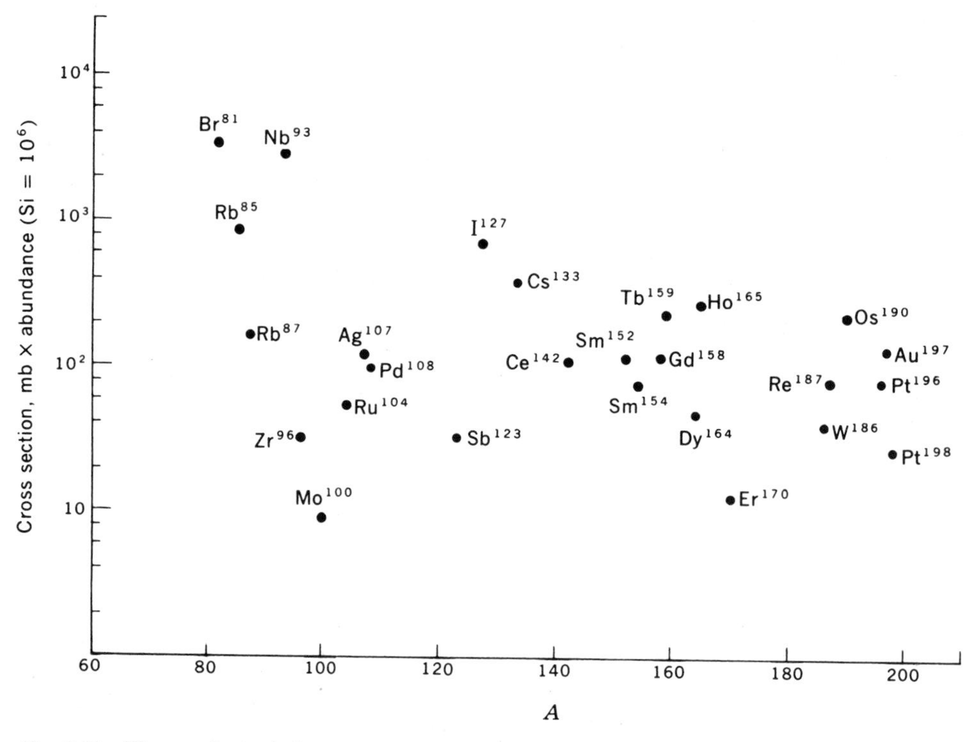

Fig. 7-21 The product of the neutron-capture cross section times the nuclide abundance for r-process nuclei. The irregularity in the product σN for these nuclei is expected and shows that the smooth variation found in Fig. 7-20 is not wholly accidental.

Two things are convenient: (1) because each abundance is directly proportional to the origin number of Fe^{56} seed nuclei, we can define the solution in terms of the abundance per initial seed nucleus; and (2) inasmuch as the product σN is expected to be smoothly varying, we choose that product rather than N itself as the set of dependent functions. Then

$$\bar{\psi}_k(s) \equiv \frac{\sigma_k \bar{N}_k(s)}{N_1(0)} = \frac{1}{\left(\dfrac{s}{\sigma_k}+1\right)\left(\dfrac{s}{\sigma_{k-1}}+1\right)\cdots\left(\dfrac{s}{\sigma_1}+1\right)} \tag{7-46}$$

It is apparent that $\bar{\psi}_k(s)$, and hence $\psi_k(\tau)$, depends symmetrically on all the cross sections σ_1 up to σ_k. The solution depends upon the magnitudes of these numbers but not on the order in which they occur.

The solution $\psi_k(\tau)$ is obtained by inverting the Laplace transform:

$$\psi_k(\tau) = \frac{1}{2\pi i}\int_{-i\infty}^{i\infty} e^{s\tau}\bar{\psi}_k(s)\, ds \tag{7-47}$$

This integral is easily evaluated in the standard way: perform a contour integration by closing the path over an infinite half circle in the left-hand half of the s plane. The contribution from this half circle vanishes, so that Eq. (7-47) becomes equal to the sum of the residues at each of the first-order poles $s_k = -1/\sigma_k$.

Problem 7-21: Evaluate the residues to show that

$$\psi_k(\tau) = \sum_{i=1}^{k} C_{ki} e^{-\sigma_i \tau} \tag{7-48}$$

$$C_{ki} = \frac{\sigma_1\sigma_2\sigma_3 \cdots \sigma_{k-1}\sigma_k}{(\sigma_k - \sigma_i)(\sigma_{k-1} - \sigma_i) \cdots (\sigma_2 - \sigma_i)(\sigma_1 - \sigma_i)} \tag{7-49}$$

where the factor $1/(\sigma_i - \sigma_i)$ is omitted in C_{ki}.

Problem 7-22: Confirm by substitution in the original differential equation that Eq. (7-48) is a solution. With some algebraic manipulation one may also show that the boundary conditions are satisfied.

This exact solution has been used in the study of radioactive decay chains, for which the differential equations are of the same form, and it is useful for small values of k in the s process, but it encounters severe numerical difficulties at large atomic weight. One is interested in values of k up to 150 in the region of Pb, in which case Eq. (7-48) becomes a sum of up to 150 terms. The terms differ greatly in order of magnitude, moreover, but none can be discarded because of the near cancellation of many of them; i.e., the answer is a small number compared to the magnitudes of most of the terms in the sum. Such a large sum is very difficult to evaluate, even with the best of electronic computers. It has therefore been necessary to find an approximate solution that characterizes the exact solution in considerable degree. That solution bears an analogy to the solution of the problem in which all cross sections are equal.

Problem 7-23: Consider an idealized s process in which all the cross sections are equal: $\sigma_k = \sigma$. Find the solution in that case. If you cannot solve the problem, show that this answer works:

$$\psi_k(\tau) = \frac{\sigma}{(k-1)!} (\sigma\tau)^{k-1} e^{-\sigma\tau} \qquad \text{const } \sigma \tag{7-50}$$

Equation (7-50) is a Poisson-like distribution in k with a maximum at $k_{\max} = \sigma\tau + 1$. As the neutron exposure τ increases, the maximum in the distribution moves to larger values of k. As the abundance distribution moves out in k, the abundance at maximum decreases, and the width of the abundance distribution increases because of the random nature of the capture process.

Problem 7-24: With the aid of Sterling's formula for $(k-1)!$, show that the maximum abundance in the distribution for constant σ decreases as it moves out in k according to

$$N_{k,\max} \approx \frac{1}{(2\pi\sigma\tau)^{\frac{1}{2}}}$$

Since the total number of nuclei is normalized to unity, the width of the distribution must increase like

$$\Delta k \approx (2\pi\sigma\tau)^{\frac{1}{2}}$$

The approximate solution for the general problem is constructed in analogy to the constant-cross-section example. The Laplace transform of Eq. (7-50) is

$$L\psi_k(\tau) = \psi_k(s) = \frac{1}{\left(\dfrac{s}{\sigma} + 1\right)^k} \qquad \text{const } \sigma \tag{7-51}$$

which is to be compared to Eq. (7-46) for the general case. The approximate solution is found by choosing for each k values of two numbers λ_k and m_k such that

$$\frac{1}{\left(\dfrac{s}{\lambda_k} + 1\right)^{m_k}} \approx \frac{1}{\left(\dfrac{s}{\sigma_k} + 1\right)\left(\dfrac{s}{\sigma_{k-1}} + 1\right) \cdots \left(\dfrac{s}{\sigma_1} + 1\right)} \tag{7-52}$$

for small values of s. The details of this approximation[1] lead to the choice

$$m_k = \frac{\left(\sum_{i=1}^{k} \dfrac{1}{\sigma_i}\right)^2}{\sum_{i=1}^{k} \left(\dfrac{1}{\sigma_i}\right)^2} \qquad \lambda_k = \frac{\sum_{i=1}^{k} \dfrac{1}{\sigma_i}}{\sum_{i=1}^{k} \left(\dfrac{1}{\sigma_i}\right)^2} \tag{7-53}$$

Then the approximate solution is

$$\psi_k(\tau) \approx \lambda_k \frac{(\lambda_k\tau)^{m_k-1}}{\Gamma(m_k)} e^{-\lambda_k\tau} \tag{7-54}$$

This choice for m_k and λ_k has the property that the first three moments of $\psi_k(\tau)$ with respect to τ for the approximate solution are equal to those for the exact solution. The values in each case are

$$\int_0^\infty \psi_k(\tau)\, d\tau = 1 \tag{7-55a}$$

$$\int_0^\infty \tau\psi_k(\tau)\, d\tau = \sum_{i=1}^{k} \frac{1}{\sigma_i} \tag{7-55b}$$

$$\int_0^\infty \tau^2\psi_k(\tau)\, d\tau = \left(\sum_{i=1}^{k} \frac{1}{\sigma_i}\right)^2 + \sum_{i=1}^{k} \left(\frac{1}{\sigma_i}\right)^2 \tag{7-55c}$$

Each function $\psi_k(\tau)$ with $k > 1$ begins at zero, rises to a single maximum at some value $\tau_{k,\max}$, and falls exponentially to zero at large τ. The equality of these first three moments between the exact and the approximate solutions has a simple interpretation in each case. (1) The normalized area ensures that all the seed nuclei, in this case *one* by normalization, pass through each value of k for some

[1] As described by Clayton, Fowler, Hull, and Zimmerman, *op cit.*

value of τ. (2) That the area centroid about the $\tau = 0$ axis is conserved ensures, when coupled with Eq. (7-55*a*) and the fact that $\psi_k(\tau)$ has a single maximum, that the maximum for both solutions is near

$$\tau_{k,\max} \approx \sum_{i=1}^{k} \frac{1}{\sigma_i}$$

For the approximate solution one may easily show that

$$\tau_{k,\max} = \sum_{i=1}^{k} \frac{1}{\sigma_i} - \frac{1}{\lambda_k}$$

(3) The equality of moments of inertia about the τ axis coupled with the two previous equalities ensures that the width of the single maximum is correct in first order. The advantage of the approximate solution is that it is easy to use numerically once the sums leading to m_k and λ_k have been performed. Experience has shown that the application of Eq. (7-54) to analysis of the *s* process leads generally to errors of no more than 10 percent. Such errors are small compared to the numerical uncertainty in experimental σN products. What one tries to obtain ultimately is the distribution of neutron exposures leading to observed abundances, and the small error of this approximate solution leads to negligible uncertainty for that problem.

The approximation is not a good one for small values of τ, however. In this case *small* is taken to mean $\tau \ll \tau_{k,\max}$, that is, exposures so small that ψ_k is very small compared to the value it achieves as τ approaches $\tau_{k,\max}$. That the approximation breaks down for small τ can be seen from the asymptotic behavior:

$$\lim_{\tau \to 0} \psi_k(\tau) = \begin{cases} C\tau^{k-1} & \text{exact} \\ C'\tau^{m_k-1} & \text{approximate} \end{cases} \tag{7-56}$$

But because m_k is generally considerably smaller than k for large k, the exact solution is much smaller than the approximate one for very small values of τ. For most applications this error has proved unimportant, but for problems in which one needs the solution for small τ, a power-series solution in τ should be generated.

The results of calculations of this type have considerable significance for nucleosynthesis. Figure 7-22 shows the value of ψ_A as a function of A for several different values of the neutron exposure of the iron. Each curve is labeled by the number of neutrons captured per initial iron nucleus rather than by the parameter τ. There is, of course, a definite relationship between the number of neutrons captured n_c and the neutron exposure τ, a relationship that depends upon the values of the cross sections in the capture chain. This relationship is displayed in Fig. 7-23 for the particular values of the neutron-capture cross sections near 30 kev used by Clayton et al. One should note also that the relationship of n_c to τ depends upon the temperature of the *s* process. Generally the values of the thermally averaged neutron-capture cross sections are decreas-

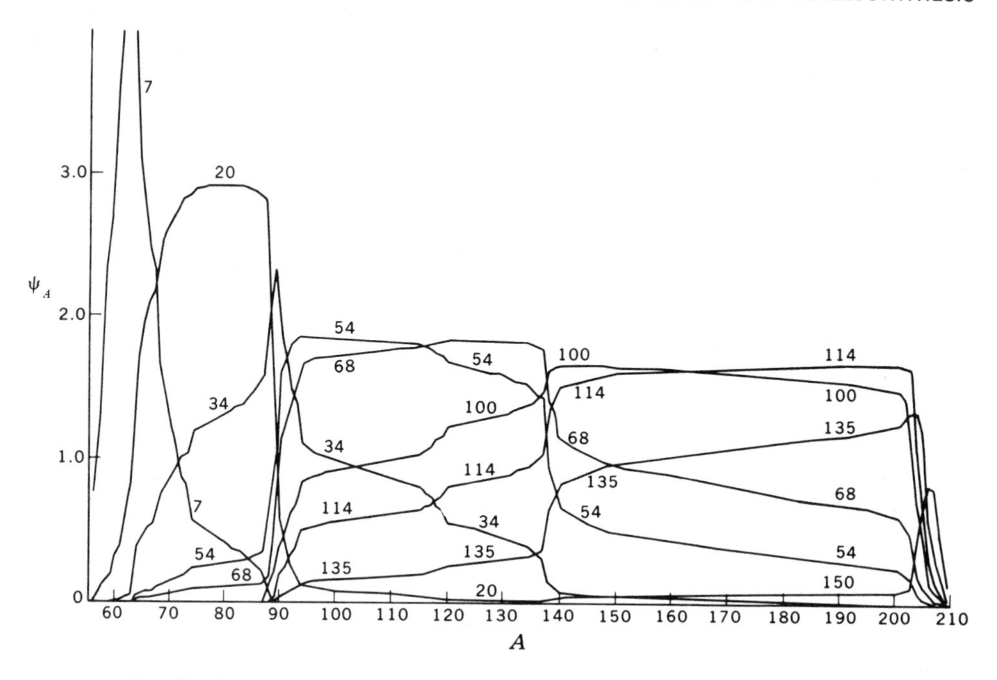

Fig. 7-22 The distributions $\psi_A = \sigma_A N_A$ for differing levels of neutron irradiation. Each curve is labeled by the parameter n_c, which is the average number of neutrons captured per initial iron seed nucleus. [*D. D. Clayton, W. A. Fowler, T. E. Hull, and B. A. Zimmerman, Ann. Phys.*, **12**:331 (1961).]

ing functions of the temperature, so that a greater neutron flux will be required at high temperatures to produce a given number of captures than will be required at low temperatures. But the abundance distribution resulting from a given number of captures as illustrated in Fig. 7-22 is nearly insensitive to the temperature, so that the parameter n_c has somewhat more generality than the parameter τ. None of the abundance distributions of Fig. 7-22 resembles in its entirety the observed *s*-process distribution for the solar system. Indeed, Fig. 7-20, which was plotted on a logarithmic scale, decreases by several orders of magnitude as A increases, although there are two relatively flat regions, whereas the maxima in the distributions of Fig. 7-22 decrease only slowly as the distribution moves to larger atomic weights. It is natural, therefore, to regard the observed solar-system distribution of *s*-process abundances as being a superposition of differing amounts of iron seed exposed to differing integrated neutron fluxes. Let the number of iron seed nuclei per 10^6 Si atoms exposed to an integrated flux τ in the interval $d\tau$ be represented by $\rho(\tau)\, d\tau$. Then because of the normalization chosen for the functions ψ, we have that the total *s*-process abundances should be given by

$$\sigma_A N_A = \int_0^\infty \rho(\tau)\psi_A(\tau)\, d\tau \tag{7-57}$$

Comparison of Figs. 7-20 and 7-22 shows that a general feature of the superposition must be that the number of seed nuclei exposed must rapidly decrease with increasing exposure. Such a situation seems to be consistent with the idea that the neutrons are liberated by nuclear reactions in stars. Because the density and temperature differ from point to point within a given star, and because the stars that have provided the sites for nucleosynthesis span a range of masses and compositions, the efficiency of neutron emission can hardly be expected to be constant. Whatever the astrophysical conditions that have produced the maximum s-process neutron flux, there must be many more similar conditions that have resulted in weaker ones. And the majority of the iron nuclei swept up into a newly formed star may be exposed to no appreciable neutron flux at all during the evolution of that star. On very general grounds, then, one expects a rapidly falling form for $\rho(\tau)$. This conclusion is even more evident if one considers the question whether $\rho(\tau)$ is zero for exposures greater than the maximum that can be encountered within a single star. For such exposures $\rho(\tau)$ is nonzero only by

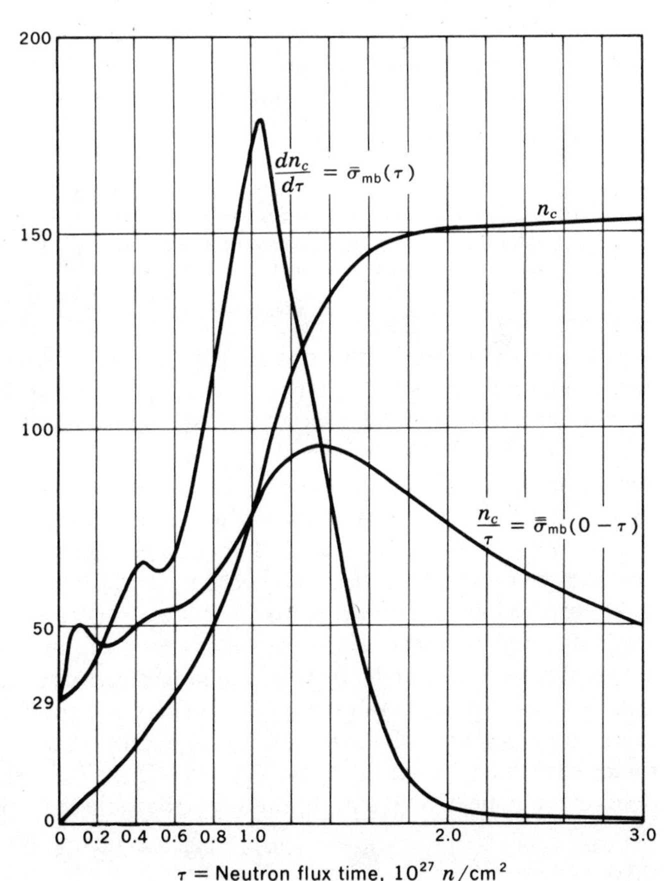

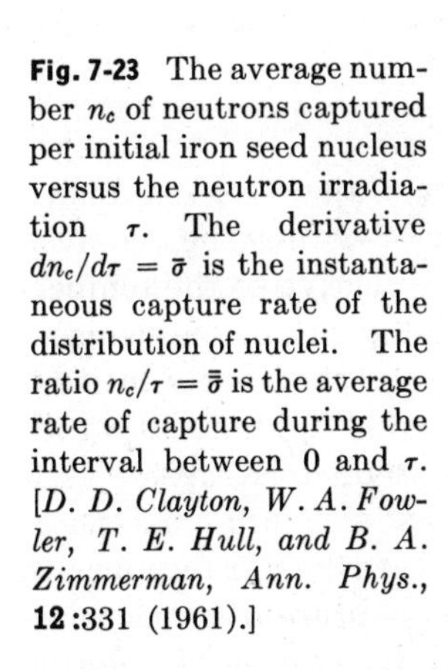
Fig. 7-23 The average number n_c of neutrons captured per initial iron seed nucleus versus the neutron irradiation τ. The derivative $dn_c/d\tau = \bar{\sigma}$ is the instantaneous capture rate of the distribution of nuclei. The ratio $n_c/\tau = \bar{\bar{\sigma}}$ is the average rate of capture during the interval between 0 and τ. [*D. D. Clayton, W. A. Fowler, T. E. Hull, and B. A. Zimmerman, Ann. Phys.*, **12**:331 (1961).]

virtue of the iron seed's having been through more than one star in its *s*-process history; but the amount of mass that has been through N, say, stars during its history must surely be a rapidly decreasing function of N.

An appealing form for investigation of the nature of superpositions is

$$\rho(\tau) = Ge^{-\tau/\tau_0} \tag{7-58}$$

where G is a normalizing constant and τ_0 is a parameter to be determined that reflects how rapidly the exposure distribution falls off. The integral in Eq. (7-57) can be performed for both the exact solution and the approximate solution for ψ_A. If the exact Eq. (7-48) is used, however, the resulting series still has the troublesome difficulties in numerical evaluation. The approximate solution Eq. (7-54) yields an integral that is easy to evaluate and reliably accurate for this purpose. With that approximation we have

$$\begin{aligned}\sigma_A N_A &\approx \int_0^\infty Ge^{-\tau/\tau_0}\lambda_A \frac{(\lambda_A\tau)^{m_A-1}}{\Gamma(m_A)} e^{-\lambda_A\tau}\, d\tau \\ &= G\left(\frac{\lambda_A\tau_0}{\lambda_A\tau_0 + 1}\right)^{m_A}\end{aligned} \tag{7-59}$$

The solid curve through the points of Fig. 7-20 is the result of evaluating Eq. (7-59) for the exposure distribution characterized by the parameter $G = 10^4$ and $\tau_0 = 0.17 \times 10^{27}\ \text{cm}^{-2}$.[1]

Problem 7-25: Using the approximate solution for ψ_A, evaluate the integral in Eq. (7-57) for the power-law flux distribution $\rho(\tau) = G'\tau^{-n}$. The abundance distributions resulting from this form of $\rho(\tau)$ fall much more steeply at the small values of A, a feature that will improve the fit in Fig. 7-20 at the small values of A, but fall less steeply at large values of A. Can you do the integral for a mixed distribution $\rho(\tau) = G''\tau^{-n} \exp(-\tau/\tau_0)$?

The fact that such simple assumptions give a caricature of the observed abundance distribution as satisfactory as the one shown in Fig. 7-20 may be taken to be a confirmation of the correctness of the *s*-process idea. The outstanding feature of the distribution is the alternation of relatively flat regions with rapidly decreasing ones. The local approximation, where σN is taken to be constant over a limited range of atomic weights, is a particularly good approximation in the regions $100 < A < 135$ and $145 < A < 200$. The product is expected to decrease relatively rapidly, however, near $A = 90$, near $A = 140$, and near $A = 208$. These rapid drops are apparent in the individual distributions in Fig. 7-22, and they are caused by the very small neutron-capture cross sections associated with the nuclei having magic neutron numbers (see Fig. 7-18). It is therefore quite natural to expect the observed σN products to decrease rapidly in these same regions. It is important to realize, however, that the abundances themselves have peaks in these regions. The cross sections drop much more precipitously near the magic numbers than the σN products do, and so the

[1] The shape of the distribution for other choices for $\rho(\tau)$ can be found in Seeger, Fowler, and Clayton, *loc cit.*

abundances N must show peaks in the magic-neutron nuclei. From a historical point of view, it was the existence of abundance peaks at these magic nuclei that motivated their explanation in terms of the neutron-capture mechanism.

Of course, the curve in Fig. 7-20 does not fit the points perfectly, nor is there any reason why it should. The experimental errors in the σN products can easily amount to a factor of 2 and perhaps more. The problems inherent in the determination of the relative abundances of two different elements are severe and constitute major modern research fields in both geochemistry and astronomy, and only a few of the neutron-capture cross sections have yet been measured with the ultimate techniques warranted by this problem. The situation improves yearly thanks to the large effort expended in this field by a large number of researchers, but the correct answers are obtained only with great difficulty. Even with the uncertainty of the present numbers, however, it appears that the exponential distribution does not yield enough abundance for values of A near 65, so that the correct form of $\rho(\tau)$ probably is greater than the exponential for small values of τ. It is also doubtful whether any ledge-precipice structure is warranted by the present data near $A = 85$ to 90. In fact, a smoothly falling curve would seem to pass through the points nicely. The details of this drop-off and the one in the Pb isotopes have been analyzed by Clayton,[1] and the reader is referred to these papers for further discussion of such points. But the generally satisfactory fit in Fig. 7-20 seems to indicate that the theory is correct.

The extraction of $\rho(\tau)$ from the macroscopic features of the σN curve provides an important datum for the theory of the chemical evolution of the galaxy. From the results on the solar system, for example, we may say that *of all the iron-group nuclei destined for incorporation into the solar nebula* 4.7 *billion years ago, a number of them equal to*

$$[\rho(\tau)\,d\tau]_{\text{solar system}} \approx 10^4 \exp\left(-\frac{\tau}{0.17}\right) d\tau \text{ per } 10^6 \text{ Si atoms} \tag{7-60}$$

had at some time been exposed to an integrated neutron flux τ *in the interval* $d\tau$. It is probably sensible to assume that the composition of the primitive solar nebula was characteristic of the average composition of interstellar gas in the galaxy at that time. If that is the case, Eq. (7-60) is a reasonable representation of the efficiency with which the earlier generations of stars within the galaxy had been able to s-process material and reinject it into space. In this way the rates of birth and evolution of stars are coupled with the efficiency of the neutron-liberating reactions within individual stars.

We must digress to establish an important point here. The composition of interstellar gas out of which new stars form reflects in some way a mixture of the by-products of countless stars. The composition has been homogenized by interstellar mixing to the point where it reflects the average rate of nucleosynthesis up to that time. There are, however, cases of individual stars with surface compositions so far from the average that we must conclude that their surfaces show

[1] *J. Geophys. Res.*, **69**:5081 (1965); *Astrophys. J.*, **148**:69 (1967).

the results of nucleosynthesis within their own interiors. One must suppose in such cases that either material from the interior mixes to the surface or, alternatively, a large amount of mass has been lost, leaving previously interior regions exposed. Either mechanism seems well within the range of possibilities for evolved stars. Historically the most dramatic such incident was the discovery by Merrill in 1952 of lines of the element technetium in the spectra of S-type stars. Since all isotopes of technetium are radioactive, its presence in the surface has been taken as proof of the processing of newly synthesized material to the surface. Technetium, in fact, does lie on the *s*-process chain, and the spectra of those S-type stars also seem to show an overabundance of those magic-neutron nuclei, viz., Zr, Ba, which are produced in overabundance by the *s* process. Sufficient abundance data have been obtained for many stars to permit plotting σN curves for their *s*-process elements. Unfortunately, the spectral lines do not distinguish between isotopes, so that one must make some assumptions in going from the abundance of an element in a stellar surface to the abundance of an isotope of that element. One certainly *does not expect* the isotopic composition of the elements in the surfaces of such stars to be identical to their isotopic composition in the solar system, but by making self-consistent assumptions regarding the distribution of the elemental enrichment among the isotopes of the element, it becomes possible to estimate *s*-process abundances. Figure 7-24 shows the estimated σN diagrams for six stars, including the sun. The bottom curve is from the old metal-deficient subdwarf γ Pavonis. Inasmuch as subdwarfs are not believed to show the results of internal nucleosynthesis on their surfaces, this curve probably reflects the low heavy-element concentration at the place and time long ago when this subdwarf formed. The discontinuity near $A = 140$ appears greater for this subdwarf than for the sun, which seems to suggest that the $\rho(\tau)$ appropriate to the gas from which it condensed not only has a generally smaller value to account for the general underabundance but is also a more steeply decreasing function of τ. The two barium stars HD 83548 and HD 116713 and the two CH stars HD 26 and HD 201626, on the other hand, are overabundant relative to the solar system in the heavy elements. These evolved stars are so rich compared to main-sequence stars that they are believed to have exposed the products of their interior. The CH stars in particular are old population II evolved stars, and their overabundance greatly exceeds the population II main-sequence counterparts. Interestingly enough, the discontinuity near $A = 140$ is for these stars less than for the solar system, which implies that the appropriate $\rho(\tau)$ is flatter. *In these cases, however, the* $\rho(\tau)$ *is to be interpreted as the result of the operation of the s process within the star itself,* in contrast to the interpretation for γ Pavonis and for the sun. The reader will note with interest the similarity between these σN curves and the family of curves resulting from different choices of τ_0, as illustrated in Fig. 7-25.

The question of where the neutrons come from is as old as the *s*-process idea itself and still largely unsolved. The difficulty is that neutrons are not released by the main line of nuclear reactions until a fairly late stage, carbon burning, is

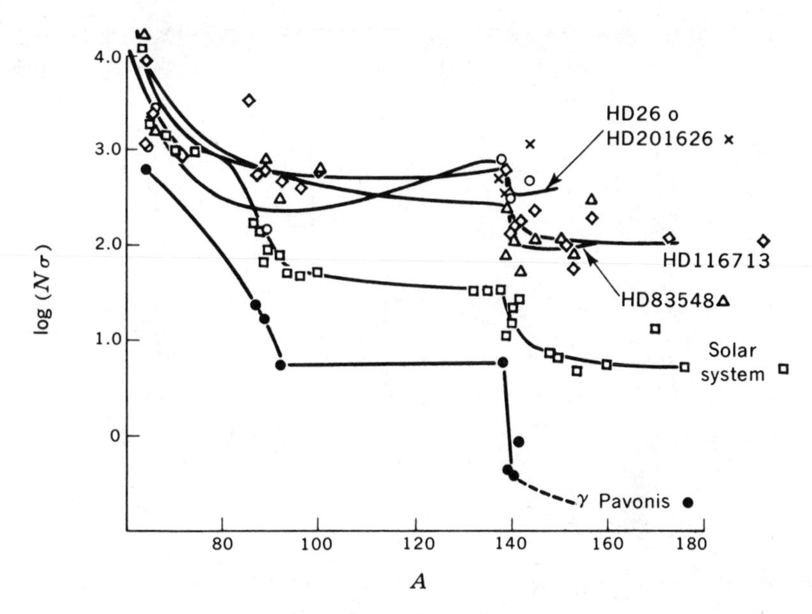

Fig. 7-24 The σN s-process curve as estimated for six different stars. Other stellar systems do not have the same abundances as those in the solar system, but the s-process correlation seems to be preserved with a change in shape. [*I. J. Danziger, Astrophys. J.*, **143:** 527 (1966). *By permission of The University of Chicago Press. Copyright* 1966 *by The University of Chicago.*]

reached. Reference to the section of Chap. 5 on carbon burning will illustrate the ways in which neutrons can be liberated in substantial amounts in that environment. The secondary reactions that liberate neutrons during helium burning may be even more interesting because they should be more common, according to the following argument. The products of helium burning, viz., C^{12}, O^{16}, are about one order of magnitude more abundant than the products of carbon burning, viz., Mg^{24}. It follows that roughly 10 times as many iron seed nuclei have been through the helium-burning environment of a star as have been through the carbon-burning environment. Thus the efficiency of neutron sources in helium burning will require careful investigation.

The operation of the CNO cycle during hydrogen burning leaves calculable amounts of C^{13} and O^{17} behind for the onset of helium burning, and to a calculable extent some Ne^{21} is made during hydrogen exhaustion at higher temperatures by the reaction $Ne^{20}(p,\gamma)Na^{21}(\beta^+\nu)Ne^{21}$. All three of these nuclei liberate neutrons during helium burning by exothermic interactions with alpha particles:

$C^{13}(\alpha,n)O^{16}$

$O^{17}(\alpha,n)Ne^{20}$

$Ne^{21}(\alpha,n)Mg^{24}$

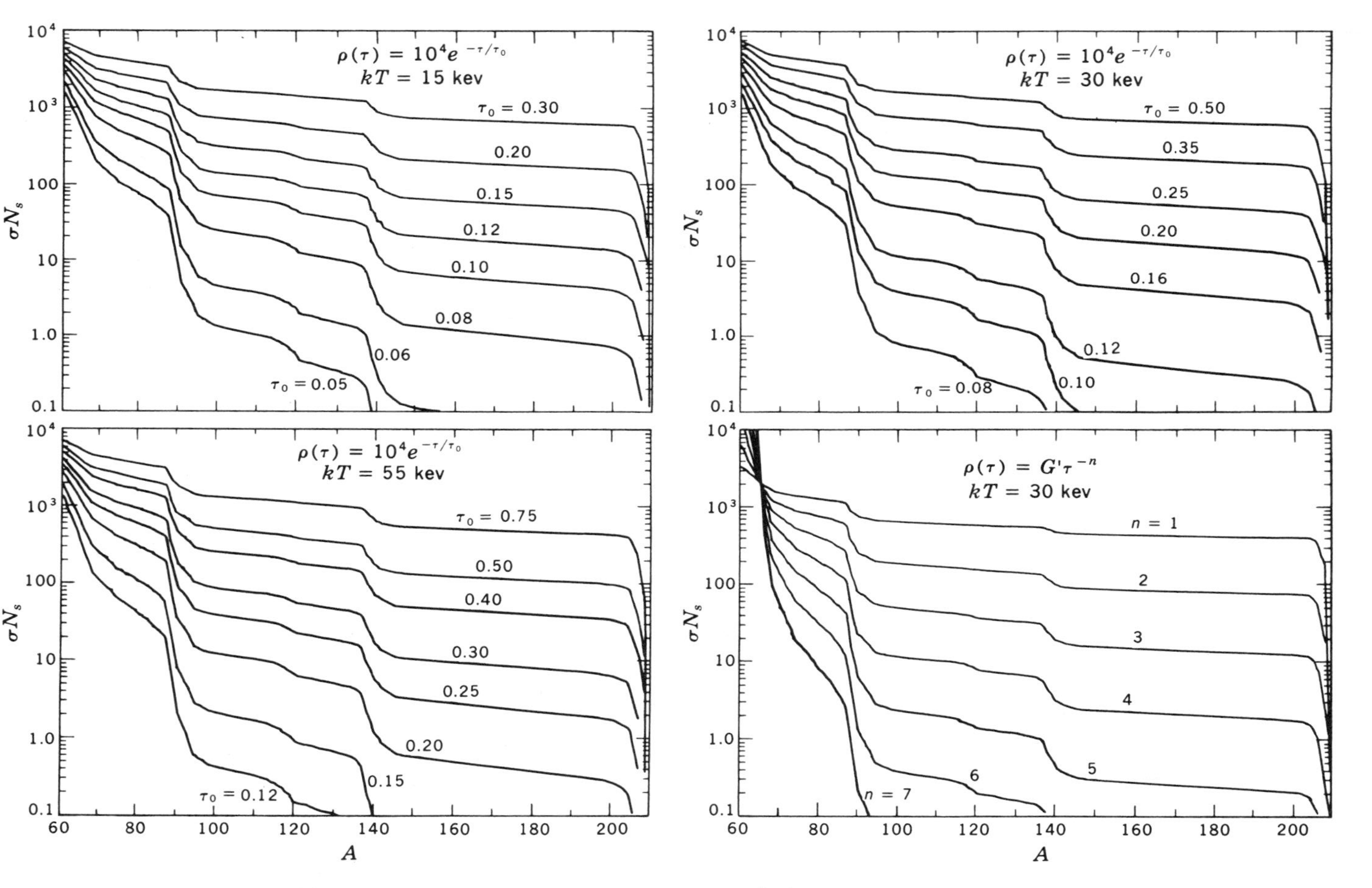

Fig. 7-25 The σN_s curves resulting from several distributions of neutron irradiations. The first three are the results of exponential distributions of neutron irradiations, as computed from Eq. (7-59). They show that the shapes of the distributions are almost the same at differing temperatures, although the value of the parameter τ_0 is temperature-dependent. The fourth figure shows several power-law distributions of neutron irradiations, each curve being normalized to the same value at $A = 65$. [*P. A. Seeger, W. A. Fowler, and D. D. Clayton, Astrophys. J. Suppl.*, **11**:121 (1965). *By permission of The University of Chicago Press. Copyright* 1965 *by The University of Chicago.*]

The efficiency of these neutron sources can be calculated only within the context of an evolving model of a star. Such a program has not yet been carried out in detail, and its necessity illustrates once again the unity of the subjects of stellar evolution and nucleosynthesis. At the present time it appears that the $O^{17}(\alpha,n)Ne^{20}$ reaction is the most promising of these three, simply because O^{17} seems to be left over in greater abundance than the other two.

Another promising possibility is afforded by the reaction $Ne^{22}(\alpha,n)Mg^{25}$. The Ne^{22} is created in great abundance by the helium burning itself. The CNO nuclei were transmuted almost entirely into N^{14} during the operation of the CNO cycle, and the N^{14} rather easily captures two alpha particles during the progress of helium burning, ending thereby as Ne^{22}. If the temperature becomes as great as 2×10^8 °K during the last stages of helium burning, the neutron-liberating reaction on Ne^{22} can proceed. Once again we see the necessity of using detailed models of evolving stars to evaluate the neutron source.[1] The advantage of this source is that the average ratio of abundance of CNO nuclei to iron-group nuclei is of order 100, so that up to 100 free neutrons can be made available per iron nucleus. Much more work remains to be done to ascertain whether the $\rho(\tau)$ observed in various objects can be easily provided for by a normal sequence of events in stellar evolution. One must, for various stellar masses and compositions, compute for each mass zone both the rate of neutron liberation and the lifetime of a free neutron, the product of which yields the free-neutron density. By following the free-neutron density in time, the exposure parameter τ can be calculated. Then for application to the *s*-process problem one must be alert to the fact that because the cross sections generally decrease with temperature, the effectiveness of a given value of the integrated flux τ in producing neutron captures varies with temperature. But it is possible to normalize all fluxes to the same temperature by a simple scaling law[2]

$$\tau(kT) \approx \left(\frac{kT}{30 \text{ kev}}\right)^{0.7} \tau(30 \text{ kev}) \tag{7-61}$$

which is accurate enough for the extraction of $\rho(\tau)$. The concept of such a scaling law allows successive irradiations at differing temperatures to be compounded.

In conclusion we must return to the basic assumption of the *s* process, that neutron-capture rates are slow compared to beta-decay rates. In some cases the radioactive nuclei produced have unusually long half-lives, and by examining

[1] H. Reeves, *Astrophys. J.*, **146**:447 (1966) has given a good discussion of the physics of neutron emission. J. Peters, Ph.D. thesis, Indiana University, Bloomington, Ind.,1967 has evaluated the efficiency of the Ne^{22} source in core helium burning in Iben's models. Interesting variants that require mixing between hydrogen and helium zones have been analyzed by G. R. Caughlan and W. A. Fowler, *Astrophys. J.*, **139**:1180 (1964), and in a different context by R. H. Sanders, *Astrophys. J.*, **150**:971 (1968).

[2] D. D. Clayton, Distribution of Neutron-source Strengths for the *s*-process, in W. D. Arnette, C. J. Hansen, J. W. Truran, and A. G. W. Cameron (eds.), "Nucleosynthesis," Gordon and Breach, Science Publishers, Inc., New York, 1968.

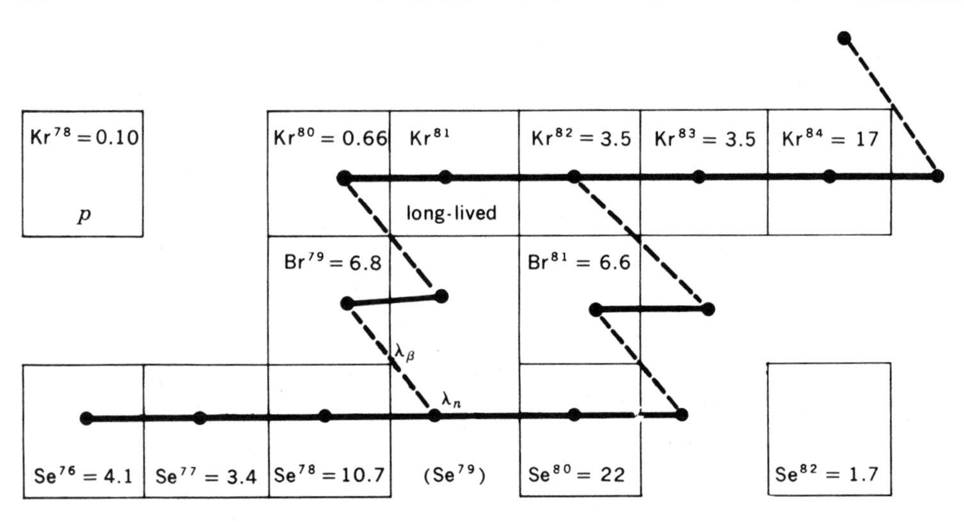

Fig. 7-26 The s-process path through selenium, bromine, and krypton. An interesting branch between neutron capture and beta decay occurs at Se^{79}, which has a laboratory half-life of 6.5×10^4 years. Both Kr^{80} and Kr^{82} are shielded from r-process production, by Se^{80} and Se^{82}, respectively. The ratio of s-process current through Kr^{80} to that through Kr^{82} is equal to the ratio of $\lambda_\beta(Se^{79})$ to $\lambda_\beta(Se^{79}) + \lambda_n(Se^{79})$. The abundance of each nucleus per 10^6 silicon atoms in the solar system is indicated.

such cases it should be possible to learn something of the average neutron-capture rate by estimating the branching between the two alternatives. One of the more useful of such cases is illustrated in Fig. 7-26. The element krypton has two s-only isotopes, and it is apparent that only a fraction of the s-process current passes through Kr^{80}, whereas all of it passes through Kr^{82}. If the p abundances are ignored, which seems justifiable in light of the small abundance of Kr^{78} and other p nuclei not pictured, we have approximately

$$\frac{\sigma(Kr^{80})Kr^{80}}{\sigma(Kr^{82})Kr^{82}} \approx \frac{\lambda_\beta(Se^{79})}{\lambda_\beta(Se^{79}) + \lambda_n(Se^{79})} \tag{7-62}$$

On the basis of nuclear systematics one expects $\sigma(Kr^{80})$ to be about twice as great as $\sigma(Kr^{82})$. This expectation can be coupled with the known abundances to estimate the branching ratio at Se^{79}.

Problem 7-26: Assume that the p contributions to each isotope of krypton are identical and calculate the ratio of λ_n to λ_β for Se^{79}.
Ans: $\lambda_n \approx 2\lambda_\beta$.

It would appear that from the results of this problem one would know the value of $\lambda_n(Se^{79})$ and hence that a measurement of $\sigma(Se^{79})$ would yield the strength of the neutron flux in the s process. Unfortunately life is not so simple. For one thing such a calculation yields only the average flux of the *s* process, although that number in itself would be a highly desirable result. A more essential complication is that the beta-decay lifetime of Se^{79} is not a known constant but instead depends upon the temperature. The ground state is found in the laboratory to have a half-life equal to 6.5×10^4 years. But Se^{79} has an excited state at only 96 kev of excitation, and when the decay rate of that thermally populated excited state is included in the total rate, one finds that the half-life becomes

$$\tau_{\frac{1}{2}}(\mathrm{Se}^{79}) \approx \frac{6.5 \times 10^4 \text{ years}}{1 + 6.47 \times 10^5 \exp(-11.1/T_8)} \tag{7-63}$$

which has a value of only 34 years at $T = 2 \times 10^8$ °K. So although the argument regarding the branching seems to be a sound one, arguments of this type usually are temperature-dependent. The hope for this kind of argument is that by carefully examining all such branches one can converge to the proper circumstances for the *s* process. General agreement has not been reached on the results of this kind of research, but it seems that $kT = 30$ kev is probably within a factor of 2 of the most common temperature.

In summary it may be said that the *s* process works very well.[1] It is one of our most successful theories, and its future for studies of the chemical evolution of the galaxy appears bright.

DETAILS OF THE *r* PROCESS

Once the class of *s*-process nuclei has been isolated, it becomes possible to extract the yield required of the *r* process. The majority of the heavy nuclei have both *s*- and *r*-process components in their makeup. By demanding that the *s*-process component be such that σN_s fall on the smooth curve of Fig. 7-20 one can attribute the remainder of the abundance to the *r* process. That is, in the spirit of Eq. (7-27) and Prob. 7-18 one calculates

$$N_r(Z,A) \approx N(Z,A) - \frac{f(A)}{\sigma(Z,A)} \tag{7-64}$$

where $f(A)$ represents whatever smooth function seems best to characterize the σN curve for the *s* process. In performing such a subtraction one clearly needs values of the neutron-capture cross sections. Most of them have not been measured, but nuclear systematics usually provides a reasonably reliable estimate of their value. A theoretical estimate for a cross section is, however, only a poor substitute for a carefully measured value, and one would ultimately like to have a measured value for every cross section on the *s*-process chain. It should also be noticed that because of numerical uncertainties, the subtraction will have little significance when N_r turns out to be a small difference between two larger numbers.

[1] The experimental evidence has been analyzed in R. L. Macklin and J. H. Gibbons, Quantitative Tests of *s*-process Stellar Nucleosynthesis for Solar System Material, *Astrophys. J.*, **149**:577 (1967).

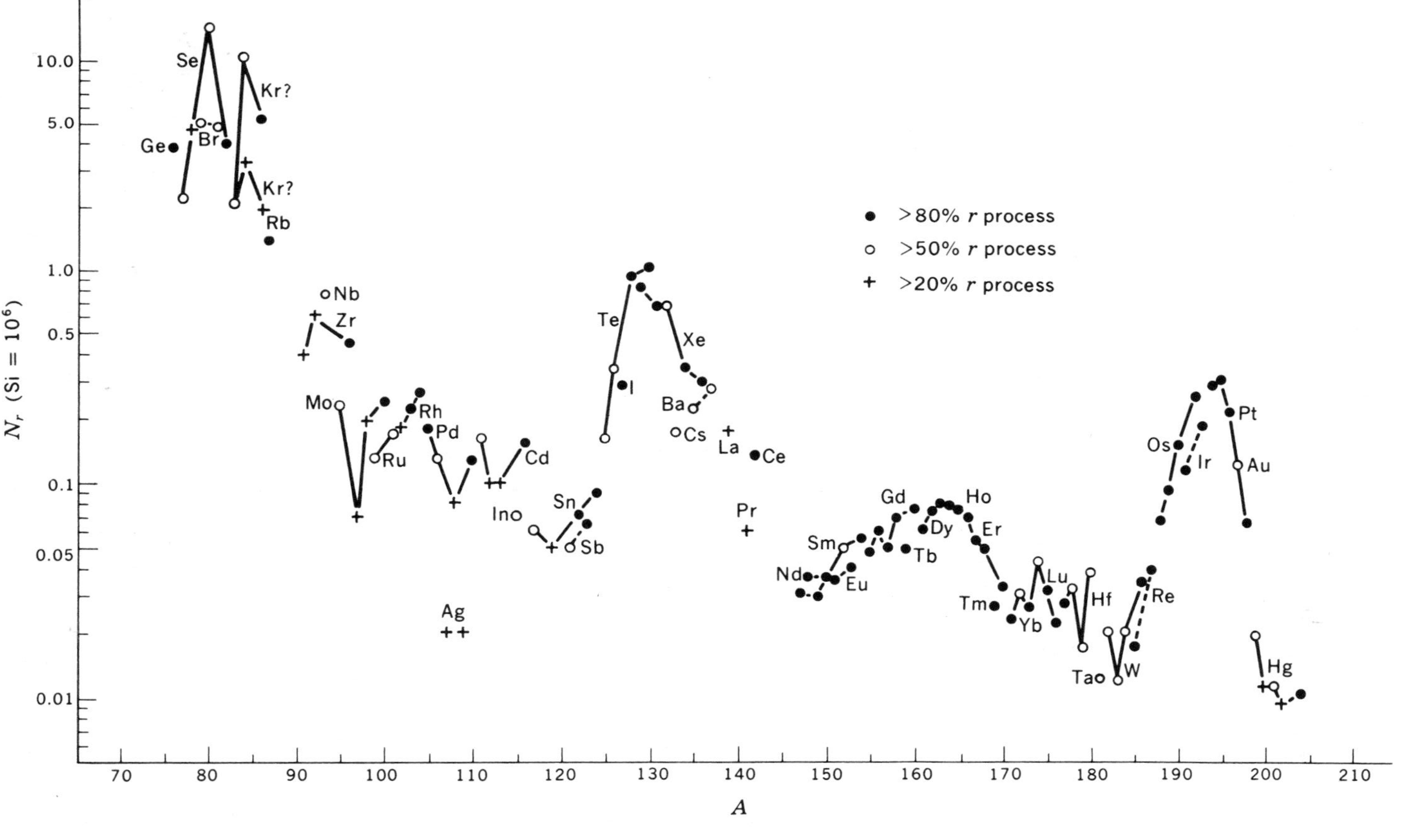

Fig. 7-27 The solar-system r-process abundances. The r-process abundances have been determined by subtracting the s-process contribution from the total abundance of the nucleus. Isotopes of a given element are joined by lines, solid lines for even Z and dashed lines for odd Z. The most inportant characteristics are the three main abundance peaks and the broad hump in the region of the rare earths. [*P. A. Seeger, W. A. Fowler, and D. D. Clayton, Astrophys. J. Suppl.,* **11**:121 (1965). *By permission of The University of Chicago Press. Copyright* 1965 *by The University of Chicago.*]

The estimated r-process abundances are plotted in Fig. 7-27 on the Si $= 10^6$ scale. Three different classes of points are indicated, based upon the relative size of the s-process subtraction. No point is plotted for which N_r is less than 20 percent of the total isotopic abundance. As a result numerous gaps appear in the figure at the location of the s-process peaks. Isotopes of the same element are connected by lines in the figure to illustrate the fact that changes in the *elemental* abundances cannot in general change the major features. The major peaks in the r-process abundances are fairly well established by the large slope in the abundances of the isotopes of the elements in the peaks. For example, the isotopic composition of Te and Xe strongly indicates an abundance peak near $A = 130$, and small uncertainties in the abundances of Te and Xe affect only the size of the peak. The outstanding features of this figure are those abundance peaks at $A = 195$, $A = 130$, and probably $A = 80$. It is the existence of these peaks, each about 10 units of atomic weight less than the s-process peaks near $A = 208$, $A = 140$, and $A = 90$, that leads to the identification of the r process.

It seems natural to assume that these r-process abundance peaks are also due to magic-neutron-shell closure at $N = 50$, 82, and 126 but that the nuclei formed are so far to the neutron-rich (or alternatively *proton-poor*) side of beta stability that the atomic number Z of the magic-neutron nuclei is about 10 units less than it is along the valley of beta stability, where the s process flows. After rapid ejection into space these nuclei would then undergo a series of beta decays ending at the most neutron-rich stable isobar, and the neutron-rich abundance peak would be preserved in the form of the observed peaks near $A = 195$, $A = 130$, and $A = 80$. It is essential that the reader achieve clear understanding of this primary idea of the r process, viz., the observed r-process abundance peaks will be attributed to abundance peaks of neutron-magic progenitors in very neutron-rich matter. This assumption is not so arbitrary as it might at first appear, because there are at least two physical ways of achieving the desired abundance peaks in this way, whereas it has proved impossible to attribute the peaks to any other physical idea.

The first possibility for producing such peaks lies with the extension of the ideas of nuclear statistical equilibrium to very high density. We have discussed earlier the fact that at temperatures of a few billion degrees the nuclear equilibrium favors the iron peak at the densities encountered in the cores of presupernova stars (say $10^4 < \rho < 10^8$). Heavier nuclei are not established with significant abundance at these densities because the photodisintegration rates are too great, but the situation changes rather remarkably at very high densities, say $\rho > 10^{10}$, such as may be encountered in the collapse of the supernova core, and it changes for two fundamental reasons: (1) on very general grounds one sees from the statistical relations that $N(A + 1)/N(A)$ is proportional to the density, so that very high density will shift the equilibrium toward more massive nuclei, and (2) the exclusion principle will lead to such large values of the Fermi energy for the electron gas that the electrons will be endoergically captured by nuclei, thus driving matter far to the neutron-rich side. Both effects are in the direction required for the phenomenon sought.

Problem 7-27: Estimate the electron density required in a degenerate gas to force the conversion of free protons to neutrons by electron capture.
Ans: $n_e > 10^{31}$ cm^{-3}.

Calculations of this phenomenon are illustrated in Figs. 7-28 and 7-29. In Fig. 7-28 we see that the abundance peak has shifted to nuclei having $N = 50$ in the neighborhood of $A = 80$ at $\rho = 1.87 \times 10^{10}$. If the density is increased still further, a considerable component breaks through to nuclei with $N = 82$ and $120 < A < 130$, as illustrated at $\rho = 1.78 \times 10^{11}$ in Fig. 7-29. These two exam-

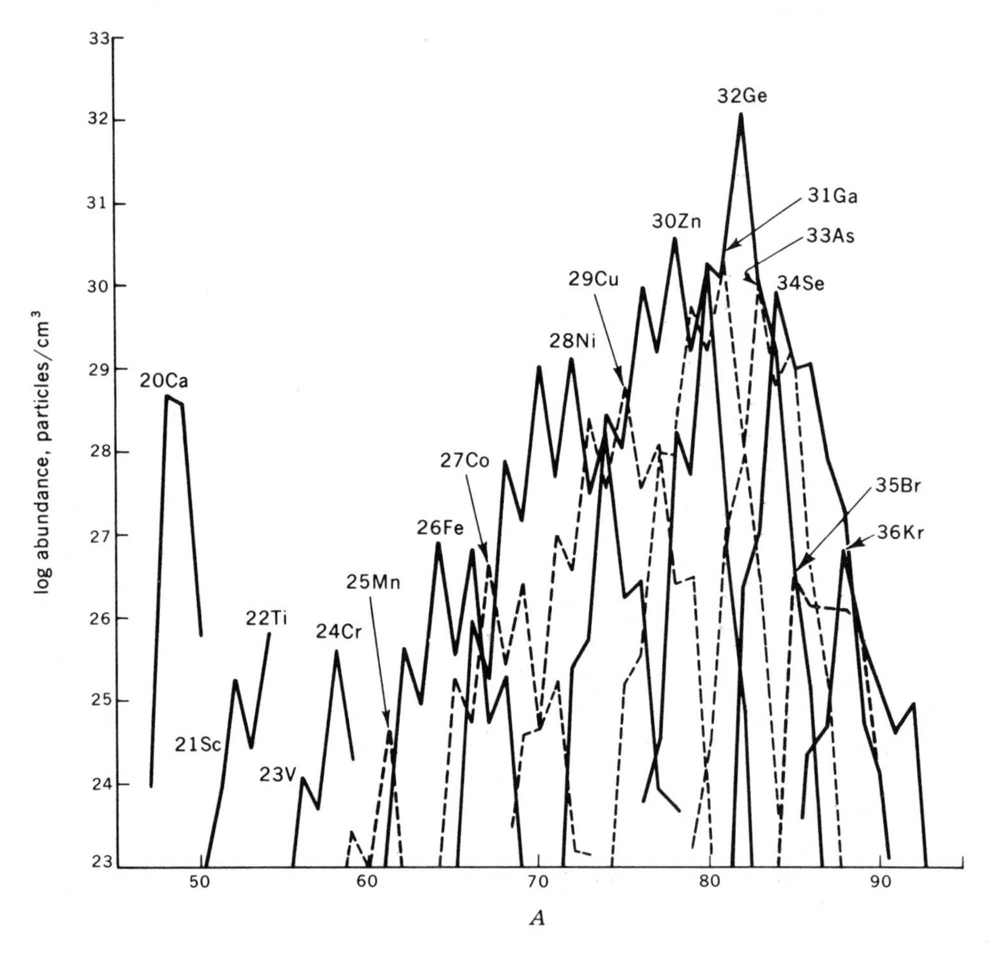

Fig. 7-28 The composition of matter in nuclear statistical equilibrium at $T = 5 \times 10^9$ °K and $\rho = 1.870 \times 10^{10}$ g/cm^3. The electron Fermi energy is $E_F = 10$ Mev at this density, and electron capture has driven matter to the neutron-rich side of the valley of beta stability. The abundance peak bears gross similarities to the first r-process abundance peak. [*S. Tsuruta and A. G. W. Cameron, Can. J. Phys.*, **43**:2056 (1965).]

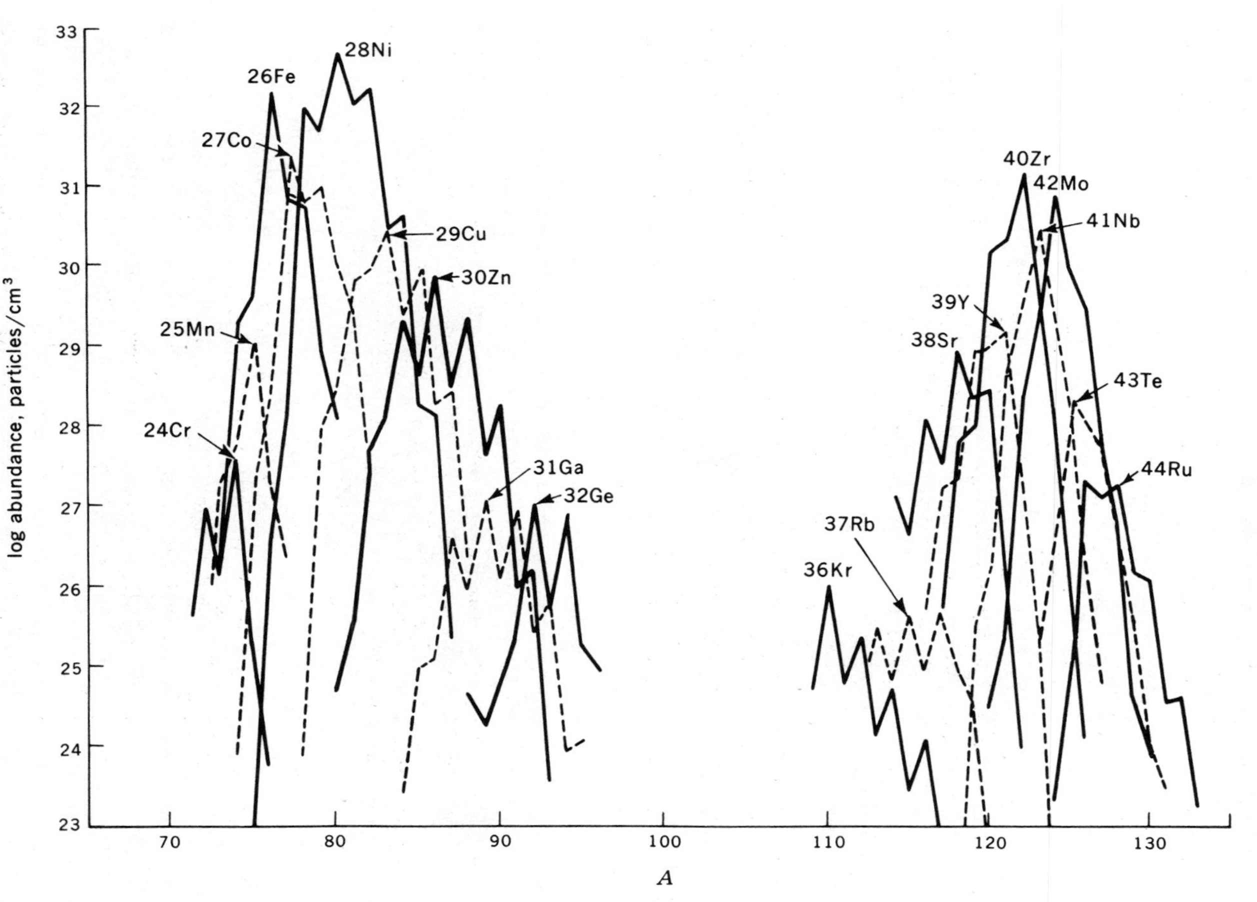

Fig. 7-29 The composition of matter in nuclear statistical equilibrium at $T = 5 \times 10^9$ °K and $\rho = 1.776 \times 10^{11}$ g/cm^3, where the electron Fermi energy is $E_F = 20$ Mev. The matter is more neutron-rich than in Fig. 7-28, and the two abundance peaks produced bear gross similarities to the first two r-process peaks. [*S. Tsuruta and A. G. W. Cameron, Can. J. Phys.*, **43**:2056 (1965).]

ples demonstrate the possibility of producing the appropriate neutron-rich abundance peaks in this way, but very little quantitative work has been done in trying thus to account for the details of r-process abundances. There is a very large knowledge gap concerning how such nuclei can be injected into the interstellar medium. The conditions illustrated are quite commensurate with the possibilities in the imploded supernova core, but as the material reexpands explosively, why is it that the nuclear equilibrium does not shift to that appropriate for lower densities, viz., back down into the iron-group peak? The preservation of the parent peaks would seem to require such rapid cooling that the photoneutron rates would be inadequate to drive the peaks back down to iron in the natural time scale of the explosion. The uncertainties surrounding these possibilities highlight once again the crucial role of the supernova in nucleosynthesis. Very detailed and correct models are required of the implosion-explosion phenomenon in order to be able to assess the nature of the nuclear debris.

Before passing on to the second r-process mechanism, however, we wish to point out an uncertainty of these nuclear calculations that is common to both mechanisms. The calculation of the abundances of neutron-rich nuclei invariably involves the binding energy of those nuclei, but these binding energies are not known from laboratory experiments. The nuclei involved cannot be produced in the laboratory. It is, therefore, always necessary to use a semiempirical formula for the nuclear binding. These formulas are obtained by applying the theories of nuclear binding to the class of known binding energies in an attempt to obtain a formula that can be extrapolated into regions of unknown nuclei. This problem is a very difficult one, and no general agreement has been reached regarding the best procedures to adopt. The associated uncertainties plague every discussion of r-process nuclei.

The standard form for the binding energy of nuclei is

$$B_0(Z,A) = \left(\alpha - \frac{\gamma}{A^{\frac{1}{3}}}\right) A - \left(\beta - \frac{\eta}{A^{\frac{1}{3}}}\right) \frac{I^2 + 4|I|}{A} + \delta\left(\pm \frac{1}{A^{\frac{1}{2}}}, 0\right) - 0.800 \frac{Z^2}{A^{\frac{1}{3}}}\left(1 - \frac{0.76361}{Z^{\frac{2}{3}}}\right)\left(1 - \frac{1.9605}{A^{\frac{2}{3}}}\right) \qquad (7\text{-}65)$$

where $I = N - Z$ is the neutron excess. The five parameters α, β, γ, η, and δ are commonly determined by a least-squares fit to the known binding energies of more than 800 nuclei. The form of these terms is suggested by nuclear theory, and each one has a simple intuitive meaning.[1] The term αA represents the approximate fact that the binding energy per nucleon is roughly constant, especially when it is reduced by the number of unsaturated nuclear interactions at "the surface of the nucleus," which is proportional to $\gamma A^{\frac{2}{3}}$. The fact that the radius scales as $A^{\frac{1}{3}}$ again appears in the last term, which is an approximation to the coulomb energy. In the absence of coulomb effects, the maximum binding would occur for $Z \approx N$, because only under those conditions can all nucleons be

[1] A discussion of these terms may be found in M. A. Preston, "Physics of the Nucleus," chap. 6, Addison-Wesley Publishing Company, Inc., Reading, Mass., 1962.

assigned to the lowest available single-particle energy eigenstates. The second term therefore reflects the quadratic decrease in binding necessitated by the exclusion principle when the numbers of protons and neutrons are unequal. The δ term reflects a pairing energy or symmetry energy and is taken as positive for even-even nuclei, zero for odd A, and negative for odd-odd nuclei. Elaboration of these points is easily obtainable from textbooks on nuclear physics.

This standard mass law will not successfully reproduce the abundance features attributed to the r nuclei, and the reason for this is very simple. The postulated models for the production of r nuclei depend upon the abrupt changes of binding energy at the magic-nucleon-shell closures, but Eq. (7-65) takes no account of the empirically known discontinuities at those nucleon numbers. This shortcoming can be cured by adding such a term between each pair of magic numbers in Z and N which is, for example, quadratic and has a value zero at each end of the shell. Thus one might choose

$$B(Z,A) \approx B_0(Z,A) - \zeta_j \frac{(N_{j+1} - N)(N - N_j)}{(N_{j+1} - N_j)^2} - \zeta_k \frac{(Z_{k+1} - Z)(Z - Z_k)}{(Z_{k+1} - Z_k)^2} \qquad (7\text{-}66)$$

where the magic numbers N_j and Z_k are at 20, 28, 50, 82, 126, and 184. These ζ coefficients have been tabulated by Seeger et al,[1] who also added terms due to increased stability associated with nuclear deformation in certain mass ranges. Such attempts are typical of the lengths to which one must go to handle this awkward problem.[2]

Problem 7-28: Calculate the binding energy of a neutron in (Z,A) as the difference of the two appropriate binding energies of the form Eq. (7-66).

A more common conceptual approach to the production of r nuclei[3] has the advantage of not requiring such inordinately high densities, requiring instead only a higher neutron density than is ordinarily encountered in advanced stages of stellar evolution. The idea uses a flow concept rather than a nuclear-equilibrium concept and is similar to an s process except that the neutron flux is presumed to be so great that neutron capture is rapid (hence the mnemonic r) compared to beta-decay rates. Consider the fate of a heavy nucleus placed in such an intense neutron flux, which was estimated by Eq. (7-31) to require neutron densities of order $n_n \approx 10^{23}$ cm^{-3}. Such a nucleus would capture neutrons rapidly, one after another, until the neutron binding energy becomes so low that it can capture no more. The nucleus would presumably then wait until beta decay increases its charge, whereupon it would again capture as many neutrons as would be consistent with neutron binding. This repetitive sequence of events would lead to a *waiting point* for each charge Z at which beta decay

[1] Seeger, Fowler, and Clayton, *loc. cit.*

[2] See also A. G. W. Cameron and R. M. Elkin, *Can. J. Phys.*, **43**:1288 (1965).

[3] Devised by Burbidge, Burbidge, Fowler, and Hoyle, *loc. cit.*

must occur before the capture chain can proceed. It is a simple matter to see that under such conditions the abundance at each charge Z is governed by the simultaneous set of differential equations

$$\frac{d}{dt} n_Z(t) = \lambda_{Z-1} n_{Z-1}(t) - \lambda_Z n_Z(t) \tag{7-67}$$

where λ_Z is the beta-decay rate at the waiting point for charge Z.

Problem 7-29: Compare these equations with those of the s process. Show that they have the same formal solutions that were found for the s process. Write the explicit solutions if Fe^{56} is the seed for the capture chain. In this case, however, the abundances vary inversely with λ_Z rather than with the neutron-capture cross sections.

We see from Eq. (7-67) and from Prob. 7-29 that the abundance n_Z at the waiting point for charge Z varies inversely with the beta-decay rate at the waiting point. If the slowest beta-decay rates occur for neutron-magic nuclei, and if the waiting points occur at neutron-magic nuclei with the proper neutron excess, a successful r-process model can be constructed in this way. Before exploring the model, however, we must refine it in several important aspects.

The first question to be answered is why it is that the sequence of rapid neutron captures is halted. As neutrons are added, the general trend is that the neutron binding energy decreases, and eventually such heavy isotopes may be created that the neutron binding has fallen to zero. Before that point is reached, however, the capture chain is halted by the photoejection of neutrons by the thermal photons. When the neutron binding is reduced to the point where the lifetime against photoejection of a neutron is less than the lifetime against neutron capture, the sequence of captures will grind to a halt.

Problem 7-30: Show that the ratio of the rate per nucleus of neutron capture by (Z,A) to the rate per nucleus of photoneutron emission from $(Z, A+1)$ is given by

$$\log \frac{\lambda_n(Z,A)}{\lambda_\gamma(Z, A+1)} = \log \frac{G(Z, A+1)}{G(Z,A)} + \log n_n - 34.0749 - \tfrac{3}{2} \log \frac{A}{A+1} T_9 + \frac{5.040}{T_9} Q_n(Z, A+1) \tag{7-68}$$

where $Q_n(Z, A+1)$ is the binding energy in Mev of a neutron in the nucleus $(Z, A+1)$:

$$Q_n(Z, A+1) = B(Z, A+1) - B(Z,A) \tag{7-69}$$

Problem 7-31: For the conditions $n_n = 10^{24}\ \text{cm}^{-3}$ and $T_9 = 1$, how small must the neutron binding energy be in order that $\lambda_\gamma(Z, A+1) > \lambda_n(Z,A)$?
Ans: $Q_n(Z, A+1) < 2$ Mev.

Neutrons are less tightly bound in an odd-N nucleus than in neighboring even-N isotopes, so that the capture chain will tend to stop in some even-N nucleus to which the binding of the next odd neutron would be so small that it would be rapidly reejected. Thus the waiting point for each element will be identified with the lightest isotope having the property that $\lambda_\gamma(A + 1)$ is greater than $\lambda_n(A)$. From Eq. (7-68) it is apparent, therefore, that the location of the waiting points depends upon the neutron density and the temperature. The greater the neutron density, the more neutron-rich the waiting points. The greater the temperature, the less neutron-rich the waiting points. The conditions $n_n \approx 10^{24}$ and $T_9 \approx 1$ are typical of the ones commonly discussed for the r process, and they lead to waiting-point binding energies of order $Q_n(Z, A + 1) \approx 2$ Mev. The neutron-capture path of the r process under these conditions is shown as the shaded band in Fig. 7-30. This band is well to the neutron-rich side of the stable isotopes represented by black dots. The capture path moves slowly upward in Z as it races outward in N until a magic number of neutrons is encountered. The extra stability of the magic neutron shell was able to maintain the neutron binding at a high level as shell closure was approached. The next neutron after shell closure has very small binding energy in such neutron-rich matter, however, and is quickly photoejected. Thus magic neutron configurations will definitely be expected to be waiting points. These neutron-magic waiting points have longer than normal beta-decay lifetimes, moreover, and so they will build up to relatively large abundance in the flow.

After the beta decay of a neutron-magic nucleus, it captures only one more neutron before it is again neutron-magic. At the shell closures, therefore, a sequence of waiting points is encountered at the same neutron number until, after several beta decays, the flow is sufficiently close to the region of stable nuclei for the binding of the next neutron to become sufficiently great to allow the path to "break through" the magic value of N. These series of magic-N beta decays are clearly visible in Fig. 7-30, particularly at the values $N = 82$ and $N = 126$. Because the beta-decay rates are progressively smaller for these nuclei, the r-process progenitors have abundance peaks at these places on the path. Extrapolation along a line of constant A shows that these progenitors will ultimately decay to stable abundance peaks near $A = 130$ and $A = 195$, in good agreement with the requirements of the process.

Not all the nuclei *waiting* at a given value of Z will be concentrated into a single isotope. The binding energies change sufficiently slowly on the average for the inverse neutron captures and photodisintegrations to spread the concentration over a small range of isotopes. This will, in fact, establish a local equilibrium with the neutron bath and the photon bath. Thus the concentration at the waiting point is described by an abundance distribution function $p(Z,A)$, defined as the probability for element Z that nuclei wait in the form of isotope A and having the property therefore that $\sum_A p(Z,A) = 1$. Of course, the fact that not all r processes will have occurred at identical values of n_n and T_9 will spread the distribution even more.

Because there will always be some free protons in the bath, the rate of the (p,n) reaction must be added to the rate of the beta decay at the waiting points. Nor is it clear, without a detailed model of the event, whether $\lambda_{p,n}$ or λ_β will

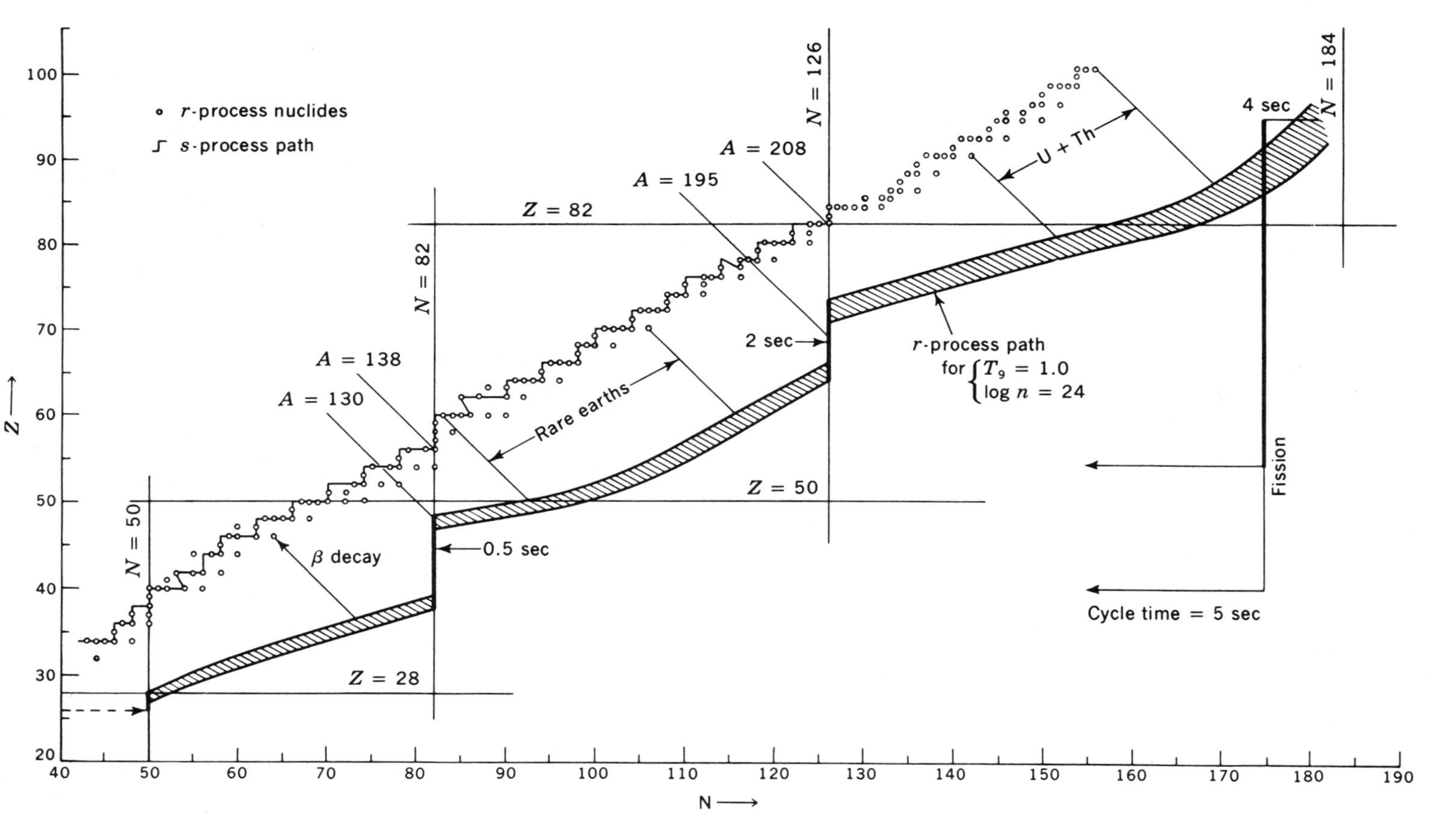

Fig. 7-30 Neutron-capture paths for the *s* process and the *r* process. The *s* process follows a path in the NZ plane along the line of beta stability. The neutron-rich progenitors to the stable *r*-process nuclei, which are here shown as small circles, are formed in a band in the neutron-rich area of the NZ plane, such as the shaded area shown here. This *r*-process path was calculated for the case $T_9 = 1.0$ and $\log n_n = 24$. After the synthesizing event the nuclei in this band beta-decay to the stable *r*-process nuclei. The abundance peaks at $A = 80$, 130, and 195 are attributed to abundance peaks in the neutron-rich progenitors having $N = 50$, 82, and 126. Neutron capture flows upward from the lower left-hand corner along the shaded band until neutron-induced fission occurs near $A = 270$. [*P. A. Seeger, W. A. Fowler, and D. D. Clayton, Astrophys. J. Suppl.*, **11**:121 (1965). *By permission of The University of Chicago Press. Copyright* 1965 *by The University of Chicago.*]

dominate as the major charge-increasing reaction. It may even be necessary in some circumstances to include (α,n) and (α,p) reactions, which also increase the nuclear charge, but this possibility has not been investigated. The function λ_Z introduced in Eq. (7-67) must at least be generalized to include all the charge-increasing reactions:

$$\lambda_Z = \sum_A p(Z,A)[\lambda_\beta(Z,A) + \lambda_{p,n}(Z,A) + \cdots] \tag{7-70}$$

Then the solution proceeds as before. The beta-decay rates show strong shell effects of the required type, because for decays with large end-point energies $\lambda_\beta \propto W^5$, where the energy W released in the decay is also computed from the mass law used for the problem. As the succession of waiting points at a closed shell drives the nucleus nearer to the region of beta stability, the energy W decreases, so that the lifetime, and hence the abundance, increases. This sequence is of the type to augment the abundance peaks. It is not clear whether the competitive (p,n) reactions have the same effect, at least to the same degree. Proton capture by a neutron-magic nucleus will proceed through a lower density of resonances than will the nonmagic nuclei, so that the rate of (p,n) reactions should be diminished at closed neutron shells. This effect is again in the proper direction, but its magnitude is uncertain. It seems that either type of charge-increasing reaction can successfully lead to an *r* process of this flow type.

The termination of the *r* process is considerably more dramatic than that of the *s* process and places added emphasis upon the role of nuclear systematics. It is impossible to build ever-larger nuclei because the coulomb energy, which is proportional to Z^2, becomes so great that the nuclei fission spontaneously. Indications from the systematics of nuclear fission are that it will occur for nuclei in this very neutron-rich region when the charge Z reaches 94. This occurs near $N = 175$ in Fig. 7-30. In the fission the heavy nucleus splits into two nuclei, with Z near 40 and 54, thereby returning two seed nuclei early into the capture chain. The number of heavy nuclei is doubled in the time required for the chain to build an average fission fragment back up to fissionable size. Large abundances of heavy nuclei can be built up in this way in a matter of seconds if the neutron supply holds out. For short times the fission can be ignored in favor of a simple flow problem toward greater atomic weights, as outlined in Prob. 7-29. For long times the abundances assume steady-state proportions such that each grows exponentially as nuclei pass around the fission cycle. The short-time solutions for three different values of the time are shown in Fig. 7-31. The conditions $n_n = 10^{24}\ \text{cm}^{-3}$ and $T_9 = 1$ are the same as those mentioned earlier, and the times are computed with the assumption that beta decays are the dominant charge-increasing reactions. It should be emphasized that the neutron captures are so rapid under these circumstances that no significant time is spent in moving out to the waiting points. The time is entirely due to the length of time the nuclei must wait before the charge is increased, either by beta decay or by (p,n) reactions. Any or all of these assumptions could be badly in error for nature's *r* proc-

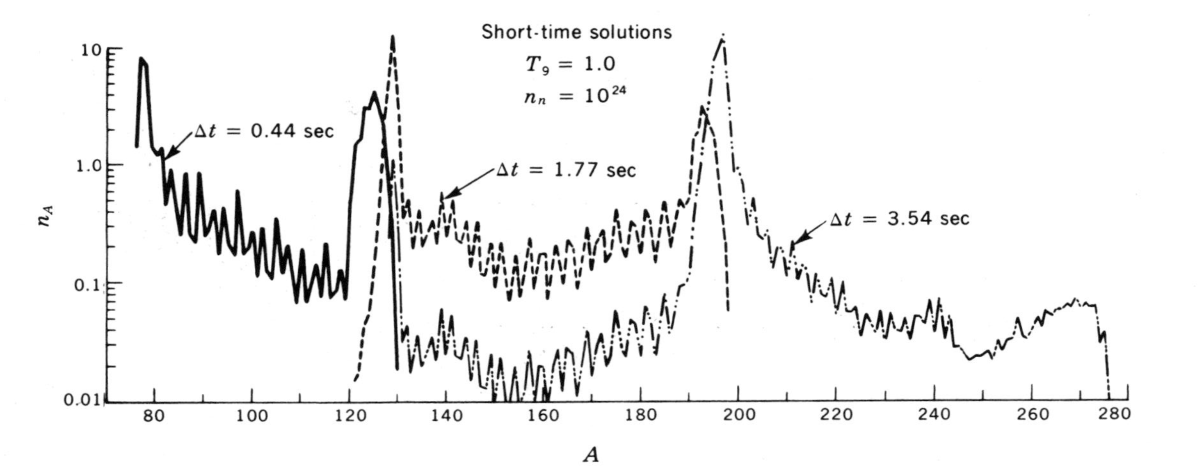

Fig. 7-31 Calculated r-process abundances resulting from iron seed placed in a neutron flux characterized by $n_n = 10^{24}$ and $T_9 = 1$ for three different time intervals. Abundance peaks near $A = 80$, 130, and 195 build up at the neutron-magic waiting points shown in Fig. 7-30. The times are computed with the assumption that beta decays are the major charge-increasing reactions. Neutron-induced fission terminates the distributions near $A = 275$. Under these conditions a significant number of nuclei have been processed from Fe to the transuranic region in times of order 3 sec. [*P. A. Seeger, W. A. Fowler, and D. D. Clayton, Astrophys. J. Suppl.*, **11**:121 (1965). *By permission of The University of Chicago Press. Copyright* 1965 *by The University of Chicago.*]

ess, but it is clear that one does obtain in this way the abundance peaks near $A = 80$, 130, and 195. Changing n_n or T_9 (or both) changes slightly the location of the peaks and changes rather drastically the time required for the process, but the possibility of obtaining the *r*-process abundances in this way is evident.

The steady-state abundance distribution obtained for long times is shown in Fig. 7-32. Fission terminates the process near $A = 276$, and the material cycles in again at points just below the *r*-process abundance peaks at $A = 130$ and $A = 195$. The time required for a given nucleus to pass through the full fission cycle is 4.9 sec under the conditions shown. This time again assumes that beta decays are the effective charge-increasing reactions.

Problem 7-32: Assume that the time t_{cycle} required for the fission cycle is determined by the beta-decay rates. From general principles, how do you expect that time to vary with n_n and T_9? Explain.
Ans: $(\partial t/\partial n_n) < 0$, $(\partial t/\partial T_9) > 0$.

Unfortunately, the *r* process does not have direct confirmation in the spectra of stars that is as convincing as that for the *s* process. The major evidence for the existence of an *r* process remains the existence of the abundance peaks on the small *A* side of the *s*-process peaks and the natural existence of uranium and thorium. These latter nuclei are known to be synthesizable by some natural mechanism simply by virtue of their existence in solar material, the implications of which will be outlined in the next section. No other mechanism seems able to synthesize transuranic abundances. The most natural site for the *r* process seems to be the supernova, so that a convincing confirmation in principle of the location of the process would be the demonstration from spectra that supernova remnants are rich in *r*-process heavy elements. Unfortunately the nuclei that are largely due to the *r* process have, with the exception of the rare earths, very unfavorable optical transitions and are thus difficult to detect in spectra of any type. It would be very encouraging to see at least one convincing case of an astronomical object that is definitely overabundant in *r* nuclei but not in *s* nuclei. Considerable attention has been given to a proposal[1] that the nearly exponential decline of supernova light curves reflects the power released from the spontaneous fission of Cf^{254} and possibly of other transuranic nuclei. This phenomenon has been often quoted as an observational indication of *r* process in supernovas, but the proposal is subject to many doubts. A most striking test will be the detection of line gamma rays emanating from the very heavy radioactive elements, but their detection will be difficult[2] at best, even if they exist in supernova remnants in the quantities required by the californium hypothesis.

If the flow mechanism of the *r* process is the one that occurs in astrophysics, it is not yet clear where the process occurs. It is no trivial matter to find neu-

[1] Burbidge, Burbidge, Fowler, and Hoyle, *loc. cit.*

[2] D. D. Clayton and W. L. Craddock, *Astrophys. J.*, **142**:189 (1965).

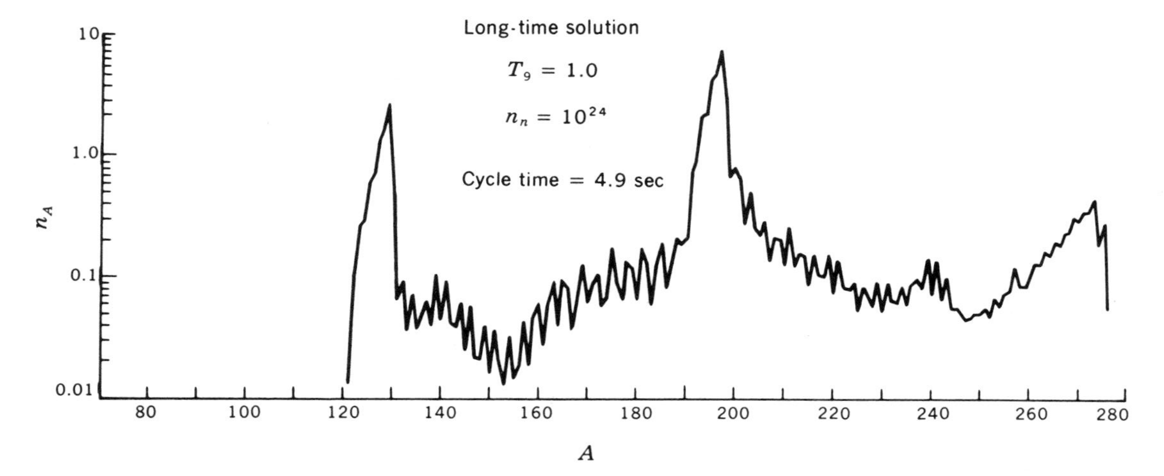

Fig. 7-32 Calculated r-process abundances after long times. The nuclei have been driven around a fission cycle until the abundance distribution becomes characteristic of a steady state. The abundance of each nucleus grows exponentially with an e time of 4.9 sec. This time, calculated on the basis of beta-decay rates, depends upon the neutron density and the temperature. [*P. A. Seeger, W. A. Fowler, and D. D. Clayton, Astrophys. J. Suppl.,* **11**:121 (1965). *By permission of The University of Chicago Press. Copyright* 1965 *by The University of Chicago.*]

tron densities great enough for the process to occur at temperatures small enough for heavy nuclei to be formed at all. The most persistent suggestions have again centered on the supernova. After the core collapse is halted, a shock wave propagates outward and raises the temperature of each layer momentarily to a value of several billion degrees. Neutrons may be liberated rapidly during this shock and then rapidly recaptured as the temperature quickly falls back down, in a matter of a few seconds. Such a site is intellectually appealing, but early estimates of the free-neutron densities encountered have not been very encouraging. A much more copious supply of neutrons exists in the imploded core, leading to the suggestion that the r process occurs as this matter is rapidly expanded from its compressed state. The problem with this idea is that one must rapidly establish enough medium-weight seed nuclei for the r process to build on from the hot gas of neutrons, protons, and alpha particles. Quantitative evaluation of this idea has not been made because once again one needs an accurate dynamical model of the supernova explosion upon which to base the calculation. A third idea has been the thermal runaways associated with flash phenomena in a degenerate electron gas. Further special requirements are needed to obtain a sufficient supply of neutrons in the gas, however, and it is not clear whether these requirements can be met realistically.

It can be stated with some confidence, however, that the proper interpretation of the origin of the r nuclei will have considerable astrophysical significance. The unusual demands placed by nuclear physics upon their circumstances of synthesis will certainly put their origin within an unusual and special event. The attempt to satisfy these nuclear demands within a natural astrophysical context is an exciting adventure.

In concluding the discussion of the mechanisms of heavy-element nucleosynthesis, we must make a brief mention of the p nuclei. Their low abundance attests to the relative infrequence of the site responsible for them, and perhaps also explains the relatively small amount of work that has been done on the problem of their origin. Suffice it to say that the idea of Burbidge et al. seems to be the correct one, that the p process is similar to the r process except that it is the proton flux which is so intense that rapid-proton capture occurs.

NUCLEAR COSMOCHRONOLOGY

About this time, Rutherford, walking in the Campus with a small black rock in his hand, met the Professor of Geology; "Adams," he said, "how old is the earth supposed to be?" The answer was that the various methods led to an estimate of 100 million years. "I know," said Rutherford quietly, "that this piece of pitchblende is 700 million years old."

This was the first occasion when so large a value was given, based too on evidence of a reliable character; for Rutherford had determined the amount of uranium and radium in the rock, calculated the annual output of alpha particles, was confident that these were helium, measured the amount of helium in the rock and by simple division found the period during which the rock had existed in a compacted form. He was the pioneer in this method and his large value surprised and delighted both geologists and biologists.[1]

So says Eve in his biography of Ernest Rutherford, whose pioneering work on natural radioactivity led to the science of nuclear cosmochronology. Rutherford's age for the piece of pitchblende was soon exceeded by many other minerals, and today there is abundant evidence pointing to major solidification of the earth and the meteorites about 4.5 to 4.6 billion years ago. The nature of this evidence and the geochemical problems associated with the formation of the solar system constitute a fascinating story.[2] But for the science of nucleosynthesis the essential question is that of the origin of the natural radioactive nuclei. To what extent is it possible to use the abundances of the radioactive nuclei and their daughters as they existed at the time of the formation of the solar system to learn something of the previous history of nucleosynthesis? In this final section we shall outline the conceptual context of this investigation.

All the relevant observations have been made on solar-system material, so that any implications for the history of nucleosynthesis within the galaxy are necessarily limited to apply to the history of only that material which was destined for incorporation into the primitive solar nebula. The composition of the solar nebula may or may not have been the average composition of gas in the galactic disk 4.6×10^9 years ago. To the extent that the two compositions were equal, the history of material incorporated into the solar system may be identified with the average nuclear history of the galactic disk, but it behooves one to remember that there may be some differences. The galaxy seems to rotate and mix in a time on the order of 2×10^8 years, so that if the galactic material were slowly enriched by stellar debris over time periods much longer than that, one could expect the galactic composition to be nearly uniform. But the abundances of radioactive nuclei with half-lives on the order of 2×10^8 years or less may be expected to reflect the local activity of synthesizing events, probably supernovas, in the recent history of the region.

We cannot expect to see the effects of radioactivity in meteorites except for those species having half-lives long enough to survive the period of time between the isolation of the solar nebula from interstellar gas and the solidification of the meteorites. It is generally believed that about 10^8 years, within a factor of 10 either way, is required to form solid objects from the gravitationally contracting nebula of the protostar, so that one does not expect to find any remnant of radioactivities having $\tau_{1/2} < 10^7$ years in the meteorites. The number of longer-lived species is not large, and of them only six show promising possibilities for

[1] A. S. Eve, "Rutherford," The Macmillan Company, New York, 1939.

[2] For an absorbing review of meteorite ages, see E. Anders, Meteorite Ages, in B. Middlehurst and G. Kuiper (eds.), "The Moon, Meteorites, and Comets," vol. 4 of "The Solar System," The University of Chicago Press, Chicago, 1963.

application to this problem. Their half-lives in increasing order are

$$\tau_{\frac{1}{2}}(I^{129}) = 16.9 \times 10^6 \text{ yr} \qquad \tau_{\frac{1}{2}}(Pu^{244}) = 82 \times 10^6 \text{ yr}$$

$$\tau_{\frac{1}{2}}(U^{235}) = 7.13 \times 10^8 \text{ yr} \qquad \tau_{\frac{1}{2}}(U^{238}) = 4.51 \times 10^9 \text{ yr}$$

$$\tau_{\frac{1}{2}}(Th^{232}) = 1.39 \times 10^{10} \text{ yr} \qquad \tau_{\frac{1}{2}}(Re^{187}) \approx 4 \times 10^{10} \text{ yr}$$

It is with the abundances of these nuclei and their daughters that nuclear cosmochronology is concerned.

These nuclei divide naturally into two groups. The half lives of the first two are so short that no detectable quantities of I^{129} and Pu^{244} exist today in the solar system. Their presence initially in the solar system can be inferred only by the detection of anomalies in the abundances of nuclei to which they decay. They are called *extinct radioactivities*. The other four nuclei are found to occur naturally today in the earth and in meteorites. In analyzing their implications for cosmochronology one may use both their abundances and the abundances of their daughters, at least in principle. In actual fact one is able to use only their abundances themselves, because all four nuclei decay to three isotopes of Pb, and the circumstances surrounding the nucleosynthesis of Pb are so complicated that it is not clear what portion of their abundances may be attributed to the radioactive-decay chains.[1]

These two classes of radioactivities probe somewhat different epochs in the history of nucleosynthesis, for which a model of the following type is generally adopted. It is assumed that nucleosynthesis has occurred continuously over a period of time T roughly equal to the age of the galaxy, or at least of the galactic disk. That is, T is a period of something like 1 to 10 billion years preceding the isolation of the solar nebula. Because of the large number of synthesizing events that are homogenized by galactic mixing over large times, it is commonly supposed that the rate of this continuous nucleosynthesis is either constant over its duration T or has decreased monotonically with time. In the popular models the duration T is followed by a time interval ΔT, commonly believed to be on the order of 10^8 years, that represents the time required for the solidification of the solid bodies within the solar system. During this interval ΔT all the radioactivities decay exponentially. The extinct radioactivities are measured by the amounts of xenon gas, to which they both decay, trapped within the meteorites. Obviously then, one measures the abundances of the extinct radioactivities at the time of meteoritic solidification rather than at the time of the isolation of the solar nebula. The two abundances are related by the laws of exponential decay over the interval ΔT. To explore the history of nucleosynthesis, however, one requires the abundances at the end of the duration period T of the continuous nucleosynthesis, and thus one needs to know the decay interval ΔT. The following generalizations should be correct.

[1] But see D. D. Clayton, *Astrophys. J.*, **139**:637 (1964) for a thorough discussion of the way in which this may be done with sufficiently accurate measurements of the abundances of Pb and U and of the neutron-capture cross sections of the Pb isotopes.

(1) The extinct radioactivities are most sensitive to the amounts of nucleosynthesis near the end of the duration T and, in addition, are strongly influenced by the length of the decay interval ΔT. They are not good probes of the beginnings of nucleosynthesis because their half-lives are so short that none of the early production remains in the initial solar nebula. Several workers, particularly Cameron and Kuroda, have emphasized that the discrete nature of the final synthesizing events should be retained, especially if the sun was formed in a stellar association wherein one or more local supernovas may have contributed a substantial amount of fresh radioactivity just before the condensation of the solar nebula.

(2) The long-lived radioactivities, U^{238}, Th^{232}, *and* Re^{187}, *receive substantial contributions from the entire duration of nucleosynthesis, and their abundances are practically independent of the decay interval ΔT.* These features make them good probes of the duration of nucleosynthesis or, as one sometimes says, of the age of the elements. But the fact that their abundances change negligibly over the decay interval ΔT makes them poor probes of that interval.

(3) U^{235} *is an intermediate case.*

Problem 7-33: Suppose that the elements were synthesized at a constant rate. Show that after long times the abundance of each radioactive species is equal to the number synthesized during one mean lifetime $\tau = \tau_{\frac{1}{2}} \ln 2$ of the nucleus. Go back over the preceding discussion with this result in mind.

There is one enormous coincidence among these six nuclei; viz., each is almost entirely due to the r process. Each of the four heavy transbismuth nuclei is entirely due to the r process, for the slow processes cannot climb the hill of transbismuth radioactivities. Reference to Fig. 7-27 and to the corresponding calculation represented in Fig. 7-32 shows that I^{129} and Re^{187} are respectively in the $N = 82$ and $N = 126$ r-process peaks, and almost all these nuclei ever produced are believed to have resulted from r-process events. It will be clear, therefore, that nuclear cosmochronology probes only the history of r-process nucleosynthesis. Inasmuch as the site of the r process is probably the supernova, these investigations probably explore the rate of supernova explosions in the history of the galaxy.

Figures 7-30 to 7-32 show the r process terminated by neutron-induced fission near $A = 270$. There is considerable uncertainty about the largest masses produced in the r process, but that uncertainty is not crucial to the production of the transuranic nuclei, for the following reason. As these very heavy nuclei begin their chains of beta decay after the synthesizing event, the nuclear charge increases, and as it does, the nuclei fission spontaneously. The rate of spontaneous fission is most strongly dependent upon the ratio of the coulomb energy ($E_c \propto Z^2/A^{\frac{1}{3}}$) to the surface energy ($E_S \propto A^{\frac{2}{3}}$); that is, when the *fission parameter* Z^2/A becomes too large, the nucleus fissions spontaneously. For the heaviest nuclei produced in the r process this fate seems unavoidable as they decay toward the valley of beta stability. Only for values of atomic weight less than $A \approx 256$

does it appear possible that the isobaric decay chain reaches beta stability without fissioning and also begins alpha decay more rapidly than spontaneous fission.

The sequence of progenitors that ultimately decay to Th^{232}, the so-called $4n$ group of transuranic nuclides, is shown in Fig. 7-33. Material produced at $A = 256$ undergoes spontaneous fission 96 percent of the time when the decay chain reaches the beta-stable nucleus Fm^{256}. Nuclei produced at $A = 252$, $A = 248$, and $A = 244$, however, predominantly decay by alpha emission and hence tend to pile up at Pu^{244} at times greater than 10^6 years. After times long compared to 10^8 years, however, the Pu^{244} has decayed primarily to Th^{232}, joining, as it does so, the nuclei produced by the r process at $A = 240$, 236, and 232. From the study of such nuclear data one sees that for relatively short periods Pu^{244} exists by virtue of its three r-process progenitors plus a small 4 percent contribution from $A = 256$, but for relatively long times it joins the three progenitors of Th^{232} to give it six progenitors in all. A similar section of the transuranic nuclide chart for the $4n + 1$ series shows five progenitors for Np^{237}, whose 2.1×10^6 year half-life is sufficiently short for it to decay unnoticed to stable Bi^{209}. The $4n + 2$ series reveals three progenitors plus about 10 percent of a fourth one that end as U^{238}, whereas the $4n + 3$ series shows a grand total of at least six progenitors ending as U^{235}. The systematic details of the r process suggest that each of these progenitors should be synthesized in roughly equal amounts because there is no shell structure nearby. One of the attempts at a more detailed calculation of the expected abundances is shown in Table 7-2. The progenitors of Pu^{244} are included as the final four entries in the Th^{232} column, and the numbers in parentheses indicate the fraction of the decays by alpha emission, for these are the only decays resulting in a residual abundance for the

Table 7-2 Calculated r-process abundances for parents of the radioactive series (Si = 10^6)†

$Th^{232}(4n)$		$Np^{237}(4n+1)$		$U^{238}(4n+2)$		$U^{235}(4n+3)$	
A	*Yield*	*A*	*Yield*	*A*	*Yield*	*A*	*Yield*
232	0.049					235	0.054
236	0.056	237	0.059	238	0.061	239	0.065
240	0.066	241	0.065	242	0.059	243	0.051
244	0.043	245	0.038	246	0.036	247	0.032
248(0.89)‡	0.029	249	0.033	250(0.10)	0.004	251	0.036
252(0.86)	0.033	253	0.040			255	0.047
256(0.04)	0.002					259	?
Total	0.278	Total	0.235	Total	0.160	Total	0.285
× 1.10 = 0.30		× 0.90 = 0.21		× 1.10 = 0.18		× 0.90 = 0.26	

† Taken from P. A. Seeger, W. A. Fowler, and D. D. Clayton, *Astrophys. J. Suppl.*, **11**:121 (1965). By permission of The University of Chicago Press. Copyright 1965 by The University of Chicago.

‡ Numbers in parentheses indicate the fraction of the decays by alpha omission.

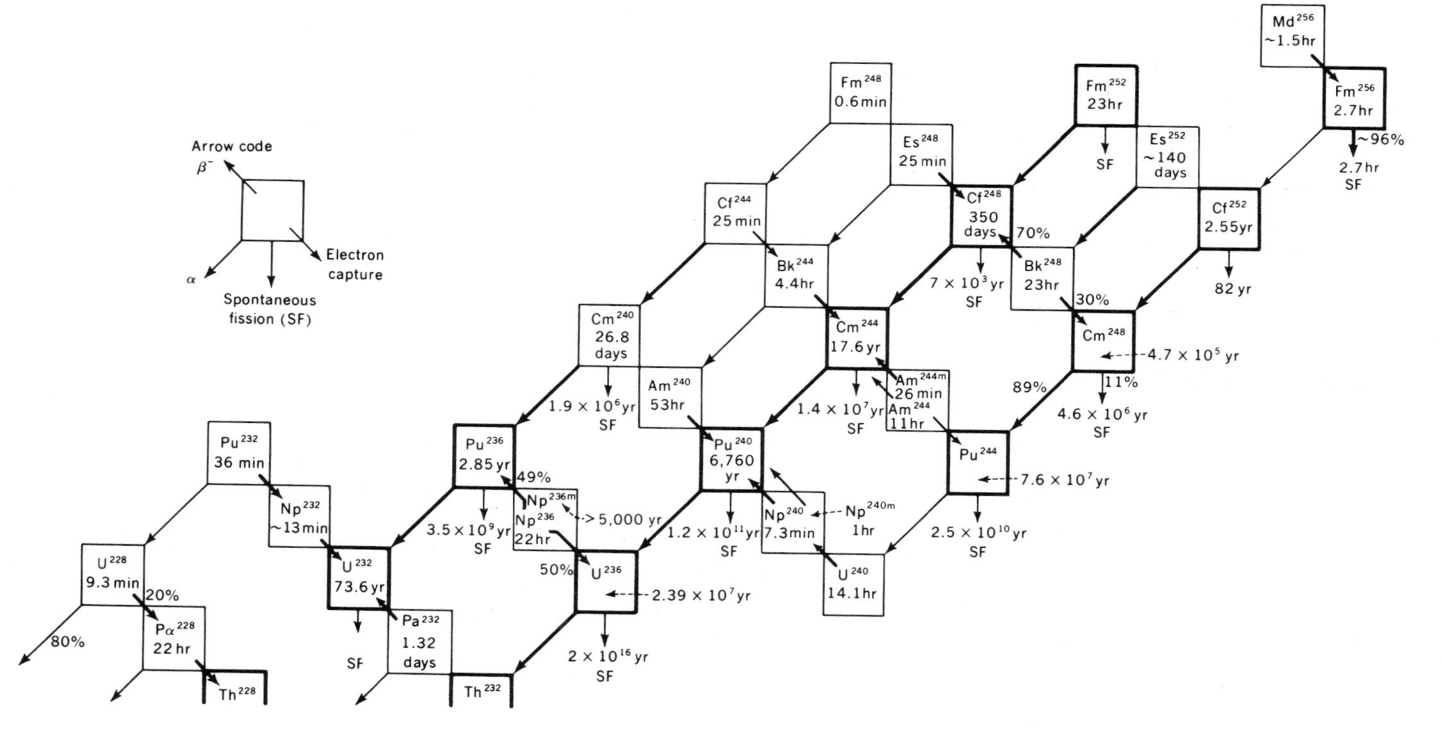

Fig. 7-33 Genetic relationships of the $4n$ group of transuranium nuclides. Decay modes of each nuclide are indicated by arrow code. A heavy arrow indicates 10 percent or greater branching by that decay mode. A light arrow indicates slight branching. Heavily outlined squares indicate beta stability. From such data it can be seen that Th^{232} has a total of 5.79 r-process parents, 2.79 of which are temporarily arrested at extinct Pu^{244}. [*Earl K. Hyde, Isadore Perlman, and Glenn T. Seaborg, "The Nuclear Properties of the Heavy Elements," vol. 2, "Detailed Radioactivity Properties,"* Copyright 1964. *Reprinted by permission of Prentice-Hall, Inc., Englewood Cliffs, N.J.*]

daughter. The numbers were normalized in such a way that the yield below $A = 200$ of Fig. 7-32 matches the observed r-process abundances on the Si $= 10^6$ scale, and so the entries represent an estimate of the total number of the progenitors ever produced on the Si $= 10^6$ scale. There are independent indications of an odd-even effect of about 20 percent that was not included in the calculations, so that the even abundances have been augmented by 10 percent and the odd abundances diminished by the same amount. This table may be used as a guide to r-process production ratios when dealing with problems of nuclear cosmochronology.

Let us now see how these ideas can be used to compute something of the history of nucleosynthesis. Consider first the relative abundances of U^{235} and U^{238}, a ratio that is very well known because it has not been subject to alteration by geochemical reactions in the processes of meteorite formation. (The chemical properties of U^{235} and U^{238} are, of course, essentially identical.) The present ratio is $(U^{235}/U^{238})_{\text{today}} = 0.00723$. With the aid of the exponential decay laws one can compute the abundance ratio at the time of formation of the solar system.

Problem 7-34: Compute the value of the uranium-isotope ratio 4.6×10^9 years ago.
Ans: $(U^{235}/U^{238}) = 0.31$.

It is clear that if all the uranium were produced in a single event, called *sudden synthesis*, and if one knew the ratio in which the two isotopes were produced, one could compute the time of nucleosynthesis by extrapolating the ratio backward in time according to the exponential laws until its value equals the production ratio. From Table 7-2 the production ratio R of U^{235} and U^{238} is expected to be

$$\frac{\text{Production U}^{235}}{\text{Production U}^{238}} = R \approx \frac{0.26}{0.18} = 1.45$$

$$\text{Without odd-even correction } R \approx 1.8 \qquad (7\text{-}71)$$

$$\text{Progenitor ratio} = R = \frac{6}{3.1} = 1.93$$

which would have been reached by purely exponential decay almost 7 billion years ago.

On the other hand, one hardly expects all of nucleosynthesis to have occurred in a single event, so that we return to the model of continuous nucleosynthesis lasting for a duration T. It has been popular to characterize the rate of continuous nucleosynthesis by an exponential time dependence in which the rate of supernovas decreases as $\exp(-\Lambda t)$ over the interval T. Differing values of the parameter Λ generate physically distinct cases. The value $\Lambda = 0$ corresponds to a constant rate of nucleosynthesis, and the value $\Lambda \to \infty$ corresponds to sudden synthesis, since it places all of nucleosynthesis at the initial time. With such a model it is not difficult to calculate the ratio $(U^{235}/U^{238})_T$ applying at the end of the period of nucleosynthesis.

Problem 7-35: Show that

$$\left(\frac{U_{235}}{U_{238}}\right)_T = R\frac{\Lambda - \lambda_{238}}{\Lambda - \lambda_{235}}\frac{\exp\left[(\Lambda - \lambda_{235})T\right] - 1}{\exp\left[(\Lambda - \lambda_{238})T\right] - 1} \tag{7-72}$$

and demonstrate that this expression has the proper limiting values for the cases of sudden synthesis and constant rate of synthesis.

Any model of the rate of nucleosynthesis characterized by values of Λ and T can also be labeled by T and the percentage of the initial rate of nucleosynthesis still occurring at the end of the duration T. To see the results of such an exercise, suppose we ignore the small decay interval ΔT required for the solidification of the meteorites and search for all models that result in the ratio $(U^{235}/U^{238})_T = 0.31$ at the end of the period of nucleosynthesis. These models are displayed in Fig. 7-34,

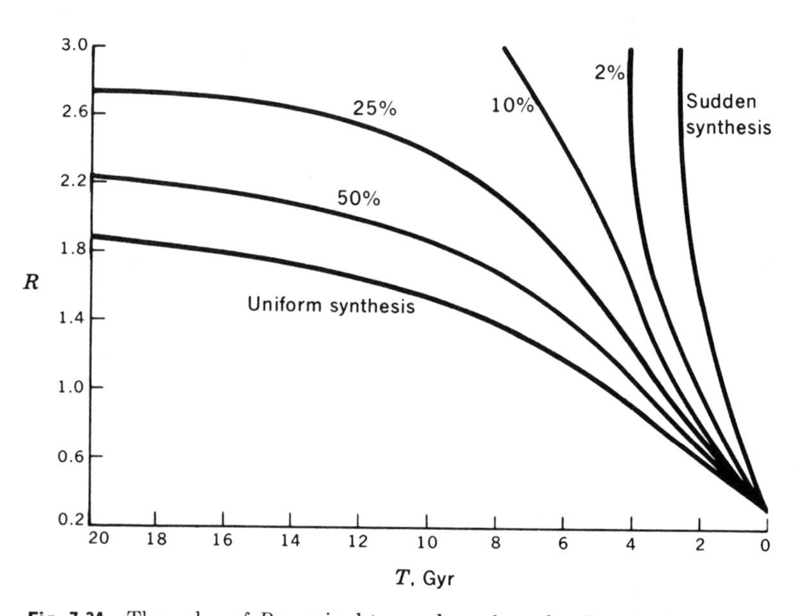

Fig. 7-34 The value of R required to produce the value $U^{235}/U^{238} = 0.31$ at the time of formation of the solar system as a function of the time of the beginning of galactic nucleosynthesis. That time T is measured back from the formation of the solar system. R is the production ratio of U^{235}/U^{238} in r-process explosions; its value is assumed to be in the range $1.4 < R < 1.8$. The relationship is demonstrated for exponential models of the rate of enrichment due to galactic nucleosynthesis of material destined for incorporation into the solar system, each curve being labeled by the percentage of the initial rate of enrichment still occurring at the time of solar-system formation, which is placed at zero on the abscissa. For nucleosynthesis rates proportional to $\exp(-\Lambda t)$ over an interval T the fraction of the initial rate at the time of solar formation is $f = \exp(-\Lambda T)$. [*D. D. Clayton, Astrophys. J.*, **139**:637 (1964). *By permission of The University of Chicago Press. Copyright* 1964 *by The University of Chicago.*]

where each curve is labeled by the steepness in the decrease in the rate of nucleosynthesis. The time is measured backward, from the time of formation of the solar system, not from today. For the expected values of R near 1.5, we see that the models of continuous nucleosynthesis give much older elements than the sudden-synthesis model does. Times for the beginning of the period of nucleosynthesis go back as far as about 10 billion years before the solar system formed for the model with constant rate of synthesis, whereas the sudden synthesis would have occurred slightly more than 2 billion years before the solar system. In fact, the results obtainable are so variable that when coupled with a liberal uncertainty in the production ratio, say $R = 1.6 \pm 0.3$, the uranium isotope ratio is *by itself* unable to define the onset of nucleosynthesis.

The logical thing to do is to augment the possibilities from uranium synthesis with those of another long-lived decay. The Th^{232}/U^{238} ratio is the most attractive insofar as the half-lives are concerned. The estimated production ratio is, from Table 7-2,

$$\frac{\text{Production Th}^{232}}{\text{Production U}^{238}} \approx \frac{0.30}{0.18} = 1.7 \pm 0.3 \tag{7-73}$$

where the error simply indicates an allowable uncertainty that would not do violence to the theory. The problem with this technique, as with many others, rests in the uncertainty of the measured abundance ratio. Unlike the two isotopes of uranium, Th^{232} and U^{238} have different chemical properties. As a result, their abundance ratio is variable. Because the chemical history of the formation process for the meteorites is still unknown, one is not sure which measured abundance ratio should be regarded as most representative of the solar system as a whole.

Estimates of the present Th/U ratio range from values somewhat greater than 4 to somewhat less than 3. A popular value has been Th/U = 3.8. It is again a simple matter to show that if that abundance is correct today, the value 4.7×10^9 years ago was $(Th^{232}/U^{238})_T \approx 2.33$. This number is still greater than the anticipated production ratio, so that one must again build a family of models of nucleosynthesis to explore the range of possibilities. Figure 7-35 shows the family of exponential models of nucleosynthesis for both of the abundance ratios discussed so far. Although the curves are not easily concordant either for sudden synthesis or for uniform synthesis, it does seem that the abundance ratios are consistent with a model of nucleosynthesis that began about 11 billion years ago and decreased in rate to about one-third of its original rate at the time the solar system was formed. Both calculations have appreciable uncertainties, but the point to be made here is that the diligent demand for concordance between two decay schemes sets the chronological model much more explicitly than the constraint of a single decay scheme does.

In making any analysis of this kind, one must distinguish between the absolute rate of nucleosynthesis in the galaxy and the rate of nucleosynthesis of nuclei that are ultimately to be incorporated into the solar nebula. It is the latter that

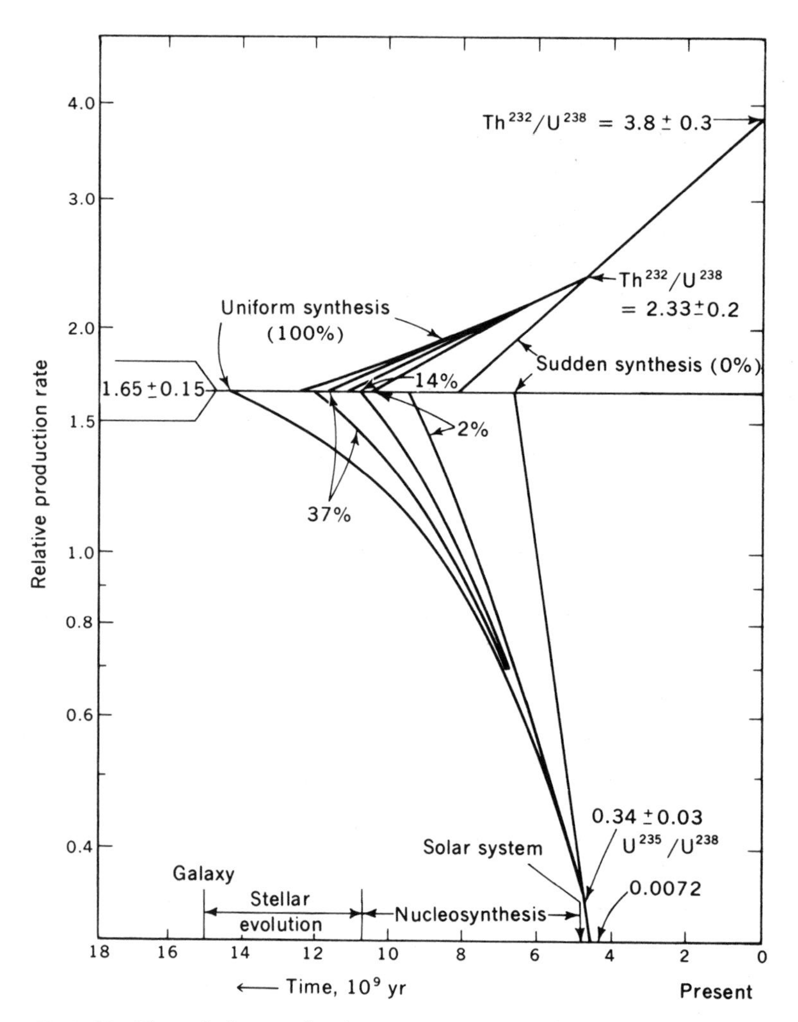

Fig.7 -35 The relative production rate of Th^{232} and U^{238} necessary to give the abundance ratio $Th^{232}/U^{238} = 2.33$ at an epoch 4.7×10^9 years ago is shown in the upper curves, whereas the lower curves, which are the same as those of Fig. 7-34, show the relative production rate of U^{235} and U^{238} necessary to give the ratio $U^{235}/U^{238} = 0.34$ at the same epoch. Coincidentally it happens that both production ratios are expected to lie somewhere in the band 1.65 ± 0.15, so that concordance may be sought in the form of intersections within this band of production ratios of curves corresponding to the same exponential rate of enrichment of solar material. The curves are labeled in terms of the present rate of enrichment of interstellar gas (per unit mass of gas) relative to the original rate of enrichment of interstellar gas (per unit mass of gas then). The best agreement falls near 12×10^9 years ago, but the uncertainties are great. [*After W. A. Fowler and F. Hoyle, Ann. Phys.*, **10**:280 (1960).]

is measurable by the radioactive chronologies. A large amount of early synthesis may have been mixed into a large mass of interstellar gas which may presently be bound within stellar interiors. Such contributions to the solar heavy elements require devaluation when compared with later synthesis that may have been mixed with a less massive interstellar medium.

The Re^{187} beta decay to stable Os^{187} is one of the most important decays for nuclear cosmochronology. It must be treated in a different manner than the U and Th decays, however. Its 40-billion-year half-life is so long that only a modest fraction of all the Re^{187} ever produced has decayed. This fact has the consequence that no useful information can be obtained from the ratio of the Re^{187} abundance to that of one of the other radioactivities. One must instead measure the fraction of the Re^{187} that has decayed over the internal T by measuring the extent of the enrichment of the daughter nucleus Os^{187}. Normally this would be a nearly impossible task, but the vicissitudes of nuclear structure have taken a very favorable turn in this region of the nuclear chart, a turn that allows one to compute the fraction of the Os^{187} abundance that is the result of cosmoradiogenic decay of Re^{187}. The relevant portion of the chart of nuclides is shown in Fig. 7-36. Both Os^{186} and Os^{187} are shielded from r-process production, by

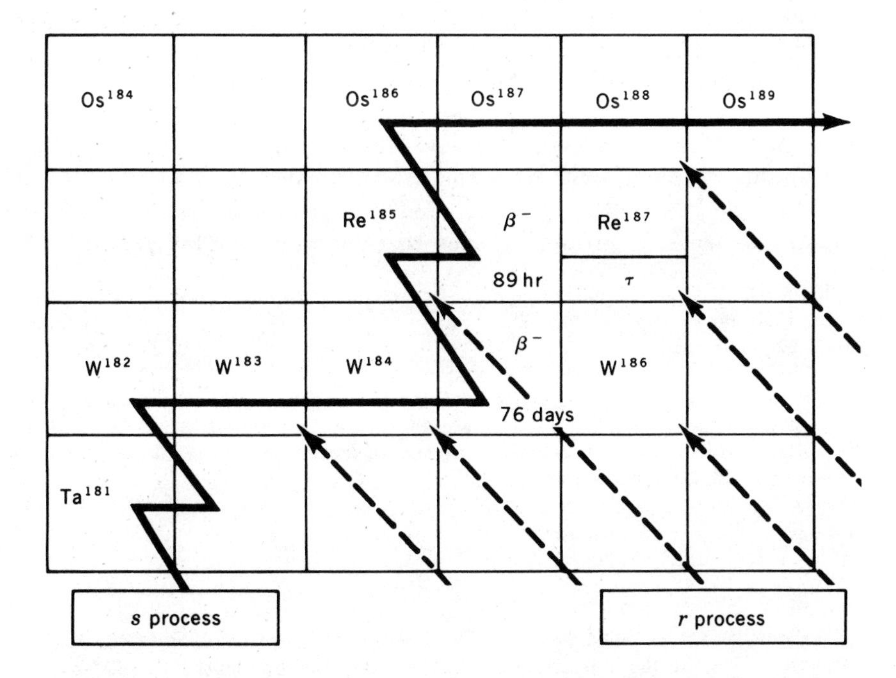

Fig. 7-36 The synthesis of Re and Os. The s-process current passes through both Os^{186} and Os^{187}, which are shielded from r-process production by W^{186} and Re^{187}, respectively. Over galactic time scales, however, a portion of the long-lived Re^{187} will decay to Os^{187}. [*D. D. Clayton, Astrophys. J.*, **139**:637 (1964). *By permission of The University of Chicago Press. Copyright* 1964 *by The University of Chicago.*]

W^{186} and Re^{187}, respectively. Because osmium lies on such a flat portion of the s-process curve, we would have equal values of σN for Os^{186} and Os^{187} were it not for the fact that Os^{187} has been augmented by the decay of Re^{187}. That is, the s-process portion of the Os^{187} abundance is equal to

$$Os_s^{187} = \frac{\sigma(186)}{\sigma(187)} Os^{186} \tag{7-74}$$

where Os^{186} represents the total abundance of s-only Os^{186}. (In principle one must make allowance for the p-process contributions, but they are again very small, as indicated by the abundance $Os^{184} \approx 0.012 Os^{186}$.)

Problem 7-36: Define "cosmoradiogenic Os^{187}" as the abundance of Os^{187} due to Re^{187} decay at the time the solar system formed, and designate it by Os_c^{187}. Show that its value is given by

$$\frac{Os_c^{187}}{Re^{187}} = \frac{Os^{187}/Os - [\sigma(186)/\sigma(187)]Os^{186}/Os}{Re^{187}/Re} \frac{Os}{Re} \tag{7-75}$$

where all abundance ratios are at the time of solar-system formation. Their values are $Os^{187}/Os = 0.0132$, $Os^{186}/Os = 0.0159$, $Re^{187}/Re = 0.65$.

One of the ratios requiring great caution is that of the abundances of the two elements osmium and rhenium, but careful analysis of different phases of iron meteorites has fairly well established its value as $Os/Re = 11.3$. Nature has again been kind in that Os and Re are similar in their geochemical properties and have not been strongly fractionated from each other. The major outstanding problem is the measurement of the neutron-capture cross sections, which will be measurable with high accuracy whenever sufficiently large samples of isotopically pure Os^{186} and Os^{187} have been prepared. There seems little doubt from the systematics of neutron capture, however, that one will find $\sigma(186)/\sigma(187) = 0.4 \pm 0.1$.

Problem 7-37: If the estimates made above are correct, what percentage of the Re^{187} decayed before the solar system was formed?
Ans: 12 percent.

Problem 7-38: Using the same exponential models of the history of nucleosynthesis that were used in the uranium discussion, calculate the ratio of cosmoradiogenic Os^{187} to the residual Re^{187} that will have been produced during the nucleosynthesis interval T in which the rate of r-process production decreases as $\exp(-\Lambda t)$.
Ans:

$$\left(\frac{Os_c^{187}}{Re^{187}}\right)_T = \left[\frac{\Lambda - \lambda_{187}}{\Lambda} e^{\lambda_{187}T} \frac{1 - \exp(-\Lambda T)}{1 - \exp(-\Lambda + \lambda_{187})T}\right] - 1 \tag{7-76}$$

Equation (7-76) has been plotted in Fig. 7-37 as a function of the ratio of the duration T to the Re^{187} half-life. There it can be seen that if $\sigma(186)/\sigma(187) = 0.4$, as assumed for prob. 7-37, the interval of time between the beginning of nucleosynthesis and the formation of the solar system varies from about $0.16\tau_{1/2}$ for sudden synthesis to $0.33\tau_{1/2}$ for uniform synthesis. Taking the time of solar-system formation to be 4.6×10^9 years ago and the half-life to be $\tau_{1/2} = 40 \times 10^9$ years,

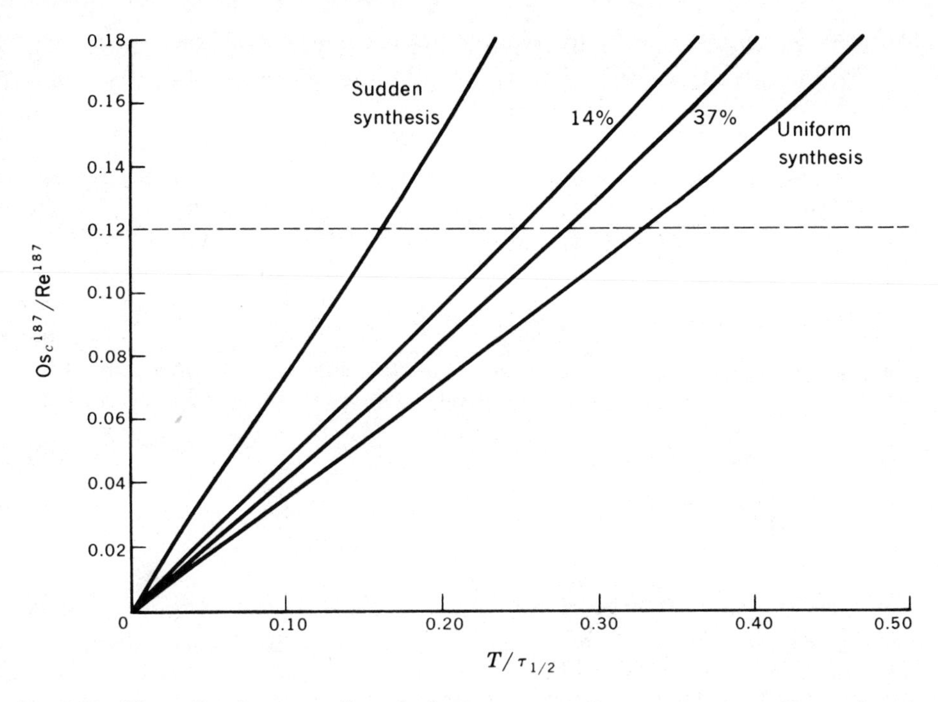

Fig. 7-37 The ratio of cosmoradiogenic Os^{187} to parent Re^{187} at the time of formation of the solar system calculated for various exponential models of the rate of enrichment due to *r*-process nucleosynthesis of material destined for the solar nebula. The abscissa measures the commencement time of that enrichment in terms of the Re^{187} half-life. It is measured backward from the time of formation of the solar system. The curves are labeled by the ratio of the rate of enrichment of solar material when the sun formed to the initial rate of enrichment. The dashed horizontal line corresponds to the value $Os_c^{187}/Re^{187} = 0.12$, which applies if the ratio of neutron-capture cross sec ions is near the expected value $\sigma(186)/\sigma(187) = 0.4$. [*D. D. Clayton, Astrophys. J.*, **139**:637 (1964). *By permission of The University of Chicago Press. Copyright* 1964 *by The University of Chicago.*]

this estimate would place the beginning of galactic nucleosynthesis at a time 11×10^9 years ago for sudden synthesis and 18×10^9 years ago for uniform synthesis. Although this range of chronologies differs somewhat from those shown in Fig. 7-35, it should be understood that no disagreement really exists, because all methods have sources of uncertainty. But it does seem that the Re^{187} chronology generally indicates an older galaxy than the other radioactive decays.[1] This may well become the most important of the decays for its cosmological implications, so that it is clear that very careful measurements are needed of these three uncertain quantities: Os/Re, $\sigma(186)/\sigma(187)$, and $\tau_{\frac{1}{2}}$.

[1] For a discussion of the degree of concordance within the framework of these models the reader should consult D. D. Clayton, *Science,* **143**:1281 (1964). That paper also defines a very sensible format for making the comparison within which the effect of errors is shown explicitly.

Study of the extinct radioactivities abounds with complications. Because they are extinct, they can be detected only by the effects they leave behind. The first discovered and most celebrated of such effects are the anomalies in the isotopic composition of xenon gas trapped in the meteorites.[1] The so-called *special anomaly*, a large overabundance of Xe^{129} in the trapped gas, has been interpreted as the product of the beta decay of 16.9-million-year I^{129}, which is presumed to have been incorporated into the lattice structure of the meteorites before it decayed. There are anomalies of smaller magnitude at most of the other of the many isotopes of xenon. Called the *general anomaly*, these overabundances are interpreted as being produced as fission fragments, and the only parent that seems to fit the necessary fission yields is the spontaneous fission of Pu^{244}. A later experimental measurement than the one shown in Fig. 7-33 determines the ratio of the rate of spontaneous fission of Pu^{244} to the rate for alpha decay to be

$$\frac{\lambda_f(\mathrm{Pu}^{244})}{\lambda_\alpha(\mathrm{Pu}^{244})} = \frac{1.03 \times 10^{-11}\ \mathrm{yr}^{-1}}{8.45 \times 10^{-9}\ \mathrm{yr}^{-1}} = 1.2 \times 10^{-3}$$

Even with this small branching ratio it turns out that the fission yields, generally measured by the size of the Xe^{136} anomaly, are great enough to account for the general anomalies. The data are highly variable, however, so that it is not possible to state any firm value for the initial abundances of these extinct nuclei. For example, the ratio of Xe^{136} from fission to U^{238} is found in three different types of meteorites to be

$$\frac{(\mathrm{Xe}^{136})_{\mathrm{fission}}}{\mathrm{U}^{238}} \approx \begin{cases} 0.075 \times 10^{-5} & \text{achondrite} \\ 0.5 \times 10^{-5} & \text{ordinary chondrite} \\ 3.8 \times 10^{-5} & \text{carbonaceous chondrite} \end{cases}$$

It is not clear whether these meteorites formed at differing times, or whether their Pu^{244}/U^{238} ratio differs because of chemical fractionation, or both. The last of these three values is so large, however, that if taken at face value it would imply a ratio $(Pu^{244}/U^{238})_0 \approx 0.26$ at the time the meteorite solidified. This is far more Pu^{244} than can be expected on a model of continuous nucleosynthesis. On the other hand, the value $(Pu^{244}/U^{238})_0 = 5 \times 10^{-3}$ in the achondrite at the time of its solidification is consistent with continuous galactic nucleosynthesis plus a decay interval on the order of 10^8 years to allow for the solidification of the meteorites. This abundance ratio is also consistent with the density of spontaneous-fission tracks detectable in these minerals. By this technique[2] the ratio $(Pu^{244}/U^{238})_0$ at the time of solidification has been estimated to be 3.1×10^{-3} and 11.4×10^{-3} for the meteorites Toluca and Moore County, respectively. It is very difficult to interpret these numbers at the present time. The initial iodine

[1] For a good exposition see J. H. Reynolds, Xenology, *J. Geophys. Res.*, **68**:2939 (1963). For more details of later developments see C. Merrihue, *ibid.*, **71**:263 (1963), and J. H. Reynolds, Isotopic Abundance Anomalies in the Solar System, in E. Segré (ed.), *Ann. Rev. of Nuclear Science*, vol. 17, Annual Reviews Inc., Palo Alto, Calif., 1967.

[2] R. L. Fleischer, P. B. Price, and R. M. Walker, *J. Geophys. Res.*, **70**:2703 (1965).

ratio inferred from the Xe^{129} special anomaly

$$\left(\frac{I^{129}}{I^{127}}\right)_0 \approx 10^{-4}$$

seems also to be consistent with continuous galactic nucleosynthesis provided shorter decay intervals, say $\Delta T \approx 50 \times 10^6$ years, are allowed for the solidification.

Problem 7-39: Assuming that the duration of nucleosynthesis was several billion years or more and that its rate was constant, what ratio $(Pu^{244}/U^{235})_T$ would have remained at the end of nucleosynthesis? You may also wish to compute the corresponding solidification interval ΔT required to give $(Pu^{244}/U^{238})_0 = 5 \times 10^{-3}$.
Ans: $(Pu^{244}/U^{235})_T = 0.051$.

It seems likely that further laboratory work will solve the problem of the extinct radioactivities in the sense that the abundance ratios at the end of nucleosynthesis and the formation times required for the various classes of meteorites will be determined. Only after a convincing concordance of this type has been struck will it be possible to evaluate fully the implications for the history of nucleosynthesis. However, it seems only appropriate to point out some of the major issues that may ultimately be resolved by these investigations.

Fowler, Greenstein, and Hoyle constructed a model[1] of the early solar system in which spallation reactions and associated neutron capture by elements in planetesimals in the early solar system resulted in the observed terrestrial abundances of the rare light elements. Those authors also point out that spallation and neutron capture by isotopes of tellurium can result, at least in part, in the xenon anomalies. We may hope that further study of the extinct radioactivities will clarify the relevance of this model to the problems of nucleosynthesis. The origin of the rare light elements is still a mystery. This model cannot, of course, produce any Pu^{244}.

Several authors[2] have suggested that the material in the primitive solar nebula should have had a larger portion of "last-minute" nucleosynthesis than would be expected on a continuous model. Their reasoning is that stars tend to form in associations in which the massive members may evolve to supernovas in a time shorter than is required for the low-mass stars like the sun to form. Perhaps one should expect one or more supernovas to have exploded nearby, mixing their fresh radioactivities into the gas that would soon condense into the solar system. Such a model will be strongly suggested if Pu^{244}/U^{238} ratios are consistently and unambiguously found to have larger values than uniform synthesis can provide. Such a model, constructed essentially to account for the extinct radioactivities, would necessarily have larger implications for the history of nucleosynthesis in the galaxy. If it accounted also for a sizable portion of our supply of U^{235}, the history of the galaxy inferred from the remainder would be older.

[1] *Geophys. J.*, **6**:148 (1962).

[2] The most thorough discussion of the idea has been made by A. G. W. Cameron, *Icarus*, **1**:1 (1962).

One must be optimistic for the future of nuclear cosmochronology. The old methods continually improve, and occasionally new methods are discovered. Perhaps more long-lived radioactivities can be made to yield useful results. The natural radioactivities are the most convincing proof of the truth of nucleosynthesis, just as the existence of stars is the most convincing proof that interstellar gas may somehow collapse into a star. Perhaps our notions of the sites of nucleosynthesis are all wrong, but wherever the truth lies, one can be sure that nuclear cosmochronology must provide a consistent factual framework. Those who wrestle with this problem are grateful for the small peculiarities of nuclear structure that have provided these marvelous clues to history.

INDEX